Structure and Function of the Circulation
Volume 1

A Continuation Order Plan is available for this series. A continuation order will bring delivery of each new volume immediately upon publication. Volumes are billed only upon actual shipment. For further information please contact the publisher.

Structure and Function of the Circulation
Volume 1

Edited by
Colin J. Schwartz, M.D., F.R.A.C.P.
*Professor of Pathology, University of Texas Health Science Center,
and Member of Scientific Staff, Southwest Foundation for Research
& Education, San Antonio, Texas*

Nicholas T. Werthessen, Ph.D.
*Bioscientist, Office of Naval Research, Boston, Massachusetts,
and Senior Research Associate, Brown University, Providence, Rhode Island*

and

Stewart Wolf, M.D.
*Vice President, Medical Affairs, St. Luke's Hospital, Bethlehem, Pennsylvania,
and Professor of Medicine, Temple University, Philadelphia, Pennsylvania,
and Director, Totts Gap Institute, Bangor, Pennsylvania*

PLENUM PRESS · NEW YORK AND LONDON

Library of Congress Cataloging in Publication Data

Main entry under title:

Structure and function of the circulation.

Includes index.
1. Blood—Circulation. I. Schwartz, Colin John, 1931- . II. Werthessen, Nicholas Theodore, 1911- . III. Wolf, Stewart George, 1914-
QP102.S75   599'.01'1   79-9413
ISBN 0-306-40278-5 (v. 1)

© 1980 Plenum Press, New York
A Division of Plenum Publishing Corporation
227 West 17th Street, New York, N.Y. 10011

All rights reserved

No part of this book may be reproduced, stored in a retrieval system, or transmitted, in any form or by any means, electronic, mechanical, photocopying, microfilming, recording, or otherwise, without written permission from the publisher

Printed in the United States of America

FOREWORD

In order to produce a superior scholarly treatise in biomedical science, three important conditions need to be met. First, the subject needs to be of recognized importance and preferably one in which a sizeable volume of new knowledge has been added recently. Second, it needs to be quite evident that the field involved requires much more up-to-date coverage than it has received and third, the choice of the editors and in turn the authors needs to be recognized as outstanding.

This major treatise fills these criteria in an admirable way. There are few who would deny the importance of knowledge concerning the circulatory system. This all pervasive system is the route by which virtually all of the cells and tissues of the body receive their nutrition and it is the major route by which metabolic waste products are carried away. Furthermore, the diseases that involve the circulatory system are, by far, the underlying causes of death and morbidity in the largest number of Americans, Western Europeans and several other populations of industrialized nations. Not only is atherosclerosis-induced-ischemic disease of the heart, brain and extremities widespread in these populations but venous occlusive disease also takes a great toll from phlebothrombosis, pulmonary embolism, etc.

These disease processes and their serious consequences would not merit a treatise like this were it not for the rather remarkable advances that have occurred recently in our understanding, especially of the lesions of the most common and serious disorder of the arteries, namely atheroscalerosis. Similarly the general understanding of the intimate morphology and function, i.e. the cell biology and the molecular biology of the normal vascular channels has also advanced rapidly in the past decade. All of this recent progress makes it a very opportune time to develop a treatise of this type.

The major factor in any scholarly work is the quality of the scholarship of the major contributors to these volumes. In this instance the choice of the chief editor could not have been better.

Colin J. Schwartz has been an active investigator of almost all
aspects of vascular structure and function in Australia, England,
Canada and now in the USA. His contributions have included work
almost equally divided between studies of structure and function.
Furthermore he has delved deeply into investigations that have,
in turn, been concerned with lymph and lymphatics, with veins and
for many years with arteries. Furthermore he has carried into all
of these inquiries a careful and critical approach and a probing
intellect. These qualities as well as the perspective of a modern
pathologist enlightened by almost a quarter of a century of study
of the vascular system have helped him to shape these volumes. He
has developed an outstanding outline of subjects for the series.
He has chosen excellent authors for each topic and he has been able
to convince each of these capable scientists to develop an
authoritative chapter in the field in which he or she works. This
is a great accomplishment. The result is a scholarly work that
should serve as a standard reference for those interested in the
vascular system for many years - although hopefully it will be
revised and reprinted when further progress in the field makes
that desirable. At present it is the only comprehensive review of
modern knowledge of this important system.

                                                Robert W. Wissler, Ph.D., M.D.
                                                Distinguished Service Professor
                                                  of Pathology
                                                The University of Chicago
                                                March, 1979

PREFACE

The "Structure and Function of the Circulation" has been divided into three volumes, of which this is the first. The need for a series such as this became apparent during the scientific sessions of both the first Lindau Conference in April of 1970 and the second conference held in Heidelberg in October of 1973. With the urging of my good friends Dr. K. T. Lee and N. T. Werthessen at an exploratory meeting in Boston, plans for this large and ambitious undertaking began to crystallize. Development of the three volumes was further catalyzed by the encouragement and helpful criticism of a number of colleagues at a subsequent meeting hosted by Dr. Stewart Wolf at a Totts Gap, Pennsylvania.

The Lindau Conferences and many subsequent discussions clearly identified a need to collate in a comprehensive and scholarly manner existing knowledge concerning the structure and biology of the circulation in health and the implications of this knowledge in achieving or facilitating a better understanding of basic disease mechanisms. Emphasis has been directed primarily to the many facets of "normal" arterial, venous, and lymphatic structure and function. The three volumes cover a broad spectrum ranging from gross and comparative vascular anatomy to the basic ultrastructure and cellular biology of the components of the circulation. Additionally, three chapters cover the historical evolution of our understanding of the circulation, from Greco-Roman times to the era of Sir Thomas Lewis, father of contemporary cardiovascular physiology, and to the present.

It is my hope that this interwoven collection of essays by scholars from around the world will serve as a useful reference source for all students of cardiovascular disease, be they involved in medical or surgical cardiology, or the basic cardiovascular sciences. Hopefully, the "Structure and Function of the Circulation" will also provide a substative scientific infrastructure upon which cardiovascular research might continue to grow and flourish.

As the matrix outlining the content and scope of these volumes was under development, I became acutely aware of many areas where our information base remains inadequate. Some such areas include the broad field of comparative physiology and biology of the circulation, and the control or regulation of normal vascular endothelial pinocytosis. Other essentially uncharted areas which illustrate further this deficiency include topics such as the nature of the interactions between arterial endothelium and medial smooth muscle, the physiological role of endothelial contractile filaments, arterial phagocytic functions, and the determinants of endothelial anti-thrombotic properties.

In developing the three volumes which comprise the "Structure and Function of the Circulation" I am deeply indebted to Dr. Werthessen who has served as managing editor. Mr. Seymour Weingarten, formerly of Plenum Press, and Dr. Mark Altschule of Boston have also provided much help in the development of these volumes, as have the staff at Plenum Press. Last but not least, I wish to thank the authors, not only for their excellent contributions, but also for their forbearance with the foibles of the editor.

Colin J. Schwartz, M.D., F.R.A.C.P.
F.R.C.P.A., F.R.C.P. (C)., F.R.C. Path.
Professor of Pathology
The University of Texas Health Science
Center at San Antonio, Texas, USA

CONTENTS

The Arteries in Greco-Roman Medicine   1
    C.R.S. Harris

Embryology of the Human Arterial System
        (Arteriogenesis) 21
    W. Pallie

Functional Morphology of Arteries During
        Fetal and Post-natal Development 95
    W.W. Meyer, S.Z. Walsh, and J. Lind

Abdominal Visceral Circulation in Man 381
    E.A. Edwards

Arterial Circulation of the Extremities 425
    H. Haimovici

Biology of the Collateral Circulation 487
    D.E. Strandness, Jr.

Measurement of Blood Pressure, Blood Flow,
        and Resistance to Blood Flow in
        the Systemic Circulation 537
    J. Ludbrook

Regulation of Arterial Blood Flow, Pressure,
        and Resistance in the Systemic
        Circulation 587
    J. Ludbrook

The Anatomy of the Renal Circulation 631
    K. Solez and R.H. Heptinstall

The Renal Circulation: Physiology and Hor-
        monal Control 661
    K. Solez and R.H. Heptinstall

The Innervation of Arteries 729
    G. Burnstock, J.H. Chamley, and
    G.R. Campbell

The Blood Supply to Nerves 769
    W. Pallie

Index 805

THE ARTERIES IN GRECO-ROMAN MEDICINE

C. R. S. Harris, MA.D.Phil. (Oxford) Ph.D. (Princeton and Adelaide). D.Litt.Hum. (Lake Erie College)
Sometime Reader in the History of Medicine, University of Adelaide, South Australia.
Retirement Address: Rockhouse, Wheatley, Oxford, England

It is not coincidence that the original meaning of the Greek word "arteria" was windpipe, long before the discovery of the two different kinds of blood vessels had been made. The use of the same word for two organs belonging to different physiological systems was no accident. It points to one of the great errors in the early history of physiology, the notion that the heart was the focus of the respiratory, as well as the center of the vascular system. Early medicine has always been associated with many crude and fantastic superstitions. The astonishing feature of the medicine of ancient Greece from the 6th century B.C. onward was its complete divorce from this type of thinking, though in the folk medicine of the classical period, there was plenty of it.

Another feature of ancient Greek medicine was that by the end of the third century B.C. a vast volume of anatomical and physiological knowledge had been acquired. Empedocles seems to have been the first of the Greek thinkers to attempt to work out a connection between blood movement and respiration. Aristotle explains that he attempted to account for the in-and-outward movement of the air and the movement of the blood inward and outward from the surface of the body in terms of the strictly physical process of the water-clock.

From simple notions like these, the early Greek doctors had to interpret their observations of the actual behavior of human bodies in sickness and in health. By the end of the fourth century B.C., they apparently had a good description of the location of the main blood vessels. The focus of the system was the heart, as is shown by Aristotle's account, which makes no distinction between arteries and veins. Aristotle makes an inexplicable error in attributing three ventricles to the heart of the larger mammals and shows no signs of any knowledge of the cardiac valves.

Aristotle's conceptions of the vascular system show an enormous improvement on that of one of the most famous of the earlier treatises of the "Hippocratic" corpus, that on The Sacred Disease (i.e., epilepsy). It regards the head as the starting point of the main blood vessels. The descriptions in the various "Hippocratic" treatises of the blood vessels show so many contradictions and inconsistencies that we need not pursue them here.

Before Alexandrian days, considerable progress had been made in the knowledge of the vascular system, even before any Greek doctor had become acquainted with the interior of the human body at first hand. The main topographical features of the two great systems of vessels connected to the two-ventricled heart had been discovered before the end of the fourth century. The "thick" or "hollow" vein (the vena cava) and the aorta, each with its own branch-network permeating the body and the limbs, as well as the three main "trees" of the bronchial system and the arteries and the veins in the lungs were described, as were the general location of the principal organs inside the human body, the stomach, the intestines, the spleen, the kidneys, and the liver, and their connection with the portal system.

At the beginning of the third century B.C., if not earlier, Praxagoras (the younger?) found that the two separate great trunks of blood vessels--still all called veins--were two quite different kinds of tubes. The distinction between arteries and veins was thus established. Herophilus showed that the coat of the arteries is about six times as thick as that of the veins. Furthermore, in a corpse the veins collapse if emptied of blood, while the arteries, like the bronchial tubes, do not. Praxagoras insisted that the arteries contained no blood but only air. He declared that when removed from a body, they would continue for some time to pulsate. In this error of the bloodless arteries, he was followed by one of the great anatomical discoverers of history, Erasistratus, who first worked out in detail the correct anatomy of the heart and discovered the valves and the way they worked. Unlike Herophilus, he refused to regard the auricles as part of the heart, looking on them as merely the terminal appendages of the great blood vessels, the vena cava, and the pulmonary veins.

This mistaken idea of Praxagoras, more than any other single factor, prevented the ancients from arriving at the discovery of the circulation of the blood. But absurd as it might seem to us, the theory was founded on true, though incorrectly interpreted, observation. Praxagoras, when he cut open the chest of the animal he dissected, may well have found that the main arteries were neither collapsed nor filled with blood, but emptied and full of air.

When death takes place, the vasoconstrictor center governing the arteries is strongly stimulated by the increase of carbon dioxide and lack of oxygen.† Some explanation of the inconsistency of these findings has been given by the Swedish pathologist Dr. Fåhreus, who made a number of autopsies at least twenty-four hours after death. He says that on introducing a cannula connected with a micrometer into the carotid or the femoral artery, he found a negative pressure varying from a few millimeters to one decimeter. If you now begin the autopsy and open the thoracic cage, the pressure immediately rises to zero owing to the absorption of air by the arteries. He believes that the explanation of this negative pressure is the following: pressure in the arteries becomes negative because after death part of the arterial system dilates. This dilation cannot take place in the big arteries; it must therefore be the little arteries that dilate and draw a certain quantity of blood from the big arteries. This phenomenon varies with the age of the person. The height of the negative pressure increases sharply from the ages of 18 to 70 and then remains relatively constant. Nor is it instantaneous at the moment of death. This may perhaps explain why the author of De corde found the ventricle, but not the aorta empty.

It is natural to suppose that the pulse had been observed by doctors before the distinction between the two vascular systems had been discovered. Galen tells us that the ancient physicians applied the term pulse (σφυμος) only to those movements of the blood vessel which were sufficiently pronounced to be perceptible to patients or to cause pain like those in inflamed parts, and never to the movement in the healthy parts. Aristotle, like Plato, was aware of the normal beating of the heart which he distinguished from the type of throbbing which he calls *"leaping,"* due to a pathological state of the blood. He also concluded that the movement or throbbing of the blood vessels is due to the heart. *"All the veins,"* he tells us in the De respiratione, *"beat at the same time, since they are connected with the heart. They beat more strongly in the young than in the old."*

In contrast both to Erasistratus and to Galen, Herophilus regarded the auricles as forming a part of the heart proper, as modern anatomists do. He saw quite clearly that the two great vessels connecting the heart with the lungs were of different kinds, that the pulmonary artery was an artery par excellence, and that the pulmonary

---

†See Abel, K., Die Lehre vom Blutkreislauf in Corpus Hippocraticum in Hermes LXXXVI, pp. 192-219, also the paper delivered by Dr. Fåhreus, entitled "Empty Arteries" delivered to the Fifteenth International Congress of the History of Medicine held in Madrid in 1957.

vein was a vein.† None of his actual descriptions either of the heart or the vessels leading into and out of it have survived, but from the accounts of his doctrines given by Galen and the Anonymous Londoner, we can be sure that he considered that both systems contained blood: the veins that derived from nourishment, and the arteries, both blood and pneuma. Herophilus seems to have known, too, that arterial blood differed from venous.

Less fortunate was his description of the wonderful arterial network, which it is rather tempting to regard as an over-imaginative reduplication of the circle of Willis.†† This organ which appears to have been transferred by Herophilus to the human brain from a structure most easily recognizable in the ox††† and a number of other domestic animals,* was to play a most important role in Galen's account of the physiology of the brain and the nervous system. It is clear that he thought the arterial system contained both blood and pneuma, and that he accepted the idea already noted in Aristotle, that the function of respiration was to cool the innate heat in the heart.

Erasistratus, who dissected the human cadaver, gave so accurate a description of the heart and its valves that Galen could not improve on it. He accepted Praxagoras' theory of the empty arteries, and like Harvey, he postulated rather than observed the connections between the arterial and venous capillaries.**

He seems to have regarded the heart as a kind of double pump, the right side distributing blood and the left pneuma to all tissues of the body. This simple scheme, as Galen was not slow to point out, simply would not work, as blood from the right ventricle can

---

†De usu partium, VI 10 (K.III, 445) and Rufus of Ephesus, De nom. part, ed. Dareberg-Rulle 162. Contemporary Greek practice was to name all the vessels connected with the right ventricle veins, and all those connected with the left ventricle, arteries. This led to a good deal of terminological confusion, since from time immemorial, the windpipe and its branches had been called arteries before the identification of any of the blood vessels. Once the three different sets of vessels in the lung had been differentiated, the bronchial tubes were called the "rough" arteries, the pulmonary veins, the vein-like or smooth arteries, and the pulmonary arteries, the artery-like veins.
††C. M. Goss "The precision of Galen's anatomical descriptions compared with Galenism," Anat. Rec. (1965), 152.
†††Cf. Siegel, Galen's System of Physiology, pp. 109ff.
*S. Sisson & J. D. Grossman, The Anatomy of Domestic Animals (Philadelphia 1948), pp. 626 & 732.
**See Galen, De venae sectione, 2 (K.XI, 153).

only be pumped into the lungs through the pulmonary artery. Moreover, how does it explain the fact that if you wound or prick an artery, blood always jets out?

If the arteries contain only air, the work of distributing the blood to all the tissues must fall upon the right ventricle of the heart. But this can only pump it via the pulmonary artery into the lungs. How then can it get into the vena cava which is obviously the main stem of the venous system? Moreover, wounded arteries bleed, a fact which Erasistratus himself does not attempt to deny.

There is no reason to doubt that Erasistratus was acquainted with the fact that all muscles contain veins, arteries, and nerves; veins to provide them with food, arteries to provide them with pneuma, the breath of life, and nerves to endow them with the power of contraction.

The purpose of respiration, Erasistratus maintained, is to fill the arteries, which in this case must surely be not the bronchial tubes but the blood vessels of the same name.†

Erasistratus did realize that the expansion of the arteries or pulse wave caused by the systole and diastole of the left ventricle was a purely mechanical phenomenon. It was not due to any peculiar vital force or power of pulsation, or possessed by the arteries as such, quite independently of the heart's movements. The air drawn into the heart via the lungs became by some unexplained principle, the vital spirit. It was distributed by the heart and the arterial system throughout the body. It was this that made the body alive.

His doctrine of the triple skein of vein, nerve, and artery also enabled him to provide a plausible theory of muscular action.

Celsus has nothing new to tell us about the arteries except that, unlike Erasistratus, he accepted the fact that they contain blood, that they do not collapse when empty, and that they do not heal, indeed they cause the most violent hemorrhage. He is also acquainted with the anastomosis of the arteries and the veins. He does not appear to be at all precise about the distinction between arteries and veins, and in many places talks about throbbing veins.

As late as the 2nd century A.D. there were still doctors who believed that the arteries contained not blood but air. Galen thought it worth while demonstrating experimentally the falsity of this assumption, though this had been recognized by most medical practitioners and schools. Even after the anatomy of the heart and the working of its valves had been accurately described, there was little or no

---
†Galen, 1 (K.IV, 471).

agreement about what actually went on in the heart and the great vessels.

It is to Galen that we must look for the gathering together of all the currents of ancient medicine still existing in his day, and the combination of what he believed to be valid in them into a carefully articulated system.

So far as the arteries are concerned, Galen distinguishes them from the veins by the fact that they have two (or three?) coats, whereas the veins have only one. "The arteries," he tells us, "have two intrinsic coats, the outer [tunica aventitia] like that of the vein, the inner [tunica media] about five times as thick and harder. It consists of transverse fibres. The outer coat, like that of the veins, has longitudinal fibres, some slightly oblique, but none transverse. The inner thick hard tunic of the arteries has a woven sort of membrane on its inner surface, which can be seen in the large vessels. Some regard it as a third coat [tunica intima].†

He follows Erasistratus in regarding the heart as consisting only of the two ventricles. He, therefore, also adopts his rather awkward terminology for the great vessels connecting the heart with the lung.

It was inevitable that in his descriptions of the human vascular system he should make mistakes by attributing to the human body features he had noticed in the animals he dissected, chiefly ungulates and simians. His description of the arterial system contains mis-attributions from both these sources, but it is not always easy to be certain which. On certain occasions Galen does tell us from which animal his description has been taken. In Book XIII of the Manual of Dissection entitled "On the veins and arteries," having described the opening of the peritoneal cavity and the greater and lesser omenta, he proceeds to describe the portal system, pointing out that the detailed vascular arrangement is not the same in all animals.

"But all animals have one point in common--the veins which attach themselves and grow into the omentum, the spleen, the stomach, and the whole of the intestines come off from the vein which originates

---

†De Anat. Admin. VII Cap 5. Mrs. May commenting on a similar passage to that quoted, taken from the De usu partium VI 11 seems to have overlooked the description given in the Anat. Admin. when she remarks in her note 46 to p. 304 of her translation, Galen, On the usefulness of the parts of the body (Cornell University Press, 1968), that Galen overlooked the tunica intima.

at the porta hepatis. Accordingly we shall first begin with a summary of what there is to see of that in the apes. For this animal is more like mankind than are the remaining animals."

In the XIIIth book of the <u>Manual of Dissection,</u> Galen gives the following description of the aorta and its ascending and descending branches:

"I will start from the heart which...is the starting point for all the arteries in the body. I say that from the heart springs out an artery in the same manner as does the trunk of a tree from the earth. And this artery is the largest of all the arteries in the body... And from it these latter have their origin. First of all it divides itself into two tremendous portions which resemble the large limbs of a tree. Then each of these two divides itself into others, and it keeps on dividing frequently in this way, until all its subdivisions come to an end...just as the division of the branches of a tree finds its end in small twigs...Now on the great artery, before it breaks up into divisions, you can see two offshoots like that which emerges from the seed of a plant when it germinates. And of these two, you see that one is wider and longer than the other, and you find that it encircles the whole heart in a ring [left coronary artery], *and surrounds it at those positions at which its two cavities* [ventricles] *unite with one another and adjoin each other* [descending branch]. *But the other offshoot* [right coronary artery], *which is narrower and shorter, distributes itself, especially to those portions of the heart of which Aristotle holds the view that they are a third ventricle..."* (trs. Duckworth, pp. 172f.)

"*After the two offshoots* [the coronary arteries], he tells us: *"you see the two mighty arteries into which the aorta divides itself. Of the two, the broader is the descending branch, the narrower the ascending."* His description of the former runs as follows:

<u>The artery which goes to the lower parts of the body</u> *"mounts on the fifth thoracic vertebra in the place upon which there lies also the vein* [azygos] *which nourishes the lower thoracic region. And this artery travels with this vein and distributes itself together with it, and branches* [posterior intercostals] *sprout off from it which go to the inter-costal spaces just like the twigs of that vein. And from these arteries branches go outward, which travel with the veins to the outer thoracic regions. For their origin is beside that of the veins, and they distribute themselves similarly to those muscles to which the veins are distributed...Now when this main artery* [aorta] *pierces the diaphragm, to this, on its course, it often gives off branches of considerable size, and sometimes small branches. In the same way a branch to the left side...springs off from it, after it has passed beyond the diaphragm, at its first entry into the regions found below the thorax and within the abdomen in front*

of the kidneys. After this branch come two arteries and these are those...of which I said that they extend to and distribute themselves in the liver, spleen and stomach and in the whole mesentery with the exception of a small portion. These arteries are not coupled in pairs, but one comes off after the other. They have their origin upon the anterior part of the main artery. But all arteries coming after them have a paired origin. Of these, two very large ones go to the kidneys. And after them come two others whose origin is paired, like the origin of the veins on every single vertebra.

"With regard to the unpaired artery which has no fellow and which arises below the kidneys [inferior mesenteric]...it breaks up in the lower mesentery. Accordingly there is no need for a long account...because of the two following considerations. In the first place although there go to the peritoneum numerous veins of arachnoid or capillary tenuity, springing from many veins but mainly from those which go to the testes, with no single one of these veins do we find an accompanying artery; although we have investigated and sought it out in all animals especially in those of large size [oxen, horses, asses, mules and even elephants]." (trs. Duckworth.)

Galen now tells us he has mentioned this because some anatomists have erred in assuming that the arteries which encircle the urinary bladder connect with the aorta.

"For they do not reach to that artery but only to the obliquely directed ones [iliac arteries] which branch off from the great artery and to the lower limbs passing across the broad bone, the sacrum. From these branches break off which subsequently hang freely without entering any one of the structures lying in this place. And you will find that there are always two of these branches. On rare occasions I have found a third delicate twig barely visible [middle sacral artery]. These branches split, divide, and break up upon all the structures which overlie the broad bone. But the two arteries which result from the division of the artery lying upon the vertebrae extend as far as the upper ends of the thighs, after passing beyond the lower ends of the muscles around the joint [deep circumflex iliac arteries]. But after they have mounted the upper ends of the thighs they distribute themselves completely in all the subdivisions of the lower limbs." (trs. Duckworth.)

Another description of the descending aorta is given in the tenth chapter of the 16th Book of the De usu partium. Here Galen describes nature's wisdom in devising the location of the vessels which pass through the diaphragm, the vena cava [inferior] the esophagus and the great artery [aorta] as follows:

"Passing through the lower parts of the thorax, the largest artery ...sends off outgrowths [aa intercostales] on each side to the

regions of the intercostal muscles...the greatest portion [rami anteriores]...branches into the thorax. For it was neither safer nor shorter to bring arteries to these or to the diaphragm from any other source...Certainly it was better for the stomach, spleen and liver...to get them only from this largest artery as soon as it arrives in the parts below the diaphragm. From this same region the artery distributed to all the intestines [a mesenterica superior] arises, because the summit of the mesentery is close by, and not only the artery but also the vein [v. mesenterica superior] and nerve [plexus coeliacus] had to begin at this point to branch into all the coils of the intestines. Again, since the kidneys come next, a pair of very large arteries [aa renales] is implanted in them... The branch going...to the right kidney has its point of origin higher than that...going to the left kidney, because the position of the kidneys themselves is not the same...

"The outgrowths [aa spermaticae or ovaricae internae - for Galen regarded the ovaries as the female testicles] come next after those to the kidneys...the one arising from the left side always certainly receives something from the one leading to the kidney and sometimes even avails itself of this source exclusively...the one on the right always arises directly from the great artery, receiving in addition a contribution from the outgrowth going to the kidney...

"The next outgrowths from the great artery are those to the epigastric muscles [aa lumbales]...Along the entire course of the great artery, beginning at the fifth thoracic vertebra and extending over the whole spine, there are certain [posterior] branches of small vessels [aa intercostales, lumbales] inserted into the spinal medulla; these are generally divided into two parts and send one not inconsiderable portion back to the spinal muscles. They penetrate within the bones [the vertebrae] where these come together and where the nerves issue from within, and there are two outgrowths at each junction because there are two apertures, one on the right side of the spine, the other on the left. All these very numerous pairs of small arteries are found along the whole length of the spine. There is the same number of them as the nerves issuing from the spinal medulla and along with the veins they penetrate the thin membrane [the pia mater] surrounding the spinal medulla. Moreover, at each offshoot of an artery, the one that is their stem, so to speak, and descends along the middle of the spine, decreases in size, like the trunks of trees after branches have grown out..." (trs. May.)

With the details of the vascular arrangements of the four limbs we have no space to concern ourselves. We therefore pass on to his description of the thoracic aorta. This is described in the ninth chapter of the tenth book of the Manuel on Dissection in the following passages:

"Look immediately at that small twig which is single, unpaired, and lies beneath the lung [left bronchial artery] at the place where the aorta joins the vertebral column...Over and above the twig...you see the great artery dividing itself...Now I will point out to you here the distribution of that artery which makes its way upwards...this also, as soon as it branches off from the artery which turns downwards, divides itself. It is not yet supported by anything, but hangs suspended in the cavity of the thorax. One of its two subdivisions, that is the greater of the two [brachiocephalic trunk] inclines towards the right-hand side and passes on until it reaches the meeting places of the two clavicles with the sternum [suprasternal notch]. This place is called the throttle that is, the hollow of the neck. The other subdivision [left subclavian artery] travels to the left axilla. Should you wish to dissect whichever of these two portions you like, you can see, with regard to the arteries branching off from it, that so long as they stay within the thorax, they continu to divide up like those veins we mentioned a short while ago. But when they come outside the thorax, then notice the manner in which, at their first emergence, the veins immediately separate themselves from the arteries and turn to the region of the skin. For the parts above the thorax you will nowhere find an artery without a vein, but you will find many veins without arteries. Indeed why should we mention small veins, when we find veins of the greatest size, I mean the superficial jugular veins without arteries? These latter arise at the place at which, as soon as the greater jugular veins reach it, there sprout from them those other veins of which you have already heard an account...Afterwards those veins which lie externally to it encircle the clavicle at their first encounter with it. They are the ones...which...after they have mounted upwards to the neck and have united and combined themselves with the other branches from the deep jugular veins, are almost all of them accompanied by arteries...Of those veins which are connected to the muscles, there is no single one without an artery...The artery...that ...comes obliquely upwards to the hollow of the neck, when it meets with the vena cava, follows its example by dividing itself into two, so that there arise from it the [two] arteries which are called the carotids...These then pass upwards alongside the two jugular veins until they reach the skull bone. But, as soon as they get near to this, they part company [with the veins] and penetrate the cranial bone as far as the brain. For their entry [vein and artery] does not take place in a single foramen but in two. The jugular veins make their way through a separate foramen peculiar to them, other carotid arteries through another foramen pecular to them, lying in front of it, I mean higher up. The foremen of the veins is common to both of them and to the components of the sixth pair of nerves arising from the brain [Nn glossopharyngeus, vagus, accessorius]. And the foramen of the arteries belongs to these, together with the nerves combined with them, throughout the whole of the neck [sympathetic trunk]...But from the ends of the arteries, after they have *pierced the skull, immediately upon their first entry and passage into its cavity, there proceeds the so-called 'reticular network' ...Out from the rete mirabile, again there ascend two arteries which encircle the brain, together with the delicate membrane [pia mater] after the fashion of a girdle, just as do the veins as we have described...in the anatomy of the brain...*

"*But as for the artery which goes to the left shoulder, to the axilla and the arm, no branch at all comes from it, but it distributes itself with the veins which are found in these parts. Further, it does not distribute itself with all these veins, but only with those which nourish the muscles. For the superficial veins beneath the skin are all without accompanying arteries.*" (trs. Duckworth.)

Another description of the thoracic branch of the aorta is given in the De usu Partium XVI.

In Chapter I Galen says:

"*...In the free space in the left side of the thorax the largest of the arteries* [the aorta]*...first emerges obliquely, but then immediately divides, the larger part being supported down along the spine and the smaller passing up to the clavicle. Here nature distributes one part of it* [a subclavia sinistra] *to the scapula, the arm, the left side of the neck, and whatever other parts are situated in the region: the other part* [truncus communis] *she extends up along the sternum and divides again into two unequal parts, making one on the left which is smaller a carotid artery* [a carotis communis sinistra] *and extending obliquely, the larger one on the right* [aa anonyma and subclavia dextra]*, from which, as it advances a little way many outgrowths are given off. For a certain artery* [a thoracalis suprema] *goes to the upper part of the thorax; another* [a choracica interna]*, descends along the sternum to the right breast; and before these, to be sure, the right carotid* [a carotis communis dextra] *grows off rising steeply. Then the oblique remainder of the artery* [a subclavia dextra]*, after reaching the place where the first rib grows out, is distributed to the scapula, the arm and the parts on the right side of the neck.*" (trs. May.)

"*The one on the left could not possibly be bent at the place where the left common carotid first grows out, for the part of the great artery* [truncus communis, a. anonyma] *which ascends along the sternum, and from which the carotid is split off almost straight, being* [only] *slightly inclined towards the right side of the whole thorax. The other offshoot* [a subclavia sinistra] *of the ascending artery, the offshoot that goes to the left scapula and arm, has nearly the same position; for this too as a whole is almost straight being* [only] *slightly inclined toward the left arm. There remains for the nerve then, one available turning point, the stem itself of the* largest artery wonderfully constructed for the nerve to use, not only because of its size, but also because of its strength and position." (trs. May.)

The description of the thoracic aorta given in Chapter 10 is as follows:

"There is a certain very large vessel [the aorta] which grows out from the left ventricle of the heart like a trunk and is distributed into the whole body. Immediately after it grows out from the heart, this very large vessel divides into two parts, one of which bends down along the spine in order to send arteries to all the parts below, whereas the other passes up to the head to furnish branches of vessels to all parts above the heart...the division is an unequal one because there are more of the animal's parts below the heart than above, and the descending portion of the artery is as much larger than the portion ascending to the throat as the number of the lower parts exceed that of the parts above...

"The descending portion of it is fixed at a point just opposite the place where it grows out, inclining neither to one side nor to the other, but going by the straightest...route it mounts upon the 5th thoracic vertebra. The ascending portion, as soon as it has grown out, immediately gives off a certain part of itself [a subclavia sinistra] which extends to the left scapula and axilla. This carried upon the lung and supported by [mediastinal] membranes passes up as far as the first rib without dividing...In that region accordingly it sends one part [truncus costocervicalis] of itself to the first inter-costal spaces, and another [a thoracis interna] extends beneath the whole sternum to the hypochondrium and breast, and a third [a vertebralis] goes to the cervical part of the spinal medulla, passing through the apertures of six vertebrae after sending off offshoots along the way to the muscles in the vicinity. The remainder of this artery is distributed to the entire left arm and to the scapula. The other, larger part [truncus communis] of the whole ascending artery, attaching itself as soon as possible to the bone at the middle of the chest, passes straight up to the throat from the place where it is given off." (trs. May.)

In another passage in the same chapter, he says:

"Examine closely the region also where the different parts of the artery first mount upon the bones; for you will see not only that a bone has been prepared as a bulwark and foundation for each part but also that...a membrane has been placed under...the descending aorta, that the cartilage which lines the inner [ventral] parts of the vertebrae has been prepared like a soft bedding for it, and that the other vessel [truncus communis] which ascends to the throat, has a very large, soft gland [the thymus] placed beneath it like a couch...

With the vena cava passing up below and with the oesophagus and vein [v azygos]...passing down from above, it was not fitting to overlook the safety of these parts...First, though [the Creator] could have attached the oesophagus to the sternum, and the vena cava to the spine, he did the opposite. For the spine is closer to the oesophagus than the sternum is, and the sternum is nearer the vena cava than the spine is, since the oesophagus, from its beginning, is carried through the whole neck by being mounted on the vertebrae, and the vessel which passes up from the right auricle of the heart, and is continuous with the [inferior] vena cava...is near the sternum. And it was better to make the bone closer to each part its bulwark, rather than to...bring a suspended vessel back through the whole breadth of the thorax to the opposite side. Then, too, another advantage has resulted for each of the parts from this position...The oesophagus goes in a straight line with the stomach lying upon the spine and receiving it and this is not compelled to leave the thorax through the middle of the diaphragm, where there is already a necessary aperture giving passage to the vena cava, and when the vein reaches the throat and is associated with the artery from the heart, it readily obtains a suitable situation. The artery also is at once kept in such a position that as the vessels pass through the neck and divide, the arteries are in the depths and the veins lie upon them.

"Furthermore it was not only by locating the oesophagus, the artery [aorta] and the vein [v azygos], nourishing the lower parts of the thorax, upon the spine and by extending the vena cava beneath the sternum, that Nature arrived at the best results; for she also avoided placing the oesophagus, artery, and vein one upon another, or putting the oesophagus in the middle, and the artery at the side. Rather, she extended the artery along the middle of the vertebrae and the oesophagus alongside it. For the artery obtained a position much safer than that of the oesophagus, as its importance for life is the greatest." (trs. May.)

After describing the abdominal aorta and its branches, Galen returns (in Chapter 11) to the upper portion and connections of the great artery, which is distributed from the heart to the neck, scapulae, arms, face, and the whole head: *"like the descending artery,"* he tells us, *"as it passes through the thorax it sends off branches to the intercostal muscles, the spinal medulla, and the parts outside the thorax, and in addition to these there are outgrowths which go to the breasts...and others going to the scapulae and the arms.*

"On each side one artery [a. carotis communis], the part remaining after these outgrowths, passes up to the head and all parts of the face and neck, are interwoven with branches of these vessels. The spinal muscles receive outgrowths from the vessels branching to the scapula, and from these vessels as soon as they emerge from the

thorax into the neck, offshoots [aa vertebrales] *pass through the lateral apertures of each of the first six vertebrae as far as the head. In fact the artery is no longer extended along the vertebrae themselves...for it is most necessary that the muscles drawing the head forward should be placed there. Moreover the oesophagus and the rough artery* [trachea] *in front of it must be placed upon the vertebrae...Nature...pierced each outgrowth* [transverse process of the vertebra] *with a regular round hole and made paths for the vessels out of the rows of apertures--the intervals between the apertures where the nerves emerge from the spinal medulla are not large--the small offshoots* [rami spirales] *from the artery penetrate into the spinal medulla...When the vessel* [a vertebralis] *ascending to the head has passed the first vertebra it divides into two terminal branches one of which grows inward to the hinder part of the encephalon; the other is distributed to the muscles surrounding the joint of the head and is attached to the ends of the vessels located in the thin meniux* [pia mater]. *Outgrowths from the vessel* [a superscapularis?] *of the scapulae are interwoven through the superficial muscles and the skin."* (trs. May.)

After dealing with the arterial arrangements of the axilla and the arms, Galen proceeds in Chapter 12 to give a brief description of *"the remaining pair of arteries"* called carotid which, buried deep in the neck, go straight up to the head:

"Each of the [common] *carotid arteries divides into two parts, one* [a carotis interna] *going more toward the back, the other* [a carotis externa] *more toward the front, and each of these parts again divides into two. The one branch* [a facialis] *from the anterior part arrives at the tongue* [a ingualis] *and the inner muscles of the lower jaw* [a submetalis], *and the other* [the main stem of the external carotid], *though placed nearer the surface than this first branch, nevertheless is also covered by large glands* [gl parotis] *and passes up* [a temporalis superficialis] *in front of the ears as far as the temporal muscle. Here it branches and its posterior part* [the parietal branch] *goes up to the crown of the head, the ends of the vessels on the left side of the head anastomising freely with those of the other side, and the ends of the inner* [deep] *vessels anastomising with the outer* [superficial] *ones.*

"The other part [a carotis interna]...*also divides into two branches, very large, but of unequal size, the smaller one* [a condyloidea] *goes up to the rear into the base of the parencephalis* [cerebellum], *being received by a large aperture* [the hypoglossal canal] *situated at the lower end of the lambdoidal suture. The other* [the main stem of the internal carotid] *passes up anterior to this* [first branch] *through the aperture* [the carotid canal] *in the petrous bone and goes to the retiform plexus."* (trs. May.)

This arterial structure, which does not exist at all in man but does exist in ungulates, is a very complicated network which occupies the same region as the circle of Willis does in man. Its chief source is the internal carotid artery. In ruminants it also receives branches from the vertebral and condyloid arteries--but not so in the pig.

In the 4th chapter of the ninth book of the De usu partium, this arterial network is described as follows:

"It encircles the gland [hypophysis] itself and extends far to the rear, for nearly the whole base of the encephalon has this plexus lying beneath it. It is not a simple network but [looks] as if you had taken several fishermen's nets and superimposed them. It is characteristic of this net of Nature's, however, that the meshes of one layer are always attached to those of another, and it is impossible to remove any one of them alone; for, one after another, the rest follow the one you are removing, because they are all attached to one another successively...Nature appropriated as the material for this wonderful network the greatest part [a carotis interna] of the arteries ascending from the heart to the head. Small branches are given by these arteries to the neck, the face, and the external parts of the head. All the rest of them, as straight as they were formed at the beginning, pass up through the thorax and neck to the head and are received there comfortably by a part of the cranium which is pierced through [by the carotid canal]...the thick meninx [dura mater] too was about to receive them and had already been pierced through along the line of their invasion, and all these things gave the impression that the arteries were making haste to reach the encephalon. But this was not the case, for when they have passed beyond the cranium, in the space between it and the thick meninx, they are first divided into very small slender arteries, and then they are interwoven and pass through one another, some towards the front of the head, some towards the back, and others to the left and right, giving the...impression that they have forgotten the route to the encephalon. However...as roots combine to form a trunk so from these many arteries there arises another pair of arteries [aa carotides cerebrales] equal to the pair that passed upwards in the beginning, and so these now enter the encephalon through the performations in the thick meninx." (trs. May.)

We have no space to examine in detail Galen's more rather than less accurate topography of the arterial arrangements of the limbs. Nor can we examine his equally competent account of the topography of the venous system. The important thing to remember is that he knew that all muscles contain arteries, veins, and nerves. He had a very fair idea of the blood vessels serving the principal organs like the liver, the lungs, the spleen, and the stomach and intestines, the pudenda, and some of the larger glands. It remains to examine his conception of the arterial system, first in its relation to the venous system, secondly to the respiratory, and thirdly to the nervous system. Throughout the body, Galen and many of his predecessors had observed that veins are, in most regions, almost always accompanied by arteries. In his great Manual on Dissection he devotes a chapter (in the 13th book) to enumerating the exceptions to this rule. He postulated that the terminal branches of the trunk of the arterial tree branching off from the aorta were all connected with the terminal branches of the venous tree branching out of the vena cava by invisible capillaries. But he refused to acknowledge the fact that the right ventricle of the heart was the starting point of the venous system. For physiological rather than anatomical reasons, he insisted that the trunk of the venous tree was the hepatic vein connecting the convex part of the liver with the heart. This doctrine was not just an amiable eccentricity of a hyper-intellectual physician. It was shared also by another most distinguished Graeco-Roman physician, Aretaeus of Cappadocia, whose description of diabetes and cardiac collapse are among the finest in Ancient Medicine which have survived. The aorta springs directly from the left ventricle and then immediately divides into its two great branches, but the hepatic vein is the trunk of the venous tree, rooted in the organ where blood is produced. This, just before reaching the heart, divides into the two main branches of the vena cava, vc superior and inferior. At the division of these branches there is a sort of backwater, as it were, which constitutes the left auricle and terminates in the tricuspid valve which opens into the right ventricle. In this way he contrives to make a large proportion--indeed probably the largest--of the blood made in the liver bypass the heart altogether and to distribute itself to the whole of the body except the lungs. At diastole the right ventricle draws blood into itself, some of which at systole it propels into the lungs and some directly into the left ventricle.

Galen knew that there are three "trees" of vessels in the lung --the bronchial system, the pulmonary arteries, and the pulmonary veins. He is not aware of the existence of the alveoli, but he postulates the interconnection of the terminal branches of all these three "trees." The terminal bronchiole is connected with the terminal pulmonary arteriole as well as with the terminal pulmonary venule. He never explains the exact nature of their function, but he is forced to assume that the terminals of the bronchioles have a smaller lumen than those of the arterioles.† This is because the terminal "twigs" of the pulmonary arteries are inserted into rough arteries, the bronchioles, not the smooth arteries [pulmonary veins].

---
†De usu partium, VII 9.

Galen devised a simple experiment to prove that the arteries contained blood. He tightened an artery in two places and then divided it between the two ligatures--blood of course came out of it.†

Though Galen will not accept the observation, attributed to Paraxagoras, that an excised artery will go on pulsating, he insists, nevertheless, that its power of pulsation, though derived from the heart, is transmitted by the heart to the arterial coats, and is something quite independent, belonging to the arteries themselves. And the power which works both the beating of the heart and the movements of the arteries, is the vital and not the psychic power. Its instrument, therefore, is the vital not the psychic pneuma. A man's heart can go on beating, thus keeping him alive, though he may have lost consciousness for days, or weeks, or even months, provided that the heart and the arteries continue to function. The function of breathing, and arterial pulsation, besides cooling the heart, produce the following effect on the innate heat: they ventilate its hearth (the left ventricle) by causing a complete evacuation of smoky wastes. This double action is called "safeguarding," SOTEERIA, or preservation. We must, I think, in spite of some of the passages to the contrary, continue to believe that Galen probably did admit that some of the air taken in by the lungs does, in fact, find its way via the left ventricle of the heart into the arterial system. For the rate of respiration is normally much slower than that of the systole and diastole of the heart, so that it is difficult to see how the air breathed into the heart by the lungs at diastole of the left ventricle, or at any rate some of it, can fail to be expelled at systole.

How did Galen come to believe that such a structure as the <u>rete mirabile</u> existed in them at all? He could certainly not have found anything like it in the ape--with the brains of which he must have been quite intimately acquainted. He tells us that it was first described by Herophilus, who had, according to unimpeachable tradition, dissected human bodies. He must, therefore, have been prepared to accept Herophilus' attribution of this structure to the human brain. Moreover, Galen in the course of his dissections of various animals, had no doubt dissected many oxen, a species in which, as Sisson and Grossman have demonstrated,†† this anatomical feature is most prominent in the form of a network intercalated in the course of an artery, not only by the middle meningeal artery which enters the cranial cavity through the <u>foramen ovale</u>, but also by *"branches which take the place of the internal carotid."* These enter the cranial cavity through the foramen <u>orbito-rotundum</u>; they *"concur with the*

---

†This experiment Galen described in a special treatise, <u>An in arterits natura sanguis contineatur</u>, 6 (K.IV, 724).
††S. Sisson and J. D. Grossman, <u>The Anatomy of Domestic Animals</u> (Philadelphia, 1948).

branches of the occipital, vertebral, middle-meningeal and condyloid arteries in the formation of an extensive rete mirabile cerebis on the cranial floor, around the sella turcica."

The veins, some large and some small, which pass upward into the diploic areas of the cranium and the membrane of the pericranium go downward into the underlying thin membrane, the pia mater. The folds [sinus transversi] of the membrane that conduct the blood come together at the crown of the head into a sort of cistern which Herophilus used to call the wine vat--the celebrated torcular Herophili. From this point, as from an acropolis, they send out conduits to all parts lying beneath, some leading from the torcular into the whole parencephalus via the occipital sinus ramifying, like the irrigation conduits in a garden; others [sinus occipitalis] send down then to the parts beneath. Nature did not entrust the remainder of the blood to a single vein, but constructed the [sinus rectus] from the part of the dura mater that extends anteriorly, with many outlets. When this aqueduct approaches near the middle (third) ventricle, there come off from it the large veins [v cerebris magna, vv cerebrae internae] to be distributed in the choroid plexuses.

Galen's actual description given in the ninth book of the De usu partium, Chapter 2, runs as follows:

"For the whole encephalon [brain] is interwoven with intricately divided arteries, many of whose branches end in its ventricles, just as many of the veins do that descend from the crown of the head. Coming from the opposite direction, they encounter the arteries and are distributed, as the arteries are, into all parts of the encephalon, both into the ventricles themselves and the other parts as well. Now just as many arteries and veins extend to the stomach and intestines and pore out bile, phlegm, and other such humours into the free space outside themselves [the cavities of the alimentary canal], *while retaining within themselves the blood; and vital pneuma, so in the same way the veins expel residues into the ventricles of the encephalon while retaining the blood; but the arteries most of all breathe forth the* [vital] *pneuma; for they come up from the parts below, whereas the veins descend into the encephalon from the crown of the head. Nature having marvellously made this provision too, in order that the substances escaping from this orifice may penetrate the whole encephalon. As long as they are contained in the vessels themselves, these substances travel with them into all parts of the body, but when they have once escaped from the vessels, each moves according to its own proper weight, the thin, light material passing up and the thick, heavy material down.* (trs. May.)

We pass now to the consideration of the most obvious characteristic activity of the arterial system, namely the pulse.

The first Greek physician to work out a classification of the different kinds of pulses was Herophilus. He was also the first to recognize arterial pulsation, not just as a symptom of inflammation, but as a normal physiological process.

Herophilus, unlike Praxagoras, did not believe that arteries could pulsate all on their own. He insisted, as Galen did after him, that the power of pulsation comes to the arteries from the heart, but was not, as Erasistratus had assumed, merely the purely mechanical effect of the expulsion on systole of the contents of the left ventricle.

Erasistratus appears to have been less interested in the pulse than Herophilus. Unlike Herophilus, he refused to attribute to the arteries any expansive or contractive power of their own. They expand only because pneuma is forced into them by the left ventricle, the quantity and density of the pneuma contributing to the strength of the pulse. Moreover, he taught that the expansion of the different portions of the artery was not simultaneous but successive. Erasistratus was perhaps the only physician of ancient Greece who had any adequate conception of the strength of the power of the heart's contraction.

If the ancients knew very little about heart disease, there are two vascular disorders with which they were acquainted, to both of which they applied surgical treatment. One of these was varicose veins and the other aneurysm. Galen was acquainted only with aneurysms resulting either from an injury or a wound. He never seems to have observed any other cases, which may seem to us less strange when we remember, first, that he never dissected the human body, and secondly, that until recently by far the commonest cause of aneurysm was syphilis. (Another argument which would seem to support the view that this disease was unknown in the ancient world.) Galen describes aneurysm as follows:

"it is a morbid dilation of an artery which occurs as the result of a wound, when the skin covering the vessel heals and forms a scar, but the wound in the artery itself remains and does not cicatrise completely or become sealed up with flesh. The condition is diagnosed from the beatings of the arteries concerned. When pressure is applied the swelling disappears completely, the whole of the blood which caused it being pushed back into the artery.†

He also tells us how he succeeded in healing an arterial wound, made by a careless young doctor who had cut an artery instead of a vein, without causing an aneurysm. He explains how death can result

---

†De tumoribus praeter naturam II (K. VII, 725).

from a wound in the artery lying under the vein on the inner side of the elbow, which was often opened in phlebotomy. He also describes one case of "gangrene" arising as the result of the premature imposition of a ligature, as well as other deaths resulting from an attempt to operate upon the aneurysm itself. All operations on aneurysm entail the interruption of the blood flow by ligature. But even if the artery is one of the larger ones, the wound in it can be healed without forming an aneurysm, if the artery be completely divided, the ends on both sides of the cut being pulled, the one upward and the other downward. In this way not only the formation of an aneurysm, but also the danger of hemorrhage can also be avoided.

Many of Galen's contemporaries maintained that it was impossible to heal a wound in any artery, but Galen discovered that this was not entirely true since he himself had succeeded in healing a wounded artery. He approximated the edges of the wound in the artery and then bound the wounded vessel with a plaster consisting of a sponge soaked in hemostatic drugs. This was kept over the wound for three days and when removed on the fourth, revealed the wound perfectly sealed up. A fresh application of drugs was used, and the cut healed perfectly without producing any aneurysm.

Galen explains that the healing is like that of a broken bone--bone is also originally produced from parental semen, which is irreplacable. Its healing cannot be accomplished by joining the fractured ends of the lost bone-tissue, and getting nature to fill up the crack with new. Nature, however, is not without resources. She produces a glue-like secretion in both cases which unites the opposed edges of the wound or fracture.

EMBRYOLOGY OF THE HUMAN ARTERIAL SYSTEM (ARTERIOGENESIS)

W. Pallie, M.B., D. Phil. (Oxon)

Chairman and Professor, Department of Anatomy, McMaster
University Medical Centre, 1200 Main Street West, Hamilton,
Ontario, Canada, L8S 4J9

In phylogenetic terms, the circulatory system evolved in response to the metabolic activities generated by a progressively enlarging metazoan that on reaching a critical cell mass, required physical pathways of ingress and egress for nutrients and metabolic products respectively. Without such the "mileau interne" could not be kept stabilized.

The first three weeks of life of the human zygote (Hauser and Corner, 1957) is sustained in the absence of a contained circulatory system. By the end of this time it has grown to be a trilaminar blastocyst of 5 mm diameter (the embryonic disc itself being about 3 mm in length). Prior to implantation (7th post-fertilization day), nurture is provided by fluids of the fallopian tubes and uterine cavity. Thereafter, the progressive erosion of the maternal endometrium or decidua provides fluid uterine debris, and is assisted later by maternal blood juxtaposition for the establishing chorionic vesicle. During this time the heart has begun to develop, commencing in the third week as have the "blood islands" (of Pander) in various regions. But though the heart has begun to beat at the beginning of the fourth week (22nd day), unidirectional pulsatile flow and circulation commences only toward the end of the fourth week (26th day). Thus, the human embryo reaches a maximal length of 4 to 5 mm before a functional circulation comes into being (Streeter, 1942).

The fifth week embryo, of over 5 mm maximal length, has a primitive heart with pulsatile contraction and vascular circuits in the yolk sac, body stalk (to chorion), the branchial region and the rest of the embryo. The chorionic sac itself is about 2 cm in diameter at this time.

With the increasing complexity of structural development of the embryo during the fetal period, the fetal blood vessels develop pari passu. Continuous adaptations in them reflect the modifications that become necessary, together with the recapitulation of phylogenetic patterns residual in human ontogeny. Throughout, the circulation must go on incessantly, as no stoppage can be provided during alterations, i.e., adjustments must be made coincident with circulatory function.

The substantive fetal circulation and cardiac development is accomplished by the 8th week. Thereafter, adaptations to fetal growth continue until birth. At this time, the heart in relative proportions is large. It remains so to cater to extraembryonic tissues (placenta) from which circulatory relief is only obtained at birth. A further point of consequence in growth adjustments is the descent of the heart from the neck (at the 4th embryonic segment) to the thorax (to the 17th embryonic segment). This is an embryonic migration it shares with the developing septum transversum, the major component of the adult diaphragm. Postnatal adjustments become the final chapter in major (and critical) adaptations in the pattern of the circulation. In these are involved the shutdown of the ductus arteriosus, the ductus venosus, and the umbilical vessels to establish what will become the postnatal and adult pattern for the circulation.

A chronological listing of embryonic stages and landmarks of cardiovascular development has been put together (Table 1). Steeter's horizons (Heuser and Corner, 1957; Streeter, 1942, 1945, 1948, 1949, 1951) have been adapted for the early weeks. It is useful to recall the words of Streeter in regard to the difficulty with early categorizations ". . . *the term is used to emphasize the importance of thinking of the embryo as a living organism which in its time takes on many guises, always progressing from the smaller and simpler to the larger and more complex.*" We must also remember the approximation in size, for in regard to very young embryos it has been said that "*simple statements of length are practically meaningless*" (Bartelmez and Blount, 1954).

Functionally significant stages in the development of the circulatory system may be identified as follows:

1. Pre-circulatory system stage (3 weeks duration).
2. Primordial circulation (5 weeks duration).
3. Development of substantive fetal circulation (30 to 32 weeks duration).
4. The fetal circulation and birth adjustments ("shut downs" or "transitional circulatory adjustments" in the first postnatal weeks).
5. Neonatal establishment (6 months and postnatally).
6. Substantive "adult" pattern.

TABLE 1

CHRONOLOGY OF PRENATAL DEVELOPMENT (For the C.V.S.)

Descriptive	Somite	Horizon (Streeter)	Size	Weeks	Ovulation Age (Days)	Cardio-vascular System
1 celled ovum		I		FIRST		
Segmenting zygote		II				
Free blastocyst		III				
Implanting blastocyst		IV				
Implanted, but avillous		V		SECOND		
Primitive villi, distinct yolk sac		VI				
Branched villi, embryonic axis defined		VII				
Hensen's node and primitive groove		VIII		THIRD	18	Blood islands appear
Neural folds, elongated rotochord	1	IX	disc 1-2 mm		20 to 21	2 endothelial tubes, myocardial cells

TABLE 1 (cont.)

Descriptive	Somite	Horizon (Streeter)	Size	Weeks	Ovulation Age (Days)	Cardio-vascular System
	4	X		THIRD		1 Heart tube -- atria, ventricle and bulbus cordis. Myocardial contractions
Neural tube closing	7		1.5-2 mm (5 mm)	THIRD	22 ± 1	Bulbo-ventricular loop -- external cardiac prominence.
Head folds	12			THIRD		Yolk sac haemopoiesis and angiogenesis.
	13	XI	3 mm	FOURTH		Intra embryonic haemopoiesis "ebb and flow" circulation Pharyngeal arches begin (1st aortic arch)
	20			FOURTH		
Arm buds appear	21	XII		FOURTH	26	Vascularization of neural tube.
	29		3.5 mm (15-20 mm)	FOURTH		Tubula chambered heart Venous system develop- ments - cardinal, vitelline, hepatic "Through" circulation (i.e., heart contracts blood flows)

TABLE 1 (cont.)

Descriptive	Somite	Horizon (Streeter)	Size		Ovulation Age (Days)	Cardio-vascular System
	30 (total ±42)	XIII	4-5 mm (20-30 mm)	FOURTH	28	Axial limb vessels begin Primitive vascular plan of paired vessels & undivided heart tube - aortic arches and - cardinal veins - vitelline and body stalk circulation
		XIV	5.5 - 7 mm (30-38 mm)	FIFTH	28-30	Cardiac septation begins with Septum primum Body mesenchyme and blood vessel haemopoesis Interventricular septum
		XV	7-9 mm		31-32	Hyaloid artery enters lentiretinal space through retinal fissure Foramen secundum Endocardial cushions appear
		XVI	7-13 mm		33 ± 1	3 ventral aortic branches coelias and mesenteric are single trunks

TABLE 1 (cont.)

Horizon (Streeter)	Size	wks.	Ovulation Age (Days)	Cardio-vascular System
XVII	11-14 mm	FIFTH	35 ± 1	Bulbus incorporated into right ventricle Heart achieves definitive form externally
XVIII	12-17 mm		37 ± 1	Perilental blood vessels are prominent
XIX	16-19 mm	SIXTH	39	Endocardial cushions fused together Liver haemopoiesis
XX	18-23 mm		42	Interventricular foramen closed
XXI	22-24 mm	SEVENTH	43	Aortic arched transformed Septum secundum appears
XXII	23-28 mm		45	Bergmeister's papilla is present in some optic discs

TABLE 1 (cont.)

Horizon (Streeter)	Size	wks.	Ovulation Age (Days)	Cardio-vascular System
XXIII	27-31 mm	SEVENTH	49	Stem of pulmonary vein absorbed into left atrium. Cardiac septation completed.
	35 mm	EIGHTH	56	Main vascular channels established as in adult. Specialization of conducting system of heart.
	40	NINTH	63	Lymph system develops. Spleen, thymus and lymph gland haemopoiesis.
	45	10th	70	
	50	11th	77	
	60	12th	84	Layers of blood vessel wall formed. Red blood cells become non-nucleate.

4th month – bone marrow haemopoiesis

TABLE 1

Wt. in gms	Size (C.R.)	Age* Mths.
11-135	61-100	4
50-350	101-150	5
260-800	151-200	6
490-1200	201-260	7
1030-2850	261-320	8
2250-3200	321-390	9
2750-3800	391-450	10

*Based on previous reports, in Hamilton and Mossman's Human Embryology, 1972.

EMBRYOLOGY OF THE HUMAN ARTERIAL SYSTEM

1. THE PRE-CIRCULATORY SYSTEM PERIOD (PRE-SOMITIC STAGES, STREETER HORIZONS 1 TO IX)

During this period, the conceptus passes through phases of sustenance as follows:

a. The zygote - by deutoplasm (or yolk) [Streeter horizon I].
b. The morula - by deutoplasm and fluid environment [Streeter horizon II and III].
c. The implanted blastocyst - by decidual cells and maternal venous sinuses [Streeter horizons IV and V].
d. The chorionic sac and embryo - by "blood islands" that forge links through intra and extraembryonic mesenchyme connecting villi via the body stalk with a vitelline system [Streeter horizon VI].

The earliest primordia of the blood vascular system are cell clusters ("blood islands") arising in the yolk sac between the splanchnic mesoderm and entoderm. The "angioblastic theory" held that all vasculogenesis" proceeded from these specific islands alone (His, 1900, Sabin, 1917). This has been disclaimed by the "local origin" theory, which proposed that mesenchyme throughout the embryo had the potential to produce angioblastic tissue (Huntingdon, 1914, McClure, 1921) and experimental proof was generated to substantiate this effectively (Reagan, 1915; Jolly, 1940).

Angiogenetic tissue or "blood islands" develop as isolated cellular cords ("angioblastic cord") (Bremer, 1914) which canalize ("angioblastic cysts") in the second week and become confluent by their lumina to form networks that extend further by endothelial budding and linkages with other islands of development [Streeter horizon VII to IX - the end of the pre-somitic period].

A scheme of angiogenesis may, thus be represented as follows:

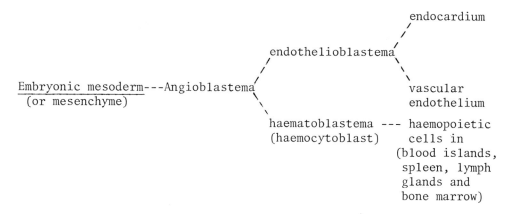

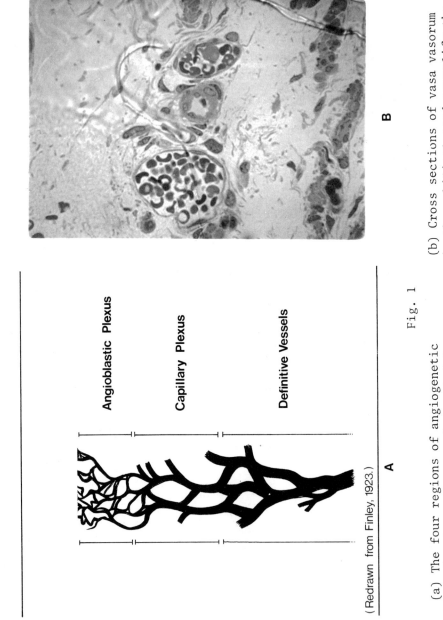

Fig. 1

(a) The four regions of angiogenetic growth progression are demarcated in the diagram.

(b) Cross sections of vasa vasorum in the umbilical cord exemplify the sprouting capillary zone.

At this stage, the developing heart excepted, the entire primitive vascular system consists of simple endothelium forming tubular plexuses. When circulation occurs, new vessels arise as outgrowths of these pre-existing vessels (Arnold, 1871; Clark and Clark, 1909; Clark et al., 1931; and Sabin, 1922). How this may occur has been suggested; tips of endothelial processes may force their way into and through differentiating tissue (Fig. 1); vessels may lengthen by true endothelial division and sprouting, or the tips of a growing plexus may exert enough stimulus to adjacent mesenchymal cells to induce them to become angioblasts (Finley, 1923). In this connection it is worth recalling that similar endothelial sprouting behavior was seen in vivo in rabbit ear chambers (Clark et al., 1931), in the tail of frog larvae (Clark, 1918) and in solid capillary buds of the vasa vasorum (Feinberg, 1971).

Fetal hemopoiesis begins within these islands in the developing lumina of the vessels in the embryonic mesenchyme of the yolk sac, body stalk, and body wall in the 3rd to 6th weeks. At the end of this period regional nets of angioblastic tissue have appeared--vitelline in the yolk sac, umbilical in the body stalk area, and paired endothelial tubes in the cardiogenic region. The cardiogenic area (a horseshoe shaped area lying in the forward tip of the anterior zone of the embryonic disc) appears on the 18th to 19th day (Streeter, 1942). It consists of a plate of mesoderm. The "cardiogenic plate" is situated ventral to the mesodermal horseshoe shaped primordial pericardial cavity that is continuous posteriorly with the extra embryonic coelom on each side (Fig. 3).

The cardiogenic plate area transforms into "two heart cords" side by side. Canalization of these (Streeter, 1942, 1945) produces the two "primitive endocardial heart tubes" (Fig. 2). The bilateral origin and migration of cardiogenic cells by labeling techniques in the chick have been described (Rosenquist and de Haan, 1966).

2. THE PRIMORDIAL CIRCULATION (Commencing in the 4th week - Streeter Horizon X)

[A] The Undivided Heart Tube - (And Paired Symmetrical Vessels)

The primordial circulation is established in the span of one week by angiogenesis and cardiogenesis during the fourth week. The two endocardial tubes begin to fuse at the cranial (arterial) end first, and fusion progresses caudally as the lateral folds develop (Figs. 2 and 3). The tubes link up with paired embryonic blood vessels already formed in situ, in the yolk sac mesoderm (vitelline vessels), the body stalk (umbilical vessels), and body wall mesenchyme (aortic arteries and cardinal veins) (Fig. 3). "Blood islands" are the solid mesenchymal masses in the yolk sac and body wall mesenchyme, and are the precursors that transform into canalized tubes lined by flat endothelial cells that contain angioblasts

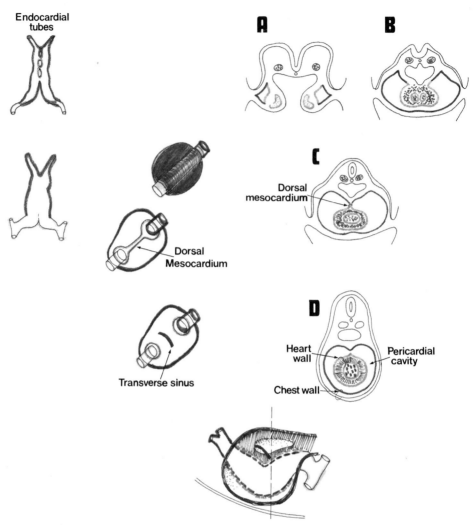

Fig. 2 - <u>The Early Heart Tubes</u>. On the left are illustrations of the bilateral origin of the longitudinal tubes. Successive pictures illustrate the dorsal mesocardium and its absorption to produce the septum transversum. The letter-bearing illustrations are cross-sections. A - The bilateral symmetry of heart tube and coelom is shown. B - The approaching fusion of the tubes, the contribution of the coelomic layer to the serous pericardium and the form of the coelomic cavity is shown. C - The dorsal mesocardium. D - The heart wall components, pericardium and chest wall formation - at a level indicated by the dotted line on the lowermost picture.

which produce embryonic blood cells within the lumen. By the 22nd day (bar the sinus venosus) is a single tube. During the early part of this (4th) week (Streeter horizons X and XI) i.e., the early somite stages, the myocardial cells contract or the primitive heart has begun to beat. The vascular system is elsewhere a mesh of simple endothelial tubes, and within it an ebb and flow type of circulation (extrapolating from the chick) exists. The fluid within these vessels contains relatively few cells. In the words of Streeter (1942), "... *the embryo is adequately maintained by a simple system of endothelial channels that cruise through the deeper lying tissues, and that in these channels an almost cell-free fluid ebbs and flows owing to the action of the pulsations of the primitive myocardium . . .*"; this provides "*over and above the diffusion already secured by the services of the coelomic fluid that irrigates the various areas of the coelomic tract.*"

The development of the "head fold" (24th day) reverses the heart beneath the embryonic disc--a rotation of 180° in the saggital plane (Fig. 3). Thence, the heart becomes caudal to the buccopharyngeal membrane, (the buccopharyngeal membrane is itself, now caudoventral to the head and brain) and ventral to the foregut. The cardiogenic plate, originally ventral to the pericardial cavity lies dorsal, and now a dorsal mesocardium exists (true for all mammals) and none is present ventrally, i.e., as the heart tube elongates it bends and gradually sinks into the dorsal wall of the pericardial cavity, as it does so, becoming suspended by the dorsal mesocardium (Fig. 2).

The primitive tubular heart wall thickens. The external layer i.e., splanchnic mesoderm becomes the myoepicardium (myoepicardial mantle), the inner cells ("cardiac jelly") generating the myocardium (cardiac muscle cells), in which contractions begin by the 26th day; ultrasonic records of contraction have been obtained on the 32nd day (Kratochwil and Eisenkut, 1967).

The heart tube is slung in the pericardial cavity at arterial and venous ends by the dorsal mesocardium. The mesocardium fenestrates at the 10 somite (Streeter horizon X) stage and is completely absorbed by the 16 somite stage (Streeter horizon XI). This resorption produces the transverse sinus of pericardium, separating the arterial from the venous end of the heart (28th day, Streeter horizon XIII) (Fig. 2).

At the end of the first month, the heart is now a simple muscular tube (as primitive type adult hearts are) and the vascular system consists of paired, symmetrical vessels. At this stage within the embryo "provisional capillary plexuses" are present. They are the sites where major vessels will be formed (Aeby, 1868; Evans, 1909). This plexus is generated selectively by influences that are "angiotactic," chemical, metabolic or by "growth activity." The

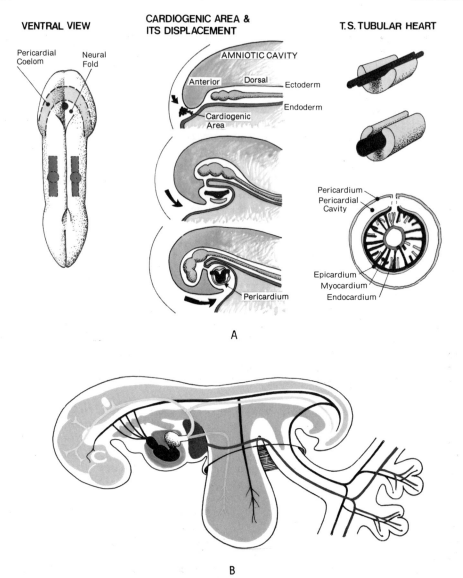

Fig. 3 - <u>Early cardiac development.</u> A - 1. The ventral view illustrates the relationship of the "cardiogenic area" to the other structures in the embryonic disc. 2. The relationships in A 1. are illustrated in the first instance and the reversal of these by "head folding" is shown in stages. 3. This series of illustrations shows the tubular heart development in relation to the pericardial cavity. B - The basic foetal channels are shown: cardinal veins; umbilical veins; vitelline veins; corresponding arteries. The branchial arch arteries are also shown.

neural tube is the first structure to be vascularized. It is followed by the (pre)-muscular and (pre)-cartilaginous areas.

Definitive vessels are selected out of these capillary plexuses. But what governs the choice is only understood in general terms. That heredity has influence is emphasized by the fact that large vessel development may occur for a time in the absence of a circulation (Clark, 1918; Knower, 1907). That physical forces also contribute has been shown by other evidence (Thoma, 1893; Mall, 1906; Clark, 1918; Woollard and Harpman, 1937).

The length of the heart increases between its point of anchorage at the ventral aortic root in front and the great veins and septum transversum behind (Fig. 4). Meanwhile between these two points the dorsal mesocardium has disappeared. Regions recognizable are:

1. Truncus arteriosus (or ventral aorta)
2. Bulbus cordis
3. Ventricle
4. Atrium   (Right horn ·
5. Sinus Venosus   (
                   (Left horn

Such an undivided heart composed of essentially four undivided chambers is seen in fishes. Since all the blood is pumped through the gills for oxygenation and carried thence to the tissues (Figs. 5 and 6), no separation of blood streams and, thus, no septation is required in fishes. Man recapitulates his phylogeny briefly and moves on (Fig. 5).

3. DEVELOPMENT OF SUBSTANTIVE FETAL CIRCULATION

Septation of the Heart

The primitive circulation (in the human in the fourth week of intrauterine life) is a single circuit generated by a single ventricle discharging via a truncus arteriosus. This arterial conduit (aortic sac or the ventral aorta) pumps blood through bilaterally located gills (of which primitive fish have seven). In the human, six pairs of gill arch arteries develop and, passing dorsally, enter the paired dorsal aorta (Fig. 5).

Fig. 4.- The heart tube

    The tubular heart and its appearances as it transforms in antero-posterior view and corresponding lateral perspectives.

    The separate parts, the manner of folding to make the venous chamber posterior and to produce the transverse and oblique sinus of pericardium is depicted. (See p.i of color insert).

Fig. 5. The comparable form of the branchial arch region in fish to that of embryonic stages in man is illustrated. However, completed gill clefts in the fish with oxygenated water circulating through them (blue arrows) do not develop in man, who instead develops a respiratory anlage for pulmonary function. The fish has seven gill arches (not all shown here) while man has six, all of which are not equally well developed; indeed, they do not appear simultaneously at any stage, so that the diagram shows a composited representation in man. (See p.i of color insert).

With the evolution of lungs as the necessary substitute for gills, a separate vascular circuit was developed for this organ. It is the pulmonary or "lesser circulation." It carries de-oxygenated blood to the lungs and returns aerated blood to the heart. This system annexes the 6th arch arteries for this purpose (Fig. 6). The remainder selectively evolves into the systemic or "greater circulation."

This separation of pulmonary and systemic circulations takes place in stages. The morphological positioning of the heart is claimed to produce spiraling blood streams (de Vries and Saunders, 1962) and experimental work appears to support the proposition that hemodynamic factors contribute to further septation here (Rychter, 1959). Partial septation of the truncus arteriosus by a "spiral valve" ("aorticopulmonary valve") directs blood from the right side to the pulmonary artery and from the left to the aorta. In the fetus, systemic venous blood is delivered to the right side of the common atrium and pulmonary venous blood to the left. If this were a postnatal situation, the aorta would contain oxygenated blood and the pulmonary artery mainly deoxygenated blood, with admixture occurring in the common atrium--an amphibian type of circulation.

Incipient septation of the ventricle by a partial "interventricular septum" is the next evolutionary step. It is the stage seen in adult reptilia. In postnatal mammals and birds, septation into right and left sides is complete through all the chambers. Ontogeny is a recapitulation of the undivided condition described in lesser forms. Thus, the embryonic stages of human cardiac development pass through these stages of incomplete septation:

(a) The stage of a single tubular four-chambered (sinus venosus, atrium, ventricle and conus arteriosus) heart with no septation resembles the condition found in fish. All the blood reaching the heart is "venous" (deoxygenated) and a "gill circulation" exists. When "lungs" develop, separate streams are required.
(b) Partial septation occurs. Atrial septa and conus valve represent the Amphibian stage.
(c) The interventricular septation next evolved is the Rep-

tilian stage. In both (b) and (c) double aortic arches exist.
(d) The prenatal fetal circulation in man is one of fetal "septation with communication" i.e., valved closure in the arterial chambers.
(e) Complete septation is found in avians that possess a right-sided aortic arch (i.e., oxygenated blood diverted to right side), while in adult mammals, completed septation is associated with a left-sided aorta (i.e., oxygenated blood is on the left side).

The prenatal human condition is one of "septation with communication," as indicated above in (d). Several septa develop and make contributions in varying degree and in combinations (Fig. 7), to right and left separation of each chamber of the heart. These septa are: (Figs. 8, 11, 12 and 13).

```
                    (-- septum spurium
     Atrial         (-- septum primum (5th week)
                    (-- septum secundum (6 - 7th week)
```

Atrioventricular -- the dorsal and ventral endocardial cushions
    Canal   (appear 5th week - fuse 6th week)

Ventricle -- the (muscular) interventricular septum (5th week)

Conus and Truncus -- the trunco-conal septum, spiral valve or
    arteriosus   aortico pulmonary septum.

Septation in various parts of the heart progresses simultaneously from its commencement in the 5th week (30th day, Streeter horizon XIV to the 7th week). However, to get a better notion of the items involved, chronological sequences will be disregarded here and convenient categories chosen instead.

(1) The Atrioventricular canal -- This short segment of the cardiac tube is a strategically important region for, as we shall see, septation here contributes to adjacent parts, namely, the atrial and ventricular septa and valves as well.

The atrioventricular canal is flattened and would present as a transverse opening from atrium to ventricle. Two endocardial cushions, dorsal and ventral respectively, arise (5th week) as outgrowths of the endocardium in the canal. In the sixth week these meet and fuse to form what has been called (uncommonly) the "septum intermedium."

(2) The Atrial Septa (septum primum and septum secundum) (Figs. 8 & 14) -- The common atrial chamber is compressed between the bulboventricular portion ventrally and the sinus venosus (and septum transversum) dorsally. This is an external foreshadowing of internal atrial septation.

Fig. 6 A comparison of branchial circulations in fish (top) and foetal man (bottom) illustrates the relegation of the sixth arch arteries to pulmonary function in man.
Venous deoxygenated blood (esp. superior vena caval blood) passing into the right heart (blue arrow) enters the sixth arch arteries that connect with the lung. In utero, most of this is bypassed via the ductus arteriosus (dotted) into the descending aorta.
The admixed venous return to the right atrium (caval and umbilical) is shunted to the left atrium via the foramen ovale and passed from the left side of the heart to the other arches that enlarge to become the carotid and subclavian streams. (See p.i of color insert).

Fig. 7A, B & C - <u>Septation of the heart</u>. Column 1 illustrations are in the coronal plane, Column 2 is a model rotated into an oblique plane to lead into Column 3 which is a lateral view of stages in cardiac septation. As arranged, horizontal rows of illustrations represent the same stage in development with connecting lines demonstrating identical landmarks.
The top row shows the ostium primum in relation to the septum primum (yellow). The developing anterior (brown) and posterior (green) endocardial cushions approach one another in the atrio-ventricular canal zone. The interventricular septum has begun to grow; the intercommunications between the ventricles and the atria are widely open.
The middle row represents an intermediate stage. The ostium secundum of the atrial septum is developed as the septum primum joins infusion with the fused endocardial cushions of the A-V canal. The A-V canals are then distinct entities (row 2). The primitive aorta is also becoming sub-divided by the "aortico-pulmonary" septum.
The bottom row is a representation of the finished state of cardiac septation, the septum secundum (blue) is thick and the thin septum primum (yellow) behaves as a flap valve against it, thereby preventing left to right blood flow. The upper thin protion of the interventricular septum, i.e., the membraneous part, is at the junction of the aorticopulmonary, endocardial cushions and interventricular septa from all of which it is composited in development. (See p.ii of color insert).

Fig. 8 - <u>Atrial septa</u>. Supplements the stages of atrial development shown in Fig. 7, indicating the components explicitly. (See p.ii of color insert).

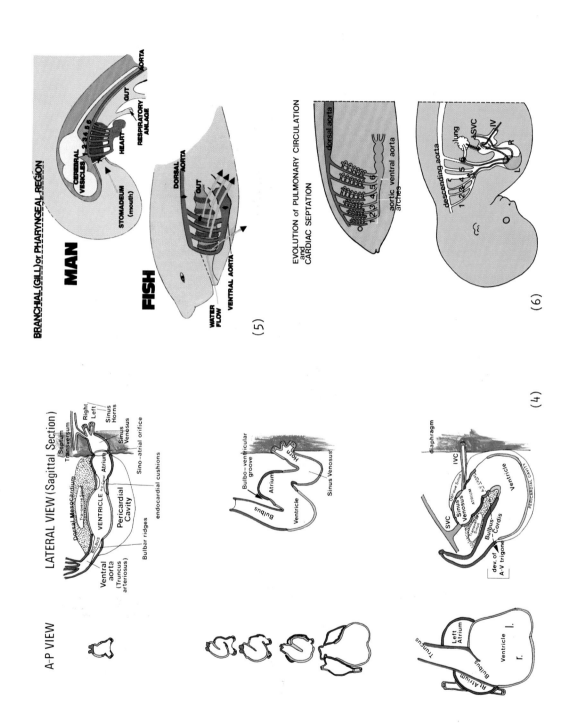

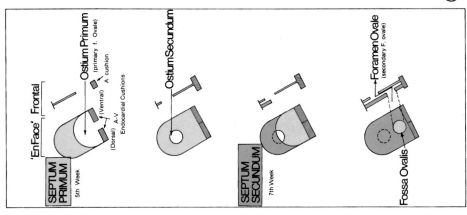
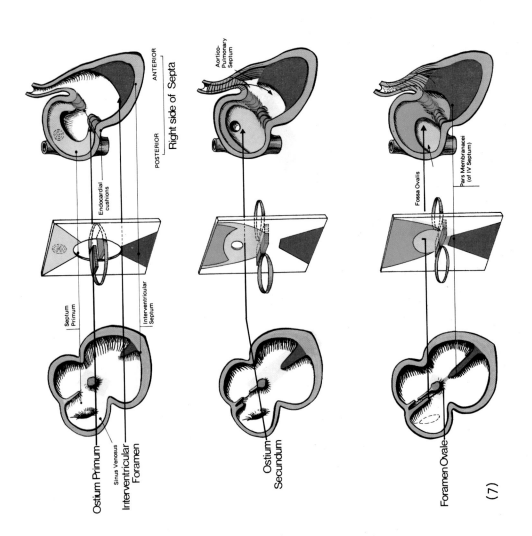

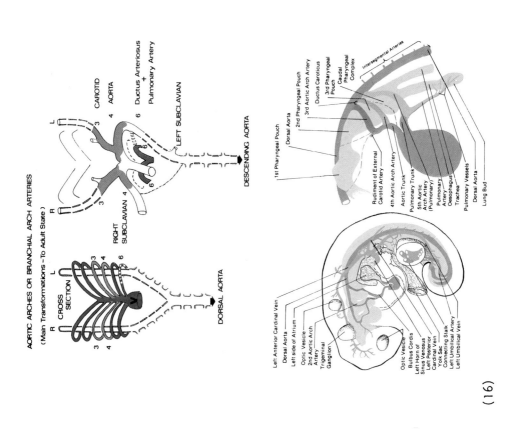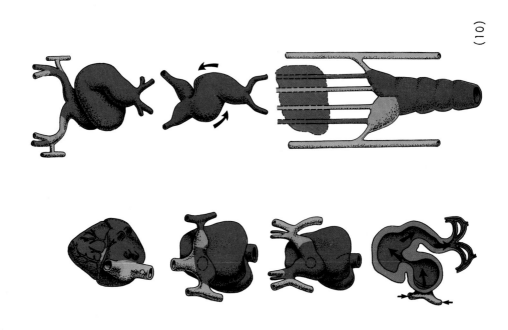

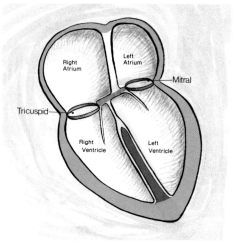 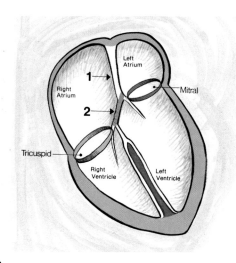

(11)

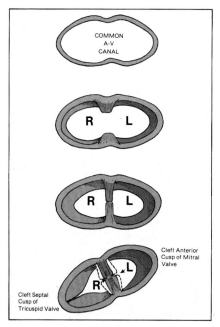 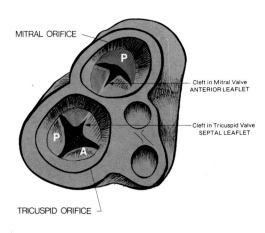

(12)

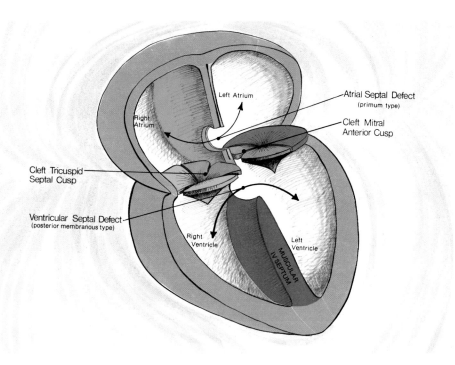

(13)

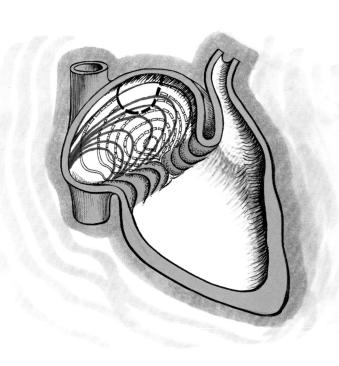

(14)

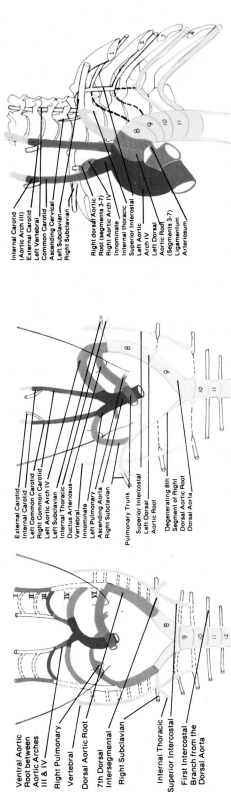
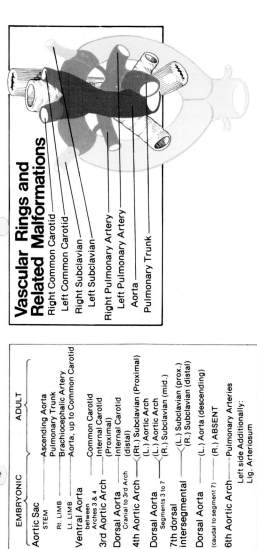

Vascular Rings and Related Malformations

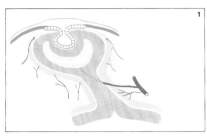
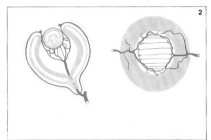
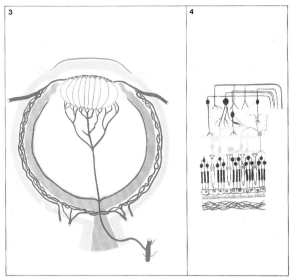
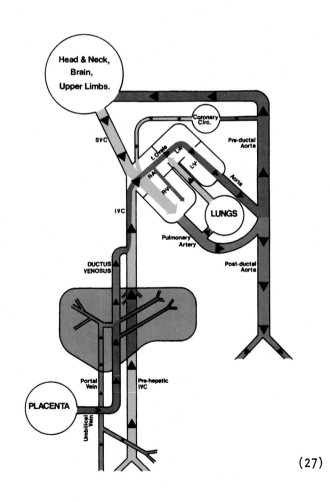

(23)

(27)

Upper Limb-Vascular Embryology (1)

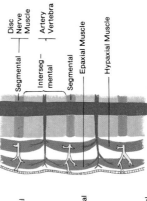
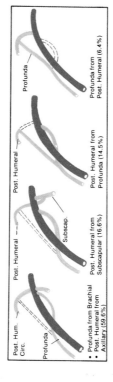

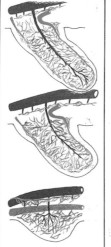
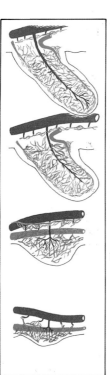

1. Superficial Brachial Artery
2. Superficial Antebrachial Artery (Originating from Superficial Brachial)
3. Superficial Brachial (from Brachial)
4. Radial originating from Superficial Brachial
5. Superficial Antebrachial Artery
6. Anastomosis (Superficial Brachial to Brachial)
7. Radial from a persistant Median Artery
8. Ulnar Artery from Superficial Antebrachial
9. Superficial Median Artery

- Profunda from Brachial
- Post. Humeral from Axillary (59.6%)

Post. Humeral from Subscapular (16.6%)

Post. Humeral from Profunda (14.5%)

Profunda from Post. Humeral (6.4%)

6th Intersegmental Artery

8th Intersegmental Artery

9th Intersegmental Artery

(25)

(24)

(a) In the fifth week an endocardial ingrowth commences in the mid-dorsocephalic wall of the atrium and assumes a sickle or crescentic shape as it grows toward the fusing dorsal and ventral endocardial cushions below. This is the "septum primum" (Fig. 8). At this stage the gap between the lower end of the septum primum and endocardial cushions is called the "primary foramen ovale" ("ostium primum" or "interatrial foramen I").

(b) Before the septum primum reaches and fuses with the endocardial cushions (themselves fused together), a gap occurs in the upper part of the septum. This perforation above mid-atrial level is called the "ostium secundum" (or "interatrial foramen II"). Thus, interatrial communication (right to left) is maintained uninterruptedly in utero.

(c) At the same time (now in the seventh week) a second and thicker septum, the "septum secundum" grows on the right side of the first septum in the "interseptovalvular space" i.e., between the site of origin of septum primum and the attachment of the left venous valve. It too is sickle shaped and grows down far enough to overlap the foramen ovale, but not far enough to reach the level of the endocardial cushions except ventrally. This septum being a firm one, while the septum primum overlying it is thin, a valve arrangement now exists--the "valve of the (secondary) foramen ovale." The mechanism allows right to left flow but not the reverse. Further, the orientation of the sickle-shaped free margin of the septum secundum is such that it overrides the inferior vena cava orifice, causing much of the inferior vena cava's blood to flow into the foramen directly, while permitting some admixture by subdividing the inferior vena cava stream in the right atrium. The part of the primary septum fused with the endocardial cushion constitutes the floor of a fossa--"the fossa ovalis." The anterior edge of the septum secundum curving posteriorly along the atrioventricular cushions produces the margin of the fossa ovalis or the "annulus ovalis." The interatrial septum, thus, has the following contributions:

1. Growth of septum primum
2. Fusion of dorsal and ventral endocardial cushions
3. Growth of septum secundum

(3) Transformation of the Sinus Venosus (sinus horn and related vessels) (Figs. 9 and 10) -- On each side, each sinus horn receives blood from the veins of the corresponding side as follows:

- The common cardinal vein (duct of Cuvier) - the fused trunk of the anterior and posterior cardinal veins draining the respective regions of the body wall of the embryo;
- The umbilical vein - draining the placenta and,
- The vitelline (omphalomesenteric) vein - draining the yolk sac.

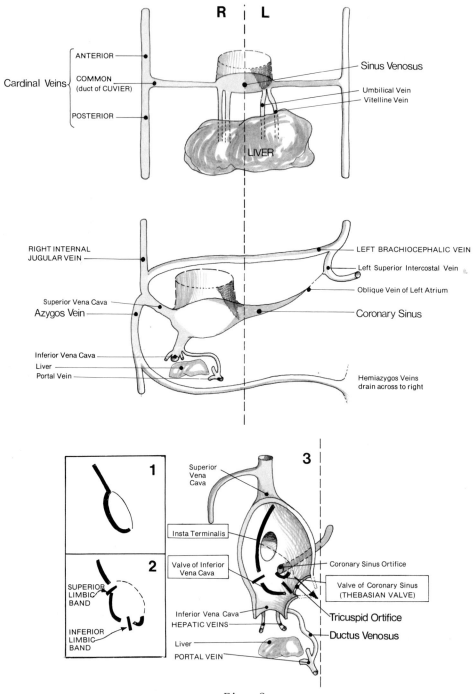

Fig. 9

← Fig. 9 - Development of right atrium and its veins

    The manner in which development biases the right side is shown. The left brachiocephalic vein crosses to the right across the midline; the left horn of the sinus venosus (yellow) becomes the coronary sinus opening into the right atrium and the heimazygos takes a transverse course across the midline to drain into the azygos vein (a right-sided structure) draining into the superior vena cava.
    The right atrial components are labeled in the bottom figure. The inset illustrates the sub-division of the embryonic valve of the vena cava.

Left Horn of Sinus Venosus

    At or near the left horn of the sinus venosus, all the vessels disappear. This reduced horn remains reduced as the coronary sinus which is ultimately absorbed into the right atrium into which it opens in the adult. A small part of the left common cardinal vein persists as a tributary of the coronary sinus, namely, the oblique vein (of Marshall) of the left atrium.

Right Horn of Sinus Venosus

    This enlarges and is incorporated into the right atrium. The root of the superior vena cava is derived from the common cardinal vein. The rest of the superior vana cava and azygos veins represent anterior and posterior cardinal veins respectively. Where the right horn opened into the atrium the "valve of the superior vena cava" existed. The fused cephalic part of this valve is the "septum spurium." The left part of the valve disappears. The right, crossed by two muscular "linked sub-endocardial bands," remains as the crista terminalis, the intervenous tubercle and the (inferior) valve of the coronary sinus.

    The horns of the sinus venosus start out embedded in the septum transversum and work free into the pericardial cavity. The horns

---

Fig. 10 - External cardiac transformations

    The symmetrical pattern of the veins draining into the horns of the sinus venosus (one horn in green) is shown, followed by the rotatory pattern of the bulbo-cardiac loop. The top right picture is a lateral view cut away and below this three successive stages of cardic transformation as seen in posterior view. Note the left horn of the sinus venosus (green) and the inferior vena cava (purple). (See p.iii of color insert).

were horizontal at first, but the caudal migration of the heart from the 4th to 17th segment pulls the common cardinal veins (and the horns) posteriorly and vertically. These vessels throw up folds of mesoderm in the lateral walls of the pleuropericardial cavity ("pleuropericardial folds") that, moving medially from a lateral position, shut off the pericardial cavity in front, from the pleural cavity behind. Since the left-sided veins dwindle rapidly, pericardial defects are commoner on the left side. The rearrangement of hepatic veins in the septum transversum causes the "ductus venosus" to open into the right side (into the inferior vena cava). A summary of the origin of the great thoracic veins is indicative of these changes.

Right Side	Left Side
Rt. brachiocephalic - (cephalic part of ant. cardinal)	Left brachiocephalic - (cross anastomosis L -- R ant. cardinal
Superior vena cava	Left superior intercostal - (terminal part of L. ant. cardinal)
Upper ½ - (terminal part of Rt. ant. cardinal)	Oblique vein of left atrium - (left common cardinal)
Lower ½ - (Rt. common cardinal)	Coronary sinus - (left horn of sinus venosus)
Azygos vein cephalic part - (Rt. post. cardinal)	
Inferior vena cava (terminal part of Rt. vitelline vein)	

(4) <u>The Right Atrium</u> -- Blood returned in the great veins is shunted across the midline from left to right; (cranially) the left brachiocephalic vein drains across to the right common cardinal vein, (caudally) the left hemiazygos veins have cross anastomosis with the right azygos vein (Figs. 9 and 10). The net effect is to enlarge the primitive right atrium which "takes up" the sinus venosus and its horns (both right and left) into its wall. The smooth-walled posterior part of the adult right atrium represents this absorption, the crista terminalis and the rough-walled (by contours of the musculi pectinati) part of the atrium representing the primitive right atrial component.

The sinus venosus opening into the atrium (to become part of the right atrium) has valves to prevent reflux during atrial contraction --a "septum spurium" diverging around the original orifice into right and left sinus valves. The right valve of the sinus venosus is cut across by muscular superior and inferior limbic subendocardial bands (Fig. 9). The septum spurium and upper end of this valve produces the adult "crista terminalis" represented externally as the "sulcus terminalis of the right atrium." The superior limbic band remains as the faintly marked "intervenous tubercle." The segment between the bands is the valve of the inferior vena cava, and the lowermost segment of the right valve is what becomes of the valve of the coronary sinus. All of these are poorly represented and non-functional in the adult heart. What remains of the left valve is incorporated into the interatrial septum or disappears.

The liver developing in the septum transversum has a host of venous transformations to affect. In the outcome, the terminal part of the right vitelline vein becomes the proximal part of the inferior vena cava, i.e., the "hepatocardiac channel" (Fig. 15). The right umbilical vein disappears while the left umbilical vein persists, and in its proximal part develops a channel connecting it to the right vitelline vein that becomes the hepatic part of the inferior vena cava. This channel that by-passes the liver i.e., from left branch of portal vein to inferior vena cava in the prenatal state, is the "ductus venosus." The development of the inferior vena cava itself is complex and is composed of a number of embryological components (Fig. 15).

(5) <u>The Left Primitive Atrium and the Pulmonary Veins</u> -- The respiratory system or tree is developed by evagination of the cephalic foregut floor. Thus, the respiratory anlage and lungs budding from it are drained by a widely connecting plexus of veins of the foregut.

The substantive pulmonary vein is a single outgrowth of the primitive left atrial chamber that grows out in the dorsal mesocardium to reach and connect with the lung veins which it takes over. Absorption of the walls of the pulmonary vein continues into the left atrium until the common pulmonary vein and its two successive dichotomizing branches i.e., four orifices, are incorporated into and constitute the major part of the adult left atrium. The small rough-walled left atrial appendage (or auricle) is all that remains of the original primitive left atrium.

In summary, the origin of each of the atrial components (Fig. 14) is as follows:

Right atrium   = Rt. horn of sinus venosus
                  Rt. primitive atrium (auricular appendage)
                  Rt. atrioventricular canal

Atrial septum   = septum secundum
                  septum primum
                  endocardial cushions

              Left atrium   =   pulmonary veins
                                  primitive left atrium (auricular appendage)
                                  left atrioventricular canal

The atrioventricular valves develop in three stages in human hearts
(Odgers, 1939):   (Figs. 12 and 13).

   1. Composed of endocardial cushion tissue and muscle derived
      from the ventricular trabeculae (11.2 to 23 mm stage).
   2. Increasing muscular component (28.5 to 61 mm stage).
   3. Becoming entirely collagenous (85 to 210 mm stage). The
      occasional presence of auricular type muscle is entirely
      capricious.

   Since the heart rotates, the right cusp of the tricuspid valve
in the embryo becomes the inferior or posterior cusp in the adult,
while the septal cusp becomes the medial one. In the mitral valve,
the right embryonic cusp becomes anterior.

6. <u>The Development of the Ventricles</u>

   Ventricular septation, which has multiple component parts, is
critical to establishing normal circulation (Fig. 11). The right
and left ventricles, correctly formed, have to be connected to the
right and left atria on the venous side and to the pulmonary and
aortic trunks at the arterial end respectively. It means that the
following have to be correctly matched: rotation of the cardiac
tube to produce its external form and proper apposition of the following interventricular septum components:

   a. Atria ventricular endocardial cushions (dorsal and ventral)
   b. Muscular interventricular septum and
   c. Bulbar septa (right and left), (continuous with the "spiral
      aortico-pulmonary septum").

<u>External Form of Ventricles and Rotation</u> (Fig. 10).

   To understand this, the following developmental events and
points are significant:

   1. The "bulbo-ventricular loop" is the ventralmost part of
      the heart tube when from a "U" shape it becomes "S" shaped.
      This twist is said to be caused by rapid growth of the tube,
      restricted (from length-wise extension) at the cranial end
      by the chest wall and brachial arches and caudally by the
      septum transversum.

   2. The proximal part of the bulbus cordis becomes absorbed
      into the common ventricle (especially the right ventricle).

This ultimately becomes the "out-flow tracts" of the right and left ventricles, i.e.,

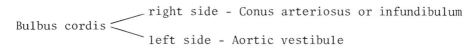

Bulbus cordis
- right side - Conus arteriosus or infundibulum
- left side - Aortic vestibule

3. Although in the adult heart the ventricles appear side by side and contract simultaneously, embryologically these are first placed tandem-wise along the heart tube, the right ventricle nearer the conus arteriosus. (Thus, embryologically, the left ventricle contracts before the right in "peristaltic" fashion.)

The region between the atria and ventricles narrows, producing atrioventricular canals and ultimately becomes free of myocardium by connective tissue ingrowth from the epicardium. At the same time the conducting system of the heart (including the bundle of His) is developing (10 mm stage) (Walls, 1947).

Ventricular Septation

a. Muscular interventricular septum

This develops in the fifth week at the apex of the ventricular chamber and grows in crescentic form toward the atrioventricular cushions. It is suggested that the muscular interventricular septum arises as a result of differential and progressive enlargement of the ventricles (van Mierop et al., 1967 and 1963). In this stage there is a fetal "interventricular foramen" between the crescentic margin of the interventricular septum and the lower end of the atrioventricular endocardial cushions (that proceed to fuse).

The manner of closure of the small fetal foramen that exists at the upper end of the ventricle is interesting. It is (on account of the U form of the bulbo ventricular loop) a junctional area for the ventricles, atrioventricular canals, and the bulbus cordis. Closure is accomplished by septa originating in each of those sections of the tube: i.e.,

1. the muscular interventricular septum
2. the endocardial cushions
3. the conus ridges.

Thus, each of these contribute, or extend to form the "membranous part of the interventricular septum" in the final closure (end of seventh week).

These illustrations (Figures 11, 12, 13 and 14) depict derivations and relationships of various cardiac septa. They can be found on pages iv and v of the color insert.

Fig. 11 - The oblique orientation of the endocardiac cushions contribution to septation is shown to illustrate how the lower portion septates the left ventricle from the right atrium (label 2).

Fig. 12 - By revealing the minimal failure of endocardial cushion defect (cleft tricuspid trimitral valve cusps) clarifies its contributions in the A-V canal zone to these structures.

Fig. 13 - This figure indicates in simplified fashion the location of defects in septation that embryology explains directly.

Fig. 14 - This figure illustrates step-wise growth in each of the three septa, the septum primum (in yellow) from above downward, the septum secundum (in blue) toward the inferior vena cava orifice, i.e., down and backward, and the endocardial cushions (posterior in green, anterior in brown) toward each other in the atrioventricular junction.

---

b. The fusion of the endocardial cushions with the primitive interventricular septum

The dorsal and ventral endocardial cushions fuse with each other to produce a median septum dividing the single A-V canals into right and left canals. Additionally, this fused septum elongates so that the right margin is pushed downward; it is this margin that is excavated to produce the septal cusp of the tricuspid valve. Thus, the left margin of the fused endocardial cushions lying higher up produces the anterior cusp of the mitral valve (Fig. 11).

The atrial septa fuse with the endocardial cushion near the mitral cusp development (i.e., at the left part). The result of this is that the membranous part of the interventricular septum here is really an "artrioventricular partition" (i.e., right atrium and left ventricle) (Fig. 11). This membrane is derived from extensions of tissue from the right side of the fused endocardial cushions (Duckworth, 1967).

c. The right and left conus ridges

These appear to continue the downward growth of the spiral aorticopulmonary septum and may be referred to as the trunco-conal ridges (or bulbo-conal septum). Some authorities (Kramer, 1942; De Vries and Saunders, 1962) distinguish the "conus ridges" from the "aorticopulmonary ridges," the dividing line being the attachment of the semilunar valves. In explanation of the origin of these septa, it is stated that asynchronous contractions of the left and right ventricles generate two streams of blood (5th week) and endothelial ridges arise to separate them.

The bulbar (truncus) ridges are right and left, and when they fuse in the 8th week, they separate the fourth and sixth aortic arches from one another. These bulbar ridges run in the truncus arteriosus spirally to reach the conus and root of ventricles to become the "conus ridges." Together the trunco conal ridges, or septum, spiral through 180° and completely separate two channels --

1. Aortic (connecting with left ventricle)
2. Pulmonary (connecting with right ventricle).

i.e., they constitute an "aorticopulmonary spiral septum." When this separation is complete, accessory subendocardial folds grow in to produce the semilunar valves at the truno-conal junction. These "valve swellings" hollow out their walled "pockets" or cusps, three each to circumscribe the aortic and pulmonary trunks respectively.

Left Ventricle

In utero, the inferior vena cava contains oxygenated blood received in it via the umbilical vein, admixed in part with venous blood from the lower half of the body. This it delivers to the right atrium, and the "valve of the inferior vena cava" (in the fetus) shunts part of this fetus blood to the left atrium (in turn the supply for the left ventricle) via the foramen ovale. The arrangement transfers the most "nutrition" to the heart, upper extremity and the head (especially the brain) and may be the explanation of the sturdier development of the upper body of the newborn as compared with the lower trunk and legs.

The left ventricle receives very little venous return from the lungs except in the last few weeks of pregnancy. Thus, a shunt from right to left in the atria is what keeps the left ventricle well developed. Premature closure of the foramen ovale leads to left ventricular atresia.

The myocardium develops as a meshwork of muscle into which, on the ventricular surface, the endocardial layer invades the interstices. Thus, trabeculae carnae, papillary muscles and chordae tendinae all represent varying degrees of excavation of the ventricular walls. The aortic vestible is the incorporated portion of the bulbus cordis.

Right Ventricle

As in the left ventricle; trabeculae carnae, papillary muscles and chordae tendinae develop. However, the size of this chamber depends on the "fixation" of the atrioventricular junction. If "atrialization" is excessive (i.e., at the expense of the right ventricle), the right ventricle is undersized. The infundibulum represents the incorporated part of the bulbus cordis.

Conduction tissues of the myocardium

Purkinje cells or "conducting tissue" (as distinct from contractile cells) are a separate differentiation in the subendocardial tissues of the fetus. The sinoauricular node differentiates in the 10 mm stage but is histologically clearly distinguishable only at birth (Walls, 1947). The atrioventricular node and bundle of His are recognizable in the 7 to 8 mm stage (Wall, 1947) but only in second month embryos. The divisions of the bundle are conspicuous by three months. Innervation of the nodal tissues has been seen in the fifth week of development (Yamauchi, 1965).

ARTERIAL EMBRYOLOGY

The changes that produce the substantive fetal peripheral circulation are conveniently dealt with in three sections:

1. The transformation of aortic arches (and cerebral circulation)
2. Development of the (dorsal) aorta and determination of vascular regions
3. Embryology of limb vasculature.

1. Transformation of Aortic Arches

Early cardiogenesis and angiogenesis of the first four weeks has traced out the manner in which the symmetrical vascular system and undivided heart tube develops and becomes a single functional circuit of blood flow. The single fused "ventral aorta," dilated and shortened to form the aortic sac, connects with the paired dorsal aorta through the first aortic (branchial) arch, a condition found by the 14 somite stage of development (23rd day). Thereafter, five other pairs of aortic arches appear in succession, the earlier ones (1 and 2) largely disappearing before the latter ones (6 and 5) appear, so that at any time no more than three aortic arch pairs are present. In fact, in man the 5th pair is of very brief transitory existence. Each aortic arch artery passes round, dorsal to the gut and trachea, from the ventral aorta to the dorsal aorta (Figs. 15 and 16).

This branchial phase of development lasts until the 12 mm CR stage (a duration of 22 days since the first arch appeared) and the pattern of flow from the ventrally placed aortic sac to the dorsal aorta becomes a succession of changes determined by the stage of arch development and retrogression approximately as follows:

	Aortic Arch	CR Length	Week
1.	1st	3 mm	4th
2.	1st and 2nd		
3.	2nd and 3rd	5 mm	5th
4.	3rd and 4th =(?5)		
5.	3rd, 4th and 6th	7 mm	6th

By the end of the branchial phase, the 3rd, 4th and 6th arch arteries are well developed, and each arising from the aortic sac, terminates in the dorsal aorta of the corresponding side. These are persistent vessels which become considerably modified in the post branchial period in utero to establish the final postnatal pattern of the "great" arteries. ("The post branchial phase" arbitrarily commences with interruption of pulmonary arch attachment.)

While the branchial phase of aortic arch development is in progress, the bulbospiral septum has divided the bulbus cordis and its distal part (the "aorticopulmonary septum") extends this partition into the aortic sac. This causes right ventricular blood to enter the partitioned dorsal part of the aortic sac which leads into the 6th aortic arches that enter the lungs. (The 6th arch is therefore the "pulmonary.") The ventral partition of the aortic sac becomes the ascending aorta.

The further evolution of the aortic arches are illustrated in diagrams (Figs. 17 and 18). Details and disagreements are recorded in the literature (Congdon, 1922; Barry, 1951; Adams, 1957; and Moffat, 1959). Some specific changes are singled out below:

The stem of the aortic sac is subdivided by the bulbospiral septum to produce the ascending aorta and the pulmonary trunk in such a manner that the pulmonary trunk is in communication with the 6th arch and the aorta with the 4th arch (Fig. 6)

The cranial part of the aortic sac between the 6th arch attachments becomes drawn out cranially into right and left limbs, or extensions, as the neck elongates. The right limb becomes the brachiocephalic artery and the left limb, the proximal part of the aortic arch up to the origin of the left common carotid.

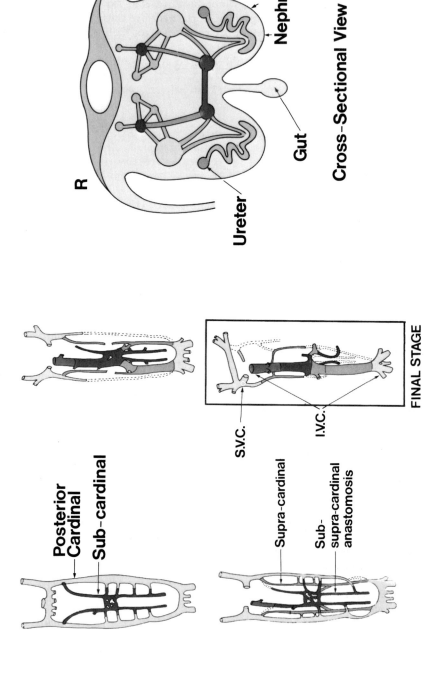

Fig. 15. The development of veins, particularly in the caudal portion of the body, is complex. Main channels form in an embryonic stage and have several interconnections as illustrated in the cross-sectional view. The general pattern of development that contributes to the final stage is represented by the four color-coded pictures.

EMBRYOLOGY OF THE HUMAN ARTERIAL SYSTEM

The ventral aorta between attachment of the 3rd and 4th arches forms the common carotids which become stretched with neck elongation. These caudal displacements and elongations follow the "relative growth" rates of all structures in the embryonic neck region.

Aortic arch arteries - 1st and 2nd

These largely disappear. As they were generated from a network of vessels, so do they regress into a plexiform network that fades out (Lewis, 1903; Bremer, 1914; and Evans, 1909). A small portion of the 2nd arch attached to the dorsal aorta (which at this segment contributes to the internal carotid), becomes the stem of the stapedial (caroticotympanic) artery. The segment of dorsal aorta connecting the 1st and 2nd aortic arch arteries contributes to the internal carotid artery.

The 3rd aortic arch arteries become the proximal segment of the internal carotid on each side respectively. The more distal part of the adult internal carotid is formed by the dorsal aorta cranial to the attachment of the 3rd aortic arch artery. The external carotid sprouts from the attachment of the 3rd arch to the ventral aorta,

Fig. 16 - The upper half of this figure shows the type plan branchial artery system (on the left) and its transformation to the adult human (on the right). The third arterial arch (pink) becomes "carotid," the left fourth arch (orange) becomes the aorta, while the right fourth becomes the subclavian. The pulmonary arteries are sixth arch derivations. The lower half of this figure indicates these specializations in situ in the fetus of this stage. (See p.iii of color insert).

Fig. 17 - It is possible to trace the embryological transformation of each segment of the embryological arterial arch pattern into the adult condition. The color-coded stages in the figure do this, and are derived from the work of Barry (1951). Malformations of this system can also be traced to incriminate particular segments and vascular ring anomalies and related malformation have been analyzed along these lines (Stewart et al., 1964). (See p.vi of color insert).

probably linking up with remnants of the 1st and 2nd arteries (Congdon, 1922); it is carried cranialward by the migration of the 1st and 2nd arch musculature that it supplies.

The 4th aortic arch arteries differ in further development on either side. On the right side this forms the proximal segment of the right subclavian. On the left side it contributes to the arch of the aorta between the left common carotid and left subclavian arteries.

The left subclavian is produced by the 7th dorsal intersegmented artery. The arch of the aorta just proximal to the left subclavian is formed by the dorsal aorta between the attachments of the 3rd and 7th intersegmented arteries. This portion of the dorsal aorta is considerably shortened by relative non-growth, so that the left subclavian has "migrated" in a cephalic direction, from a position one segment caudal to the unfused aorta, past the pulmonary arch, up the aortic arch to its adult position. This seems to be achieved by rearrangement of cells around the orifice, spreading apart above and closing below, and reflects the fetal plasticity of endothelium (Heuser, 1923).

The right subclavian is produced by the same segment of dorsal aorta just referred to i.e., the segmented portion 3 to 7 on the right. The more distal part of the right subclavian is a segment of the 7th right intersegmental artery. Thus, the right subclavian has three embryonic segments:

    Proximally   - the 4th right aortic arch
    Intermediate - the right dorsal aorta (between segmental
                    arteries)
    Distally   - the left 7th intersegmental artery.

Dorsal aorta beyond the 7th segmental artery

The right and left dorsal aortae fuse caudal to the 9th segment. The dorsal aorta related to segments 8 and 9 disappear on the right side, and with it, the attachment of the 6th aortic arch (pulmonary) to the dorsal aorta on the right side.

On the left all the above persist, producing a descending aorta on the left side connected to the heart. The connection between this vessel and the 6th arch on the left side constitutes the "ductuc arteriosus," the fetal by-pass of the pulmonary circuit.

Dorsal aorta between arches 3 and 4 - this is referred to as the "ductus caroticus" and disappears on both sides. Growth and elongation of the neck, accompanied by its flexion, makes the aortic arches crowd together ventrally and stretch apart dorsally, this latter encouraging the disappearance of the dorsal part of attenuation.

The 6th arch - the proximal parts produce the right and left branches of the pulmonary artery. The distal parts behave differently on the two sides, on the left side retaining its connection with the dorsal aorta (i.e., the ductus arterious or "ductus Botalli"). On the right side this is lost with the deletion of the right dorsal aorta and the displacement of residual parts of it into the right subclavian that becomes a vessel in the neck.

The aorta in the adult is thus comprised of:

- the ventral aorta (between the 4th and 6th arches)
- the left 4th aortic arch
- dorsal aorta (left side 3-7 (8) intersegmental) unfused left
- fused dorsal aortae (caudal to segment 8)

The pulmonary artery - The trunk is subdivided from the aortic sac which connects with the 6th aortic arches, which become the right and left banches of the pulmonary trunk.

Aortic arch formation occurs through the period of the embryo from 4 to 16 mm and anomalies arising from aortic arch development in this period have been extensively classified (Stewart, Kincaid and Edwards, 1964). Double aortic arches, right aortic arch, aberrant (retro-oesophageal right subclavian and patent left or right) ductus arteriosus, as well as coarctation of the aorta are the more common types, in regard to all of which there is extensive literature.

2. Development of the (Dorsal) Aorta (excepting branchial developments) (Fig. 18)

The aortae and their development in the branchial region has been outlined. The fact that right and left dorsal aortae had fused in the midline caudal to the 9th segment to form a single dorsal aorta has been identified. This is achieved stepwise. At the 23 somite stage (Streeter horizon XII, day 26), the fusion extends from the 10th to the 16th segment. The umbilical arteries, at this stage, continue to separate dorsal aorta into the body stalk. The original vessel that does this is overshadowed by the ventral branch (which in the trunk region becomes the major vessel) that supplies the flank and ventral body wall. At this stage, the original stem of the dorsal intersegmental artery has a (smaller) dorsal and (larger) ventral ramus or branch. Later, the fusion of the dorsal aorta extends to the 23rd body segment (= L4) and represents the terminal point of the adult abdominal aorta. In the embryo, a connection from the dorsal aorta develops at this point and becomes the dorsal root of the umbilical artery. The original ventral root origin of the artery then disappears. This latter is probably the 5th lumbar intersegmental artery which is implicated with the internal iliac artery. The umbilical artery ultimately appears to arise from this vessel. The dorsal aorta, yet more caudally, runs on as the attenuated extension of the fused dorsal aortae, represented by the adult median sacral artery.

Branches of the embryonic dorsal aorta (Figs. 18 and 19)

1. Dorsal somatic arteries

The dorsal aortae (right and left) and their fused caudal median part give off intersegmental arteries to supply the spinal cord and somitic structures, 30 such pairs arise. Those that arise from the unfused portion all disappear, except for the 7th intersegmental which forms parts of the subclavian arteries.

Each intersegmental artery supplies branches to somitic structures, the vertebral column and the spinal cord. These horizontally oriented branches are linked up by several longitudinal anastomoses: ventral somatic, precostal, postcostal, and post-transverse.

In the cervical region - The seven pairs of intersegmental arteries and their branches are linked by a series of longitudinal anastomoses (Fig. 18) in relation to the vertebrae, namely: precostal

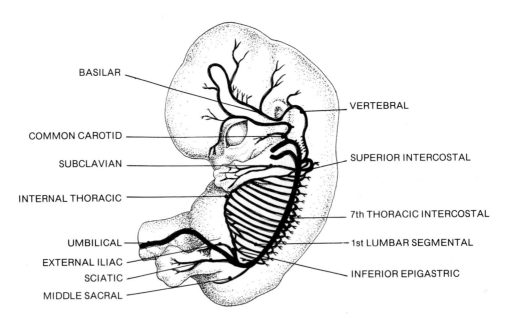

Fig. 18. Arterial development in a seven-week human embryo is shown.

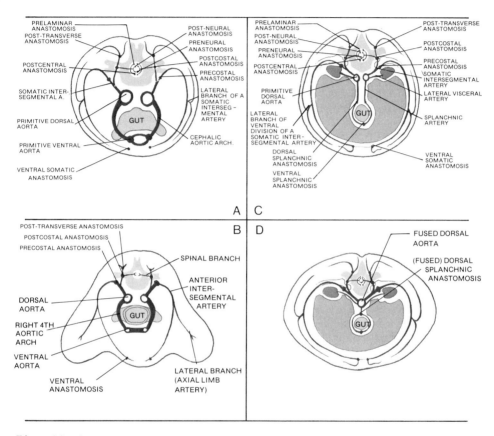

Fig. 19. A. Demonstrates the anastomoses and intersegmental arteries in the cervical region as seen in cross-section.
B. The arrangement of vessels at the level of the limbs is shown in cross-section.
C. and D. Show the arrangement of segmental and intersegmental arteries at an early stage in C and later in D.

(on the ventral branch); postcostal, and post-transverse (on the dorsal branch), and four others within the vertebral canal: postcentral, prelaminar and preneural and postneural in relation to the spinal cord. Of the seven pairs only the 7th intersegmental artery persists, connected to the aorta.

On the left side, the subclavian artery, from its commencement to the origin of the vertebral artery, is formed by this connection. More distally, up the first rib, the ventral branch of this 7th intersegmental artery is the primordium of the subclavian artery. Beyond the first rib it is the much enlarged lateral branch (of this ventral branch) that continues the subclavian trunk into the limb appendage. (Note that seventh vertebra is 7th somite, while the first rib is of 8th somite origin.)

On the right side, except for the fact that the proximal part of the right subclavian is a part of the fourth aortic arch, the distal portions of it are similar in derivation to that on the left.

It is possible from a comparison of adult vascular relationships and those indicated embryonic anastomosis (Fig. 19) to recognize that: (a) the vertebral artery is usually derived in its first part from the dorsal branch of the 7th intersegmental; the second part is the postcostal anastomosis; the third part is the spinal branch of the first intersegmental artery; and the fourth part of the preneural anastomosis.

(b) The thyrocervical trunk and ascending cervical artery is a precostal anastomosis derivation.
(c) Spinal branches of the 2nd part of the vertebral artery are already represented in the embryonic pattern and these anastomoses within the canal remain in the adult.
(d) The costocervical trunk arises from precostal anastomoses caudal to the subclavian, and the superior intercostal artery giving off the 1st and 2nd posterior intercostals, derive from the ventral rami of segmental arteries here.

In the thoracic region - The embryonic pattern of the intersegmental arteries is largely retained (Figs. 18 and 19). Again, the adult derivatives and their embryonic origins are readily discerned by comparing relationships. Thus, the intercostals are clearly ventral rami of intersegmental arteries, and the ventral somatic anastomosis is the column from internal thoracic artery into the superior and inferior epigastric anastomotic channel in the adult.

In the lumbar region - The lumbar arteries are again embryonic intersegmental arteries continued forward as their ventral rami.

In the sacral region - The further caudal fusion of the dorsal aorta produces the median or middle sacral artery.

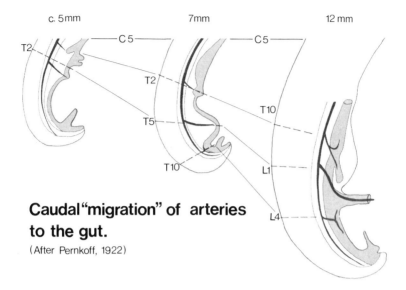

Caudal "migration" of arteries to the gut.
(After Pernkoff, 1922)

Fig. 20. The points of origin of the ventral arterial branches of the aorta to the gut migrate caudal in relation to body segments. This is demonstrated in the diagram through the 5 mm, 7 mm and 12 mm stages.

In the third month, in relation to the lower end of the median artery, arterial venous anastomoses channel through a mass of tissue, the cells of which are claimed to be smooth muscle or postembryonal angioblasts (Krompecher, 1940) in a study of fetal and neonatal material. This is the anococcygeal body in the adult.

2. <u>Visceral branches</u> (Figs. 19 and 20)

These are imperfectly segmental branches which are numerous at first and become reduced. They are categorized into (a) ventral splanchnic or visceral branches, and (b) lateral splanchnic or visceral branches.

(a) Ventral (splanchnic) branches - At first these are a series of paired multiple vitelline arteries to the yolk sac and umbilical arteries to the allantois. As the dorsal aortae fuse together, single ventral vessels appear, apparently by fusion. Their numbers are further reduced until by the fifth week of development only three major ventral vessels remain attached to the aorta, namely, the coeliac, superior mesenteric and inferior mesenteric arteries. The umbilical arteries being further displaced caudo-laterally remain as original ventral paired arteries. They now arise from secondary connections to intersegmental arteries, thereby losing their aortic connection, and bearing a dorsal relationship to the ureter. The umbilical arteries now run from the internal iliac artery. Then, after the development of the ventral abdominal wall, they run in this structure, becoming the lateral umbilical ligament on their obliteration after birth.

The three substantive ventral branches all descend several segmental levels--the coeliac artery 13 segments, the superior mesenteric 11 segments, and the inferior mesenteric 3 segments. The manner of this caudal "migration" is explained by one or both of two means: unequal growth of the dorsal as compared to the ventral wall of the aorta, and/or secondly, by successive attachment to new caudal roots and subsequent atrophy of the older cranial attachment. Pernkopf's (1922) studies of series of embryos demonstrated intermediate stages through the presence of multiple obliquely placed roots appearing in transitions demanded by the second method of migration (Fig. 20).

The superior mesenteric artery for a time retained its connection ventrally beyond the gut wall into the extraembryonic yolk sac. This has been named the "omphalomesenteric artery" at this stage. Occasionally, a persisting fibrous strand representing this portion beyond the gut and reaching to the umbilicus remains in the adult. Indeed, a persistant omphalo-mesenteric artery causing intestinal obstruction and gangrene of the associated intestinal loop (Meckel's diverticulum), has been reported (Manning & McLaughlin, 1947).

The bronchial and oesophageal branches of the aorta represent minor ventral branches of the embryo that also persist in the adult.

(b) The lateral splanchnic branches are irregular multiple branches for the benefit of the intermediate cell mass originally, and are therefore devoted to the pro-meso and metanephros, the gonads and suprarenal glands. Many involute, leaving the suprarenals (including the inferior phrenic), renal and gonadal arteries to represent the final condition.

3. Arterial development and its determinants

As a rule, a provisional capillary plexus is established by "angiotactic" influences in the embryo. It is through it, in stages, that various regions and tissues become vascularized. This primitive endothelial plexus is _self_ differentiated independently of the circulation (Evans, 1909; Sabin, 1920; Hughes, 1935). Within the plexus definitive vessels are selectively enhanced, but the specific determinant of selection of main channels in these plexus to produce arteries or veins remains uncertain.

Genetic factors as well as hemodynamic considerations are cited as determinants. Mechanical influences--rate and direction of flow and pressures and pulses of the blood stream have been identified. For example, when segments of veins are transplanted into arteries, the wall of the transplant increases in thickness (Fischer, 1908; Fischer and Schmieden, 1909). In tissue culture, the need for mechanical tension within the media appears necessary to propagate and maintain tension-bearing elements such as elastin and collagen fibers.

Thoma's (1893) histomechanical principles have provided the basis of these considerations with but minor modification. These principles are as follows:

(1) The increase in size of the lumen of a vessel, which is proportional to growth in surface of the vessel wall, is dependent upon the rate of the blood flow.
(2) The growth in thickness of a vessel wall is proportional to (dependent upon) the tension in the wall, which itself is determined by the diameter of the vessel lumen and by the blood pressure.
(3) Increase of blood pressure above a certain limit (which is defined by the metabolism of the particular tissue) leads to the new formation of capillaries.
(4) Increase in length of a vessel is governed by the tension exerted in a longitudinal direction by adjacent extravascular structures.

However, as Hughes (1943) warns, it is hazardous to extrapolate from postembryonic situations to those in embryonic development. Further, the development of differentiated arteries and veins belongs to the "functional period of development" (Huxley and de Beer, 1934), as witnessed by in vivo observations (Clark, 1918). Again, some exceptional patterns of development occur in various regions. Venous sprouting (as against arterial) is claimed in the retinal vessels in cat and human (Michaelson, 1948). In the kidney, after the initial ingrowth of branches of renal artery and vein forming the vasa recta, the glomeruli arise in situ from angiogenic masses proliferated from the visceral layer of Bowman's capsule, and these then link up (Lewis, 1958). As an exception to the usual primitive capillary plexus that precedes the laying down of main vessels, the dorsal intersegmental branches of the aorta arise as definitive single stems from the aorta.

When histogenesis begins in the single layered capillary tube, mesenchymal cells resembling fibroblasts become apposed to the endothelium externally (Clark and Clark, 1925), and the general mesenchyme adjacent continues to be the the source of the succeeding

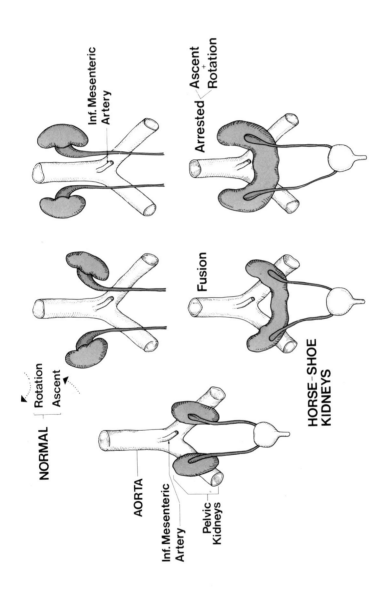

Fig. 21. The kidneys and their relationship to blood vessels is an interesting embryological transformation.
The "ascent" of the kidney from the pelvis changes its relationship to the aorta--from being below the bifurcation to reach up along it to the level of L2 vertebra. In this process, it receives blood vessels in succession from the primitive aorta that it discards in setps as it acquires its substantive renal artery. This "ascent" is dramatized by the failure of the horse-shoe kidney to ascend all the way, on account of the inferior mesenteric artery that halts the process by getting in the way.

vessel coats. This development proceeds in stages that have been described (Benninghoff, 1930; Hughes, 1943). By the 4th intrauterine month of development, vessels with three layers of tunic are seen and vasa vasorum have appeared (Massloff, 1914). In the 5th intrauterine month, the external elastic lamina separating the media from the adventitia appears.

Vessels established early in the embryo may later be modified to produce adult patterns. Aberrations of these processes may produce anomalous patterns. Thus, fusion of arteries are seen in the production of the basilar artery from the two vertebrals. Cross anastomosis and take-over is exemplified in the left brachiocephalic vein. Fusion and absorption of segments of vessels (or "longitudinal annexation") is exemplified in the inferior vena cava development and in the development of the popliteal segment of the adult arterial trunk of supply to the lower limb. "Migration" of the level of origin of an artery is well exemplified in the coeliac and mesenteric arteries and is discussed in that section. The headward migration of the kidney and the arrest of this process in "horse-shoe kidney" by the inferior mesenteric artery is a special case (Fig. 21). Thus, anomalies in their turn result from abnormal fusions, persistence of a channel usually discarded, or by the deletion of one normally retained.

<u>Venous valves</u> appear in the 4th month and are at full quota by the end of the 6th month (Kampmeier and Birch, 1927). An interesting unconfirmed "law" states that *"the interval between valves in veins of adults is some multiple of 7 mm for the lower limbs and 5.5 mm for the upper* (Bardeleben's law). *It is felt that muscle contraction and gravity are responsible"* for valves but their "causation" has yet to be clarified.

Branching is one of the aspects of arterial patterns that is conspicious. On the physiological principle of minimum work applied to the angle of branching arteries, theoretical patterns have been evolved (Fig. 22) (Murray, 1926). Thus, the minimum theoretical angle for a simple bifurcation is 75° and the daughter vessel size 79.4% of the parent trunk. Thoma's figures (1921-2) have been cited to match the theoretical expectation with measurements, and these do (Thompson, 1959) as the table indicates:

Internal Diameter of Abdominal Arteries (R. Thoma 1921-22)
(in millimeters)
(4 subjects)

Abdominal aorta	15	12	14.1	13.9
Right common iliac	10.8	8.8	10.4	8.6
Left common iliac	10.7	8.6	9.5	10.0
Mean	10.8	8.7	10.0	9.3
iliac : Aorta	0.72%	0.73%	0.71%	0.67%

It also turns out that the total cross sectional area of the daughter vessels in such a division (as in other simple dichotomous branching) is a 1.7 to 1.2 times increase of cross section on the parent trunk.

The angles of branching determine the caliber of the offshoot and have been theoretically deduced (Murray, 1926) (Fig. 22). This has been compared with actual measurements at the division of the branchial artery (Turner, 1959) and his results from 54 specimens are tabulated below:

	Radius - Range (in centimeters)	Mean	Theoretical* Angle (Mean)	Observed Angle (mean)
Brachial artery	(0.245-0.0988)	0.1851		
Ulnar	(0.206-0.0643)	0.1476	36.52)	) 38.63
Radial	(0.197-0.059)	0.1282	38.72)	) (standard error 2.407)

* The theoretical angle θ calculated on the basis of

$Cos\theta = \frac{r1}{R}$, R and r, being the ratio of the parent and branch vessels respectively.

Thus, Turner (1959) concludes that on the basis of optimal angle for minimum loss of pressure, as well as on the basis that flow is proportional to cross sectional areas, there is close correspondence between predicted and observed mean values.

The vertebro-basilar junction has also been examined in this manner, and the fit is similar (Turner, 1957). Thus, the basilar artery has a mean diameter of 0.1780 (range 0.0887-0.2863), while the right vertebral is 0.1316 (range 0.0456.0.2287) as against the left which is 0.1286 (range 0.0418-0.2343). The observed angle of 64° is a reasonable fit to theoretical deductions.

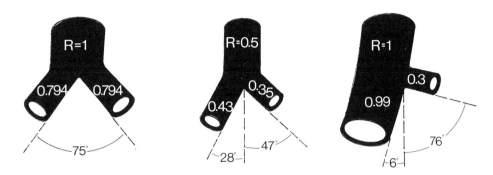

Fig. 22. At an arterial bifurcation the angles of branching of each daughter vessel is theoretically determined on hydrodynamic principles on the basis of its caliber. The diagram (after Murray, 1926), illustrates this pattern which is matched in many arterial bifurcations that exist.

There are quanitifications of branching in segments of circulations that are available. These are presented in table form (Tables (2) and (3)--(3) for the blood vessels in the bat's wing (Wiedeman, 1963); (2) for the mesenteric artery of the dog (Mall, modified by Schleier).

Vascular geometry of this kind is not specifically available for the human. It has been speculated that individual differences of vessel caliber to significant organs e.g., adrenals or pituitary glands may be responsible for predelictions to particular pathological states (Levine, 1964).

While 1 μ change of diameter in a 5 μ capillary is claimed to produce a 80% change in flow if all else remains constant, in large arteries a considerable degree of "critical stenosis" appears to be necessary to reduce the blood flow significantly to the specific region. Thus, an 80% reduction appears to be required in the dog iliac artery (May, de Weese and Rob, 1963), 70% in the abdominal aorta (Haimovici, 1954) and 45% to 55% in the aortic arch (Gupte and Wiggers, 1955). In converse terms, the cross section must be reduced to 20% of normal to compromise flow (May et al., 1963; Linddrom, 1950). Of course, there is continued growth of vessels throughout development in utero and into adulthood. Calibers of a few vessels as they grow with age have been presented in another chapter.

TABLE 2

Dimensions of Blood Vessels in the Bat's Wing

(after Wiedeman, 1963)

Vessel	Average length mm	Average Diameter μ	Average number of branches	Number of vessels	Total cross-sec area μ²	Capacity mm³ x 10⁻³	Percent of capacity
Artery	17.0	52.6	12.3	1	2,263*	38.4	10.1
Small artery	3.5	19.0	9.7	12.3	4,144	14.4	3.8
Arteriole	0.95	7.0	4.6	119.3	5,101	4.7	1.2
Capillary	0.23	3.7	3.1#	548.7	6,548	1.5	0.39
Post-capillary venule	0.21	7.3		1,727.0	78,233	16.4	4.3
Venule	1.0	21.0	5.0	345.4	127,995	127.9	33.7
Small vein	3.4	37.0	14.1	24.5	27,885	94.7	25.0
Vein	16.6	76.2	24.5	1	4,882	81.0	21.4

*Average of individual cross-sectional areas
#Calculated

TABLE 3

Branching of Mesenteric Artery of Dog*

	Total number	Radius (cm.)	Total cross-sectional area (sq.cm.)	Length (cm.)	Mean velocity (cm./sec.)	Pressure fall (mm. Hg)	Volumes (% of total)
1. Mesenteric artery	1	0.15	0.07	6.0	16.80	0.8	2.6
Main branches	45	0.05	0.12	4.5	10.10	3.2	3.4
End branches	15	0.03	0.13	3.91	9.30	7.4	3.2
2. Short and long intestinal branches	1,899	0.0068	0.20	1.42	5.80	23.5	1.7
Last branches	26,640	0.0025	0.57	0.11	2.10	7.2	0.4
Branches to villi	328,500	0.00155	2.48	0.15	0.48	5.4	2.3
Arteries of villi	1,051,000	0.00122	4.18	0.20	0.28	8.1	5.3
3. Capillaries of villi	47,300,000	0.00040	23.78	0.04	0.05	2.4 / 58.0	5.9 / 24.8
4. Veins at base of villi	2,102,400	0.00132	11.59	0.10	0.10	1.03	7.2
5. Veins before entering submucosa	131,400	0.00375	5.80	0.10	0.20	0.29	3.6
Last branches of submucosa	18,000	0.0064	2.32	0.15	0.51	0.37	2.2
Last branches of short intestinal veins	28,800	0.0032	0.93	1.1	1.30	2.50	6.4
Long and short intestinal veins	1,899	0.0138	0.84	1.42	1.40	1.40	7.4
6. Last branches of mesenteric vein	45	0.075	0.79	3.91	1.50	0.22	19.2
Branches of mesenteric vein	15	0.12	0.67	4.5	1.70	0.07	18.7
Mesenteric vein	1	0.3	0.28	6.0	4.2	0.05 / 5.93	10.5 / 75.2

*After Mall; modified from Schleier.

The umbilical vessels are unique, and interesting measurements have been made of them (Chacko and Reynolds, 1954). Thus, constriction of the artery may reduce its length by 17%, and as compared to the distended state, the wall thickness may increase 6 to 7 times (the ductus arteriosus appears to show a similar pattern of behavior). As a measure of growth in length, the umbilical cord at 3 months is 6 cm, growing at full term to be equal to the length of the fetus itself at that time, and averaging 54 cm. The length at term, however, may vary from 18 cm to 122 cm (Walker and Pye, 1960). The cord shows spiral tension due to fetal movement or to unequal growth of the umbilical fetal vessels, but specific causal factors determining this are uncertain.

Vascular development and histogenesis in particular tissues

Each tissue or organ has a specific pattern of histogenesis, and where mesenchyme is incorporated, the vascular development associated with this is often characteristic of the developmental organization of the tissue elements. In this section, a few interesting examples are dealt with to illustrate such patterns.

The retinal vasculature is of particular significance on account of the sensitivity of these vessels to neonatal oxygen therapy that may result in retrolental fibroplasia as a sequel.

The eye develops from three elements (Fig. 23);--the optic stalk that converts into a "cup" that leads back in a hyaloid groove along the optic stalk (neurectodermal derivative); the lens vesicle that is invaginated from the "skin" surface (ectodermal derivative) and soon loses contact with the surface; thirdly, the mesoderm containing blood vessels which, besides forming the outer coats of the eyeball, also invades the neurectodermal component. These structures arise early in development. The optic stalk in the 4th intrauterine week is quickly converted into the "optic cup" and the hyaloid artery then enters the "optic" ("choroidal" or "hyaloid") fissure on the inferior aspect of the optic stalk (30th day); at this time the optic cup has also induced the lens vesicle.

In the 5th intrauterine week there are three systems developing (Fig. 23):

(1) the choroidal capillary net - surrounding the optic cup. This develops from the surrounding vaso-active mesoderm and is later enhanced by anterior, posterior, and long ciliary arteries.
(2) the hyaloid system - supplying the interior of the eye including the primary vitreous and reaching the lens vesicle to supply its vascular tunic.

Fig. 23 - The arterial development of the eye

The optic vesicle (from ectoderm) and the optic cup (neurectoderm from forebrain) come together in a supporting bed of mesenchyme that contains blood vessels. The "central artery of the retinal" begins to follow the path of the optic cup (1).

In (2) the central artery continues through to become the "hyaloid artery" tunneling the vitreous body to reach the lens where it anastomoses circumferentially with vessels derived from the collagenous coats of the eyeball. These systems are elaborated somewhat, but in a later stage the interconnections have petered out (3). In the pre-natal stage, the hyaloid artery is also eliminated so that the patent central artery of the retina stops short at the rear of the vitreous at the optic disc shere it branches out into the retina.

In-set (4) illustrates the pattern of supply of retinal layers, leaving outer layers of the retina avascular, between this superficial invasion by retinal artery capillaries and the externally located vascular plexus (choriocapillary layer) in the choroid coat of the eyeball.

(See p.vii of color insert)

(3) the primitive annular system - of vessels derived from the hyaloid artery in part and from the anterior ciliary arteries around the rim of the optic cup.

The "tunica vasculosa lentis" is a mesodermal and vascular layer that invests the developing lens, the anterior portion of which is called the pupillary membrane. This membrane separates the anterior and posterior chambers and contains delicate vascular loops maximally developed in the 5th intrauterine month. Resorption of the pupillary membrane and its vascular loops begins in the 7th intrauterine month and is complete in the 8th month, only traces remaining at term. The optic ("choroidal") fissure closes in the 6th to 7th weeks, so that the central artery of the retina with its companion vein is contained centrally in the anterior part of the optic nerve.

The chorio-capillary layer is developed synchronously with the pigmentation of the outer (pigment) layer of the optic cup. Like the pia arachnoid, of which it is a forward continuation, it is highly vascular by the 6th intrauterine week. Early in the 5th intrauterine week (6 mm stage) an interesting transient vascularization of the optic cup occurs. Fine capillaries invade the retinal layer as they do the rest of the neural tube at this time, reflecting the exceptionally active growth here. Very shortly after (end of 5th week or 7 mm stage), degeneration of these vessels occurs rapidly, and they are absent from the retina until the 5th intrauterine month (100 mm stage). This transient vascularity is a feature of the mammalian retina (Mann, 1928). In contrast, in the case of the cerebral vessels, this first vasculature goes on to become the definitive supply of the cerebral cortex. During this period, retinal cytodifferentiation proceeds from the posterior pole (13 mm stage) to reach the equator (20 mm) and the ora serrata (65 mm).

In the fifth month, formative vacularization of the retina commences and proceeds peripherally, the temporal retina being the last part to be accomplished. This may proceed to occur asymmetrically on the two sides; this correlates interestingly with the fact that retrolental fibroplasia is found in 90% of cases in the temporal quadrant near the ora serrata and may be unilateral. It is claimed that the presence of oxygen inhibits the development of capillarity in the retina; in fact, capillary free zones exist around the retinal arteries, and increased oxygen is said to enter these areas (Ashton, 1957). However, this does not appear to be self limited and is affected by O_2 diffusion from the chorio-capillaries, since it ceases to occur if the retina is detached (Ashton, 1957). We must recall here, that in the neonatal and adult retina, only the inner one-third is supplied by the retinal vessels. The outer two-thirds of the retina is devoid of any vessels, and diffusion from without, from the chorio-capillaris, must traverse the membrane of Bruch to reach the retinal pigment layer at first.

Pari passu with vascularization, functional histomorphology is established in the retina. All the layers are distinct by the 6th intrauterine month and the retina is light sensitive in the 7th intrauterine month. However, neonatal development is important. Thus, perception evolves in the first year of life, and in keeping with this, the fovea completes development (in the 13th month, Michaelson, 1948; Cogan, 1963). Color vision is achieved yet later (2nd year).

The retinal arteries are of interest for the further reason that they may reflect cervical artery stenosis (Wright, 1974), as indeed they do the general arterial status. With these considerations in mind, Parr and Spears (1974) have reported the caliber of retinal arteries expressed as the equivalent width of the central artery of the retina.

The nervous system is vascularized very early in development, and vessels from the pia enter the cortex, penetrating it radically. Cross connections between these build up a series of arcades that occupy the thickness of the developing cortex (Lewis, 1957). While long corticomedullary vessels do penetrate into the white matter, a relatively avascular junction exists between this cortical capillary plexus and the medullary one derived from central vessels that penetrate the base of the brain and proceed toward the surface. There is a uniform capillary plexus that gradually becomes elaborated and differentiated as the histogenesis of the neural tube progresses. In addition, capillary density increases. Thus, estimates of capillary lengths in the 3 to 4 month fetus in utero have been given as 20 mm/mm^3, increasing to 128-167 mm/mm^2 at term, (Niemineva and Tervila, 1953). Further, capillary loop counts in various regions demonstrate angioarchitectonic specialization in utero (Blinkov and Glezer, 1968).

The muscles arising from paraxial mesoderm are in close proximity to intersegmental arteries. These invade the pre-muscle masses and are the first tissues to be vascularized shortly after the nervous system. Thus, myotomes carry segmental nerves and intersegmental arteries with them in their subsequent migrations (Mall, 1898; Evans, 1912). Specific patterns of development in muscle vascularization are contained in the literature (Spalteholz, 1888; Campbell and Pennefether, 1919; Le Gros Clark and Bloomfield, 1945; Bloomfield, 1945).

The heart muscle is first pervaded by endocardial channels in situ in the developing myocardium and giving rise to a labyrinthine sinusoidal system in the developing trabeculae of muscle. The coronary arteries bud out of the aortic sinus, the left first, and then the right coronary (Licata, 1962).

Cutaneous vessels - in the head region have been studied (Finley, 1923). In their development, four zones have been identified (Fig.

1). An innermost vascular net of open blood vessels leading to a zone of solid vascular networks awaiting canalization, a zone of active endothelial sprouting following this, and in the outermost part the avascular zone awaiting invasion. The "hair pin" papillary patterns evolve gradually, but these are not unchanging even in the adult (Ryan, 1967).

Cartilage - is free of blood vessels in the adult, being nourished by perichondial vessels. In fetal cartilage, however, cartilage canals of vessels traverse developing cartilage. The general pattern of cartilage canals is established in the 7th intrauterine week (C.R. 22 cm) (Brookes, 1958). In the mandible, vascular canals develop in the 5th month (130 mm C.R.) and are present until the 3rd year of life (Blackwood, 1965). Human fetal diaphyseal cortex is arterialized from the medulla; a periosteal arterial supply is absent, but capillaries of the two systems do anastomose (Brookes, 1958).

Vasa vasorum in the aorta - have been assessed in neonates, at which time the ascending aorta and arch show greatest density, falling off at the end of the first year of life. The reverse appears true for the descending aorta (Clarke, 1965). It has been claimed that successively the outer, middle, and inner thirds of the aortic media are vascularized in the 4th, 10th and 13th years; however; in quantitation, the abdominal aorta has the greatest vascularity of any part by the 15th year (Clarke, 1965).

Pulmonary arteries in the lung undergo changes which in part reflect the nonfunctional intrauterine state and attendant changes at and after birth (Civin and Edwards, 1951; O'Neal et al., 1957 and Rosen et al., 1957). In utero, the arteries of the lung are relatively thick, small, and muscular--an adaptation that restricts pulmonary blood flow in favor of diverting it via the main pulmonary artery via the ductus arteriosus and the aorta. In the last trimester, in utero the arteries are intralobular structures lined by thick endothelial cells with an ill-defined internal elastic lamina appearing six months after birth to separate it from the tunica media. After birth, the lumen enlarges as does the overall external diameter, elastic laminae become evident in six months after birth, and elastic fibers appear in the media as well. This process continues even up to 20 years of age. Arterioles are clearly recognizable after the 3rd month of life. Thus, while the thick-walled arteries in utero are transformed into normal thin walled adult proportions in the first year of life (Valenzuela et al., 1954; Civin and Edwards, 1951), their ultimate transformations are incomplete until 20 years later (Spencer, 1962).

3. Embryology of Limb Vasculature

Development of the vasculature of the upper limb (Figs. 24 and 25)

The upper limb buds appear at the end of the fourth intrauterine week and the vascular plexuses associated with four to five intersegmental branches of the dorsal aorta extend into the limb bud. Through this preliminary plexiform capillary arrangement, the 7th intersegmental artery becomes established as the primitive axial artery (the subclavian--axillary-- brachial--interosseons trunk), and marginal veins return blood from the limb bud to originate what will become "cephalic" and "basilic" venous channels (Woollard, 1922).

In the next stage, the median artery develops as a branch from the axial artery and connects with the capillary plexus in the terminal part of the limb bud. The interosseons artery (the distal part of the primitive axial trunk) regresses, and the median artery is now the main contributor to the digital plexus.

In the next development (6th week), the ulnar artery arises more proximally from the axial vessel, and establishing connection with the median artery, forms the carpal arch that provides the digital arteries for the developing fingers.

In the 7th embryonic week, a third channel develops yet more proximally from the axial vessel. This is the superficial brachial (Senior, 1926), or the superficial axillary (Singer, 1933a) artery, which runs on the medial side of the arm and diagonally in the forearm from ulnar to radial side to terminate on the dorsum of the wrist supplying the thumb and index.

By the 8th week the adult pattern is establishing itself. The median artery regresses to become the companion artery of the median nerve (which in the adult is represented as the branch of the anterior interosseons artery to the median nerve).

The distal part of the superficial brachial artery, having established connection with the axial vessel at the forearm, becomes the substantive radial artery of the adult.

The ulnar artery remains as the branch from the axial vessel, and with the radial forms the carpal arches in the hand. The profunda brachii is a later development from the brachial segment of the axis artery. However, it is claimed that the profunda represents the original axial artery and that the rotation of the limb is what made the brachial artery the dominant vessel (Strandness, 1969).

In summary, the proximal part of the axial artery becomes the subclavian--axillary--brachial trunk, while distally it is represented in the adult by the anterior interosseons and the "arteria nervi mediana." The radial, ulnar and profunda branches are the later major developments that are represented in the adult.

Variations

1. The usual intersegmental vessel that enlarges to become the axial artery of the upper limb bud plexus is the 7th intersegmental. However, other segmental vessels may take over this function in whole or part (Miller, 1939) when the relationship of the brachial plexus nerve roots becomes correspondingly anomalous. These are described in detail, and some of the characteristic relationships are redrawn (Fig. 25) from that report (Miller, 1939).

2. The axial artery, according to Singer (1933a) is represented by two embryonic channels superficial and deep axillary arteries. The deep vessel is the normal persistant adult artery and is responsible for separating the cords of the brachial plexus in the usual way, the arteries lying between the lateral (C5, 6, 7) and medial cords (C8 and T1). If the superficial axillary artery persists instead, the brachial plexus cords are undivided and may be fused together as a single cord as has been recorded (Singer, 1933b; Hasan and Narayan, 1964). Precise relationships of intersegmental arteries to other segmental structures has been clarified by Prader (1947).

3. The profunda humeral artery is established later, though the network of vessels spiraling the humerus exist early in limb vascularization. Variations at this level are common and have been examined by Grant (1962) in 235 specimens. Fig. 25 is redrawn from this paper and gives the incidence recorded for these variants. Several other variants of the axillary arteries are described. Thus, the axillary artery may have 6 to 11 branches (Huelke, 1959).

4. The "superficial brachial" artery running superficial to the median nerve may persist in the adult. This vessel may also give rise to a "high origin" of the radial, or more rarely, the ulnar artery (Keen, 1961). The danger of superficial vessels in the cubital area has been noted (Hazlett, 1949; Ganon, 1966), and Ganon records an incidence of 24% in cases of gangrene resulting from accidental intra-arterial injection. Wankoff (1962) from an extensive study has derived a scheme to account for the many variations that had been reported (Charles et al., 1931; Charles, 1894; De Gains and Swartley, 1928; McCormack, L. J., et al., 1953; Misra, 1955). Fig. 25 is redrawn from Wankoff's report to list the variations and identify their embryological determinations.

The radial and ulnar arteries are rarely entirely absent, but if this does occur, the anterior interosseous artery substitutes for them (Chaatrapati, 1964).

EMBRYOLOGY OF THE HUMAN ARTERIAL SYSTEM

5. The brachial artery may occasionally be contained in a bony supracondylar foramen (of the humerus), a condition normal in the cat.

6. The deep palmar arch is completely represented in 97% of cases, while the superficial is complete in 80%, and this latter has five types of formation.

Common Anomalies of Upper Limb Arteries   (modified from Keen, 1961)

		% incidence in 142 subjects
1.	Subscapular artery from 3rd part of axillary	71%
2.	Common trunk for subscapular and post. humeral circumflex	31%
3.	"High origin" of subscapular artery	29%
4.	Profunda brachii from 3rd part of axillary	26%
5.	"Superficial brachial artery" (all types)	12.3%
6.	Strong median artery reaching palm of hand	9.5%
7.	Profunda brachii from circ. humeral (passing behind terres major)	7%
8.	Profunda brachii arising from post. circ. humeral	6%
9.	"High origin" of radial	5.9%
10.	Post. circ. humeral passes below teres major and ascends behind it.	2.8%
11.	"High origin" of ulnar	2.8%
12.	"Superficial radial"	1%

Figs. 24 and 25 - These illustrate developmental aspects of upper limb vascularization. The axial artery [24(2)] is indicated in red and the secondary branches and plexuses in gray. Embryological anomalies [24(2)] show the abnormal vessel development indicated. The arteries are "intersegmental" structures and the seventh intersegmental is usually the axial vessel. If adjacent ones take over this major role their relation to the brachial plexus reveals this anomalous development. (See p.viii of color insert).

The dorsal carpal arch has been classified into six types. Details of these terminal anastomoses in 650 limbs are on record. (Coleman and Anson, 1961).

<u>Veins</u> - Marginal veins appear on the radial (preaxial) and ulnar (postaxial) sides of the limb bud (6th week). The ulnar marginal vein becomes the basilic--axillary--subclavian trunk. The subclavian vein originally opens into the post cardinal vein, but with caudal cardiac migration, has moved to open into the precardinal vein.

The radial marginal vein is less developed, until it gives rise to the cephalic vein that originally opens into the external jugular vein. Later, it transfers drainage to an intraclavicular position, into the axillary vein.

Development of the lower limb vessels (Fig. 26)

The umbilical circulation must develop very early, and therefore in the second intrauterine week the umbilical arteries arise from the unfused dorsal aorta on each side and run to the body stalk medial to the (pro/meso) nephric duct.

When the dorsal aortae fuse in the midline, a new vessel presumptively derived from the 20th intersegmental (5th lumbar) artery connects the caudal end of the aorta to the umbilical artery. This lies dorsal to the nephric duct, and with its appearance the original ventral segment disappears.

With the development of the hind limb band, an axial vessel arising from the proximal portion of this secondary umbilical artery passes into the plexus in the limb band (Senior, 1919). This vessel has been called the "sciatic artery" or "ischiadic artery" (9 mm stage).

In the next stage, a new vessel arises from the middle of the new dorsal root of the primitive umbilical artery. This is the commencing external iliac-femoral trunk of the adult. This vessel extends into the leg, establishing several connections with the axial vessel (Senior, 1919 and 1920), producing several branches to become the major arterial supply of the limb. It takes over the digital vessels in the foot as the axial vessel recedes. The anterior tibial artery, developed as a longitudinal section of anastomoses with the axial artery and its uppermost connection, while the popliteal vessel becomes the adult arrangement. In the foot, it terminates as the dorsalis pedis artery and supplies this area.

Thus, the axial vessel is represented by the inferior gluteal artery, the artery to the sciatic nerve (a branch of the former artery), a proximal segment of the popliteal artery and a part of the peroneal artery). The obturator artery arises late in development

EMBRYOLOGY OF THE HUMAN ARTERIAL SYSTEM

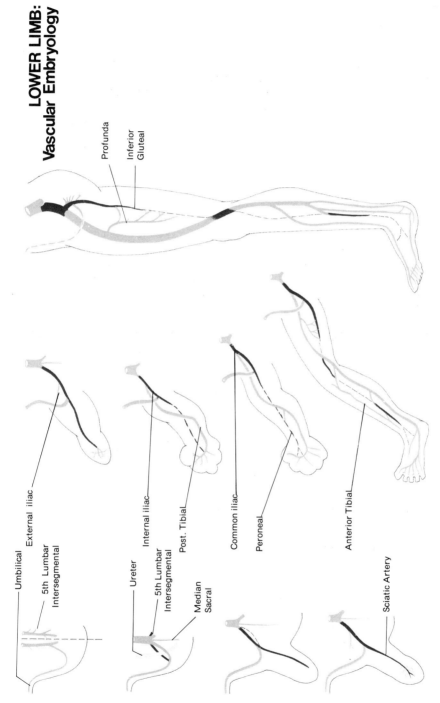

Fig. 26. Vascular embryology of the lower limb - The axial vessel is marked in red.

as a plexus situated on the medial side of the thigh. By the third month, the adult pattern is established.

The venous pattern is developed as in the upper extremity upon the pre- and postaxial marginal veins that drain the limb band. The connection between these is established to become the dorsal venous arch. Postaxially, this is drained by the short saphenous vein that continued proximally as the popliteal, inferior gluteal, and internal iliac veins into the embryonic posterior cardinal vein.

In later development, the short saphenous vein, established as a superficial vein, ends by draining into the popliteal vein in the adult.

The preaxial vein becomes the great saphenous vein continued deeply into the femoral vein at the saphenous opening. In the 20-30 mm embryo, an anastomotic development takes over the drainage of the thigh into the long saphenous from the short saphenous. Thus, as a variation, arrest at any of these stages may cause a persistent short saphenous drainage directly up the thigh into the inferior gluteal vein, or a situation where the short vein connects anastomotically with the long saphenous.

The venae comitantes of the deep arteries are secondary developments connecting with the popliteal and femoral venous segments.

Anomalies on Lower Limb Arteries

(after Keen, 1961)

		% Frequency in 140 subjects
1.	Dorsalis pedis predominant in forming plantar arch	48%
2.	Both circumflex arteries from profunda fenions	42%
3.	Dorsalis pedis and lateral plantar equal in size	35%
4.	Medial circumflex from femoral	31%
5.	Lateral circumflex from femoral	20%
6.	Lateral plantar is dominant in forming plantar arch	17%
7.	Both circumflex from femoral	7%
8.	High division of popliteal	5%

Anomalies on Lower Limb Arteries (cont.)

(after Keen, 1961)

		% Frequency in 140 subjects
9.	Peroneal artery replaces dorsalis pedis	5%
10.	Trifurcation of popliteal	4.3%
11.	Absent medial circumflex	2.5%
12.	Peroneal replaces post tibial	2.5%

4. The Fetal Circulation and Birth Adjustment

The substantive fetal circulation is modified in uterine life by hemodynamic changes in its component parts. However, the general pattern of this circulation is maintained in utero. In the fetal circulation (Figs. 28 and 29) the blood is so distributed that the right ventricular output is about twice that of the left, and admixed blood from both ventricles and atria are passed around the general body circulation, in particular to the lower half of the body. Thus, in order to express the magnitude and capacity of the various critical fetal pathways involved, the "combined ventricular output" is used to assess percentage flow in the respective circuits.

The placenta receives an increasing blood supply via the umbilical arteries, from 17% to 33% from 10 (50 g. wt.) to 20 weeks (150 g. wt.) of gestation respectively in the human fetus (Rudolph et al., 1971). These figures were sensitive to pH changes and to the effects of hysterotomy, and a figure of 49% of the combined ventricular output in a 250 g. fetus has been recorded (Rudolph et al., 1971). In the fetal lamb near term, the figure is up to 36% to 42% of the combined ventricular output. It is a fair estimate that the umbilical vein at term returns rather less than half of the combined ventricular output, and the O_2 saturation of the contained blood is about 80%.

Inferior vena cava venous return

64% to 75% of the combined ventricular output is returned to the right atrium via the inferior vena cava, the higher figure being reached as early as the 20th week in the human fetus.

The portal circulation accounts for 5.5% to 9.2% of the combined ventricular output, i.e., the amount delivered by the gut arteries and carried thence via the portal veins.

The infra hepatic vena cava returns about 30% to 40% of the combined ventricular output. The total return of the inferior vena cava to the right atrium (64% to 75% of combined ventricular output) is broken down as follows:

Portal vein (5.5% to 9.2%)   )
) via ductus venosus or hepatic
Umbilical vein (17% to 33%)   )   veins

Infra hepatic inferior vena cava (30% to 40%)

The Fetal Hepatic Circulation

The single persisting left umbilical vein, connecting with the left branch of the portal vein, supplies the left lobe of the liver in the main with highly oxygenated placental blood. The main portal vein likewise distributes blood to the right lobe preferentially. Though in fact admixed blood from the two sources (portal and umbilical) reaches both lobes, the left lobe is favored in utero with a better oxygen supply. Indeed, some degenerative changes have been demonstrated in the right lobe of the liver before birth (Gruenwald, 1955).

The "ductus venosus" that by-passes the liver circulation, by connecting the left branch of the portal vein to the suprahepatic inferior vena cava, shunts an amount of blood ranging from 20% to 80% of the combined umbilical and portal returns (25% to 40% of the combined ventricular output). It is unclear what determines this wide range of umbilico-portal blood shunt. The portion is independent of total umbilical venous return, is not related to fetal age, and appears to adapt its caliber passively to changes in venous flow; in this latter regard, the ductus is not particularly sensitive to PO_2, PCO_2 or pH and only mildly constricted by large doses of norepinephrine. It is possible that different fetuses have varying hepatic resistences, or some anatomical variant of the umbilical vein may account for differences (Rudolph, 1971). However, a smooth muscle sphincter has been identified at the junction of ductus venosus and umbilical vein, and a vagal innervation of the ductus has been claimed. Further, it is suggested that when umbilical vein pressure rises, as when the uterus contracts, this sphincter diverts blood into the liver to protect the fetal heart from direct overload (Lind et al., 1965).

Superior Vena Cava

This vein empties the venous return from the upper half of the body (including brain, head and neck, upper limbs, and upper half of thoracic walls) to the right atrium and accounts for 20% to 25% of the combined ventricular output return.

Atrial Circulations

The right atrium directly receives all the fetal venous return except that of the pulmonary veins (7% of combined ventricular output). This consists of the following:

Inferior vena cava return - 64% to 75%

Superior vena cava return - 23% to 32%

Coronary sinus return - 3.4%

and negligible returns from "venae cordis minimae."

It has long identified (Kilian, 1826) that the inferior vena cava stream of blood entering the right atrium divides into two streams, since the "crista dividens" margin of the foramen ovale straddles the inferior vena cava orifice. Supposedly, the layer stream (the "via sinistra") passes through the foramen ovale into the left atrium. While in the early fetal stages, higher volumes pass in the via sinistra, lower figures were obtained as fetal growth progressed (range 74% to 14%) (Rudolph, 1971). The "via dextra" component of the inferior vena cava blood passes into the right artium, and admixed with superior vena cava and coronary venous blood, passes into the right ventricle. The superior vena cava stream itself, as a percentage of total cardiac output, falls steadily with increasing fetal age (Rudolph, 1971). From the right atrium, two exits are available:

1. Through foramen ovale to left atrium ("via sinistra") - 33% combined ventricular output. Most of this is derived from the inferior vena cava stream (25% to 28% combined ventricular output) that is, the portion partly deflected by the valves.
2. Through the tricuspid valve into the right atrium ("via dextra") - 66% combined ventricular output. Nearly all the superior vena cava stream (22% to 25% combined ventricular output) passes along this course. In addition, 38% to 42% combined ventricular output from the inferior vena cava is added to this, as well as the small (3% to 4% conbined ventricular output) coronary sinus return.

Left (side of the) heart circulation

The left atrium receives about 33% of the combined ventricular output made up as follows:

1. Through "via sinistra" (largely inferior vena cava blood) - 25% to 28% combined ventricular output.
2. Direct pulmonary vein return - 5% to 10% combined ventricular output.

Since inferior vena cava blood contains the higher proportion of oxygenated (umbilical vein blood - 80% O_2 saturation) blood, this averages out at about 65% O_2 saturation. Passing into the left ventricle, two-thirds of this blood (22% combined ventricular output) is thence distributed to the upper half of the body via the aorta and the aortic arch. The remainder (10%) continues beyond the arch part of the isthmus into the descending aorta.

Right (side of the) heart circulation

The right ventricle receives 66% of the combined ventricular output representing superior vena cava blood (22% to 25% combined ventricular output), part of the inferior vena cava return (28% to 42% combined ventricular output) and that from the coronary sinus (3% to 4% combined ventricular output).

The pulmonary trunk in the fetus leads directly to the ductus arteriosus which carries most of the blood (55% to 60% of the combined ventricular output) into the descending aorta, and a smaller portion (7%) into the lungs via the pulmonary arteries. In utero, there is an increase in pulmonary blood flow as the number of respiratory units increase during fetal growth. Even though individual arteries are developing more media muscle, the vascular resistance decreases with this pulmonary growth. Goldberg found that low O_2 environment in the fetal rodent produces a thicker media and vice versa (Rudolph, 1974).

The descending aorta - This receives two contributions to make up 65% to 70% of the combined ventricular output it receives, namely from the ductus arteriosus via the right heart or pulmonary trunk (55% to 60%) and rather less (10% combined ventricular output) through the aortic isthmus from the aortic arch stream of the left side.

The measurement in previable human fetuses by the injection of 50 μ labeled microspheres, and calculated from the radioactivity in organs thereafter, have produced interesting results (Rudolph et al., 1971). The lower half of the body has a considerably larger portion of the output (81% to 54%) and the proportion in this for the placenta increases with age from 17% to 33%. The gut supply rose from 5.5% to 9.2%, while the kidney supply fell from 6.5% to 3.2%. These measurements were made in fetuses ranging from 64 g. to 225 g. The blood pressure was recorded in them at systolic/diastolic = 33/22 (to 46/27). Rudolph (1971) states that *"there were no statistically significant changes with growth in the percentage of cardiac output distributed to the brain, heart, adrenals and spleen ..."* Averages for brain flow = 14%, myocardium 2.6%, and adrenal 5% of the cardiac output in each case.

The distribution in summary is approximately as follows:

- gut - 5.5% to 9.2% combined ventricular output
- umbilical arteries - 17% to 33% combined ventricular output
- other branches of aorta - 30% to 40% combined ventricular output

  (kidney - 6.5% to 3.2%).

In Summary

The right ventricle via the pulmonary trunk ejects 66% of the combined ventricular output
- 59% passes via ductus arteriosus into descending aorta
- 7% enters pulmonary circulation

The left ventricle via the aorta ejects 33% of the combined ventricular output
- 20-22% passes into brain, head, trunk and upper limbs
- 10% passes via aortic isthmus into descending aorta.

In the fetus, the diameters of the vessels are in rough proportion to the proportion of blood flow these transmit:

Ascending aorta	- 31%		(Inferior vena cava	- 69%
Pulmonary trunk	- 66%	Veins	(Superior vena cava	- 21%
Descending aorta	- 59%		(Pulmonary veins	- 7%
Aortic isthmus	- 10%			

The fetus lives in a watery environment, recapitulating its phylogenetic history. Of further interest is the turnover of water and watery constitutents in utero (Fig. 25).

BIRTH ADJUSTMENTS (Figs. 26 and 27)

Birth is attended by dramatic events, the most critical of which is the <u>act of respiration</u> which replaces a positive fluid pressure with a negative air pressure in the lung alveoli. Indeed, within minutes postnatally, the lungs are fully expanded (Lind Stern and Wegelius, 1969). Thus, by sharp lowering of peripheral vascular lung resistance (estimated at 80% reduction), the pulmonary blood flow is opened up rapidly and left atrial pressure rises; diversion of blood into the lungs is accompanied by <u>closure of the ductus arteriosus.</u>

Fig. 27 - Is a diagrammatic illustration of the fetal circualtion.
  The inferior vena cava - returns 65 to 75% of the venous
  blood. Of this, 20 to 80% is by-passed via the ductus
  venosus. The left half of the liver receives most of the
  umbilical vein blood.
  The arterial flow to the gut equals 5.5 to 9.2%, the kidney
  6.5 to 3.2% and the placenta 17 to 33% of the combined ventricular output. 30 to 40% of the combined ventricular output returns via the inferior vena cava.
  The superior vena cava - returns 23 to 32% of the venous
  blood, while the pulmonary vein (7%) and the coronary sinus
  (3.4%) return a small percentage of the combined ventricular
  output.
  The right ventricle - receives 66% of the venous return
  which is made up of superior vena cava 23%, inferior vena
  cava 40% and carries 3% approximately.
  50 to 60% of the combined ventricular output is shunted
  from the pulmonary artery to the post ductal aorta and the
  smaller fraction goes to the lungs.
  The left ventricle - receives 33% of the venous return and
  of this 22% goes to the upper limb and head. (See p.vii of
  color insert).

The placental circulation at term accounts for about 40% of the total cardiac output. During the second stage of labor, when the placenta is still attached to the fetus, contraction of the umbilical vessel walls cuts down the placental blood flow. If three minutes are allowed to elapse before the cord is tied, some 200 cc of blood returns to the fetus from the placenta before shutdown. The removal of the source of this flow of blood reduces right atrial venous return, and the pressure in this chamber falls.

We have a rise in left atrial pressure on one hand and a fall in right atrial pressure on the other. This closes the foramen ovale by compressing the primum flap of the septum (lying on the left) against the septum secundum. Thus, mechanical pressure effects result in functional closure with the first breath. This results from increased pulmonary venous return to the left atrium and diminished inferior vena cava return from cessation of placental circulation.

The closing down of the placental circulation reduces the output of the descending aorta by some 40%, and this produces a marked rise in systemic vascular resistance. The inferior vena cava blood flow is also reduced, and the ductus venosus flow is diminished to become negligible in a short time.

Thus, at three sites, vascular shutdowns play a critical part --namely at the umbilical vessels, the ductus venosus. The structure of these vessels shows considerable smooth muscle and an innervation (Boyd, 1941; Baron, 1942; Noback, Anderson and Cooper, 1951; Holmes, 1958), making it possible that a nervous reflex may initiate closure in them. It has been shown that a rise in PO_2 closes the ductus arteriosus and vice versa (Oberhansli-Weiss et al., 1972) The ductus venosus appears to be only mildly constricted by large doses of norepinephrine and to be insensitive to PO_2 and pH: it is affected passively by pressure changes and the umbilical vein shut-down must affect it directly. (McMurphy et al., 1972). The part played by vasoactive substances such as bradykinin and prostoglandins is uncertain.

5+6 - NEONATAL DEVELOPMENT AND THE ESTABLISHMENT OF THE ADULT CIRCULATION

At birth, the fetal circulation containing by-passes and a placental circulation to allow the ventricles to work "in parallel," has been converted into two systems (systemic and pulmonary) connected "in series" and a placental circulation has been replaced by the opened-up pulmonary circuit. Shutdown in the ductus arteriosus and venosus has been "physiological" as it has been in the intraabdominal part of the umbilical vein, after the umbilical cord has been tied and sectioned.

The ductus arteriosus before closure, diverted 55% to 60% of the combined ventricular output away from the pulmonary circulation, acting as physiological right-left shunt. At term, it is about 10 mm in diameter and is shut "physiologically" or "functionally" 10 to 15 hours after birth. The stimulus to close appears to be the PO_2 rise (60 mm to 95 mm O_2) after birth, for this effect is reversible, and the ductus may open if the PO_2 is developmentally immature and accounts for less efficient closure in these cases. While it does finally close, it may take 10 weeks or more postnatally to do so. (Danilowicz et al., 1966). Normally, anatomical closure is complete in 10 to 21 days. At two weeks postnatally, the ductus is said to be partially open in 65% of infants (Christie, 1963). The neonatal blood pressure falls to 2/3rd the value at birth in the first 24 hours and rises back slowly. The reversibility of ductus arteriosus closure in this early period provides a potentially functional "parallel" arrangement of systemic and pulmonary circulations that can accommodate dynamic circulatory changes in the immediate postnatal period. There is first a progressive decrease in responsiveness to PO_2 changes, after which, in a few days small intimal hemorrhages, thromboses, and fibrosis are seen in the intima and the media as well. Ultrastructural features that have been described as anticipating closure actively are (a) subendothelial vacuolization, (b) extension of smooth muscle cells through internal elastic lamina, (c) interruption of elastic laminae, and (d) distended endoplasmic reticulum (Jones et al., 1969). The actual blood supply of the ductus wall has not been unequivocally demonstrated in man. The muscle of the wall is spirally arranged and thicker than that of the adjacent aorta and pulmonary arteries, so that contraction both shortens and narrows its lumen. The intima, too, is four to five times thicker than that of the neighboring vessel walls and consists of loose areolar tissue with a mucimoid appearance. "Idiopathic" potency of the ductus has been blamed on the absence of muscle in it (e.g., in utero fibrosis), congenital factors and a lack of sympathetic innervation.

The ductus constriction is most prominent initially on the pulmonary side of the ductus. A small projection from the descending aorta at the ductus attachment is seen angiographically for several postnatal weeks and is called the "ductus ampulla." The site of this ampulla is 5 to 10 mm distal to the origin of the left subclavian from the aorta and corresponds to the point of attachment of the ductus arteriosus previously. The ductus arteriosus and its postnatal behavior is of obvious significance to the status of the circulation. Detailed considerations of this and its relation to congenital circulatory anomalies are contained in a recent monograph (Cassels, 1973).

The pulmonary trunk and its immediate branches increase in size with postnatal increase of flow at the expense of the ductus, so that these become the mainstream instead (Fig. 27).

The aortic isthmus is influenced by changes that are undergone by the ductus arteriosus. Thus, after birth the normal aortic isthmus is about 25% narrower than that of the descending aorta. Coarctation of the aorta takes various forms, and the narrowed (or even interrupted) segment, though usually "juxtaductal," may be "preductal" or "postductal" and be very localized or extend some length on the aorta. Thus, the site rarely may be between brachiocephalic and left common carotid arteries, or between the left common carotid and the left subclavian. The usually stenotic segment is the "aortic isthmus" between the left subclavian and the ductus; this may extend into the left subclavian or be restricted to the aorta opposite the ductus alone. Many suggested explanations are proffered in regard to coarctation. Abnormal development of the third, fourth or fifth left brachial arch arteries and segments between them are a possible interpretation. It has also been proposed that extension of some contractile tissue from the ductus to the adjacent aorta may implicate that segment with birth contraction as in the ductus itself and produce coarctation (Skoda et al., 1893). Altered fetal hemodynamics may be responsible for the production of coarctation, and instances of juxtaductal narrowing be associated with the existence of a posterolateral aortic shelf opposite the attachment of the ductus (Heyman, Spitznas and Rudolph, 1971).

The ductus venosus closes physiologically at birth with the shutdown of the umbilical vein postnatally. However, for the first few (3-7) days after birth, indicator substances appear to shunt rapidly to the right atrium via this route. In fact, the ductus can be negotiated by a soft catheter via the umbilical vein during this time. After this time, it degenerates to become the ligamentum venosum. It is of interest, however, that the obliterated umbilical vein can be cut down on and cannulated surgically even in the adult (Gonzalez, 1959; Boyley, et al., 1967 and Braastad et al., 1967). The umbilical arteries degenerate, and in the abdominal wall are represented as obliterated cords--the medial umbilical ligaments, attached to the distal part of the internal iliac artery or its superior vesical branch.

The Foramen Ovale and Atria

The dimination of the umbilical venous return from the placenta at birth markedly reduces inferior vena cava return to the right atrium. This lowers the right atrial pressure, and the lowered pressure of the inferior vena cava stream meeting the crista dividens reduces the force that previously helped to open the foramen ovale. Simultaneously the increased pulmonary blood flow with ventilation of the lungs at birth increased venous return to the left atrium. The combined effect is the reversal of pressure gradients in the right versus the left atrium, the latter becoming 1 to 2 mm Hg higher than the right atrium postnatally. However, though this pressure difference is sufficient to close the flap value of the foramen ovale in the early postnatal period, this is not sustained, and shunting (R → L) occurs from time to time. Thus, the pressure tends to be higher in the right than the left atrium during systole. Further, a rise in pulmonary vascular resistance such as occurs in breath-holding or crying may raise right heart pressures sufficiently to open the foramen ovale. This reversed flow has been demonstrated angiographically in the first postnatal week (Lind. et al., 1969) and is a cause for some cyanotic episodes associated with crying in early neonatal life. Firm septal closure usually takes 2 to 3 months though this may be prolonged to 6 months, progressing from below upward. An opening patent to a small probe is retained in 15 to 20% of adult hearts.

Right Ventricle vs Left Ventricle

While right ventricular output is about one third greater than the left, at birth this is adjusted by a raised left ventricular output (due to increased left atrial pulmonary return) and a reduced ventricular output (smaller inferior vena cava return to right atrium) so that they are equalized.

In the newborn, physiological pulmonary hypertension (cp. older child or adult) exists and levels off rapidly in the first two weeks to reach mature levels only 6 months to 1 year later. The vessels of the pulmonary arterial bed decrease in relative media thickness as their lumina increase and pari passu the pulmonary blood pressure levels off.

At birth, the ventricles are of equal weight and thickness. Later, with increasing left ventricular load and reduced pulmonary vascular resistance, the left ventricle enlarges and thickens. The histological changes into adult myocardium are conspicuous as well with reduction in perimysial connective tissue and relative reduction of nuclear substance so that sarcomere numbers are increased per limit value.

Thus, in this period several obliterations are completed--the ductus arteriosus becomes the ligamentum arteriosum, the ductus venosus becomes the ligamentum venosum, the intra-abdominal part of the umbilical vein becomes the ligamentum teres and the umbilical arteries in the abdominal wall are now the medial umbilical ligaments The foramen ovale, too, is usually sealed off and in passing it is worth recalling that fetal hemoglobin is removed gradually, a fact of consequence to tissue oxygen delivery.

Interventricular septum

This septum is usually complete long before birth, but if ventricular septal defects do exist at birth, it is known that they often undergo spontaneous closure after birth, even in childhood (Glancy and Roberts, 1967; Evans et al., 1960). Defects in the muscular septum are ovoid in infancy and elongate with age, a process that may help growth closure of the defect.

REFERENCES

Adams, W.E. (1957) "On the possible homologies of the occipital arteries in mammals, with some remarks on the phylogeny and certain anomalies of the subclavian and carotid arteries," Arch. Anat., 29, 90-113.

Aely, C. (1868) "Der Bau des Menschlichen Kopers," pp. 1003, Liepzig, F. Voges (cited by Aely, 1963).

Arnold, J. (1871) "Experimentelle Untersuchumgen uber die Entuicklun der Blut Kapillaren," Virchow, Arch. Path. Anat., 53, 70-92.

Ashton, N. (1957) "Retinal vascularisation in health and disease," Amer. J. Opthal., 44, 4, Pt. 11, 7-17.

Barron, D.H. (1942) "The sphincter of the ductus arteriosus," Anat. Rec., 82, 389.

Barry, A. (1951) "The aortic arch derivatives in the human adult," Anat. Rec., 111, 221-238.

Bartelmez, G.W. and M.P. Blount (1954) "The formation of neural crest from the primary optic vesicle in man," Carn. Inst. Wash. Publ., 603. Contr. Embryol., 35, 55-71.

Bayley, J.H. and O.G. Cabbalhaes (1964) "The umbilical vein in the adult, diagnosis, treatment and research," Amer. Surg., 30, 56.

Benninghoff, A. (1930) in Millendorf's "Handluch der microskipischer Anatomie des Menschen," bei von Mollendorf Bd. 6, 112-232, Berlin, Springer.

Blackwood, H.J.J. (1965) "Vascularisation of the condylar cartilage of the human mandible," J. Anat., 99, 551-563.

Blinkov, S.M. and I.I. Glezer (1968) "The human brain in figures and tables," Basic Books Inc., Plenum Press, New York.

Boyd, J.D. (1941) "The nerve supply of the mammalian ductus arteriosus," J. Anat. (Lond.), 75, 457-468.

Braastad, F.W.; R.E. Condon and F. Gyorkey (1967) "The umbilical vein, surgical anatomy in the normal adult," Arch. Surg., 95, 948.

Bremer, J.L. (1914) "The earliest blood vessels in man," Amer. J. Anat., 16, 447-476.

Brookes, M. (1958) "The vascularisation of long bones in the human foetus," J. Anat., 92, 261-267.

Cassels, D.E. (1973) "The ductus arteriosus," C. C. Thomas.

Chaatrapati, D.(1964) "Absence of the radial artery," Indian. J. Med. Sci., 18, 462-465.

Chacko, A.W. and S.R.M. Reynolds (1954) "Architecture of distended and non-distended human umbilical cord tissues, with special reference to the arteries and veins," Contrib. Embryol. Carne Inot., 35, 135-150.

Charles, C.M.; L. Penn; H.F. Holden; R.A. Miller and E.B. Alvis (1931) "The origin of the deep brachial arteries in American white and in American negro males," Anat. Rec., 50, 299.

Charles. J.J. (1894) "A case of absence of the radial artery," J. Anat. & Physiol., 29, 449.

Christie, A. (1930) "The normal closing time of the foramen ovale and the ductus arteriosus: an anatomic and statistical study," Amer. J. Dis. Child., 40, 323.

Christie, G.A. (1963) "The development of the limbus fossa ovalis in the human heart - a new septum," J. Anat., 97, 45-54.

Civin, W.K. and J.E. Edwards (1951) "The postnatal structural changes in the intrapulmonary arteries and arterioles," Arch. Path., 51, 192.

Clark, E.R. (1918) "Studies on the growth of blood - vessels in the tail of the frog larva - by observation and experiment on the living animal," Amer. J. Anat., 23, 37-88.

Clark, E.R. and E.L. Clark (1909) "Observations on living, growing lymphatics in the tail of the frog larva," Anat. Rec., 3, 183-198.

Clark E.R. and E.L. Clark (1925) "The development of adventital (Rouget) cells in the blood capillaries of amphibian larvae," Amer. J. Anat., 35, 238-264.

Clark, E.R.; W.J. Hitschler; H.T. Kirly-Smith; R.O. Rex and J.H. Smith (1931) "General observations on the ingrowth of new blood vessels into standardised chambers in the rabbit's ear," Anat. Rec., 50, 129-167.

Clarke, J.A. (1965) "An x-ray microscopic study of the postnatal development of the vasa vasorum in the human aorta," J. Anat., 99, 877-889.

Cogan, D.G. (1963) "Development and senescence of the human retinal vasculature," Trans. Opthal. Soc., U. K., 83, 465-489.

Coleman, S.S. and B.J. Anson (1961) "Arterial patterns in the hand based upon a study of 650 specimens," Surg. Gynaec. Obstet., 113, 409-425.

Congdon, E.D. (1922) "Transformation of the aortic arch system during the development of the human embryo," Carnegie. Contrib. Embryol., 14(68), 47-110.

Danilowicz, D.; A.M. Rudolph and J.I.E. Hoffman (1966) "Delayed closure of the ductus arteriosus in premature infants" Pediatrics, 37, 74.

De Gains, C.F. and W.B. Swartley (1926) "The axillary artery in white and negro stocks," Amer. J. Anat., 41, 353.

De Vries, P.A. and J.B. Saunders (1962) "Development of ventricles and spinal outflow tract in human heart," Contr. Embryol. Carnegie. Inst., 37(256), 87-114.

Evans, H.M. (1995) "On the development of the aorta, cardinal and umbilical veins, and other blood vessels of embryos from capillaries," Anat. Rec., 3, 498-518.

Evans, H.M. (1912) "Development of the vascular system" in Manual of Human Embryology, Eds. F. Keibel & F. P. Mall, J. B. Lippincott Co., Philadelphia. Vol. 2, pp 570-623.

Evans, J.R.; R.D. Rowe and J.D. Keith (1960) "Spontaneous closure of ventricular septal defects," Circulation, 22, 1044.

Feinberg, R.J. (1971) "Solid capillary buds of the vasa vasorum," Surgery, 70(5), 730-735 and 751-757.

Finley, E.B. (1923) "Development of the subcutaneous vascular plexus in the head of the human embryo," Carnegie. Contr. Embryol., 14(71), 155-162.

Fischer, B. (1906) Verh. ges. dtsch. Naturf. Koln., 2, 38 (quoted by Hughes, 1943).

Fischer, B. and V. Schmieden (1909) Frankfurt Z. path. B III. (quoted by Hughes, A. F. W., 1935 and 1943).

Ganon, R. (1966) "Superficial arteries of the cubital fossa with reference to accidental intra-arterial injections," Canad. J. Surg., 9, 57-65.

Glancy, D.L. and W.C. Roberts (1967) "Complete spontaneous closure of ventricular septal defect."

Gonzoles, C.O. (1959) "Portography; a preliminary report of a new technique via the umbilical vein," Clin. Proc. Child. Hosp., 15, 120.

Grant, J.C.B. (1962) "An atlas of Anatomy," Baltimore, William & Wilkins Co.

Gupta, T.C. and C. Wiggers (1951) "Basic haemodynamic changes produced by aortic coarctation of different degrees," Circulation, 3, 17.
Hamiovici, H. (1954) "Stenosing arterial thrombosis," Surgery, 36, 1075, Amer. J. Med., 43, 846.
Hamilton, W.J. and H.W. Mossman (1972) "Human embryology," Williams and Wilkins Co., Baltimore.
Hasan, M. and D. Narayan (1964) "A single cord human brachial plexus,J. Anat. Soc. India, 13, 103.
Hazlett, J.W. (1949) "The superficial ulnar artery with reference to accidental intra-arterial injections," J. Canad. Med. Assn. 61, 289-293.
Heuser, C.H. (1923) "The branchial vessels and their derivatives in the pig," Carnegie. Contr. Embryol., 15(77), 121-139.
Heuser, C.H. and G.W. Corner (1957) "Developmental horizons in human embryos," Carnegie, Contr. Embryol., 36(244), 31-39.
His, W. (1900) "Lecthoblast and angioblast der Wirbeltiere," Abhandl. math-naturio Kl. K. Sachs. Ges., 22. (quoted by Arcy, 1963).
Holmes, R.L. (1958) "Some features of the ductus arteriosus," J. Anat. (Lond.), 92, 304-309.
Huelka, D.F. (1959) "Variations in the origin of the branches of the axillary artery," Anat. Rec., 135, 33-41.
Hughes, A.F.W. (1935) "Studies on the area vasculosa of the embryo chick," J. Anat., 70, 76-122.
Hughes, A.F.W. (1943) "The histogensis of the arteries of the chick embryo," J. Anat., 77, 266-287.
Huxley, J. and G.R. de Beer (1934) "The elements of experimental embryology," Cambridge.
Jolly, J. (1940) "Recherches sur la formation du systeme vasculaire de l'embryo," Arch. d'Anat. Micr., 35, 295-361.
Jones, M.; M.V. Barron and M.W. Wheat (1969) "An ultrastructural evaluation of the closure of the ductus arteriosus in rats," Surgery, 66, 891.
Kampmeier, O.F. and C.L. Birch (1927) "The origin and development of the venous valves," Amer. J. Anat., 38, 451-499.
Keen, J.A. (1961) "A study of the arterial variation in the limb, with special reference to symmetry of vascular patterns," Amer. J. Anat., 108, 245-261.
Knower, H.M. (1907) "Effects of early removal of the heart and arrest of the circulation on the development of frog embryos," Anat. Rec., 1, 161-165.
Kramer, T.C. (1942) "Partitioning of truncus and conus and formation of membraneous portion of the interventricular septum in human heart," Amer. J. Anat., 71, 343-370.
Kratochwill, A. and D. Eisenhut (1967) "The earliest proof of foetal heart action by means of ultrasonics" in "Intrauterine danger to the foetus," Ed. Horsky & Stembera, Excepta Med. Foundation, Amsterdam.

Krompecher, S. (1940) "Die Gefaswandentwicklung in Kausalhistogenetischer und vergleichend functionellar Darstellung," Zeit. Anat. u. Entw., 64, 96-275.
Levine, S.A. (1964) "A neglected and promising kind of anatomic research," Circulation, 29(3), 325-327.
Lewis, O.J. (1958) "The development of blood vessels of the metanephros" J. Anat., 92, 84-97.
Licate, R. (1962) in "Blood vessels and lymphatics," Ed. D. I. Abrainson, New York, Academic Press.
Lind, J.; L. Stern and C. Weglius (1964) "The human foetal and neonatal circulation," C. C. Thomas, Springfield, Ill.
Linddrom, A. (1950) "Arteriosclerosis and arterial thrombosis in the lower limb -- a Roentgenological study," Acta Radio., 33, Suppl., 80.
Mall, F.P. (1906) "A study of the structural unit of the liver" Amer. J. Anat., 5, 227-309.
Mann, I. (1928) "Developmental abnormalities of the eye," Cambridge Univ. Press., London.
Manning, V.R. and E.F. McLaughlin (1947) "Persistant omphalomesenteric artery causing intestinal obstruction and gangrene of Meckel's diverticulum," Ann. Surg., 126, 358-365.
Massloff, M.S. (1914) "Zur Frage uber die Eutwickling der grossen Gefasse beim menschlichen embryo," Arch. mikr. nat., 84, 351-368.
May, A.G.; J.A. de Weese and C.G. Rob (1963) "Hemodynamic effects of arterial stenosis," Surgery, 53, 513-524.
McClure, C.F.W. (1921) "The endothelium problem," Anat. Rec., 22, 219-237.
McCormack. L.J.; E.W. Cauldwell and B.J. Anson (1953) "Brachial and antebrachial arterial patterns: a study of 750 extremities," Surg. Gynaec. Obstet., 96, 43-54.
McMurphy, D.M.; M.A. Heymann; A.M. Rudolph and K.L. Melmon (1972) "Developmental changes in constriction of the ductus arteriosus: Responses to oxygen and vasoactive substances in the isolated ductus arteriosus of the foetal lamb," Pediat. Res., 6, 231.
Michaelson, I.C. (1948) "Vascular morphogenesis in the retina of the cat," J. Anat., 82, 167-174.
Miller, R.A. (1939) "Observations upon the arrangement of the axillary artery and brachial plexus," Amer. J. Anat. 64, 143-163.
Misra, B.D. (1955) "The arteria mediana," J. Anat. Soc. India, 4, 48.
Moffat, D.B. (1959) "Developmental changes in the aortic arch system of the rat" Amer. J. Anat., 105, 1-35.
Murray, C.D. (1926) "The physiological principle of minimum work applied to the angle of branching of arteries," J. Gen. Physiol., 9, 385-841.
Niemineva, K. and L. Tenila (1953) "On the capillary bed of the human fetal cerebellar hemispheres," Acta. Anat., 19, 204-209.

Noback, G.J.; F.D. Anderson and W.G. Cooper (1951) "On the presence of nerve tissue in the media of the human ductus arteriosus," Anat. Rec., 109, 331.

Oberhansli-Weiss, I.; M.A. Heymann; A.M. Rudolph and K.L. Melmon (1972) "The pattern and mechanisms of response of the ductus arteriosus and umbilical artery to oxygen," Pediat. Res., 6, 693.

Odgers, P.N.B. (1939) "The development of the arterioventricular valves in man," J. Anat., 73, 643-657.

O'Neal, R.M.; W.A. Thomas; K.T. Lee and E.R. Rabin (1957) "Alveolar walls in mitral stenosis," Circulation, 15, 64.

Parr, J.C. and G.F.S. Spears (1974) "General calibre of retinal arteries expressed as the equivalent width of the centre arteries of the retina," Amer. J. Opth., 77(4), 472-477.

Parr, J.C. and G.F.S. Spears (1974) "Mathematical relationships between the width of a retinal artery and the width of its branches," Amer. J. Opthal., 77, 478-483.

Pernkopf, E. (1922) "Die Entnicklung der Form des Magendarm kanales biem Menschen," Z. Anat. Entwicklungsgesch., 64, 96-275.

Parder, A. (1947) "Die Entwicklung der Form Magendarmkanales biem Menschehichen Keimling," Acta. Anat., 3, 115-152.

Reagan, F.R. (1915) "Vascular phenomena in fragments of embryonic bodies completely isolated from yolk-sac endoderm," Anat. Rec., 9, 329-341.

Rosen, L.; D.H. Bowden and I. Uchida (1957) "Structural changes in pulmonary arteries in first year of life," Arch, Path., 63, 316.

Rosenguist, G.C. and R.L. de Haan (1966) "Migration of precardiac cells in the chick embryo: A radioamtographic study" Contr. Embryol. Carnegie., 38(263), 113-121.

Rudolph, A.M.; M.A. Heyman; K.A.W. Teramok; C.T. Barrett and N. C R. Raiha (1971) "Studies on the circulation of the previable human foetus," Pediat. Res., 5, 452-465.

Rychter, Z. (1959) "Cerni soustara zauidku Kunete," Ceskoslov. Morfol., 7, 1-20 (cited by de Vries & Saunders, 1962).

Sabin, F.R. (1917) "Origin and development of the primitive vessels of the chick and of the pig," Carnegie Contr. Embryol., 6, 61-124.

Sabin, F.R. (1920) "Studies on the origin of blood vessels and red blood cells as seen in the living blastoderm of chicks during the second day of incubation," Carnegie Contr. Embryol., 9, 213-262.

Sabin, F.R. (1922) "Direct growth of veins by sprouting," Carnegie Contr. Embryol, 14, 1-10.

Senior, H.D. (1920) "The development of the human femoral artery, a connection," Anat. Rec., 17, 271-279.

Senior, H.D. (1926) "A note on the development of the radial artery," Anat. Rec., 32, 220.

Senior, H.D. (1919) "An interpretation of recorded arterial anomalies of the human leg and foot," J. Anat., 53, 130.

Singer, E. (1933a) "Embryological pattern persisting in the arteries of the arm," Anat. Rec., 55, 403-410.
Singer, E. (1933b) "Human brachial plexus united into a single cord: Description and interpretation," Anat. Rec., 55, 411-419.
Skoda, et al., quoted by Hochhaus, H. (1893) "Uber das offenbleiben des ductus botalli," Deutsch. Arch. Klin. Med., 51, 1-10.
Spencer, H. (1962) "The structure of the pulmonary arteries during intra and extrauterine life," in "Chronic pulmonary hypertension."
Stewart, J.R.; O.W. Kincaid and J.R. Edwards (1964) "An atlas of vascular rings and related malformations of the aortic arch system," Springfield, Thomas.
Strandness, D.E. (Jr.) (1969) "Peripheral arterial disease -- a physiological approach," Little, Brown & Co., Boston.
Streeter, G.L. (1942) "Developmental horizons in human embryos," Contrib. Embryol. Carnegia Inst., 30, 211-245.
------------ (1945) "Developmental horizons in human embryos," Contrib. Embryol. Carnegie Inst., 31, 27-63.
------------ (1948) "Developmental horizons in human embryos," Contrib. Embryol. Carnegie Inst., 32 133-203.
------------ (1949) "Developmental horizons in human embryos," Contrib. Embryol. Carnegie Inst., 33, 149-167.
------------ (1951) "Developmental horizons in human embryos," Contrib. Embryol. Carnegie Inst., 34, 165-196.
Thoma, R. (1893) "Untersuchungen uber die Histogenise and Histomechnik des Gefassystems," Stutgart, F. Enke.
Thompson, D.W. (1959) "On growth and form," Cambridge University Press.
Turner, R.S. (1957) "A comparison of theoretical with observed angles between the vertebral arteries at their junction to form the basilar," Anat. Rec., 129, 243-253.
------------ (1959) "The angle of origin of the ulnar artery," Anat. Rec. 134, 761-767.
Van Mierop, L.H.S.; R.D. Alley; H.W. Kausel and A. Stranahan (1962) "The anatomy and embryology of endocardial cushion defects," J. Thorac. Cardiovasc. Surg. 43, 71.
Van Mierop, L.H.S., et al. (1963) "Pathogenesis of transportation complexes I. Embryology of the ventricles and great arteries," Amer. J. Cardiol., 12, 216-225.
Walker, C.W. and B. G. Pye (1960 "The length of the human umbilical cord in a statistical report," Brit. Med. J., 20 (5172), 546-548.
Walls, E.W. (1947) "The development of the specialized conducting tissue of the human heart," J. Anat. (Lond.), 81, 93-110.
Wankoff, W. (1962) "Uber einige gesetzma zigkeiten beider variabitat der artenen der ober extremitat," Anat. Anz., 111, 216-240.
Wiedeman, M.P. (1963) "Dimensions of blood vessels from distributing artery to collecting vein," Circ. Res., 12, 375-378.

Wollard, H.H. (1922) "The development of the principal arterial stems in the forelimb of the pig," Carnegie Contr. Embryol., 14, 139-154.

Wollard, H.H. and J.A. Harpman (1937) "The relation between size of an artery and the capillary bed in the embryo," J. Anat., 72, 18-22.

Wright, E.S. (1974) "Detection of cervical arterial stenosis during routine eye examination," Amer. J. Opthal., 77, 483-490.

Yamanchi, A. (1965) "Electron microscopic observations on the development of S-A and A-V nodal tissues in the human heart," Z. Anat. Entnickf., 124, 562-587.

FUNCTIONAL MORPHOLOGY OF HUMAN ARTERIES DURING FETAL AND POSTNATAL DEVELOPMENT

Wladimir W. Meyer,[1] M.D., S. Zoe Walsh,[2] M.D., and John Lind[3]

[1] Professor and Chairman of the Department of Pediatric Pathology, University of Mainz, Mainz, West Germany
[2] Senior Research Investigator, Karolinska Hospital and Wenner-Gren Research Laboratory, Stockholm, Sweden
[3] Professor Emeritus, Department of Pediatrics, Karolinska Hospital and Wenner-Gren Research Laboratory, Stockholm, Sweden

> "...development of structure and development of function go hand in hand. And if the function cannot be subserved without the development of the structure, equally the stimulus of the function is necessary for the proper maturation of the structure." (Liley, 1972)

1. GENERAL CONSIDERATIONS

The equilibrium between the distending force of the blood pressure and the arterial wall is determined essentially by a simple law of mechanics, the <u>law of Laplace.</u> According to this law, the total force or tension (T) in the vessel's wall represents the product of the radius of the vessel (r) and the blood pressure (p), $T = r \cdot p$. The law of Laplace may be used to estimate and compare the tension produced by blood pressure in vessels of different size and thereby determine the increase in functional load on arteries. Since tension increases not only with blood pressure but also with the radius of the vessel, the highest tension is presumably produced in the wall of the ascending aorta where tension amounts to about 200,000 dynes/cm, i.e., 200 g/cm. In the vena cava, which also has a wide lumen but is subject to a much lower blood pressure, the total wall tension is also high, about 20,000 dynes/cm, i.e., 20 g/cm, whereas in the capillary wall, tension is very low - 16 dynes/cm, i.e., 16 mg/cm. Since the radius is small, the thin wall of a capillary can withstand the distending force of capillary blood pressure (F 25 mm Hg.). It should be borne in mind, however, that such estimates are mere approximations because human arteries have curves and numerous branches and are capable of active constriction and rapid alterations in caliber in more distal segments.

Function of the Main Structural Components of the Arterial Wall

Elastic tissue, smooth musculature and collagen are the three main components of the arterial wall that cope with "transmural" pressure, i.e., the distending force produced by blood pressure on the arterial wall.

Elastic tissue is present in all vessels except capillaries and arterio-venous anastomoses but its amount varies in individual parts of the arterial tree. The main function of elastic tissue in the arterial wall is "to produce elastic tension automatically and without expenditure of biochemical energy" (Burton, 1972) and, in this way, resist the distending force of the blood pressure. By virtue of these physical characteristics, which are due to reversible distensibility of elastic tissue, a part of the stroke volume propelled during systole in the aorta is temporarily accommodated in the proximal segment of the arterial tree, i.e., mainly in the aorta, and is then transferred more distally during the following diastole (Böger and Wezler, 1936; Wezler and Böger, 1937, 1939; Wezler and Sinn, 1953; Bader, 1963). By this means the highly pulsatile blood flow produced by the left ventricle is transformed into more continuous flow. On the other hand, the unstretched elastic structures behave like a spring and resist compression of the arterial wall. In this way, they prevent the irreversible closure of the vessel which would otherwise occur with increasingly active tone of the smooth musculature in vessels deprived of or poor in elastic fibers (see p.143).*

Collagen fibers show much lower distensibility than elastic fibers and their modules of elasticity are hundreds of times higher than the modules of elastic tissue (Burton, 1972). Collagenous networks, however, are loosely arranged and mostly unstretched in arterial walls subjected to normal blood pressure. They begin to exert their tension first when the arterial wall is exposed to an increasing degree of stretch. With further widening of the artery,

---

*This and following references to other parts of the text provide specific examples.

the stiff collagenous fibers resist further stretch and prevent overdistension of the arterial tube.

The function of smooth musculature is to produce, in addition to the elastic tension of the arterial wall, an active tension resulting from a graded state of contraction, i.e., active tone. The actual tension that smooth muscle can exert was initially underestimated. According to Somlyo and Somlyo (1964), the active contractile force that can be produced by arterial smooth musculature varies between 1.5 and 2.5 x 10^6 dynes/cm^2. These data accord with more recent results reported by Dobrin and Rovick (1969). Even in elastic arteries, smooth musculature represents an important component of the arterial wall. In the thoracic aorta, for example, the amount of smooth musculature is estimated to range from 25 to 35 percent (McDonald, 1974). Thus, the force produced by smooth musculature at maximal contraction is supposed to be 8 - 9 x 10^5 dynes/cm^2 in the aortic wall and is threfore higher than the strain to which the elastic tissue is exposed. Hence, even when smooth muscle is contracting less than maximally, the tension in the wall may be doubled and the diameter of larger elastic arteries may actually decrease (McDonald, 1974). In this connection it is of interest that in newborn piglets the main pulmonary artery decreases in caliber in response to hypoxia (Rowe and Mehrizi, 1968). The addition of contractile tension in parallel to elastic tension increases the elastic modules and makes the arterial wall stiffer (Gerova and Gero, 1967; McDonald, 1974).

The endothelial lining probably plays a very small role in total elasticity of the vessel because very little force is required to deform endothelial cells (Burton, 1972).

The admixture of elastic, collagenous and muscular structural components is different in individual arterial provinces. In general, the amount of elastic tissue is greatest in the proximal segment of the arterial tree, i.e., in large arteries subjected to the highest transmural pressure. The media of these arteries, i.e., aorta, brachiocephalic, subclavian, common carotid, common iliac, and proximal segments of the renal, mesenteric and celiac arteries, consists to a considerable extent of elastic membranes. They are therefore called elastic arteries.

In short transitional segments, the number of elastic membranes in the media diminishes relatively abruptly, becoming evident first in the outer layer of the media and then rapidly throughout the rest of the media. Shortly beyond the transitional area, the media of large, medium-sized and small arteries consists mainly of smooth musculature with only tiny elastic fibers. These arteries are called muscular arteries.

The transition from an elastic to a muscular segment occurs at the following levels: immediately above the carotid sinus (p. 149), in the upper segment of the brachial artery (p. 208), at the origin of the external iliac artery (p. 213) and in the proximal segment of the celiac, mesenteric and renal arteries. All smaller arterial branches that originate from larger elastic trunks are muscular arteries, and the transition to this structural pattern occurs at the orifices of the branching arteries. For instance, the intercostal and lumbar arteries assume a muscular structural pattern immediately at their origin from the aorta.

An <u>intermediate structural pattern</u> is seen in the trunk and main branches of the pulmonary artery. As in the elastic arteries, the media consists of alternating layers of smooth musculature and elastic membranes, but the amount of the musculature is much greater than in the aorta. The elastic and muscular elements also show a distinctive pattern of connection not seen in other human muscular or elastic arteries (p. 253). The proximal segments of the coronary arteries represent muscular arteries during fetal development, but later develop a special structural pattern which cannot be regarded as characteristic of either of the usual two types (p. 111).

Types of Musculo-elastic Junctions in the Arterial Wall

In human elastic arteries smooth muscle cells are located in the interlamellar spaces, i.e., between the concentrically arranged elastic membranes. Their ends are attached to the opposite surfaces of both adjacent membranes (Fig. 1A). Thus, it may be assumed that at least in a distended arterial wall the elastic and muscular elements are arranged in parallel (Benninghoff, 1930).

Another type of connection is seen in the aorta of larger animals, e.g., in cows. In the outer aortic media larger groups of muscle bundles alternate more or less regularly with layers of strong, concentrically arranged elastic membranes, i.e., they are arranged in series (Figs. 1B, 2).

A similar arrangement of elastic and muscular elements becomes visible during growth in tangential sections of the human pulmonary trunk. In this vessel, star-like elastic membranes in the media are connected by numerous smooth muscle cells. Thus, the elastic and muscular elements are also arranged in series (Fig. 3) (p. 253).

In the human inferior vena cava and in the trunk of the portal vein, longitudinal smooth muscle bundles form elastic tendons similar to those seen in the aorta of animals (Figs. 2, 4). In veins, the tendons connect the muscle bundles not only with the elastic networks which parallel the musculature, but they also insert in junctions of vertically and horizontally arranged fibers from which

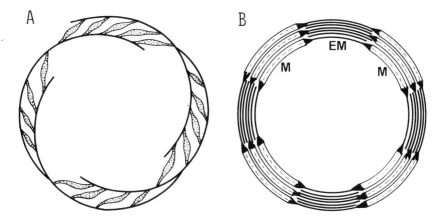

Fig. 1. Diagram showing two types of musculo-elastic junctions. (A) In the human aorta, the muscle cells (arrows) are located between the elastic membranes and their ends are attached to the opposite surfaces of the adjacent membranes. (B) In the cow aorta, groups of muscle bundles (M) alternate with layers of thick elastic membranes (EM). See Fig. 2. Modified from Benninghoff (1930).

the muscular tension can be transmitted to circular or oblique elastic structural components of the wall. In this way the longitudinal musculature may influence the size of the venous lumen. Similar junctions of elastic and muscular elements may also exist in human arteries.

The connections between smooth musculature and elastic components of the arterial wall ensure coordinated function of both structural components, graded vasomotor control and favor a stable equilibrium between the distending force of the blood pressure and the tension of the arterial wall (Burton, 1972). With different states of contraction of smooth musculature, i.e., with different tone, the degree of distention of elastic fibers and membranes may be changed, even at the same blood pressure, and the arterial lumen adjusted to the blood flow directed to the supplied area. According to Burton (1972), without elastic tissue, stable equilibrium between blood pressure and the arterial wall is unthinkable. Vessels poor in elastic tissue show an unstable equilibrium and are regarded as "critical vessels" that tend to be either distended or irreversibly closed (see p. 79).

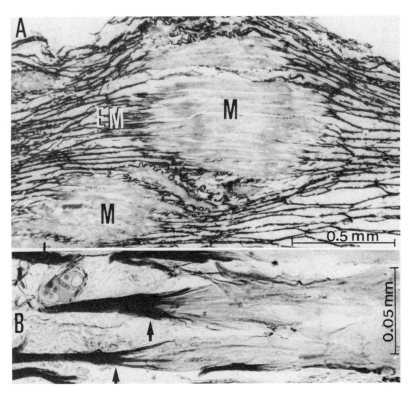

Fig. 2. Cross-section of the aorta of a cow. (A) Groups of muscle bundles (M, unstained) alternate with layers of strong elastic membranes (EM) along the circumference.
(B) Higher magnification shows that the elastic tendons (arrows) connect the muscle bundles with the elastic membranes. Orcein elastic stain.

Increase in Total Wall Tension with Growth and Associated Structural Changes

With growth, the radius of the arteries increases considerably. The internal diameter of the ascending aorta, for instance, increases from 6 mm in newborns to 22 - 25 mm in young adults. A comparable enlargement in diameter also occurs in other arterial provinces. Blood pressure likewise increases. Consequently, total wall tension (calculated by determining the product of the radius and blood pressure) augments considerably with growth, i.e., from a circumferential tension in the ascending aorta of 3,000 - 4,000 dynes/cm in the newborn to 2000,000 dynes/cm in the adolescent.

The increase in total wall tension is compensated for by an appropriate increase in wall thickness and strengthening of individual structural components of the arterial wall. In the aorta, for instance, the number and thickness of lamellar units increase significantly (see p. 81). In most arterial provinces, however, growth does not consist solely of a numerical increase in fibers, membranes, and smooth muscle cells. In many arteries profound remodelling of the original structure is a continuous process that accelerates after birth and results in formation of additional structures that are not present in earlier development (Gillman, 1968).

The most conspicuous microscopic changes that occur with growth are seen in the innermost layer of the arterial wall, i.e., the layer

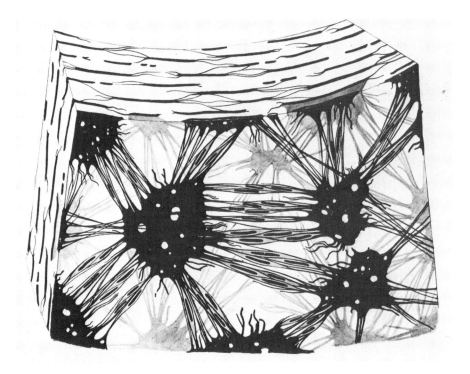

Fig. 3. Diagram of a fully-developed pulmonary trunk in a young adult. Note that star-like elastic membranes are interconnected by fine elastic networks and smooth muscle bundles. These bundles probably alter the tension of the membranes and networks by varying their tone and may play a role in adaptation of the pulmonary artery to the existing circulatory workload. From Meyer and Richter, 1956.

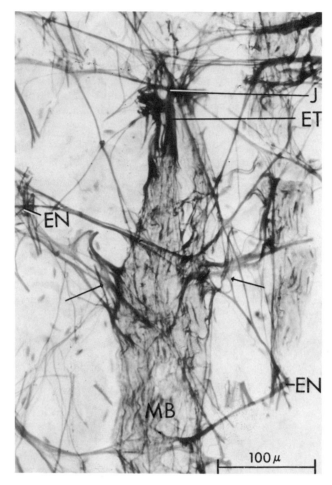

Fig. 4. A longitudinal muscle bundle (MB) terminates in an elastic tendon (ET) which inserts in a junction (J) of circularly and obliquely oriented elastic fibers. Other elastic tendons (arrows) originate from the surface of the same muscle bundle and join adjacent elastic networks (EN). Tangential frozen section of longitudinal musculature of the portal vein. 33-year-old woman. Orcein and hematoxylin stain. From Meyer and Kleibsch, 1964.

subjected to the greatest increase in the distending force of blood pressure. The chief finding is the <u>formation of secondary elastic sheets</u> between the endothelial lining and the primary internal elastic membrane that forms during fetal life. These sheets usually develop in large and medium-sized muscular arteries. They initially consist of tightly packed fine longitudinal elastic fibers but later assume the more homogeneous appearance of an elastic lamella. Progressive "splitting" of the internal elastic membrane has often been regarded as a process leading to the formation of secondary elastic intimal structures. Such splitting, however, probably never occurs and it seems more likely that the primary internal elastic membrane represents a matrix from which a part of the secondary elastic sheets arise. In earlier stages of development, the luminal aspect of the internal elastic membrane sometimes appears granular and the secondary longitudinal elastic fibers seem to be formed in close proximity to the membrane by "budding" of its substrate. However, the new subendothelial elastic fibers and sheets probably also develop independently of the substrate of the internal elastic membrane.

Secondary elastic structures become thicker with growth. They seem to take over increasingly the static and dynamic functions of the primary internal elastic membrane which gradually becomes displaced into the deeper part of the intima.

The secondary elastic sheets and membranes stain more intensely than the comparatively pale primary internal elastic membrane with the usual elastic stains, particularly Weigert's resorcin fuchsin. In most arteries which become the site of early calcific deposits, e.g., in the iliac arteries, calcium appears first in the primary internal elastic membrane, but not in the secondary elastic sheets, at least during infancy and childhood. Hence, the secondary elastic structures that form during growth probably have a different chemical composition and possibly other physical characteristics than the primary membranes.

The formation of secondary membranous intimal elastic structures is commonly associated with proliferation of longitudinal smooth muscle cells between the primary and the secondary elastic membranes and development of the longitudinal musculo-elastic intimal layer along the inner aspect of the media (pp. 119, 134). The most pronounced development of axially arranged structural components is seen in coronary arteries during postnatal growth (pp. 110, 133). The increasing longitudinal shearing forces from transmission of the pulse wave are probably responsible for the appearance and further increase in longitudinal structural elements in the inner layer of arteries during growth. However, since the parameters of the physical forces acting upon the arterial wall in the individual arterial segments are still not exactly known, the interpretation of the structural changes occurring with growth is difficult and remains a matter of conjecture.

At the same time as new intimal structures develop along the endothelial lining, numerous gaps appear in the adjacent primary internal elastic membrane during fetal development and after birth. Since these gaps are especially prominent in arteries that show accelerated growth during fetal development, i.e., in the iliac arteries (p. 190), they also probably represent a type of structural adaptation to the increasing functional load upon the enlarging arterial tube.

Numerous circular gaps in the primary internal elastic membrane also appear during postnatal growth in large and medium-sized muscular arteries, particularly in extremity arteries (p. 230). They develop first in larger proximal arterial segments. In the femoral artery, for instance, the gaps are already present in the first year of life. The early development of circular membrane gaps is probably closely related to marked longitudinal growth of the extremity arteries and additional longitudinal stretch consequent to flexion and extension of the extremities.

With growth, considerable remodeling occurs at bifurcations, branchings, and curved arterial segments and results in the formation of cushions which protrude into the lumen. They are present around the orifices of branching arteries in all vascular provinces and have been described in detail in the coronary and cerebral arteries (pp. 129, 180) (Figs. 72, 73). The cushions consist of networks of elastic fibers, elastic sheets, and smooth muscle cells which form ring-like structures around the orifices. The primary internal elastic membrane is usually interrupted beneath the cushions. The circular arrangement of the elastic fibers and muscle cells probably represents a structural accommodation to the increased circumferential tension to which the orifices are presumably subjected. It is uncertain why the primary internal elastic membrane is interrupted at these sites, but it seems possible that the amount of membrane is inadequate.

Profound structural changes occur in similar cushions that develop along the concave walls of curved arterial segments, e.g., in the carotid siphon. They initially consist of elastic tissue but, with growth, come to contain increasing amounts of ground substance and collagen (pp. 167-173). The cushions at arterial branchings, bifurcations, and along the concave parts of the curvatures as well as the musculo-elastic layer of the thickened intima in some straight arterial segments often become the site of early lipid infiltration and atherosclerotic lesions (pp. 174-178). This suggests that the additional functional load responsible for the appearance and subsequent differentiation of these structures may also favor the development of pathological changes at the same sites of the arterial tree.

Early calcifications, like lipid deposits, also selectively appear in specialized structures that develop in the arterial wall

FUNCTIONAL MORPHOLOGY OF ARTERIES 105

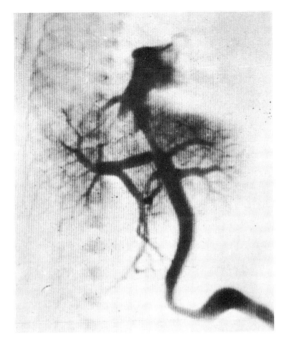

Fig. 5. Angiographic demonstration of the ductus venosus and the portal sinus in a 12-week-old fetus. Note that the portal trunk is of smaller caliber than its branches and that the angle of the portal trunk is directed somewhat to the right. Hence, most of the blood is probably directed to the right branch of the portal sinus. From Lind et al., 1964.

during growth. When an appropriate gross staining technique is used, the location, form, and extent of the affected predisposed structural components become visible along the arterial luminal surface. In this way, demonstration of early lesions contributes to better knowledge of normal arterial structures. Some patterns of early calcification and lipid deposits that appear in the arteries during growth are therefore included in various parts of this chapter (see pp. 138-140, 173-174, 184-187, 230-232, 236-238).

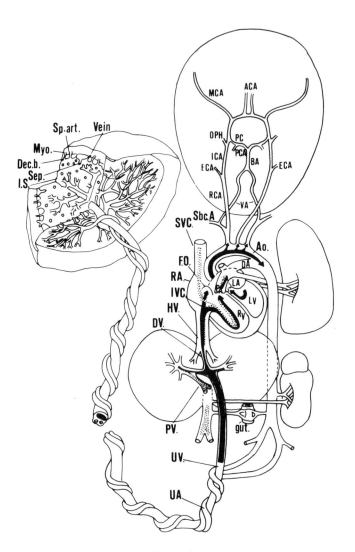

Fig. 6.

← Fig. 6. Schematic diagram showing the course of the fetal circulation as well as the relatively large size of the brain and placenta and the considerable length of the umbilical cord. ACA - anterior cerebral artery; AC - anterior communicating artery; MCA - middle cerebral artery; PC - posterior communicating artery; OPH - opthalmic artery; PCA - posterior cerebral artery; ICA - internal carotid artery; BA - basilar artery; RCA - right carotid artery; VA - vertebral artery; ECA - external carotid artery; Sbc.A. - subclavian artery; SVC - superior vena cava; RA - right atrium; FO - foramen ovale; RV - right centricle; PA - pulmonary artery; LA - left atrium; LV - left ventricle; DA - ductus arteriosus; Ao - aorta; UV - umbilical vein; PV - portal vein; DV - ductus venosus; HV - hepatic vein; IVC - inferior vena cava; UA - umbilical artery; Vein - uterine vein; Sp.art. - spiral arteriole; Myometrium; dec.b. - decidua basalis; Septa - Septum; I.S. - intervillous space. From Walsh, Meyer and Lind, 1974.

2. ARTERIES IN THE FETAL AND NEONATAL CIRCULATION

Course of the Fetal Circulation

Knowledge of the structural and functional characteristics of the fetal circulation is important not only for proper interpretation of the adaptive changes occurring in the arterial system during fetal life, but also for better understanding of remodeling of arterial structures in different vascular provinces shortly after birth and during further growth and development. Cineangiographic and microsphere studies in the fetal lamb and in the human fetus have provided much information about the course of the fetal circulation and relative size of vessels (Barclay et al., 1944; Lind and Wegelius, 1954; Rudolph and Heymann, 1970). These studies show that at midterm, arterial blood from the placenta enters the umbilical vein, which has a larger diameter than the descending aorta, and then passes into the umbilical recess where it divides into two almost equal streams. One stream enters the ductus venosus, a direct continuation of the umbilical recess and a by-pass around the liver, whereas the other enters at a right angle the left and then the right branch of the portal vein before perfusing the liver sinusoids and entering the liver veins (Figs. 5, 6). The exact amounts passing through these two routes at different times during gestation are not known but it is often assumed to be about 50%. Both streams reunite at the inferior vena cava, which carries relatively well oxygenated blood from the umbilical vein, less well oxygenated blood from the liver and relatively poorly oxygenated blood from the lower half of the body.

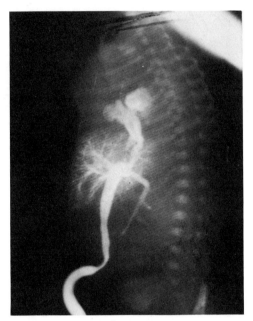
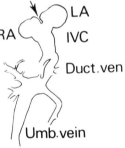

Fig. 7. Angiographic demonstration of the crista dividens. The contrast medium is injected into the umbilical vein. It then enters the umbilical recess, ductus venosus and inferior vena cava. At the crista dividens, it splits into two streams. Fetus of about 14 weeks gestation. From Walsh, Meyer and Lind, 1974.

On entering the heart, the blood is again divided into two streams by the crista dividens, i.e., the free edge of the interatrial septum which extends posteriorly almost to the limits of the atrial cavity (Fig. 7). One stream enters the left atrium directly via the foramen ovale, thus by-passing the lungs, and is joined by a small amount of less saturated blood from the lungs. From here it passes to the left ventricle and it is then pumped into the ascending aorta to be distributed mainly to the coronary arteries and aortic arch vessels, thereby supplying blood to the heart, the comparatively large head, and the upper body.

The path taken by this stream from the inferior vena cava through the foramen ovale and left side of the heart to the upper part of the body is referred to as the via sinistra (Figs. 8, 9).

The other stream of blood from the inferior vena cava continues to the right atrium where some mixing occurs with relatively poorly oxygenated blood returning via the superior vena cava and the coronary sinus. From here it enters the right ventricle and it is then pumped into the pulmonary artery. Since the pulmonary vascular bed offers relatively higher resistance than the systemic vascular bed, and pulmonary arterial pressure is higher than systemic pressure, the blood largely by-passes the lungs by entering the ductus arteriosus, a continuation of the main pulmonary artery (Fig. 10). The remainder enters the comparatively small pulmonary branches. It then passes into the descending aorta where it joins the small residue from the left ventricle which has not gone to the head. Hence little blood enters the aortic isthmus, the segment of the aorta between the origin of the left subclavian artery and the entry of the ductus arteriosus, and it remains of comparatively small caliber (pp. 87, 105).

The path taken by the stream of blood from the inferior vena cava through the right side of the heart and to the lower part of the body is referred to as the via dextra.

Thus both the via sinistra and the via dextra serve as bypasses around the lungs. Most studies in the anesthetized exteriorized fetal lamb suggest that slightly more than 50% of the blood from the inferior vena cava takes the path of the via sinistra and the remainder the via dextra.

Studies performed by injection of radionuclide-labeled 15 µ microspheres in the near term fetal lamb in utero, at three to five days after surgery, however, show that only about one-third of the blood from the inferior vena cava (26% of the combined ventricular output of 500 ml/kg/min) traverses the foramen ovale and that the major portion of right ventricular output enters the ductus arteriosus (60% of the combined output) (Rudolph and Heymann, 1970). In the human previable fetus (weighing 50 to 300 g), about 40% of the

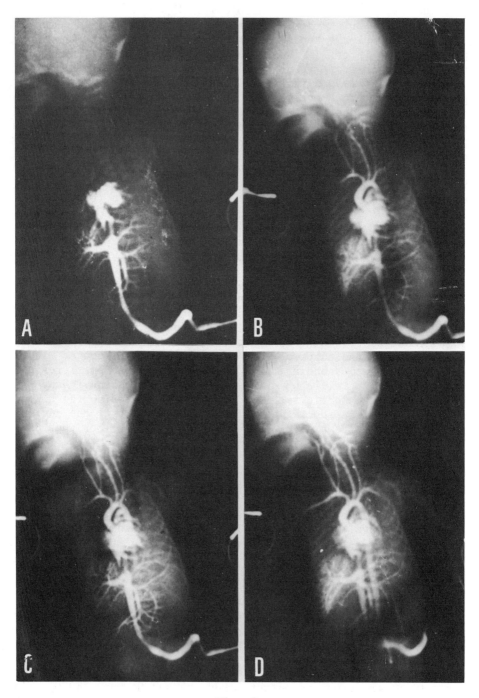

Fig. 8.

combined ventricular output (vs 50 to 60% in the fetal lamb) is returned for oxygenation to the low vascular resistant placenta. It is uncertain whether the amount changes during gestation, but in the sheep and goat a greater part of the blood volume is in the placenta early in gestation, while later in gestation the situation is reversed. This is presumably due largely to the considerably greater increase in vascular bed of the fetus than of its placenta during the latter part of pregnancy.

Since arterial pressure increases during fetal life (Mott, 1961) and much of the flow to the lower part of the body and long umbilical arteries derives mainly from right ventricular output, the work load of the right heart should become greater than that of the left. This concept accords with autopsy findings showing equal thickness of the right and left ventricular myocardium until midgestation followed by relatively increased thickness of the right ventricular myocardium. It also accords with ECG studies which show an upright R wave in VI from early gestation onward, usually a sign of relative right ventricular dominance (Emery and MacDonald, 1960; Recavarren and Arias-Stella, 1964; Hort, 1966; Walsh, Meyer and Lind, 1974).

Fetal Foramen Ovale

The foramen ovale is an obliquely elongated cleft between the inferior vena cava and the left atrium which forms at about six weeks gestation when the ostium primum is closing (Fig. 11). The foramen ovale constitutes the end of the inferior vena cava which is divided at its termination by the interatrial septum, a protruding fold of endocardial tissue known as the crista dividens (Fig. 7.)

← Fig. 8. The via sinistra seen in left anterior oblique projection. Contrast medium injected into the umbilical vein of a human fetus of about 15 weeks gestational age.
(A) The contrast medium opacifies the liver vessels, ductus venosus, inferior vena cava and the cardiac chambers.
(B) in the following systole, the ascending aorta, pulmonary trunk and upper part of the descending aorta become visualized.
(C) The aortic arch vessels, ductus arteriosus and left pulmonary artery are now well visualized.
(D) During ventricular systole, the aorta shows a slight increase in caliber. The descending aorta and umbilical arteries are now opacified but the medium has become diluted by blood from the descending aorta. From Walsh, Meyer and Lind, 1974.

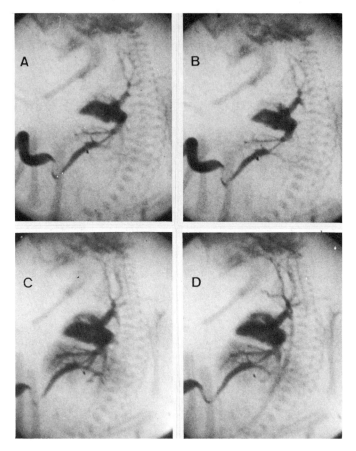

Fig. 9. The via sinistra seen in right anterior oblique projection. Contrast medium injected into the umbilical vein of a human fetus of about 12 weeks gestational age.
(A) Contrast medium injected into the umbilical vein. The contrast medium has entered the ductus venosus and hepatic vessels and outlines the heart, ascending aorta and aortic arch vessels.
(B) During ventricular systole, as evidenced by the increasing size of the atrial appendage, only the uppermost part of the descending aorta is opacified. This indicates that the medium is being diluted by blood flow from the ductus arteriosus.
(C) Two seconds later, the right ventricle is also contrast filled, as judged by filling of the infundibulum of the right ventricle.
(D) In the following systole, the entire descending aorta appears because of contrast filling from the right ventricle.

The foramen is quite large in early gestation but does not grow to the same extent as the septum. Hence, the size of its functional outlet decreases (Fig. 11E). Throughout gestation, the cross-sectional area of the foramen is less than that of the inferior vena cava but the relationship changes. Thus, in pre-term infants it measures one-half the area of the inferior vena cava, whereas in term infants, it measures slightly more than one-third (Patten et al., 1929-30). It may, therefore, be assumed that a comparatively smaller proportion of blood from the inferior vena cava by-passes the lungs by way of the foramen ovale--i.e., takes the via sinistra in the latter part of gestation. At term, the foramen measures 8 mm in diameter (Patten, 1953).

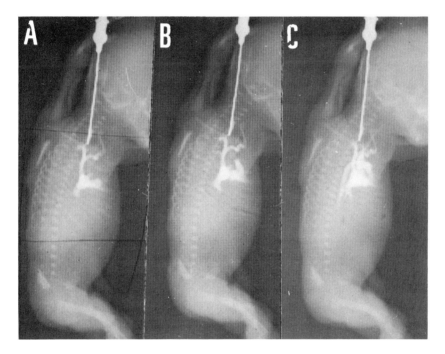

Fig. 10. The via dextra. Contrast medium injected into the jugular vein.
(A) The superior vena cava, right atrium and right ventricle are outlined.
(B) During right atrial systole, the right ventricle and its infundibulum are well filled.
(C) During ventricular systole, the pulmonary artery trunk. ductus arteriosus and descending aorta become opacified. Right anterior oblique projection. Fetus of about 12 weeks gestation. From Lind et al., 1964.

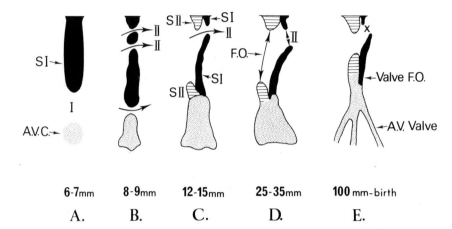

Fig. 11. Developmental stages in formation of the septum primum, septum secundum and foramen ovale.
(A) The septum primum (SI) grows down toward the atrioventricular canal (A.V.C.), the later site of the A.V.C. cushion, to close the ostium primum (I).
(B) Before reaching this, a new opening forms in it, the ostium secundum (II), which is also known as the foramen ovale (F.O.).
(C, D, E) While the septum secundum (SII) develops and the ostium primum closes, blood continues to pass from right-to-left through this opening, which gradually becomes smaller and whose functional outlet size is indicated (S). After birth, pressure in the left atrium rises and exceeds that in the right atrium. The valve of the foramen ovale (or SI) then shuts off return flow. Later it closes structurally when fibrous adhesions form. Based on a drawing by Patten, 1968. From Walsh, Meyer and Lind, 1974.

During fetal life, the foramen and the ductus venosus constitute the main routes for direct passage of blood from the placenta to the brain. In studies quantitating blood flow in the fetal lamb in utero, only about one-third of the inferior vena caval return is estimated to cross the foramen into the left atrium (Heymann et al., 1973), but since the size of the brain of the fetal lamb relative to body weight is considerably less than that of the brain of the human fetus, these values probably cannot be extrapolated to the human. The foramen also probably plays a role in normal development and function of the left heart because pulmonary venous return is small and only about 3% of superior vena caval blood crosses the foramen (Rudolph and Heymann, 1967). Moreover, premature reduction of the size of the foramen is associated with reduction in the capacity and muscular development of the left ventricular wall (Lev and Killian, 1942).

Toward the end of gestation, more blood successively enters the pulmonary circulation and the size of the functional outlet of the foramen ovale decreases. Left atrial pressures rise and the valve approximates the limbus of the foramen. This complex pattern of change requires normal progression of development of the pulmonary vascular bed, proper development of an adequately large septum secundum to cover the ostium secundum, as well as multiple functional adaptations concerned with the relation between pulmonary and systematic vascular resistance and pressures and control mechanisms. It is thus not surprising that in about one of every five fetuses and newborns, the edge of the septum secundum does not completely overlap the rather large ostium secondum (Morison, 1952).

With the reduction in size of the foramen ovale, more blood is obliged to enter the right heart--i.e., take the via dextra. Presumably as a result of this increased hemodynamic lead, the weight and capacity of the right atrium exceeds that of the left, and right heart valves are slightly wider than those of the left heart (Müller, 1883; Merkel and Witt, 1955; Schulz and Giordano, 1962). Other differences between the left and right heart chambers are also present (Table 1). Thus, in a sense the size of the foremen ovale also plays a role in structural changes in the right heart.

Before birth, pressures in the right atrium are believed to be only slightly higher than those in the left, i.e., in the fetal lamb mean pressures in the right and left atrium are 3.5 mm Hg and 2.5 mm Hg, respectively, and the shape of the pulse pressure curves is different in both chambers (Assali et al., 1968). Since the lower part of the unresorbed septum primum is located in the left atrium and rests loosely over the foramen, it acts as a one-way valve (valvula foraminis ovalis) only permitting flow from right-to-left (Fig. 11). Hence, when the pressure becomes higher in the left than in the right atrium, flow ceases if the valve is competent.

The valve is composed almost entirely of cardiac muscle during fetal life and therefore cannot be regarded as an entirely passive structure unaffected by changes in the cardiac cycle. Cineangiographic studies in the lamb and human fetus also attest to the capacity of the foramen ovale to constrict during atrial systole (Walsh, Meyer and Lind, 1974).

TABLE 1.

Summary of differences between the right and left heart chambers in the fetus, newborn and infant

	Right vs Left Atrium	Right vs Left Ventricle
Fetus	Weight of muscle mass greater throughout gestation, particularly toward term and in males (Müller, 1883). Increase in size greater toward end of gestation (Merkel and Witt, 1955).	Weight greater after fetus is viable (Müller, 1883; Merkel and Witt, 1955; Emery and MacDonald, 1960). Valves larger (Schulz and Giordano, 1962). Myofibrils and Mitochondria more rapid development (Schulz and Giordano, 1962). Muscle fiber diameter greater at term (Hort, 1955). Ventricular capacity greater in fetuses weighing 500 to 2500 g (Shanklin, 1959).
Newborn and Infant	Weight of muscle mass greater but difference less marked. No difference after 2 months (Müller, 1883).	Valves larger (Schulz and Giordano, 1962). "Atrophy" after birth (Müller, 1883; Boellard, 1952; Hort, 1955; Schulz and Giordano, 1962; Recavarren and Arias-Stella, 1964). Weight increase slower (Recavarren and Arias-Stella, 1964; Emery and Mithal, (1960). External surface greater (Hort, 1966). Ventricular volume--same at birth, double at one month (left ventricle unchanged) (Kyrieleis, 1963). Ventricular capacity greater (Hifflesheim and Robin, 1864). Muscle fiber diameter becomes relatively less (Hort, 1955). Rate of doubling of muscle fiber slower (Linzbach, 1952; Hort, 1953). Myocardial blood flow decreases (Increases by 50 percent in left ventricle) (Filer, 1966). RNA concentration does not increase after birth (does in left ventricle) (Gluck et al., 1964). Slower development of adult or H type isozyme of lactic dehydrogenase (6 months later than left ventricle) (Filer, 1966).

From Walsh, in prep.

Control Mechanisms During Stress

In the human fetus, the activity of various reflexes is assumed on the basis of studies in various other species, on the isolated human fetal heart, studies of reflexes in pre-term infants after birth, and in term infants during labor and delivery. The data indicate that some degree of autonomic nervous control is established early in gestation before the fetus becomes viable and that the heart of the mature fetus is under autonomic nervous control. Chemoreceptor and baroreceptor reflexes are also well developed at birth. Reflex control of the circulation becomes important in the event of an acute reduction in oxygen supply and/or increase in demand. It is then that a variety of mechanisms permit the fetus to maintain preferential perfusion of the heart, brain, adrenals, and placenta at the expense of other organs. These include the distinctive pattern of the fetal circulation, increasing control of the autonomic nervous system, active cardiovascular reflexes, and high cardiac outputs. This selective vasoconstriction in various vascular beds seems to be accompanied by an increase in blood pressure and fall in heart rate (Elsner et al., 1970; Cosmi et al., 1973) and is especially severe in conditions associated with acidosis (Rudolph and Yuan, 1965; Campbell et al., 1967).

The effects of this selective "circulatory ischemia" on various organs in the human fetus are not known. Recent data suggest that hypoxemia may cause degeneration of the intramural ganglion cells in the distal colon and thereby result in Hirschsprung's disease (Smith, 1968; Touloukian and Duncan, 1975). Hence a systemic stress in utero may alter neorogenic function without producing other evidence of tissue injury. More severe damage to blood supply in the latter part of gestation may cause congenital anomalies. Thus Louw and Barnard (1975) ligated a branch of the mesentery artery in a puppy nine days before term during cesarean section. The puppy continued to term, and after spontaneous delivery, was found to have atresia of the gut with no fibrous remnant between the two ends.

As far as can be judged, anoxic survival depends primarily on the amount of glycogen in the heart, though not on the amount in the liver, and therefore on the rate of anaerobic metabolism (Dawes et al., 1963). Myocardial cells contain large amounts of glycogen throughout gestation (about 30 to 40 mg/g in the cardiac ventricles of most species, including man) and the stores increase at term (Shelley, 1961). This would suggest that the enzymes necessary for glycogen synthesis appear very early.

Umbilical Blood Flow

Umbilical blood flow is determined by vascular resistance and the pressure gradient which drives blood from the descending aorta

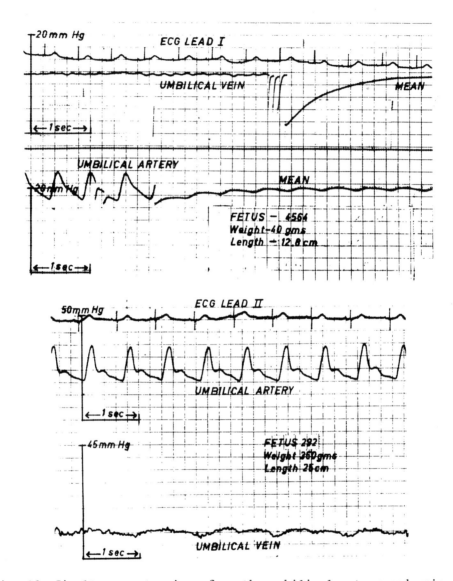

Fig. 12. Simultaneous tracings from the umbilical artery and vein.
A - Human fetus in situ, weight 40 g, length 12.8 cm.
B - Human fetus in situ, weight 360 g, length 25 cm. Note comparatively low umbilical venous pressure and slowly rising umbilical arterial pressure. From Walsh, Meyer and Lind, 1974.

to the placenta and back to the heart. In the fetal lamb the pressure drop takes place across the extensive capillary bed of the fetal villi in the placenta (about 35 mm Hg). In this species, umbilical blood flow appears to increase with both gestational age and weight, initially largely because of a fall in vascular resistance, subsequently because of the gradual rise in fetal arterial pressure. The rise in systemic blood pressure is attributed partly to increasing activity of sympathetic vasoconstrictor tone. In the lamb, umbilical venous pressure tends to remain at about 10 to 12 mm Hg, whereas in the human, umbilical venous pressure at term is somewhat higher, about 25 mm Hg (Fig. 12). In the opinion of Dawes (1968), umbilical blood flow can and does alter in response to small changes in arterial blood pressure but umbilical vascular resistance is low and invariate.

Reported values for umbilical blood flow in the fetal lamb and human differ and the differences are probably largely methodological. In one oft-quoted study in the anesthetized fetal lamb, umbilical blood falls from 230 ml/kg/min at about 90 days gestation to about 170 ml/kg/min at term (Dawes and Mott, 1964), while in another study in the unanesthetized fetal lamb in utero, umbilical blood flow is about 240 ml/kg/min at term (Rudolph and Heymann, 1970). In contrast, reported values for umbilical blood flow in the human fetus are considerably lower than in the lamb. Assali et al. (1960), for instance, noted a mean flow of 110 ml/kg/min (with the cuff electromagnetic flow-meter) at 10 to 28 weeks gestation, while Stembera et al. (1965) reported a value of 75 ml/kg/min (with local thermodilution) between 35 and 42 weeks gestation, immediately after delivery. Although there are almost no other data available it seems likely that total flow approximates 150 ml/kg/min or even slightly more in the average human fetus at term (Dawes, 1968). Judging from some very carefully carried out studies, umbilical blood flow as a percentage of the combined ventricular output seems to be lower in the human fetus than in the lamb fetus.

Rudolph and Heymann (1970) believe that whereas total umbilical blood flow/kg body weight does not change during gestation, umbilical blood flow represents a progressively smaller proportion of total cardiac output, falling from about 50% of cardiac output early in gestation to about 40% at term. This is ascribed to a fall in total body resistance in relation to placental resistance.

In previable human fetuses weighing 75 to 650 g, fetal O_2 consumption is about 4 ml/min/kg (Assali et al., 1960). Maternal O_2 inhalation does not increase this value unless the fetus is hypoxic (Bartels, 1970).

Cessation of The Placental Circulation

Numerous studies show that unless the umbilical cord is clamped immediately at delivery, at which time the placenta contains approximately one-third the total blood volume (about 125 ml blood), the placental circulation will continue at a rapidly declining rate until placental separation (two-four minutes after delivery) (Fig. 13). During this period, oxygen tension, oxygen saturation, carbon dioxide tension, pH and base excess (in the umbilical vein) will remain largely unchanged despite simultaneous onset of respiration (Engström et al., 1966). The rate of transfusion, though not the amount, is hastened by maternal administration of ergot derivatives. The effects of maternal anesthesia, parity, and other maternal factors are largely unknown. The position of the infant vis-a-vis the level of the vulva will cause the infant to receive less blood during a particular interval if held above (e.g., at the time of cesarean section) or to receive more blood if held below the vulva (Gunther, 1967; Yao and Lind, 1969), but the time of onset of respiration will probably not affect ultimate values (Yao et al., 1968), as postulated by Redmond et al. (1965). Štembera et al. (1964), using thermodilution, found that 100 to 150 ml of blood, which corresponds to about 15 ml O_2, is transferred to the infant in 30 seconds. If clamping is delayed, between 80 and 100 ml/kg of blood is shifted from the placenta to the infant within 3 minutes. In an average-sized adult, this would amount to transfusion of 5 - 6 L. of blood.

Whole blood viscosity increases with hemoconcentration and a decrease in pH. Viscosity is doubled at a hematocrit of 70% (Burnard, 1966), which is not much greater than in infants with delayed clamping of the cord. Some evidence suggests that an elevated blood viscosity and hematocrit are better tolerated by the newborn infant than the adult, possibly because of higher suspension stability of the blood consequent to lower levels of fibrinogen and high molecular weight proteins (Dintenfass, 1971). Although some data suggest that blood volume after stripping of the cord is about the same as after delayed ligation of the cord (Usher et al., 1963), very vigorous stripping may cause transient respiratory distress and cyanosis consequent to acute expansion of blood volume. In this situation the foramen and ductus arteriosus presumably act as escape outlets and permit right-to-left shunting, thereby preventing acute right ventricular overload. Electrocardiographic changes have also been noted, such as increased P amplitude, prolonged P duration, and P-R intervals, lower R/S ratio in V1, because of low amplitude R wave, and delayed T wave inversion in V1. These findings are ascribed to a delay in fall of pulmonary artery pressure and closure of the ductus arteriosus (Walsh, 1969).

Placental separation occurs at different times, depending on a variety of factors including maternal analgesia, anesthesia, oxytocics, and mode of delivery of the infant. In one study, separation

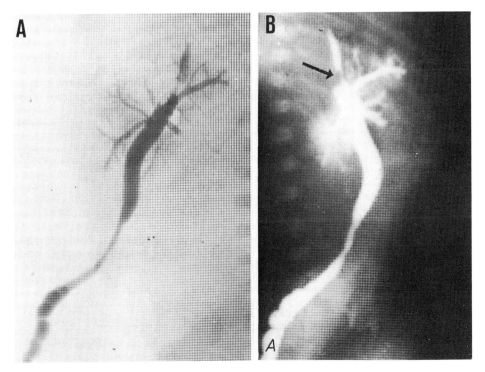

Fig. 13. (A) Demonstration of early constrictions of the umbilical vein in a human fetus of about 16 weeks gestation. Angiogram in left anterior oblique projection. The contrast medium has been injected into the umbilical vein about one minute after cesarean section. Note the multiple beaded constrictions in the umbilical vein. The ductus venosus has also constricted and is thread-like at its origin from the umbilical recess.
(B) Functional closure of the ductus venosus and umbilical vein at birth. 2 ml contrast medium have been injected into the umbilical vein of a full-term newborn infant about two minutes after vaginal delivery. The extra abdominal umbilical vein shows multiple beaded constrictions (A) whereas the ductus venosus (arrow) is almost completely constricted at its origin from the umbilical recess. From Lind et al., 1964.

was shown to be impeded by immediate clamping of the cord (Walsh, 1968) as five of 59 mothers (9 percent) required manual removal of placental secundines while none of 58 mothers of infants with delayed clamping of the cord developed this complication. Similar findings have since been independently reported by Botha (1968) and again noted in a larger number of infants by Walsh. This is not surprising since the completeness with which the placenta is separated is determined both by how much of the subplacental area of the uterine wall is reduced as well as by the speed with which this is accomplished and a larger residue of blood (after early clamping) should therefore hinder this process.

Structure and Closure Mechanism of Umbilical Vessels at Birth

There is still no agreement on the structure of the umbilical vessels and especially on the arrangement of the smooth muscle cells and the functional significance of individual layers of the umbilical artery during postnatal closure (v. Hayek, 1935; Goerttler, 1951; Roach, 1973; Rudolph and Heymann, 1974). For this reason the structure and closure mechanism of umbilical vessels were studied in 10 cm long segments of the umbilical cord. The central parts of the cords were clamped with Bunce's double hemostat (Bunce, 1961, 1974), a method which permits umbilical vessels in clamped segments to remain filled with blood during fixation with little or no change in size or shape because of simultaneous closure of both branches. The cords were clamped from five to 120 seconds after delivery and it was therefore possible to study structural changes during different stages of contraction (Meyer et al., in press).

Umbilical Vein

The umbilical vein and arteries have some structural features in common which may account for the fact that postnatal closure of the lumina is effected by a similar rearrangement of structural components. Since the structure of the umbilical vein is simpler than that of the umbilical artery, various structural characteristics essential for interpretation of the closure mechanism are more easily demonstrated and understood by first discussing the umbilical vein. Presentation of the findings in the umbilical vein is also justified because the vein carries "arterial" blood during fetal life.

The structural similarities between the umbilical vein and arteries are presumably largely due to the relatively high blood pressure to which the umbilical vein is exposed during fetal development. In the few studies available in the human infant at term, umbilical venous pressure is about 25 mm Hg immediately after delivery (pp. 23, 25). This is the highest blood pressure which has been recorded in any venous channel of the human circulation under normal conditions. In the umbilical artery, the pressure is probably more than three times higher than in the umbilical vein at term. However, as we have seen, the functional load upon the vascular wall is not only determined by the blood pressure but also by the radius of the vascular tube. In double-clamped umbilical cords that are clamped almost immediately after delivery, the fully patent umbilical vein has an internal mean diameter of 7 mm. Hence the total circumferential tension is about 12,000 dynes/cm in the umbilical vein. This is determined by multiplying the blood pressure (25 mm Hg) by 1,330 dynes/cm^2 (= 1 mm Hg), which is equal to 33,250 dynes/cm^2, by the radius ($\simeq$ 0.35 cm). Since the radius of each fully patent umbilical artery is considerably smaller than that of the vein (0.12 - 0.15 cm), the total wall tension in the arterial wall, despite the higher blood pressure ($\simeq$ 75 mm Hg) only approximates 15,000 dynes/cm, i.e., it is in the same range as in the umbilical vein. Alterations in pressure and lumen size of arteries, however, occur during stress and no doubt significantly affect these calculations.

Structure of the widely patent umbilical vein. The wall of the umbilical vein consists mainly of smooth musculature which is circularly arranged and therefore better able to cope with the distending force of the blood pressure. Since the umbilical vein contains only small amounts of elastic tissue, which in other vessels withstands the distending forces, the equilibrium between the blood pressure and the venous wall must be maintained mainly by the circular smooth musculature.

The regular circular pattern of the musculature is well seen on longitudinal and cross-sections of the fully patent vein. In longitudinal cryostat sections, which have been immersed in Feyrter's staining solution immediately after sectioning and show no artefacts due to drying, the venous wall consists of tightly packed cross-sectioned muscle cells surrounded by metachromatic ground substance (Fig. 14). In a few places the musculature falls apart in longitudinal rows of cross-sectioned muscle cells (Fig. 15). Some striking artefacts are, however, produced by drying sections on the mount to achieve better adhesiveness before staining. The venous musculature then falls apart into parallel, longitudinally arranged bands (Fig. 16). With drying and shrinking of the intercellular matrix between the rows of muscle cells, the fragments lose their initial vertical position on the mount and tip over. They are then seen from one side and look like bands that parallel the axis of the vein (Fig. 17). These bands obviously represent fragments of thin cylindrical muscular sheets which are concentrically arranged in the venous wall and constitute its main structural component. The fragments of muscular sheets are also evident in tangential sections of the venous wall (Fig. 18). During preparation of sections, muscle cells remain held together in sheets but artefactual splits occur between individual band-like fragments. This would suggest that cylindrical muscular sheets are only loosely connected by a plastic semi-fluid ground substance.

In tangential sections of the luminal layer of the vein, axially arranged oval-shaped nuclei of the endothelial cells are clearly visible (Fig. 19). Beneath the endothelial lining, there are only circularly oriented muscle cells with typical rod-shaped nuclei. No other cellular elements are present between the endothelium and the musculature of the vein. On cross-sections of the fully patent veins, a sheet of tiny longitudinal or obliquely oriented elastic fibers is present beneath the endothelial lining. A true internal elastic membrane which would resist contraction and closure of the lumen after birth does not exist in the umbilical vein. However, with contraction, individual subendothelial elastic fibers come closer together and form a more compact elastic sheet which closely resembles an internal elastic membrane (Fig. 20).

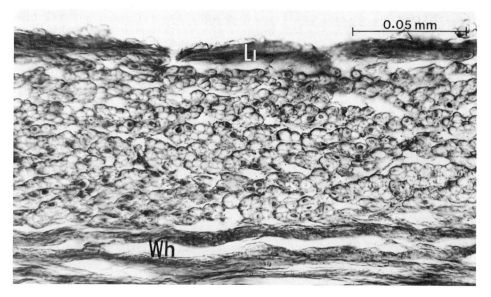

Fig. 14. Longitudinal cryostat section of the wall of the fully patent umbilical vein. The venous wall consists of tightly packed, cross-sectioned, muscle cells surrounded by tiny sheets of metachromatic ground substance. The black rings around the muscle cells stain red in the section. The thin luminal layer (Ll) contains comparatively large amounts of metachromatic material (black) and a few longitudinally arranged cells beneath the endothelial lining. Wharton's jelly (Wh) consists of longitudinally arranged lamellae of metachromatic ground substance. L - lumen. Feyrter's thionine staining method.

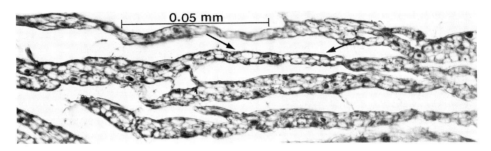

Fig. 15. In this longitudinal section of the umbilica vein, the ven-venous musculature falls into rows of cross-sectioned muscle cells (arrows). The rows represent fragments of cylindrical muscular sheets which constitute the main structural component of the wall. The sheets are tightly held together (see Figs. 16, 17, 18). Cryostat section of a fully patent vein from an early double-clamped umbilical cord. Feyrter's thionine staining method.

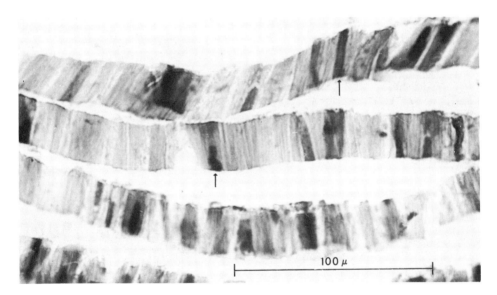

Fig. 16. The smooth musculature of the umbilical vein which has not contracted, falls apart into individual bands after longitudinal section in a cryostat. These bands are parts of the interconnecting muscular sheets, the main structural element of the vein. Nuclear fragments can be seen (arrows). Frozen section. From Walsh, Meyer and Lind, 1974.

Venous structure during contraction and closure. Closure of
the umbilical vein is initiated by multiple localized contractions
which can be seen in utero during cesarean section or appear immediately after delivery. In a 10 cm long segment of the umbilical
cord, up to four or five contractions develop in the umbilical vein
within the first 15 seconds after delivery (Moinian et al., 1969).
The contractions look like transverse grooves on the cord surface
and correspond to indentations in the vascular wall in longitudinally
opened vessels. The indentations, which were formerly believed to
be preformed valves (Hoboken's valves), evolve into diaphragmatic
structures which narrow the lumen considerably. With double clamping of the cord at different times after birth, it is possible to
follow the various stages of contraction that lead to closure.

In localized contractions, the smooth musculature retains its
circular orientation. Hence, in longitudinal sections the diaphragmatic structures narrowing the lumen consist of tightly packed cross-sectioned muscle cells separated by thin sheets of metachromatic
ground substance (Figs. 21 - 23). With contraction, the original
thin muscular sheets become transformed by folding into strong muscular bundles. This rearrangement is distinctly seen in longitudinal sections, particularly in the outer parts of the contracting
venous wall, i.e., beneath the involved cord surface (Fig. 23). The
newly-formed muscular bundles are loosely arranged here and individual bundles have distinct outlines. Remnants of the original muscular sheets may still be visible between the bundles.

The numerous localized contractions that form after delivery
must serve to considerably diminish the amount of blood flowing
through the umbilical vessels and thereby contribute to a fall in
blood pressure. With the decrease in blood pressure, the active

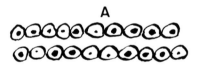
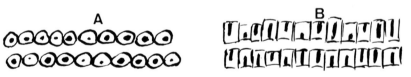

Fig. 17. Diagram showing longitudinal sections of muscular sheets
when drying is prevented (A) and when drying occurs on the
mount. (B) Before drying, fragments of cross-sectioned
muscle cells retain their vertical position. After drying,
fragments tip over and become visible from one side.

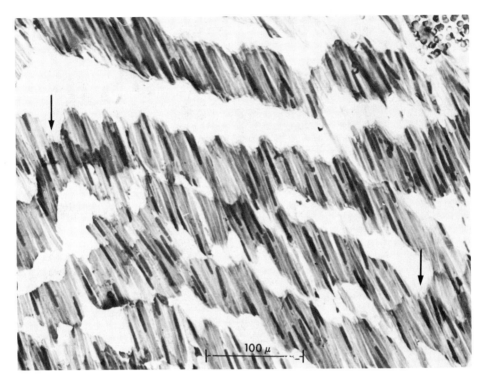

Fig. 18. On tangential section, it is easier to see the interconnections between the muscular sheets (↓) and the numerous nuclei in their full length. From Walsh, Meyer and Lind, 1974.

tension of the smooth musculature should overcome the reduced distending force acting upon the venous wall and with further contracttion, narrow and ultimately close the larger venous segments. The tortuosity of the umbilical vein which accompanies contraction may also contribute to complete closure of the lumen (Fig. 22).

Umbilical Artery

Structure of the fully patent artery. In umbilical cords which are double-clamped within the first seconds after delivery, the mean internal diameter of each umbilical artery measures 2.4 mm. In most early clamped cords, i.e., clamped within 5 seconds, however, the arteries are frequently wider--about 3 mm in diamter. Consequently, the calculated cross-sectional luminal area of both early clamped umbilical arteries (which is determined by multiplying the square of

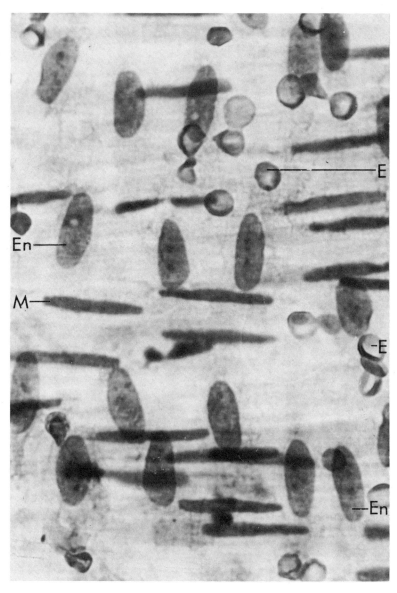

Fig. 19. Normal structure of the endothelium of the umbilical vein on tangential section. Note the oval-shaped nuclei of the endothelial cells (En), scattered erythrocytes (E) and cigarette-shaped nuclei of the muscle cells (M). Oil immersion (x 1000). From Walsh, Meyer and Lind, 1974.

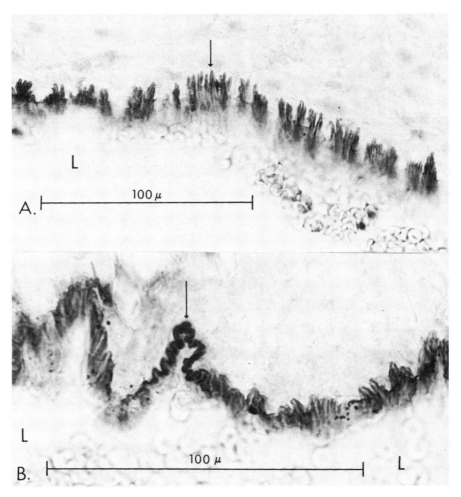

Fig. 20. (A) Subendothelial elastic layer of a partly contracted umbilical vein. This layer consists of longitudinal or slightly oblique elastic fibers, stumps of which are seen standing upright on tangential section. A few erythrocytes are present in the lumen (L) (Cryostat, orcein elastic stain).
(B) Subendothelial elastic layer of a more contracted segment of the umbilical vein. The longitudinal elastic fibers are closer together. In one area (↓) they are cross-sectioned and simulate an internal elastic membrane (Cryostat section, orcein elastic stain). From Walsh, Meyer and Lind, 1974.

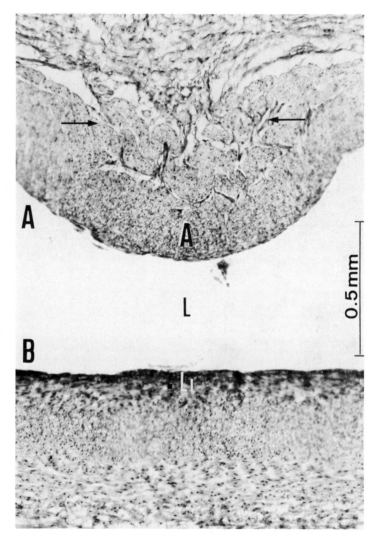

Fig. 21. Early phase of a localized venous contraction shown on longitudinal section from a double-clamped umbilical cord. The contracted part of the wall (A) nearest the cord surface protrudes into the lumen (L) and consists of tightly packed cross-sectioned bundles which are loosely arranged in the outer layers (areas between arrows) beneath Wharton's jelly (Wh). (See Fig. 23 which shows higher magnification.) The opposite wall (B) does not participate in the contraction and has a straight inner contour. The inner layer (LI) is rich in metachromatic ground substance (black). Wharton's jelly is seen beneath the musculature. Cryostat section. Feyrter's thionine staining method.

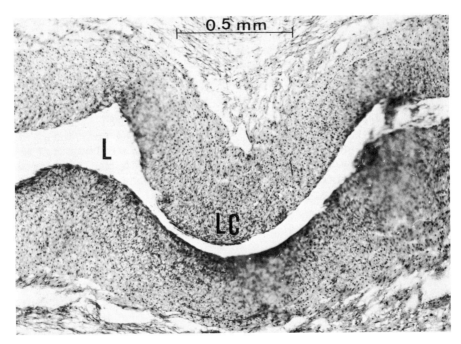

Fig. 22. Ultimate occlusion of an umbilical vein in the area of a former localized contraction (LC). Longitudinal cryostat section. A later stage of venous contraction than that shown in Fig. 21. The tortuosity of the vein that occurs with occlusion and which is evident in this figure may also contribute to ultimate occlusion of the lumen (L). Wh - Wharton's jelly. Feyrter's thionine staining method.

the radius by Π (Π = 3.14) measures 9.0 - 14.0 mm^2 while the cross-sectional luminal area of the umbilical vein, which has a diameter of 6.8 - 7 mm, measures 34.2 - 38.5 mm^2.

In cryostat <u>longitudinal sections</u> of the arterial wall immersed in Feyrter's staining solution immediately after sectioning, artefacts due to drying are eliminated and the wall of the umbilical artery is seen to consist of two distinctly different layers (Fig. 24). The structure of the <u>outer</u> arterial layer is identical to that of the muscular layer of the umbilical vein. It consists of tightly packed, circularly arranged muscle cells, cross-sections of which are well delineated from each other by thin, deeply stained sheets of metachromatic ground substance (Fig. 24A). As in the sections of the umbilical vein, longitudinal rows of muscle cells can often be distinguished (Fig. 25A). In the <u>inner</u> arterial layer, abundant metachromatic ground substance constitutes a prominent structural component (Fig. 24). With light microscopy, only a few cells embedded in this substance display the typical structural features of smooth muscle cells. Almost no nuclei show a definite axial orientation which should be present if the cells of this layer were arranged longitudinally.

In cryostat longitudinal sections dried before staining to ensure better adhesion to the mount, the musculature of the outer arterial layer falls apart into numerous parallel, longitudinally arranged, band-like fragments (Fig. 25B). As in the umbilical vein, these fragments obviously represent parts of thin cylindrical muscular sheets which are concentrically arranged in the outer arterial layers and constitute its main structural component. Since most nuclei in the bands are arranged perpendicular to the direction of

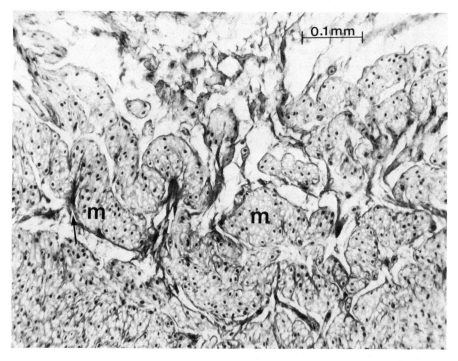

Fig. 23. Cross-sectioned muscle bundles (m) partly delineated by sheets of loose metachromatic connective tissue that stain black (arrows) are seen in the outer layer of the contracted part of the venous wall (see Fig. 21 which shows lower magnification). The muscle bundles form by contraction of the sheets which are visible in the wall of the fully patent umbilical vein (see Figs. 16 and 18). Wh - Wharton's jelly. Feyrter's thionine staining method.

FUNCTIONAL MORPHOLOGY OF ARTERIES 133

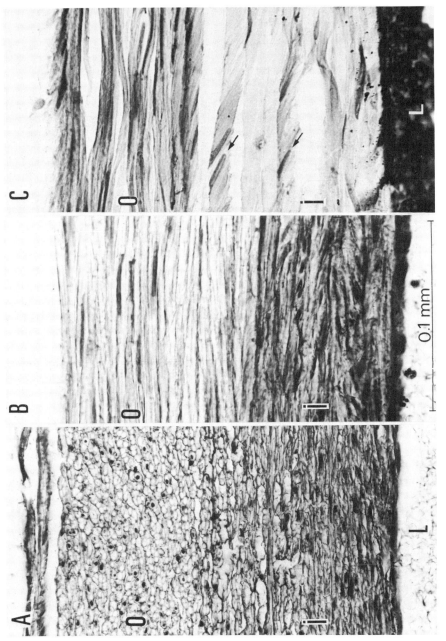

Fig. 24. (Caption follows on next page).

Fig. 24 (p. 133) - Structure of the fully patent umbilical artery in cryostat longitudinal (A) and cross-sections (B,C). In the longitudinal section (A) stained with Feyrter's thionine technique before drying, the outer muscular arterial layer (0) consists of tightly packed muscle cells, cross-sections of which are distinctly outlined. The internal arterial layer (i) includes large amounts of red-stained metachromatic ground substance which appears black in the microphotograph. In the cross-section (B), the outer layer shows circularly arranged muscle cells with cigarette-shaped nuclei. In the internal arterial layer (i) the nuclei are obliquely arranged. No longitudinal muscular layer is seen. (C) Cryostat cross-section stained with hematoxylin and eosin after drying. In the band-like fragments of the internal arterial layer, the nuclei are obliquely arranged (arrows) which would suggest that the cells have a spiral course in this layer. L - lumen. The thickness of both arterial layers is approximately equal.

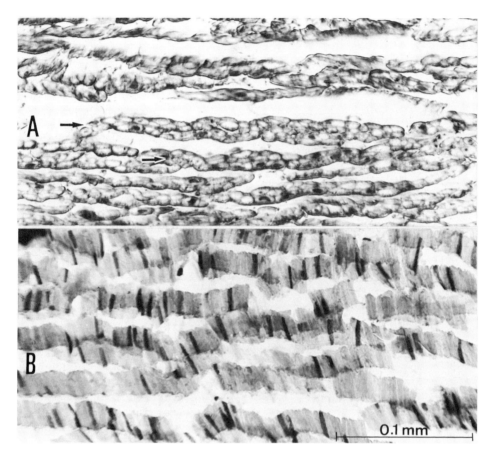

Fig. 25. In the longitudinal cryostat section from the fully patent umbilical artery (A), the musculature often falls apart in longitudinal rows of muscle cells (arrows). Feyrter's thionine stain. In the longitudinal sections which have been dried before staining (B), the rows take the form of band-like fragments. Hematoxylin-phloxine-tartrazine stain.

flow, the muscle cells in the outer arterial layer probably have a circular course.

In cryostat cross-sections of the fully patent umbilical artery, the outer arterial layer consists of circularly arranged muscular cells with long, rod-shaped nuclei (Fig. 24B, C). In the inner arterial layer, the nuclei appear to be obliquely arranged (Fig. 24C) which would suggest that the cells in this layer mainly take a helical course. The fine elastic fibers and the networks of reticulin

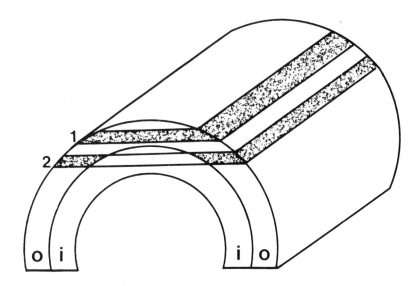

Fog. 26. Schematic diagram of part of a fully patent umbilical artery showing the outer (o) and inner (i) layers. Serial sections from below the surface contain elements from the outer layer (1) whereas sections from a deeper level (2) contain elements from the outer layer in the lateral portions and from the inner layer in the middle portion of the section.

fibers probably follow the predominant orientation of the cells in this layer.

Examination of patent arterial segments on <u>longitudinal serial tangential sections</u> and proceeding from the adventitia to the lumen provides additional insight into the structure of the arterial wall and the arrangement of its components (Fig. 26). In tangential sections from the outermost layers, the muscle cells are parallel and arranged perpendicularly to the direction of flow. A spiral, i.e., a helical arrangement of muscle cells is nowhere discernible (Fig. 27A). Near the lumen, elements of the internal layer become visible (Fig. 26). In comparison with the well-defined regular structural pattern of the outer circular muscular layer, the cells of the inner layer are irregularly and loosely arranged. They have a sparse cytoplasm and poorly defined outlines (Fig. 28). Their nuclei mostly

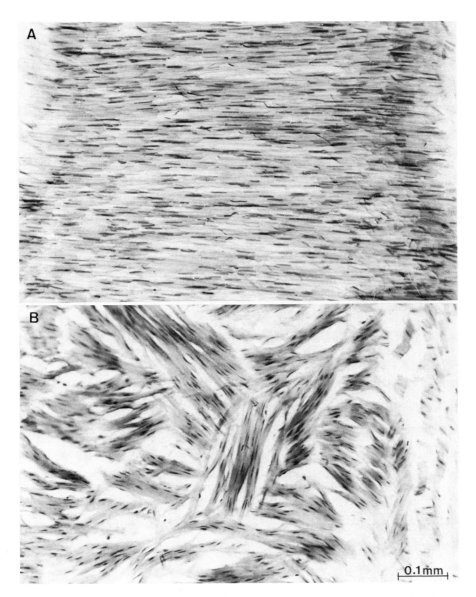

Fig. 27. (A) In an axial tangential section from a fully patent umbilical artery, the nuclei of muscle cells in the outer muscular layer have a circular course and parallel each other. (B) In an axial tangential section from a contracted umbilical artery, this layer has undergone a profound rearrangement of its original structural pattern. The muscle cells now form bundles which have a predominantly oblique course. Arrows show axis of the vessel. Cryostat sections, hematoxylin and eosin.

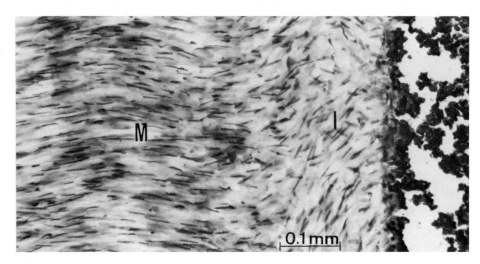

Fig. 28. The outer muscular (M) and the inner (I) layers are shown in a tangential section from a fully patent umbilical artery. Whereas the muscular layer (M) shows parallel arranged, tightly packed, intensely stained muscle cells, the inner layer (I) consists of irregularly arranged cells with a sparse cytoplasm and poorly defined outlines. L - arterial lumen partly filled with erythrocytes. Cryostat section. Phloxine-tartrazine stain.

have an oblique course and often cross each other at an angle of 90°. Only the nuclei of the endothelial lining are aligned along the axis of the artery in regular fashion.

Structural changes in the umbilical artery during contraction and closure. Closure of the umbilical arteries begins with the appearance of up to 15 localized contractions in both arteries within the first 10 seconds after delivery (Moinian et al., 1969). The location of the arterial contractions shows no definite relation to the location of the venous contractions in the cord. However, the number of venous and arterial contractions may be increased in the same cord segment. Along the cord the contractions look like transverse grooves (Fig. 29A). Since the contractions mainly involve the sectors of the umbilical artery nearest to the surface of the cord, the narrowing lumen shifts away from the cord surface, i.e., toward the center of the cord.

At low magnification (Fig. 29B), a localized contraction is seen to consist largely of a thickened outer layer of dark tissue beneath whitish Wharton's jelly; the inner layer forms a thin light stripe near the lumen. Longitudinal section (Fig. 30A) confirms the impression that the crest narrowing the lumen is mainly formed by contracted circular musculature which now consists of tightly packed bundles. These bundles, like those in the umbilical vein, develop by folding of the cylindrical sheets which are seen in adjacent non-contracted portions of the arterial wall. Cross-sections at sites of localized contractions also show that the main part of the contraction is formed by a contracted outer layer of circular musculature. The inner layer is moderately wide and of about equal thickness around the circumference. It, therefore, seems to play little or no role in narrowing the lumen at the sites of localized contractions (Fig. 30B).

As in the umbilical vein, localized contractions are followed by ultimate closure of larger arterial segments. Structural components in both arterial layers then become completely rearranged.

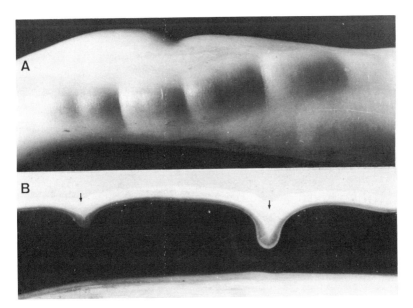

Fig. 29. (A) Localized contractions in an umbilical artery take the form of transverse grooves along the cord surface. x 2.
(B) In an umbilical artery which has been cut along the axis, the contractions take the form of indentations (arrow) which considerably narrow the arterial lumen. They consist mainly of circular musculature (darker portion of contraction). The inner layer (lighter portion) is thin. x 10. (See Fig. 32).

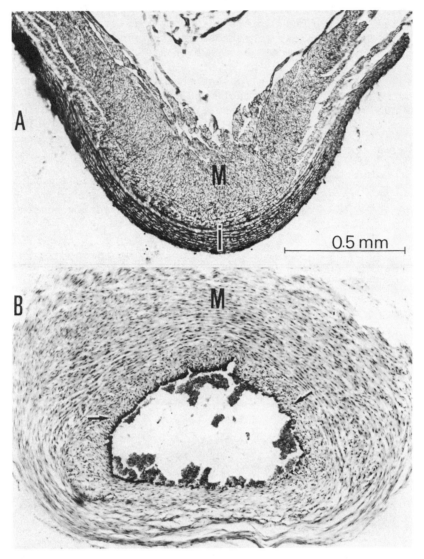

Fig. 30. Localized arterial contractions of the umbilical artery shown in longitudinal (A) and cross (B) sections consist mainly of contracted circular musculature (M). The inner arterial layer (i) forms a relatively small part of the contraction. In the microphotograph of the cryostat section stained with Feyrter's thionine technique (A), the metachromatic ground substance of the inner arterial layer (i) appears black. In the cross-section of the contraction (B), stained with hematoxylin and eosin, the arrows point to the inner arterial layer.

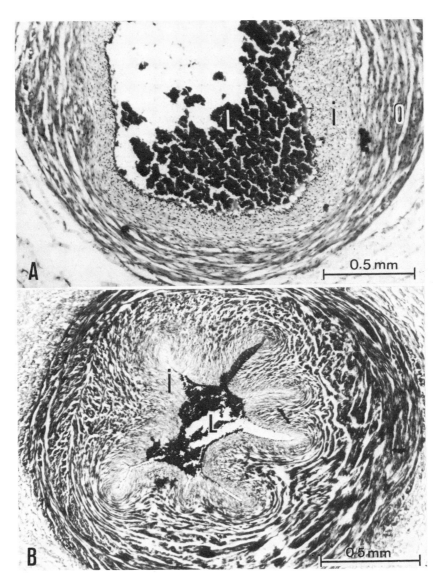

Fig. 31. (A) Cross-section of a partly contracted umbilical artery. In contrast to the intensely stained circular musculature of the outer arterial layer (O), the inner layer (i) shows loosely arranged cells embedded in a lightly stained intercellular substance. The lumen (L) contains blood. Hematoxylin-phloxine-tartrazine stain. (B) In the fully contracted umbilical artery, the folded, but not much thickened inner layer (i) fills the narrowed lumen (L). The rearranged muscular bundles of the outer layer partly assume an oblique course. Heidenhain's azocarmine-aniline blue stain.

In the outer layer, the muscular sheets, demonstrated in the fully patent artery, become transformed into muscular bundles which assume a different course. Most bundles now seem to be obliquely arranged although a few show a circular orientation (Figs. 27B, 31).

Contraction and transformation of muscular sheets into muscle bundles results in superimposition and displacement of an increasing number of smooth muscle cells in the outer arterial layer toward the lumen. Considerable thickening of the wall ensues and the tension, i.e., the distending forces to which the individual muscle cells are exposed in the fully patent artery, decline further with the decrease in volume of blood flow and blood pressure. With the reduction in functional load, the forces which counteract contraction of the musculature diminish, and ultimate irreversible closure of umbilical arteries becomes possible.

With progressive contraction of the outer muscular arterial layer, the inner layer becomes increasingly folded and the initially round arterial lumen becomes stellate on cross-section (Fig 31B). The folds protruding into the lumen, however, consist not only of the displaced inner layer but also of groups of numerous strong muscle bundles from the outer layer which have shifted toward the narrowing lumen. The folded inner layer is now about as thick at the peaks as in the depressions between the folds. This would suggest that, as in the localized contractions, the inner layer is rather passively shifted and folded into the narrowing lumen and does not actively participate in occlusion of the artery.

The considerable structural differences between the outer and inner layers that are seen in the fully patent vessel (Fig. 24) become more pronounced after contraction (Fig. 31). The outer arterial layer consists of intensely stained contracted muscle bundles, whereas the inner layer is composed mainly of loosely arranged cells with sparse cytoplasm. The structural differences between both layers become apparent particularly when the smooth musculature is deeply stained, e.g., by the phloxin-tartrazin or Heidenhain's azocarmine aniline blue staining methods.

Electron microscopic examination. Spiteri et al. (1966) found no difference between the ultrastructure of smooth muscle cells in the human umbilical artery and that of smooth muscle cells in other human and animal muscular arteries. Electron microscopic examination of our material, however, shows that only the cells of the outer arterial layer display the typical structural features of fully differentiated smooth muscle cells which are described in detail by Spiteri et al. Most cells are contracted and their large central nuclei therefore have many deep indentations. The cytoplasmic organelles, among them, in particular, some mitochondria, a few cisternae of the rough endoplasmic reticulum and accumulations of glycogen particles, are confined to a relatively small perinuclear zone. Most of the

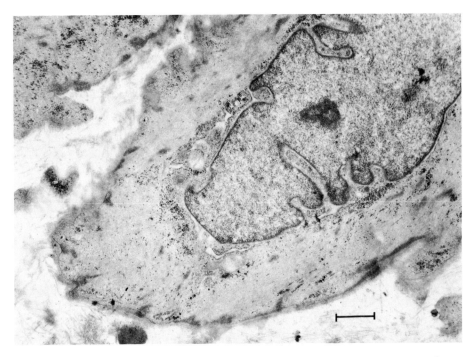

Fig. 32. Electron micrograph of a portion of the inner well of an umbilical artery showing a cell with small bundles of contractile filaments. Vascular lumen above. x 15, 300.

cytoplasm, however, is occupied by large numbers of thin (about 50 Å) filaments aligned in parallel with irregular dense patches, i.e., the Z-line-equivalents, between them. Many pinocytotic vesicles are present beneath the cell membrane which is covered by a thin continuous basement membrane.

The cells of the inner layer of the arterial wall have a different structure since only a few small bundles of 50 Å filaments are present (Fig. 32) and most of the cytoplasm has numerous organelles with a well-developed, dilated, rough endoplasmic reticulum and prominent Golgi apparatus (Fig. 33). There are only a few pinocytotic vesicles and the basement membranes of the cells are discontinuous and have large gaps. Electron microscopy, therefore, confirms the impression derived from light microscopy that there are structural differences between the cells in the outer and inner layers of the umbilical artery. The reasons for the discrepancy

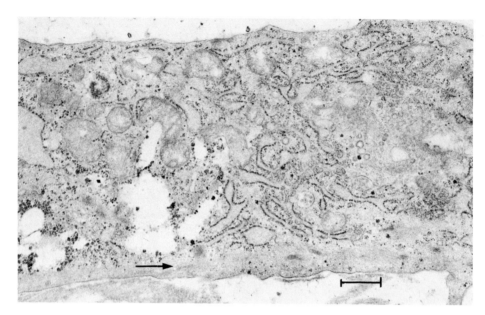

Fig. 33. Inner zone of the umbilical artery. Portion of a cell rich in organelles but with only small filament bundles near the cell membrane (→). x 13,000.

between our findings and those of Spiteri et al. are not apparent but may be methodological.

The endothelial lining of the vascular lumen is formed by a dense layer of overlapping cells connected by membrane junctions which are not to be regarded as well-differentiated endothelial cells because they are similar, if not identical, to the cells in the inner layer of the wall. There is, therefore, no real barrier to transfer or transport of substances in either direction.

In conclusion, closure of umbilical arteries is due mainly to contraction and rearrangement of the outer circular muscular layer. The inner layer probably serves as a plastic tissue component that fills in the narrowing lumen and contributes to the later stages of closure. The cells of the inner layer have a spiral course and are poorly differentiated smooth muscle cells which can only play a very minor role in vascular contraction. A well differentiated longitudinal muscular layer does not exist in human umbilical arteries. Thus, our findings do not support the concept put forward by Roach (1973) and based on findings in lambs which postulates that closure of umbilical arteries results from contraction of a longitudinal muscular layer which allegedly represents the main structural element in these vessels. Even the "valves" of Hoboken, i.e., the localized contractions of the umbilical arteries, are produced, according to Roach (1973), *"by longitudinal muscle bellies, i.e., by contraction of the longitudinal muscle which we have demonstrated to be essential for closure of large vessels."* This conclusion is inconsistent with our findings which show that localized contractions of umbilical arteries are formed mainly by contracting circular musculature.

The same is true of localized venous contractions. The wall of the umbilical vein consists of circularly arranged smooth muscle cells and an endothelial lining. Consequently, localized venous contractions are exclusively formed by contracted and rearranged elements of the circular venous musculature.

Thus, the circular musculature of the umbilical vessels alone seems to be capable of narrowing the lumen considerably even when the vascular tube is still exposed to an approximately normal and relatively high blood pressure.

Roach (1973) also states that ultimate closure of umbilical arteries is achieved *"by protrusion of the inner longitudinal layer of muscle into the lumen."* In her opinion, contraction of longitudinal musculature should decrease the size of the lumen more readily than contraction of circular musculature because the cross-sectional area of the contracting arterial tube increases in proportion to the degree of shortening of the arterial segment. In her studies the umbilical arteries of lambs delivered by cesarean section shorten on contraction by 23 - 32 percent of their initial length. This shortening is associated with a comparable increase in the cross-sectional area of the arterial wall and no decrease in external diameter occurs. In human cords, we found only a 10 percent shortening after complete closure and emptying of the umbilical vessels and the external diameter of the arteries decreased by one-half to two-thirds during the transition from a widely patent to a completely occluded vessel. If closure of the lumen were due to contraction of longitudinal musculature and shortening of the vessel, as postulated by Roach, a decrease in the external diameter should not occur (Roach, 1973: page 144, Fig. 1). Species differences, however, cannot be excluded and may explain the differences in our findings.

REGULATION OF UMBILICAL BLOOD FLOW

It is difficult to reconcile the results of in vitro and in vivo studies concerning reactivity of umbilical blood vessels. This is partly because experiments have often not been performed under physiological conditions.

In vitro, umbilical vessels respond by reacting to numerous factors. Thus, vasodilatation has been reported in response to high CO_2 and low O_2 tensions. Vasoconstriction has been induced by physical factors, such as exposure to cold and tactile stimulation. A variety of substances, including 5-hydroxytryptamine (5-HT), bradykinin, prostaglandins E_2 and $F_2\alpha$, potassium chloride, oxytocin (Altura et al., 1972; Eltherington et al., 1968; Hillier and Karim, 1968) and high oxygen tension have induced contraction. Excessively high concentrations of O_2 have sometimes been used (e.g., 95-100% O_2 producing a PO_2 of 400 mm Hg or more) (Panigel, 1962).

Methodological differences may also significantly influence in vitro results because although longitudinal and spiral strips of the human umbilical artery show no difference in qualitative response, spiral strips and circular rings contract more strongly than longitudinal strips (Altura et al., 1972; Hillier and Karim, 1968), possibly because the longitudinal muscle is comparatively weak. This in vitro difference likewise does not support the hypothesis proposed by Roach (1973) that longitudinal muscles are important for closure of umbilical vessels after birth (see pp. 48-50).

In vivo, umbilical vessels show almost no direct local effect following large changes in fetal arterial PO_2 from 15 to 65 mm Hg. Severe asphyxia produces vasoconstriction but this presumably occurs in response to release of catecholamines from the fetal adrenals. The state of "maturity" of the fetus seems to influence the response since no such effect occurs in the immature fetal lamb, possibly because smaller amounts of catecholamines are liberated.

The relative importance of these various factors on closure in vivo is difficult to evaluate. Recent studies on longitudinal and spiral strips of the human umbilical artery suggest that it takes at least 10 minutes for muscle tone to increase after a significant rise in PO_2 (Tuvemo and Strandberg, 1975). This would suggest that oxygen may not be important since umbilical vessel closure occurs within the first minutes of life whereas the PO_2 takes 8 - 10 minutes or more to rise from 25 - 35 mm Hg (at birth) to 55 - 75 mm Hg (Engström et al., 1966; Tunell, 1974). The level of the PO_2, however, affects the contractile response to prostaglandin (PGE_2), bradykinin and 5-HT (Nair and Dyan, 1973) in the sense that a higher PO_2 mildly enhances the response (Lewis, 1968). A rise in PO_2, however, has a considerably greater enhancing effect on the contractile response of the ductus arteriosus (Oberhansli-Weiss et al., 1972). Recent studies by Jonsson et al. (1976) suggest that local biosynthesis of prostaglandins is very active in the human umbilical cord after delivery and that they are present in high enough concentrations to stimulate the smooth muscles of the arteries of the umbilical cord.

Nerve Supply

The presence of neural elements in the extra-abdominal part of the cord in the human is disputed. Spivack (1943) reports that no such innervation is present in the human infant or guinea pig, except possibly at the skin edge, a view which is shared by Hollingsworth (1974), after studies using electrical stimulation and histochemical methods. Zaitev (1959), however, has published illustrations showing such elements at a distance of about 10 cm beyond the umbilicus. Nadkarni (1970), using both the light and electron microscope, likewise found structures that he believed were myelinated nerves within the smooth muscle cells in human cords, but found no nerve fibers with the light microscope after silver staining. Lachenmayer (1971) was also unable to find adrenergic nerves in this segment, Ellison (1971), however, used a thiocholine technique, and found some acetylcholinesterase positive fibers which extended a distance of up to 15 cm but concluded that the fibers were probably sensory.

The presence of neural elements in the intra-abdominal part of the umbilical vein has been clearly established (Spivack, 1943). Adrenergic nerves increase in number in the direction of the ductus venosus (Ehinger et al., 1968) and are most numerous at the site of origin of the ductus where there is also an increase in the number of smooth muscle cells in the wall. Lachenmayer (1971) likewise found a rich adrenergic innervation of the intra-abdominal part of the umbilical arteries. Nerve fibers from the celiac plexus and anterior and posterior vagal trunks are also present at the junction of the umbilical vein with the ductus venosus.

Neonatal Circulation

Multiple changes occur in the immediate hours after birth. They ensure a successful transition from intrauterine to extrauterine life. Systemic arterial pressure does not change immediately after the first breath despite the abrupt cessation of the placental circulation, the increase in flow through the pulmonary vasculature, and the large change in blood volume consequent to delayed clamping of the cord. The pressure starts to fall circa (10 to 15 mm Hg) after 10 to 15 minutes. It then slowly starts to rise again after one to four hours of age. Infants with immediate clamping of the cord, e.g., following cesarean section or with nuchal cord, of low birth weight and with severe hypoxia, have lower initial pressures.

Cardiac output is two to three times greater per unit body weight during the first two days of life than in the adult. This may, in part, be due to the high metabolic rate in the postnatal period. But there are difficulties inherent in conclusions drawn from such measurements because of shunts through the ductus arteriosus, foramen ovale and bronchial vessels.

On the first day of life, systemic blood flow is low and mean systemic vascular resistance is higher than is found later in normal children. On the second day, systemic flow increases primarily because of closure of the ductus arteriosus. Since flow increases relatively more than systemic arterial pressure during the same period, the systemic vascular resistance must fall apparently to about two-thirds of its initial value.

In studies of peripheral blood flow in the healthy newborn, (using venous occlusion plethysmography), mean values fall from 8 ml/100 ml/min of cardiac output to as low as 0.4 ml/100 ml/min a few hours later and then rise again to slightly lower than initial values (Celander, 1960). Comparatively speaking, peripheral blood flow is greater and resistance to flow less in infants of normal birth weight than in the adult and the difference is greater in infants of low birth weight (Celander, 1966). It is somewhat dubious, however, whether such comparison is very meaningful since the newborn infant (and especially the pre-term infant) has a greater proportion of skin and considerably smaller amount of muscle than the adult (18% vs 45% as observed in the foot and calf) (Celander, 1966). It is interesting in this connection that hypovolemia such as that induced by a 15% reduction of blood volume during exchange transfusion causes an increase in peripheral resistance which is considerably greater than the calculated increase in resistance to flow in the ascending aorta during identical volume depletion (Wallgren and Lind, 1967). Thus the need for caution in the interpretation of such data should be obvious.

Heart volume increases during the first 15 to 30 minutes of life. It then decreases during the rest of the first week. The causes of changes in heart volume in the neonate are obviously complex but they cannot be due to differences in muscle mass. Tiisala et al. (1966) have shown that there is a direct relation between heart volume and blood volume at birth, hence early clamped infants have a smaller heart volume than late clamped infants. There is also a direct relation between the amplitude of QRS deflections in various leads and duration of various intervals of the ECG and heart volume (Walsh, 1966). Infants with asphyxia and pulmonary atelectasis at birth also show a greater increase in heart size during the first days than healthy infants (Martin and Freidell, 1952; Burnard and James, 1961). Closure of fetal channels and accompanying alterations in the balance of resistance and pressure in the pulmonary and systemic circuits must likewise be important in this regard.

In infants born at term, right ventricular mass may exceed left ventricular mass by as much as 25%, especially in the male, and the relative thickness of its wall is greater than generally accepted as normal in adults (Hort, 1955; Emery and MacDonald, 1960; Schulz and Giordano, 1962). After birth, right ventricular mass decreases by about 20% of its weight at birth and at six months of age it has only regained its weight at birth. In contrast, left ventricular mass increases rapidly after birth and by the beginning of the fourth month, adult ventricular weight relationships are achieved. During this period, fiber diameter seemingly decreases in the right and increases in the left ventricle. In accordance with these findings, other changes have also been noted--such as alterations in RNA concentration, rate of doubling of muscle fiber nuclei, rate of development of the adult or H type isozyme of lactic dehydrogenase, differences in vascularity, etc. (Table 1, p. 22) (Gross, 1921: Linzbach, 1952; Gluck et al., 1964; Filer, 1966). Although total myocardial blood flow and flow per unit weight reportedly rise markedly in the left and fall in the right ventricle (Yuan, Heymann and Rudolph, 1966), the difference in vascularity of the ventricular walls is disputed. Gross' (1921) illustrations, however, seem convincing.

The electrocardiogram also mirrors the multiple changes accompanying extrauterine adaptation, such as differences in blood volume consequent to early and late clamping of the cord, asphyxia, etc., and has been shown in a series of studies to alter in a distinct manner after birth. Walsh has summarized her findings in two papers (1975[a,b]). Changes in the neonatal period have also been reported in the phonocardiogram but, so far as can be judged at present, the information obtained with this technique is of comparatively little value in longitudinal studies after birth (Fig. 34).

Information on cardiovascular and respiratory reflexes in the newborn are beyond the scope of this section and the reader is therefore referred to Stawe (in press) and studies by Dawes and co-workers, which have been summarized by Dawes (1968), among others.

Neonatal Foramen Ovale

Closure of the foramen ovale occurs in two stages: functionally when the valve is held closed by hemodynamic changes and anatomically when the valve becomes adherent to the edges of the orifice. Functional closure is induced by two events: aeration of the lungs and cessation of the placental circulation.

With the first breath, pulmonary vascular resistance decreases and pulmonary venous flow to the left atrium increases, hence pressure relationships in the atria reverse. During the first minutes, the pressure (in the human infant) may be only about 1 mm Hg higher in the left than in the right atrium but after the first 30 minutes of life, the pressure in the left atrium is always greater than in the right. The mean pressure gradient remains about 4 mm Hg between 30 minutes and 14 hours of age (the same as in the adult) and then falls to 2 mm Hg at 24 hours of age. During the first hour of life, the difference in pressure gradient is not affected by the size of

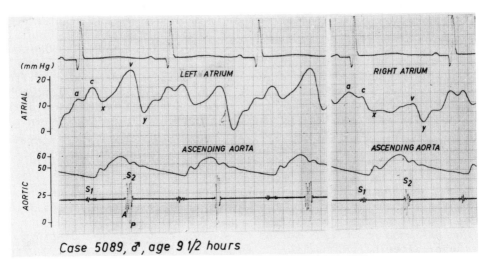

Fig. 34. Relation between atrial and aortic pressures to the electrocardiogram and the phonocardiogram in a healthy term newborn infant. Note the differences in contour between the left and right atrial curves (Courtesy of Drs. R. A. Arcilla and J. Lind). From Walsh and Lind in Stave, 1970.

the placental transfusion though the level of the pressures is. Hence, early clamping is associated with comparatively low pressures (mean pressure in the left and right atrium: 4 mm Hg and 0 mm Hg, respectively) and late clamping and stripping of the cord with comparatively high pressure (mean pressures in the left and right atrium: 9.5 mm Hg and 5 mm Hg, respectively) (Arcilla et al., 1966). In both groups of infants the configuration of the left atrial curve is characterized by tall "v" waves with rapid ascending and descending limbs and comparatively high "x" levels (a contour reminiscent of mitral regurgitation) whereas that of the right atrial curve is characterized by low "v" waves and "x" levels (Fig. 36). Hence it is easy to distinguish the location of the catheter.

With cessation of the placental circulation, umbilical blood flow ceases, inferior vena caval pressure falls and the inferior vena caval-left atrial pressure difference is reversed. This pressure change is observed even if ventilation does not occur (Dawes, Mott and Widdicombe, 1955). The valve now shuts incompletely and thereby serves to balance atrial intake and plays a role in the

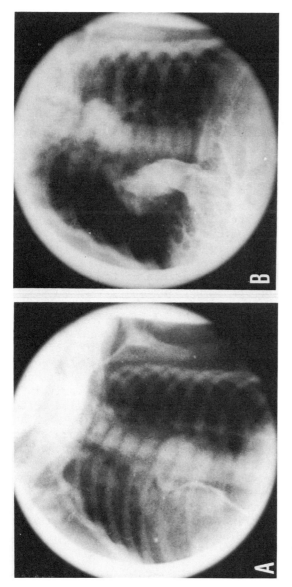

Fig. 35. Functional closure of the foramen ovale after the first cry.
(A) Injection of 2 ml contrast medium in the umbilical vein before the first cry. Most of the medium is directed into the left atrium which is faintly opacified.
(B) After the first cry, another injection of 2 ml contrast medium is administered. The medium now outlines the right heart and pulmonary artery. No right-to-left shunt is seen. In-between frontal and left anterior oblique projection.

smooth transition from intrauterine to extrauterine life. The foramen may close functionally with the first vigorous cry consequent to aeration of the lungs and cessation of umbilical blood flow (Fig. 35). More commonly, however, an insignificant right-to-left shunt through the foramen may persist during the first days of life (Lind and Wegelius, 1954; Stahlman et al., 1962) increasing with any condition causing an increase in pulmonary vascular resistance or interfering with pulmonary ventilation (crying, straining, Valsalva) (Fig. 36). Significant left-to-right shunting (with pulmonary:systemic flow ratios of as much as 2:1) has been noted in healthy infants but its frequency is not known. Left-to-right shunting of no physiological significance sometimes occurs in infants asphyxiated at birth (James, Burnard and Rowe, 1961), may persist for several months (Hoffman, Danilowicz and Rudolph, 1965) and then gradually disappears.

The valve of the foramen ovale becomes increasingly plastered against the septum secundum as left atrial return rises; thereby favoring anatomical closure. Some differences in the time of closure and persistence of left-to-right shunting through the foramen presumably arise from anatomic variations in the size of the valve e.g., excessively large opening, multiple performations, etc., and/or disturbances in hemodynamic relations between the left and right heart. Various congenital malformations of the heart, accompanied by an elevation of right atrial pressure, e.g., pulmonary stenosis (or atresia) with an intact ventricular septum, tricuspid atresia, will tend to hinder closure; likewise disturbances in the pulmonary circuit accompanied by increasing pulmonary vascular resistance and pressure may also tend to re-open the channel. Conversely, congenital malformations of the heart accompanied by an elevation of left atrial pressure, e.g., severe obstructive lesions in the left heart, may accelerate closure of the valve. If this occurs early in intrauterine life, it may be associated with the so-called left heart syndrome, a term used to describe a group of anomalies with an obstructive lesion on the left side of the heart, hypoplasia of the left ventricle, a small aorta, hypertrophy of the right ventricle and a large pulmonary artery (Noonan and Nadas, 1958). If it occurs late in gestation, left heart development may be normal (Naeye and Blanc, 1964). Premature closure of the foramen ovale--i.e., a foramen with a diameter of less than 2 mm at a few days of age (Lev et al., 1963), is not all that uncommon since it has been reported in 9/1500 pediatric autopsies (Naeye and Blanc, 1964).

3. DUCTUS ARTERIOSUS

Introduction

The ductus arteriosus is derived from the distal portion of the sixth left aortic arch. At term, it represents a wide arterial

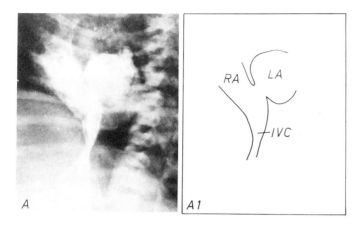

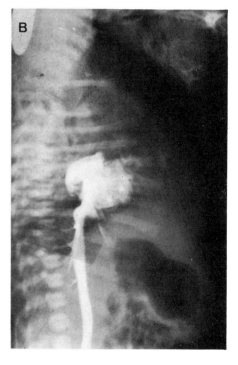

Fig. 36. Angiographic demonstration of a right-to-left shunt through the foramen ovale in a healthy 18-hour-old crying newborn infant.
(A) In left anterior oblique projection, a considerable amount of contrast medium is shunted from right-to-left through the foramen ovale whose upper rim is well seen. From Lind et al., 1964.
(B) In right anterior projection, the stream of the shunt follows the upper wall of the left atrium.

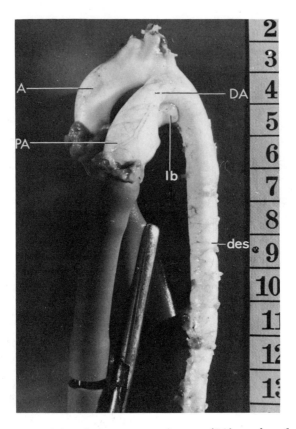

Fig. 37. The aorta (A), ductus arteriosus (DA) and pulmonary artery (PA) from a term infant have been filled with saline at a pressure of 100 mm Hg after ligation of all branches of the great vessels. There is no visible demarcation between these vessels and the take-off of the duct. The trunk of the pulmonary artery may be slightly overdistended. lb - ligated left main branch of pulmonary artery, des - descending aorta. Centimeter scale on right. From Simon, 1959.

segment which connects the main pulmonary trunk with the upper thoracic aorta (Fig. 37). The duct constitutes the direct continuation of the pulmonary trunk. Since it has nearly the same caliber as the trunk, no outer demarcation between these two vessels is visible. The upper part of the duct forms a slight curve in a dorsal direction and then enters the lower medial wall of the aortic arch just below the aortic isthmus at an angle of about 30° (range 25° - 37°) (Mancini, 1951). The aortic orifice of the duct is oval-shaped and its upper edge forms the "crista reuniens" which protrudes into the vascular lumen but plays no role in postnatal closure of the duct. The upper part of this protrusion has a structure similar to that of the aortic wall and is rich in elastic tissue, whereas the lower aspect has a structure similar to that of the duct and is rich in smooth musculature (Fig. 38).

The duct has an external diameter of 0.5 - 0.6 cm and a length of 1.25 cm at birth (Walkoff, 1869; Bruce et al., 1964). Angiographic studies in the human fetus suggest that the internal diameter of the duct is slightly less than that of the two vessels it connects (Fig. 39) (Lind and Wegelius, 1954). This is presumably because a part of the right ventricular output is diverted into both main pulmonary arterial branches before entering the duct. The duct probably appears narrower at autopsy than it is in vivo, because of postmortem contraction. Such contraction can be expected to be greater in a muscular than in an elastic artery. But even after postmortem distension of the vessel, however, the duct does not become as large as the pulmonary trunk (Fig. 37).

According to Blanc and Nodot (1957), the ratio of the internal circumference of the duct, at its midportion, to that of either the left common carotid or left subclavian artery is normally less than 1.25 during the first two days of life. Larger ducts are seen with premature closure of the foramen ovale and coarctation of the aorta. Long and tortuous ducts occur with other forms of congenital heart disease, e.g., with atresia of the pulmonary ostium.

The wall of the duct is supplied by blood from the left coronary and first intercostal arteries (Prichard, 1968) and it has many vasa vasorum and capillaries in the adventitia and outer part of the media but none in the inner media and intima. The adventitia also contains numerous nerves which are partly afferent (Boyd, 1940; Holmes, 1958). Present at term are sympathetic nerves from the lower cervical and upper thoracic portion of the sympathetic trunk, as well as vagal fibers derived directly or by way of the superficial plexus (Allen, 1955). There is also evidence in the human fetus of adrenergic and cholinergic receptor function and specific adrenergic nerve fibers that penetrate deeply into the smooth muscle layer of the media (Boréus et al., 1969).

FUNCTIONAL ASPECTS

Fetal Ductus Arteriosus

In the fetus, the ductus arteriosus is a widely patent channel that serves as a by-pass around the lungs for blood from the right heart. Since pulmonary arterial pressure is slightly higher than systemic pressure and pulmonary vascular resistance is considerably higher than systemic resistance, the major portion of blood in the pulmonary artery is shunted from right-to-left through the duct and

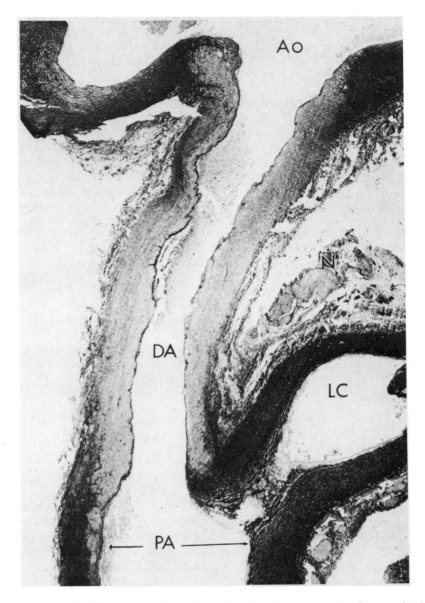

Fig. 38. Longitudinal section through the ductus arteriosus (DA) of a full-size newborn infant. The wall of the duct contains few elastic fibers and much muscular tissue which does not stain with this technique. The aorta (Ao) and the trunk of the pulmonary artery (PA) and its left branch (lc) are largely composed of dark staining elastic elements. Nerves (N). (x10). Orcein elastic stain. From Walsh, Meyer and Lind, 1974.

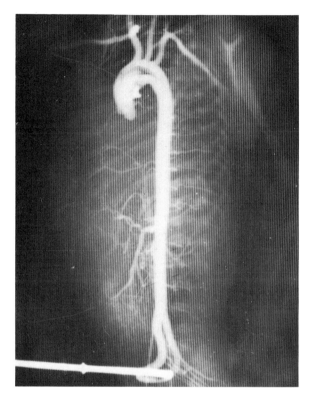

Fig. 39. Human fetus about 19 weeks gestation. Retrograde aortography. Contrast medium injected into an umbilical artery with good filling of the descending aorta, aortic arch, cephalic vessels and ductus arteriosus. From Lind et al., 1964.

returned via the descending aorta to the placenta for gaseous exchange (Figs. 6, 10, 39). The amount of blood entering the duct in the fetal lamb in utero, at least three to five days after surgery, may constitute 60% of the combined output, as estimated in studies using injection of radionucleide labeled 15 μm microspheres (Rudolph and Heymann, 1970). Values vary, depending on the method used, the conditions of the animal, and the species. There are, moreover, wide individual variations which depend on the various resistances in the hemodynamic system.

Since the wall of the duct is muscular, it is capable of reducing the size of its lumen and cannot be regarded as a passive structure incapable of exerting control on blood flow. Intermittent narrowing of the duct will oblige more blood to enter the pulmonary branches and thereby favor development of the pulmonary vascular bed, augment right ventricular work, and possibly cause the smooth muscle mass in the pulmonary arterioles to increase if it occurs often enough. It is well known that during fetal stress, the amount of blood flowing through the duct increases and that this helps maintain high placental flow. It is achieved by further vasoconstriction of pulmonary vessels in response to asphyxia and catecholamines.

Animal studies in vivo and in vitro show that there is a linear correlation between ductal blood flow and PO_2 --i.e., when the blood PO_2 rises, flow decreases (Assali et al., 1963). The effect of O_2 appears to be exerted on the smooth muscle cells themselves which become stronger toward term, and not on some intermediate cell type (Kovalçik, 1963; Fay, 1971). The effect cannot be elicited in the isolated human duct until _after_ midterm (McMurphy and Boréus, 1971).

Adrenaline, noradrenaline, and acetylcholine produce ductal constriction in the isolated duct of fetal lambs and guinea pigs, an effect which can be blocked by dibenamine and atropine without altering the response to oxygen (Born et al., 1956; Kovalçik, 1963). The human fetal duct shows prompt contraction following administration of noradrenaline and acetylcholine. Its adrenergic nerve terminals resemble those of the adult. The latter finding has been interpreted as evidence of earlier maturation of the fibers and a role in postnatal closure of the duct (Boréus et al., 1969). Extreme asphyxia seems to induce ductal constriction by release of sympathetic amines (Born et al., 1956). There is some disagreement about the effect of catecholamines, however, and especially noradrenaline, because of differences in reported results. In the opinion of Panigel (1962), the changes in ductal flow are probably due to general hemodynamic effects.

Neonatal Ductus Arteriosus

Closure of the ductus arteriosus normally occurs in two stages. First rapid functional constriction develops over a period of hours and then anatomical obliteration follows over a period of weeks, months, and occasionally requires years.

Various studies show that the fetal direction of flow, from right-to-left, may continue during the first 60 minutes of life (James, 1959; Saling, 1960) and persists only if pulmonary vascular resistance is increased (e.g., by crying or hypoxia). Reversal of the shunt then ensues, although flow may be bidirectional for as long as six hours. By the age of 15 hours, flow through the ductus has ceased or is no longer physiologically significant (Moss et al., 1963) (Fig. 40). Both the direction of flow and its magnitude depend on multiple factors, such as the fluctuating relationship between resistance and pressures in the pulmonary and systemic circuits, the degree of constriction of the duct, and the cardiac output. A left-to-right shunt may persist after this age in conditions associated with hypoxemia, such as the respiratory distress syndrome, pulmonary atelectasis, birth, and residence at high altitude, as well as after premature delivery. Persistence of a left-to-right shunt will slow the rate of decline of pulmonary vascular resistance and pressure. Recirculation of blood through the lungs may then raise the oxygen content which in turn may act as a stimulus to closure. Conversely, a <u>left-to-right shunt may be eliminated</u> by administration of 100% oxygen during the first eight hours of life. Studies in dogs indicate that the ability to tolerate a left-to-right shunt depends on the presence of normal sympathetic nervous system activity and of catecholamine release (Rudolph et al., 1963). In this species, normal catecholamine levels improve ventricular contractility, they lower left ventricular end-diastolic pressure and favor normal pulmonary vessel maturation, whereas low catecholamine levels have the opposite effect.

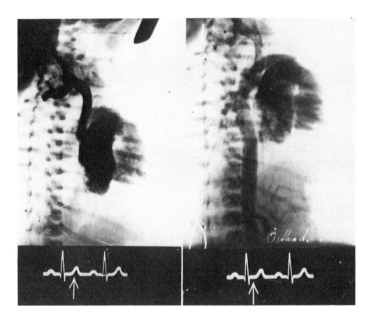

Fig. 40. Angiogram showing a transient right-to-left shunt through the ductus arteriosus in a six-hour-old term infant with cyanotic extremities.
(A) The superior vena cava and right atrium are well opacified whereas the right ventricle and pulmonary artery are relatively poorly outlined.
(B) In the following ventricular systole, the pulmonary artery, ductus arteriosus and descending aorta become well filled.

Most authors agree that the duct constricts at birth, primarily in response to a high PO_2, but the exact mechanism is not understood. It has been suggested by Oberhansli-Weiss et al. (1972), based on in vitro studies, that the oxygen response is mediated by acetylcholine and blocked by atropine. Others are of the opinion, after studies in the guinea pig, that oxygen triggers constriction of ductal smooth muscle by increasing the rate of oxidative phosphorylation consequent to a direct effect on a terminal cytochrome component (Fay, 1973). In the fetal lamb it has been shown that although the duct is rich in sympathetic nerve supply and substantial amounts of norepinephrine are present in increasing concentration from the aortic to the pulmonary end (Ikeda et al., 1972), the oxygen response is not dependent on the presence of extrinsic nerves.

Experimental studies are difficult to assess, at least partly because of the sensitivity of the duct to a wide variety of stimuli. Thus, constriction of the duct can be induced by normal breathing on delivery; lung inflation with a gas mixture containing oxygen; mechanical or electrical stimulation of the duct or various nerves; infusion of sympathetic amines in the presence of a constant arterial saturation; bradykinin and prostaglandin synthetase inhibitor (indomethacin). As with other studies of this kind, species differences undoubtedly exist--e.g., a smaller diameter duct will tend to close more easily while a larger diameter duct will tend to close with greater difficulty.

Neither of the great vessels constricts after birth because they are primarily elastic arteries. Some ductal muscle may become incorporated at the site of the insertion of the duct in the ventral wall of the aorta and this tissue, in various laboratory animals, responds to O_2 in the same manner as the duct. In the opinion of Wielenga and Denkmeijer (1968), when the amount of this muscle is greater than normal, it may lead to coarctation of the aorta.

In children with cyanotic heart disease, closure of the duct may occur at the expected time even though survival depends on persistent patency of the duct--e.g., pulmonary and tricuspid atresia. In both of these conditions the diameter of the duct is markedly reduced because in fetal life there is no pulmonary outflow from the right ventricle and blood flows from the venae cavae and atria into the left ventricle and out the aorta. Thus, the pulmonary circulation is supplied by blood flowing from left-to-right from the descending aorta through the ductus arteriosus. The duct, therefore, conducts only about 10% of the combined ventricular output rather than the 50 or 60% it normally transports. Closure presumably occurs because the PO_2 after birth is higher than before, or because of release of circulating vasoactive substances in the severely hypoxemic infant or greater facility of a smaller diameter duct to occlude. It is interesting in this connection that infusion of prostaglandin E_1 (PGE), a smooth muscle relaxant, has recently been shown to effectively dilate the ductus arteriosus in ten infants with pulmonary atresia (Heymann and Rudolph, 1977).

In other lesions of the heart, the duct may be larger than normal--e.g., in aortic atresia. It may then fail to close completely because of reduced sensitivity of the smooth muscle to PO_2 or inability of a larger diameter duct to completely occlude. In this condition, there is no normal escape of blood from the left side of the heart and the valve of the foramen ovale may herniate into the right atrium. A channel then forms which allows passage of blood from the left to the right atrium. Pulmonary venous obstruction may develop since this opening is often small.

MORPHOLOGY

The ductus arteriosus is a muscular artery that joins two large elastic arteries, i.e., the pulmonary trunk and the aorta. Since all three vessels are subjected to approximately the same blood pressure during fetal development, the differences in structure between these vessels must be determined by other factors influencing morphogenesis. A muscular arterial segment has the advantage of being able to adjust lumen size to functional demand, and in this manner, influence the distribution of right ventricular output between the aorta and the main branches of the pulmonary artery. With partial constriction of the duct, more blood will be diverted to the lungs, and this will favor development of the pulmonary vasculature, especially in the latter part of gestation when the wall has presumably become stronger. After birth, a muscular wall may be a necessity because only a muscular vessel can contract and occlude completely in a short time. A large amount of unstretched elastic tissue, such as that found in elastic arteries, will resist closure and even make it impossible.

Microscopic Structure

On longitudinal sections, the <u>media</u> of the duct consists of a thick middle layer of cross-sectioned, circularly arranged muscle cells and of thin inner and outer layers of longitudinally arranged muscle cells (Walkoff, 1869; Linzenmeier, 1914; Harris and Heath, 1962; Heath, 1969). Unlike the peripheral muscular arteries, the muscle cells are not tightly packed together, but form thin concentric sheets that alternate with prominent layers of mucoid ground substance (Figs. 41, 42). These muscular sheets resemble interconnecting cylinders of decreasing size.

Thus, the structure of the media has much in common with the structure of the umbilical vein and the outer muscular layer of the umbilical artery (pp. 28-29, 32-38). As in the umbilical vessels, the abundant ground substance probably helps the muscular sheets to

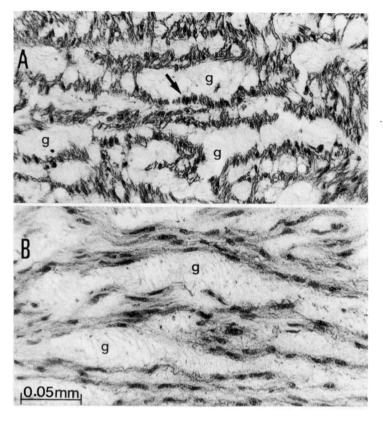

Fig. 41. (A) Longitudinal section showing the media of the duct with rows of cross-sectioned smooth muscle cells (arrows) that alternate with layers of abundant ground substance (g). (B) In a cross-section of the duct, the muscle bundles separated by accumulations of ground substance are seen in longitudinal view.

shift with contraction and ensures rapid postnatal narrowing and closure of the lumen. A reduction in the amount of this semi-fluid substance frequently occurs in the respiratory distress syndrome, and may favor a delay in closure of the duct with persistence of a shunt between the pulmonary artery and aorta.

At term, the intima of the duct is about one-fourth as thick as the media. Some intimal thickening is present, a feature which is often indicative in the newborn of an accommodation to a reduced

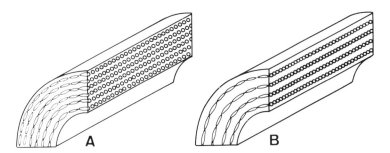

Fig. 42. Diagram showing different arrangements of smooth musculature in the media of a peripheral muscular artery (A) and in the media of the ductus arteriosus (B). In the peripheral artery, the circularly arranged smooth muscle cells are tightly packed together. In the media of the duct, sheets of muscle cells are separated by large amounts of ground substance.

Fig. 43. Luminal surface of a patent duct from a full-term newborn infant. The duct was opened longitudinally along its ventral upper wall. The troughs between the " intimal mounds" are filled with blood and appear gray-black in the photograph. Note the small size of the orifices of the main pulmonary branches (arrows) in relation to the caliber of the pulmonary trunk (P). A - descending thoracic aorta. Millimeter scale on right.

functional requirement. Hence this finding may reflect diminished blood flow in the latter part of gestation. Multiple small elevations, known as "intimal mounds," are visible grossly along the luminal surface of the duct (Fig. 43). They permit the folded intima to be easily distinguished from the smooth surface of the adjacent pulmonary trunk and the aorta. The mounds may be the result of intimal proliferation, but since they are only seen in retracted ducti, it is possible that they are the result of postmortal retraction which presumably becomes more pronounced toward the end of gestation. Microscopically, they consist of loose connective tissue rich in ground substance.

Closure Mechanism

Closure of the duct is mainly effected by contraction of the well-developed smooth musculature. Rapid and complete closure of the wide lumen of the human duct probably occurs only if the underlying structure of the wall permits profound rearrangement, i.e., by shifting of the muscular sheets or muscular bundles that form with contraction of the original muscular sheets. Some degree of dissociation of pre-existing connections between individual layers of the ductal wall also appears necessary for smooth and rapid displacement of these layers toward the ductal lumen.

In 1935, von Hayek suggested that the muscle cells of the media have a helical course, i.e., they are spirally arranged in right and leftward directions. Von Hayek postulated that closure could only occur with a helical arrangement of the musculature, because with a circular arrangement the cells would have to shorten to about one-seventh of their length, which is impossible. Using whole body freezing, Hörnblad (1967, 1969) and Hörnblad and Larsson (1967) studied the process of ductal closure in small animals (guinea pig, rabbit, rat, mouse). They found that complete occlusion of the lumen is achieved by contraction and central displacement of the media; the intima contributed little to this process.

In the much larger human duct, both folding and central displacement of the intima appear to be essential for closure. The intramural shearing forces that might resist central displacement of the intima are abolished by the dissociation that occurs at the intimal-medial junction. Before birth, the ground substance in the media is distributed evenly between the muscle sheets (Figs. 41, 42). On delivery, large amounts of mucoid ground substance accumulate in the inner media. The accumulations often look like cystic spaces which measure as much as 0.3 mm in diameter (Fig. 44). The "cysts" frequently alternate with bizarre "muscle nodules," i.e., whorls of contracted, concentrically arranged, tightly packed smooth muscle cells (Fig. 45). Some of the nodules may include small foci of necrosis and scattered leucocytes. Ground substance is no longer seen

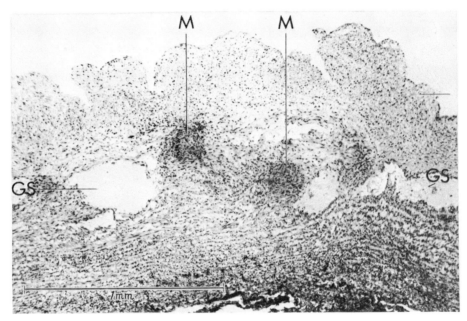

Fig. 44. Longitudinal section of partly contracted duct with intimal folds, dissociation of the intimal-medial junction, cyst-like accumulations of ground substance (GS) and whorl-like muscular nodules (arrows). Full-term infant. Hematoxylin and eosin. From Meyer and Simon, 1960.

(Fig. 46). Thus, the findings contrast sharply with the normal structure seen before closure (Meyer and Simon, 1960).

The cyst-like accumulations of ground substance seem to result from a redistribution of this substance rather than from an absolute increase. With dissociation and contraction of the media, the muscle bundles become torn from their interconnections and form nodules. The ground substance then becomes extruded from the contracting parts of the musculature and accumulates in the cysts that alternate with muscular nodules. The intima becomes undermined during this process (Fig. 47). The connections between the intima and inner media loosen, and the shearing forces which otherwise would resist intimal folding become abolished. Although these preparatory changes are not always striking or may even be absent, they nevertheless provide insight into the process of closure. Since they are often absent in premature infants, maturation of ductal structure must be important for their development and, consequently, for rapid postnatal closure of the ductus.

The outer two-thirds of the media of the human duct shows no signs of dissociation. Hence the outer media retains its ability to contract and produce occlusion of the duct.

In small animals the architecture of the smooth musculature seems to be adequate for ductal closure without any perceptible loosening of pre-existing connections between the individual structural components of the arterial wall. This is probably because shearing forces are small in ducts of small caliber.

Secondary Changes

The dissociation of the inner medial layer and the undermining of the intimal-medial junction often result in disruption of the intima. Tears appear between the intimal folds and usually become covered by superficial thrombi rich in fibrin. They commonly result in small intramural hematomas (Fig. 48). Blood seeps into the

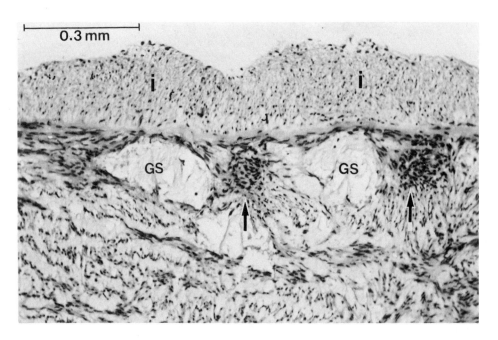

Fig. 45. Longitudinal section of partly contracted ductus arteriosus showing loss of the usual structural pattern of the media. Accumulations of ground substance (GS) alternate with whorl-like nodules of contracted musculature (M) in the inner layer of the media beneath the thickened intima (i). Hematoxylin and eosin. Two-day-old infant. From Meyer and Simon, 1960.

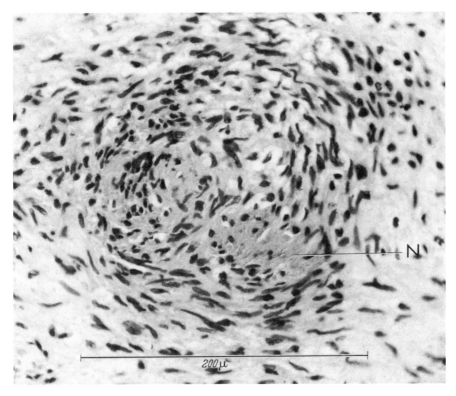

Fig. 46. Higher magnification of muscle whorl showing necrobiotic changes (N), many pyknotic nuclei but only a few leukocytes. From Meyer and Simon, 1960.

subintimal accumulations of the ground substance which then becomes filled with erythrocytes. Larger dissecting intramural hematomas may occur which then resemble the substrate of idiopathic cystic medionecrosis or experimental lathyrism. Thus, some changes that normally occur in the duct may be interpreted as the physiological counterpart of some pathological changes. The term used for describing the destructive lesions accompanying ductal closure, i.e., "physiological angiomalacia" (Meyer and Simon, 1960), therefore seems justified. Such tears and hematomas may be induced by intermittent contractions of the duct. With further dissociation of the inner wall, the changes may become more severe.

Contraction of the duct is followed by <u>necrosis</u> of the middle layer (Gräper, 1921; Kennedy and Clark, 1941; Blumenthal, 1947). It involves the entire circumference of the duct (Fig. 49) and probably arises from decreased perfusion of the narrowed lumen. The increased thickness of the contracted wall may also considerably impair transmural transport of vital nutritional substances. Thus, necrosis may be due to pathogenetic factors similar to those responsible for the necroses that are often seen in an arterial wall thickened by severe atherosclerosis (Blumenthal, 1947).

Despite the frequent occurrence of intimal tears and flat fibrin plaques, <u>occlusive thrombosis</u> is seldom found.

Organic Closure

The intimal folds come together in the center of the contracting duct. The fibrin plaques along the luminal surface and other superficial lesions involving endothelial destruction may favor final adhesion of the folds by stimulating proliferation of the intimal connective tissue and thus produce organic closure. The aortic and

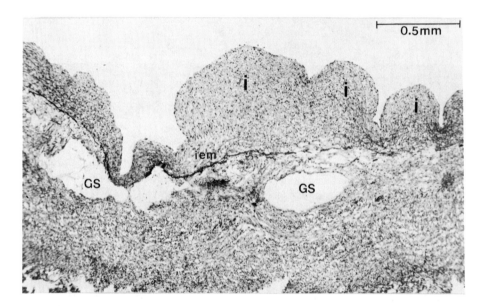

Fig. 47. Intimal folds (i) now overlie the markedly dissociated intimal-medial junction with cyst-like accumulations of ground-substance (GS) and loosening of adjacent parts of the media. iem - internal elastic membrane. Longitudinal section of the ductus arteriosus of a stillborn infant. Orcein elastic stain. From Meyer and Simon, 1960.

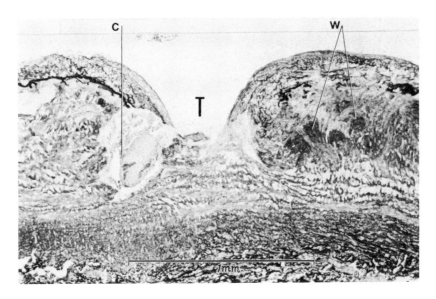

Fig. 48. Tear in the thickened intima (T) and the inner media (arrow) associated with hemorrhagic infiltration of the adjacent intimal and medial tissues containing a cyst-like accumulation of ground substance (C). Whorls of elastic fibers (w) formed by dissociated elastic networks of the inner ductal media are seen on the right. Two-day-old newborn infant. The internal elastic membrane (iem) is well outlined on the left, but fragmented on the right. Longitudinal section. Orcein elastic stain. From Meyer and Simon, 1960.

pulmonary ostia become cone-shaped. By the age of four months, only small indentations remain at both ends of the obliterated duct.

The duct probably ceases to function as a vascular channel during the first two days of life consequent to contraction (Mitchell, 1957; Wilson, 1958). Definite and complete organic closure, however, requires several weeks. Partial patency of the duct appears to exist in two-thirds of normal hearts at two weeks, in 12 percent at eight weeks, and in one percent at one year (Christie, 1930). According to Elsässer (1842), organic closure ensues within six weeks of birth. Gundobin (1912), however, found that in about 50 percent of cases, the duct was closed within 4 - 8 weeks of birth.

In conclusion, functional narrowing of the duct and elimination of its function ensues shortly after birth but contraction is reversible during the first hours of life. Preparatory dissociation of

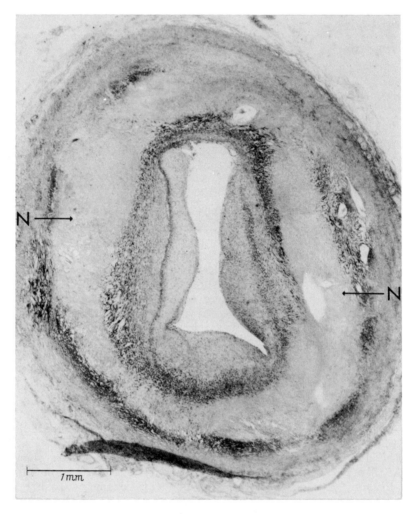

Fig. 49. Cross-section of a narrowed ductus arteriosus from a three-week-old infant showing extensive necrosis (N) in the media. The intima is thickened and partly folded. PAS stain. From Meyer and Simon, 1960.

the intimal-medial junction favors intermittent narrowing of the duct and rapid final closure by intimal folds. There are no special structures, such as those described by Fay and Travill (1967) in rabbits, which can close the lumen of the human duct. Although the data concerning organic closure are contradictory, it can be assumed that final obliteration of the duct occurs in most normal hearts between the second and fourth weeks of life (Jager and Wolleman, 1942).

PATHOLOGICAL PATENCY

Chuaqui et al. (1977) studied the patent duct in 13 children aged 4 months to 9 years. Most specimens were obtained at surgery. The length of the persistent duct was between 4 and 6 mm, and the internal diameter was 3 - 5 mm. Microscopically, three stages of transition from a muscular to an elastic artery were found. The earliest changes were confied to the intima and were characterized mainly by hyperplasia of elastic lamellar intimal structures ("stage I"). In other cases, an increase in elastic tissue was also present in the media ("stage II"). In a more advanced stage, the amount of elastic fibers and lamellae increased further in the media and the duct showed the structural pattern of an elastic artery ("stage III"). A definite correlation between the microscopic findings and the clinical picture could not be established. In view of these findings the authors concluded that some individual factors must influence remodeling of the ductal wall.

Persistent patency of the duct also results in enlargement of the pulmonary arterial branches. Although patency is not necessarily associated with higher intrapulmonary pressure, if the diastolic pressure in the pulmonary artery exceeds 40 mm Hg, histological evidence of pulmonary vascular disease will invariably be present at surgery (Bor and Valach, 1959). Pronounced atherosclerosis in the pulmonary trunk and its branches is frequently observed.

4. STRUCTURAL FEATURES FAVORING CLOSURE OF TEMPORARY FETAL VASCULAR CHANNELS

The structure of vascular channels that cease to function after birth must be adequate to transport increasing amounts of blood during fetal development and yet capable of closing rapidly and completely after delivery. Both the umbilical artery and the ductus arteriosus are muscular arteries, which would suggest that both functions may best be served by smooth musculature. A muscular wall may even be a necessity because rapid and complete closure of both of these arteries can be achieved only by active contraction of a vascular wall that has to withstand a relatively high blood pressure.

The structure of the musculature must also be capable of rapid and profound rearrangement with shifting of muscle cells and other structural components in order to narrow and occlude the lumen. As has been shown in various sections, the smooth musculature of the ductus arteriosus, umbilical arteries and vein, differs from that of persistent fetal or adult muscular arteries. In temporary vessels, smooth muscle cells form cylindrical muscular sheets which are separated from each other by an abundant semi-fluid plastic ground substance. This substance permits the muscular sheets to glide over each other, an important prerequisite for profound and rapid remodeling of the vascular wall that leads to closure.

Poor development of elastic and collagenous tissue is another characteristic structural feature of the duct and umbilical vessels. The umbilical vein has tiny subendothelial elastic fibers that may simulate an internal elastic membrane after the vessel contracts. Similarly, the umbilical artery has numerous fine elastic fibers in its inner layer but no coarse elastic fibers or internal elastic membrane. The ductus arteriosus has a well-developed internal elastic membrane and some elastic tissue in the part of the outer media that is continuous with the elastic networks of the outer aortic and pulmonary arterial walls. The rest of the media consists mainly of sheets of smooth musculature and is poor in elastic tissue.

Absence of differentiated elastic structures may also favor rapid closure of the lumen because unstretched elastic tissue resists further contraction of smooth musculature and closure of the lumen. Collagenous networks likewise probably hinder complete contraction and are absent in the umbilical vessels.

The ductus venosus is the only temporary vascular channel which has well-developed elastic networks in its wall. However, functional closure of the ductus venosus seems to be a relatively passive process that follows the rapid fall in blood pressure in the umbilical recess after cessation of umbilical blood flow. Hence, the relatively poorly developed smooth musculature in the wall of the ductus venosus, unlike that of the ductus arteriosus and umbilical vessels, does not need to overcome the continuous distending force of a higher blood pressure during closure. Functional closure, therefore, seems to be achieved merely by simple contraction of the orifice (Meyer and Lind, 1966).

Some distinctive preparatory changes appear to be necessary for contraction of the ductus arteriosus. They include dissociation of the intimal-medial junction, fragmentation of the internal elastic membrane and formation of cyst-like accumulations of ground substance which interrupt the innermost layer of the media. These preparatory changes obviously diminish the shearing forces which would otherwise resist contraction of the wall and folding of the intima. It is uncertain whether metabolic, hormonal or other factors are involved in the preparatory structural changes and dissociation that are associated with contraction.

No similar preparatory changes have been described in the umbilical vessels, although it is possible that biochemical changes may alter the ground substance and make the smooth muscle cells more sensitive to different physical and humoral stimuli. Since closure of the umbilical vessels is initiated by localized contractions, certain areas of the wall may become more sensitive at term. Hence the sites of localized contractions may be predetermined by structural or metabolic characteristics of the affected vascular segments.

Equilibrium between the walls of the umbilical vessels, which have only a few elastic elements, and blood pressure must be maintained largely or even solely by active tone of the smooth musculature. This is unlike the situation in persistent blood vessels that always have a considerable amount of elastic tissue in their walls. Elastic tissue continuously resists the distending force of blood pressure with minimal or no expenditure of energy and prevents overdistension of the vascular tube. Elastic tissue also prevents complete critical, i.e., irreversible, closure of vessels which may occur if active muscle tone rises above a critical value or if active muscle tone remains unchanged but blood pressure falls below a critical value. Since the extra-abdominal umbilical vessels consist mainly of smooth musculature, probably lack nerves 10 cm beyond the skin-umbilical junction and show only poorly developed elastic structures, they may be regarded as "critical vessels" in the sense used by Burton (1972), i.e., the equilibrium between blood pressure and active vasomotor tone is unstable and complete ("critical") closure may easily occur.

Theoretically, the equilibrium between transmural pressure and active vasomotor tone may be disturbed in umbilical vessels even during intrauterine life, for example, by a knot in the umbilical cord with consequent reduction in blood pressure. On the other hand, a local increase in vasomotor tone may be induced by the appearance of some vasoactive substance(s) in fetal blood during the latter part of gestation which conceivably, even in the presence of a normal blood pressure, may stimulate localized contractions and thereby endanger the fetal circulation. Pharmacodynamic studies on isolated umbilical arteries show that oxytocin has an enhancing effect on responses to common spasmogens, such as acetylcholine and 5 - hydroxytryptamine (Yao et al., 1975). Various other drugs may have a similar effect in vivo.

On the other hand, different mechanisms, e.g., a rapid increase in fetal heart rate, redistribution of blood with a larger amount returning to the placenta, responses by the autonomic nervous system and carotid bodies, etc., may well offset any transient change since the fetus seems well equipped with the homeostatic mechanisms necessary for maintenance of a satisfactory internal environment.

5. AORTA

The aorta is classified as an elastic artery, since its wall consists mainly of elastic membranes. Because of its elastic properties, i.e., reversible distensibility, the aorta distends during ventricular ejection and accommodates about half of the stroke volume, an amount which is transferred during the following diastole to the peripheral vasculature. This characteristic, known as the Windkessel effect, ensures transformation of the highly pulsatile flow produced by the left ventricle into more continuous streaming of blood to the distributing arteries (Böger and Wezler, 1936; Wezler and Sinn, 1953) (p. 2).

The aorta, and particularly the arch, is exposed to greater rhythmical distension and higher blood pressure because of its more central location. In the arch, blood flow is deflected through an angle of 180°, and a considerable volume of blood is diverted into its three large branches. These hemodynamic features probably place an additional functional load upon the wall and influence structural differentiation in this segment.

DEVELOPMENT

Only three aortic arches and the ventral and dorsal aortic roots are important in the formation of the adult vessels. After the first and the second aortic arches involute, the head of the embryo comes to be supplied by remnants of the third arch and the body of remnants of the fourth arch. Outgrowths of the remnants of the sixth aortic arch form the pulmonary artery.

According to Dalith (1961, 1962, 1964, 1968, 1972), the series of involutions that normally occur during development of the thoracic aorta and its large branches leave persistent structural alterations in the walls of the adult arteries at the sites of obliteration of the temporary vessels. The structural alterations include circumscribed thinning of the media, changes in microscopic architecture and focal thickening of the adventitia. They may resemble the alterations occurring in the aortic wall at the site of the obliterated entry of the ductus arteriosus. Dalith found six areas along the aortic arch and brachiocephalic arteries which he considered to be sites of developmental structural deficiency, i.e., "weak spots" in the arterial wall. Since the sites of developmental defects found in young adults correspond to the location of calcified atherosclerotic plaques in the elderly, Dalith concluded that the weak spots which arise during embryonic development may be important for the location and subsequent development of atherosclerotic lesions. Such lesions are more commonly found in the abdominal than in the thoracic aorta, a finding which Dalith (1964) ascribes to the large number of mesonephric vessels which subsequently obliterate.

FUNCTIONAL MORPHOLOGY OF ARTERIES

CHANGES IN MICROSCOPIC STRUCTURE WITH GROWTH

Introduction

Media. The aortic media consists of numerous layers of strong, concentrically arranged, elastic lamellae which alternate with interlamellar zones. The zones are formed by sheets of smooth musculature containing fine elastic fibers and ground substance rich in mucopolysaccharides. One lamella and an adjacent interlamellar zone form a <u>lamellar unit</u>. There are about 40 lamellar units at birth (Grünstein, 1896) and up to 52 in young adults (Knieriem and Heuber, 1970). The increase in number is accompanied by an increase in thickness of the lamellar units from 8.2 µm at four years of age to 10.6 µm in young adults. <u>Wolinsky</u> and <u>Glagow</u> (1967) showed that the number of lamellar units in the media of the adult mammalian aorta is almost proportional to the radius regardless of the species or variations in the wall thickness. Hence, the average tension per lamellar unit of the media is remarkably constant regardless of the radius or the species.

No similar relation exists between the increasing caliber of the growing human aorta and the number of interlamellar units in different age groups. The increase in number of interlamellar units with growth is far exceeded by the increase in diameter and, consequently, by the estimated increase in total wall tension. Thus, in the growing human, adaptation to a higher functional load seems to be achieved by structural differentiation rather than by numerical increase of structural components. This permits the highest volume distensibility at the end of growth and thereby ensures the greatest adaptation to functional demands in young adults (see p. 101).

Internal layer. Intima. Intimal thickening with accumulation of ground substance rich in mucopolysaccharides begins during fetal development. In human fetuses it has been found below the entry of the duct, along the lesser curvature of the ascending aorta and around the orifices of the aortic branches (Hörnblad, 1970). The further differentiation of the aortic intima is characterized by development of a "musculo-elastic layer." Thoma (1883, 1921) and Jores (1903, 1924) showed that this layer is already present at birth. It appears between two prominent elastic lamellae and consists of loose networks of elastic fibers and smooth muscle cells. During the first years of life, a second layer of the intima, known as the "hyperplastic elastic layer," develops internal to the musculo-elastic layer. It consists mainly of concentrically arranged elastic fibers and lamellae. Toward the end of the third decade, a third layer, i.e., the "connective tissue intimal layer," develops internal to the hyperplastic elastic layer. It is formed mainly by loose collagenous networks and ground substance with a variable number of cells. These findings have since been confirmed by more recent investigations (Prior and Jones, 1952; Movat et al., 1958).

The findings presented in this chapter show that the sequence of structural changes described by Jores (1924) represents only one of several similar developmental patterns occurring in the luminal layer with growth. Moreover, the structural changes observed during growth differ considerably in individual aortic segments.

In the ascending aorta of infants and children, the luminal layer consists mainly of prominent longitudinal elastic structural components and smooth muscle cells showing a variable course. Thus, the structure of the luminal layer differs from that of the media and it is not separated from the media by a limiting elastic lamella or an internal elastic membrane. It, therefore, cannot be classified as an intima although it may resemble an intima in some parts of the circumference and may even have a similar function. Since it is doubtful that the luminal layer arises from the intima, the term "internal layer" seems preferable.

A fragmented limiting elastic lamella formed by longitudinal elastic fibers in the deeper stratum of the internal layer, i.e., at the border with the media, appears first in the descending thoracic aorta and does not assume the characteristics of an internal elastic membrane until it reaches the abdominal aorta. With the appearance of a distinct limiting elastic lamella, the intima becomes separated from the media in these aortic segments.

Since the internal layer of the proximal aorta and the intima of the descending aorta become the predominant substrates of lipid deposition early in life, and consequently the site of early atherosclerotic lesions, the development of the microscopic structure of this layer deserves particular attention. For this reason, much of the following section is devoted to describing changes in the internal layer.

1. Fetuses

In fetuses, the retracted aorta appears conspicuously thick-walled (Fig. 50). In mid-gestation, the ratio of the internal diameter to the wall thickness is about 3:1 to 4:1 in the ascending aortic segment near the origin of the brachiocephalic trunk; it increases distally to about 5:1 to 6:1 just above the aortic bifurcation.

Ascending aorta. The internal elastic lamella is well-developed and extends around most of the circumference, apart from the area immediately adjacent to the trunk of the pulmonary artery (Fig. 51). It is considerably thicker than the lamellae in the media, is reduplicated in many areas, and has a grainy structure on cross-section produced by striations which can be seen on tangential section. Hence the lamella may be formed by the confluence of longitudinal subendothelial elastic fibers and differs considerably from the

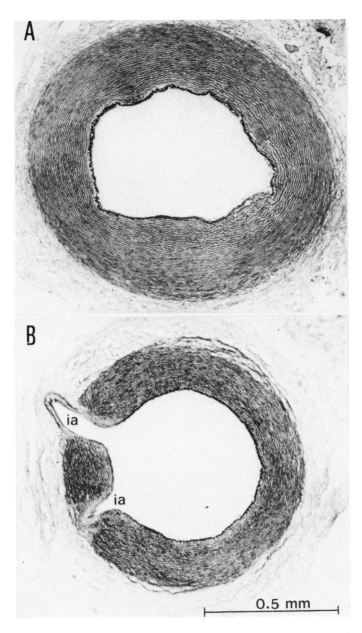

Fig. 50. Cryostat cross-sections of the ascending (A) and thoracic descending aorta (B) in a 15-week-old fetus weighing 62 g. Note the great thickness of the wall of the aorta and relatively large caliber of the intercostal arteries (ia). Weigert's resorcin fuchsin. Case J.N. 16694/75.

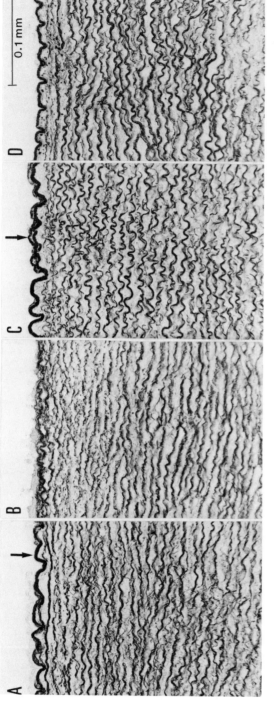

Fig. 51. Cross-sections of the aorta from a fetus of about 22 weeks gestational age with a body weight of 565 g. (A) In the ascending aorta, a prominent subendothelial elastic lamella (arrow) is seen. (B) It is replaced by networks of fine axially subendothelial elastic fibers in the part of the circumference adjacent to the pulmonary trunk. (C) In the descending aorta, a sheet of fine longitudinal fibers (arrow) covers a strong endothelial elastic lamella. (D) The subendothelial elastic sheet becomes thinner. The thickness of the medial lamellae and of the lamellar units is approximately equal in all segments. Cryostat section. Weigert's resorcin fuchsin. Case J.Nr. 30563/76. Same magnification in all figures.

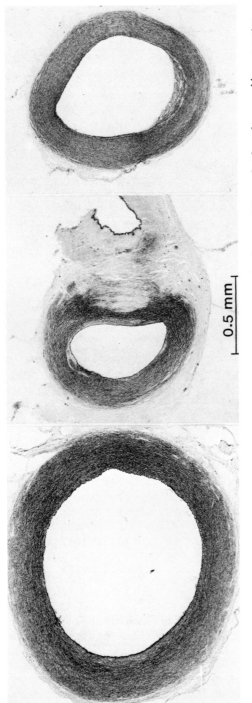

Fig. 52. Cross-sections of the ascending aorta (A), aortic isthmus (B) and the descending aorta just below the ductal orifice (C) from a fetus with a body weight of 565 g. Note the relatively narrow lumen of the isthmus. The wall of the isthmus adjacent to the duct (D) is partly penetrated by ductal smooth musculature. (asterix). In the descending aorta (C), a prominent luminal crescent-shaped cushion is seen (arrow) (see Fig. 53). Note also the considerable thickness of the aortic wall in all three segments. Cryostat sections. Weigert's resorcin fuchsin. Case J.Nr. 30565/76.

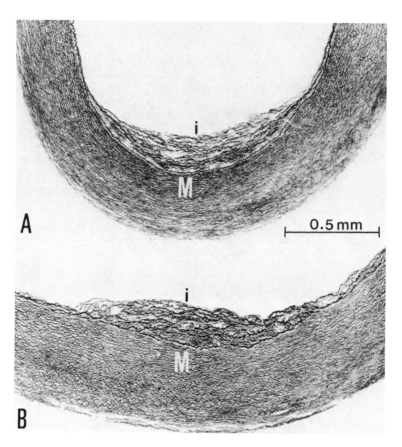

Fig. 53. (A) Thick intimal cushion (i) in the fetal thoracic aorta below the entry of the duct. The cushion consists of concentric elastic fibers and sheets and is distinctly separated from the media (M) by a prominent elastic lamella. (B) In newborn infants an identical intimal cushion is usually present below the entry of the duct in the descending aorta. Cryostat section. Weigert's resorcin fuchsin.

homogeneous primary internal elastic membrane in large muscular arteries. It stains deeply with Weigert's resorcin fuchsin.

Two or three inner medial lamellae near the internal elastic lamella appear fragmented and also seem to consist of longitudinal elastic fibers. Only a few longitudinally arranged smooth muscle cells are present between them.

In the sectors nearest to the wall of the pulmonary trunk, i.e., in about one-sixth of the circumference, the internal elastic lamella is replaced by two to four concentric superimposed elastic sheets formed by fine longitudinal elastic fibers.

Aortic isthmus. The isthmus has a considerably smaller lumen than the adjacent portions of the arch and the descending aorta (Fig. 52). The internal elastic lamella is prominent and resembles that in the ascending aorta. It is reduplicated in some sectors and replaced by two or three thin elastic sheets at the site of entry of the duct.

Descending aorta below the orifice of the ductus arteriosus. A large protruding crescent-shaped intimal cushion is usually present along the anterior or antero-medial aortic wall (Fig. 53A). It is nearly as thick as the adjacent media and consists of several layers of concentric elastic sheets, circularly arranged muscle cells and abundant metachromatic ground substance. Small circumscribed accumulations of this substance are present at the base of the cushion, i.e., along the inner aspect of the internal elastic lamella which separates the cushion from the media.

Lower thoracic aorta. The internal elastic lamella is well-developed in most sectors. It is often covered by an additional thin secondary elastic sheet. In a few sectors, it is replaced by two or three concentric sheets of fine longitudinal elastic fibers.

Abdominal aorta. The internal elastic lamella is nearly continuous. It has the same structural pattern that is found in more proximal segments but appears a little thinner. A fine secondary elastic sheet covers the internal elastic lamella in a few sectors. Flat intimal cushions are present around the orifices of larger branches; no lamella is evident at these sites.

In conclusion, the fetal aorta is a thick-walled vessel that has a thin intima and a well-differentiated internal elastic lamella in most segments. The structure of the lamella differs considerably from the internal elastic membrane in muscular arteries. A fine secondary elastic sheet along the inner aspect of the internal elastic lamella is present in only a few sectors of the lower descending thoracic aorta and abdominal aorta.

2. Newborn Infants

The aorta is thick-walled, and in this respect, resembles the fetal aorta. The ratio of the internal diameter to wall thickness is greater than in the fetus and is about 6:1 in the ascending aorta near the origin of the brachiocephalic trunk.

TABLE 2.

Aortic Measurements in the Human During Growth

Parameter	Age	Measurement	Source
Wet weight	newborn 20 years	1.1 g 21 g	- Simon and Meyer, 1958
Length			
from summit of aortic arch to bifurcation (measured on aortograms)	100-400 mm fetus newborn	1/4-1/5 total length 125 mm	- Solbiati, 1958
from origin of left subclavian artery to bifurcation	newborn	110 mm at filling pressure of 10 mm 122 mm at filling pressure of 100 mm	- Simon and Meyer, 1958
Site of aortic bifurcation (on aortograms)	200 mm fetus	level with 4th lumbar vertebra (as in adult)	- Solbiati, 1958
Diameter (mean internal)			
aortic outlet isthmus	newborn	6.0 mm 4.3 mm	- Patten, 1930
descending aorta		5.75 mm	- Solbiati, 1958
thoracic aorta	100 mm fetus 200 mm fetus	1 mm 2.5 mm	
	newborn 6 years young adult	5 mm 9 mm 14 mm	- Dragendorff, 1930; Knieriem and Hueber, 1970
Volume (mean)	newborn	7 ml at pressure of 60 mm Hg 8 ml at pressure of 100 mm Hg	- Simon and
	10 years	45 ml at pressure of 80 mm Hg 80 ml at pressure of 120 mm Hg	Meyer, 1958;
	adult	120 ml at pressure of 80 mm Hg 160 ml at pressure of 120 mm Hg	Meyer, 1964
Wall thickness	newborn adult	0.66 mm 1.20 mm	- Grünstein, 1896

Ascending aorta. The subendothelial elastic lamella is less prominent in newborn infants than in fetuses and is often replaced by layers of fine sheets of longitudinal elastic fibers (Fig. 54A).

Aortic isthmus. The lumen is still considerably narrower than that of the adjacent proximal and distal aortic segments (Table 2). The internal elastic lamella is prominent in only a small part of the circumference.

Descending thoracic aorta. As in the fetal aorta, a prominent crescent-shaped intimal cushion is present along the antero-medial or dorso-medial wall below the orifice of the duct. It is about one-fourth to one-third as thick as the underlying media (Fig. 53B).

Lower thoracic aorta. A rather prominent subendothelial elastic sheet, formed by coalescent longitudinal elastic fibers, is seen in most parts of the circumference. It is usually located just beneath the endothelial lining or is covered by another thinner elastic sheet of similar structure (Fig. 54B).

Abdominal aorta. The same elastic sheet is well seen in this vascular segment. Fragments of thin secondary elastic sheets are present in only a few sectors along the luminal aspect (Fig. 54C).

Abdominal aorta above the bifurcation. The subendothelial elastic sheet becomes homogeneous and increasingly assumes the structural characteristics of a typical internal elastic membrane. It is mostly continuous but becomes completely interrupted beneath the extensions of the dividing crest of the aortic bifurcation as well as at the crest itself. In the lateral parts of the extensions, the membrane seems to fall apart into numerous fine elastic sheets which fan out between the adjacent structures (Fig. 55).

In conclusion, the aorta of the newborn, like that of the fetus, is still a thick-walled vessel with a narrow isthmus. A well-developed intimal cushion is visible below the ductal orifice. A prominent subendothelial elastic lamella, formed by strong coalescent longitudinal elastic fibers, develops in the lower descending thoracic aorta and extends along the circumference of the abdominal aorta. It is covered by a few thin secondary elastic sheets in some sectors. Toward the bifurcation, the subendothelial elastic lamella assumes the structural characteristics of a typical internal elastic membrane. At the bifurcation, it is largely continuous except in the area of the crest and its extensions.

3. *Infants at the Age of Five to Six Months*

Ascending aorta. During the first month of life, a thick internal layer develops in the ascending aorta, which is especially

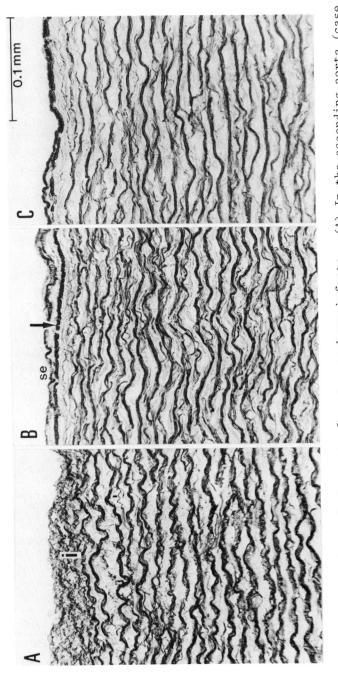

Fig. 54. Cross-sections of the aorta from two newborn infants. (A) In the ascending aorta (case 1 K/77), an inner layer (i), consisting of fine longitudinally arranged elastic fibers, has developed and replaced the compact subendothelial elastic lamella present in fetuses (see Fig. 51). (B) In the lower thoracic aorta, a thin inner layer is separated from the media by a prominent elastic lamella which is formed by strong longitudinal elastic fibers, cross-sections of which are distinguishable (arrow). A thin subendothelial elastic sheet forms the luminal border of the inner layer (se). (C) In the middle abdominal aorta, a prominent subendothelial lamella is presnt. It consists of strong coalescent longitudinal elastic fibers. (B, C: case 20 K/77). Cryostat section. Weigert's resorcin fuchsin. Same magnification in all figures.

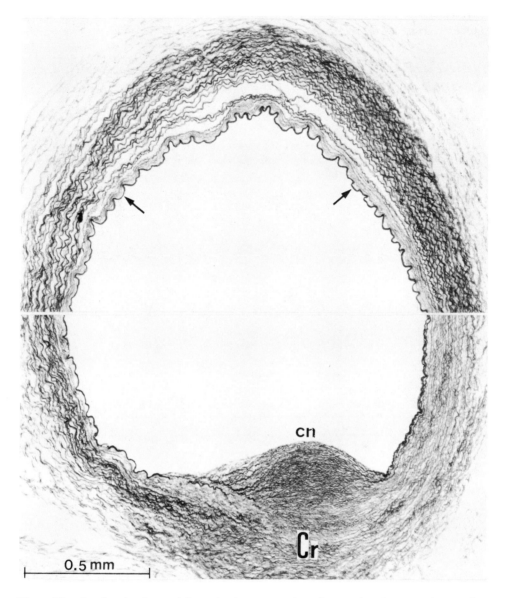

Fig. 55. An intimal cushion (cn) protrudes into the lumen along the lateral aspect of the dividing crest of the aortic bifurcation (Cr) in a retracted specimen. It consists of densely packed fine elastic networks. No internal elastic membrane is present beneath the cushion but it is present along the opposite lateral wall of the bifurcation (arrows), i.e., along the wall of the uppermost portion of the common iliac artery. Newborn infant. Case 20 K/77. Weigert's resorcin fuchsin.

prominent proximal to the origin of the brachiocephalic trunk (Fig. 56A). Along the inner concave wall of the aortic arch, it may comprise up to one-third of the total wall thickness. It consists of a deeper stratum formed by tightly packed, longitudinal elastic structures, and a luminal stratum of looser structure, which consists mainly of circular elastic fibers embedded in ground substance and interspersed with a few anastomosing cells. In some parts of the circumference, these cells may contain moderate amounts of fine lipid droplets. In other sectors, the luminal stratum may be absent. No limiting elastic structures are seen between the strata or between the deeper stratum and the media. The subendothelial elastic lamella is poorly developed.

Descending aorta below the obliterated orifice of the duct. A prominent crescent-shaped intimal cushion is present along the antero-medial wall. It may be much thicker than in newborn infants and consists of a dense network of fine concentric elastic fibers embedded in ground substance and interspersed with a few anastomosing cells. In the remaining part of the wall, the thickened internal layer appears to be an extension of the cushion and shows a similar structural pattern.

Middle segment of descending thoracic aorta. The intima is considerably thickened around the entire circumference, and along the dorsal and ventral walls comprises up to one-fourth of the total wall thickness. It consists predominantly of circular elastic fibers interspersed with a moderate number of cells and small accumulations of ground substance (Fig. 56B). A fragmented internal lamella, formed by prominent longitudinal elastic fibers, is now visible beneath the thickened intima in large parts of the circumference at this level.

Lower thoracic aorta. The intima shows a similar structure but gradually becomes thinner distally; it comprises up to one-ninth of the total wall thickness. The fragmented internal elastic lamella extends throughout most of the circumference.

Upper abdominal aorta. A moderately thickened intima is present along the dorsal wall. Its elastic networks are finer than those in more proximal segments. The internal elastic lamella is prominent but remains discontinuous (Fig. 56C).

Lower abdominal aorta. Intimal thickening is confined to the dorsal wall. The internal elastic lamella is well-developed, nearly continuous, and has the more homogeneous aspect of an internal elastic membrane.

Abdominal aorta above the bifurcation. The internal elastic lamella is prominent and continuous but is completely interrupted in the area of the protruding proximal extensions of the dividing

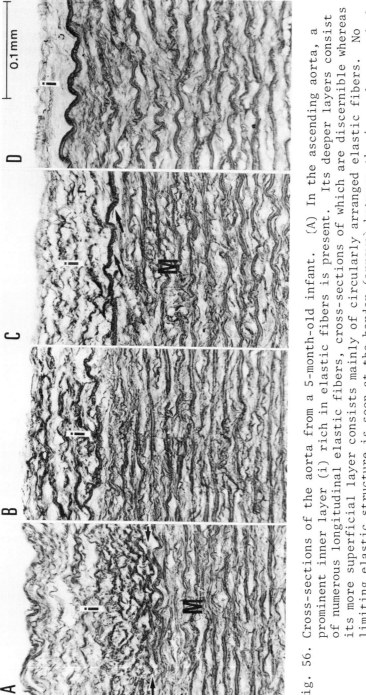

Fig. 56. Cross-sections of the aorta from a 5-month-old infant. (A) In the ascending aorta, a prominent inner layer (i) rich in elastic fibers is present. Its deeper layers consist of numerous longitudinal elastic fibers, cross-sections of which are discernible whereas its more superficial layer consists mainly of circularly arranged elastic fibers. No limiting elastic structure is seen at the border (arrows) between the inner layer and the media (M). (B) In the middle segment of the descending thoracic aorta, the inner layer is also prominent but is usually not as thick as in the upper descending aorta. (C) In the upper abdominal aorta, a discontinuous elastic sheet (arrows) appears between the inner layer (i) and the media (M). In the abdominal aorta (D), it becomes more prominent and the inner layer (i) is now thin and poor in elastic elements. Cryostat sections. Weigert's resorcin fuchsin. Case 3K/77. Same magnification in all figures.

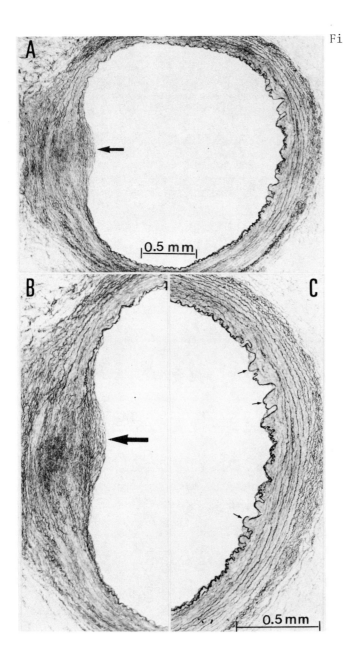

Fig. 57. (A) Cross-section of the most proximal segment of the common iliac artery immediately below the aortic bifurcation. (B) and (C) show the medial and lateral walls, respectively, at higher magnification. The extension of the cushion along the lateral surface of the dividing crest (Fig. 55) is seen along the medial wall of the common iliac artery (arrows). The internal elastic membrane is replaced at this site by sheets of superimposed elastic fibers (B). A well-developed internal elastic membrane is present elsewhere. (C) The membrane in the lateral wall (small arrows) shows a coarse irregular wavy pattern which is probably produced by calcific deposits. 5-month-old infant. Case 3K/77. Cryostat section. Weigert's resorcin fuchsin.

crest along the ventral and dorsal walls. The intima is slightly thickened immediately above the extensions; elsewhere it is thin.

The prominent internal elastic membrane in the lower abdominal aorta extends below the bifurcation into the lateral walls of the common iliac arteries where it constitutes the predominant substrate for early calcification (Fig. 57) (see p. 190). No internal elastic membrane is present along the medial walls of the common iliac arteries which represent extensions of the dividing crest. Neither the dividing crest nor its extensions develop calcifications during infancy.

In conclusion, during the first months of life, the internal layer of the ascending aorta develops an increasingly complex structural pattern and this layer may be regarded as an intima although no limiting elastic lamella is present. The intimal cushion below the obliterated ductal orifice remains prominent. The internal layer has become thicker in the descending thoracic aorta but it becomes thinner along the abdominal aorta. A discontinuous internal elastic lamella appears in the middle thoracic aorta and becomes prominent in the abdominal aorta.

4. Children at the Age of One Year

Ascending aorta. A distinct internal layer comprising one-sixth of the total wall thickness and consisting mainly of longitudinally arranged elastic and muscular elements extends over a large part of the circumference. No internal elastic lamella or similar limiting elastic structure is present along the border with the media. The endothelium rests on a sheet of tightly packed longitudinal elastic fibers. It does not differ significantly from the adjacent elastic sheets.

Isthmus. The aortic lumen has not yet attained the same size as the adjacent proximal and descending aortic segments. An inner layer of axially and circularly arranged elastic sheets is present but is less prominent than in the ascending aorta.

Aorta below the obliterated orifice of the duct. A large protruding intimal cushion extends 1 - 1 1/2 cm below the orifice. It may be twice as thick as the aortic media and consists mainly of fine longitudinal elastic fibers that often form circular sheets. The periphery of the cushion extends into a well-developed internal aortic layer which comprises one-sixth to one-fourth of the total wall thickness.

Lower descending thoracic aorta. In some sectors a stretched limiting elastic lamella is present and separates the still prominent inner layer from the media.

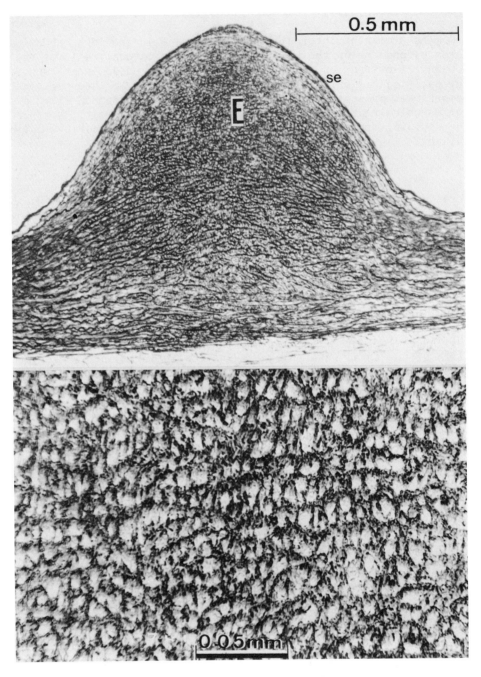

Fig. 58.

Abdominal aorta. The internal layer becomes thinner and increasingly well separated from the media by a fragmented internal elastic lamella. Thus, the inner layer has now assumed the characteristics of a moderately thickened intima rich in longitudinal elastic elements.

The extensions of the bifurcation's crest consist of tightly packed, longitudinally arranged, muscle cells and numerous fine elastic fibers which show the same axial arrangement (Fig. 58). Toward the apex of the crest, the smooth muscle cells and the elastic fibers increasingly assume a horizontal course and become the main structural component of the central part of the dividing crest. In the lateral parts of the crest and its extensions, the smooth musculature and the elastic fibers retain their axial course, extend distally into the medial wall of the proximal segments of the common iliac arteries and form oval-shaped thickenings there. In retracted arteries, these thickenings look like cushions protruding into the lumen (Figs. 55, 57).

Abdominal aorta at the bifurcation. Increasingly prominent secondary elastic sheets appear beneath the endothelial lining and over the fragmented primary internal elastic membrane. The fragments of this membrane are seen mainly in the lateral walls of the bifurcation and in the adjacent sectors of the lateral walls of the common iliac arteries. Secondary elastic sheets also cover the crest and its extensions. Slight intimal thickening is occasionally present along the lateral aspects of the dividing crest.

In conclusion, by the age of one year, a prominent internal layer consisting mainly of longitudinal structural components has developed in the ascending thoracic aorta. The lumen of the isthmus remains narrower than that of the adjoining aortic segments. A large protruding intimal cushion is present below the obliterated orifice. It extends into the descending thoracic aorta. A prominent internal layer extends around the rest of the circumference and along the lower thoracic aorta. It becomes thinner in the abdominal aorta where it resembles an intima since it is separated from the media by a fragmented elastic sheet.

← Fig. 58. Extension of the dividing crest (E) of the aortic bifurcation shown in cross-sections at two different magnifications. The extension consists of tightly packed, longitudinally arranged muscle cells and fine accompanying elastic fibers which surround the cells. The subendothelial elastic layer (se) is formed by sheets of coalescent elastic fibers. Case RM 14.1.77. Eighteen-month-old infant. Cryostat section. Weigert's resorcin fuchsin.

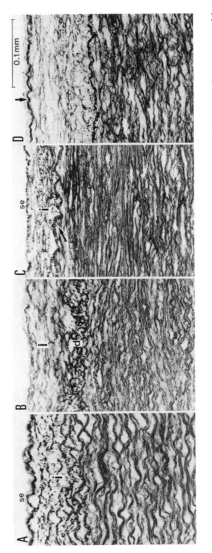

Fig. 59. Cross-sections of the aorta from an eleven-year-old child. (A) In the ascending aorta, the prominent inner layer (i) now includes more smooth musculature (unstained) between the elastic fibers which are mainly longitudinally disposed. Strong coalescent elastic fibers from a prominent subendothelial elastic sheet (se). No limiting elastic lamella is present between the inner layer (i) and the media (M). (B) In the lower thoracic aorta, the inner layer has two strata. The thinner deeper stratum (d) consists mainly of cross-sectioned longitudinal elastic fibers whereas the luminal stratum (l) consists mainly of fine concentrically arranged elastic fibers. No limiting elastic structure is present between the inner layer (i) and the media (M). A prominent subendothelial elastic lamella is also absent at this level. In the upper abdominal aorta (C), cross-sectioned longitudinal elastic fibers form a discrete subendothelial elastic sheet (se). The inner layer (i) is separated from the media by a discontinuous elastic lamella (arrow). (D) In the middle abdominal aorta, the inner layer has become more prominent and is separated from the endothelium by a subendothelial elastic sheet of strong coalescent elastic fibers (arrow) and from the media by another elastic sheet. Note the different structural pattern of the inner layer in the individual aortic segments, particularly in the ascending aorta and in the abdominal aorta. The differences in the arrangement of the medial lamella may be partly due to different stages of postmortal retraction. Cryostat section. Weigert's resorcin fuchsin. Case 13 K/77. Same magnification in all figures.

5. Children at the Age of 10 - 16 Years

Ascending aorta. A distinct internal layer comprising up to one-tenth of the total wall thickness is present along the entire circumference but appears slightly thicker along the outer convex wall of the aortic arch (Fig. 59A). The layer consists mainly of longitudinally arranged elastic fibers which often form discontinuous concentric sheets. Longitudinally arranged smooth muscle cells are seen between the elastic networks, but there are considerably fewer cells in the internal layer than in the interlamellar spaces of the media. There is no internal elastic membrane or other limiting elastic structure between the internal layer and the media. The internal layer is separated from the lumen by a subendothelial sheet of coalesced longitudinal elastic fibers. The thickness of the sheet varies along the circumference but increases with growth.

Upper descending thoracic aorta. The internal aortic layer is present along the entire circumference. It is poorer in cells than the underlying media. In a few sectors of the concave wall of the arch, two strata in the internal layer may be discerned: a deeper longitudinal musculo-elastic layer and a luminal layer consisting mainly of fine circularly disposed elastic fibers.

Lower descending thoracic aorta. The ventral wall is thicker than the dorsal wall and the internal layer is more prominent along the thicker ventral wall. The internal layer is formed by a deeper stratum of longitudinally arranged elastic fibers and muscle cells and a luminal stratum consisting mainly of circularly disposed elastic fibers and muscle cells (Fig. 59B). A limiting elastic sheet, formed by coalesced longitudinal elastic fibers, is seen in a few sectors between the media and the internal layer. Beneath the endothelium only a thin elastic sheet of longitudinal elastic fibers is present.

Upper abdominal aorta above the origin of the celiac trunk. As at proximal levels, the aortic wall is thicker in the ventral part of the circumference but the internal layer may be more prominent along the thinner dorsal wall. Strong coalesced longitudinal elastic fibers form a largely discontinuous limiting elastic sheet which separates the internal layer from the media along most of the circumference (Fig. 59C). The luminal stratum of the internal layer consists mainly of fine circularly arranged elastic fibers. The muscle cells parallel the fibers. A subendothelial elastic sheet formed by longitudinal elastic fibers is seen along the luminal aspect of the internal layer. It is thicker in adolescents.

The inner layer is slightly thickened in areas of lipid infiltration where circularly arranged cells contain numerous fine fat droplets. Lipid inclusions are also seen in some muscle cells in the underlying media.

Middle and lower abdominal aorta. The internal layer is thicker than in the more proximal aortic segment (Fig. 59D), and in adolescents, it may amount to one-sixth to one-third of the total wall thickness. The internal layer is markedly thickened along the dorsal and lateral aortic walls above the bifurcation. It extends into the lateral walls of the proximal segments of the common iliac arteries.

The two main strata of the internal layer gradually become distinct. The deeper stratum is rich in longitudinal and circular elastic fibers which are more or less densely interspersed with axially oriented smooth muscle cells and collagen networks. Thus, the structural pattern of this stratum approximately corresponds to a "longitudinal musculo-elastic layer." A fragmented limiting elastic lamella is seen between this layer and the media. Along the abdominal aorta, the lamella becomes more prominent, and in adolescents, assumes the characteristics of an internal elastic membrane. Toward the bifurcation it becomes even more fragmented with age. In some sectors, only a few fragments of the membrane can still be seen.

The luminal stratum of the internal layer is poor in cells but rich in ground substance containing numerous fine elastic fibers and discrete collagenous networks. A sheet of strong, deeply stained, longitudinal elastic fibers is present between the luminal stratum and the musculo-elastic layer. No prominent elastic limiting structure is visible beneath the endothelial lining.

As in younger age groups, the lateral surfaces of the dividing crest of the aortic bifurcation protrude slightly into the lumen. They are now covered by a hyperplastic elastic layer of variable thickness. Strong elastic lamellae, which are probably derived from coalescent subendothelial longitudinal elastic networks, are seen beneath this layer and at the carina of the crest.

In conclusion, at the age of 10 - 16 years, the internal layer is well-developed and shows varying structure, but consists mainly of longitudinally arranged structural components. The thickness of the layer increases moderately in the middle and lower thoracic aorta and still more in the abdominal aorta. In the latter segment, the internal layer amounts to one-sixth to one-third of the total wall thickness. The increase in thickness is associated with a change in structural pattern involving the formation of two different strata.

The inner layer represents the main substrate for early lipid deposition. At first, this is acompanied by only a slight increase in thickness, and in the area of lipid deposition, nearly all cells become laden with numerous lipid droplets.

No limiting elastic structure is seen between the internal layer and the media in most of the thoracic aorta. In the lower thoracic aorta, however, there is a largely discontinuous elastic sheet composed of coalescent longitudinal elastic fibers. In the abdominal aorta, the sheet becomes more prominent but remains discontinuous.

A prominent subendothelial elastic sheet, formed by networks of longitudinal or oblique elastic fibers of variable thickness, is present in adolescents along the thoracic aorta. In the abdominal aorta, this layer becomes increasingly covered by an additional luminal stratum and displaced to a deeper part of the inner layer. No prominent elastic structure is seen beneath the endothelial lining along the abdominal aorta of adolescents.

CHANGES IN ELASTIC DISTENSIBILITY

The Windkessel function of the aorta, i.e., its storage capacity, depends not only on the elastic properties of the wall but also on the volume. With the same increase in pressure, a larger aorta will accommodate more fluid than a smaller one because less stretch will occur. With age, the wall of the aorta shows a decrease in elastic distensibility and the tube becomes dilated. The latter offsets the rigidity of the wall. Hence, the storage capacity of the aorta, i.e., its Windkessel function, does not decrease as much as might be expected.

The storage capacity of the aorta may be estimated postmortem. This is done by taking the entire aorta and ligating all of the branches (Simon and Meyer, 1958; Simon, 1959; Meyer 1964a and b) (Fig. 60). The increase in volume is determined by increasing the filling pressure in different pressure ranges. The data are then used to plot curves (Fig. 61) and the relative increase in volume with increasing pressure, i.e., the <u>relative volume distensibility</u> (rvd) may be calculated for different ages (rvd = $\frac{\Delta v \cdot 100}{V}$) (Fig. 62). This method permits comparison of distensibility of aortas of different size, i.e., at different stages of development.

In newborn infants and in 7- to 15-month old infants, the pressure volume curves of the aorta are much steeper than later in life. This is chiefly due to the fact that the initial aortic volume and the absolute increases in volume with increasing pressures are small in comparison to those of larger aortas in older age groups (Fig. 61).

Relative volume distensibility in newborn infants and 7- to 15-month-old infants is similar to that in 40-year-old men (Fig. 62). In the latter age group, the elastic properties have already decreased markedly. Thus, at birth and during the first year of life, the aorta is comparatively rigid, a finding which is supported by the faster pulse wave velocity in newborns and infants than in young adults (Bolt, 1948; Graser, 1953) because pulse wave velocity varies directly with the elasticity modulus of the arterial wall, i.e., with its rigidity. Consequently, the faster pulse wave in newborns presumably reflects the rigidity of the aorta in early infancy.

The aorta also retracts less after excision in infants than in older children and adolescents (Scheel, 1908). This finding may likewise be interpreted as indicating greater rigidity of the aorta in infancy, because retraction is directly proportional to the elastic properties of the arterial tube, and the greater the elastic properties, i.e., the reverse distensibility of the arteries, the greater the degree of retraction.

With growth, the storage capacity increases continuously and reaches its highest values at the age of 16 to 19 years (Fig. 62).

Aorta. Closing Remarks

The aorta is a thick-walled vessel in the fetus and the newborn infant, and the ratio of the internal diameter to wall thickness in

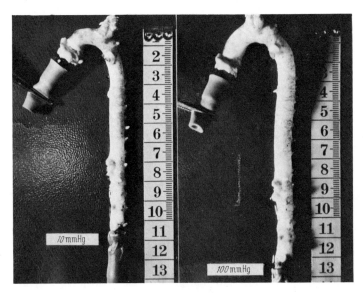

Fig. 60. Technique used for determination of postmorten volume distensibility. Cannulae are inserted in both ends of the aorta and all branches are ligated. A considerable increase in volume and length occurs when the aorta is filled with saline at a pressure of 100 mm Hg. Stillborn infant. From Meyer, 1964 a.

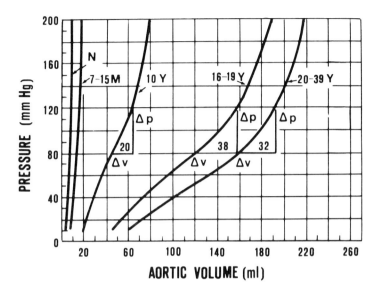

Fig. 61. Curves showing pressure-volume relations in different age groups. Each curve is based on the mean volumes (ml) obtained at different pressures in a number of aortas. The volume increases (Δv) in the physiological pressure ranges (Δp) are indicated for aortas from individuals more than 10 years of age. The sharp slopes of the curves in newborns and infants reflect the small absolute increase in initial volume and low volume distensibility of the aorta. (N-newborn, M-months, Y-years). Modified from Meyer, 1964 a.

the upper descending thoracic aorta is about 4:1 in the fetus and 6:1 in the newborn infant. The ratio between the total volume of the distended aorta and its weight is 3.2:1 in the newborn infant and increases with age to a maximum value of 6:1 at 16 - 19 years of age (Simon and Meyer, 1958). The ratio then decreases with age. In other words, when growth ceases, only a minimum of wall tissue seems to be needed to maintain the Windkessel function of the aorta as compared to earlier and later ages (Table 2, p. 88).

The greater rigidity of the aorta in newborns and infants than found in adolescents must be related to the comparatively greater

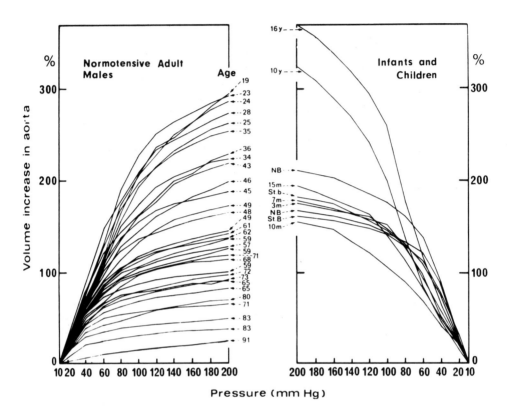

Fig. 62. Relative volume distensibility (or percent relative volume increase) of the aorta with rising filling pressure (initial volume at 10 mm Hg). Volume increases in infants and newborns are plotted in reverse manner (on the right) for the sake of comparison. Values for relative volume distensibility in infants less than 15 months of age are approximately the same as in adults in the fifth decade. The striking increase in relative volume distensibility in adolescence is mainly the result of an increase in volume since the wall of the aorta becomes stiffer during the second decade. From Meyer, 1964 a.

thickness of the wall in the newborn. It may also be related to some hemodynamic characteristics of the fetal circulation. For instance, the unusually long arterial segment between the aorta and the placenta may require a higher kinetic energy from the heart. From a theoretical point of view, a stiffer Windkessel may be needed to store more energy between contractions of the right ventricle, since during fetal life it is the right ventricle with its comparatively thick wall which supplies blood to the descending aorta and placenta (p. 13).

The aortic isthmus deserves particular attention. It is considerably narrower than the adjoining aortic segments in the fetus and newborn infant (Fig. 52). In retracted specimens, the entry of the duct into the wall of the aorta seems to contribute substantially to narrowing of the isthmus, but this may be an artifact. The isthmus not only has a narrower lumen but it also has a thinner wall than the adjacent segments of the aorta. These structural features support the view that the amount of blood entering the isthmus during fetal life is small. These anatomical findings have been amply demonstrated by Patten (1953) and in cineangiographic studies in the lamb and human fetus (Barclay et al., 1944; Lind and Wegelius, 1954; Dawes and coworkers, summarized in Dawes, 1968) (Fig. 63). Hence, during fetal development the isthmus in a sense is the site of division of the systemic circulation into two anatomical and functional parts (p. 15). This division of the aorta may also account for differences in structural differentiation within the vessel.

Considerable changes occur in the internal layer during postnatal growth. In the ascending aorta, this layer tends to have a variable structure and consists mainly of longitudinally disposed elastic elements. In newborn infants, it is somewhat poor in smooth muscle cells but they become more prominent during the first months of life. At the age of one year, the internal layer extends over a large part of the circumference and amounts to one-sixth of the total wall thickness. It continues to be prominent during further growth but becomes thinner in relation to total wall thickness. No internal elastic membrane or other limiting elastic structure is seen at the border of this layer with lamellar units of the media along the ascending and most of the descending thoracic aorta. For this reason, the classification of this layer as a derivative of the intima seems unsatisfactory (see pp. 135-136).

A discontinuous limiting elastic lamella formed by coalescent longitudinal elastic fibers develops between the internal layer and the media in the lower thoracic aorta at the age of two to five months. Along the abdominal aorta, the lamella becomes more prominent and assumes the characteristics of a homogeneous internal elastic membrane.

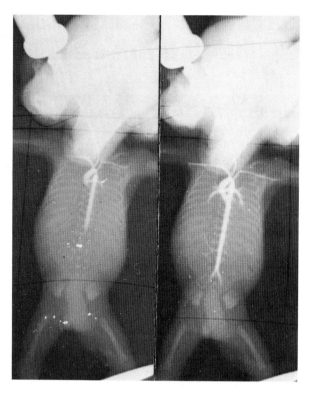

Fig. 63. Fetus of about 12 weeks gestation. LAO projection. 2 ml contrast medium injected into a carotid artery. The ascending aorta, arch, descending aorta, and the "physiological coarctation," i.e., the aortic isthmus, are seen (A). Note filling of both pulmonary artery branches in (B). From Lind et al., 1974.

In the fetus and newborn infant, prominent cushions in the internal layer are generally present both proximal and distal to the ductal orifice. They probably represent a structural accommodation to some hemodynamic factors related to the large blood volume entering the aorta via the duct. These cushions persist several months and are still prominent at the age of one year.

Diffuse thickening of the intima develops in the middle and lower thoracic aorta during the first months of life. By the age of one year, the intima is one-fifth to one-fourth as thick as the media. Its thickness decreases along the lower thoracic aorta and, in the abdominal aorta, it is only moderately thick in part of the circumference.

The thickening of the internal layer in the thoracic aorta during the first months of life may be related to some of the changes in hemodynamics in this segment that follow closure of the duct. Thoma (1883, 1921) observed diffuse intimal thickening throughout the descending aorta, i.e., between the obliterated entry of the duct and the aortic bifurcation, in the first months of extrauterine life. He ascribed this to the decrease in blood volume transported by the descending thoracic and abdominal aorta after cessation of the placental circulation. We could only partly confirm Thoma's findings since no significant intimal thickening was found in the abdominal aorta during the first months of life.

Later, the internal layer remains prominent in the thoracic aorta, and it becomes thicker in the abdominal aorta. At the age of 10 - 11 years, the intima of the abdominal aorta comprises up to one-fifth of the total wall thickness and is thicker than in more proximal segments. The intima is thickest above the bifurcation, particularly along the dorsal and lateral walls. During adolescence, intimal thickening increases in the abdominal aorta, and two different strata can be distinguished.

<u>Lipid deposits</u> in the human aorta are initially limited to the preformed internal layer or the aortic intima. At first only a few cells contain droplets but later most cells contain numerous droplets. No thickening of the internal layer occurs in the earlier stages although thickening is common later.

6. CORONARY ARTERIES

DEVELOPMENT

At about six weeks of gestation, solid out-pouchings arise at the root of the aorta from the lining of the aortic sinuses. With growth outward through the epicardium, these precursors of the coronary arteries achieve round, oval, or elliptic lumens. The left coronary artery appears by the end of the seventh week of gestation and it is followed almost immediately thereafter by the right coronary artery. The anterior descending branch, posterior descending branch, and left anterior circumflex branch then appear. By the beginning of the twelfth week of gestation, all major coronary branches are present. By mid-term, the coronary vessels seem to be comparatively wide on retrograde angiocardiography (Fig. 64).

ANATOMY

The left coronary trunk arises perpendicularly above the left sinus of Valsalva, forms a sharp curve immediately below the orifice and continues downward and to the right along the aortic wall. Its axis forms an angle with the aortic axis, i.e., it is inclined to

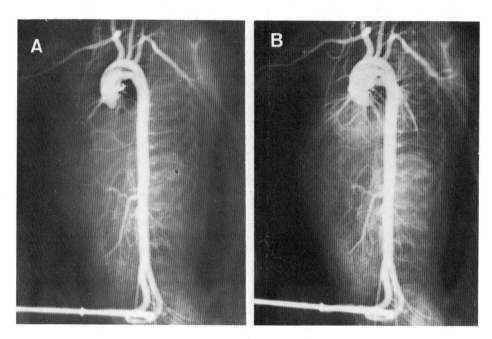

Fig. 64. Retrograde aortography in a human fetus of about 19 weeks gestation. The contrast medium has been injected in a retrograde direction in an umbilical artery. (A) The internal iliac arteries, aorta and its branches are well filled. Note good filling in the celiac trunk and faint outlines of the intercostal arteries. The coronary arteries are beginning to opacify. The ductus arteriosus is well seen and has about the same width as the aortic arch. Left anterior oblique projection. (B) The pulmonary arteries and the coronary arteries have now become well opacified. The contrast medium is now beginning to outline the myocardium. Note the thread-like caliber of the external iliac arteries and the striking difference between the caliber of the cerebral and other branches of the aorta.

the right. Thus, the bloodstream in the left coronary trunk is directed downward and in an opposite direction to that of the bloodstream in the ascending aorta. The short left coronary trunk divides into an anterior descending and an anterior circumflex branch (Fig. 65).

The proximal portion of the right coronary trunk arises obliquely above the corresponding sinus Valsalva, forms a flat curve immediately behind the orifice and proceeds perpendicular to the axis

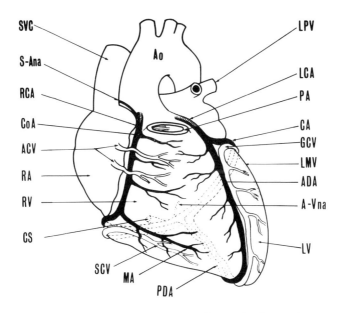

Fig. 65. Diagram showing the distribution of the main coronary arteries (black) and some of the large coronary veins of the newborn heart. SVC-superior vena cava; S-Ana-sinoatrial nodal artery; RCA-right coronary artery; CoA-conus artery; ACV-anterior cardiac veins; RA-right atrium; RV-right ventricle; CS-coronary sinus; SCV-small cardiac vein; MA-marginal artery; PDA-posterior descending artery; Ao-aorta; LPV-left pulmonary vein; LCA-left coronary artery; PA-pulmonary artery; CA-circumflex artery; GCV-great cardiac vein; LMV-left marginal vein; ADA-anterior descending artery; A-Vna-atrio-ventricular nodal artery; LV-left ventricle. From Walsh, Meyer and Lind, 1974.

of the ascending aorta behind the base of the pulmonary trunk and in front of and below the right auricular appendage.

The patterns of the heart vasculature have been discussed in detail by Gross (1921), Schlesinger (1940), James (1961) and Winterscheid (1969).

DEVELOPMENT OF MICROSCOPIC STRUCTURE

Introduction

Pronounced individual variations in the pattern of distribution of the coronary arteries and associated differences in the relative size of the main coronary branches must be kept in mind when assessing structural changes in the coronary arteries before and after birth. For example, differences in size and development of the right coronary artery and the left circumflex branch are common in cases of so-called right and left preponderance, (James, 1961; Winterscheid, 1969). With "right preponderance," the posterior circumflex branch of the right coronary artery extends to the "crux" of the heart and forms the posterior descending artery which proceeds along the posterior interventricular sulcus. In these hearts, the right coronary artery supplies blood not only to the right ventricle but also to the adjacent portion of the dorsal wall of the left ventricle. With "left preponderance," a part of the dorsal wall of the right ventricle adjacent to the left ventricle is supplied by the posterior circumflex branch of the left coronary artery. The left circumflex branch is better developed than in those hearts with "right preponderance" wherei the right coronary artery apparently dominates. But in both patterns, the left coronary artery is the main source of arterial blood supply to most of the heart's musculature. Thus, the terms right and left preponderance refer only to the variation in the degree of vascular supply of the dorsal part of the heart.

Winterscheid's (1969) statement that *"variation is the hallmark of the coronary blood vessels"* is also true of their microscopic structure. This is so because the structural pattern of the arterial tube depends greatly on vascular caliber, i.e., on the radius of the tube. Structural variations are also evident at different levels of the same segment and even in the same cross-section since opposite arterial sectors seldom have uniform structure or thickness. Differences in the same vessel probably arise in part from curves which subject individual sectors to different hemodynamic loads and considerably influence structural differentiation during growth. Differences between both coronary arteries likewise may be ascribed to differences in curves since the left trunk takes a sharper curve below its orifice than does the trunk of the right coronary artery. Therefore, the description of *"the developmental changes in the coronary arteries must of necessity be one representing the average of a large number of sections"* (Gross et al., 1934). Since structural variability is common, even at the same level, illustrations of opposite sectors of the walls of the coronary arteries are shown in figures.

Some differences in thickness of the individual arterial sectors may be due to different degrees of postmortal retraction of the individual parts of the arterial tube. These may be reduced by perfusing the coronary arteries during fixation under appropriate pressure.

The microscopic structure of coronary trunks and their main branches differs considerably from that of other arteries of comparable size (Wolkoff, 1923; 1929; Ehrlich et al., 1931; Gross et al., 1934). The internal elastic membrane, which usually separates the intima from the media, is present in only a part of the circumference of the left coronary trunk of pre-term and term newborn infants. It completely disappears during growth. The longitudinal musculo-elastic bundles (which develop along the inner aspect of the circular musculature of the trunk during gestation) form a prominent longitudinal musculo-elastic layer in the first year of life. Since it seems unlikely that this layer is derived solely from the intima, it should not be classified as an intimal layer (p.   ). Similar difficulties arise in the interpretation and terminology of individual layers of other main proximal coronary branches, particularly of the left descending branch. The view that the border between the intima and media disappears and an "intermediary layer" of longitudinal musculature forms (Gross et al. (1934), among others) reflects the difficulty in dividing the arterial layers of the main coronary branches into intima and media. The inconsistency of such a classification becomes obvious in the description and interpretation of the complex developmental changes that occur in the proximal segments of the coronary arteries during fetal and postnatal development. We have, therefore, chosen to avoid the terms "intima" and "media" in this section and refer to two layers, an internal and an external layer. The <u>internal layer</u> consists of an outer longitudinal musculo-elastic layer and an inner elastic layer which is covered by an endothelial lining; the <u>external layer</u> consists of the circular musculature and the adventitia.

<u>Left Coronary Trunk</u>

1. <u>Pre-term infants</u>. In viable infants with a gestational age of about seven months, average body weight of 1000 g and body length of 27 cm, the most proximal portion of the trunk has a thick wall consisting of circularly arranged elastic membranes and networks of elastic fibers that seem to be continuous with the elastic structures of the aorta (Fig. 66A).

In cross-sections taken immediately below this level but still in the vicinity of the aorta, the wall of the coronary trunk rapidly

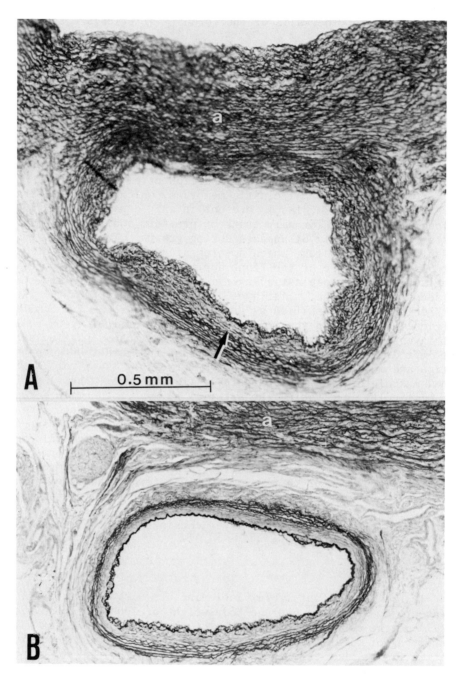

Fig. 66.

becomes thinner. A band of circular musculature appears and the trunk increasingly assumes the structural features of a muscular artery. The thick elastic layer, which is present internal to the circular musculature at more proximal levels, becomes thinner and disappears. Numerous cross-sectioned, longitudinally arranged, smooth muscle cells and fine elastic fibers now appear along the internal border of the circular musculature. In the outer sector of the wall of the trunk, i.e., farthest from the aorta, these longitudinal muscle cells and elastic fibers usually form an almost continuous internal layer which constitutes about one-half of the entire wall thickness (Fig. 66B, 67B). A well-differentiated internal elastic membrane is present in only a few sectors of the wall of the trunk nearest the aorta (Fig. 67A).

In serial sections taken immediately below this level, i.e., from the middle segment of the trunk, the internal elastic membrane extends over larger parts of the circumference. It is prominent along the inner wall of the trunk, i.e., nearest the aorta, which mainly consists of circular musculature. The opposite wall does not have an internal elastic membrane. Instead the internal layer is formed by longitudinal musculature and longitudinally arranged elastic fibers. In some areas, the longitudinal structural elements form small cushions that protrude into the lumen.

In the following segment, the structural pattern is typical of a muscular artery with a well-differentiated internal elastic membrane and circular smooth musculature around the entire circumference. In a few small sectors, longitudinally arranged smooth muscle bundles are seen along the inner aspect of the vessel.

← Fig. 66. The structural pattern of the left coronary trunk from a pre-term infant differs considerably over a very short distance. (A) In the most proximal portion, i.e., near the origin of the vessel, the wall of the trunk consists mainly of circularly arranged elastic sheets and membranes that seem to be continuous with the elastic structures of the aorta (a). Small bundles of smooth musculature (arrow) are seen in the wall of the trunk farthest from the aorta only. No internal elastic membrane is present at this level.
(B) In the cross-section taken immediately below this level, the wall of the trunk is considerably thinner and shows a well-developed circular musculature (unstained, light). The internal elastic membrane is seen in the sectors nearest the aorta (above, a). In the opposite sectors, i.e., in the wall farthest from the aorta, fine longitudinal muscle bundles and elastic fibers form the internal layer of the wall. Cryostat sections. Weigert's resorcin fuchsin. Body weight 1000 g. Case 24 K/76.

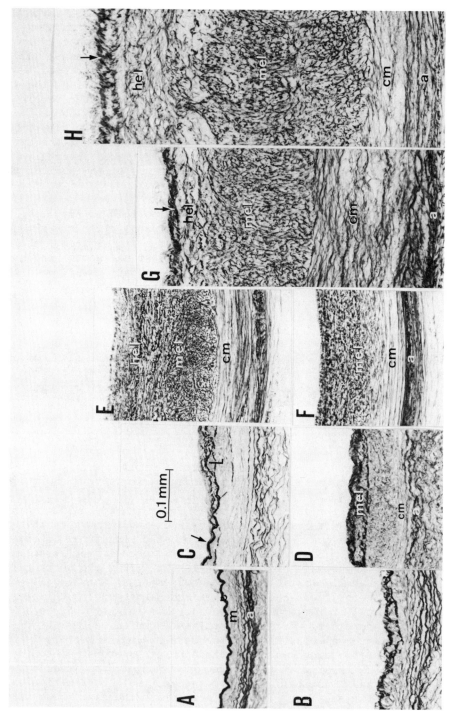

Fig. 67.

Fig. 67. Series of figures showing the structure of the left coronary trunk at different ages. (A) In the fetus, the inner wall of the trunk, i.e., that nearest the aortic wall, consists mainly of circular musculature (m), a prominent internal elastic membrane and circularly arranged elastic networks in the adventitia (a).
(B) In the opposite outer sector, the internal layer of the arterial wall is formed by longitudinally arranged structural components, i.e., muscle cells and elastic sheets consisting of longitudinal elastic fibers. The adventitia is rich in elastic fibers. Weigert's resorcin fuchsin section. Case 24 K/76.
(C) and (D) show the structure of the outer and inner walls of the trunk at birth. The outer wall (D), i.e., farthest the aortic wall shows a thick internal musculo-elastic layer (mel). The adjacent wall consists mainly of longitudinally arranged smooth muscle cells and fine elastic fibers. The circular musculature (cm) forms only a thin layer near the adventitia (a). Along the inner wall (C), i.e., nearest the aortic wall, the longitudinal structural components (1) are also prominent but do not extend along the entire luminal layer. The internal elastic membrane (arrow) is seen in some places. Case 29 K/76.
(E, F) At the age of two years, the wall of the left coronary trunk consists mainly of a longitudinal musculo-elastic layer (mel) and a hyperplastic elastic layer (hel) which are poorly separated from each other. The circular musculature (cm) constitutes a relatively small part of the vascular wall which is well separated from the adventitia (a). (E) and (F) show opposite sectors of the trunk. The coronary arteries were fixed under a pressure of about 80 cm formalin solution. Case 81 K/76.
(G, H) At the age of 17 years, the longitudinal musculo-elastic layer (mel) and the hyperplastic elastic layer (hel) have increased further in thickness and constitute the major structural component of the arterial wall. The circular musculature (cm) and adventitia (a), which consists of circular elastic fibers, form the outer layers of the wall of the trunk. A distinct subendothelial elastic layer (arrows) is present near the lumen. There is no internal elastic membrane. (G) and (H) show opposite sectors of the same cross-section. Case RM 135/76. Cryostat frozen sections. Weigert's resorcin fuchsin. The magnification is the same in all figures.

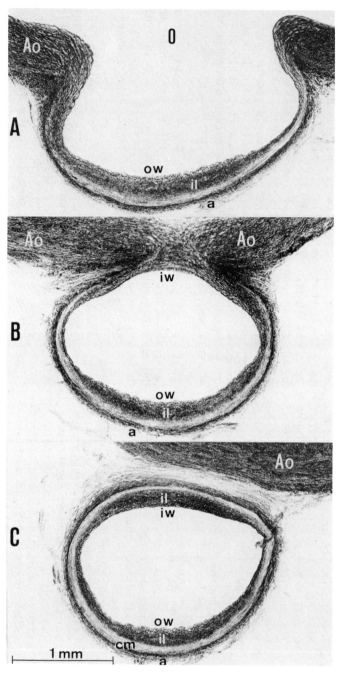

Fig. 68.

← Fig.68. The structure of the proximal segment of the left coronary
trunk from a two-month-old infant (Case 19 K/77). The
coronaries were fixed in formalin under a filling pressure
of about 10 mm Hg.
(A) At the orifice (o), the outer wall of the trunk (OW)
is formed mainly by a prominent internal layer (il) which
consists of dark staining longitudinal musculo-elastic
bundles. A relatively thin layer of light unstained circular musculature is seen between the internal layer and
the adventitia (a).
(B) In the next section, the internal layer (il) remains
prominent along the outer wall (OW) of the trunk. i.e.,
farthest from the aorta (Ao). A - adventitia.
(C) In the following cross-section taken immediately below
(B), the internal layer is thick along the outer and the
opposite inner walls of the trunk, i.e., nearest the aorta.
A small part of the aortic wall is seen at the right (Ao).
The lightly stained circular musculature (cm) is well seen
around the circumference but is comparatively thin. a -
adventitia. Cryostat sections. Weigert's resorcin fuchsin.

Just above the division of the trunk, the longitudinally arranged smooth muscle bundles again become more prominent, particularly along the extensions of the dividing crest. In these sectors,
the internal elastic membrane is replaced by sheets of longitudinal
elastic fibers which also surround the longitudinal smooth muscle
bundles.

2. Term newborn infants. The structure of the most proximal
thick-walled segment of the left coronary trunk resembles that seen
in pre-term infants and consists of concentrically arranged elastic
membranes. Just below this level, the circular musculature becomes
visible and extends throughout the entire circumference. In serial
sections, prominent bundles of longitudinal muscle cells appear
along the inner border of the circular musculature (Fig. 67C, D).
As in pre-term infants, they form a more or less continuous layer
along the outer wall of the trunk, i.e., farthest from the aorta,
and consititute one-half or more of the entire wall thickness. No
limiting elastic sheets or membranes are present between this layer
and the circular musculature. In sectors with longitudinal musculature, the subendothelial elastic layer is formed by concentric
elastic sheets consisting of longitudinal elastic fibers. Bundles
of longitudinally arranged muscle cells are also present between
these sheets. A more or less continuous internal elastic membrane
is visible in sectors without longitudinal muscle bundles, particularly along the posterior wall of the trunk, i.e., nearest the
aorta.

Comparison of cross-sections of the trunk immediately above and at its division into the descending and circumflex coronary branches shows that prominent, longitudinal, musculo-elastic structures extend from the trunk into the proximal part of the left descending branch.

3. One- to three-year-old children. A continuous layer of circular musculature is distinctly visible around the orifice of the trunk (Fig. 68A). It is covered by a thick internal layer which consists mainly of longitudinally arranged musculo-elastic bundles (Figs. 67E, F, 68). This layer extends distally and at lower levels constitutes the major structural component of the part of the circumference farthest from the aortic wall, i.e., covered by the epicardium. The opposite wall, i.e., nearest to the aortic wall, and just below the orifice consists of circularly arranged, elastic sheets which parallel the membranes in the wall of the aorta (Fig. 68B). At lower levels, this wall is formed mainly by longitudinally arranged structural components and its structure resembles that of the opposite outer wall located below the epicardium (Fig. 68C, 69). The adventitia is prominent along the entire circumference and rich in strong circular elastic sheets.

4. Adolescents and young adults. As in younger age groups, the wall of the most proximal segment of the trunk just below the orifice is continuous with the aortic wall and consists of numerous tightly packed elastic membranes that parallel the adjacent aortic membranes in the wall nearest the aorta. At lower levels, it consists of two main layers: an internal and an external layer (Fig. 67G, H).

The internal layer is the most prominent layer in the wall of the trunk. It is mainly formed by a well-differentiated longitudinal musculo-elastic layer covered by a thick elastic layer consisting of concentric elastic sheets and circular elastic fibers. Thick longitudinal or oblique elastic fibers form a prominent subenthelial elastic sheet near the lumen (Fig. 67G, H, arrows). A thin layer of loose connective tissue develops along the inner aspect of this sheet and often extends over a large portion of the circumference. In some areas it becomes relatively thick and it may then become the site of lipid deposits. No internal elastic membrane is present at any level of the trunk.

The external layer is formed by a circular band of smooth musculature that contains an increasing number of elastic fibers which parallel the muscle cells. It is therefore less clearly separated from the surrounding adventitia than in children.

The total thickness of the trunk wall varies considerably in individual sectors of the same cross-section (Fig. 67G, H). It is uncertain to what extent these differences are due to differences in postmortal retraction.

In summary, development of the <u>left coronary trunk</u> is characterized by the following findings:

1. <u>Fetuses and pre-term infants</u> with a body weight of about 1000 g. The wall of the trunk is typical of a muscular artery and consists mainly of circular musculature. A layer of fine, longitudinal, musculo-elastic bundles is commonly present along the luminal aspect of the vessel and especially in the outer part of the circumference, i.e., farthest from the aorta. A more or less continuous internal elastic membrane is present in most parts of the wall. In sectors with an internal layer of longitudinal musculature, the membrane is replaced by sheets of fine longitudinal elastic fibers which are also present between the muscle bundles.

2. <u>Term newborn infants</u>. The wall of the trunk displays similar structural characteristics. In many sectors, particularly in the outer wall, numerous longitudinal muscle bundles and elastic fibers form an additional layer along the inner aspect of the circular musculature. This layer represents the precursor of a prominent musculo-elastic layer that develops during growth.

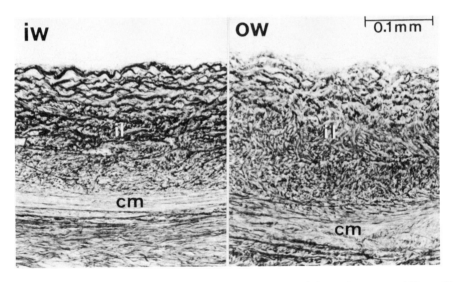

Fig. 69. Higher magnifications of the inner (I) and outer (O) walls of the left coronary trunk shown in Fig. 68C. The internal layer (il) is rich in elastic elements and constitutes the major component of the arterial wall. The circular musculature (cm) forms only a very thin layer in the inner and a somewhat thicker layer in the outer wall. Resorcin fuchsin.

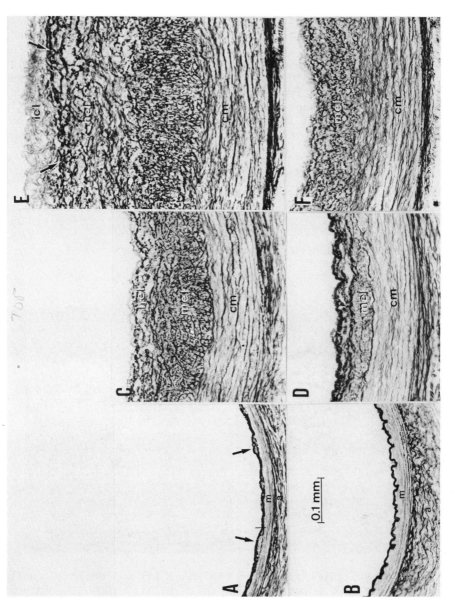

Fig. 70.

Fig. 70. Shows the microscopic structure of opposite sectors of the left anterior descending coronary branch in a newborn infant (A, B), a two-year-old child (C, D), and in a 17-year-old adolescent (E, F).
In the newborn infant (A, B), the branch has the structure of a muscular artery with a well-developed circular musculature (m). In numerous sectors, longitudinal muscle bundles of variable thickness (arrows) delineated by sheets of longitudinal elastic fibers are present in the luminal layer. The adventitia (a) is rich in rather loose elastic networks. Case 87 K/76.
In the two-year-old child (C, D), the longitudinal musculo-elastic layer (mel) and the hyperplastic elastic layer (hel) represent the main structural components in most parts of the circumference (C). In the other sectors (D), the circular musculature (cm) still constitutes the main structural element and in some areas it is separated by a sheet of longitudinal elastic fibers (arrows) from the superimposed inner layer. Case 46 K/77.
In the 17-year-old adolescent (E, F), the wall of the branch is formed mainly by longitudinal structural components, i.e., the longitudinal musculo-elastic layer (mel) and the hyperplastic elastic layer (hel). The circular musculature (cm) represents a minor structural component in most of the circumference. Numerous stronger elastic sheets (arrows) are seen below the thin intimal connective tissue layer (icl), which contains a few fat-laden cells, near the lumen. Case RM 13.5.76. Cryostat frozen sections. Weigert's resorcin fuchsin. Same magnification in all figures.

3. *One- to three-year-old children.* The wall of the left coronary trunk consists mainly of a thick longitudinal musculo-elastic layer and a prominent luminal elastic layer. In most sectors, the circular musculature forms a comparatively thin external layer and comprises one-fourth to one-third of the total wall thickness. No internal elastic membrane is present.

4. *Adolescents.* The wall of the trunk consists of five main layers. The longitudinal musculo-elastic layer and the more internal elastic layer remain the major structural components of the trunk and they can often be easily distinguished. A prominent subendothelial elastic sheet is present near the lumen. It develops during the second decade and consists of coarse, tightly packed, longitudinal, elastic fibers. A thin connective tissue layer appears along the inner aspect of this sheet. The circular musculature and the adventitia form the external layers. No internal elastic membrane has developed in any portion of the trunk.

Hence, structural changes in the left coronary trunk after birth consist primarily of increasing differentiation and thickening of longitudinal structural components.

Left Anterior Descending Branch

1. *Fetuses and pre-term infants.* The wall of the proximal segment of the branch immediately below its origin consists mainly of circularly arranged musculature. In some sectors, well-developed longitudinal muscle bundles are seen internal to the circular musculature. The bundles are separated from the lumen by sheets of longitudinal elastic fibers instead of an internal elastic membrane. The latter is well-developed in the rest of the circumference. The adventitial elastic networks are fairly prominent.

In subsequent sections, the longitudinally oriented muscle cells decrease or disappear. Only small "intimal cushions," consisting of longitudinal muscle bundles, persist and occasionally protrude into the lumen. The sheets of elastic fibers which surround the muscle bundles may resemble a "duplicated" internal elastic membrane.

2. *Term newborn infants.* As in fetuses, the circular musculature represents the main structural component of the wall of the left anterior descending branch (Fig. 69A, B). A longitudinal musculo-elastic layer internal to the circular layer is present in many sectors. In some areas, it may be as thick as the external circular musculature. No elastic sheets or fragments of an internal elastic membrane are visible at the junction of both layers. In sectors that include longitudinal musculature, the subendothelial elastic layer consists of superimposed concentric sheets of longitudinal

elastic fibers. In sectors that consist solely of circular musculature, i.e., have no longitudinal muscle bundles, a well-differentiated internal elastic membrane is clearly seen.

In more distal sections, the left descending branch increasingly assumes the typical structural pattern of a muscular artery although longitudinal musculo-elastic structures are still evident in some sectors. They often form cushions that protrude slightly into the lumen in retracted arteries but they flatten under pressure. The internal elastic membrane extends over an ever greater portion of the circumference.

3. <u>One- to three-year-old children</u>. The circular smooth musculature is well-developed throughout the circumference but it now constitutes only about one-half of the wall thickness in the proximal portion of the branch. A thick, longitudinal, musculo-elastic layer has developed along most of the circumference internal to the circular musculature. This layer is covered by a prominent elastic layer consisting of concentric elastic sheets of mostly longitudinal elastic fibers (Fig. 70C, D). The individual layers of the branch are poorly distinguished from each other in most parts of the circumference. Only the border between the circular musculature and adventitia, which is rich in elastic membranes, is clearly visible. No true internal elastic membrane is present in the proximal portion of the branch.

4. <u>Adolescents and young adults</u>. The longitudinal musculoelastic layer and the superimposed elastic layer represent the major structural components of the proximal segment of the branch (Fig. 70E, F). Along the thicker inner wall of the branch, i.e., nearest the myocardium, both layers form up to two-thirds of the entire wall thickness. In a large part of the circumference, they are poorly separated from each other. Cross-sections of prominent elastic sheets, consisting of tightly packed, strong, longitudinal or oblique elastic fibers, are present near the lumen in most of the circumference. No internal elastic membrane is seen. A thin luminal connective tissue layer, formed by loose collagenous fibers and ground substance, is present. In some areas, it is thickened and includes intra- and extracellular lipid deposits.

Structural features of the <u>left anterior descending branch</u> during various stages of development may be summarized as follows:

1. <u>Pre-term infants</u>. The wall of the proximal segment of the branch consists mainly of circular musculature. Along the inner aspect of this layer, there are usually some well developed longitudinal muscle bundles. A well-differentiated internal elastic membrane is confined to areas that consist solely of circular musculature.

2. **Term newborn infants.** The circular musculature remains the predominant structural component of the branch. In some sectors of the most proximal segment, a longitudinal musculo-elastic layer as thick as the external circular musculature is present. An internal elastic membrane is seen in sectors consisting solely of circular musculature but not in sectors with circular and longitudinal musculature. At lower levels, the structural pattern becomes increasingly typical of a muscular artery with an internal elastic membrane that extends throughout most of the circumference.

3. **One- to three-year-old children.** There is a considerable increase of longitudinally arranged, musculo-elastic structures which in some sectors form up to two-thirds of the total wall thickness.

4. **Adolescents and young adults.** The wall of the branch consists mainly of a prominent longitudinal, musculo-elastic layer and a superimposed elastic layer of variable thickness. No significant change in structural pattern occurs in the left anterior descending branch between one and three years of age to adolescence, apart from the increase in size and development of a connective tissue intimal layer that may contain lipid deposits.

Left Anterior Circumflex Branch

1. **Fetuses and pre-term infants.** The structure of this branch is typical of a muscular artery with circular musculature and a well-differentiated internal elastic membrane. Only a few small longitudinal musculo-elastic bundles are occasionally seen along the inner aspect of the circular musculature.

2. **Term newborn infants.** Little change in structure has taken place during the last weeks of gestation. Longitudinal musculo-elastic structures are often present but they are usually confined to sites of branchings.

3. **One- to three-year-old children.** In the proximal segment of the branch, the internal layer has developed and accounts for one-fourth to one-third of the entire wall thickness. It consists mainly of concentric elastic sheets formed by coalescent longitudinal elastic fibers. An internal elastic membrane with numerous small gaps separates the internal layer from the outer layer of circular musculature, except in areas with "cushions." These consist of longitudinal musculo-elastic bundles and protrude slightly into the lumen.

4. **Adolescents and young adults.** With growth, the internal elastic layer becomes more prominent and, in some places, it may be three times thicker than the outer layer of circular musculature, i.e., nearly as thick as in the right coronary artery (see p. 129).

However, the thickness of the internal layer varies considerably around the circumference, and in some sectors, the internal layer may be even thinner than the underlying circular musculature. As in other proximal coronary segments, a sheet of densely packed, coarse, longitudinal, elastic fibers develops near the lumen. Intimal proliferation does not occur. A fragmented internal elastic membrane is present between the internal layer and circular musculature in only a part of the circumference. It is absent in those sectors where the internal layer is thickest. Longitudinal musculo-elastic bundles are often seen between the membrane and circular musculature.

In summary, development of the <u>left anterior circumflex branch</u> is characterized by the following findings:

1. <u>Pre-term infants</u>. This branch represents a muscular artery. It has a well-differentiated internal elastic membrane and a few longitudinal musculo-elastic bundles internal to the circular musculature.

2. <u>Term newborn infants</u>. The branch retains the structure of a muscular artery but the longitudinal musculo-elastic bundles are probably more numerous.

3. <u>One- to three-year-old children</u>. The internal layer shows approximately the same structural pattern as in other proximal coronary segments and extends throughout most of the circumference. It is well delineated from the circular musculature by a fragmented internal elastic membrane.

4. <u>Adolescents and young adults</u>. The internal arterial layer has become the main structural component in most of the circumference. In some sectors it is still clearly distinguished from the circular musculature by a fragmented internal elastic membrane.

<u>Right Coronary Trunk</u>

1. <u>Fetuses and pre-term infants</u>. At the orifice, the wall of the artery is thick and consists mainly of circular elastic membranes interspersed with numerous longitudinal and oblique elastic fibers in more internal layers. The adventitial elastic networks are continuous with the adventitial networks of the aortic wall. A band of smooth musculature appears first beneath a thick layer of elastic membranes and is best seen in axial sections from the most proximal portion of the trunk.

In the following cross-sections, the elastic layer thins and the wall of the trunk assumes the structure of a muscular artery with an almost continuous internal elastic membrane encircled by a thick layer of circular smooth musculature and an adventitia of

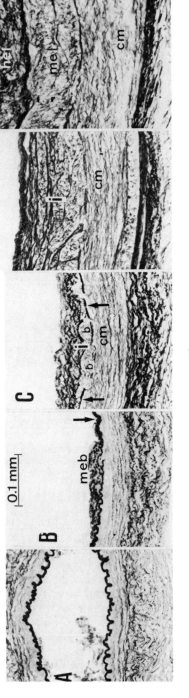

Fig. 71.

Fig. 71. Series of figures showing the development of the microscopic structure of the proximal portion of the right coronary artery.
(A) In a pre-term infant, the trunk has the structural pattern of a muscular artery and a well-developed internal elastic membrane. Case 24 K/76.
(B) In a term newborn infant, the internal elastic membrane (arrow) is replaced in some sectors by longitudinal musculo-elastic bundles (meb). Case 87 K/76.
(C) By the age of one year, a prominent internal layer (i) consisting of longitudinal muscle bundles (b) and elastic sheets has developed. In some sectors it is about as thick as the adjacent circular musculature (cm). Fragments of the internal elastic membrane (arrows) are seen at the border between these layers. Case 56 K/76.
(D, E) At the age of three years, the internal layer (i) has become thicker as evidenced by these cross-sections of opposite sectors of the trunk. In some sectors (D), cross-sections of axially arranged smooth muscle bundles (arrows) are seen below the fragments of the internal elastic membrane. Case 88 K/76.
(F, G) At the age of 17 years, the longitudinal musculo-elastic layer (mel) and the super-imposed hyperplastic elastic layer (hel) constitute the major structural components of the arterial wall. The circular musculature (cm) is rich in elastic fibers which parallel muscle cells. Same magnification in all figures.

coarse collagenous networks intermingled with numerous fine elastic fibers (Fig. 71A). Small musculo-elastic cushions commonly protrude into the lumen.

The transition from an elastic to a muscular artery is less abrupt when both the right coronary artery and the conus artery have a common orifice or when the conus artery emerges from the proximal portion of the right coronary trunk. With these variants, the inner elastic layer of the proximal segment is continuous with the prominent musculo-elastic cushions that are present at sites of branching, and particularly at the dividing crest.

2. <u>Term newborn infants</u>. The most proximal part of the right coronary artery is thick-walled, and as in fetuses, consists mainly of concentric elastic membranes which are continuous with the aortic wall. Longitudinal sections show that circular smooth musculature is also present at the site of origin, and in its vicinity, the trunk assumes the structure of a muscular artery. At this level the internal elastic membrane is continuous in most parts of the circumference. In a number of places the membrane is replaced by several concentric sheets of longitudinal elastic fibers (Fig. 71B). Small longitudinal muscular bundles are visible between the elastic sheets. Some of the bundles are separated from the circular musculature by fragments of an internal elastic membrane. However, no limiting elastic structures are present between the larger longitudinal musculo-elastic bundles and the circular muscular layer.

In hearts with a common orifice for the right and conus artery or a conus artery arising from the most proximal portion of the right coronary artery, both vessels assume the typical structural features of muscular arteries distal to the orifice or to the site of branching, a feature also present in the fetus.

3. <u>One- to three-year-old children</u>. As in younger and particularly in older age groups, a considerable change in structure occurs along the proximal segment of the right coronary trunk. In the most proximal portion, i.e., immediately distal to the orifice, the wall of the trunk is rich in elastic elements, and a prominent internal longitudinal musculo-elastic layer represents the major structural component of the arterial wall. (Fig. 72A). Only 2 - 4 mm distally, the internal layer becomes thin and the circular musculature forms the main part of the wall (Fig. 72B). Fragments of the internal elastic membrane or elastic sheets are present along the border of the internal layer with the circular musculature. No limiting elastic structures, however, are seen in many sectors, a finding described as "border disappearance" (Gross et al., 1934).

This structural pattern persists in more distal parts of the vessel where the separation of the internal elastic layer from the longitudinal or circular musculature by an elastic sheet or fragments of an internal elastic membrane becomes more distinct. The limiting elastic sheet, however, may not be present in the parts of the circumference with a prominent continuous longitudinal musculo-elastic layer.

4. <u>Adolescents and young adults</u>. The proximal segment of the right coronary trunk consists of an internal and an external layer. The internal layer is formed mainly by longitudinally arranged musculo-elastic bundles and a superimposed hyperplastic elastic layer of varying thickness (Fig. 71F, G). The internal layer is two to three times thicker than the circular muscular layer which, together with the adventitia, forms the external arterial layer. No prominent elastic sheets or membrane separate the circular musculature from the internal layer. However, fragments of concentric elastic sheets or of a homogeneous elastic membrane appear commonly between the individual layers in cross-sections taken some millimeters distal to the orifice.

The innermost elastic layer consists of a network of circular and longitudinal elastic fibers and sheets. Near the lumen it is formed by stronger concentric sheets of rather loosely arranged, coarse, longitudinal, elastic fibers. A thin layer of loose intimal connective tissue extends throughout the circumference. It is considerably thickened in the most proximal part of the artery where it frequently contains lipid deposits.

Structural changes in the wall of the <u>right coronary artery</u> may be summarized as follows:

1. <u>Fetuses and term newborn infants</u>. Near the orifice, the wall of the vessel rapidly assumes the structural pattern of a muscular artery with a prominent layer of circular musculature and an almost continuous internal elastic membrane. Small longitudinal muscle bundles and elastic sheets are present along the inner aspect of the circular musculature.

2. <u>One- to three-year-old children</u>. The number of concentrically arranged elastic sheets and longitudinal muscle bundles increases. This results in formation of a distinct internal layer which is partly separated from the external layer of circular musculature by elastic sheets or fragments of an internal elastic membrane.

3. <u>Adolescents and young adults</u>. The well-differentiated longitudinal musculo-elastic layer and the superimposed elastic layer become the main structural components of the proximal segment of the right coronary artery, whereas the circular muscular layer now constitutes only one-third to one-half of the total wall thickness.

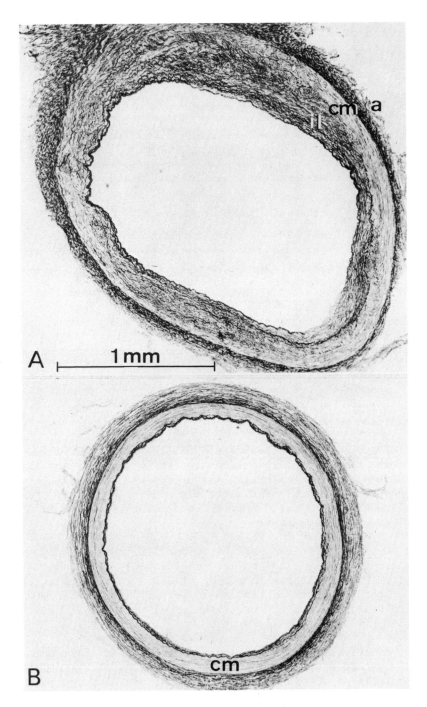

Fig. 72.

←Fig. 72. Pronounced structural differences are seen over a very short distance in the coronary wall of the proximal segment of the right coronary artery after it has been fixed under a filling pressure of about 10 mm Hg. (A) In the most proximal portion, i.e., near the origin of the vessel, a thick internal layer (ie) consisting of longitudinally and obliquely arranged structures and rich in dark-stained elastic fibers constitutes the main component of the wall. The unstained circular musculature (cm) forms a relatively thin external layer that is well delineated by the adventitia (a) which is rich in elastic networks. (B) One-to-two millimeters distal to the origin of the conus artery, the well-developed circular musculature (cm) represents the main structural component of the wall. The internal layer is thin and consists of fine elastic sheets and small longitudinal muscle bundles. Two-and-a-half-year-old boy. Case 46 K/77. Cryostat sections. Weigert's resorcin fuchsin.

Thus, as in the left coronary trunk and its main branches, development of the right coronary artery is characterized mainly by growth and development of a prominent internal layer consisting of longitudinally arranged structural components, particularly of the longitudinal musculo-elastic layer, and a superimposed elastic layer.

"Intimal Cushions"

Intimal cushions are commonly present at branchings and bifurcations of coronary arteries and have been described by many investigators (Wolkoff, 1923; Ehrlich et al., 1931; Gross et al., 1934; Minkowski, 1947; Lober, 1953; Schornagel, 1956; Moon, 1957; Neufeld et al., 1962; Levene, 1965; Jaffé et al., 1971)(Fig. 72). At branchings, the internal elastic membrane is replaced by numerous fine elastic sheets that alternate with smooth muscle cells (Fig. 73). In microscopic sections which show the branching artery in longitudinal view, the muscle cells and elastic fibers that form the sheets are seen in cross-section. Hence, they probably encircle the orifice of the branch, and the cushions represent cross-sections of ring-like structures that protrude in the lumen. Their raised cushion-like appearance is probably due to postmortal arterial retraction, because they flatten in arteries perfused during fixation under appropriate pressure.

Cushions may be regarded as a normal feature of arterial structure. They probably permit the vessel to adapt to the higher circumferential tension and additional hemodynamic load at sites of branchings. Later in life, however, the cushions become the sites of predilection for lipid infiltration and atherosclerotic lesions.

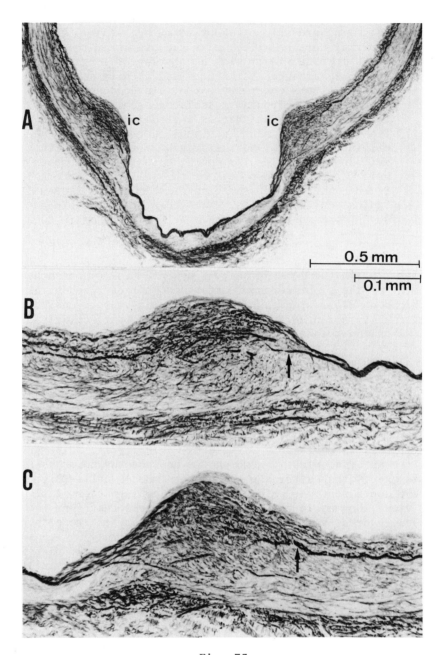

Fig. 73.

←Fig. 73. (A) Intimal cushions (ic) at the orifice of a branch of the right coronary artery in a 30-month-old boy. (B) and (C) show the cushions at higher magnification. They consist of dense networks of fine elastic fibers and sheets which are seen in cross-section and seem to form circular structures around the orifice of the branch. The internal elastic membrane (arrows) is interrupted below the cushions. Case 46 K/77. Cryostat section. Weigert's resorcin fuchsin.

In newborns, such intimal thickenings have also been found in other areas (Dock, 1946; Fangman and Hellwig, 1947; Lober, 1953; Schornagel, 1956; Vlodaver et al., 1969; Jaffé et al., 1971). A more rapid increase in intimal thickening has been reported in ethnic groups susceptible to coronary heart disease (Vlodaver et al., 1969). Structures described as intimal thickenings may represent, in part, elements of the internal longitudinal arterial layer, i.e., the normal structural component of the developing coronary trunks. With stronger postmortal arterial retraction, parts of this layer may protrude in the lumen and create the impression of intimal cushions.

Closing Remarks

The increase in longitudinally arranged muscular and elastic elements in the coronary trunks and their main branches during postnatal development results in formation of a prominent internal layer which is not seen in other arteries of comparable size (Wolkoff, 1923; 1924; Ehrlich et al., 1931; Gross et al., 1934; Jaffé et al., 1971). In adolescents and young adults, the internal layer tends to be considerably thicker than the circular musculature which, together with the surrounding adventitia, forms the external layer.

Thickening of the internal layer is apparent first in infants and is greatest in the trunk of the left coronary artery. It then extends to the left anterior descending branch. Involvement of the right coronary trunk usually occurs more slowly than in the proximal segments of the left anterior descending coronary branch but at about the same rate as in the anterior circumflex branch. Accelerated growth of a luminal connective tissue layer in the right coronary artery does not occur during childhood and adolescence although it has been frequently observed in young adults (Gross et al., 1934). In general, differentiation and remodeling of coronary arteries during growth is more pronounced in proximal segments, whereas distal branches retain the initial structural pattern of muscular arteries.

The longitudinal musculo-elastic layer, which constitutes the major component of the internal arterial layer, arises from small, longitudinally arranged, muscle bundles that first appear in the fetus and newborn along the inner aspect of the circular musculature.

The musculo-elastic bundles become visible on both sides of the internal elastic membrane, as well as in small gaps in the membrane (Fig. 74). The bundles are usually well delineated from the circular musculature by a sheet of fine longitudinal elastic fibers. Another sheet of elastic fibers is seen along the inner aspect of the bundles. Both sheets converge and appear to be continuous with the border of the interrupted internal elastic membrane. These findings have often been interpreted as "splitting" or reduplication of the membrane. Formation of new elastic sheets around developing longitudinal muscular elements, however, seems a more likely explanation since it is hard to believe that a thick, homogeneous, fully differentiated internal elastic membrane can split and form two new concentric membranes. In infancy and childhood, the internal elastic membrane is seldom homogeneous and seems to consist of tightly packed, longitudinal, elastic fibers. In arteries perfused during fixation under appropriate pressure, individual elastic fibers forming the internal elastic membrane can easily be distinguished on crosssections. In earlier stages of development, ingrowth of smooth muscle cells into the sheets that still consist of loosely arranged elastic fibers is possible. With consolidation of the elastic sheets around the newly-formed muscle bundles, these limiting elastic structures may later appear to be homogeneous elastic membranes. Thus, as in other arteries, limiting sheets of longitudinal elastic fibers obviously represent precursors of mature elastic membranes.

During growth, the internal elastic membrane becomes increasingly fragmented in large proximal segments of the coronary arteries. Numerous gaps then become visible, presumably at sites where the elastic fibers forming the membrane are more loosely packed. At the same time, smooth muscle bundles gradually fill the gaps. As the arterial tube enlarges, the internal elastic membrane disappears completely consequent to the profound remodeling of the arterial wall that accompanies growth. Hence, the initial distinct outlines of the intima and media evident in fetal life are slowly lost, a process described by Gross et al. (1934) as "border disappearance" (Fig. 74).

It remains uncertain whether the longitudinal musculo-elastic layer and the adjacent luminal elastic layer of the coronary trunks and their main branches are derived from the intima or from the inner part of the media. To our knowledge there is no convincing evidence that both layers are derived solely from the intima. The development of the longitudinal musculo-elastic layer on both sides of the original primary internal elastic membrane in fetuses and infants (Fig. 74) suggests that it may be derived from both the intima and the media. The layers of the coronary artery that develop internal to the circular musculature should therefore not be classified as the coronary's intima because there is no limiting elastic membrane such as that found at the intimal-medial junction in other arteries along the inner aspect of the circular musculature. In our opinion,

Fig. 74. Early stage of development of longitudinal musculo-elastic bundles along the intimal-medial junction of the left coronary trunk. Roundish cross-sections of muscle cells (unstained, arrows) are seen on both sides of the fragmented internal elastic membrane. The originally distinct border between the intima and the media is no longer visible ("border disappearance"). Newborn infant. Case 87 K/76. Weigert's resorcin fuchsin.

the usual subdivision of the arterial coats, therefore, does not seem applicable to the trunks and main branches of the coronary arteries and one should distinguish only between the internal and external layers of the coronary wall. An advantage of this classification is that the border between the circular and longitudinal smooth musculature is always distinctly visible even in the absence of a limiting elastic sheet.

Interpretation of the internal arterial layer as the coronary's intima may also be misleading because it implies early, and in comparison with other arteries, extremely pronounced development of the intima which allegedly favors the development of early atherosclerotic lesions. Gross et al. (1934) found that the "intima" of the left anterior descending branch was often twice as thick as the "media" by the end of the second decade. It is true that the coronary's intima becomes thickened earlier than in other arteries. However, not all arterial layers that appear early internal to the circular musculature can be regarded to be the result of excessive intimal proliferation.

Distinctive structural features of the coronary trunks and their main branches obviously develop in response to various hemodynamic factors. Highly pulsatile streaming to which both coronary arteries are subjected may be associated with increased shearing forces which, in turn, may play a role in the development of longitudinal structures. Since marked differences in the dynamics of blood flow exist between the left and right coronary arteries (Mansfield et al., personal communication, 1976), the differences in structure and predisposition to atherosclerosis in both arteries, particularly in the trunk and main branches, may also be determined in part by a different functional load.

EARLY PATHOLOGICAL CHANGES IN INFANCY

Premature coronary atherosclerotic lesions with prominent intimal thickening and narrowing of the lumen are associated with familial hyperlipoproteinemia type II (Frederickson et al., 1967). These lesions have also been observed in children dying of chronic glomerulonephritis, a condition associated with hyperlipemia and hypercholesterinemia (Fig. 75). In these children, the hypertension accompanying the nephritis also probably contributes to development of atherosclerotic lesions.

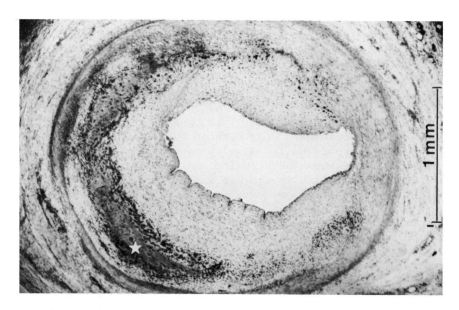

Fig. 75. Stenotic atherosclerotic plaque in the right coronary artery of a 6-year-old girl who died of chronic glomerulonephritis. Considerable narrowing of the lumen is present because of intimal proliferation with deposition of a large amount of lipids (black) and extensive necrosis in the deeper layers (asterix). During the course of the illness, the child's blood pressure increased and serum lipids rose to 700 mg %. Fettrot and Weigert's resorcin fuchsin stain. Case 768/75.

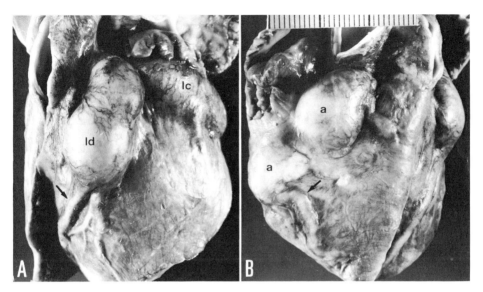

Fig. 76. Multiple large aneurysms are present along the course of the coronary arteries in a four-month-old infant who died of a severe illness believed to be the mucocutaneous lymph node syndrome.
(A) Note the 2.5 x 1.5 cm aneurysm in the left anterior descending branch (ld) and the aneurysm in the left circumflex branch (LC). (From Aterman et al., 1977).
(B) Two aneurysms (a) are seen in the right coronary artery. The coronary branches distally to the aneurysms are diffusely thickened and partly tortuous (arrows). The lumena of the aneurysms are filled with partly organized thrombi. Case 132/75.

Intimal thickening in coronary arteries has also been experimentally produced by feeding animals an atherogenic diet, e.g., in the rat (Wissler et al., 1953, 1954), in the rabbit (Krylow, 1916; Wolkoff, 1930; Sinizina, 1964), in the chick (Pick and Katz, 1965), and in the rhesus monkey (Taylor et al., 1963; Taylor, 1965).

Unlike the iliac arteries and carotid siphon, the internal elastic membrane of the coronary arteries does not calcify in infancy and childhood. Total wall tension is lower in smaller coronaries than in the large iliac arteries or carotid siphon. Consequently the functional load on the internal elastic membrane is also lower, which may be an important factor preventing development of calcific deposits. Moreover, the internal elastic membrane gradually disappears entirely from the wall of the main coronary trunks and becomes replaced by secondary, newly formed elastic structures which, as in other arteries, show no affinity for calcium in infancy.

In a few cases, however, pronounced idiopathic calcification has been reported in infancy (Brown and Richter, 1941; Stryker, 1946; Menten and Fetterman, 1948; Bickel and Janssen, 1963; Traisman et al., 1965). The calcium deposits then appear in the inner- and middle-thirds of the media of medium-sized and large coronary branches. Pronounced intimal proliferation follows and causes narrowing of the lumen. Multiple myocardial infarctions or myocardial fibrosis may then develop. Hypercalcemia of unknown origin, in some cases probably due to an inborn error of calcium metabolism (Traisman et al., 1956) or to an inborn hypersensitivity to Vitamin D (Seelig, 1969), may be involved. Grasso and Selye (1962) produced identical lesions in the rabbit by administering dihydrotachysterol, which was used as a sensitizer, and then injecting ferrous saccharate, which acted as a challenger. Interestingly, this calciphylactic phenomenon (Selye, 1962) resulted in the same clinical picture seen in children with idiopathic calcinosis, i.e., multiple myocardial infarctions and heart failure.

The greater susceptibility of growing coronary arteries to pathogenetic factors may be responsible for the severe changes observed in the mucocutaneous lymph node syndrome (MCLS). Many cases have recently been reported in Japan, and it seems to be a distinct clinical entity (Kawasaki, 1974; Mortimer, 1976). One to two percent of children who develop this condition show symptoms of coronary disease (Kawasaki, 1974; Kato, 1975). In cases coming to autopsy, the coronary arteries seem to be selectively affected. The lesions usually consist of large protruding aneurysms along the main coronary stems (Fig. 76), the lumina of which are partly or completely occluded by thrombi. Most of the wall of the aneurysms is formed by connective tissue and shows evidence of thrombus incorporation and organization (Aterman et al., 1977). Fibrinoid lesions are not common. The relation of MCLS to polyarteritis nodosa remains to be elucidated. Heart failure with or without myocardial infarction is the usual cause of death.

7. CAROTID ARTERIES

The common carotid arteries and the brachiocephalic trunk, from which the right carotid artery originates, are elastic arteries and belong to the central portion of the systemic circulation. The function of this segment of the arterial system is not only to distribute blood to the individual parts of the body, but to transform pulsatile blood flow into a more continuous streaming directed to the head and brain. Since baroreceptors are located at the carotid bifurcation, this segment also plays an important role in maintenance and regulation of systemic blood pressure (Fig. 77).

FUNCTIONAL MORPHOLOGY OF ARTERIES 233

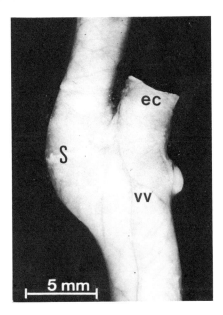

Fig. 77. The carotid bifurcation of an 18-month-old infant has been dissected free and fixed under a pressure of about 50 cm formalin. The protruding sinus area (s) is distinctly visible. cc-common carotid artery, ic-internal carotid artery, ec-external carotid artery, vv-vasa vasorum (veins).

The amount of blood transported through the carotid arteries appears to undergo considerable changes during intrauterine and early extrauterine life. The head is comparatively larger in the embryo, fetus, and newborn infant than in the adult. Thus, at 12 to 13 weeks of gestation, the head of the embryo amounts to about 45 percent of total body weight (23 - 30 g), while at term it comprises about 20 percent of total body weight (3.500 g). In contrast, in the adult it is only 3 - 5 percent of total body weight. Hence, the relative amount of left ventricular output transported to the brain must be considerably greater in the fetus and newborn infant than in the adult. This is reflected by the relatively large caliber of the aortic arch vessels in the newborn infant as compared with the adult. The differences in relative size can be demonstrated more precisely by distending and fixing the arch and its main branches under appropriate intravascular pressure and then photographing it at different magnifications until the infant and adult aortas seem to be the same size (Walsh, Meyer, and Lind, 1974).

EARLY LIPID DEPOSITS AND THEIR RELATION TO PREFORMED ARTERIAL STRUCTURES

Common Carotid Artery

The common carotid artery and its bifurcation represent common sites of pronounced atherosclerotic lesions (Benecke, 1922; Hultquist 1942; Samuel, 1956; Hutchinson and Yates, 1957; Schwartz and Mitchell 1961, 1962; Fisher et al., 1965; Javid et al., 1970; Solberg and Eggen, 1971) which, in adults, often produce narrowing and even occlusion of the lumen. The precursors of these lesions, i.e., lipid streaks, appear in the predisposed segments of the carotid artery early, i.e., toward the end of the first decade or in the second decade of life. Their shape, location, and extent are largely determined by structures which form in the inner layer of the arterial wall during postnatal development and show a greater affinity for lipids. Gross demonstration of early lipid deposits is, therefore, helpful for assessing the extent of these specialized structures along the arterial luminal surface.

In the trunk of the common carotid artery, fatty streaks develop initially on the antero-medial wall just below the origin of the external carotid artery. They then extend proximally and, in young adults, form a characteristic triangular deposit which is narrower in the proximal segment of the trunk (Fig. 78). The opposite dorso-lateral wall does not become involved for a long time. This suggests that either some structural features of the predisposed antero-medial wall or some rheological or other factors may favor lipid deposition.

Examination of arterial rings and cross-sections of the common carotid artery at different levels reveals conspicuous structural differences between predisposed and resistant arterial sectors (Meyer and Noll, 1974). The differences become evident during the first decade and pronounced during the second decade of life. In the predisposed antero-medial wall, a well-delineated musculo-elastic layer consisting of longitudinally arranged muscle cells and fine elastic fibers develops along the inner aspect of the internal elastic membrane during the first few years of life. The membrane may be continuous but is more often interrupted by numerous presumably longitudinally oriented gaps. In a recent study of 15 children 3 to 13 years of age, 11 already had a distinct musculo-elastic layer in the predisposed arterial sector. In five of these eleven children, aged 3 - 7 years, the common carotid artery was still free of gross lipid deposits. In the older children, small scattered lesions were predominantly located in the upper portion of the carotid trunk but structural differences between the opposite sectors were also conspicuous at arterial levels without lipid deposits. Even with more pronounced lesions in the young adult, large parts of the

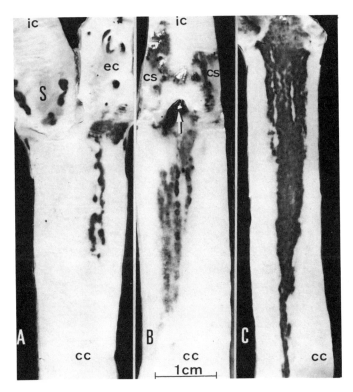

Fig. 78. Various stages in the development of gross patterns of lipid deposits in the common carotid artery.
(A) Lipid deposits (black) initially form along the length of the vessel below the origin of the external carotid artery (ec) and around the periphery of the carotid sinus (s). i.c. - internal carotid artery, cc - common carotid artery. 20-year-old man. Case 1243/73.
(B) Deposits then become confluent along the antero-medial wall of the common carotid artery below the origin of the external carotid artery (arrow) and extend around the central part of the sinus. In this specimen the central sinus area (cs) has been opened longitudinally and divided into two parts which extend to the borders of the vessel. Note also the lipid-free carina of the carotid bifurcation (arrow). 34-year-old man. Case 227/73.
(C) With more advanced lesions, lipid infiltrates the antero-medial wall and assumes a characteristic triangular form. 31-year-old man. Case 1027/72. Fettrot VII B. From Meyer and Noll, 1974.

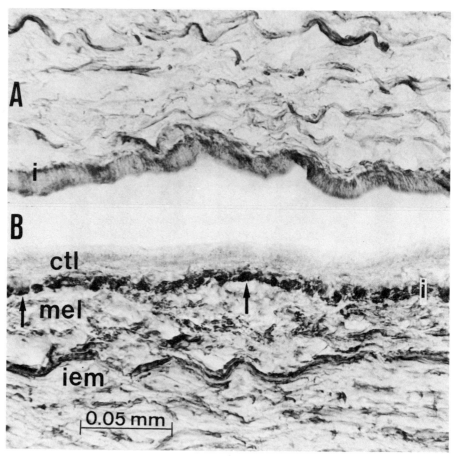

Fig. 79. Structural differences between the dorso-lateral (A) and antero-medial (B) walls of the common carotid artery develop in early childhood. They are well seen on a cross-section of an arterial ring taken from the middle of the common carotid artery from a 20-year-old man. Cause of death: traffic accident. (A) In the dorso-lateral wall, the intima (I) is thin. It consists mainly of a subendothelial layer formed by tightly packed elastic fibers. No internal elastic membrane is seen. (B) In the antero-medial wall, the intima (I) is thickened and has a well-developed musculo-elastic layer (mel) between the internal elastic membrane (iem) and the compact sheet of strong longitudinal elastic fibers (arrows). These fibers are seen in cross-section below the thin connective tissue layer (ctl) of the intima. No lipids are present at this level. Cryostat frozen section. Weigert's resorcin-fuchsin. Case 1145/71. From Meyer and Noll, 1974.

musculo-elastic layer on both sides of the infiltrated area or between individual lipid accumulations remained free of lipid deposits. Thus, development of a musculo-elastic layer cannot be the result of lipid infiltration.

With growth, the musculo-elastic intimal layer becomes thicker. Toward the end of the first decade, a compact sheet of tightly packed, longitudinally arranged, strong elastic fibers develops near the arterial lumen separating the musculo-elastic layer from the loose subendothelial connective tissue (Fig. 79). Both the musculo-elastic and the connective tissue layers result in moderate intimal thickening in the predisposed sectors, whereas the intima of the opposite resistant arterial wall remains thin until early adulthood.

The structural differences between predisposed and resistant arterial sectors seem to appear earlier and are usually more pronounced in the upper part of the common carotid artery, i.e., below the carotid bifurcation where lipid deposits first appear. At this site the musculo-elastic intimal layer extends over a larger part of the arterial circumference than in lower more proximal arterial segments. In general, the extent of the thickened and differentiated intima in the arterial tube seems to correspond to the triangular area of the luminal surface which gradually becomes the predominant substrate of lipid deposition in children and young adults.

Lipids seem to appear first in the sheet of strong longitudinal elastic fibers which develops along the medial aspect of the musculo-elastic intimal layer (Fig. 80). Lipid infiltration of these fibers has been found microscopically in common carotid arteries which do not have gross fatty streaks but show only a "blush," i.e., a pale diffuse reddish staining of the predisposed area. Lipid infiltration of the longitudinal elastic sheet is also often seen around fatty streaks and below characteristic triangular lipid deposits.

Carotid Sinus

Early development of lipid deposits in the sinus may likewise depend on early differentiation and thickening of the intima during postnatal growth. Thus, by five to seven years of age, the intima shows considerable thickening along the protruding part of the sinus. The intimal thickening consists mainly of longitudinally or obliquely arranged elastic and collagenous fibers and muscle cells. A sheet of tightly packed longitudinal elastic fibers is often present near the lumen and appears to be similar if not identical to the luminal elastic layer frequently found in the predisposed sector of the common carotid artery. Toward the end of the first decade of life, the thickened intima, still largely free of lipids, has extended over the entire sinus area and may be more than one-fourth as thick as the media (Fig. 81).

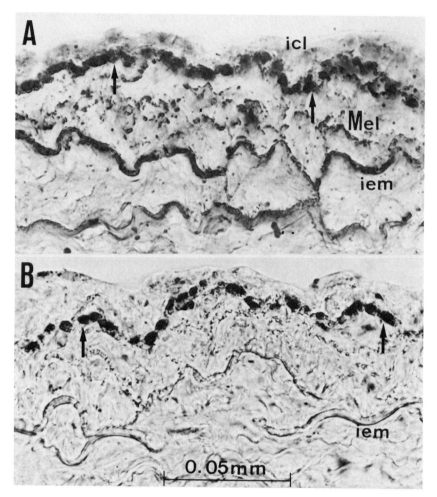

Fig. 80. Cross-sections of the intimal luminal layer of the common carotid artery showing selective deposition of lipids in the sheet of strong elastic bundles. (A) With Weigert's resorcin-fuchsin stain, cross-sections of strong elastic bundles (arrows) are distinctly seen between a thin intimal connective tissue layer (icl) and the intimal musculo-elastic layer (Mel). iem - internal elastic membrane. (B) With Fettrot, the lipid infiltrated, red-stained cross-sections of the strong elastic bundles (arrows) appear black Note the fragments of the internal elastic membrane (iem). 10-year-old boy. Case 51/74. Cause of death: traffic accident. From Meyer and Noll, 1974.

FUNCTIONAL MORPHOLOGY OF ARTERIES 239

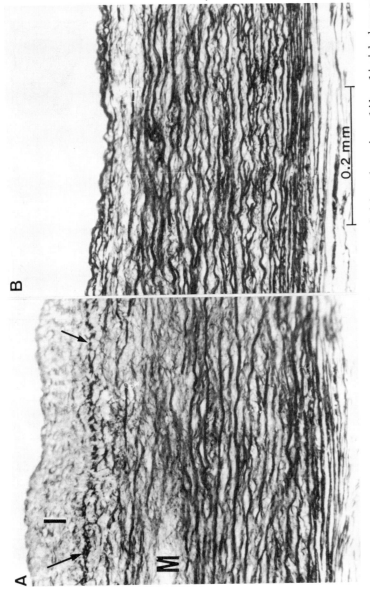

Fig. 81. Cross-sections of the protruding lipid-free area within the ring-like lipid deposit in the carotid sinus (A) and of the arterial wall external to the lipid deposit at the same level (B). (A) In the sinus, the intima (I) is about one-fourth as thick as the media (M). Arrows point to the sheet of longitudinal elastic fibers which extends along the inner aspect of the musculo-elastic intimal layer. (B) In the segment immediately external to the sinus, no intimal thickening is present. 10-year-old boy. Case 51/74. Weigert's resorcin fuchsin stain. From Meyer and Noll, 1974.

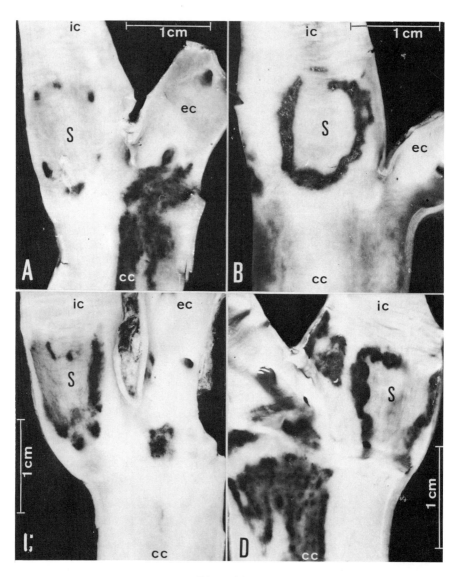

Fig. 82.

←Fig. 82. Gross patterns of early lipid deposits in the carotid sinus.
(A) Note scattered punctate lipid deposits (black) around the periphery of the carotid sinus (S). 13-year-old boy. Case RM 183/74.
(B) A ring-like lipid deposit encircles the central protruding sinus area. 10-year-old boy. Case 51/74.
(C) Extensive lipid deposits are present in the peripheral parts of the sinus. A pale diffuse "blush" is seen in the central sinus area and along the antero-medial wall of the common carotid (cc). 19-year-old woman. Case 1335/74.
(D) Lipid deposits in the periphery of the sinus from a 20-year-old man. Cause of death in all cases: traffic accidents. Fettrot 7 B. S-carotid sinus; ic-internal carotid artery; ec-external carotid artery; cc-common carotid artery. From Meyer and Noll, 1974.

The media of the sinus has the structure of an elastic artery and usually appears to be thinner than the media of the adjacent internal carotid artery, which has the structure of a muscular artery. The total thickness of the wall of the sinus and of the adjacent wall of the internal carotid artery, however, is about the same. This is because the thinness of the medial layer in the sinus is offset by the increased thickness of the intimal layer.

Although considerable intimal thickening is present in the entire sinus area, lipids tend to appear primarily along its periphery and, in the second decade of life, may form a ring-like deposit (Fig. 82). This is somewhat surprising since there are no apparent structural differences between the intima of the periphery which becomes the predominant site of early lipid deposition and the intima of the central sinus which remains lipid-free. The relation between the peculiar ring-like pattern of early lipid deposition in the sinus to penetration by nerves into the sinus periphery still needs to be elucidated.

A close relationship between the extent of lipid deposits and structure is seen at the upper border of the sinus where lipid deposits end abruptly and the border forms a straight horizontal line (Fig. 83). Microscopically, the upper border of the lipid deposits generally corresponds to the transitional zone between the elastic and muscular segments where the intima becomes gradually thinner and the media assumes the structural pattern of a muscular artery (Fig. 84). Thus, lipid deposits are limited to the portion of the elastic segment with a thickened intima and are not found in the adjacent upper muscular segment of the artery.

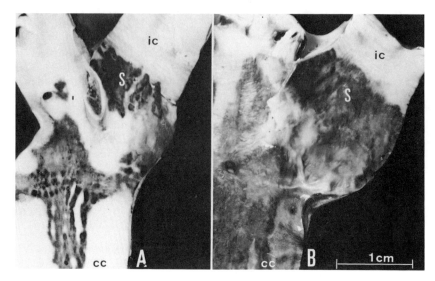

Fig. 83. Advanced lipid infiltration of the carotid sinus (S). Note sharp delineation of the lipid-infiltrated area from the lipid-free portion of the internal carotid artery (ic). Common carotid artery (cc). (A) Carotid artery from a 22-year-old woman (1140/71); (B) Carotid artery from a 46-year-old man (1126/71). From Meyer and Noll, 1974.

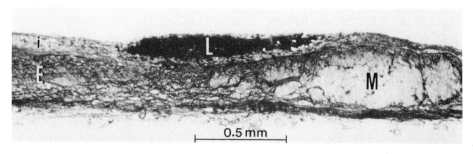

Fig. 84. Lipid deposits in the upper part of the carotid sinus (L, black) are located in the transitional zone between the proximal elastic (E) and distal muscular (M) segments and do not extend into the muscular segment even in later life. Note the pale unstained smooth musculature of the muscular arterial segment (M) and the thickened intima (i) of the sinus. Resorcin-fuchsin and Fettrot stains. 27-year-old woman. Case 118/73. From Meyer and Noll, 1974.

Closing Remarks

Postnatal development of the common carotid artery and the carotid sinus is closely associated with differentiation of specialized structural components which show an increased affinity for lipids and determine the site, shape, and extent of early lipid deposits. It is noteworthy that medial calcification, independent of intimal lipid deposits, also preferentially affects the antero-medial wall of the common carotid artery and extends over approximately the same triangular area as the lipid deposits (Meyer, 1975) (Fig. 85). The appearance of intimal and medial lesions in the same portion of the arterial tube strongly suggests that a common factor, e.g., an increased hemodynamic load, may be responsible for their development.

As in many other arterial segments, lipids appear first in the longitudinal musculo-elastic layer of the carotid trunk, particularly in its prominent luminal sheet of newly-formed strong longitudinal elastic fibers. This layer probably represents a structure which copes with increasing longitudinal tension or shearing forces to which the intima of the common carotid artery becomes exposed during transmission of the pulse wave. Early selective accumulation of lipids in this layer may be regarded as a morphological manifestation of a local metabolic disorder arising in an additional intimal structure which develops as a structural accommodation to an increased mechanical strain.

8. CEREBRAL ARTERIES

The brain is supplied by two pairs of arteries: the internal carotid and vertebral arteries (Fig. 86).

The internal carotid artery arises from the carotid bifurcation, and its cervical portion extends vertically to the base of the skull. It is a typical muscular artery with a well-developed internal elastic membrane that has numerous transverse gaps. The cervical portion of the internal carotid artery enters the carotid canal and forms a tortuous arterial segment, known as the carotid siphon, at the base of the skull. After penetrating the dura, the internal carotid artery joins the circle of Willis and mainly supplies the anterior and middle cerebral arteries.

The vertebral artery arises from the superior aspect of the proximal segment of the subclavian artery. It enters the costo-transverse foramen in the transverse process of the 6th cervical vertebra and extends vertically upward through the foramina of the 5th to the 2nd vertebra (Fig. 86). Thus, the artery is encased in bone in alternate segments.

The vertebral artery, like the internal carotid artery, forms a siphon before entering the cranial cavity. The curves of the

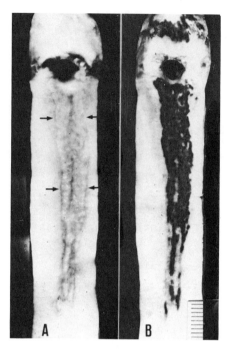

Fig. 85. (A) A grayish strip of medial calcification (arrows) is seen after the von Kossa reaction along the antero-medial wall of the common carotid artery from a 36-year-old man. (B) It extends over approximately the same area as the lipid deposits after subsequent gross lipid staining. From Meyer, 1977.

vertebral siphon, unlike those of the carotid siphon, are prominent early in gestation (Fig. 87) and are not well-protected by bone. Hence the vertebral siphon is exposed to over-distension and compression during lateral rotation of the head. Its tortuous contour may serve to protect the artery from excessive movements. The curved shape of both arterial siphons, however, probably damps the strong pulse wave which might otherwise have disturbing effects on the brain once the sutures and fontanelles close.

Above the perforation of the dura, both vertebral arteries unite to form the basilar artery which joins the circle of Willis. The cerebellar arteries and the posterior cerebral artery are the main branches supplied by the basilar artery.

A distinctive alternating pattern of lipid infiltration occurs in the vertebral artery of adults (Meyer, 1964c). It affects the segments of the vertebral artery that are located between the osseous

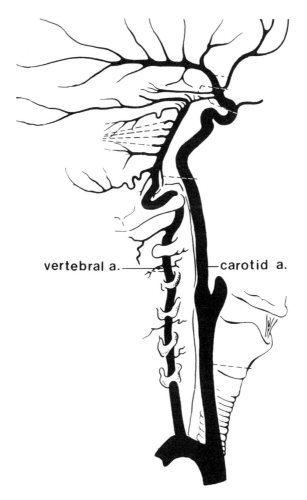

Fig. 86. Diagram showing the carotid and vertebral arteries and their intra-cranial ramifications. Note the course of the vertebral artery through the costotransversal foramina. From Lindenberg, 1957.

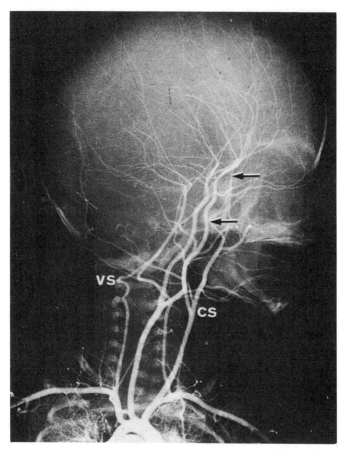

Fig. 87. Angiogram of the cerebral arteries of a 27 cm long and 15 cm CRL fetus. The curvatures of both carotid siphons (between the arrows) are still flat. (x 1 1/4). CS-left carotid sinus; VS-siphon of the right vertebral artery. KS II/3 59. From Meyer and Lind, 1972.

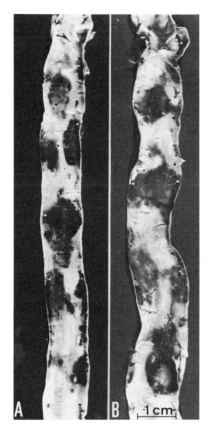

Fig. 88. A and B. Rhythmical pattern of lipid deposits in the vertebral arteries of a 54-year-old woman (A) and of a 78-year-old man (B). The red-stained deposits appear black in the photograph. In older age groups the lesions primarily occur in the dilated segments between the costotransverse foramina.

rings and not encased in bone (Fig. 88). It is uncertain whether there are any structural differences between vessel segments protected by bone and those that are not and, if so, whether they favor the alternate arrangement of lipid deposits. This arrangement may also be related to differences in elasticity of the surrounding tissues at alternating levels and their different capacity to damp the recoil of arterial pulsations.

FUNCTIONAL ASPECTS

Cerebral Blood Flow

No organ except the heart is more dependent upon its supply of blood than the brain.

Before birth, the distinctive pattern of the fetal circulation ensures somewhat better oxygenation of the brain and heart than of the lower part of the body by permitting relatively direct passage of "well-oxygenated" blood from the placenta to these organs through the via sinistra (Figs. 6, 8, 9, p. 13, 15, 16). The measured difference in O_2 saturation between the branches of the ascending and descending aorta is not very great in the mature fetal lamb and monkey since it averages 4 to 9 percent (Dawes, 1968). Preferential supply of oxygen to the brain and heart is also maintained during moderate hypoxemia, hypercapnia, or a combination of both, and during acute asphyxia, e.g., that induced by temporary occlusion of the umbilical cord. Multiple mechanisms are undoubtedly involved but the evidence supports the view that there is a considerable degree of nervous control of the circulation present toward the end of gestation. Local mechanism regulating heart rate, force of contraction, and peripherical vasomotor tone, however, are probably sufficient for survival without central nervous control.

After birth, the pattern of the circulation likewise ensures a better oxygen supply to the heart and brain since, with the first breath, pulmonary vascular resistance falls and right ventricular output now flows preferentially into the pulmonary vascular system. Pulmonary venous return increases, left atrial pressure rises above right atrial pressure, and the valve over the foramen ovale closes functionally, if not anatomically (see pp. 56-58).

Some Factors Affecting Cerebral Blood Flow

The major homeostatic mechanism for regulation of cerebral blood flow appears to be the responsiveness of the cerebral circulation to changes in tension of respiratory gases. In the anesthetized mature fetal lamb, lowering the arterial PCO_2 and raising the PO_2 by ventilation with room air (instead of 7% CO_2 and 3% O_2 in nitrogen) causes a reduction in carotid, sagittal sinus and coronary flow and an increase in femoral arterial flow (Dawes, 1968). The changes in flow in the brain and heart may be due to the local effect of PO_2 and PCO_2. The low O_2 tension in fetal life itself favors an increase in cerebral blood flow and a decrease in cerebrovascular resistance (Lucas et al., 1966) and probably also plays an important role in capillary growth. Indeed, some studies suggest that capillary density is the only anatomical indicator of O_2 consumption of limited regions of the brain so it is not surprising that children residing at high altitude or exposed to chronic hypoxia due to the presence of cardiac defects show an increase in capillary density. Cerebral blood flow is twice as rapid in the fetus as in the adult (in rabbits and rats) and falls by 10 to 45% shortly after birth. The rate of decline appears to parallel the level of the PCO_2.

Not all studies support this view, at least in the rat, since Dahlquist (1976) found no significant difference in mean cerebral blood flow between infant (20-day-old) and adult (3-month-old) animals anesthetized with barbiturate anesthesia which is known to reduce cerebral blood flow in the adult rat (Goldman and Sapirstein, 1973). The discrepancy between these findings and those of other investigators remains to be elucidated. In dogs, blood flow rises two- to three-fold at different times in gray and white matter. In white matter, blood flow reaches a peak at about two weeks and then gradually falls to levels which may even be below those found at birth. In gray matter, blood flow reaches a peak at about six weeks and remains at this level until maturity. In Kennedy's (1970) opinion, the postnatal increase in flow is probably favored by the rise in arterial blood pressure (which would tend to increase perfusion pressure) and fall in hemoglobin with consequent reduction in viscosity (which would decrease cerebro-vascular resistance and increase the flow).

Other factors also influence cerebral blood flow, but their importance during gestation and in the newborn remains to be elucidated. For example, rapid hemorrhage induces a fall in the rate of cerebral blood flow (Barker, 1966), a condition which may arise in association with anomalies of insertion of the placenta (e.g., placenta praevia) and patterns of development of the placental vessels which make them susceptible to damage. It is uncertain whether rapid transfusion has the opposite effect on cerebral blood flow, but blood volume increases by 25 to 50 - 60% in five minutes in full-size infants, if the cord is not clamped until after pulsations have ceased (Usher et al., 1963).

However, adults with polycythemia vera show the lowest values for cerebral blood flow and their blood viscosity lies within the same range as that of newborn infants (i.e., late clamped infants). Severe polycythemia is also encountered in the plethoric twin in the twin transfusion syndrome; it results from vascular communications in the monochorial type of placenta in which one twin transfuses the other, usually by an artery-to-artery or artery-to-vein anastomosis. Vein-to-vein anastomoses are uncommon. The organs (and body size) of the plethoric twin are considerably larger than those of the anemic twin--hence considerable differences in organ blood flow may be assumed to exist.

A rise in intracranial pressure also induces a fall in the rate of cerebral blood flow (Kety et al., 1948), and although no significant change in pressure presumably occurs while the membranes are intact, vaginal delivery after rupture of the membranes causes a rise in intracranial pressure and a fall in cerebral blood flow. The greater the decrease in blood flow, the greater the likelihood that vascular lesions will be produced. These tend to occur in white matter because of differences in vascularity and blood flow rates in different regions of the brain which place this part of the nervous system at a disadvantage.

The rate of cerebral blood flow in the human fetus is unknown, but in lambs cerebral blood flow increases from 50 ml/100 gm/min during gestation to 100/ml/100 gm/min at term. Since the lamb's brain constitutes only about 1.5% of total body weight and the human infant's brain is about 12-13% of total body weight, it may be assumed that the rate of cerebral blood flow in the human is greater.

Cerebral Oxygen Consumption

As the rate of cerebral oxygen consumption has proved difficult to measure, only a few determinations are available. In the rat, cerebral oxygen consumption is relatively low at birth (30-50% of adult values per unit of wet weight). It increases markedly at about one week of age and almost reaches adult values by three weeks of age (Fazekas et al., 1941; Mandel et al., 1957). Garfunkel et al. (1954) found that human infants have lower values than young children, but they made only a few determinations, and in any case, children have a rate of 5.2 ml/100 gm/min which may correspond to 50 percent of a five-year-old child's basal oxygen consumption (Kennedy and Sokoloff, 1957). The rate in the adult is lower, (3.5 ml/100 gm/min (Kety, 1950)), yet the rate of cerebral oxygen consumption in the adult human brain appears to be among the highest of any organ in the body. The rate of cerebral oxygen consumption in the fetal lamb at term is estimated to be about 2.7 ml/100 gm/min, and judging from available studies, it remains relatively constant even during large changes in maternal or fetal blood gas tensions since oxygen extraction compensates for increases or decreases in cerebral blood flow. Cephalic oxygen consumption falls only during a sharp reduction in cerebral blood flow, such as that observed immediately after birth, as the cerebral arteriovenous oxygen content difference remains unchanged.

ANATOMIC-ROENTGENOGRAPHIC CORRELATIONS

Carotid Siphon

The curves of the siphon at mid-gestation are flat and its course is more vertical than in children and young adults (Meyer and Lind, 1972; Kier, 1974) (Figs. 87, 91). A cluster of relatively large branches originates from the cavernous portion of the siphon in fetuses of 24 - 28 weeks of gestation (Kier, 1974). The largest branch emerges from the infero-lateral aspect, the "persistent primitive maxillary artery" (De La Torre and Netsky, 1960). Its branches anastomose with branches from the external carotid artery. In adult man it persists as the lateral meningeal branch of the internal carotid artery.

Circle of Willis

Seydel (1964) studied the circle of Willis in 98 human fetuses and neonates. The internal carotid arteries were symmetrical in 97 cases and an anomaly was present in only one: an absent left communicating artery. The segments of the arterial circle are equal in fetuses of four months' or less gestation, but some differences in size of the various segments appear in the latter part of gestation and in the newborn (Lazorthes et al., 1971). In most instances the asymmetry involves the posterior communicating arteries. Similar results have been reported by Kier (1974), who found a relative bilateral symmetry during the second trimester and observed no anomalies. In contrast, anomalies occur in about 50 percent of adults and are commoner in the posterior part of the circle (Padget, 1944; Stephens and Stillwell, 1969). According to Lazorthes et al., asymmetry and anomalies of the circle do not arise during fetal development, but result from lifelong hemodynamic factors such as compression of the carotid and vertebral arteries by movements of the head and neck. Some of these physiological factors may become permanent as a result of arteriosclerotic changes. There is some question whether this is the sole or even the major mechanism causing asymmetry. The mechanisms are, no doubt, complex and must be related to a variety of factors such as the degree of development of anastomoses of cerebral vessels, flow patterns, pulse waves, closure of sutures and fontanels, etc.

Middle Cerebral Artery

The development of arteries arising from the circle of Willis is closely related to growth and differentiation of individual parts of the brain. This is particularly true of the middle cerebral artery (Kier, 1974). In mammals this artery is enlarged to the extent that it appears to be the main trunk of the cranial division of the internal carotid artery. The enlargement is obviously due to the anatomical and functional dominance of the cerebral hemispheres, a characteristic feature of the evolutionary changes of the mammalian brain which becomes evident in the early stages of embryonic development.

In the human, numerous primitive sulci, which are infoldings of the entire thickness of the wall of the hemispheric vesicle, form along the surface of the rapidly-growing hemispheres during the second and third months of gestation. The sylvian, hippo-campal, parieto-occipital and calcarine sulci remain, but the other sulci

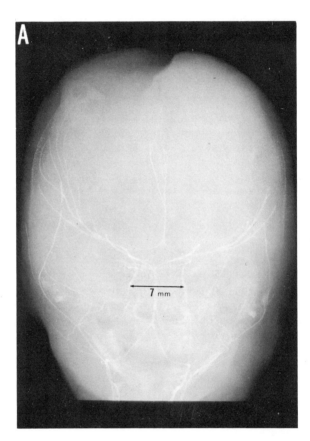

Fig. 89. Frontal roentgenogram of the skulls of injected fetuses. Their age are: (A) 13 weeks; (B) 20 weeks; (C) 28 weeks. As the convolutions develop, the branches of the anterior, middle and posterior cerebral arteries gradually change from a rectilinear to a tortuous pattern. The proximal segments of the anterior and middle cerebral arteries become more horizontal as the brain increases in size. From Kier, 1974.

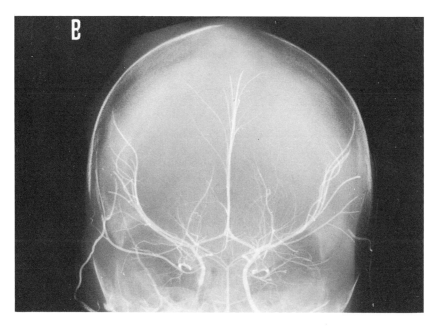

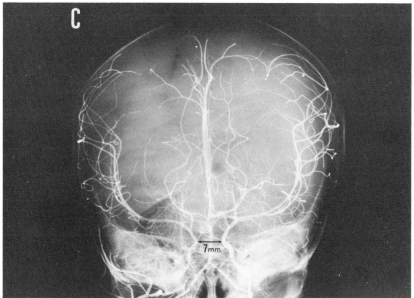

Fig. 89B, C.

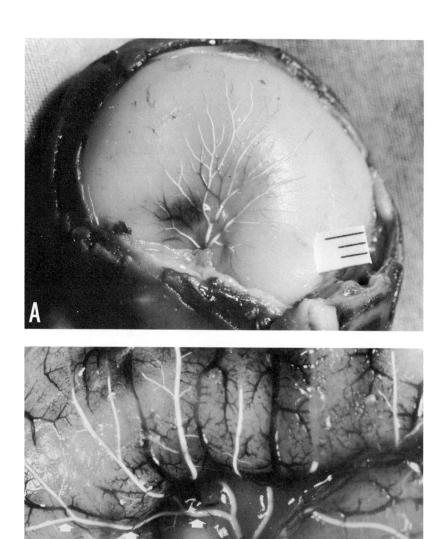

Fig. 90A, B.

disappear during the fourth month, presumably because of more rapid growth of the cranium. The more rapidly enlarging frontal, parietal and temporal areas of the cortical mantle (the opercula) grow over the insula and overhang it to form the sylvian fossa. Permanent sulci develop, which unlike the early sulci, are only depressions of the surface. Most of the principal convolutions and fissures are present before the eighth month of gestation although convolutions and sulci are still short and secondary furrows are not yet developed before term. Nevertheless, the brain of the newborn human infant possesses more convolutions than the brain of an adult orangutan (Larroche, 1966).

During rapid growth of the hemispheres, the first segment of the middle cerebral artery changes its course from an oblique upward forward to a more horizontal direction (Takuku and Suzuki, 1972). At the end of the first trimester in the frontal projection, the branches appear straight and close to the skull (Fig. 89A) whereas in the lateral projection, they show a vertical course and appear rectilinear (Fig. 90A, B; Fig. 91A, B, C). As the opercula enlarges and the sylvian fissure becomes more prominent, a new pattern of the middle cerebral artery and its branches becomes evident. Characteristic curves of the branches develop as a result of growth of the gyri (Fig. 91D) and the distance of the insular branches from the skull is seen to increase in the frontal projection (Kier, 1974) (Fig. 89C).

Anterior Cerebral Artery

With growth of the corpus callosum, a structure which exerts a major influence on the course of this vessel, the artery changes its course from vertical at the end of the first trimester to more anterior and forms a curve to pass around the enlarging genu of the corpus callosum shortly after mid-term (Figs. 91C, D). As the hemispheres enlarge, the horizontal portion of the anterior cerebral artery shifts to a lower position. The peripheral arterial branches lose their straight appearance and become more curvilinear.

← Fig. 90. Changes in the fetal middle cerebral artery result from opercular growth. (A) Lateral view of the cerebral hemisphere of a 14-week fetus. The insular, opercular and cortical segments of the injected middle cerebral artery are entirely visible. The branches are rectilinear and course vertically over the completely smooth brain surface. The dark blood vessels are veins. Millimeter scale is located over the occipital lobe. (B) Close-up view of the insular region of a 24-week-old fetus. The entire orbitofrontal artery (arrows) is visible since the frontal operculum has not yet developed. From Kier, 1974.

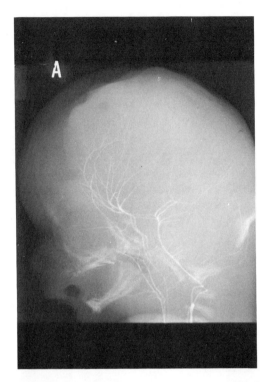

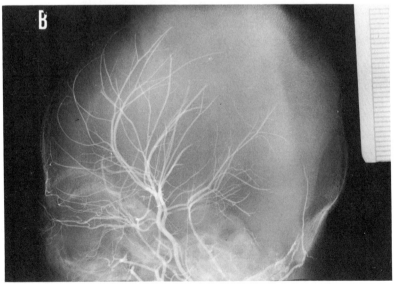

Fig. 91A, B.

FUNCTIONAL MORPHOLOGY OF ARTERIES 257

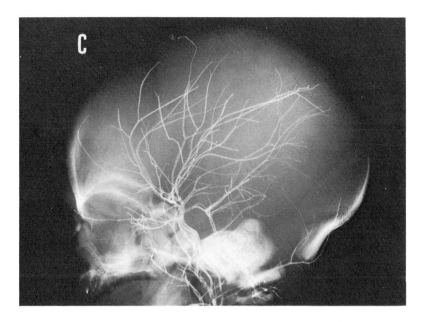

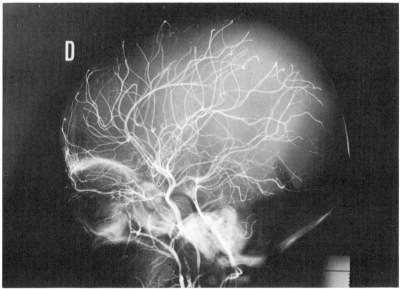

← Fig. 91. Lateral roentgenograms of the skulls of injected fetuses. Their ages are: (A) 13 weeks; (B) 20 weeks; (C) 24 weeks; (D) 28 weeks. With development of the brain, the angle of the pericallosal artery decreases and depresses the middle cerebral artery axis. Note the flat and vertical carotid siphon. From Kier, 1974.

The Posterior Cerebral Artery and The Vertebro-Basilar System

In submammals, i.e., in fish, amphibians, reptiles, and birds, the carotid arteries are the only source of blood to the brain since the basilar and spinal arteries are supplied by the carotid circulation. In these four classes of vertebrates, the internal carotid arteries divide into a cranial and a caudal division. In the human embryo, at the 4 mm stage (24 days), the caudal division of the internal carotid artery terminates in the mesencephalon. At the 5 to 6 mm stage (30 days), the caudal division anastomoses with the longitudinal neural artery at the mesencephalon and becomes the definitive posterior communicating artery that replaces the primitive trigeminal artery. The posterior cerebral artery is probably formed by a branch of the posterior communicating artery which supplies the mesencephalon and remains prominent for some time during fetal development.

In mammals, the carotid circulation is unable to meet the requirements of the fast-growing forebrain, the enlarging cerebellum, the pons and crura cerebri. Pre-existing anastomoses between the basilar and vertebral arteries, which already exist in the salamander and frog, enlarge to form the vertebral arteries that enter the skull through the foramen magnum. They appear in the order of monotremes (platipus, spiny ant-eater), the lowest sub-class of mammals.

As the cerebellar hemispheres enlarge in the human fetus, trunks of the anterior inferior and posterior inferior cerebellar arteries, arising from the basilar artery, appear. The developing tonsils displace the posterior inferior cerebellar arteries, changing their course from oblique to more vertical in the adult. The wide variations in origin, course, and territorial supply of branches arising from the vertebral and basilar arteries has been documented by Smaltino et al. (1971). According to Padget (1948), this variability relates to persistence of the primordial vascular plexus in the region of the hindbrain and to the late development of the cerebellum.

MORPHOLOGY

Carotid Siphon

The carotid siphon represents the tortuous distal segment of the internal carotid artery which is located at the base of the skull and consists of four superimposed curves (Figs. 92, 93). The first curve is formed at the level of the external opening of the carotid canal where the artery enters the bone. The artery then extends horizontally along the carotid canal in an antero-medial direction. This is the petrous part of the siphon (Fig. 93). Here the arterial tube is surrounded by the venous plexus and the dura, and it is, therefore, not in direct contact with bone. Above the internal opening of the carotid canal, the artery describes a second curve

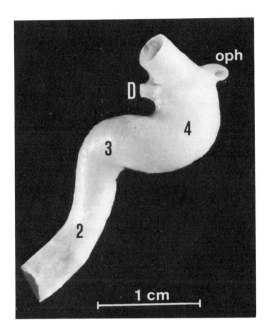

Fig. 92. Carotid siphon of a four-year-old boy which has been dissected free after fixation in formalin at a filling pressure of about 90 mm Hg. The numbers refer to the curvatures. The first curvature was removed and is not seen in the figure. The small adherent part of the adjacent dura mater (D) indicates the level of its perforation by the siphon. Oph - cross-sectioned ophthalmic artery.

upward and enters the sinus cavernosus. The cavernous part of the siphon begins here. The artery then passes vertically along the body of the sphenoid bone, and on reaching the posterior clinoid process, forms a third curve. The artery continues horizontally forward along the carotid groove of the sphenoid bone and is slightly inclined upward or downward. In the anterior portion of the carotid groove, the artery changes its direction and forms, around the latero-anterior aspect of the anterior clinoid process, a sharp fourth curve at an angle of 180°. The artery then perforates the dura, passes posteriorly beneath the optic nerve, and forms an additional curve before dividing into the anterior and middle cerebral arteries.

Development of Microscopic Structure

The tortuous shape of the carotid siphon seems to play a role in the development of conspicuous structural characteristics of the wall, and in particular, in pronounced differences in structure of opposite parts of the curves (Fig. 94). The differences arise during fetal life and appear to be closely related to the development of increasing tortuosity during the second half of gestation (p. 157).

In <u>newborn infants</u>, intimal thickenings or cushions consisting of densely packed elastic fibers are regularly present along the <u>inner</u> concave wall of the curves (Fig. 95). The rest of the circumference is encircled by a prominent continuous internal elastic membrane. At the junction with the cushion, the membrane seems to "split" into two or more lamellae. The stronger outer lamella separates the cushion from the media. The inner lamella fans out into the elastic networks of the cushion. At this age, intimal cushions are particularly prominent in the most tortuous portion of the siphon, i.e., in the third and fourth curvatures. The media consists mainly of circular musculature with numerous tiny elastic fibers. The adventitia contains fairly well-developed fine elastic networks.

<u>During the first year of life</u>, the carotid siphon enlarges considerably and shows a two-fold increase in diameter. The increase

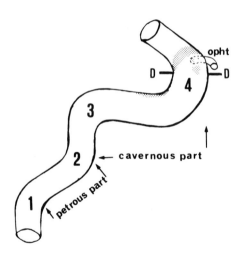

Fig. 93. Diagram of the right carotid siphon showing the sites of predilection of early calcification (dotted area) and indicating the petrous and cavernous portions of the internal carotid artery. The numbers refer to the curvatures of the siphon. D-D - dura mater, oph - origin of the ophthalmic artery. From Meyer and Lind, 1972.

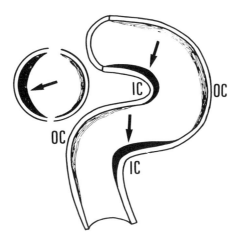

Fig. 94. Diagram showing a longitudinal section of the right carotid siphon and illustrating the structural differences between the opposite walls of the curves and the location of intimal cushions. In childhood, the subendothelial elastic layer is well developed along the outer walls (OC) of the curvatures, whereas prominent intimal cushions (arrows), consisting mainly of elastic elements and connective tissue, are present along the inner walls (IC) of the curvatures. Inset shows a cross-section of the vessel at the site of one of the arrows.

in caliber is associated with strengthening of intimal elastic structures. The subendothelial elastic layer now consists of two or more membranes, an outer homogeneous primary internal elastic membrane that separates the intima from the media, and a thicker inner secondary membrane. The intimal cushions along the <u>inner</u> walls of the curves become larger and thicker (Fig. 96). They now consist of concentric elastic sheets which are more densely packed near the lumen. These sheets are continuous with the hyperplastic secondary elastic sheets which cover the primary internal elastic membrane in the adjacent parts of the circumference. Small bundles of smooth muscle cells and some loose collagenous networks are present between the elastic sheets of the cushions. The primary internal elastic membrane, which separates the cushions from the underlying media, often looks stretched. Beneath the cushion, the media is thinner than in adjacent sectors. In the opposite sector, i.e., along the

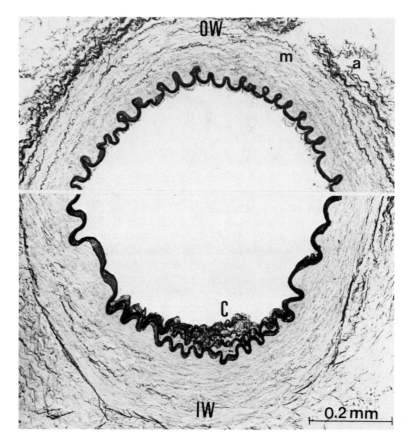

Fig. 95. Carotid siphon of a newborn term infant. Note small intimal cushion (C) consisting mainly of fine elastic fibers along the inner concave wall (IW) of the fourth curvature. In the opposite sector (above), i.e., along the outer wall (OW) of the curvature, there is only a wavy internal elastic membrane below the endothelial lining. m - media; a - adventitia. Case 46 K/76. Cryostat section. Weigert's resorcin fuchsin.

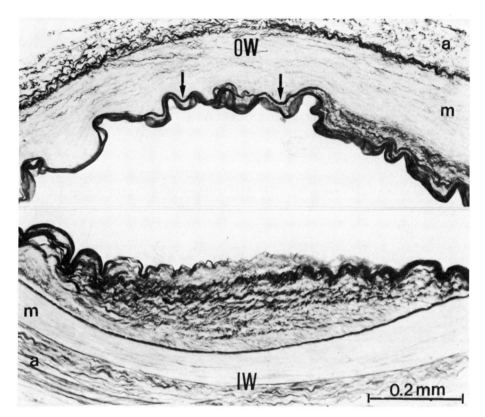

Fig. 96. Structural differences between opposite sectors of the upper fourth curvature of the carotid siphon from a one-year-old boy. Along the inner concave wall (IW) of the curve, a prominent intimal cushion (i) has developed internal to the stretched internal elastic membrane (arrows). It consists mainly of elastic sheets. The media (m) appears thin beneath the cushion. Along the outer convex wall (OW) of the curve, a thick, secondary elastic, intimal layer is seen internal to the thin internal elastic membrane (arrows). Case 28 K/76. Cryostat section. Weigert's resorcin fuchsin.

outer wall of the curves, flat cushions are also common. They consist of fine elastic networks covered by hyperplastic subendothelial elastic membranes. In children, these cushions never attain the same size and thickness as the cushions along the inner wall of the curves.

By the age of four to five years, the intimal cushions along the inner walls of the curves have become larger. The elastic sheets have coalesced to form a thick limiting membrane beneath the endothelial lining. The primary internal elastic membrane at the base of the cushion remains thin and contains small gaps. The center of the cushion contains a moderate number of elastic elements, loose collagenous networks, and ground substance. In the opposite sector,

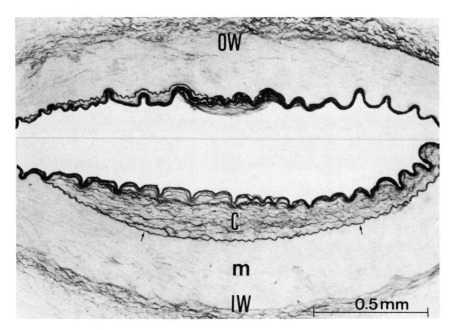

Fig. 97. Structural differences between opposite sectors of a carotid siphon from a 13-year-old boy. Along the inner concave wall (IW) of the fourth curvature (below), a prominent intimal cushion (C) rich in elastic fibers is seen internal to the fine primary internal membrane (arrows). Along the outer convex wall (OW), a thick hyperplastic secondary elastic layer is present internal to the fine wavy primary internal elastic membrane (arrows). m - media. Case 58 K/76. Cryostat section. Weigert's resorcin fuchsin.

i.e., along the outer wall of the curves, the primary internal elastic membrane is covered by a thick hyperplastic secondary elastic sheet which has assumed the homogeneous appearance of an elastic membrane. Numerous fine elastic fibers and small longitudinal muscle bundles are commonly present between both membranes.

Between 5 and 15 years of age, the intimal cushions increase further in size along the inner wall of the curvatures, and the loose collagenous tissue increasingly becomes the major structural component. The elastic networks, which were initially tightly packed, gradually become more loosely separated over the enlarging cushions. In the adjacent and opposite sectors, the secondary subendothelial elastic membrane has become three to five times thicker than the underlying primary internal elastic membrane which shows little change in structure and thickness (Fig. 97).

In young adults, the intimal cushions often show a more complex structure and consist of two superimposed layers. In cross-sections, both layers are crescent-shaped. Sometimes the luminal layer is shifted laterally, and then extends beyond the deeper component of the cushion. The deeper layer of the cushion, i.e., that near the media, is usually richer in elastic elements than the luminal layer which consists mainly of loose collagenous networks embedded in abundant ground substance. Both layers are usually distinctly separated from each other by elastic sheets or membranes. The development of an additional luminal layer in the intimal cushion may be related to dilatation and increasing tortuosity of the curved arterial segment which are already present at this age. The associated distortion of the initial curved shape of the arterial tube probably results in a change in hemodynamics which may stimulate intimal proliferation above the original intimal cushions.

Predilection Sites of Early Calcific Deposits

The strong subendothelial elastic networks in the outer convex and lateral walls of the curves show a great affinity for calcium and become a predominant substrate for gross calcifications in early life. Calcific deposits appear first along the outer wall of the upper fourth curve, the most prominent curvature of the carotid siphon where blood flow is deflected through an angle of 180°. They are grossly visualized, after the von Kossa reaction, above the orifice of the ophthalmic artery, i.e., in the part of the curve which is located just above the perforation of the dura mater by the carotid siphon (Figs. 93, 98). The deposits are either punctate or linear and often arch over a large part of the circumference. The extent and severity of involvement increases with age, but remains about the same in the left and right siphons. By three to five years of age, coarse, plaque-like calcific deposits are often present above the orifice of the ophthalmic artery. To date, gross calcific

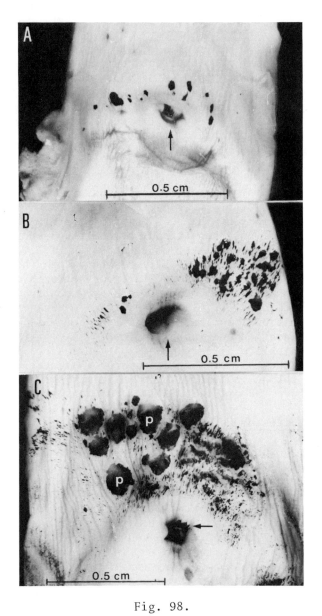

Fig. 98.

←Fig. 98. Gross demonstration of early calcifications in the carotid siphon using von Kossa's reaction. (A) Early scattered calcific deposits (black) are present above the orifice of the ophthalmic artery (arrow) in a 13-month-old child. (Case 26/70). (B) A part of the arterial circumference above and lateral to the orifice of the ophthalmic artery (arrow) is interspersed with polygonal and linear calcific deposits. Eight-year-old girl. Cause of death: traffic accident. (C) The entire circumference of the carotid siphon above the orifice of the ophthalmic artery shows extensive calcific deposits. Larger calcific plaques (F) are partly covered by grayish connective tissue. 10-year-old boy. Cause of death: accident. From Meyer and Lind, 1972.

deposits have been demonstrated at this site in all children from 1 - 16 years of age (Meyer and Lind, 1972a). Thus, selective gross calcifications of the upper fourth curve of the carotid siphon may be regarded as a conspicuous early lesion of the human arterial system. With more pronounced calcification, smaller deposits are also found below the ophthalmic orifice, i.e., in the lower part of the anterior convex wall of the siphon that extends along the carotid groove of the sphenoid bone.

Later, calcifications also appear in the more proximal portion of the siphon. In young adults they become visible around the prominent intimal cushion which develops along the inner "concave" wall of the fourth curvature and remains free of calcific deposits for years (Fig. 99).

The upper distal part of the third curvature, where blood is diverted through an angle of at least 90°, is another site of early calcification. Pronounced calcific deposits along the upper as well as the adjacent lateral and medial walls of this segment are common by the age of 25 to 35 years (Fig. 99).

Microscopically, calcific deposits seem to involve the primary internal elastic membrane first and are then limited to the crests of the membrane. In most lesions, the thick secondary elastic layer which covers the primary membrane, also becomes calcified. With further calcification the wavy pattern of the membrane is lost (Fig. 100A). Extension of calcific deposits into the underlying media seems to be an early change (Fig. 100B, C). Larger deposits often become covered by a thickened proliferated intima (Fig. 101). Microradiographs (Fig. 102) show that the pattern and extent of calcification is not altered by the von Kossa reaction although it does, however, produce better opacification of the lesions.

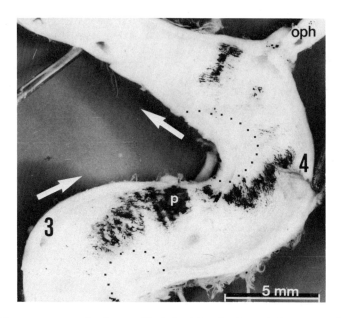

Fig. 99. Upper part of the medial half of the right carotid siphon from a 17-year-old male who died in a traffic accident. After von Kossa reaction, confluent axially arranged calcific deposits (black) are seen above the orifice of the ophthalmic artery (oph). No calcifications are present in the peak of the concave inner wall of the fourth curvature (4), i.e., in the area of slightly raised intimal cushion (dotted line in the center of the figure). Tightly packed confluent linear incrustations are seen below and in front of this cushion. Proximal to the peak of the fourth curvature, there is a large calcific plaque (p) which is partly covered by grayish connective tissue. Numerous fine, coalescent calcific deposits are seen below the plaque. Gross staining lipid shows only faint reddish tinge, i.e., early lipid infiltration limited to the area of both intimal cushions located at the inner walls of the third and fourth curvatures (inside the dotted lines). Arrow indicates the direction of blood flow. Case 43/72.

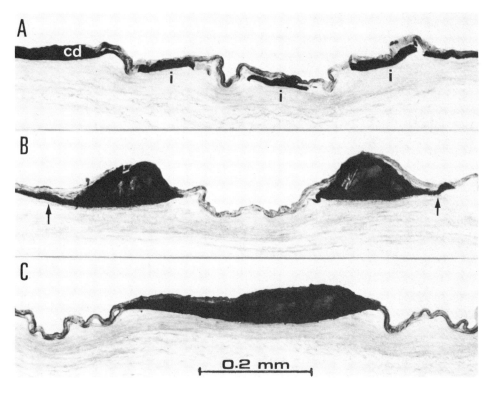

Fig. 100. Microscopic demonstration of calcific deposits in the carotid siphon of a three-year-old boy who died in a traffic accident. (A) Early calcific deposits (i, black) appear primarily in the interrupted primary internal elastic membrane although a few deposits may extend into the overlying secondary elastic sheet (cd). (B) and (C) With more extensive involvement, both layers and the adjacent media usually become involved. Along the periphery of the plaques, however, only the primary internal elastic membrane may be calcified (arrows). Von Kossa reaction and Gomori's aldehyde fuchsin stain for elastic tissue. From Meyer and Lind, 1972.

The calcium content in even the smallest von Kossa positive lesion can also be demonstrated with Voigt's method (1957). With this technique, yellow anisotropic crystals are formed within the lesion after exposure of microscopic sections to a solution of N, N-naththalyl-hydroxylamine sodium salt (Fig. 103). Gross alizarin staining reveals patterns of calcification which are identical to those demonstrated with the von Kossa reaction in the contralateral carotid siphon. Numerous gypsum crystals appear in the deposits when concentrated sulphuric acid is placed on the microscopic sections.

Larger calcific deposits protrude into the arterial lumen which may be partly due to postmortem retraction of the arterial tube. The deposits are covered by a connective tissue layer, however, and may be raised above the adjacent flat intima. This would tend to produce marked deformation of the luminal surface even in vivo.

Lipid Deposits

Lipid deposits develop later than primary calcifications and become visible after gross lipid staining in the upper part of the siphon toward the end of the second decade. They are usually found mainly in the intimal cushions located along the inner walls of the curves. Thus, their localization initially differs entirely from

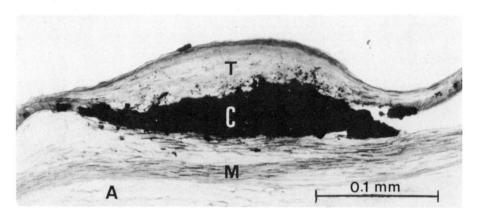

Fig. 101. At times, large calcific deposits (C) along the wall of the carotid siphon become covered by a thick layer of connective tissue (T). M - media, A - adventitia. 12-year-old boy. Cause of death: accident. Von Kossa reaction. From Meyer and Lind, 1972.

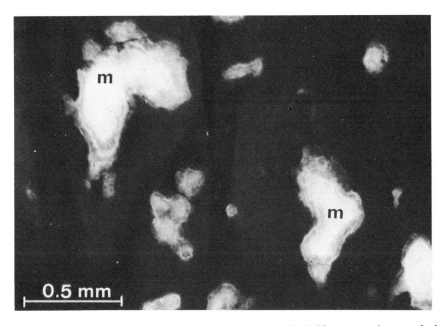

Fig. 102. Numerous calcific deposits (m) of different size and density are seen in a microradiograph of the wall of the carotid siphon using "Micro 60" (manufactured by C.H.F. Müller, Hamburg). Three-year-old boy. Case 871/71. Cause of death: traffic accident. From Meyer and Lind, 1972.

that of early calcifications. Even when pronounced, calcifications do not extend into the cushions although they may completely encircle the lipid infiltrated cushion (Fig. 99).

Intimal cushions later become the site of severe atherosclerosis. Secondary calcifications are then commonly found in the thickened and degenerated intima. Hyperplastic subendothelial elastic sheets, which initially represent the predominant substrate of primary calcifications, also become the site of pronounced lipid infiltration. Toward the end of the third and in the fourth decade of life, calcific and lipid deposits tend to involve the same portions of the siphon and seriously affect both of its upper curves.

It is uncertain whether the attachment of the siphon to the surrounding hard tissue, i.e., bone and dura, plays a role in the development of lesions in this segment. The siphon is largely surrounded by the venous sinus which probably acts as a hydraulic suspension system and damps the recoil arising with transmission of the pulse wave. During growth, the siphon is, therefore, probably sufficiently protected from rebound of the pulse wave. With subsequent increasing enlargement of the arterial tube, the distance between the artery and its hard encasement may decrease and the recoil becomes more effective. Pronounced calcifications, at least in younger age groups, however, seldom appear in the petrous portion of the siphon which passes through an osseous canal and where the influence of the surrounding bone should be most effective. In contrast, it is the upper part of the siphon, which is only partly attached to the bone but is more tortuous, that usually becomes the site of marked calcifications and atherosclerotic lesions. Likewise early calcifications of childhood develop in the uppermost tortuous segment of the siphon which is located outside bone. Thus, the curved

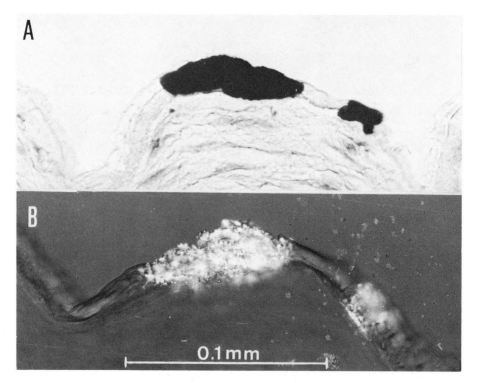

Fig. 103. Demonstration of early calcific deposits in the internal elastic layer of the carotid siphon using von Kossa's and Voigt's methods. (A) Calcific lesions stain black with von Kossa's method. (B) They become anisotropic with Voigt's method in the following cryostat sections. Two-year-old girl. From Meyer and Lind, 1972.

shape of the siphon and associated special hemodynamic features seem to be more important in the development of lesions than the presence of hard tissue to which the siphon is attached.

Cerebral Arteries

The anatomical pattern of the brain vasculature differs considerably from that of other viscera. In the spleen, kidney, and other organs the arteries ramify, after entering the hilus and fan out in arboreal fashion. In contrast, the cerebral arteries arch over the surface of the brain and ramify along the fissures and sulci. Before penetrating the brain, the arteries extend a considerable distance and show little decrease in caliber. The walls of the cerebral arteries appears thinner and more transparent than those of arteries supplying visceral organs. Unlike extracranial and meningeal arteries, they are not accompanied by veins, and venous blood is drained by superficial veins merging into the dural sinuses or via the subependymal veins that join the great vein of Galen.

The microscopic structure of the cerebral arteries differs from that of other arteries and is characterized by the following features (Binswanger and Schaxel, 1917; Thoma, 1923; Hackel, 1927; Wolkoff, 1933; Rotter et al., 1955; Stehbens, 1974):

1. The internal elastic membrane is thicker than in extracranial arteries of comparable size, but as in other arteries, it is interrupted by numerous round or oval-shaped fenestrae. The membrane in larger cerebral arteries is homogeneous but seems to be formed by axially arranged elastic fibers in smaller branches. Larger circular gaps, which are common in the internal elastic membrane of larger extracranial arteries, do not seem to occur in cerebral arteries.

2. The media appears thinner and consists of circularly disposed smooth muscle cells which probably assume a slightly spiral course (Strong, 1938). The number of muscular layers forming the media depends on the caliber of the vessels. There are fewer elastic fibers than in the media of other arteries.

3. In contrast to the extracerebral arteries, the adventitia is thin in cerebral arteries and poor in elastic fibers which are fairly evenly distributed throughout the collagenous networks. There is no external elastic membrane. The comparatively poor development of the elastic structures in the adventitia is probably due to the position of the cerebral arteries, which in contrast to the extracranial arteries, are not exposed to the additional longitudinal stretch or compression arising from movements of the body, i.e., factors which are responsible for abundant development of elastic elements in the adventitia of other arteries.

4. There are no noteworthy differences in the structure of intracranial arteries penetrating the base of the skull and cerebral arteries, apart from more marked development of the adventitia in the proximal segments.

The two most characteristic features of the cerebral arteries, i.e., well-developed internal elastic membrane and small amount of elastic tissue in the adventitia may, like the tortuous pattern of the siphons, help to damp the pulse wave and reduce its effect on brain tissue.

Intimal Pads

Intimal thickening develops at sites of bifurcations and branchings of cerebral arteries during fetal life and postnatal growth (Hackel, 1927; Wolkoff, 1933; Rotter et al., 1955; Stehbens, 1960), a finding which has also been noted in other arteries. The microscopic structure and development of these cushion-like thickenings or "pads" in the cerebral arteries has been thoroughly studied by Stehbens (1960, 1972). The "facial" and "dorsal" pads of the uppermost portion of the trunk appear first and are present at 26 weeks of gestation. "Lateral" pads, i.e., intimal thickenings located "*immediately below the site where the lateral wall curves into the proximal side of the daughter branch*" appear next, and the apex of the bifurcations is involved last. In fetuses, the pads consist predominantly of muscle cells and fine elastic fibers which seem to be longitudinally arranged with only a small amount of collagen. The pads are often separated from the media by a thinned elastic membrane and a new "incomplete" elastic sheet is present beneath the endothelium.

During the first years of life, the thickness and extent of the pads increase. By the age of seven years they have coalesced and there are no longer any individual pads. The largest intimal cushion is still present at the original site of the "facial," "dorsal," and "lateral" pads, even in youth (Stehbens, 1972). With growth, the amount of elastic tissue in the pads increases, and numerous new elastic sheets form. They are usually not homogeneous, but fibrillary. The internal elastic lamina beneath the pads is thin, stains poorly, and often has a straightened contour. Elsewhere, the internal elastic lamina is thickened and "beading" is common. The pads now contain increasing amounts of collagen and metachromatic ground substance but fewer muscle cells. In older individuals, the lateral pads become thicker and often consist of several layers which are separated by new elastic laminae. In contrast, the intima at the apex remains comparatively thin.

Like intimal cushions at bifurcations and orifices of branching arteries in other arterial provinces, the pads of the cerebral arteries may be regarded as a normal anatomical differentiation of the arterial wall. They presumably develop in response to the greater mechanical load on these parts of the arterial tube. There is no evidence that they result from a degenerative process.

Aside from intimal cushions at bifurcations and orifices of branching arteries, the intima of the cerebral arteries also becomes diffusely thickened with age. This is followed by the development of fibrosis and hyalinization associated with a decrease in cellularity and the appearance of lipophages. The latter changes are usually encountered by the fourth or fifth decades of life.

Medial Gaps

Forbus (1930) first drew attention to discontinuities or gaps in the medial coat at bifurcations of cerebral arteries. In his opinion the defects were inherited sites of mechanical weakness that might be responsible for the development of arterial aneurysms. Based on his findings, a "congenital" or "developmental" etiology of cerebral aneurysms was formulated which has not been substantiated. In a comprehensive study, Stehbens (1959, 1963, 1972) clarified the morphology, the commonest sites, the frequency, and the evolution of medial gaps with age. He showed that muscular gaps are not congenital lesions but are acquired after birth, and that the role of medial gaps is not one of prime importance in the etiology of aneurysms.

In infants, medial gaps are wedge-shaped, and according to Stehbens, the medial musculature is substituted by adventitial tissue which seems to be "invaginated" and may extend to the internal elastic membrane. At the apices of bifurcations, the collagen within the medial gaps extends longitudinally along the distal sides of the branches. There is no evidence that medial gaps themselves are areas of undue "inherent" weakness in the arterial wall. The histological appearance, whether of collagen or elastic tissue, provides no accurate indication of their tensile strength.

The apices of the bifurcations and the lateral angles are the most frequent sites of muscular gaps. At the apex, medial gaps are much more frequent in the middle cerebral artery than in the internal carotid artery, and conversely, those at lateral angles are more frequent in the latter than in the middle cerebral artery. Medial gaps also occur at the basilar bifurcation and at the forks of cerebral, cerebellar, and spinal arteries. Medial gaps have also been found in human coronary, mesenteric, renal, and splenic arteries (Forbus, 1930; Stehbens, 1959; Hassler, 1962), and in arteries of different animals (Stehbens, 1972). On the other hand, medial gaps do not seem to occur in elastic arteries (Hassler, 1963).

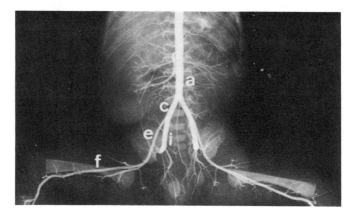

Fig. 104. Postmortem angiogram of a newborn infant. The common (C) and internal (I) iliac arteries appear wide, while the external (E) iliac and femoral (F) arteries are comparatively narrow. A - abdominal aorta. (From Meyer and Lind, 1972).

Medial gaps at the apices of forks increase in number with age. The lateral angle gaps are usually acquired after birth since none have been found in fetuses or infants less than six months of age. After this time, medial gaps are rarely, if ever, absent in human cerebral arteries (Carmichael, 1945; Stehbens, 1959). According to Stehbens (1959), the increase in frequency and size of medial gaps after birth suggests that they may be "degenerative lesions," a view which is not altogether convincing, since it is hard to believe that an early developmental change is pathological. Evidence against the probability that medial gaps are the commonest sites of cerebral aneurysms is provided by the differences in location and microscopic substrates present in early phases of aneurysmal formation (Stehbens, 1972).

9. ILIAC ARTERIES

The common and internal iliac arteries are of particular interest because they connect the aorta to the umbilical arteries in fetal life and are, therefore, subject to the full brunt of the large blood volume directed to the placenta. By dint of their more distal location from the heart, they may also be subject to higher peak pressures and peak flow velocities than more centrally located arteries (Hurthle, 1934; Wehn, 1957). These factors should tend to become of increasing importance in later gestation since intravascular pressures are increasing.

The large blood flow to which the common and internal iliac arteries are exposed during gestation results in accelerated growth and widening of the common and internal iliac arteries as compared with other fetal vessels. Angiograms of human fetuses of 14 to 20 weeks gestational age show that the internal iliac artery is considerably larger than the external iliac artery which supplies the lower limb (Fig. 104). Direct measurements reveal that the mean circumference of the internal iliac artery at term is nearly twice that of the external iliac artery, i.e., 7 and 4 mm, respectively. With cessation of the placental circulation and closure of the umbilical vessels, the difference in size begins to alter and the ratio is reversed by the end of the first decade or possibly before this (Fig. 105, Table 3).

TABLE 3.

Circumferences of Abdominal Aorta, Common, Internal, and External Iliac Arteries in Children

Age Groups	No. of Cases	Aorta (above bifurcation)	Common Iliac Artery	Internal Iliac Artery	External Iliac Artery
Newborns	10	12.9±0.2	7.7±1.0	7.2±0.7	4.1±0.4
6 dy-6 wk	7	12.4±1.6	7.5±1.4	6.2±0.4	4.4±0.9
3-8 mth	5	12.5±0.4	8.0±0.3	6.1±0.2	5.7±0.5
10-18 mth	6	14.3±0.7	8.8±0.6	6.6±0.6	7.1±0.5
3-5 yr	7	17.0±2.0	10.9±0.7	7.4±0.3	8.7±0.9
7-12 yr	6	21.4±2.2	15.6±2.3	9.3±1.0	11.3±1.7

NORMAL STRUCTURE

The common and internal iliac arteries are elastic arteries and their media, therefore, consists predominantly of elastic membranes which alternate with well-developed smooth musculature. In both vessels, the number of smooth muscle cells between elastic lamellae increases with postnatal growth but the structural pattern remains unchanged. In contrast, the external iliac artery is a muscular artery (pp. 219-221). The transition from an elastic to a muscular structural pattern occurs at the origin of the external iliac artery and can be seen grossly after postmortal contraction because transverse structures form along the luminal surface of muscular arteries (pp. 210-213).

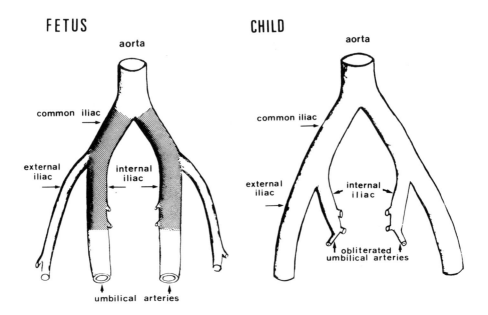

Fig. 105. Diagram showing postnatal changes in the relative size of the iliac arteries. In the fetus and newborn, the common and internal iliac arteries are comparatively large because they transport a large volume of blood to the umbilical arteries. After birth, the internal iliac artery supplies blood to the pelvis and neighboring tissues while the external iliac artery supplies an increasingly large volume of blood to the lower extremities. This leads to a reversal in the size ratio of the internal and external iliac arteries. (Dotted parts indicate the selective localization of early calcifications.) Modified from Walsh, Meyer and Lind, 1974.

STRUCTURAL FEATURES IN ILIAC ARTERIES DETECTED BY DEMONSTRATION OF EARLY CALCIFIC DEPOSITS

The rising hemodynamic load during intrauterine life and the associated more rapid increase in the caliber of the common and internal iliac arteries result in a higher total tension on the wall, since tension varies with the product of the blood pressure and the radius of the vessel (law of Laplace). The rapid increase in caliber and tension probably lead to structural changes which favor the early and selective development of calcific deposits in

FUNCTIONAL MORPHOLOGY OF ARTERIES

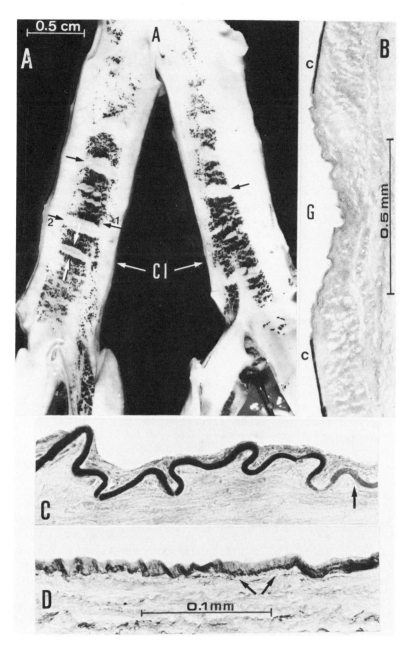

Fig. 106. (Caption follows on next page).

Fig. 106, p.279 (A) Gross demonstration of calcifications (black) in the iliac arteries of a seven-month-old infant. In the common iliac arteries (CI), which have been opened along their medial walls, calcifications are located predominantly in the middle of the luminal surface, i.e., along the lateral arterial wall. Note many calcium-free transverse strips in the midst of the calcifications (arrows) and larger calcium-free areas along both sides of the calcified parts. A - bifurcation of the aorta.
(B) Longitudinal frozen section of the part of the artery indicated by white arrows in (A). The calcium-free strip corresponds to a wide gap (G) in the primary internal elastic membrane. The calcified edges (black, c) of the interrupted membrane are seen above and below the gap. Von Kossa reaction and Gomori's aldehyde fuchsin.
(C) Cross frozen section of the calcified area indicated by arrow 1 in (A). Note pronounced calcification of the primary internal elastic membrane (black). A portion of the membrane is not calcified (arrow). Von Kossa reaction and Gomori's aldehyde fuchsin. Same magnification as in (D).
(D) Cross-section of the same artery at the level indicated by arrow 2. The subendothelial or subintimal elastic layer consists of confluent longitudinal elastic fibers, dot-like cross-sections of which are seen in the area indicated by arrows. There is no primary internal elastic membrane here or in other calcium-free areas. Seven-month-old infant. Weigert's resorcin fuchsin elastic stain. (All figures from Meyer and Ehlers, 1972.)

the common and internal iliac arteries. Such deposits have regularly been demonstrated grossly with the von Kossa reaction in most infants and children (Meyer, 1968, 1971; Meyer and Ehlers, 1972; Meyer and Lind, 1972b). Since the calcifications form distinct patterns, preformed structures of the arterial wall may be responsible for their distribution along the arterial luminal surface.

Common iliac artery. Calcific deposits tend to develop along the lateral arterial wall and, if the artery is opened along its medial aspect, scattered deposits can be seen along the central portion of the luminal surface. When the lesions are extensive, they become confluent and form a longitudinal strip with numerous well-delineated, unstained transverse bands within the calcified areas (Fig. 106). The arterial luminal surface on both sides of the longitudinal strip likewise tends to remain free of calcific deposits.

Internal iliac artery. Calcific deposits are distributed along the dorso-medial wall, which corresponds to the outer curvature of the iliac-umbilical arch. With extensive calcification, a common finding in this vascular segment, the lesions form a longitudinal strip along the dorso-medial wall that is usually well-delineated from adjacent calcium-free areas (Fig. 107). Most branches of the internal iliac artery originate from the dorso-medial wall, and their orifices are seen in the midst of black-stained incrustations (Fig. 107C).

The selective development of early calcific deposits in specific parts of the arterial tube suggests that there may be some underlying structural differences between affected and unaffected areas of the wall. Microscopic examination shows that calcific deposits in the common and internal iliac artery form only in the "primary" internal elastic membrane. This membrane develops in early gestation at the junction of the intima with the media and becomes a strong homogeneous elastic structure at birth. It evolves before new elastic sheets arise along the inner aspect of the internal elastic membrane. It, therefore, seems justified to regard the elastic structure that appears first at the intimal-medial junction as the "primary" internal elastic membrane and the later appearing elastic structures in the intima as "secondary" elastic sheets and membranes (p. 9). Whereas the primary internal elastic membrane of the common and internal iliac arteries represents the selective substrate of early calcifications, the secondary elastic membranes of these vessels do not calcify in infancy or childhood.

Microscopic examination shows that the primary internal elastic membrane does not extend throughout the entire circumference of the common and internal iliac arteries. In the common iliac artery of the newborn infant, the primary internal elastic membrane is present in the lateral wall and is absent in larger parts of the medial wall (Figs 57, 108, 109). Serial sections taken from the aortic bifurcation and the proximal segments of both common iliac arteries show that the internal elastic membrane of the aorta becomes interrupted in the area of the crest and in its proximal extensions along the ventral and dorsal aortic walls. It remains intact in the lateral aortic wall and continues from there into the lateral walls of both common iliac arteries (Figs. 55, 57). In contrast, the medial walls of the common iliac arteries, which are extensions of the membrane-free crest of the bifurcation, do not have a primary internal elastic membrane along their course. Thus, the selective substrate of early calcification is absent along the medial walls of the common iliac arteries and calcification of these walls does not occur in childhood. The same is true of transverse calcific-free bands seen along the lateral walls. These bands correspond to large transverse gaps in the internal elastic membrane that are well seen in longitudinal and tangential sections (Figs. 106B, 110). Since similar gaps in the internal elastic membrane also occur in arteries with minimal or no calcification, they cannot result from calcification but must be a normal structural component of the common iliac artery at birth.

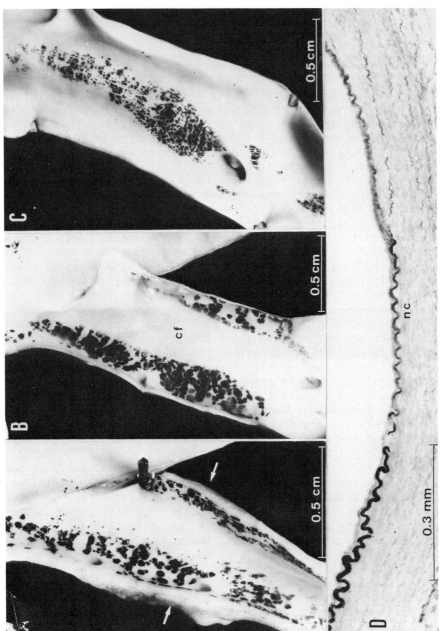

Fig. 107.

Fig. 107. Calcification patterns in the internal iliac artery at 13 months (A), 18 months (B), and 5 years (C) of age.
(A, B) In arteries opened along their dorso-medial walls, calcifications (black) appear on both sides of a longitudinal, calcium-free, unstained longitudinal area (cf) in the middle of the luminal surface. This area corresponds to the antero-lateral wall of the artery.
(C) In arteries opened along their antero-lateral walls, calcified areas are located in the middle of the luminal surface, i.e., along the dorsomedial wall.
(D) Shows a microscopic cross-section of the internal iliac artery obtained at the level of the arrows seen in (A). The primary internal elastic membrane is thickened and calcified (black) on the left, not calcified in the middle (nc) and is absent on the right. Only networks of the longitudinal, partly confluent, elastic fibers are visible in this sector. Von Kossa reaction and Gomori's aldehyde fuchsin stain for elastic tissue. (From Meyer and Ehlers, 1972.)

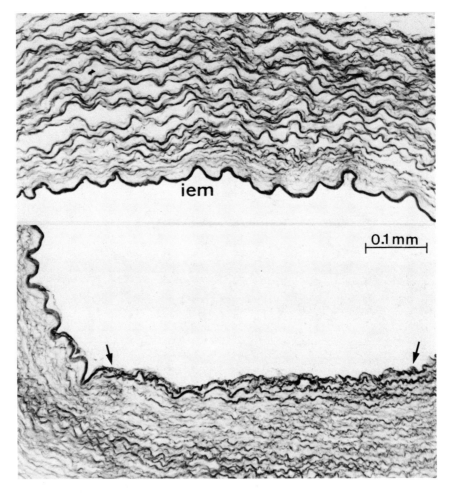

Fig. 108. Opposite sectors of the common iliac artery of a newborn infant. (A) The internal elastic membrane (iem) is complete along the lateral wall. (B) It is replaced by several fine elastic sheets along the medial wall (arrows). Case 20 K/77.

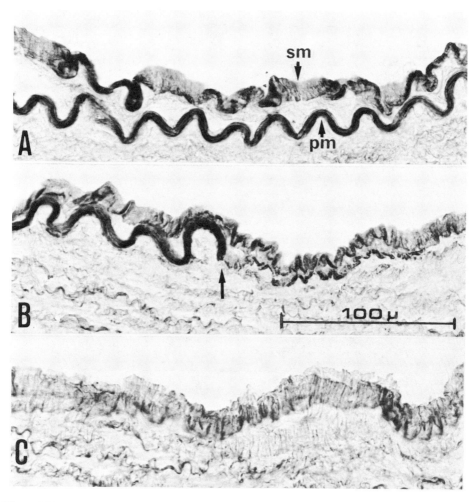

Fig. 109. Structural differences between individual sectors of the common iliac artery of a newborn infant (1030/71). All three photomicrographs were taken of the same cross-section at the same magnification. (A) The lateral sector shows a well-developed primary internal elastic membrane (pm) covered by a secondary membraneous structure (sm). (B) Here the primary internal elastic membrane ends abruptly (arrow). (C) In the medial sector, there is only a striated membranous structure, formed by coalesced longitudinally arranged fibers, at the intimal-medial junction. Weigert's resorcin fuchsin. From Meyer and Ehlers, 1972.

The site of membrane gaps can be anticipated grossly even in unstained arteries because they are usually confined to the central parts of slightly elevated transverse folds that are sometimes visible along the arterial luminal surface. The folds correspond to "spindles" in larger muscular arteries (pp. 213-219). In muscular arteries, e.g., the external iliac or femoral arteries, spindles usually develop later and only a few are present during the first year of life.

In the internal iliac artery, a well-developed and more or less continuous primary internal elastic membrane is present along the dorso-medial wall, i.e., the same portion of the wall which becomes the site of calcification. There is usually no primary internal elastic membrane in the opposite antero-lateral wall and calcific

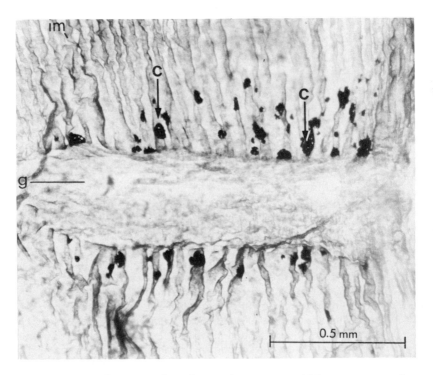

Fig. 110. Tangential section from the common iliac artery of a newborn. Punctate calcifications (c, arrows) are located near a gap (g) in the internal elastic membrane (im). The membrane has a characteristic corrugated pattern above and below the gap. Frozen section. From Meyer, 1968.

deposits do not form in this segment. Larger longitudinal musculo-elastic bundles at the intimal-medial junction which develop after birth also do not have a membrane. They often protrude considerably into the lumen and may then be visible as irregular elevations along the luminal surface.

In areas of the luminal layer of both common and internal iliac arteries without a primary internal elastic membrane, the subendothelial intimal elastic structures consist of several thin concentric elastic sheets formed by very fine, tightly interconnected, longitudinal elastic fibers. These fibers are clearly visible on cross- and tangential sections of the innermost arterial layer (Fig. 109C). They fuse together early during development and often form a membrane-like subendothelial sheet at birth. This secondary elastic sheet also develops in sectors that contain a primary membrane. Two elastic limiting structures are then seen beneath the endothelium (Fig. 109A). The secondary elastic sheet early assumes the characteristics of a homogeneous elastic membrane and is frequently as thick as the subjacent primary membrane at birth.

With growth, numerous additional new elastic sheets and networks form in the intima, i.e., between the primary internal elastic membrane and the endothelium (Fig. 111). At the age of four or five years, the newly formed secondary elastic membranes still have a longitudinal striated pattern and the primary internal elastic membrane has also retained its original more homogeneous appearance.

Absence of the internal elastic membrane in large parts of the circumference of the common and internal iliac arteries probably results from accelerated increase in diameter of these vessels during fetal life. Persistence of a continuous internal elastic membrane may not be possible with such an increase in arterial caliber. Adaptation of the membrane to growth with extension over a rapidly increasing circumference can scarcely be achieved by distention alone because the upper limit of normal distensibility would soon be surpassed and the membrane would become fragmented. On the other hand, continuous remodeling of the membrane's organic substrate without formation of larger defects seems scarcely possible in the short span of intrauterine life because the turnover of elastic tissue is slow in comparison with other constituents of the arterial wall, e.g., the mucopolysaccharides (Buddecke, 1963).

Examination of the iliac arteries in different stages of fetal development shows that in the common iliac artery a continuous internal elastic membrane encircles the entire arterial lumen by 12 to 16 weeks of gestation (Figs. 112A, 113). On serial cross-sections, only small membrane defects are present. With further development, more extensive interruptions become visible which may represent the precursors of the larger membrane gaps present at term. In contrast,

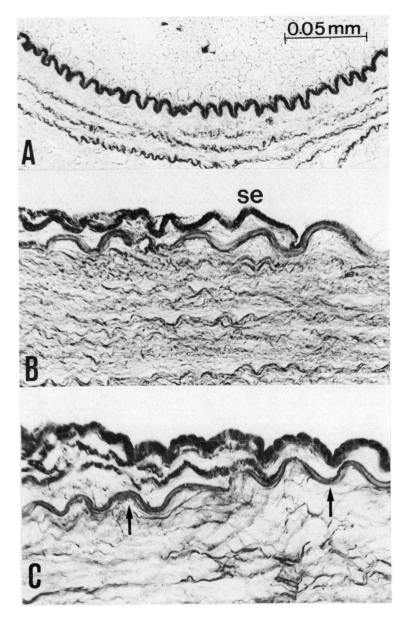

Fig. 111.

←Fig. 111 Development of intimal elastic structures in the common
        iliac artery.
        (A) Early in fetal life, only a primary internal elastic
        membrane is present along the inner aspect of the media.
        Fetus of about 15 weeks gestational age. Case 569.
        (B) At term birth the primary internal elastic membrane
        may be covered by secondary intimal elastic sheet (se)
        in some sectors. Case 847/74.
        (C) During the next few years of life, numerous secondary
        elastic sheets develop internal to the internal elastic
        membrane (arrows). Child of 6 years of age. Case 110 K/76.

the external iliac artery, which grows more slowly, retains a continuous internal elastic membrane throughout fetal life (Fig. 114). This supports the view that absence of a membrane in larger parts of the common iliac artery at birth may be the consequence of accelerated widening of this vessel during fetal development.

The findings in the internal iliac artery may be explained in the same manner, although interpretation of changes in the structure of this vessel is more difficult. In contrast to the straight tube of the common iliac artery, which has practically no branchings, the internal iliac artery is curved and several smaller arteries originate from its dorso-medial convex wall. Both the curved shape of the artery and the orifices of branching arteries in themselves produce considerable deviations from the structural pattern that is present in straight arterial segments. This is because at branchings, an additional hemodynamic load, e.g., turbulence, etc., results in formation of "intimal cushions" which may cope more adequately with the local functional load than a continuous internal elastic membrane. An additional load may also be present along the outer walls of the curves which favors the development of gaps in the membrane (pp. 233-235). In 14- to 16-week old fetuses, however, the internal elastic membrane of the internal iliac artery is almost continuous in serial cross-sections obtained from its most proximal straight segment (Fig. 112B). This indicates that in the internal iliac artery the large defects in the internal elastic membrane are not preformed but develop in the second half of gestation.

EFFECT OF SINGLE UMBILICAL ARTERY ON ILIAC ARTERIES

The close relation between hemodynamic load and structural features of the iliac arteries is evident in children with a single umbilical artery (SUA). This vascular anomaly occurs in 0.75 - 1.0% of consecutive deliveries and is frequently associated with congenital malformations (Benirschke and Bourne, 1960; Faierman, 1960; Thomas, 1961; Fujikura, 1964; Seki and Strauss, 1964; Fujikura and Froehlich, 1972). Among 20,000 umbilical cords examined by Bryan

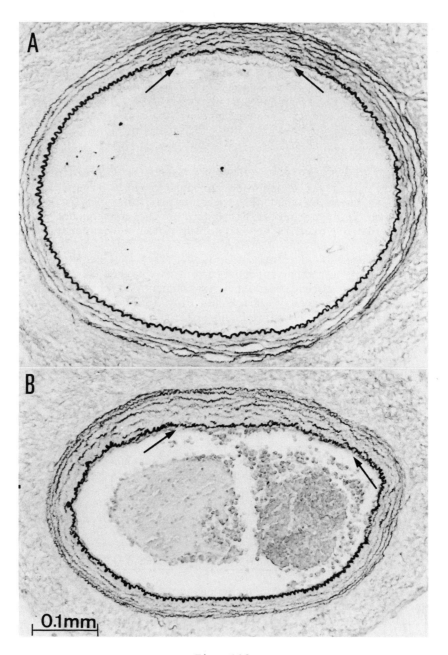

Fig. 112.

← Fig. 112 Cross-sections of the common iliac and the internal iliac arteries in a fetus of 15 weeks gestation.
(A) In the common iliac artery, the internal elastic membrane extends around most of the circumference apart from small sectors between the arrows. At this site the membrane is replaced by thin discontinuous sheets of longitudinally arranged elastic fibers which can be seen with higher magnification.
(B) In the internal iliac artery, the internal elastic membrane shows more numerous interruptions, e.g., see sector between arrows.
(A) Case 569/71, fetus body weight 84.5 g.
(B) Case 346/67, fetus body weight 73 g.

and Kohler (1974), 143 cases of SUA were found, an incidence of 0.72%. Of these cases, 83 (58%) were female and 60 (42%) male, a female:male ratio of 1.4:1. Four infants were twins and all of the co-twins had normal umbilical vessels. Twenty-five infants (17.5%) had major congenital malformations which were detected in the perinatal period. Twenty of the infants were stillborn or died within the first week of life. Only four survived beyond the neonatal period. The overall perinatal mortality associated with SUA was 17.5%. Follow-up studies of children who were born with SUA but had no congenital malformations show normal development up to the age of four years (Froehlich and Fujikura, 1972).

With a single umbilical artery, a unique hemodynamic situation arises during gestation. The entire blood volume to the placenta is transported from the abdominal aorta through the common and internal iliac arteries on only one side of the body (Fig. 115). Consequently, the common and internal iliac arteries on the side of the SUA become subject to a higher hemodynamic load during fetal life and develop a considerably larger caliber and thicker wall than those on the other side of the body (Fig. 116). In children born with a SUA and aged from one day to 3 1/2 months, the mean circumference of the common iliac artery on the side of the SUA was 8.2 mm, and on the opposite side, only 3.8 mm. The mean circumference of the internal iliac artery was 7.1 mm and 2.9 mm, respectively.

In infants with SUA, iliac arteries on the side of the SUA are larger while iliac arteries on the opposite side are smaller than vessels from infants of similar age (Meyer and Lind, 1972, 1974). Among 12 children with SUA who were examined postmortem, one infant had anomalous branching of the iliac arteries on the side of the SUA. In this case, both the external and internal iliac arteries originated from the bifurcation of the abdominal aorta (Fig. 117). In two other cases, the origins of the internal iliac arteries were located

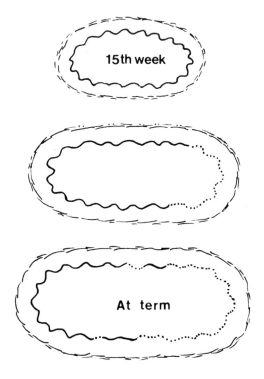

Fig. 113. Diagram showing the gradual disappearance of the internal elastic membrane from half of the circumference of the common iliac artery during fetal development. Above. At 15 weeks gestation, a wavy continuous internal elastic membrane encircles the entire lumen. Below. At term, it is absent along the medial aspect of the circumference of the vessel.

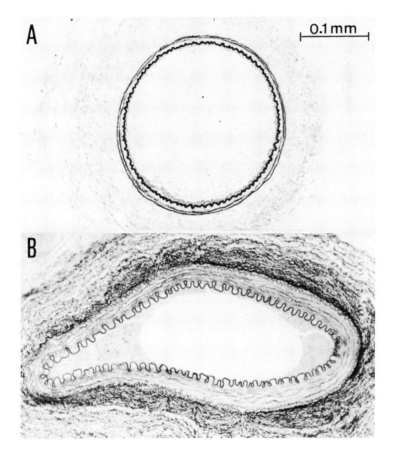

Fig. 114. Cross-sections of the external iliac artery from a fetus of 15 weeks gestational age (A), and from a newborn (B). Note the continuous internal elastic membrane.

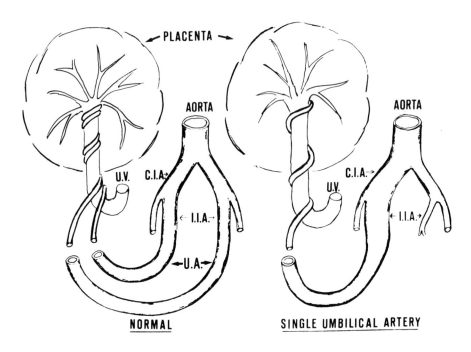

Fig. 115. Diagram of a part of the systemic arterial tree in a fetus with two umbilical arteries (left) and in a fetus with a single umbilical artery (right). Note the differences between the calibers of the iliac arteries. UA - umbilical artery, IIA - internal iliac artery, CIA - common iliac artery, UV - umbilical vein. From Meyer and Lind, 1974.

at different levels and the common iliac artery on the side of the SUA was considerably shorter than the artery on the opposite side of the body (Fig. 121).

Microscopic examination also reveals considerable structural differences between the iliac arteries on both sides of the body. In term infants, both common iliac arteries show pronounced differences in total wall thickness and in the structure of the media (Fig. 119). On the side of the SUA, the common and internal iliac arteries have the structure of elastic arteries with tightly packed interconnecting elastic membranes in the media. On the opposite side, the common and internal iliac arteries have the structure of muscular arteries with a media consisting of smooth muscle cells interspersed with a few fine elastic fibers and pronounced subendothelial elastic structures. Some of these structural differences probably develop during the latter part of gestation as evidenced by findings in a twin with a birth weight of 620 g (Fig. 118). In this fetus, the media of the common and internal iliac arteries on both sides of the body consisted of concentrically arranged elastic membranes that alternated with narrow interlamellar spaces. Thus, both arteries were elastic arteries. The wall of the small common iliac artery, which

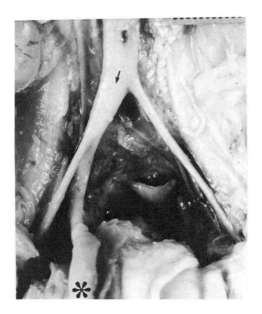

Fig. 116. Comparison of common iliac arteries in a pre-term infant with a single right umbilical artery (asterix). On the side of the single umbilical artery, the common iliac artery (arrow) is approximately four times as large as on the opposite side. The internal iliac artery is also larger on this side. On contrast, the external iliac arteries (arrows) are of an equal size. Case 183/73.

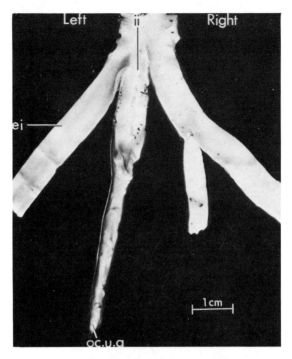

Fig. 117. Dorsal view of the luminal surface of iliac arteries from a one-year-old infant who, on autopsy, was found to have a single obliterated (closed) umbilical artery (ob. u.a.) on the left side. The left internal iliac artery (ii) originates level with the bifurcation and its circumference is greater than that of the right internal iliac artery which originates at the normal level. Punctate calcifications are grossly visible in the left internal iliac artery but not in the right. Von Kossa reaction. ei - external iliac artery. From Walsh, Meyer and Lind, 1974.

did not participate in the placental circulation, however, was only one-quarter as thick as that of the common iliac artery on the side of the SUA.

During postnatal growth, the interlamellar spaces of elastic iliac arteries on the side of the SUA become wider and include more muscle cells than are normally found in newborn infants. The smaller iliac arteries on the opposite side retain the structural pattern of muscular arteries (Fig. 120). These structural differences probably account for the significant differences in pulse wave velocity between iliac arteries on both sides of the body reported by Berry et al. (1976). They studied 18 children born with a SUA who were

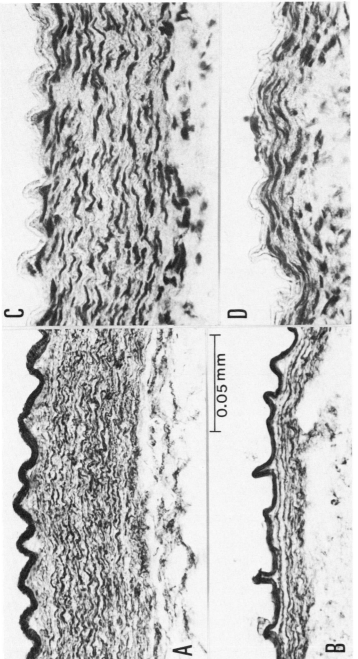

Fig. 118. Cross-sections from the right (A) and left (B) common iliac arteries of a premature twin (birth weight 620 g). Both vessels have the structural pattern of elastic arteries. On the side of the single umbilical artery, the wall of the common iliac artery (A) is approximately four times thicker than that on the opposite side (B). Cryostat frozen section. Weigert's resorcin fuchsin stain. (C) and (D) show cross-sections of the same arteries stained with hematoxylin-eosin. (A) and (B) from Meyer and Lind, 1974.

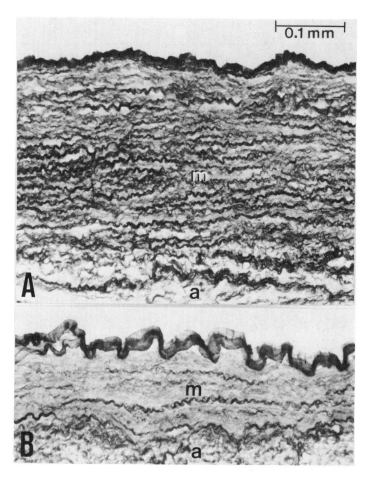

Fig. 119. Cross-sections of common iliac arteries of a term newborn infant. The common iliac artery (A) on the side of the single umbilical artery is an elastic artery with well-developed elastic sheets in its media. In contrast, the thin-walled common iliac artery on the opposite side (B) is a muscular artery with a media (m) poor in elastic networks. a - adventitia. Cryostat frozen section. Weigert's resorcin-fuchsin stain. From Meyer and Lind, 1974.

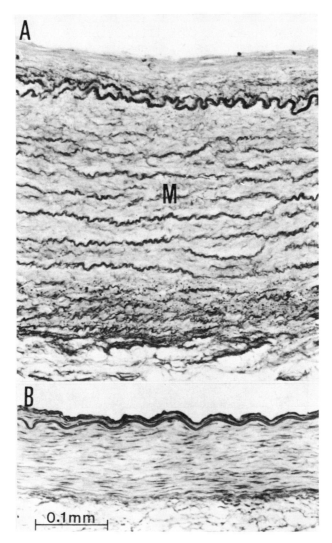

Fig. 120. Cross-sections of (A) an enlarged right common iliac artery on the side of the obliterated umbilical artery and (B) of the smaller left common iliac artery on the opposite side from an 18-month-old infant (see Fig. 122). Whereas (A) the right common iliac artery displays the microscopic structure of an elastic artery with numerous lamellar units in its media (M), (B) the left common iliac artery shows the structural pattern of a muscular artery. Thus, structural differences between the vessels are still present in later infancy. Orcein stain. From Meyer and Lind, 1974.

between five and nine years of age. They measured vessel compliance in vivo with a non-invasive method (ultrasound) and found differences in compliance values between both iliac arteries of SUA children. Compliance in this vessel was also considerably higher than in the aorta, whereas in normal children, iliac compliance is usually about the same or slightly less.

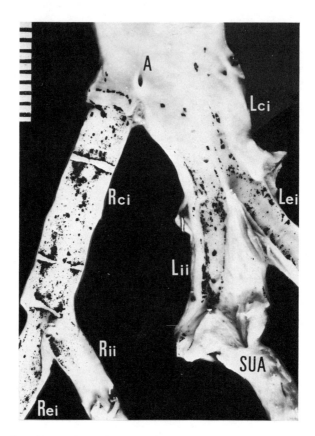

Fig. 121. Iliac arteries from a one-month-old infant with a single left umbilical artery. Note the large caliber of the common iliac artery (Lci) and that of the internal iliac artery (Lii) on the side of the obliterated single umbilical artery (SUA). Conspicuous calcifications (black) are present in both iliac arteries, but the pattern differs in the iliac arteries on the side of the single umbilical artery. In the right common iliac artery (Rci), circular calcium-free bands (arrows) are seen. Microscopically, they correspond to the gaps in the internal elastic membrane. Von Kossa reaction. Millimeter scale at left. Case 906/73. From Meyer and Lind, 1974.

Marked gross calcification of iliac arteries is common in infants with SUA. It is usually more conspicuous in the iliac arteries on the side of the SUA. In these vessels, calcific deposits appear irregularly distributed along the luminal surface (Fig. 121). In contrast, in the narrower contralateral iliac arteries, calcific deposits form a more regular circular pattern and appear as fine transverse streaks. In two cases, pairs of circular calcified bands were seen along the well-delineated, transverse, calcium-free strips (Fig. 121, right). Microscopically, the strips correspond to wide transverse gaps in the internal elastic membrane, whereas the bandlike deposits represent the calcified edges of the interrupted internal elastic membrane. The more regular distribution of calcium deposits in the narrower iliac arteries of children with SUA depends on the shape and direction of gaps in the internal elastic membrane, a characteristic structural feature of most human muscular arteries. Thus, the calcification pattern is typical of a muscular artery.

In the enlarged common iliac artery on the side of a single obliterated (closed) umbilical artery, atherosclerotic lesions have been observed in two children 18 months and 4 years of age (Meyer and Lind, 1974) (Figs. 122, 123, 124, 125). They appear to be among the earliest atherosclerotic lesions in the human arterial system. The findings suggest that local structural and hemodynamic factors are important in the development of these lesions. Since "fatty streaks" were not present elsewhere in the arterial system, a generalized metabolic disorder is probably not the cause. It seems unlikely that the lesions are the immediate consequence of the higher hemodynamic load to which arteries are subjected during fetal development and which probably favor the development of early calcifications since no lipid deposits have been detected in iliac arteries on the side of the SUA in newborns. Atherosclerotic lesions obviously appear later and are probably related to remodeling of enlarged iliac arteries, a process which starts after birth as an accommodation to diminished blood flow after cessation of the placental circulation. This structural adaptation is associated with proliferation of the intima which contributes to narrowing of the lumen. The proliferating intima may develop an increased affinity for lipids (Taylor et al., 1954; Hass et al., 1961; Cox et al., 1963), i.e., an increased "lipidosis potential," as postulated by Hass (1967), and in this way, become the site of early atherosclerotic lesions. In addition, the adaptive narrowing of the vessel may be accompanied by thickening of the media which could impair transmural transport and favor lipid deposition in the thickened intima.

10. EXTREMITY ARTERIES

The upper extremities are supplied by blood from the subclavian artery via the axillary and brachial arteries. The lower extremities are supplied by blood from the external iliac artery which

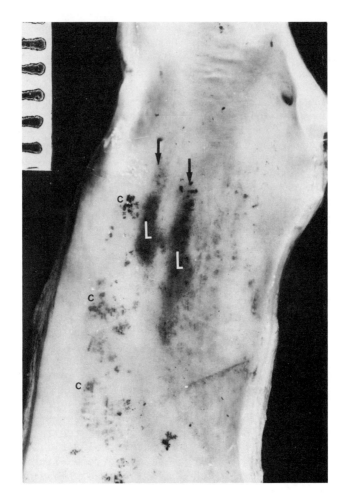

Fig. 122. Right common iliac artery on the side of the closed single umbilical artery from an 18-month-old child. After the von Kossa reaction, numerous fine calcific incrustations (c) become visible. Subsequent gross fat staining with Fettrot 7B reveals two prominent longitudinal strips of lipid infiltration (L, vertical arrows) and a more discrete reddish staining near the origin of the artery (1). The red-stained lipid infiltrated areas appear black in the photograph. Case 610/71. Millimeter scale upper left.

represents a direct continuation of the common iliac artery. Below the inguinal ligament, the external iliac artery continues distally as the femoral artery. Both the brachial artery below the orifice of its deep branch and the external iliac artery are muscular arteries.

TRANSITION FROM AN ELASTIC TO A MUSCULAR SEGMENT

In the upper extremity, the transition from an elastic to a muscular artery occurs along the brachial artery at the level of the origin of its deep branch. This can be seen grossly in excised specimens. The luminal surface of the distal muscular segment develops a distinct transverse pattern after retraction, whereas that of the proximal elastic segment remains relatively smooth (Fig. 126A).

Microscopic examination shows that the transition occurs over a short distance: 2 - 3 mm. It is characterized by an abrupt reduction in the medial elastic sheets, first in the outer media, then in the rest of this layer. In the more proximal part of the transitional zone, only the outer media is free of elastic sheets, whereas the more luminal medial layers contain extensive elastic networks which are formed mainly by circularly arranged elastic fibers. In the more distal portion of the transitional zone, the elastic elements disappear, and the media assumes the structural pattern typical of a muscular artery. It is now clearly separated from the intima by a prominent internal elastic membrane (Fig. 127).

Structural changes also occur in the adventitia. In the more proximal elastic segment, the adventitial elastic fibers are rather loosely spread throughout the abundant coarse collagenous networks. During the transition to a muscular segment, the elastic networks become more tightly packed around the media, but the collagenous networks remain prominent in the outer layer.

All arteries in the lower extremity are muscular arteries because the transition from an elastic to a muscular segment occurs in the pelvis at the origin of the external iliac artery. The transition is less abrupt than in the upper extremity because the elastic sheets disappear more gradually from the media. A considerable reduction in medial elastic sheets occurs along the common iliac artery, which belongs to the elastic arterial segment. In its proximal portion, the interlamellar spaces become wider and contain more smooth muscle cells than in the abdominal aorta, but the vessel is still an elastic artery (Rotter and Rottman, 1952). In its more distal portions, the bundles of smooth muscle cells are larger and they are accompanied by networks of finer elastic and collagenous fibers. For the most part, the bundles show a circular or oblique course, but in some places they may have an irregular course. Although smooth musculature predominates here, a typical muscular structural pattern consisting of a regular circular arrangement of densely packed smooth musculature appears first at the origin of the external iliac artery.

It seems unlikely that any significant displacement of the transitional area in a distal or a proximal direction occurs with

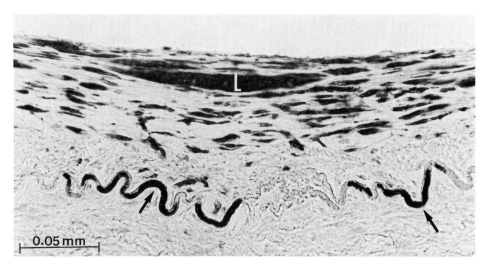

Fig. 123. Cross-section of the common iliac artery shown in Fig. 122. In the area of the prominent lipid infiltration, the intima is considerably thickened and densely interspersed with fusiform lipid-containing cells. Near the lumen, a larger accumulation of lipophages (L) is seen. The underlying internal elastic membrane is partly calcified (arrows).

growth and development of the arterial system or in later adult life. However, some individual variations in the level and extent of transitional areas in iliac arteries have been reported by Rotter and Rottman (1950).

DEVELOPMENT OF THE LUMINAL RETRACTION PATTERN IN MUSCULAR ARTERIES

During the first and second decades of life, folds appear along the luminal surface of excised and retracted muscular arteries, and in later life, they form intricate patterns. The evolution of these patterns was first described in detail by Dietrich (1930).

During the first year of life, scattered, slightly raised, whitish, transverse strips appear in larger muscular arteries, e.g., in the external iliac artery and in the proximal portion of the femoral artery near the origin of its deep femoral branch (Fig. 128). These strips represent precursors of more prominent folds which develop in increasing numbers a few years later and extend over larger areas. Toward the end of the first decade of life, the entire luminal surface of the excised and retracted large muscular arteries is usually densely interspersed with numerous fine circular folds and

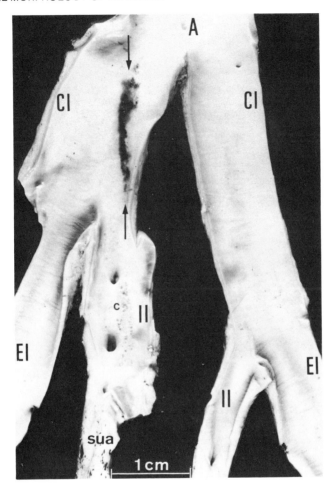

Fig. 124. A raised longitudinal atherosclerotic lesion (between arrows) is seen in the enlarged right common iliac artery on the side of the obliterated (closed) single umbilical artery (sua) from a four-year-old boy. The red stained lesion appears black in the photograph. Fine calcific incrustations (c) are seen in the right internal iliac artery. A - abdominal aorta, CI - common iliac artery, II - internal iliac artery, EI - external iliac artery. Case 116/72. Von Kossa reaction and gross fat staining with Fettrot 7B. Millimeter scale on right.

appears ruffled (Fig. 129A). Prominent spindle-shaped ridges, called <u>spindles</u> because of their shape, are also commonly seen between the numerous narrow folds. The spindles have raised, somewhat serrated borders and concave oval-shaped central parts (Fig. 130).

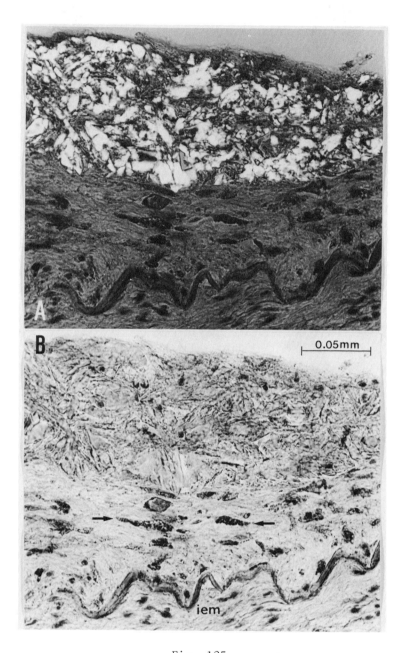

Fig. 125.

← Fig. 125 (A) In the luminal layer of the atherosclerotic plaque from the right common iliac artery (on the side of the obliterated single umbilical artery) numerous anisotropic crystals and fat droplets are seen (see Fig. 124).
(B) Same section seen in non-polarized light. The deeper intima includes numerous lipophages (arrows). iem - internal elastic membrane. Cryostat frozen section, Fettrot 7B stain.

This transverse, ruffled relief is most pronounced in large extremity arteries of adolescents and young adults, especially after blood or a mixture of blood and india ink has been smeared on the luminal surface (Fig. 129A). It only becomes clearly visible in retracted arteries, and when vascular segments are fixed under approximately physiological pressure (that assumed to have been present in vivo), the luminal surface becomes smooth and only large folds, i.e., spindles, may be distinguished. The formation of transverse folds in retracted arteries is important, nevertheless, because it is determined by preformed normal structures which are incompletely understood. Longitudinal arterial sections show that early transverse strips and larger transverse folds, i.e., spindles, correspond to gaps in the internal elastic membrane (Fig. 128B). It is possible that normal structural characteristics of the underlying media may also be involved in the formation of transverse folds along the luminal surface after retraction. Since the luminal pattern undergoes a series of changes during life with formation of complex patterns, it seems likely that profound remodeling occurs not only in the arterial luminal layer, but also in the underlying media throughout life.

In some arterial segments, e.g., in the lower femoral and in the popliteal arteries, the transverse ruffled luminal pattern is completely interrupted by longitudinal, slightly raised band-like cushions (Fig. 131). They usually include numerous orifices of branching arteries. The cushions consist mainly of longitudinally arranged musculo-elastic bundles which are separated from the subjacent media by fine elastic networks. The internal elastic membrane is usually absent in these areas. Similar cushions have been found in fetal popliteal and brachial arteries and are regarded as "stress zones" in the sense that they constitute a structural adaptation of the arterial wall to increasing longitudinal stretch (Robertson, 1960).

With development of the longitudinal structural components of the intima, particularly of the hyperplastic elastic layer, the transverse folds become covered by an increasing number of fine, slightly raised, longitudinal lines or ridges (Fig. 129B). By the fourth or fifth decades of life, the original transverse pattern may

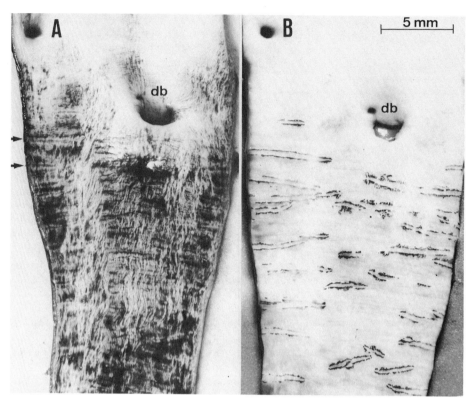

Fig. 126. (A) The luminal surface of the muscular arterial segment of the brachial artery, below the origin of its deep branch (db), has numerous transverse folds (arrows) that are partly covered by longitudinal intimal structures whereas that of the elastic arterial segment, above the orifice, is smooth. The vascular surface has been photographed after application of a mixture of blood and india ink. 55-year-old man. Cause of death: myocardial infarction. Case 720/67.

(B) Segment of the brachial artery shown in (A) after the von Kossa reaction. Numerous pairs of calcific bands (black) are seen in the muscular segment of the brachial artery that arises at the level of the orifice of the deep branch (db). The pairs of calcific bands delineate gaps in the internal elastic membrane (p. 195). No calcific deposits are present in the adjacent elastic segment above the orifice. Millimeter scale at right. From Zabka, 1968.

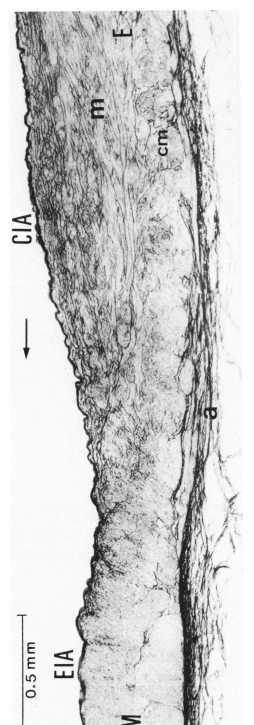

Fig. 127. Longitudinal section showing the transition from an elastic (E) to a muscular (M) structural arterial pattern at the origin of the external iliac artery (EIA). In the elastic segment, i.e., in the distal portion of the common iliac artery (CIA), the media (m) is densely interspersed with elastic sheets, which accompany longitudinally and obliquely arranged muscle bundles (grayish, unstained). In the outer medial layer, cross-sections of circularly arranged muscle bundles (cm) are visible. Toward the origin of the external iliac artery (EIA), the amount of the elastic structures decreases in the media, and the vessel assumes the structure of a muscular artery. The elastic networks of the adventitia (a) which are rather loosely spread throughout the abundant collagenous networks along the common iliac artery become more tightly packed around the media of the external iliac artery. The arrow indicates the direction of blood flow. Cryostat section, Weigert's resorcin fuchsin. 16-year-old boy. Case RM 31.1.77.

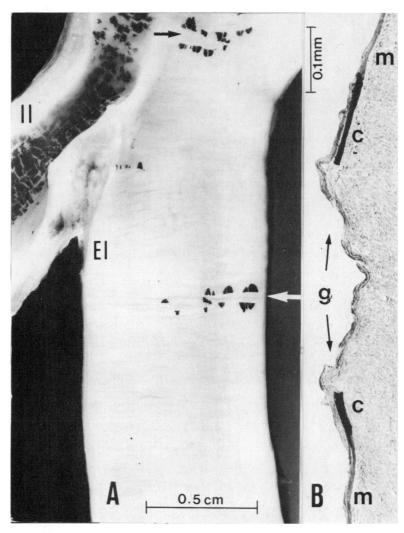

Fig. 128. (A) Fine transverse folds (arrows) of the luminal surface are seen in the left external iliac artery (EI). They are partly delineated by black-stained calcific deposits. Extensive calcifications are present in the left internal iliac artery (II). Two and a half-year-old girl. Cause of death: Waterhouse-Friderichsen syndrome. Case 632/72. (B) Microscopically, on longitudinal section, the fold seen in (A) (white arrow) corresponds to a gap (g, arrows) in the primary internal elastic membrane. Note the black-stained calcific edges of the membrane (c) and the non-calcified parts of the membrane (m) above and below this. Von Kossa reaction and Gomori's stain for elastic tissue.

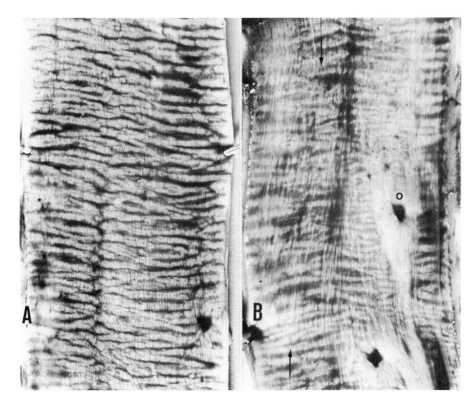

Fig. 129. Demonstration of various patterns of the luminal relief after blood and india ink are smeared along the arterial luminal surface.
(A) Numerous transverse folds covered by fine longitudinal grooves form after retraction of the femoral artery.
(B) In some areas the original transverse pattern is still discerned (arrows) but it is mostly covered by an elastic hyperplastic intimal layer consisting of numerous fine, longitudinal structures. No pattern can be distinguished around the orifices (o). Microscopic examination of these areas shows prominent cushions consisting of longitudinally arranged musculo-elastic bundles at the intimal-medial junction. 15-year-old boy. Case 735/67. (See Fig. 131.)

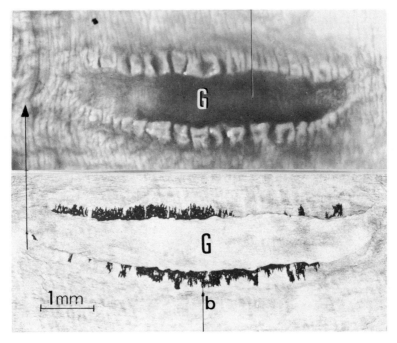

Fig. 130. (A) Prior to staining, the concave central part of an arterial spindle in the femoral artery (G) is still filled with blood. It is outlined by the serrated border of the fragmented internal elastic membrane. The corrugated longitudinal pattern of the luminal surface that is produced by circular retraction of the internal elastic membrane is absent in the membrane-free area of the central part of the spindle.
(B) The same spindle after von Kossa reaction. Larger parts of both edges of the interrupted internal elastic membrane are calcified (black) and form a pair of calcific bands (b) above and below the membrane gap (G). The arrow points in the direction of the vascular axis. 35-year-old man. Died a few hours after a traffic accident. Case 124/67. From Stelzig and Meyer, 1967.

no longer be distinct. The age at which this occurs varies greatly in individual arterial segments but, in general, it occurs first in larger muscular trunks and then in smaller branches.

In some other large arteries, particularly in the common and internal iliac arteries and especially in the trunks of the renal arteries, another change in pattern is common. In these vessels, the entire luminal surface assumes a regular wavy pattern (Fig. 132A). This "wavy folding" (Dietrich, 1930) may also occur in other muscular arteries, e.g., the femoral and splenic arteries. It is usually associated with pronounced elastic intimal hyperplasia but appears to be due mainly to profound alterations of the underlying media, i.e., particularly a reduction in muscular elements with age (Dietrich, 1930). The longitudinal and wavy luminal parterns show little change after fixation under appropriate intravascular pressure and thus do not result from arterial retraction (Heard, 1950; Meyer, 1963).

ALTERATIONS IN MICROSCOPIC STRUCTURE WITH AGE

Lower Extremity Arteries in Fetuses

The external iliac artery shows a thin intima consisting of endothelial lining only. A strong internal elastic membrane encircles the entire arterial lumen. The pronounced wavy pattern of this membrane (Fig. 133A) suggests that considerable retraction of the arterial tube occurs after death. The media consists of three to four layers of circularly arranged smooth muscle cells and numerous fine, concentrically arranged but discontinuous, wavy elastic sheets between them. Some of the sheets are formed by tightly packed longitudinal elastic fibers, cross-sections of which are occasionally discernible. A wavy external elastic membrane, which appears reduplicated in some sectors, is present along the outer aspect. The adventitia is two to three times as thick as the media. It consists mainly of a dense layer of longitudinally arranged collagen fibers. Numerous cross-sectioned, fine, longitudinal elastic fibers are seen in the midst of the adventitial collagen networks.

The femoral artery (Fig. 135A) also shows a comparatively well-differentiated internal elastic membrane. In some fetuses it appears to be less "wavy" than it is in the external iliac artery which would suggest that less postmortal retraction occurs here. Presumably for this reason the lumen of the femoral artery may appear even wider than the adjoining portion of the external iliac artery. In a few small sectors, the internal elastic membrane may be reduplicated, i.e., a second thinner lamella external to the main membrane may be present in the mid-term fetus. Between both elastic lamellae, solitary, circularly arranged muscle cells are visible. Only scattered fine elastic fibers or thin circular elastic sheets are present

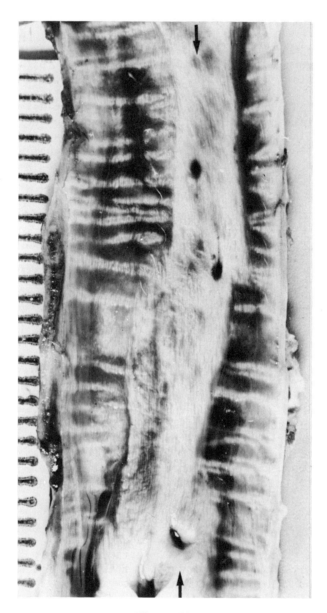

Fig. 131.

← Fig. 131 The transverse pattern of the distal femoral artery is interrupted by a longitudinal elevated strip containing several orifices (arrows) of branching arteries. The surface of the stip has a fine longitudinal ruffled pattern. Microscopically the strip consists mainly of longitudinally arranged musculo-elastic bundles. 15-year-old boy. Cause of death: traffic accident. Case 753/67. Millimeter scale on left. From Fellhofer, 1969.

in the media which consists of about three layers of muscle cells in mid-gestation. The outer aspect of the media is separated from the adventitia by a thin discontinuous elastic lamella. The adventitia is about twice as thick as the media. It consists of a network of somewhat coarse, mainly circularly arranged, collagen fibers which are interspersed with fine longitudinally arranged elastic fibers.

Lower Extremity Arteries in Newborns

As in fetuses, the intima of the external iliac artery is thin and consists of an endothelial lining that covers a strong, folded, wavy, continuous internal elastic membrane. The media is formed by 10 - 12 layers of circularly arranged smooth muscle cells interspersed with concentrically arranged fine elastic sheets and fibers which seem to separate it into circular layers in some sectors (Fig. 133B). Sheets of an intensely red-stained metachromatic ground substance are easily demonstrated along the elastic structures of the media with Feyrter's method. A well-developed but discontinuous external elastic membrane is present. The adventitia is approximately twice as thick as the media. It consists mainly of coarse collagen bundles which seem to show a different course, but are probably predominantly obliquely arranged. Collagen networks are densely interspersed with numerous longitudinally arranged elastic fibers around the outer aspect of the media.

The femoral artery displays a similar structural pattern (Fig. 135B). As seen in the fetus, its lumen seems to be somewhat larger than that of the external iliac artery, presumably because it retracts less after death. The internal elastic membrane is continuous and is covered by a thin endothelial layer. The media consists of eight to ten layers of circularly arranged muscle cells, and fine, discontinuous elastic sheets and fibers. They are less prominent than in the media of the external iliac artery. A well-developed external elastic membrane separates the media from the adventitia. Its structure is similar to that of the external iliac artery.

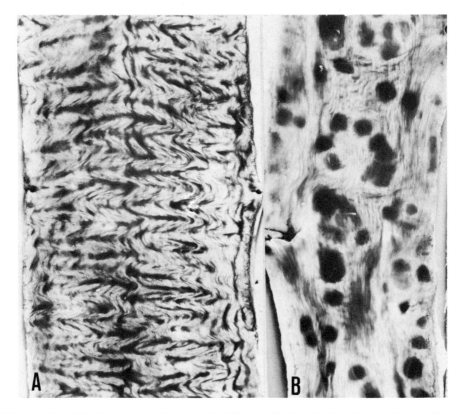

Fig. 132. (A) Pronounced wavy folding of the luminal surface of an external iliac artery. 52-year-old woman. Case 101/68. (B) Medial calcification leads to formation of numerous roundish grooves and results in distortion of the initial luminal pattern. The longitudinal strips are due to elastic intimal hyperplasia. 47-year-old woman. Case 376/68. From Fellhofer, 1969.

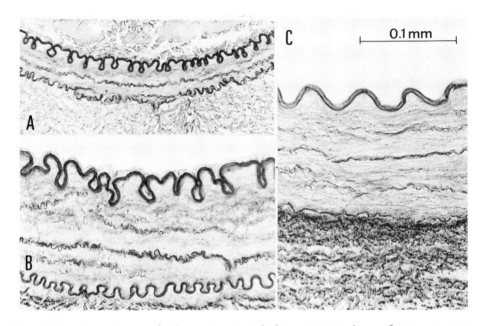

Fig. 133. Structure of the external iliac artery in a fetus, a newborn infant and a two-month-old infant.
(A) In the fetus (body weight 565 g), the internal elastic membrane shows a pronounced wavy pattern which is due to strong retraction. The media contains a few fine elastic fibers and there is a discrete external lamina along the outer aspect of the media.
(B) In the newborn, the wall has become thicker. The internal elastic membrane still has a prominent wavy pattern and is also thicker than in the fetus. The external elastic lamella is now better developed.
(C) In the two-month-old infant, the internal elastic membrane is not as strongly retracted as in the newborn and even appears somewhat stretched, as are the elastic fibers in the media. The external elastic lamina is no longer well seen. Cryostat frozen sections, Weigert's resorcin fuchsin stain. Same magnification in all three figures.

Lower Extremity Arteries at Two Months of Age

During the first two months of life, the lumen of the external iliac artery enlarges considerably and its internal diameter nearly doubles (Fig. 133C). Its wall thickens and the media now consists of approximately 16 to 20 layers of circularly arranged muscle cells with numerous fine elastic fibers between them. The adventitia consists of circularly or obliquely arranged collagen bundles and numerous cross-sectioned longitudinally arranged elastic fibers that are thicker than in the newborn. They form dense networks around the outer aspect of the media.

The proximal femoral artery is about the same size as the external iliac artery (Fig. 135C) but its medial elastic networks are composed of fewer and finer elastic fibers. Occasionally a fine secondary elastic sheet consisting of longitudinal elastic fibers is seen internal to the internal elastic membrane. The latter often shows small indentations that may represent an early stage of the complete interruption that develops later in life.

Lower Extremity Arteries at One to Two Years of Age

During the first year of life, a secondary subendothelial elastic sheet develops in the external iliac artery. It extends around the entire circumference and is formed by tightly packed longitudinal elastic fibers that give it a granular structure on cross-sections. It stains more deeply with Weigert's resorcin-fuchsin and is thinner than the homogeneous primary elastic membrane. In most sectors, the secondary elastic sheet rests immediately on the primary membrane and there is only a narrow space between them (Fig. 134A). In some sectors, however, the lamellae are separated by a small amount of ground substance and scattered muscle cells. Intimal cushions surrounded by both lamellae and containing numerous elastic fibers and muscle cells are present only near and around sites of arterial branching. The media consists of approximately 30 to 35 layers of muscle cells and is rich in fine elastic fibers which show the same circular course as the muscle cells. The wide adventitia is formed mainly of strong longitudinal elastic fibers.

The femoral artery has a similar structural pattern but the elastic networks of the media are less prominent than in the external iliac artery (Fig. 136A).

Lower Extremity Arteries at Five to Six Years of Age

The external iliac artery has a well-developed secondary subendothelial elastic lamella which is now thicker than the primary internal elastic membrane (Fig. 134B). The structural differences between the lamellae are as well seen as in younger children. The

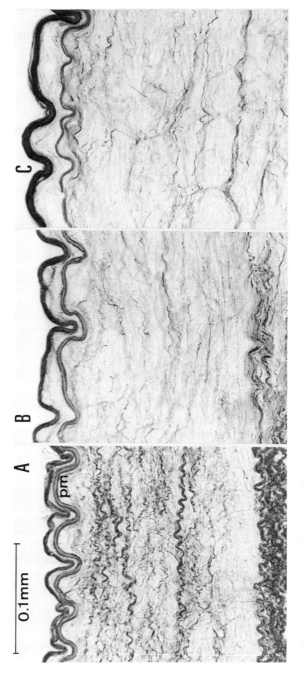

Fig. 134. Structural changes in the external iliac artery from one to 27 years of age.
(A) at one and a half years of age, a distinct secondary elastic sheet (arrows) covers the primary internal elastic membrane (pm), which has not changed since infancy. The elastic fibers in the media are prominent and the elastic sheets in the adventitia are somewhat coarser.
(B) At six years of age, a prominent, dark, deeply-stained, secondary sheet of about the same thickness as the adjacent primary internal elastic membrane has appeared. The elastic networks of the media have decreased and are thinner than in infancy. [See (A).]
(C) At 27 years of age, the thickness of the secondary elastic sheet has increased further. In contrast, the primary internal elastic membrane seems thinner now. No apparent change in medial structure has occurred. Cryostat frozen sections. Weigert's resorcin fuchsin stain.

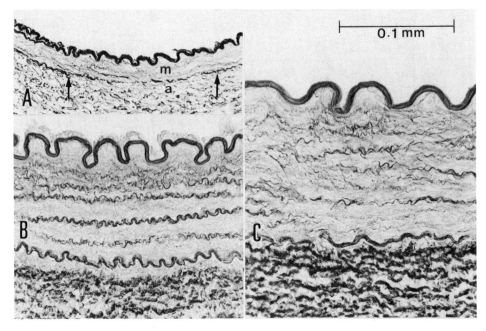

Fig. 135. The structure of the femoral artery above the origin of the deep femoral branch in a fetus (body weight 565 g), a newborn infant and a two-month-old infant.
(A) In the fetus, the wavy internal elastic membrane is prominent and may be reduplicated (left). A discrete external elastic lamella is present (arrows). m - media, a - adventitia.
(B) At birth, the arterial wall has become considerably thicker. The wavy internal elastic membrane remains prominent but stains less deeply. The media includes numerous elastic sheets and fibers. There is a distinct external elastic lamina.
(C) At the age of two months, the internal elastic membrane has become slightly thicker than at birth, but is less retracted. The medial musculature (unstained here) is interspersed with numerous circularly arranged elastic fibers which have become thinner. The adventitia is rich in elastic fibers which show an oblique or longitudinal course. Cryostat frozen sections. Weigert's resorcin fuchsin. Same magnification in all figures.

primary internal elastic membrane appears homogeneous and sharply outlined, whereas the coarse secondary lamella is granular and its contours are somewhat irregular on cross-section. It is reduplicated in a few areas. Small gaps in the primary internal elastic membrane are common, but large gaps are seen in only a few sectors. In these sectors the membrane is replaced by several concentric elastic sheets that alternate with thin rows of longitudinally arranged muscle cells.

The space between the secondary elastic lamella and the primary internal elastic membrane is clearly seen around the entire vascular circumference. It is formed by ground substance, tiny elastic fibers, and a few longitudinally disposed smooth muscle cells. The media contains a moderate amount of fine elastic fibers. The adventitial layer consists of densely packed, strong, longitudinal elastic fibers, bundles of which often penetrate into the outer media where they take an oblique or circular course.

The femoral artery displays a similar structural pattern (Fig. 136B, C). The secondary subendothelial elastic lamella, however, is less prominent than in the external iliac artery. It is about as thick as the primary membrane and closely applied to the "waves" of this membrane. The interlamellar space is narrower than in the external iliac artery. Small gaps in the internal elastic membrane are common in longitudinal sections. They correspond to the transverse strips or small spindles seen in Fig. 129A.

Lower Extremity Arteries in Young Adults (Early Third Decade)

In the external iliac artery the subendothelial elastic layer is formed by a prominent secondary elastic sheet which is now much thicker than the primary internal elastic membrane. On cross-sections of retracted arteries, the wavy pattern of this sheet differs from that of the primary elastic membrane in that each fold encloses about two folds of the primary membrane (Fig. 134C). The longitudinal elastic fibers that form the secondary subendothelial elastic sheet are now more closely packed together. Their nonhomogeneous granular appearance is still evident in cross-sections. Tangential sections show that the layer has a longitudinal striated pattern. Reduplication of the layer occurs in many sectors, but both sheets remain close together. A narrow space or layer is present between the secondary luminal elastic layer and the primary internal elastic membrane. It is filled mainly with metachromatic ground substance containing numerous tiny elastic and collagenous fibers and a few loosely arranged cells that take a different course. In some sectors, there are some prominent longitudinal muscle sheets between both elastic lamellae.

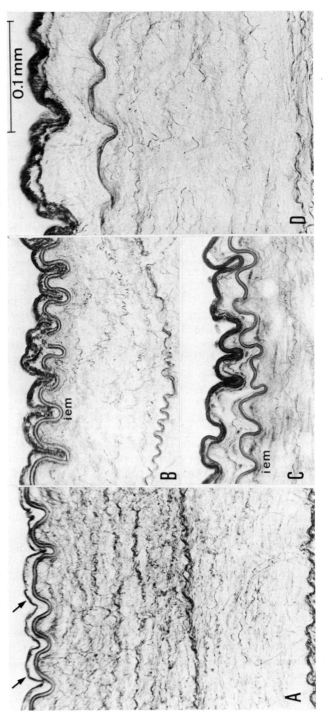

Fig. 136.

Fig. 136. Structural changes in the femoral artery from one and one-half to 27 years of age.
(A) At one and one-half years of age, the primary internal elastic membrane appears slightly thinner than at two months of age (see Fig. 135C) and is covered by a thin secondary elastic sheet (arrows). The media is now twice as thick as at two months of age (see Fig. 135C).
(B), (C) At the age of six years, the secondary elastic sheet is well developed. It does not seem to be homogeneous in some sectors (B) where it is formed by longitudinal elastic fibers, cross-sections of which are seen internal to the internal elastic membrane (iem). It has a more homogeneous appearance in other sectors (C) where it is thicker than the adjacent primary internal elastic membrane (iem). The media now contains fewer elastic fibers.
(D) at 27 years of age, the secondary internal elastic sheet is thick, partly split and not homogeneous. The space between the internal elastic structures (not stained here) is filled with smooth muscle cells and ground substance. The media contains only a few scattered fine elastic fibers. Cryostat frozen sections. Weigert's resorcin fuchsin. Same magnification in all figures.

The media near the internal elastic membrane is usually densely interspersed with fine elastic fibers that mostly parallel the circularly arranged muscle cells. Near the middle of the media the number of elastic fibers decreases considerably.

On longitudinal sections, numerous small gaps in the primary internal elastic membrane are visible in longitudinal sections (pp. 213-219, 230-232). The edges of the interrupted membrane often seem to be connected by very fine, wavy, elastic fibers. Fine elastic sheets are also present throughout the media. They frequently seem to outline round or polygonal areas of cross-sectioned muscle cells which may represent cross-sectioned interconnecting muscular bundles in the media. These areas tend to be smaller in the innermost parts and larger in the outermost parts of the media.

The adventitia consists of densely packed, strong, longitudinally arranged, elastic fibers with an increasing number of collagen fibers in the outer layers of the adventitia.

The femoral artery has a similar structural pattern at this age. In comparison with the external iliac artery, the secondary luminal elastic sheet may seem thinner and its wavy pattern may not be as coarse (Fig. 136D). The space between the elastic layers is usually narrower in the femoral artery than in the external iliac artery. Numerous small and large gaps in the primary internal elastic membrane are seen in the longitudinal arterial sections. The amount of ground substance seems to be greater below the gaps than in the adjacent media. In the third decade, the number of muscle cells beneath the gaps may decrease.

THE SYSTEM OF COMMUNICATING GAPS IN THE INTERNAL ELASTIC MEMBRANE OF MUSCULAR ARTERIES

Although large circular gaps in the internal elastic membrane were first described by Dietrich in 1930, their exact shape, size and distribution in individual arterial segments have only become appreciated with studies on gross demonstration of membrane calcifications (Meyer and Stelzig, 1967, 1968, 1969). This is because early calcific deposits form along the margins of the interrupted elastic membrane. With appropriate technique it is possible to grossly demonstrate gaps in most large muscular arteries, e.g., the femoral, brachial, and popliteal arteries, and some medium-sized muscular arteries, e.g., the cervical portion of the internal carotid artery (Figs. 137-143).

In the femoral artery, circularly arranged gaps of about 0.1 - 0.3 mm predominate (Fig. 138). They may completely encircle the

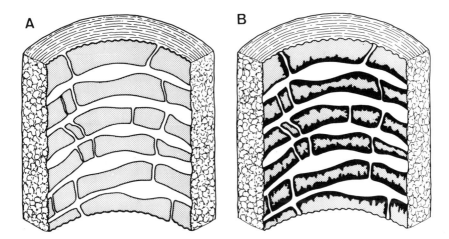

Fig. 137. (A) Diagram showing circularly-oriented gaps in the internal elastic membrane which are interconnected by numerous smaller longitudinal and oblique membrane splits.
(B) Calcifications in the internal elastic membrane develop mainly along the edges of the membrane gaps. After the von Kossa reaction, they are seen to form pairs of black bands that outline the membrane gaps. Modified from Walsh, Meyer and Lind, 1974.

circumference and are often connected by fine longitudinal and oblique gaps. Larger circular gaps are located in the central parts of more prominent spindles (Fig. 138B).

In the external iliac artery longitudinally arranged gaps predominate whereas only a few larger circular gaps are usually present (Fig. 140).

A labyrinthine network of membrane gaps is often visible above and below larger arterial branchings, e.g., in the area of the orifice of the deep femoral branch and in the most proximal portion of this branch. Above small orifices of branching arteries, the gaps seem to converge toward the orifice, but the calcific bands which outline the gaps do not continue into the branch and the area around the orifices does not calcify (Figs. 138, 142, 143). Microscopic examination shows that the primary internal elastic membrane ends at some distance from the orifices and is replaced by intimal cushions consisting of musculo-elastic bundles, elastic sheets, and dense networks of fine elastic fibers. These elastic structures, unlike the primary internal elastic membrane, are not subject to early calcification.

No comparable system of membrane gaps can be seen in the carotid siphon although scattered larger gaps sometimes occur. Larger gaps usually develop in the proximal segment of the main coronary branches (p. 110). In these arteries, however, the membrane seldom

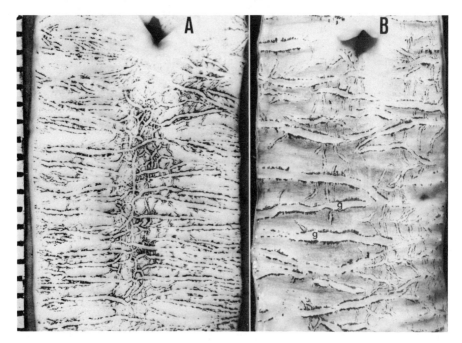

Fig. 138. (A) System of communicating membrane gaps along the luminal surface of the proximal segment of a femoral artery. Note the labyrinthine arrangement of gaps in the middle of the artery and the absence of calcific bands around the orifice of the branch (above). Millimeter scale on left. Von Kossa reaction. From Meyer, 1975.
(B) In the deep femoral branch, larger circular gaps (g) are located in the spindles. The gaps are outlined by pairs of black calcific bands. Von Kossa reaction. 47-year-old man. Case 954/74. Millimeter scale on left.

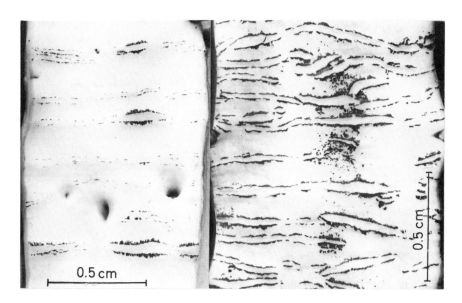

Fig. 139. Pairs of calcific bands (black) delineate preformed circular gaps in the internal elastic membrane of the brachial arteries from a 17-year-old man (left, cause of death: traffic accident) and from a 56-year-old man (right, cause of death: cerebral hemorrhage). Von Kossa stain. From Meyer, Zabka and Stelzig, 1969.

calcifies early, and for this reason, the pattern of gaps during postnatal growth cannot be demonstrated grossly.

FACTORS RELATED TO DEVELOPMENT OF GAPS IN THE INTERNAL ELASTIC MEMBRANE

The gaps in the primary internal elastic membrane appear long before the membrane edges become calcified. They can be easily demontrated in longitudinal sections of children's extremity arteries. The reason why a system of gaps develops in medium and large muscular arteries with the body's growth remains to be elucidated. Progressive stretch of the arterial tube during postnatal growth may be a factor, since the length of the leg quadruples during infancy and childhood. Even after cessation of growth, the arteries continue to be constantly subjected to longitudinal stretch, as evidenced by retraction of excised arteries to about one-third of their original length in situ. Simon (1959) compared the length of the aorta in situ and after removal from the body. He then ligated the branches and filled the aorta with saline. Under physiological pressure the aorta regained its original length as seen in situ. The same is also probably true of large muscular arteries and especially those of the extremities. Hence, stretch seems to be mainly due to intravascular pressure and not to attachment of the arterial tube to the surrounding tissue. Moreover, the arteries of the extremities are often subjected to additional stretch during flexion and extension.

Some of the differences in the pattern of gaps in various segments may reflect differences in the amount of stretch to which the arteries are exposed. In the femoral artery, circularly arranged membrane gaps predominate, whereas in the external iliac artery,

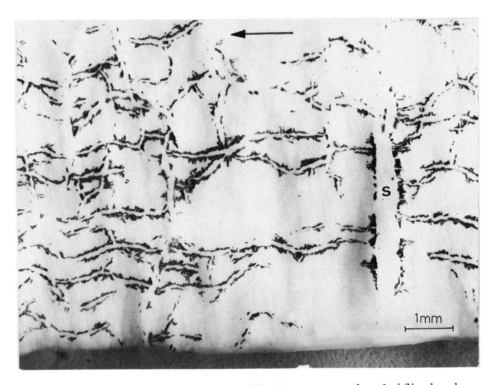

Fig. 140. Pairs of mostly longitudinally arranged calcific bands (black) are present in the middle third of the external iliac artery. They outline a system of communicating gaps in the internal elastic membrane. Only one larger transverse gap in a spindle (s) is present. The arrow indicates the direction of the arterial axis. Von Kossa reaction. 19-year-old man. Cause of death: traffic accident. Case 498/67. From Meyer and Stelzig, 1968.

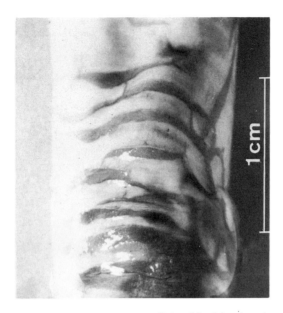

Fig. 141. Numerous transverse gaps (black) in the internal elastic membrane are seen along the luminal surface of the outer convex wall of the curved splenic artery. The outer part of the curve has been inverted for better demonstration of the entire luminal surface and now appears convex. The membrane-free parts of the luminal surface, i.e., the transverse membrane gaps, are stained by postmortally hemolyzed blood and appeared reddish in the fresh specimen (see Fig. 142). 25-year-old man. Case 402/68. Cause of death: chronic glomerulonephritis with hypertension. From Meyer and Weber, 1968.

which is located deep in the pelvis and may be less subject to stretch with movements of the body, only a few larger circular gaps are present.

In the splenic artery, which has a curved course even in young adults, large membrane gaps are present along the outer convex wall which may be exposed to a higher functional load and stretch than the inner concave parts of the curves (Meyer and Weber, 1968) (Figs. 141, 142).

POSSIBLE ROLE OF MEMBRANE GAPS IN NUTRITION OF THE ARTERIAL WALL

Since both the gaps in the internal elastic membrane and the numerous fine fenestrae (pores) in the membrane connect the luminal arterial layer to the media, it is tempting to assume that they are important for transmural transport of oxygen or nutritional substances from the vascular lumen to the media. Some observations suggest that permeability is greater in membrane-free parts of the luminal surface because, even after formalin fixation, the gaps in the membrane stain selectively with hemolyzed blood (Figs. 141, 143). They also show a reddish coloration when a mixture of fuchsin and picric acid is applied to the luminal surface. Studies in animals likewise support this view. For example, Wilens and McClusky (1954) showed that perfusion of rabbit arteries with hyperlipemic serum results in preferential deposition of lipids in the media beneath the splits in the internal elastic membrane. Similarly in dogs fed cholesterol, lipid first penetrates the media beneath the gaps in the internal elastic membrane (Fry, personal communication, 1972). In human arteries, lipids also accumulate in the media beneath the gaps.

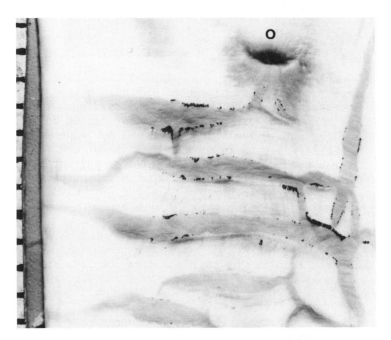

Fig. 142. Higher magnification of part of the luminal surface of the splenic artery shown in Fig. 141.
The membrane-free parts appear gray and are partly outlined by black-stained calcified edges of the internal elastic membrane. Note the membrane-free area around an orifice (o) of a branch that was also stained with hemolyzed blood. Millimeter scale on left. From Meyer and Weber, 1968.

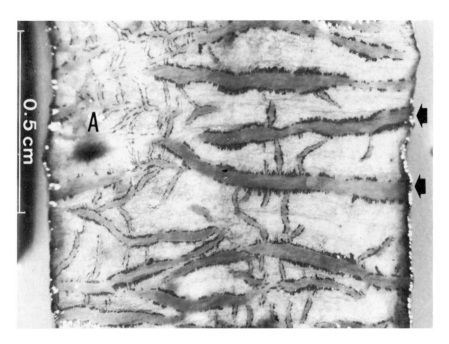

Fig. 143. System of communicating gaps in the internal elastic membrane in the deep femoral branch of a 27-year-old man. The gaps (gray) are stained with hemolyzed blood (arrows) and are outlined by pairs of calcific bands. The luminal surface around a small orifice has also been stained with hemolyzed blood (A, left). Case 967/71. Von Kossa reaction. From Meyer, 1972.

Whatever the significance of the membrane gaps and pores for nutrition of the arterial wall may be during arterial growth, the development of secondary elastic structures in the intima internal to the primary internal elastic membrane early in life probably diminishes the permeability of the arterial luminal layer. The effect must become more marked later in life because the gaps in the membrane become covered by an increasing number of intimal layers. Hence their function as nutritional channels must alter significantly.

CALCIFIC BANDS AND THEIR ROLE AS CRYSTALLIZING SITES FOR LARGER CALCIFICATIONS

In large muscular arteries, i.e., the external iliac and femoral arteries, scattered incrustations are seen grossly along the gaps in the membrane in the first year of life (Fig. 128). These deposits later extend more completely around the gaps and form pairs of "calcific bands" that outline the membrane-free parts of the luminal surface. Such bands develop in the femoral artery by the end of the first decade and are usually present in large muscular arteries of the upper and lower extremities by the end of the second decade or in the early third decade (Figs. 138A, 139A).

The calcific bands that appear during arterial growth often become the site of more extensive calcification (Meyer and Stelzig, 1967; Stelzig and Meyer, 1967). During the second decade of life, scattered, round, granular or oval-shaped, solid calcific deposits form along the incrusted edges of the internal elastic membrane and extend into the media (Figs. 144-148). Granular calcifications coalesce along the calcific bands and form larger circular aggregations in the underlying media. In young adults, these aggregations do not extend into the membrane-free parts of the wall, i.e., in the parts of the media located below the central areas of the spindles (Fig. 147). In this stage of calcification, circular calcific aggregations in the media alternate with circular calcium-free parts, a finding which is typical of Mönckeberg sclerosis.

Thus, some of the structural changes in the luminal arterial layer during growth seem to determine both the site of some early calcifications as well as the later localization of medial calcifications typical of Mönckeberg sclerosis in the adult. However, Mönckeberg sclerosis may also arise from fine medial calcifications that develop in the middle of the media and are not related to the early calcifications of the internal elastic membrane. In this type of Mönckeberg sclerosis, the alternating pattern of ring-like calcific deposits is probably determined by some structural features of the media which remain to be elucidated.

11. PULMONARY CIRCULATION

MORPHOLOGY

The trunk of the pulmonary artery and its two branches form a fork-shaped arterial segment which connects the right ventricle and the intrapulmonary arterial tree (Fig. 149). In fetuses and newborn

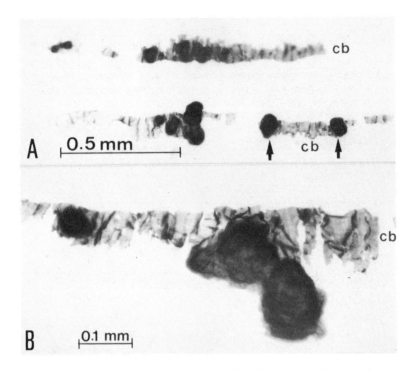

Fig. 144. (A) Dense grain-like calcific deposits (arrows) are present along the calcific bands (cb). (B) The deposits are seen to have a striated structure on higher magnification. Dawson's alizarin staining method. The calcifications are photographed in translucent light after clearing the specimen in KOH and glycerin. From Ludwig, 1976.

infants, the pulmonary fork is connected with the aortic arch by the ductus arteriosus which is the direct extension of the pulmonary trunk beyond the origin of its two main branches (Fig. 37). At term, the pulmonary trunk is a short arterial segment which measures about 1 cm in length and 6.7 mm in diameter. It is slightly wider than the aorta but its wall is nearly as thick in most of its circumference (Fig. 150). In fetuses and term newborn infants, the caliber of the main pulmonary branches is considerably smaller than that of the trunk (Fig. 43). By the age of three or four months, the diameter of each branch has increased and amounts to about three-fourths that of the trunk.

FETAL PULMONARY CIRCULATION

In the antenatal period, the lung is not an organ of gas exchange but it has a higher oxygen consumption than in the adult (Fritts et al., 1960). The high metabolism of the lung is not only due to the need for growth but also to nonventilatory functions, notably formation of a fluid the volume of which amounts to the functional residual capacity, i.e., about 30 ml/kg body weight (Adams et al., 1963). In the sheep fetus, the fluid is produced at the rate of 3 to 6 ml per hour (Goodlin and Rudolph, 1970) and it is assumed to enlarge the radii of the distal airways, thereby lowering the pressure required for the first breath of air.

In the fetus, pulmonary vascular resistance is greater than in the systemic circulation and most blood from the right ventricle bypasses the lungs by flowing from the main pulmonary artery through the ductus arteriosus to the descending aorta. A still greater reduction in the pulmonary blood flow occurs in conditions associated with fetal hypoxemia, especially in combination with metabolic or respiratory acidosis (Rudolph and Yuan, 1965; Campbell et al., 1967).

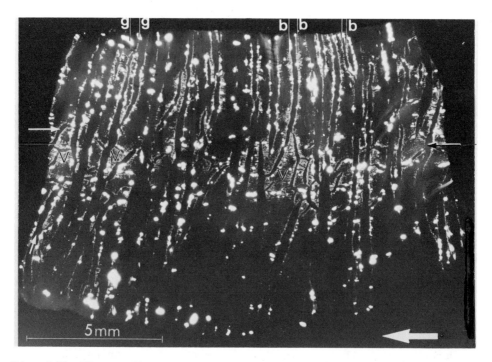

Fig. 145. Microradiograph of the luminal layer of an opened and flattened femoral artery. Numerous roundish, grain-like calcific deposits (white, g) are seen along the calcific bands (b) that outline fine, circularly arranged, membrane gaps (small arrows). The arrow (right) points in the direction of the arterial axis. 44-year-old man. Died two days after a traffic accident. From Stelzig and Meyer, 1967.

However, a high carbon dioxide tension (in the cat) has a mild vasodilating effect which slightly reduces the vasoconstricting effect of the low pH of respiratory acidosis (Viles and Shepherd, 1968). The site of the high resistance of the fetal pulmonary vascular bed is believed to be localized to precapillary muscular arteries, which according to Naeye (1961), develop a thick muscular coat during the latter part of gestation. This smooth muscle mass presumably controls the volume of flow to the lungs by diverting larger or smaller amounts of blood via the ductus arteriosus to the placenta. This view has been challenged recently by Cosmi (1975) since the volume of blood flowing to the lungs is small, i.e., less than 10% of total cardiac output. He believes that the peripheral circulation plays a far more important role in the controlling flow and cites studies in the sheep fetus which show that neural and humeral stimuli produce both cardiovascular and respiratory responses (Condorelli and Cosmi, 1972). The importance of the small muscular arteries in controlling pulmonary blood flow has also, in a sense, been challenged by the findings of Hislop (1969) since she found that although fetal pulmonary arteries are more muscular than adult pulmonary arteries, their medial thickness decreases slightly from the 12th week of gestation to term. It should be noted, however, that reported differences in medial thickness may be due to differences in the amount of medial musculature (atrophy or hypertrophy) and/or the state of contraction of the vessel (dilatation or constriction). Hence the question cannot be regarded as settled.

It is pertinent that during the third trimester, the percentage of cardiac output to the lungs increases and that flow to each 100 g of lung tissue likewise increases (from 28 to 126 ml/min/100 g) (Rudolph and Heymann, 1970). Since the increase in flow exceeds the concomitant increase in pulmonary arterial pressure, pulmonary vascular resistance falls considerably during this period, possibly because of growth and development of new vessels. At term, pulmonary blood flow per unit weight lung (in the lamb) is about the same as blood flow to the brain, is less than that to the heart, kidney, or spleen, and more than that, to the intestine (Rudolph and Heymann, (1970). The lamb's brain, however, then constitutes 1.5% of body weight, whereas the human infant's brain constitutes 12% of body weight (Widdowson, 1974).

The high pulmonary vascular resistance of the fetal lung could be the resultant of several factors acting upon it locally. A number of studies have suggested that the most important of these is the low oxygen tension (20-25 mm Hg, as compared to postnatal values of 80-100 mm Hg) and relatively high carbon dioxide tension (about 40 mm Hg) with accompanying increase in H^+ ion concentration [fall in pH of blood normally perfusing the lung (in lambs)] (Cook et al., 1963; Cassin et al., 1964) which cause vasoconstriction. Indeed, there is evidence that in the lower ranges of oxygen tension occurring in fetal pulmonary arterial blood, a small decrease in PO_2 may

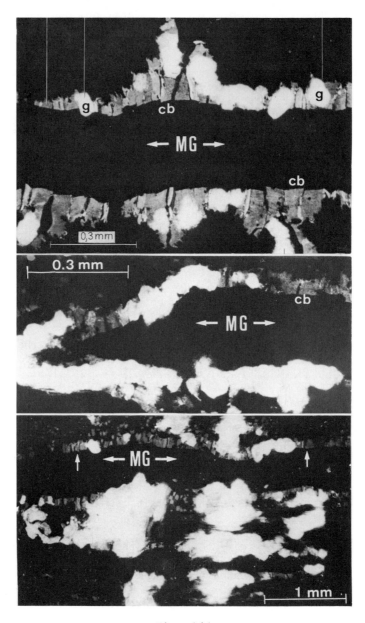

Fig. 146.

← Fig. 146 Microradiographs showing progressive calcification in the femoral artery.
(A) Solid round and oval-shaped calcific deposits (white, g) have become confluent along the calcified edges (c) of a membrane gap (MG, arrows). 47-year-old man.
(B) With more extensive calcification, the grain-like deposits along the calcific bands (cb) form larger, circularly arranged deposits (white). 44-year-old man. Died two days after a traffic accident.
(C) Still larger calcific deposits form by further apposition of calcium and confluence of smaller aggregations. The original crystallizing sites, i.e., the calcific bands, are still seen in some areas (vertical arrows). The membrane gaps (MG, arrow) remain free of calcifications. Same case as in (B). From Stelzig and Meyer, 1967.

induce a striking increase in pulmonary vasoconstriction (Rudolph and Yuan, 1965). Reflex pulmonary vasoconstriction from stimulation of aortic or cardiac chemoreceptors also occurs when arterial PO_2 falls or H^+ concentration rises (Colebatch et al., 1965). Others argue that although it is true that oxygen is of major importance in the control of pulmonary vascular resistance, isolated perfusion studies (in the cat) do not support the view that a low oxygen tension in pulmonary arterial blood produces vasoconstriction in pulmonary arteries (Duke, 1954). In their opinion, if alveolar oxygen tension is normal it is immaterial how low the oxygen tension is in pulmonary venous blood (Hoffman, 1975). They believe that alveolar hypoxia causes pulmonary vasoconstriction both before and after birth, i.e., in various disease states as well as consequent to residence at high altitude (Bergofsky and Holzman, 1967; Glazier and Murray, 1971). It is important to note that the effect of oxygen on pulmonary vessels is opposite to that on other vessels. Staub (1963) has shown with a rapid freezing technique that oxygen can diffuse from surrounding alveoli directly into the small pulmonary arteries. Locally produced chemical mediators, e.g., histamine (Hauge, 1968), are also involved since a cuff of pulmonary parenchyma surrounding the artery is necessary for demonstration of the oxygen effect (Lloyd, 1968).

Pulmonary vasoconstriction is also produced in fetal animal preparations by small doses of catecholamines with alpha-adrenergic activity (epinephrine, noraepinephrine). This is unlike the adult lung in which a response can be blocked by dibenamine (which blocks the vasoconstrictor action of these catecholamines). It is still uncertain how important they are in the intact animal. Stimulation of sympathetic nerves before the lamb is viable or the lungs can be expanded with gas causes a 75% decrease in pulmonary blood flow. Asphyxia likewise produces pulmonary vasoconstriction in immature

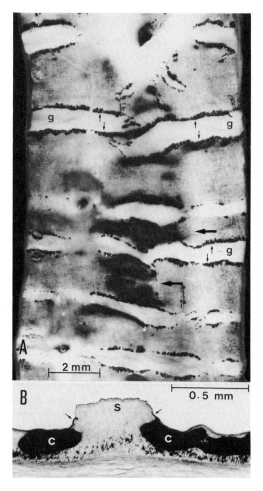

Fig. 147. (A) Pairs of black calcific bands (small arrows) delineate calcium-free membrane gaps (g) in the luminal layer of a deep femoral artery. Massive medial calcifications (horizontal arrows) are present between the membrane gaps. They coalesce with the calcific bands. 34-year-old woman. Cause of death: acute pancreatitis. Von Kossa reaction. (B) Longitudinal cryostat section of the artery shown in A. Larger black-stained calcific deposits (c) are located immediately proximal and distal to the cross-sectioned protruding central part of a spindle (s). The deposits (c) coalesce with the calcified edges of the internal elastic membrane seen on both sides of the spindle (arrows). Von Kossa reaction and Gomori's elastic stain.

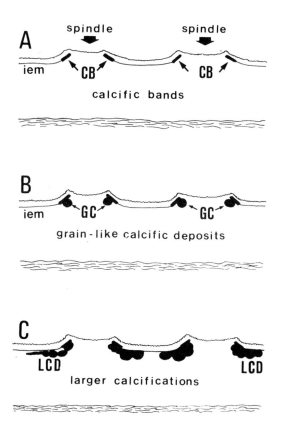

Fig. 148. Diagram showing development of calcifications in muscular arteries (longitudinal sections).
(A) Calcium deposits (CB) appear along the edges of the gaps (G) in the internal elastic membrane (IEM). At the arterial luminal surface they are seen as pairs of calcific bands (see Figs. 139-143).
(B) Small grain-like calcific deposits (GC) develop along the calcific bands and extend into the adjacent media.
(C) Further deposition and confluence of calcium deposits results in formation of larger incrustations in the media (LCD). These represent the substrate of Mönckeberg's sclerosis. I - intima, ADV - adventitia, S - spindle. Modified from Meyer, 1971b.

and mature lambs. Such vasoconstriction can be abolished by dibenamine. Serotonin, however, causes constriction of both arteries and veins. Catecholamines with beta-receptor activity (isoproterenol) produce pulmonary vasodilatation. According to Heymann et al. (1969), bradykinin is the mediator for the fall in pulmonary vascular resistance consequent to ventilation of fetal animals. The ability of this substance to lower pulmonary vascular resistance has also been demonstrated in the normal human adult. In these subjects bradykinin has little effect after inhalation of air, but if made hypoxic first (by breathing a low oxygen mixture), bradykinin causes pulmonary vasodilatation, a response which has also been noted for acetylcholine (Segel, 1970). The effects of histamine are disputed and it is uncertain whether species differences may be involved. According to some workers, histamine produces vasodilatation, especially in vessels which have previously been constricted (Dawes, 1968). According to others, it may be an important mediator of hypoxic vasoconstriction (Hauge and Melmon, 1968), but in the dog it has been shown to increase pulmonary vascular resistance by venous constriction (Glazier and Murray, 1970).

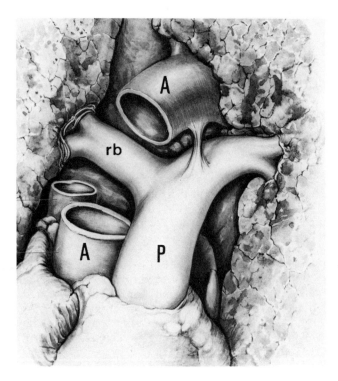

Fig. 149. The dissected pulmonary fork of an adult. A part of the ascending aorta (A) has been removed to show the right main branch (rb) of the pulmonary trunk (F).

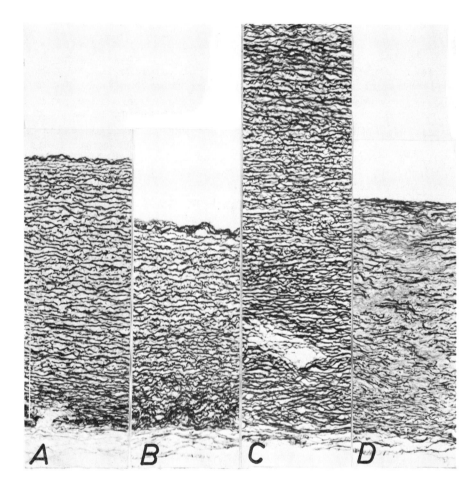

Fig. 150. Cross-sections of the aorta and pulmonary trunk from a newborn (A, B) and an infant aged 8 months (C, D). The wall of the aorta (A, C) increases markedly in thickness while retaining its structural pattern. In contrast, the wall of the pulmonary artery (B, D) shows little increase in thickness but alters considerably in structure. (See Figs. 154, 155). From Meyer and Simon, 1959.

Pulmonary vasodilatation is produced by ventilation with air, O_2 and N_2 but not by expansion of the lung with warm saline (Cassin et al., 1964; Lauer et al., 1965). This vasodilatation occurs within the parenchyma by local action and is independent of fetal blood gas composition.

AERATION

At birth, the infant must exchange a liquid-filled lung for an air-filled lung (Fig. 151). Thus, the intrauterine lung which is filled with a fluid that appears to arise from blood and alveolar lining cells must be replaced by air. This is achieved partly by compression of the infant's thoracic cage and partly by removal of fluid by the pulmonary capillaries and lymphatics. Studies in lambs indicate that pulmonary lymphatic flow increases on aeration and may account for resorption of 25% or more of pulmonary fluid (Boston et al., 1965). Lymph flow is greater in mature than in pre-term fetuses. Ultrastructural studies show that pulmonary fluid passes from the basement membrane to the interstitium of the alveolar wall and then to the peribronchial and perivascular connective tissue which contains lymphatics (Gonzalez-Crussi and Boston, 1972). It is uncertain how much of the fluid is absorbed by capillaries, but they are capable of it since available studies show that capillary hydrostatic pressure is about 12 mm Hg in the newborn. This is considerably lower than the colloid osmotic pressure (22 mm Hg) (Adams et al., 1971). Indeed, Peltonen et al. (1965) injected contrast medium into the trachea of the term lamb fetus in utero. When the lamb took the first breath of air, the contrast medium spread immediately throughout both lung fields. A few seconds later, it had almost disappeared and the pulmonary veins, left atrium, left ventricle, and aorta were outlined by the contrast media. The authors concluded that immediate increase in blood flow in the lung is a major route for transport of alveolar lung fluid.

Once fluid resorption is complete, an acellular alveolar lining layer forms over the epithelial cells which contains the components of the surfactant system. The lining is formed in part from the non-reabsorbed (surfactant) constitutents of pulmonary fluid and in part from the secretion of surfactants by type II cells. The low surface tension properties of the lining stabilizes the alveoli by minimizing their tendency to collapse and control liquid balance in order to ensure alveolar dryness (Clements, 1962); Pattle, 1965). The width of the lining varies inversely with the degree of expansion of the lung.

NEONATAL PULMONARY CIRCULATION

At birth, the placenta ceases to function and the site of gas exchange and the life of the infant is suddenly and urgently dependent upon establishment of pulmonary respiration. When air breathing

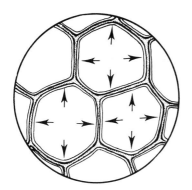 

Fig. 151. The effect of lung aeration on the pulmonary vasculature. Before birth, the lung is filled with fluid and there is no gas-fluid interphase. Pressure is therefore exerted laterally and outward on the pulmonary vessels passing between the walls of the terminal air spaces. This tends to compress the vessel and increase resistance to pulmonary flow. After birth, the lung becomes expanded by gas which creates a gas-fluid interphase. This results in a surface tension effect which tends to exert an inward pull on the walls of the terminal air spaces. The caliber of the pulmonary vessels is thereby widened and the resistance to pulmonary flow decreases. Modified from Lind et al., 1964. From Walsh, Meyer and Lind, 1974.

begins, pulmonary vascular resistance falls from a prenatal level of about 1.6 to about 0.3 mm Hg/ml/kg and pulmonary blood flow increases markedly from about 35 ml/kg/min to 160 ml/kg/min (Dawes et al., 1953) (Figs. 152, 153). At the same time, cessation of umbilical blood flow reduces the volume of blood conducted to the heart through the inferior vena cava. The greater part of the reduction in pulmonary vascular resistance is achieved by dilatation of the small pulmonary arterioles (Civin and Edwards, 1951; Dammann and Ferencz, 1956; Rosen et al., 1957; Edwards, 1957; Könn and Storb, 1960; Wagenwoort et al., 1961). The dilatation is consequent to the rise in pO_2 and fall in pCO_2 on ventilation with air. Part of the decrease in pulmonary vascular resistance results also from the

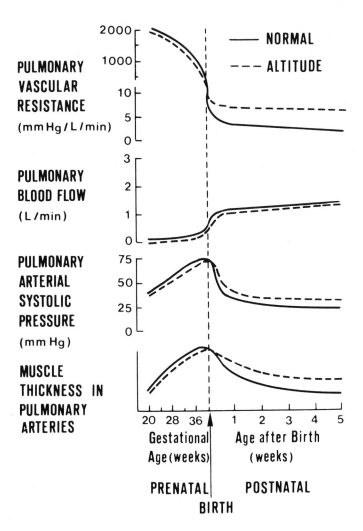

Fig. 152.

← Fig. 152. Schematic diagram of fetal and postnatal changes in pulmonary vascular resistance, pulmonary blood flow, pulmonary arterial systolic pressure, and thickness of smooth muscle in the medial layer of pulmonary arterioles. There is no known difference in pulmonary vascular development in the fetus of women living at high altitude or at sea level. Postnatally, pulmonary vascular resistance falls more slowly in infants born at high altitude. Pulmonary blood flow increases rapidly at birth in the normal infant and probably at the same rate at altitude. Pulmonary arterial pressure normally falls rapidly after birth, the fall is delayed at altitude and pressure remains slightly elevated. The muscle in the pulmonary arterioles does not regress as rapidly as normal in infants residing at high altitude. Modified from Rudolph, 1970. From Walsh, Meyer and Lind, 1974.

mechanical effect of ventilating the lung with gas (Cassin et al., 1964) consequent possibly to changes in alveolar geometry that may be produced by the establishment of a gas-liquid interface (Dawes, 1968). In other words, vessels exposed to alveolar pressure may be pulled open by the surface tension of the alveolar lining layer (Butler et al., 1962). Normal expression of lung fluid during delivery (which is maintained at a small positive pressure before birth) may also favor a decrease in pulmonary vascular resistance as may the change in intrapleural pressure (measured as intraesophageal pressure) from above atmospheric in the fetus in utero to subatmospheric (Scarpelli et al., 1967). As breathing continues and aeration improves, the rhythmic fluctuations in systemic pressure and changes in gas tension serve to effect a further decline in pulmonary vascular resistance.

REMODELING OF THE MICROSCOPIC STRUCTURE OF THE PULMONARY FORK WITH GROWTH

During fetal life, the aorta and pulmonary artery are subject to comparable pressures and have a similar, though not identical structure. During the second half of gestation, the media of both great vessels consists of overlapping elastic membranes which alternate with layers of circularly arranged smooth musculature. At birth the arrangement of the elastic membranes seems to be less regular in the pulmonary trunk than in the aorta and the membranes of the trunk contain more small gaps.

Pronounced structural remodeling begins after birth in the pulmonary trunk and its main branches. In newborns, the elastic membranes are closely packed (Fig. 154A). During the first months of life, they seem to "fragment" (Fig. 154B), a change which has been

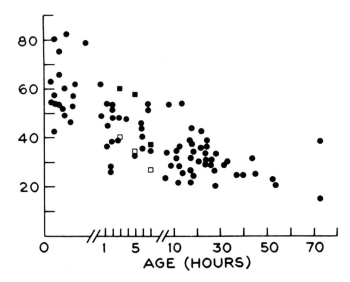

Fig. 153. Mean pulmonary arterial pressure rapidly declines after birth in normal human infants but at a somewhat faster rate when the placental transfusion is small. Average values at 3, 5, and 7 hours of age in 22 late clamped (closed squares) and 10 early clamped infants (open squares) are also shown. (● = individual values). Modified from Dawes, 1968. Based on data of Adams and Lind, 1957; Saling, 1960; Rudolph et al., 1961; Emmanouilides et al., 1964 and Arcilla et al., 1966. From Walsh, Meyer and Lind, 1974.

ascribed to "disuse atrophy" (Heath, 1959, 1969), but which has been shown on tangential sections to be due to the gradual spreading apart of preformed elastic membranes. By the age of five to six months, they form well-delineated star-like elastic sheets (Fig. 155B) (Meyer and Simon, 1959). On routine cross-sections, the membranes have stick-like shapes (Fig. 154B). The rearrangement of elastic membranes is accompanied by a steady increase in the smooth musculature which develops between the elastic membranes and becomes an essential structural component of the wall of the pulmonary trunk and its main branches.

An adult structural pattern is acquired by the age of two years. It is characterized by well-differentiated star-like elastic membranes which are distinctly seen on tangential sections (Figs. 3, 155C). The distance between the individual membranes is now greater. The membranes are connected by numerous fine elastic fibers and muscle bundles which parallel the fibers. In cross-sections, the amount of smooth musculature now appears to be much greater than in the aorta.

The structural remodeling which occurs after birth can be duplicated to some extent by overdistending the pulmonary trunk of newborn infants. This technique separates the elastic membranes of the media, and after fixation in a distended state, the individual star-like membranes can be distinguished. The microscopic structural pattern then resembles that seen in later infancy (Fig. 156).

Thus, with growth, the initial tight structure of the elastic tissue of the pulmonary trunk gradually becomes looser, a pattern that is adequate for the lower blood pressure to which the wall of the pulmonary artery is now exposed. Simultaneously, the mean ratio of wall thickness in the pulmonary trunk to that of the aorta declines from 1.0 at birth to 0.4 - 0.8 between six months and two years of age (Heath et al., 1959).

The fetal pattern of the pulmonary trunk and its branches, i.e., their dense lamellar structure, is retained, however, if pulmonary pressures remain at systolic level (Heath and Edwards, 1958). Thus, in patent ductus arteriosus accompanied by pulmonary hypertension, hyperplasia of the elastic structural components develops and the wall of the pulmonary trunk becomes as thick by the end of the first year of life as in an adult (Fig. 157).

DEVELOPMENT OF ELASTIC DISTENSIBILITY OF THE PULMONARY FORK WITH GROWTH

The elastic properties of the pulmonary trunk and its main branches can be studied in dissected specimens removed from the body (Meyer and Simon, 1959) (Fig. 158). A cannula is inserted into the proximal part of the pulmonary trunk and both main branches are ligated at the level of their first branch. In newborn infants, the ductus arteriosus is also ligated at its origin. The specimen is then filled with saline and exposed to increasing filling pressures. Curves are plotted which show changes in volume in different pressure ranges and ages (Fig. 159). On the basis of these data, the relative volume distensibility (rvd), i.e., the percentage increase in volume with pressure increment (rvd = $\frac{\Delta v \cdot 100}{V}$), is calculated. As in the aorta (p. 101), this paramenter permits comparison of elastic properties, i.e., the volume distensibility, in a growing pulmonary fork which steadily increases its absolute

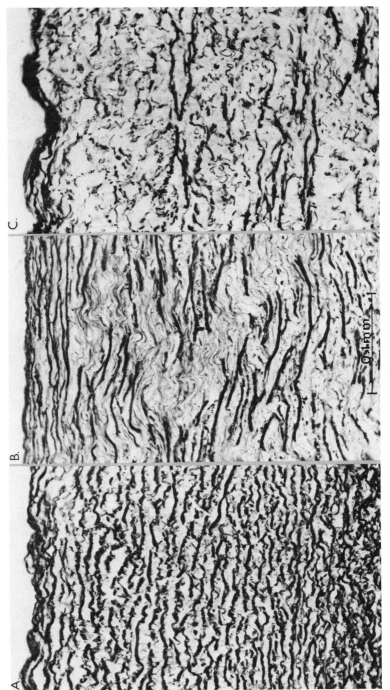

Fig. 154.

Fig. 154. Cross-sections of the pulmonary trunk from a mature stillborn (A), an 8-month-old infant (D) and a young adult (C). (Same specimens as those shown in Fig. 155). At birth (A), elastic membranes are densely packed. Later in infancy, (B), they appear fragmented and fine elastic networks are present in the interspaces. In the adult, (C), these membranes spread still further apart and much muscular tissue has developed between the elastic networks which have now assumed another direction. Near the lumen, the elastic membranes are closely packed and create the impression of an internal elastic lamella. All three cross-sections are shown at the same magnification. Hence the inner two-thirds of the wall are shown in Figs. A and B but only the inner third of the wall is shown in Fig. C.

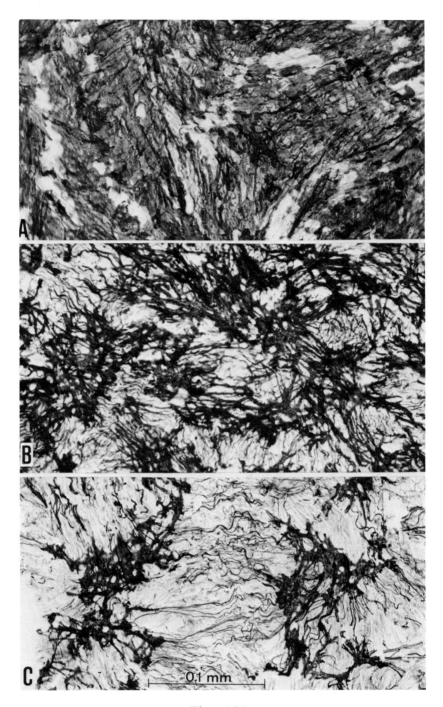

Fig. 155.

←Fig. 155. Tangential sections of the pulmonary trunk from a mature stillborn (A), 8-month-old infant (B), and an adult (C) at the same magnification. (Same specimens as those shown in Fig. 154.) Elastic membranes gradually spread apart during growth. At birth (A), they are very compactly arranged and overlap. Later in infancy (B), they spread apart and can then be seen to have a star-like form similar to that found in later life. In the adult (C), they spread further apart and numerous elastic fibers and smooth muscle cells (unstained) are present between them. Orcein stain. From Meyer and Simon, 1959.

volume. However, only the elastic properties of the elastic structural component of the arterial tube can be estimated in postmortal specimens. This is because the smooth musculature, which also increases considerably with growth in the pulmonary trunk, loses its active tone after death and becomes but a plastic material.

At term, the relative volume distensibility of the fork is high. In the pressure range of 10 - 50 mm Hg, the volume of the pulmonary fork rises by 150% and in the pressure range of 10 - 100 mm Hg by 227% (Fig. 161). Volume distensibility decreases after birth, and between two and five months of age, the relative increase in volume is only 80% and 95% in the same pressure ranges (Fig. 160). The decrease in volume distensibility reflects the change in functional load on the pulmonary fork, i.e., the drop in blood pressure, and the associated structural remodeling in the pulmonary trunk after birth. After the first year of life and during the first two decades, the relative volume distensibility of the pulmonary fork again increases progressively but not to the same high levels as at birth. Thus, postnatal development of the elastic properties of the pulmonary fork differs considerably from that of the aorta (p. 101).

The relative volume distensibility remains elevated after birth in conditions accompanied by persistent pulmonary hypertension, e.g., persistent ductus arteriosus (Meyer and Simon, 1959) (Fig. 161). In acquired pulmonary hypertension, i.e., mitral stenosis, the relative volume distensibility rises considerably above the normal level (Meyer and Schollmeyer, 1957). As with a patent ductus arteriosus, the increase is related to the pronounced hyperplasia of the elastic elements of the pulmonary fork that develops with hypertension. This illustrates the close relationship between pulmonary pressure and both volume distensibility and the arterial structural pattern.

CLOSING REMARKS

1. The cardiovascular system is the first organ system to become established. This occurs when the embryo is only a few millimeters long because it is incapable of growing without a functional

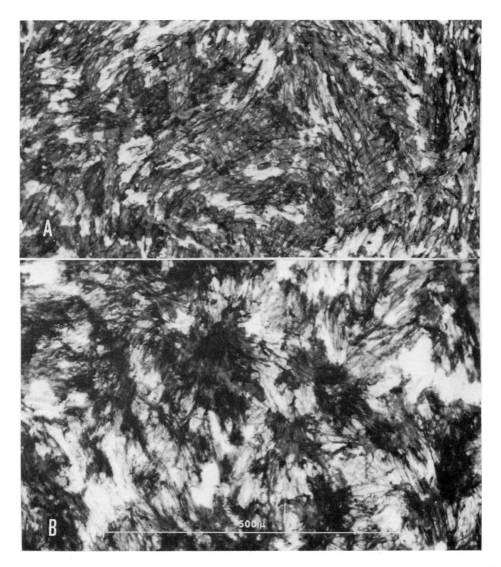

Fig. 156. Tangential sections of a non-distended (A) and a distended (B) pulmonary trunk in a newborn infant. (A) At birth, star-like membranes overlap each other and only a few areas do not contain elastic elements. (B) When the pulmonary trunk is fixed at a pressure of 60-100 mm Hg, individual star-like elastic membranes become visible and the microscopic appearance then resembles that seen in later infancy (see Fig. 155B). However, the muscular elements and fine elastic networks connecting these membranes have still not developed. Orcein stain. Both figures photographed at same magnification. From Meyer and Simon, 1959.

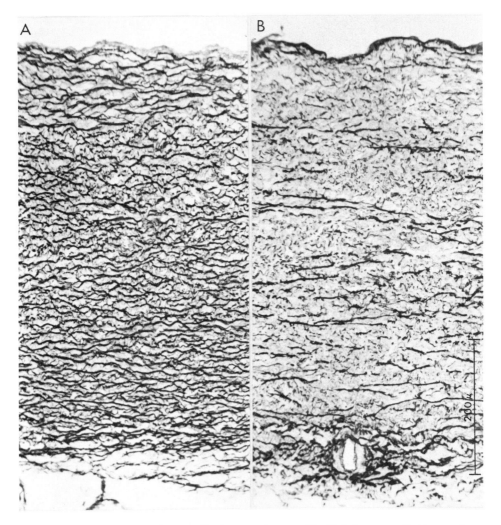

Fig. 157. Cross-sections of the entire wall thickness of the pulmonary trunk from a 4 1/2-month-old infant with a patent ductus arteriosus (A) and a healthy young adult (B). In the presence of a patent duct, the wall of the pulmonary trunk of the infant retains its fetal structural pattern and becomes as thick as the wall of the adult pulmonary artery. Below, a small artery (B) is seen in the outer part of the adult pulmonary trunk. Both figures photographed at same magnification. Orcein elastic stain. From Meyer and Simon, 1959.

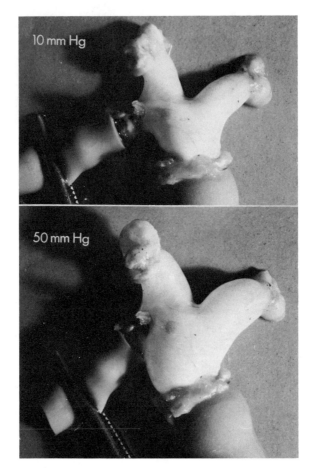

Fig. 158. Demonstration of the method used for determining the volume distensibility of the pulmonary trunk and its main branches (pulmonary fork) in a full-term newborn infant. Cannulas are inserted in the pulmonary trunk (below) and ductus arteriosus (left). Both main branches are ligated. The vessel segment increases significantly in size with increase in pressure. x 2.3 (see Figs. 159, 160, 161). From Meyer and Simon, 1959.

circulation. By the end of the second month of life, apart from not entirely adequate atrioventricular valves and a patent interatrial foramen, partition of the heart and great vessels is complete. After mid-gestation the structural changes in the heart largely reflect the remodeling process that accompanies growth in size and elaboration of the muscular pattern. Arterial growth, especially after growth, is associated with significant alterations in structural pattern and function. As in the heart, these alterations represent an adaptation to the changing hemodynamic load.

2. Normal development of the fetus and rapid postnatal adaptation depend largely on patency of temporary fetal vascular channels and their ability to close rapidly after delivery. The structure of these channels differs considerably from that of persistent fetal and adult vessels. Since three of the vessels have a muscular wall,

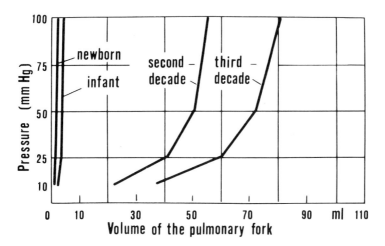

Fig. 159. Curves showing postmortal pressure-volume relations in the "pulmonary fork" at different ages. Each curve has been derived from mean volumes obtained at different pressures in a number of pulmonary arterial segments. The steepness of the slopes of the curves in newborns and infants reflects the small absolute increase in initial volume. Such curves provide no information concerning the volume distensibility of the "pulmonary fork" but the data on which they are based may be used for plotting curves of relative volume distensibility (see Fig. 160). Modified from Meyer, 1964.

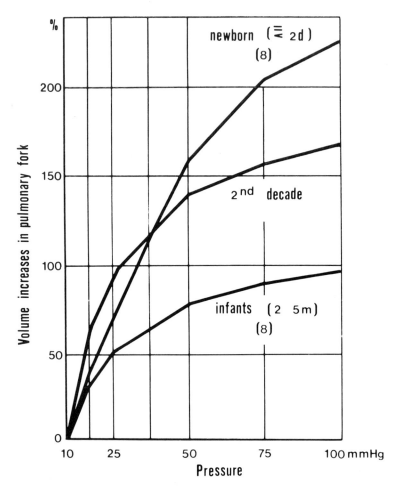

Fig. 160. Changes in relative volume distensibility of the pulmonary fork with growth. Each curve shows the mean percent increase in initial volume with increasing pressure. Note the higher distensibility of the pulmonary fork in newborn infants and its marked decrease shortly after birth. The distensibility increases again in the second decade. From Meyer and Simon, 1959.

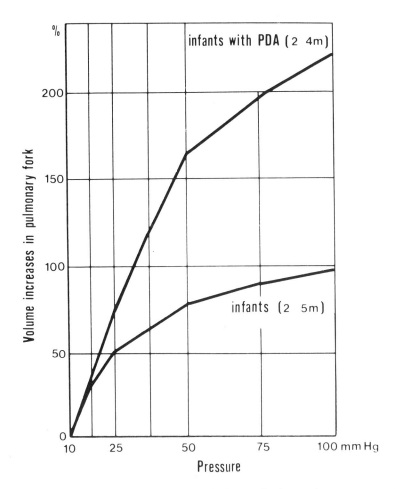

Fig. 161. Relative volume distensibility of the pulmonary trunk and its main branches in two- to five-month-old infants with no evidence of heart disease (lower curve) and in those of similar age with a persistent ductus arteriosus (upper curve). No postnatal decrease in volume distensibility occurs in infants with a patent duct (see Fig. 160). From Meyer and Simon, 1959.

i.e., the umbilical arteries and vein and the ductus arteriosus, patency and closure are presumably best achieved by smooth musculature, at least in these vessels. Various structural characteristics seem to favor rapid closure. One such characteristic feature is the presence of large amounts of metachromatic ground substance between the muscular elements. This substance probably helps the muscular elements to glide over each other and thereby ensures profound rearrangement of the musculature. Poor development of elastic and collagenous tissue is another important structural feature which favors rapid closure because large amounts of unstretched elastic and/or collagenous tissue would resist contraction and prevent closure of the vessel. Some preparatory changes that lead to dissociation of the intimal-medial junction eliminate shearing forces and also significantly contribute to complete closure of the duct. At least theoretically, poor development of elastic tissue may affect the equilibrium between the distending force of the blood pressure and the vascular wall and result in premature closure of temporary vessels, particularly of the umbilical arteries or vein, and thereby endanger the life of the fetus. It is uncertain, however, whether this occurs in man in vivo.

3. The aorta constitutes the main vessel in the systemic circulation and transforms the highly pulsatile flow produced by the left ventricle into more continuous streaming of blood to the distributing arteries. Its wall is very thick in the fetus. The ratio of the thickness of its wall to its radius is considerably higher in the newborn infant than in the adolescent and young adult. The highest reversible volume distensibility, i.e., the highest degree of accommodation to an increased functional demand, is attained about the end of the second decade, i.e., at the same age that the ratio of wall thickness to radius is lowest. Thus, postnatal differentiation of aortic structure is associated with relative thinning of the wall, particularly of the media, and development of a structural pattern which ensures the highest functional capacity with a minimum of substrate. The exact nature of the medial changes that occur with growth, however, are still obscure. As in other arteries, considerable remodeling of the internal aortic layer occurs with growth and the changes differ considerably in individual segments. The internal aortic layer seems to contain fewer cells than the media, and with early lipid deposition which is initially confined to this layer, nearly all cells become laden with lipids. It is uncertain to what extent normal metabolic activity of these cells may be impaired in early life by increasing lipid-deposition and whether, as a result, the intercellular substance of this layer may "run out of control" during growth. It seems possible that reduction of cellular control over intercellular substance may favor the development and progression of atherosclerotic lesions at sites of fatty streaking.

4. Of all human vessels, the coronary arteries demonstrate the most pronounced postnatal remodeling of structure, especially during the first years of life. In the main coronary trunks, complex longitudinal musculo-elastic structures become the major constituent of the vascular wall. This presumably occurs in response to the hemodynamic load imposed by regular contraction of the heart and pulsatile streaming of blood in these arteries. The major changes in structural pattern are encountered along the short proximal segments, i.e., immediately distal to the orifices where cross-sections differ considerably at successive levels. In retracted arteries, the elements of the internal longitudinal musculo-elastic layer may protrude into the lumen and resemble "intimal cushions." When the arteries are fixed under appropriate pressure, however, the coronary wall unfolds and the cushions flatten. The structural changes resulting from growth-bound remodeling and the variations in structure along the course of the proximal coronary segments, as well as the artifacts due to postmortal contraction, must be kept in mind when examining coronary arteries of patients with congenital heart disease. Knowledge of the histotopography of the coronary arteries and their postnatal development may be of value in studying early lipid deposition and its initial localization.

5. Rapid growth of the human brain in the fetus is accompanied by accelerated growth of the carotid and cerebral arteries. This is reflected by the caliber of aortic arch vessels in fetuses and newborn infants which is comparatively larger than in adults. The early development of lipid deposits in the common carotid artery and in the carotid sinus may be favored by accelerated growth of these arterial segments and early differentiation of specialized intimal components which presumably reflect a structural accommodation to a higher functional load.

The development of tortuous segments, i.e., the vertebral and the carotid siphons between the extracranial and intracranial cerebral arteries, probably serves to damp the pulse wave which might otherwise damage the brain tissue in its rigid bony cavity. The tortuosity of these vascular segments probably places an additional functional load upon the arterial tube and leads to differentiation of structural components which often become the sites of predilection of early calcific and lipid deposits. The calcifications result in distortion of the arterial wall and may favor the development and later progression of atherosclerotic lesions and occlusive vascular disease.

6. Remodeling of arterial structure consequent to accelerated growth occurs in the common and internal iliac arteries, vessels which connect the aorta to the placenta during intrauterine life. No similar remodeling occurs in the external iliac artery, a vessel which does not participate in the placental circulation and grows

more slowly. The distinctive structural features of the iliac arteries in infants with a single umbilical artery illustrate the importance of the relationship between functional load and structural features in growing human vessels.

7. Peculiar gaps in the internal elastic membrane appear during growth in large muscular arteries, and especially in extremity arteries. They seem to represent a structural accommodation to the longitudinal stretch to which these arteries are subjected by movements. The gaps may be important for nutrition of the arterial wall. The margins of the interrupted elastic membrane along the gaps become the site of predilection of early calcifications during growth.

8. Postnatal development of arterial structure in the pulmonary circulation seems to be related to various adaptive mechanisms that lead to a postnatal decline in pulmonary vascular resistance and pressure accompanied by closure of the duct. This is associated with the gradual development of a structural pattern which seems to be characteristic of a low pressure system and is not seen in arteries of the systemic circulation.

9. With growth and differentiation of the arteries, special structures develop in the arterial wall which become sites of early pathological changes, i.e., calcification and/or lipid deposits. These changes often increase in severity. Thus, the seeds of later occlusive disease appear to be sown in childhood and adolescence (Strong and McGill, 1969).

REFERENCES

Adams, F.H., Fujiwara, T., and Rowshan, G., 1963, The nature and origin of the fluid in the fetal lamb lung, J. Pediat. 63:881.
Adams, F.H., Yanagisawa, M., Kuzela, D., and Martinek, H., 1971, The disappearance of fetal lung fluid following birth, J. Pediat. 78:837.
Allen, F.D., 1955, The innervation of the human ductus arteriosus, Anat. Rec. 122:611.
Altura, B.M., Malaviya, D., Reich, C.F., and Orkin, L.R., 1972, Effects of vasoactive agents on isolated human umbilical arteries and veins, Am. J. Physiol. 222:345.
Arcilla, R.A., Oh, W., Lind, J., and Blankenship, W., 1966, Portal and atrial pressures in the newborn period, Acta Pediatr. Scand. 55:615.
Assali, N.S., Rauramo, L., and Peltonen, T., 1960, Measurement of uterine blood flow and uterine metabolism. VIII. Uterine and fetal blood flow and oxygen consumption in early pregnancy, Am. J. Obstet. Gynecol. 79:86.
Assali, N.S., Morris, J.A., Smith, R.W., and Manson, W.A., 1963, Studies on ductus arteriosus circulation, Circ. Res. 13:478.

Assali, N.S., Morris, J.A., and Beck, R., 1968, Cardiovascular hemodynamics in the fetal lamb before and after lung expansion, Am. J. Physiol. 208:122.

Aterman, K., Dische, R., Franke, J., Fraser, G.M., and Meyer, W.W., 1977, Aneurysms of the coronary arteries in infants and children, Virchows Arch. A. Path. Anat. Histol. 374:27.

Bader, H., 1963, The anatomy and physiology of the vascular wall, in "Handbook of Physiology" Section 2: Circulation, Vol.II (W.F. Hamilton and P. Dow, eds.), American Physiological Society, Washington, D.C.

Barclay, A.E., Franklin, K.J., and Prichard, M.M.L., 1944, The Foetal Circulation and Cardiovascular System and the Changes that they undergo at Birth, pp. 74-75, 160-161, 219, Blackwell Scientific Publications, Oxford.

Barker, J.N., 1966, Fetal and neonatal cerebral blood flow, Am. J. Physiol. 210:897.

Bartels, H.W., 1970, "Prenatal Respiration," pp. 148-150, North Holland, Amsterdam-London.

Benecke, R., 1922, Atherosklerose der A. carotis communis und ihre Bedeutung für das Verständnis der Blutsäulenformen, Frankf. Z. Path. 28:409.

Benirschke, K., and Bourne, G.L., 1960, The incidence and prognostic implication of congenital absence of one umbilical artery, Am. J. Obst. Gyn. 79:251.

Benninghoff, A., 1930, Blutgefässe und Herz, in "Handbuch der mikroskopischen Anatomie des Menschen" (W.v.Möllendorff, ed.), Vol. 6, Part 1, p. 93, Julius Springer, Berlin.

Bergofsky, E.H., and Holtzman, S., 1967, A study of the mechanisms involved in the pulmonary arterial pressure response to hypoxia, Circ. Res. 20:506.

Berry, C.L., Gosling, R.G., Laogun, A.A., and Bryan, E., 1976, Anomalous iliac compliance in children with a single umbilical artery, Brit. Heart J. 38:510.

Bickel, E., and Janssen, W., 1963, Arteriopathia calcificans infantum, Arch. f. Kinderheil. 169:274.

Binswanger, O., and Schaxel, J., 1917, Beiträge zur normalen und pathologischen Anatomie der Arterien des Gehirns, Arch. f. Psychiatrie Nervenkr. 58:141.

Blanc, W.A., and Nodot, A., 1957, An anatomical study of circulatory adjustment at birth. Normal and dilated ductus, Am. J. Dis. Child. 94:533.

Blumenthal, L.S., 1947, Pathologic significance of the ductus arteriosus, Arch. Path. 44:372.

Böger, A., and Wezler, K., 1936, Die zentrale Stellung des Windkessels im Kreislauf, Klin. Wschr. 15:1185, 1241.

Bolt, W., 1948, Die hämodynamischen Kreislaufgrössen im Kindesalter, Klin. Wschr. 26:590.

Bor, J., and Valach, V., 1959, Evaluation of pulmonary hypertension in patent ductus arteriosus, Ann. Pediatr. 193:14.

Boréus, L.O., Malmfors, T., McMurphy, D.M., and Olson, L., 1969, Demonstration of adrenergic receptor function and innervation in the ductus arteriosus of the human fetus, Acta. Physiol. Scand. 77:316.

Born, G.V.R., Dawes, G.S., Mott, J.C., and Rennick, R., 1956, The constriction of the ductus arteriosus caused by oxygen and by asphyxia in newborn lambs, J. Physiol. (Lond). 132:304.

Boston, R.W., Humphreys, P.W, Reynolds, E.O.R., and Strang, L.B., 1965, Lymph flow and clearance of liquid from the lungs of the fetal lamb, Lancet, 1965, 2:473.

Botha, M.C., 1968, The management of the umbilical cord in labor, S.A.J. Obstet. Gynecol. 6:30.

Boyd, J.D., 1940, The nerve supply of the mammalian ductus arteriosus, J. Anat. 75:457.

Brady, J.P., and Tooley, W.H., 1966, Cardiovascular and respiratory reflexes in the newborn infant, Pediatr. Clin. North. Am. 13:801.

Bruce, J., Walmsley, R., and Ross, J.A., 1964, "Manual of Surgical Anatomy," E. and S. Livingstone, Edinburgh.

Bryan, E.M., and Kohler, H.G., 1974, The missing unbilical artery, I. Prospective Study based on a maternal unit, Arch. Dis. Childh. 49:844.

Buddecke, E., 1963, Biochemische Grundlagen arterieller Verschlusskrankheiten, Verh. dtsch. Ges. Kreisl.-Forsch. 29:182.

Bunce, D.F.M., II, 1961, Structure of the distended vascular wall. IVth International Congress of Angiology (Prague), pp. 708-713.

Bunce, D.F.M., II, 1974, "Atlas of Arterial Histology," Warren H. Green, St. Louis.

Burnard, E.D., and James, L.S., 1961, Radiographic heart size in apparently healthy newborn infants - Clinical and biochemical correlations, Pediatrics 27:726.

Burnard, E.D., 1966, Influence of delivery on the circulation, in "The Heart and Circulation in the Newborn and Infant" (D.E. Cassels, ed.) p. 92, Grune & Stratton, Inc., New York.

Burton, A.C., 1972, "Physiology and Biophysics of the Circulation," Second edition, Year Book Medical Publishers, Chicago.

Butler, T., Brunderman, I., Hamilton, W.K., and Tooley, W.H., 1962, Surface tension and pulmonary vascular resistance, Fed. Proc. 21:445.

Campbell, A.G.M., Dawes, G.S., Fishman, A.P., and Hyman, A.I., 1967, Regional redistribtion of blood flow in the mature fetal lamb, Circ. Res. 21:229.

Carmichael, R., 1945, Gross defects in the muscular and elastic coats of the larger cerebral arteries, J. Path. Bact. 57:345.

Cassels, D.E., 1973, "The Ductus Arteriosus," Charles C. Thomas, Springfield, p. 309.

Cassin, S., Dawes, G.S., Mott, J.C., Ross, B.B., and Strang, L.B., 1964, The vascular resistance of the foetal and newly ventilated lung of the lamb, J. Physiol. 171:61.

Celander, O., 1960, Blood flow in the foot and calf of the newborn, Acta Paediatr. Scand. 49:488.
Celander, O., 1966, Studies of the peripheral circulation, in "The Heart and Circulation in the Newborn and Infant," (D. Cassels, ed.), Grune and Stratton, New York.
Christie, A., 1930, Normal closing time of the foramen ovale and the ductus arteriosus, Am. J. Dis. Child. 40:323.
Chuaqui, B., Piwonka, G., and Farrú, A.O., 1977, Über den Wandbau des persistierenden Ductus arteriosus, Virchows Arch. A. Path. Anat. Histol. 372:315.
Civin, W.W., and Edwards, J.E., 1951, Postnatal structural changes in the intrapulmonary arteries and arterioles, AMA Arch. Path. 51:192.
Clements, J.A., 1962, Surface phenomena in relation to pulmonary function, Physiologist 5:11.
Colebatch, H.J.H., Dawes, G.S., Goodwin, J.W., and Nadeau, R.A., 1965, The nervous control of the circulation in the foetal and newly expanded lungs of the lamb, J. Physiol. 178:544.
Condorelli, S., and Cosmi, E.V., 1972, Cardiovascular respiratory reflexes in the adult and fetal animal: Evidence for a reflex originating in skeletal muscle, in "Bases Fondamentales de l'Anesthesie et de la Réanimation Obstétricale," Vol. 4, Librairie Arnette Publ., Paris.
Cook, C.D., Drinker, P.A., Jacobson, N.H., Levison, H., and Strang, L.B., 1963, Control of pulmonary blood flow in the foetal and newly born lamb, J. Physiol. 169:10.
Cosmi, E.V., Condorelli, S., and Myers, R.E., 1973, Reanimazione del neonato di pecora dopo asfissia acuta endouterina, Acta Anes. Ital. 24:133.
Cosmi, E.V., 1975, Fetal Homeostasis, in "Pulmonary Physiology of the Fetus, Newborn and Child," (E.M. Scarpelli, ed.) pp. 61-95, Lea & Febiger, Philadelphia.
Cox, G.E., Trueheart, R.E., Kaplan, J., and Taylor, C.B., 1963, Atherosclerosis in rhesus monkeys: IV. Repair of arterial injury - an important secondary atherogenic factor, Arch. Path., 76:166.
Cross, K.W., Klaus, M., Tooley, W.H., and Weisser, K., 1960, The response of the newborn baby to inflation of the lungs, J. Physiol. (Lond.) 151:551.
Dahlquist, G., 1976, "Utilization of ketone bodies by the developing brain," Thesis, Karolinska Institute, Stockholm.
Dalith, F., 1961, Calcification of the aortic knob: its relationship to fifth and sixth embryonic aortic arches, Radiology 76:213.
Dalith, F., 1964, An embryological concept for the prevalence of atherosclerotic lesions in the abdominal aorta, J. Atheroscler. Res. 4:239.
Dalith, F., 1968, Embryological aspects of atherogenesis, Ann. N.Y. Acad. Sci. 149:865.

Dalith, F., 1971, A concept concerning genetically controlled properties of the normal arterial wall in atherosclerotic plaque formation, Lex and Scientia 8:104.

Dalith, F., and Molho, M., 1962, Calcified plaques in the brachiocephalic arteries: A concept of focal atherogenesis, J. Atheroscler. Res. 2:416.

Dammann, J.F., and Ferencz, C., 1956, The significance of the pulmonary vascular bed in congenital heart disease, Am. Heart. J. 52:7.

Dawes, G.S., 1968, "Foetal and Neonatal Physiology: A Comparative Study of the Changes at Birth," p. 167, Year Book Medical Publishers, Inc., Chicago.

Dawes, G.S., Mott, J.C., Widdicombe, J.G., and Wyatt, D.G., 1953, Changes in the lungs of the newborn lamb, J. Physiol. (Lond.) 121:141.

Dawes, G.S., Mott, J.C., and Widdicombe, J.G., 1955, Closure of the foramen ovale in newborn lambs, J. Physiol. 128:384.

Dawes, G.S., Mott, J.C., Shelley, H.J., and Stafford, A., 1963, The prolongation of survival time in asphyxiated immature fetal lambs, J. Physiol. 168:43.

Dawes, G.S., and Mott, J.C., 1964, Changes in O_2 distribution and consumption in fetal lambs with variations in umbilical blood flow, J. Physiol. 170:524.

De La Torre, E., and Netsky, M.G., 1960, Study of persistent maxillary artery in human fetus: some homologies of cranial arteries in man and dog, Am. J. Anat. 106:185.

Dietrich, K., 1930, Beiträge zur Pathologie der Arterien des Menschen, Virchows Arch. Pathol. Anat. 274:452.

Dintenfass, L., 1971, "Blood microrheology," Butterworths, London.

Dobrin, P.B., 1969, Influence of vascular smooth muscle on contractile mechanics and elasticity of arteries, Amer. J. Physiol. 217:1644.

Dock, W., 1946, The predilection of atherosclerosis for the coronary arteries, J.A.M.A. 131:875.

Dörfler, J., 1935, Ein Beiträg zur Frage der Lokalisation der Arteriosklerose der Gehirngefässe mit besonderer Berücksichtigung der Arteria carotis interna, Arch. Psych. Nervenkrh. 103:180.

Duke, H.N., 1954, The site of action of anoxia on the pulmonary blood vessels of the cat, J. Physiol. 125:373.

Edwards, J.E., 1957, Functional pathology of the pulmonary vascular tree in congenital cardiac disease, Circulation 15:164.

Ehinger, B., Gennser, G., Owman, C., Persson, H., and Sjöberg, N.O., 1968, Histochemical and pharmacological studies on amine mechanisms in the umbilical cord, umbilical vein and ductus venosus of the human fetus, Acta Physiol. Scand. 72:15.

Ehrich, W., de la Chapelle, C.E., and Cohn, A.E., 1931, Anatomical ontogeny, B. Man. I. A study of the coronary arteries, Am. J. Anat. 49:241.

Ellison, J.P., 1971, The nerves in the umbilical cord in man and the rat, Am. J. Anat. 132:53.
Elsässer, C.L., cited by Dragendorff, O., 1931, in "Handbuch der Anatomie des Kindes" (K. Peter, G. Wetzel, and F. Heiderich, eds.), Vol. II, p. 319, J. F. Bergmann, Munich.
Elsner, R., Hammond, D.D., and Parker, H.R., 1970, Circulatory responses to asphyxia in pregnant and fetal animals - A comparative study of widdell seals and sheep, Yale J. Biol. Med. 42:202.
Eltherington, L.G., Stoff, J., Hughes, T., and Melmon, K.L., 1968, Constriction of human umbilical arteries, Circ. Res. 22:747.
Emery, J.L., and MacDonald, M.S., 1960, The weight of the ventricles in the later weeks of intrauterine life, Br. Heart J. 22:563.
Engström, L., Karlberg, P., Rooth, G., and Tunell, R., 1966, The Onset of Respiration, in "A Study of Respiration and Changes in Blood Gases and Acid-base Balance," Association for the Aid of Crippled Children, New York.
Faierman, E., 1960, The significance of one umbilical artery, Arch. Dis. Childh. 35:285.
Fangman, R.J., and Hellwig, C.A., 1947, Histology of the coronary arteries in newborn infants, Am. J. Pathol. 23:901.
Fay, F.S., 1971, Guinea pig ductus arteriosus, I. Cellular and metabolic basis for oxygen sensitivity, Am. J. Physiol. 221:470.
Fay, F.S., 1973, Biochemical basis for response of ductus arteriosus to oxygen, in "Foetal and Neonatal Physiology" (K.S. Comline, K.W. Cross, G.S. Dawes, D. W. Nathanielsz, eds.) Cambridge University, Cambridge.
Fay, J.E., and Travill, A., 1967, The "valve" of the ductus arteriosus - an enigma, Can. Med. Assoc. J. 97:78.
Fazekas, J.F., Alexander, F.A.D., and Himwich, H.E., 1941, Tolerance of the newborn to anoxia, Am. J. Physiol. 134:281.
Fisher, C., Gore, E., Okabe, N., and White, P.D., 1965, Atherosclerosis of the carotid and vertebral arteries - extracranial and intracranial, J. Neuropath. Exp. Neurol. 24:455.
Forbus, W.B., 1930, On the origin of miliary aneurysms of the superficial cerebral arteries, Bull. Johns Hopkins Hosp. 47:239.
Fredrickson, D.S.R., Levy, R.I., and Lees, R.S., 1967, Fat transport in lipoproteins - an integrated approach to mechanisms and disorders, New Engl. J. Med. 276:34, 94, 148, 215, 273.
Fritts, H.W., Jr., Richards, D.W., and Cournand, A., 1960, Oxygen consumption of tissues in the human lung, Science 133:1070.
Froehlich, L.A., and Fujikura, T., 1973, Follow-up of infants with single umbilical artery, Pediatrics 52:6.
Fry, D., 1972, Personal communication.
Fujikura, T., 1964, Single umbilical artery and congenital malformations, Am. J. Obstet. Gynec. 88:829.
Fujikura, T., and Froehlich, L., 1972, A follow-up of children with single umbilical artery, Am. J. Path. 66:19.
Garfunkel. J.M., Baird, H.M., and Ziegler, J., 1954, The relationship of oxygen consumption to cerebral functional activity, J. Pediatr. 44:64.

Glazier, J.B., and Murray, J.F., 1971, Sites of pulmonary vasomotor reactivity in the dog during alveolar hypoxia and serotonin and histamine infusion, J. Clin. Invest. 50:2550.

Gluck, L., Talner, N.S., Gardner, T.H., and Kulovich, M.V., 1964, RNA concentrations in the ventricles of full term and premature rabbits following birth, Nature 202:770.

Gillman, T., 1968, On the possible roles of arterial growth, remodelling, repair, and involution in the genesis of arterial degeneration, Ann. NY Acad. Sci. 149:731.

Goerttler, K., 1951, Die Bedeutung der funktionellen Struktur der Gefässwand, I. Untersuchungen an der Nabelschnur-Arterie des Menschen, Gegenbaur's Morphologisches Jahrbuch 91:368.

Goldman, H., and Sapirstein, L.A., 1973, Brain blood flow in the conscious and anesthetized rat, Am. J. Physiol. 224:122.

Gonzales-Crussi, F., and Boston, R.W., 1972, The absorptive function of the neonatal lung, Lab. Invest. 26:114.

Goodlin, R.C., and Rudolph. A.M., 1970, Tracheal flow and function of fetuses in utero, Am. J. Obstet. Gynec. 106:597.

Graser, F., 1953, Die Pulswellengeschwindigkeit im frühen Kindesalter, Klin. Wschr. 31:816.

Gräper, L., 1921, Die anatomischen Veränderungen kurz nach der Geburt. III. Ductus Botalli, Z. Anat. Entwickl. Gesch. 61:312.

Grasso, S., and Selye, H., 1962, Calciphylaxis in relation to the humoral production of occlusive coronary lesions with infarction, J. Path. Bact. 83:495.

Gross, L., 1921, "The Blood Supply to the Heart in its Anatomical and Clinical Aspects," P.B. Hoeber, New York.

Gross, L., Epstein, E.Z., and Kugel, M.A., 1934, Histology of the coronary arteries and their branches in the human heart, Am. J. Pathol. 10:253.

Grünstein, M., 1896, Über den Bau der grösseren menschlichen Arterien in verschiedenen Altersstufen, Arch. Mikr. Anat. 47:583.

Gundobin, N.P., 1912, "Die Besonderheiten des Kindesalters," Allgemeine Medizinische Verlagsanstalt, Berlin.

Gunther, M., 1957, The transfer of blood between baby and placenta in the minutes after birth, Lancet 1:1277.

Hackel, W.M., 1928, Über den Bau und die Altersveränderungen der Gehirnarterien, Virchows Arch. Path. Anat. 266:630.

Harris, P., and Heath, D., 1962, "The Human Pulmonary Circulation," E. & S. Livingstone, Ltd., Edinburgh, p. 139.

Hass, G.M., 1967, The pathogenesis of human and experimental athero-arteriosclerosis, in "Cowdry's Arteriosclerosis, A Survey of the Problem" (H.T. Blumenthal, ed.) p. 689, 2nd ed., Charles C. Thomas, Springfield, Illinois.

Hass, G.M. Trueheart, R.E., and Hemmens, A., 1961, Experimental athero-arteriosclerosis due to calcific medial degeneration and hypercholesterolemia, Am. J. Path. 38:289.

Hassler, O., 1962, Physiological intimal cushions in the large cerebral arteries of young individuals, Acta. Pathol. Microbiol. Scand. 55:19.

Hauge, A., 1968, Role of histamine in hypoxic pulmonary hypertension in the rat, I. Blockade or potentiation of endogenous amines, kinins and ATP, Circ. Res. 22:371.

Hauge, A., and Melmon, K.L., 1968, Role of histamine in hypoxic pulmonary hypertension in the rat, II. Depletion of histamine, serotonin and catecholamines, Circ. Res. 22:385.

von Hayek, H., 1935, Der functionelle Bau der Nabelarterien und des Ductus Botalli, Z. Anat. Entwickl. Gesch. 105:15.

Heard, E., 1950, Irreducible folds in the internal elastic lamina of the renal arteries, J. Path. Bact. 62:591.

Heath, D., 1969, Pulmonary vasculature in post-natal life and pulmonary haemodynamics, in "The Anatomy of the Developing Lung" (J. Emery, ed.) p. 147, William Heinemann Medical Books Ltd., Suffolk.

Heath, D., Du Shane, I.W., Wood, E.H., and Edwards, J.E., 1959, The structure of the pulmonary trunk at different ages in cases of pulmonary hypertension and pulmonary stenosis, J. Path. Bact. 77:443.

Heath, D., and Edwards, J.E., 1958, The pathology of hypertensive and pulmonary vascular disease: a description of six grades of structural changes in the pulmonary arteries with special reference to congenital cardiac septal defects, Circulation 18:533.

Heymann, M.A., Rudolph, A.M., Nies, A.S., and Melmon, K.I., 1969, Bradykinin production associated with oxygenation of the fetal lamb, Circ. Res. 25:521.

Heymann, M.A. Creasy, R.K., and Rudolph, A.M., 1973, Quantitation of blood flow patterns in the foetal lamb in utero, in "Foetal and Neonatal Physiology" (K.S. Comline, K.W. Cross, G.S. Dawes, Nathanielsz, eds.) pp. 129-135, Proceedings of the Sir Joseph Barcroft Centenary Symposium, Cambridge, University Press, Cambridge.

Hillier, K., and Karim, S.M.M., 1968, Effects of prostaglandins E_1, E_2, $E_1\alpha$, $F_2\alpha$, on isolated human umbilical and placental blood vessels, J. Obstet. Gynecol. Br. Commonw. 75:667.

Hislop, A., 1969, The non-muscular phase of the pulmonary circulation in the child, in "Fifth International Cystic Fibrosis Conference" (D. Lawson, ed.) p. 340, C. Nicholls and Co., Ltd., London.

Hoffman, J.I.E., 1975, The normal pulmonary circulation, in "Pulmonary Physiology of the Fetus, Newborn and Child" (E.M. Scarpelli, ed.) pp. 259-272, Lea & Febiger, Philadelphia.

Hoffman, J.I.E., Danilowicz, D., and Rudolph, A.M., 1965, Hemodynamics, clinical features and course of atrial shunts in infancy, Circulation, 32 (Suppl. 2):113.

Holmes, R.L., 1958, Some features of the ductus arteriosus, J. Anat. 92:304.

Hörnblad, P.Y., 1967, Studies on closure of the ductus arteriosus, III. Species differences in closure rate and morphology, Cardiologia (Basel) 51:262.

Hörnblad, P.Y., 1969, Effect of oxygen and umbilical cord clamping on closure of the ductus arteriosus in the guinea pig and the rat. Studies on closure of the ductus arteriosus, VI, Acta. Physiol. Scand. 76:58.

Hörnblad, P.Y., 1971, Intercellular changes in the human aortic wall during pre-neonatal and early postnatal life, Anat. Anz. 128:375.

Hörnblad, P.Y., and Larsson, K.S., 1967, Studies on closure of the ductus arteriosus. I. Whole-body freezing as improvement of fixation procedures, Cardiologia (Basel) 51:231.

Hort, W., 1955, Morphologische Untersuchungen an Herzen vor, während und nach der postnatalen Kreislaufumstellung, Virchows Arch. Path. Anat. 326:458.

Hort, W., 1966, The normal heart of the fetus and its metamorphosis in the transition period, in "The Heart and Circulation in the Newborn and Infant" (D.E. Cassels, ed.) p. 210, Grune & Stratton, Inc., New York.

Hultquist, G.T., 1942, "Über Thrombose und Embolie der Arteria carotis und hierbei vorkommende Gehirnveränderungen," Gustav Fischer Verlag, Jena.

Hürthle, K., 1934, Experimentelle Beeinflussung der Höhen und Formen der Carotis- und Cruralispulsen, Pflügers Arch. 233:262.

Hutchinson, E., and Yates, P., 1957, Carotico - vertebral stenosis, Lancet 1:2.

Ikeda, M., Sonnenschein, R.R., and Masuoka, D.T., 1972, Catecholamine content and uptake of the ductus arteriosus of the fetal lamb, Experentia 28:914.

Jaffé, D., Hartroft, W.S., Manning, M., and Eleta, G., 1971, Coronary arteries in newborn children, Acta Paediatr. Scand., suppl. 219.

Jager, B.V., and Wollenman, O.J. Jr., 1942, An anatomical study of the closure of the ductus arteriosus, Amer. J. Path. 18:595.

James, L.S., 1959, "Adaptation to Extrauterine Life," 31st Ross Conference on Pediatric Research., Columbus, Ohio, Ross Laboratories, p. 28.

James, L.S., Burnard, E.D., and Rowe, R.D., 1961, Abnormal shunting through the foramen ovale after birth, (Abstr.), Am. J. Dis. Child. 102:550.

James, T.N., 1961, "Anatomy of the Coronary Arteries," P.B. Hoeber, New York.

Javid, H., Ostermiller, W.E., Hengesh, J.W., Dye, W.S., Hunter, J.A., Najafi, H., and Julian, O.C., 1970, Natural history of carotid bifurcation atheroma, Surgery 67:80.

Jonsson, C.E., Tuxemo, T., and Hamberg, M., 1976, Prostaglandin biosynthesis in the human umbilical cord, Biol. Neonate 29:162.

Jores, L., 1903, "Wesen und Entwicklung der Arteriosklerose auf Grund anatomischer und experimenteller Untersuchungen," J.F. Bergmann, Wiesbaden.

Jores, L., 1924, Arterien, in "Handbuch der Speziellen Pathologischen Anatomie und Histologie," Band II, Herz und Gefässe (F. Henke and O. Lubarsch, ed.) O. Springer, Berlin.

Kato, H., Koike, S., Yamamoto, M., Ito, Y., and Yano, E., 1975, Coronary aneurysms in infants and young children with acute febrile mucocutaneous lymph node syndrome, J. Pediatr. 86:892.

Kawasaki, T., Kosaki, F., Okawa, S., Shigematsu, J., and Yanagawa, H., 1974, A new infantile acute febrile mucocutaneous lymph node syndrome (MLNS) prevailing in Japan, Pediatrics 54:271.

Kennedy, C., Grave, G.J., Jehle, J.W., and Sokoloff, L., 1970, Blood flow to white matter during maturation of the brain, Neurology 20:613.

Kennedy, C., and Sokoloff, L., 1957, An adaptation of the nitrous oxide method to the study of the cerebral circulation in children: normal values for cerebral blood flow and cerbral metabolic rate in children, J. Clin. Invest. 36:1130.

Kennedy, J.A., and Clark, S.L., 1941, Observations on the ductus arteriosus of the guinea pig in relation to its method of closure, Anat. Rec. 79:349.

Kety, S.S., 1950, Circulation and metabolism of the human brain in health and disease, Am. J. Med. 8:205.

Kety, S.S., Shenkin, H.A., and Schmidt, C.F., 1948, The effect of increased intracranial pressure on cerebral circulation functions in man, J. Clin. Invest. 27:493.

Kier, E.L., 1974, Fetal cerebral arteries: a phylogenetic and ontogenetic study, in "Radiology of the skull and brain," Vol. II, Book 1, pp. 1089-1130, The Mosby Company, St. Louis.

Knieriem, H.J., and Hueber, A., 1970, Quantitative morphological studies of the human aorta, Beitr. Path. Anat. 140:280.

Könn, G., and Storb, R., 1960, Über den Formwandel der kleinen Lungenarterien des Menschen nach der Geburt, Beitr. Path. Anat. 123:212.

Kovalçik, V., 1963, The response of the isolated ductus arteriosus to oxygen and anoxia, J. Physiol. (Lond.) 169:185.

Krylow, D., 1916, Sur l'artériosclérose expérimentale des artéres coronaires, Compt. Rend. Soc. Biol. 79:399.

Lachenmeyer, L., 1971, Adrenergic innervation of the umbilical vessels, Z. Zellforsch. 120:120.

Larroche, J.C., 1966, The development of the central nervous system during intrauterine life, in "Human Development" (F. Falkner, ed.) W.B. Saunders, Philadelphia.

Lauer, R.M., Evans, J.A., Aoki, H., and Kittle, C.F., 1965, Factors controlling pulmonary vascular resistance in fetal lambs, J. Pediatr. 67:568.

Lazorthes, G., Gouazé, A., Santini, J., Lazorthes, V., and Laffont, J., 1971, Le modelage du polygone de Willis, Neurochirurgie 17:361.

Lev, M., and Killian, S.T., 1941, Hypoplasia of the aorta without transposition with electrocardiographic and histopathologic studies of the conduction system, Am. Heart J. 24:794.

Lev, M., Arcilla, R., Rinoldi, H.J., Licata, R.H., and Gasul, B.H., 1963, Premature narrowing or closure of the foramen ovale, Am. Heart J. 65:638.

Levene, C.I., 1956, The early lesions of atheroma in the coronary arteries, J. Path. Bact. 72:79.

Lewis, B.V., 1968, The response of isolated sheep and human unbilical arteries to oxygen and drugs, J. Obstet. Gynecol. Br. Commonw. 75:87.

Liley, A.A., 1963, Amniotic fluid, in "Modern Trends in Human Reproductive Physiology" (Carey, H.M., eds.) Butterworths, London.

Liley, A.W., 1972, The fetus as a personality, Aust. N.Z.J. Psychiatry 6:99.

Lind, J., and Wegelius, C.E., 1954, Human fetal circulation: Changes in the cardiovascular system at birth and disturbances in postnatal closure of the foramen ovale and ductus arterious, Cold Spring Harbor Sympos., Quant. Biol. 19:109.

Lind, J., Stern, L., and Wegelius, C., 1964, "Human Fetal and Neonatal Circulation," Ch. C. Thomas, Springfield.

Lindenberg, R., 1957, Die Gefässversorgung und ihre Bedeutung für Art und Ort von kreislaufbedingten Gewebsschäden und Gefässprozessen, in "Handb. Spez. Path. Anat.," 13, I. Teil, Bandteil B, pp. 1071-1164, Springer, Berlin and Heidelberg.

Linzbach, A.J., 1952, Die Anzahl der Herzmuskelkerne in normalen, überlasteten, atrophischen und mit Corhormon behandelten Herzkammern, Z. Kreisl. Forsch. 41:641.

Linzenmeier, G., 1914, Der Verschluss des Ductus arteriosus Botalli nach der Geburt des Kindes, Zeitschr. f. Geburtsh. 76:217.

Lloyd, T.C., Jr., 1968, Hypoxic pulmonary vasoconstriction: Role of perivascular tissue, J. Appl. Physiol. 25:560.

Lober, P.H., 1953, Pathogenesis of coronary sclerosis, Arch. Path. 55:357.

Louw, J.H., and Barnard, C.N., 1955, Congenital intestinal atresia, observations on its origin, Lancet II:1065.

Lucas, W., Kirschbaum, T.H., and Assali, N.S., 1966, Cephalic circulation and oxygen consumption before and after birth, Am. J. Physiol. 210:287.

Ludwig, J., 1976, "Vergleichende makroskopische Darstellung von Arteriencalcinosen mit von Kossa-Reaktion und Alizarin-Färbung," Thesis, Mainz.

Mandel, P., Bieth, R., and Weill, J.D., 1957, in: "Metabolism of the nervous system" (D. Richter, ed.), Pergamon Press, London, p. 291.

Mancini, A.J., 1951, A study of the angle formed by the ductus arteriosus with the descending thoracic aorta, Anat. Rec. 109:535.

Mansfield, P., 1976, Personal communication.

Martin, J.F., and Freidell, H.L., 1952, The roentgen findings in atelectasis of the newborn, Am. J. Roentgenol. 67:905.

McDonald, D.A., 1974, "Blood Flow in Arteries," 2nd edition, Edward Arnold, London.

McMurphy, D.M., and Boréus, L.O., 1971, Studies on the pharmacology of the perfused human fetal ductus arteriosus, Am. J. Obstet. Gynec. 109:937.

Merkel, H., and Witt, H., 1955, Die Massenverhältnisse des foetalen Herzens, Beitr. Path. Anat. 115:178.
Meyer, W.W., 1955 Über die eigenartige Beziehung des elastischen Gerüstes zur glatten Muskulatur im extrapulmonalen Abschnitt der Lungenarterie des Menschen, Z. Zellforschung 43:383.
Meyer, W.W., 1963, Über das sogenannte innere Relief der Extremitätenarterien, Verh. Dtsch. Ges. Kreislauff. 29:201.
Meyer, W.W., 1964 a, Funktionelle Morphologie des Gefässsystems im Kindesalter, Mschr. Kinderheilk. 112:37.
Meyer, W.W., 1964 b, Comment, in "Die Arterieninnenwand: Biochemie und Elastizität" (Bibliotheca Cardiologica) Fasc. 15, pp.76-85, S. Karger, Basel.
Meyer, W.W., 1964 c, Über die rhythmische Lokalisation der atherosklerotischen Herde im cervicalen Abschnitt der Vertebralarterie, Beitr. Path. Anat. 130:24.
Meyer, W.W., 1968, Calcinosis of the fetal segment of the pelvic arteries in infancy and childhood, Beitr. Path. Anat. 138:149.
Meyer, W.W., 1971 a, Arteriencalcinosen, Beziehung zum Gefässwachstum und prospektive Bedeutung, Dtsch. Med. Wschr. 96:1093.
Meyer, W.W., 1971 b, Early forms of calcinosis in human arteries, Sandorama (Basel), 1975, I: 11-16.
Meyer, W.W., 1977, The mode of calcification in atherosclerotic lesions, in "Atherosclerosis: Metabolic, Morphologic and Clinical Aspects," Advances in Experimental Medicine and Biology, Vol. 82, pp. 786-792, Plenum Press, New York.
Meyer, W.W., and Ehlers, U., 1972, Early calcification patterns of the iliac arteries and their relation to the arterial structure, Z. Zellforsch. 130:378.
Meyer, W.W., and Lind, J., 1965, Über die Struktur und den Verschluss-Mechanismus des Ductus venosus, Z. Zellforsch. 67:390.
Meyer, W.W., and Lind, J., 1966, The ductus venosus and the mechanism of its closure, Arch. Dis. Childh. 41:597.
Meyer, W.W., and Lind, J., 1972 a, Calcifications of the carotid siphon - a common finding in infancy and childhood, Arch. Dis. Childh. 47:355.
Meyer, W.W., and Lind, J., 1972 b, Calcifications of the iliac arteries in newborns and infants, Arch. Dis. Child. 47:364.
Meyer, W.W., and Lind, J., 1974, Iliac arteries in children with a single umbilical artery, Structure, calcifications, and early atherosclerotic lesions, Arch. Dis. Childh. 49:671.
Meyer, W.W., and Noll, M., 1974, Gross patterns of early lipid deposits in the carotid artery and their relation to the preformed arterial structures, Artery 1:31.
Meyer, W.W., and Richter, H., 1956, Das Gewicht der Lungenschlagader als Gradmesser der Pulmonalarteriensklerose und als morphologisches Kriterium der pulmonalen Hypertonie. Eine quantitativ-anatomische und feingewebliche Untersuchung, Virchows Arch. Path. Anat. 328:121.

Meyer, W.W., Rumpelt, H.J., Yao, A.C., and Lind, J., Structure and closure mechanism of the human umbilical artery, in press.

Meyer, W.W., and Schollmeyer, P., 1957, Die Volumendehnbarkeit und die Druck-Umfang-Beziehungen des Lungenschlagader-Windkessels in Abhängigkeit von Alter und pulmonalen Hochdruck, Klin. Wschr. 35:1070.

Meyer, W.W., and Simon E., 1959, Die phasenartige Abwandlung der Pulmonalis-Volumendehnbarkeit im Verlauf des Lebens in ihrer Beziehung zur Struktur der Arterienwand, Arch. Kreislauff. 31:95.

Meyer, W.W., and Simon, E., 1960, Die präparatorische Angiomalacie des Ductus arteriosus Botalli als Voraussetzung seiner Engstellung und als Vorbild krankhafter Arterienveränderungen, Virchows Arch. Path. Anat. 333:119.

Meyer, W.W., and Stelzig, H.H., 1967, Verkalkungsformen der inneren elastischen Membran der Beinarterien und ihre Bedeutung für die Mediaverkalkung, Virchows Arch. Path. Anat. 342:361.

Meyer, W.W., and Stelzig, H.H., 1968, Morphologie des Spaltensystems der inneren elastischen Membran muskulärer Arterien, Z. Zellforsch. 88:415.

Meyer, W.W., and Stelzig, H.H., 1969 a, A simple method for gross demonstration of calcific deposits in the arteries, Angiology 20:423.

Meyer, W.W., and Stelzig, H.H., 1969 b, Calcification patterns of the internal elastic membrane, Calcif. Tissue Res. 3:266.

Meyer, W.W., and Weber, G., 1968, Calcinose und Spaltensystem der inneren elastischen Membran der geschlängelten Milzarterien, Virchows Arch. Path. Anat. 345:292.

Meyer, W.W., Zabka, W., and Stelzig, H.H., 1969, Calcification patterns of the internal elastic membrane of the upper arm artery, Angiologica (Basel), 6:13.

Minkowski, W.L., 1947, The coronary arteries of infants, Am. J. Med. Sci. 214:623.

Mitchell, S.C., 1957, The ductus arteriosus in the neonatal period, J. Pediatr. 51:12.

Moinian, M., Meyer, W.W., and Lind, J., 1969, Diameters of umbilical cord vessels and the weight of the cord in relation to clamping time, Am. J. Obstet. Gynecol. 105:604.

Moon, H.D., 1957, Coronary arteries in fetuses, infants and juveniles, Circulation 16:263.

Morison, J.E., 1952, "Foetal and Neonatal Pathology," p. 156, Butterworth & Co., Ltd., London.

Mortimer, E.A., 1976, Mucocutaneous lymph node syndrome. The epidemiologic approach toward finding its causation, Am. J. Dis. Child. 130:593.

Moss, A.J., Emmanouilides, G., and Duffie, E.R., 1963, Closure of the ductus arteriosus in the newborn infant, Pediatrics 32:25.

Movat, H.Z., More, R.H., and Haust, M.D., 1958, The diffuse intimal thickening of the human aorta with aging, Am. J. Path. 34:1023.

Müller, W., 1883, "Die Massenverhältnisse des menschlichen Herzens," L. Voss, Hamburg und Leipzig.

Nadkarni, B.B., 1970, Innervation of the human umbilical artery. An electron microscope study, Am. J. Obstet. Gynecol. 107:303.

Naeye, R.L., 1961, Arterial changes during the perinatal period, Arch. Path. 71:121.

Naeye, R.L., and Blanc, W.A., 1964, Prenatal narrowing or closure of the foramen ovale, Circulation 30:736.

Nair, X., and Dyer, D.C., 1973, Effect of metabolic inhibitors and oxygen on responses of human umbilical arteries, Am. J. Physiol. 225:1118.

Neufeld, H.N., Wagenvoort, C.A., and Edwards, J.E., 1962, Coronary arteries in fetuses, infants, juveniles and young adults, Lab. Invest. 11:837.

Noonan, J.A., and Nadas, A.S., 1958, The hypoplastic left heart syndrome, Pediat. Clin. N. Am. 5:1029.

Oberhansli-Weiss, I., Heymann, M.A., Rudolph, A.M., and Melmon, K.L., 1972, The pattern and mechanisms of response to oxygen by the ductus arteriosus and umbilical artery, Pediat. Res. 6:693.

Padget, D.H., 1944, The circle of Willis, its embryology and anatomy, in "Intracranial arterial aneurysms" (Dandy, W.E., ed.), Comstock Publishing Co., Inc., Ithaca, N.Y.

Padget, D.H., 1948, The development of the cranial arteries in the human embryo, Contrib. Embryol. 32:205.

Panigel, M., 1963, Placental perfusion experiments, Am. J. Obstet. Gynec. 84:1664.

Patten, B.M., 1953, "Human Embryology" (2nd ed.), p. 698, Blakiston Company, New York.

Patten, B.M., Sommerfield, W.A., and Paff, G.H., 1929-1930, Functional limitations of the foramen ovale in the human foetal heart, Anat. Rec. 44:165.

Pick, R., and Katz, L.N., 1965, The morphology of experimental cholesterol- and oil-induced atherosclerosis in the chick, in "Comparative Atherosclerosis" (J.C. Roberts and R. Straus, eds.) p. 77, Harper and Row, New York.

Prichard, M.M.L., 1968, Personal communication cited by Dawes, G.S., in "Foetal and Neonatal Physiology. A comparative study of the changes at birth," p. 165, Year Book Medical Publ., Chicago.

Prior, J.T., and Jones, D.B., 1952, Structural alterations within the aortic intima in infancy and childhood, Am. J. Path. 28:937.

Recavarren, S., and Arias-Stella, J.G., 1964, Growth and development of the ventricular myocardium from birth to adult life, Br. Heart. J. 26:187.

Redmond, A., Isana, S., and Ingall, D., 1965, Relation of onset of respiration to placental transfusion, Lancet 1:283.

Roach, M., 1973, A biophysical look at the relationship of structure and function in the umbilical artery, in "Foetal and Neonatal Physiology. Proceedings of the Sir Joseph Barcroft Centenary Symposium" (K.S. Comline, K.W. Cross, G.S. Dawes, and D.W. Nathanielsz, eds.), p. 141-163, Cambridge University Press.

Robertson, R.H., 1960, Stress zones in foetal arteries, J. Clin. Pathol. 13:133.

Rosen, L., Bowden, E.A., and Uchida, J., 1957, Structural changes in pulmonary arteries in first year of life, Arch. Path. 63:316.

Rotter, W., and Rottmann, I., 1952, Über den Umbau der Gefässtrecken vom rein elastischen Typ in solche vom rein muskulären Typ in Bereich der Aorta und der Arteria iliaca communis, Arch. f. Kreislauff. 18:76.

Rotter, W., Wellmer, H.H., Hinrichs, G., and Müller, W., 1955, Zur Orthologie und Pathologie der Polsterarterien (sog. Verzweigungs- und Spornpolster) des Gehirns, Beitr. Pathol. Anat. 115:253.

Rowe, R.D., and Mehrizi, A., 1968, "The Neonate with Congenital Heart Disease," W.B. Saunders, Philadelphia, p. 57.

Rudolph, A.M., and Heymann, M.A., 1967, The circulation of the fetus in utero: Methods for studying distribution of blood flow, cardiac output and organ blood flow, Circ. Res. 21:163.

Rudolph, A.M., and Heymann, M.A., 1970, Circulatory changes during growth in the fetal lamb, Circ. Res. 26:289.

Rudolph, A.M., and Heymann, M.A., 1974, Fetal neonatal circulation and respiration, Ann. Rev. Physiol. 36:187.

Rudolph, A.M., Heymann, M.A., Teramo, K.A.W., Barret, C.T., and Raihä, N.C.R., 1971, Studies on the circulation of the previable human fetus, Pediat. Res. 5:452.

Rudolph, A.M., Mesel, E., and Levy, J.M., 1963, Epinephrine in the treatment of cardiac failure due to shunts, Circulation 28:3.

Rudolph, A.M., and Yuan, S., 1965, Response of the pulmonary vasculature to hypoxia and H^+ ion concentration changes, J. Clin. Invest. 45:399.

Saling, E., 1960, Neue Untersuchungsergebnisse über den Kreislauf des Kindes unbittelbar nach der Geburt, Arch. Gynäk. 194:287.

Samuel, K.C., 1956, Atherosclerosis and occlusion of the internal carotid artery, J. Path. Bact. 71:391.

Scarpelli, E.M., Clutario, B.C., and Taylor, F.A., 1967, Primary identification of the lung surfactant system, J. Appl. Physiol. 23:880.

Scheel, O., 1908, Gefässmessungen und Arteriosklerose, Virchows Arch. Path. Anat. 191:135.

Schlesinger, M.J., 1940, Relation of anatomic pattern to pathologic conditions of the coronary arteries, Arch. Path. 30:403.

Schornagel, H.E., 1956, Intimal thickening in the coronary arteries in infants, AMA Arch. Path. 62:427.

Schultz, D.M., and Giordano, D.H., 1962, Hearts of infants and children, Arch. Pathol. 74:464.

Schwartz, C.J., and Mitchell, J.R.A., 1961, Atheroma of the carotid and vertebral arterial system, Brit. Med. J. 2:1057.

Schwartz, C.J., and Mitchell, J.R.A., 1962, Observations on localization of arterial plaques, Circ. Res. 11:63.

Seelig, M.S., 1969, Vitamin D and cardiovascular, renal and brain damage in infancy and childhood, Ann. N.Y. Acad. Sci. 147:537.

Segel, N., Staněk, V., Joshi, R., and Singhal, S., 1970, The influence of bradykinin on the pulmonary hypoxic response in normal man, Progr. Resp. Res. 5:119.

Seki, M., and Strauss, L., 1964, Absence of one umbilical artery, Arch. Path. 78:446.

Selye, H., 1962, "Calciphylaxis," University of Chicago Press, Chicago.

Seydel, H.G., 1964, The diameters of the cerebral arteries of the human fetus, Anat. Rec. 150:79.

Sharpe, G.L., and Larsson, K.S., 1975, Studies on closure of the ductus arteriosus, X. In vivo effect of prostaglandin, Prostaglandins 9:703.

Shelley, H.J., 1961, Glycogen reserves and their changes at birth, Br. M. Bull. 17:137.

Simon, W., 1959, Untersuchungen über das Fassungsvermögen und die Volumendehnbarkeit des gesamten Aortenwindkessels beim Menschen und über Länge und Umfang des Aortenrohres in Abhängigkeit vom Aortendruk, Thesis. Marburg.

Simon, E., and Meyer, W.W., 1958, Das Volumen, die Volumendehnbarkeit und die Druck-Längen-Beziehungen des gesamten aortalen Windkessels in Abhängigkeit von Alter, Hochdruck and Atherosclerose, Klin. Wschr. 36:424.

Sinizina, T.A., 1964, "Experimental Atherosclerosis of Coronary Arteries," (in Russian), Medizina, Leningrad.

Smaltino, F., Bernini, F.P., and Elefonte, R., 1971, Normal and pathological findings of the angiographic examination of the internal auditory artery, Neuroradiology 2:216.

Smith, B., 1968, Pre- and postnatal development of the ganglion cells of the rectum and its surgical implications, J. Pediat. Surg. 3:386.

Solberg, L.A., and Eggen, D.A., 1971, Localization and sequence of development of atherosclerotic lesions in the carotid and vertebral arteries, Circulation 43:711.

Somlyo, A.P., and Somlyo, A.V., 1968, Vascular smooth muscle, Part I, Pharmacol. Rev. 20:197.

Spiteri, M., J. Nguyen H. Anh, and Panigel, M., 1966, Ultrastructure du muscle lisse des artères du cordon ombilical humain, Path. Biol. 14:348

Spivack, M., 1943, On presence or absence of nerves in umbilical blood vessels of man and guinea pig, Anat. Rec. 85:85.

Stahlman, M.T., Merrill, R.E., and Le Quire, V.S., 1962, Cardiovascular adjustments in normal newborn lambs, Am. J. Dis. Child. 104:360.

Staub, N.C., 1963, Site of action of hypoxia on the pulmonary vasculature, Fed. Proc. 22:453.

Stawe, U., editor, in press, "Physiology of the Perinatal Period," New York, NY, Plenum Pub. (second edition).

Stehbens, W.E., 1959, Medial defects of the cerebral arteries in man, J. Path. Bact. 78:179.

Stehbens, W.E., 1960, Focal intimal proliferation in the cerebral arteries, Am. J. Path. 36:289.
Stehbens, W.E., 1963, Aneurysms and anatomical variation of cerebral arteries, Arch. Path. 75:45.
Stehbens, W.E., 1972, "Pathology of the Cerebral Blood Vessels," C.V. Mosby, St. Louis.
Stehbens, W.E., 1975, Flow in glass models of arterial bifurcations and berry aneurysms at low Reynolds numbers, Quart. J. Exp. Physiol. 60:181.
Stelzig, H.H., and Meyer, W.W., 1967, Mikroradiographische und histologische Untersuchungen zur Morphogenese der Mönckeberg' schen Mediaverkalkung, Fortschr. Röntgenstrahlen 107:504.
Štembera, Z.K., Hodr, J., and Janda, J., 1965, Umbilical blood flow in healthy newborn infants during the first minutes after birth, Am. J. Obstet. Gynecol. 91:568.
Stephens, R.B., and Stilwell, 1969, "Arteries and Veins of the Human Brain," Springfield, Ill., Charles C. Thomas, Publisher.
Strong, J.P., and McGill, H.C., Jr., 1969, The pediatric aspects of atherosclerosis, J. Atherosclerosis Res., 9:251.
Stryker, W.A., 1946, Arterial calcification in infancy with special reference to the coronary arteries, Am. J. Path. 22:1007.
Takuku, A., and Sizuki, J., 1972, Cerebral angiography in children and adults with mental retardation, Dev. Med. Child. Neurol. 14:756.
Taylor, C.B., 1965, Experimentally induced arteriosclerosis in non-human primates, in "Comparative Atherosclerosis" (J.C. Roberts and R. Straus, eds.) p. 215, Harper and Row, New York and London.
Taylor, C.B., Cox, G.E., Hall-Taylor, B.J., and Nelson, L.G., 1954, Atherosclerosis in areas of vascular injury in monkeys with mild hypercholesterolemia, Circulation, 10:613.
Taylor, C.B., Patton, D.E., and Cox, G.E., 1963, Atherosclerosis in rhesus monkey. IV. Fatal myocardial infarction in a monkey fed fat and cholesterol, Arch. Path. 76:404.
Thoma, R., 1883, Über die Abhängigkeit der Bindegewebsneubildung in der Arterienintima von den mechanischen Bedingungen des Blutumlaufes. Erste Mitteilung. Die Rückwirkung des Verschlusses der Nabelarterien und des arteriösen Ganges auf die Struktur der Aortenwand, Arch. Path. Anat. Physiol. Klin. Med. 93:443.
Thoma, R. 1923, Über die Intima der Arterien, Virchows Arch. Path. Anat. 230:1.
Thoma, R., 1923, Über die Genese und die Lokalisation der Arteriosklerose, Virchows Arch. Path. Anat. 245:78.
Thomas, J., 1961, Untersuchungsergebnisse über die Aplasie einer Nabelarterie unter besonderer Berücksichtigung der Zwillingsschwangerschaft, Geburtsh. Frauenheilk. 21:984.

Tiisala, R., Tähti, E., and Lind, J., 1966, Heart volume variations during first 24 hours of life of infants with early and late clamped umbilical cord, Ann. Pediatr. Fenn. 12:151.

Touloukian, R.J., and Duncan, R., 1975, Acquired aganglionic megacolon in a premature infant: Report of a case, Pediatrics 56:459.

Traisman, H.S., Limperis, N.M., and Traisman, A.S., 1956, Myocardial infarction due to calcification of the arteries in an infant, Am. J. Dis. Child. 91:34.

Tunell, R., 1974, "Studier av den respiratoriska och metaboliska anpassningen till extrauterint liv hos nyfödda barn," Thesis, Karolinska Institute, Stockholm.

Tuvemo, T., and Strandberg, K., 1975, Effects and interactions of oxygen and prostaglandins on the tone of the isolated human umbilical artery, Uppsala J. Med. Sci. 80:131.

Usher, R., Shephard, M., and Lind, J., 1963, The blood volume of the newborn infant and placental transfusion, Acta Paediatr. Scand. 52:497.

Viles, P.H., and Sheperd, J.T., 1968, Evidence for a dilator action of carbon dioxide on the pulmonary vessels of the cat, Circ. Res. 22:325.

Vlodaver, Z., Kahn, H.A., and Neufeld, H.N., 1969, The coronary arteries in early life in three different ethnic groups, Circulation 39:541.

Voigt, G.E., 1957, Ein neuer histotopochemischer Nachweis des Calciums (mit Naphtalhydroxamsäure), Acta histochemica 4:122.

Wagenvoort, C.A., Neufeld, H.N., and Edwards, J.E., 1961, The structure of the pulmonary arterial tree in fetal and early postnatal life, Lab. Invest. 10:751.

Walkoff, F., 1869, Das Gewebe des Ductus arteriosus und die Obliteration desselben, Z. Ration. Med. 36.

Wallgren, G., and Lind, J., 1967, Quantitative studies of the human neonatal circulation. IV. Observations on the newborn infant's peripheral circulation and plasma expansion during moderate hypovolemia, Acta Paediat. Scand. (suppl. 179) 55-58.

Walsh, S.Z., 1966, The electrocardiogram in the neonate and infant, in "The Heart and Circulation in the Newborn and Infant," (Cassels, D.E., ed.), pp. 263-273, Grune and Stratton, New York.

Walsh, S.Z., 1968 a, Maternal effects of early and late clamping of the umbilical cord, Lancet 1:996.

Walsh, S.Z., 1968 b, Early versus late clamping of the cord: A comparative study of the ECG in the neonatal period, Biol. Neonate. 12:343.

Walsh, S.Z., 1969, Early clamping versus stripping of the cord: A comparative study of the electrocardiogram in the neonatal period, Br. Heart J. 31:122.

Walsh, S.Z., Meyer, W.W., and Lind, J., 1974, "The Human Fetal and Neonatal Circulation. Function and Structure," Charles C. Thomas Publisher, Springfield, Ill.

Walsh, S.Z., 1975, ECG changes during the first 5 - 6 days after birth. Some factors that may be involved, Praxis (Revue Suisse de Médicine) 64:747.

Walsh, S.Z., 1975 b, Characteristic features of the ECG of premature infants during the first year of life. With a comment on a large Q in lead III and the incidence of pattern types in right and left precordial leads, Praxis 64:754.

Wehn, P.S., 1957, Pulsatory activity of peripheral arteries, Scand. J. Clin. Lab. Invest. 9: Suppl. 30.

Wezler, K., and Böger, A., 1937, Über einen neuen Weg zur Bestimmung des absoluten Schlagvolumens des Herzens beim Menschen auf Grund der Windkesseltheorie und seine experimentelle Prüfung, Naunyn-Schmiedeberg's Arch. Exp. Path. Pharm. 184:482.

Wezler, K., and Böger, A., 1939, Die Dynamic des arteriellen Systems, Ergeb. Physiol. 41:292.

Wezler, K., and Sinn, W., 1953, "Das Strömungsgesetz des Kreislaufes," Editio Cantor, Aulendorf i. Württ.

Widdowson, E.M., 1974, Changes in body proportions and composition during growth, in "Scientific Foundations of Paediatrics" (J.A. Davis and J. Dobbing, eds.) pp. 153-163, W.M. Heinemann Medical Books, Ltd., London.

Wielenga, G., and Denkmeijer, J., 1968, Coarctation of the aorta, J. Path. Bact. 95:265.

Wilens, S.L., and McCluskey, R.T., 1954, The permeability of excised arteries and other tissues to serum lipid, Circ. Res. 2:175.

Wilson, R.R., 1958, Post-mortem observations on contraction of human ductus arteriosus, Brit. M.J. 1:810.

Winterscheid, L.C., 1969, Collateral circulation of the heart, in "Collateral Circulation in Clinical Surgery" (D.E. Strandness, ed.) pp. 93-167, W.B. Saunders Company, Philadelphia, London, Toronto.

Wissler, R.W., Eilert, M.L., Schroeder, M.A., and Cohen, L., 1954, Production of lipomatous and atheromatous arterial lesions in the albino rat, AMA Arch. Path. 57:333.

Wolinsky, H., and Glagow, S., 1967, A lamellar unit of aortic medial structure and function in mammals, Circ. Res. 20:99.

Wolkoff, K., 1923, Über die histologische Struktur der Coronararterien des menschlichen Herzens, Virchows Arch. Path. Anat. 214:42.

Wolkoff, K., 1929, Über die Atherosklerose der Coronararterien des Herzens, Beitr. Path. Anat. 82:555.

Wolkoff, K., 1930, Über experimentelle Atherosklerose der Koronararterien des Herzens, Beitr. path. Anat. 85:386.

Wolkoff, K., 1933, Über Atherosklerose der Gehirnarterien, Beitr. path. Anat. 91:515.

Yao, A.C., and Lind, J., 1969, Effect of gravity on placental transfusion, Lancet 2:505.

Yao, A., Nergårdh, A., and Borçus, L., Influence of oxytocin and meperidine on the isolated human umbilical artery, (in preparation).

Yuan, S.H., Heymann, M.A., and Rudolph, A.M., 1966, Relationship between ventricular weight, pressure and myocardial blood flow in the newborn piglet, Circulation 34:243.
Zabka, W., 1968, "Die Kalzinose der inneren elastischen Membran der Oberarmarterie in makroskopischer und mikroskopischer Darstellung," Thesis, Mainz.
Zaitev, N.D., 1959, Development of neural elements in the umbilical cord, Arkh. Anat. Gistol. Embriol. 37:81.

ACKNOWLEDGMENTS

The authors wish to thank Mrs. Christa Schmalzel for her efficient typing of the manuscript, Mr. Wilfried Meyer for help with the drawings, and Mr. Hilding Johansson for his valuable assistance. We should also like to extend our thanks to Professor Carl Wegelius, Director of the Wenner-Gren Cardiovascular Research Laboratory, Norrtulls Hospital, Stockholm, where part of this chapter was written.

ABDOMINAL VISCERAL CIRCULATION IN MAN

Edward A. Edwards, M.D.

Clinical Professor of Anatomy, Emeritus, Harvard Medical School, Surgeon Emeritus, Peter Bent Brigham Hospital; Acting Chief, Department of Rehabilitation Medicine, Veterans Administration Outpatent Clinic, Boston.

INTRODUCTION

The aorta enters the abdomen with an internal diameter of about 20 mm. transmitting an average of 3.5 liters of blood per minute. The two external iliac arteries leave the abdomen, each with an internal diameter of about 9 mm. and transmitting together about 700 ml. per minute (Hobbs and Edwards, 1963). Between these two levels, most of the spent blood has gone to the viscera of the abdomen, including the pelvis. A consideration of the anatomy of the arteries serving this perfusion must first deal with the pattern of the vessels involved, variations encountered, effects of interruption of the major pathway to each organ, the arterial collaterals available, and the adequacy of such collaterals. The branching of the visceral arteries are also of functional significance. No vessels are known in the adult human which are strictly end arteries in a morphological sense, that is, arteries which have no communication whatsoever with neighboring arteries. However, physiologically speaking, end arteries do exist, since some are connected with adjacent vessels via arterioles or capillaries, which communications may be functionally inadequate. Some of the viscera are thus known to be divided into segments whose arterial blood supply is significantly circumscribed from neighboring portions of the organ.

Footnote:

Portions of the text and several figures for this chapter are taken from Operative Anatomy: Abdomen and Pelvis. (Edwards, E.A., Malone, P.D., and McArthur, J.D., 1975), with permission of the publishers, Lea and Febiger.

I. THE ABDOMINAL AORTA AND THE INTERNAL ILIAC ARTERIES AS SOURCES OF VISCERAL SUPPLY

A. Origin of Visceral Trunks From The Aorta

The major topography for the origin of the major branches of the aorta is shown in Fig. 1. The range of variations in their height of origin are given in Table I, based upon Adachi (1928), George (1935), Caldwell and Anson (1943), and Feller and Woodburne (1961).

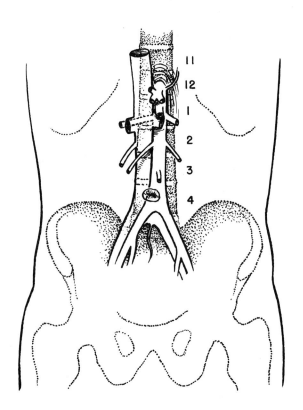

Fig. 1. Levels of branching of the abdominal aorta. (From Edwards et al., 1975.)

TABLE 1

Vertebral level of origin of major branches of the abdominal aorta.*

	Most Frequent	Range
Inferior Phrenic	TXII disc	T XI to L I disc
Celiac	T XII disc or L-I	T XI to L-I disc
Superior Mesenteric	L I	T XII to L II
Renal	L I	T XII to L II disc
Gonadal	L II	L I to L III disc
Inferior Mesenteric	L III	L II to L III disc
Common Iliac (Aortic Bifurcation)	L IV	LIII disc to LV

*From Edwards et al., 1975

The celiac and superior mesenteric arteries are combined as a celiacomesenteric trunk in about 1% of individuals. The only significant variation affecting the inferior mesenteric is its rare source for a supernumerary renal artery. It is absent with exceeding rarity, a pattern seen in marsupials (Adachi). One or more middle suprarenal arteries stem from the aorta on either side. In about half of all subjects the testicular or ovarian artery of one or both sides takes origin from a renal artery, rarely from a middle suprarenal. The gonadal artery may be doubled on either side. Variations in origin of the renal arteries will be described below.

1. Visceral Branches of the Inferior Phrenic and Lumbar Arteries

These arteries, mainly parietal in distribution, do contribute significantly to visceral structures. The inferior phrenic arteries, major sources of blood to the suprarenal glands, originate from the aorta directly in only half of all subjects. In the remainder one or both of these vessels, but especially the left, stem from the celiac or one of its components (left gastric, left or middle hepatic), or a renal artery.

The lumbar arteries give off branches to the spinal cord, irregular in size and number. The arteria radicularis magna, the largest artery of supply to the cord below its cervical portion, has a wide range of origin from the eighth thoracic to the fourth lumbar arteries. Most frequently, however, it stems from the right or left second lumbar artery. Examples of paraplegia from sacrifice of the parietal source of this artery are quoted in Edwards et al., 1972.

B. Collaterals For The Abdominal Aorta

Observations on obstruction of the upper abdominal aorta are fragmentary. Presumably supradiaphragmatic arteries--the internal thoracic and intercostals--conduct blood to the upward coursing branches of the internal iliac and the lumbar and inferior epigastric arteries, and thus to the aorta below the obstruction. In the few recorded instances of survival after occlusion of the upper aorta, including the origin of major visceral trunks, parietal arteries were found to be the major collateral pathway for the deprived viscera (Chiene, 1869; Rob, 1970).

Collaterals in obstruction of the infrarenal segment were studied by aortography and dissection by Edwards and LeMay (1955). In these cases, the superior rectal artery is the single largest vessel reaching the iliac systems (Fig. 2). When the inferior mesenteric is occluded at its take-off, flow through the superior rectal comes from the much enlarged marginal artery of the colon (Fig. 12), thus remotely from the superior mesenteric. Many smaller vessels send branches downward to the iliac arteries in instances of

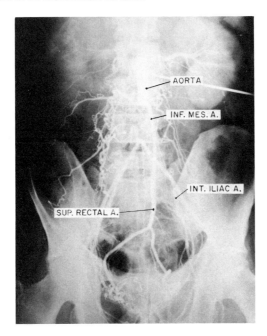

Fig. 2. Collaterals in localized obstruction of the abdominal aorta. (From Edwards, 1957.)

aortic or common iliac artery obstruction. These include the lower intercostal and lumbar arteries and the adipose capsular branches of the renals. The superior to inferior epigastric pathway, important in aortic coarctation, does not seem significant in arteriosclerosis.

C. Visceral Branches Of The Internal Iliac Arteries: The Umbilical Artery

The internal iliac artery divides into a smaller posterior and larger anterior division, the latter including the origin of the umbilical artery (Fig. 3). In roughly half of all subjects, the posterior division gives off the superior gluteal, iliolumbar, and lateral sacral arteries. In others the inferior gluteal, and uncommonly the internal pudendal as well, also stem from the posterior division.

The anterior division has both parietal and visceral branches. The parietal are the inferior gluteal and the obturator. With extreme rarity, the inferior gluteal may be the major artery of the lower limb, with the popliteal its direct continuation, as in the early embryo (Adachi, 1928; Bower et al., 1977).

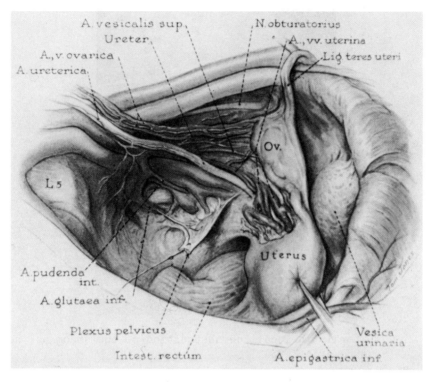

Fig. 3a. Blood vessels and nerves on the side wall of the female pelvis.
I. Arteries and veins beneath the peritoneum. The course of the ureter is clearly shown.

The umbilical artery is the largest of the several visceral branches of the anterior division of the fetus, and its proximal portion remains patent after birth to supply some of the pelvic organs. The fetal umbilical artery appears to be the direct continuation of the common iliac; indeed its diameter is twice that of the external iliac, and the other branches of the internal iliac seem but twigs of the umbilical. As measured in the lamb fetus (Rudolph, 1969), 52% of the cardiac output passes through the two umbilical arteries early in gestation, and 40% near term. The physiology of the fetal and neonatal circulation is extensively reviewed by Young (1963). The two umbilical arteries establish a gross communication between them in over 95% of cases, about 1 cm. from the placenta (Benirschke, 1965). In his atlas, Hyrtl (1870) shows that in some cases there is a long coalescence of the two umbilical arteries reaching toward the umbilicus for a considerable distance. No communication has been noted between the two vessels within the fetal abdomen. Uncommonly, one umbilical artery is absent. In these

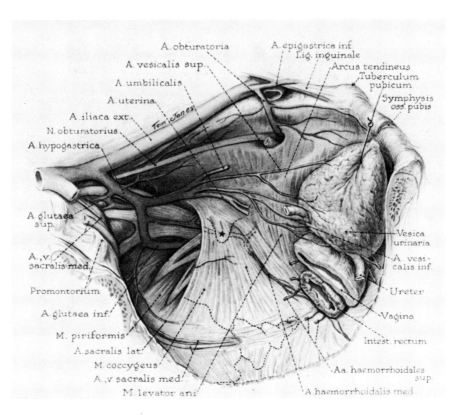

Fig. 3b. Blood vessels and nerves on the side wall of the female pelvis.
II. Arteries and nerves of deeper dissection. The ovarian artery has been removed, and the uterine artery divided. The obturator artery has an aberrant origin from the inferior epigastric. Its course lateral to the external iliac vein is unusual. Arteries to the bladder, vagina, and terminal rectum are shown. The middle rectal (hemorrhoidal) stems from the internal pudendal in this subject. The asterisk marks the position of the ischial spine. Posteriorly the major bulk of the sciatic nerve is being formed by the union of the lumbosacral trunk (L-4,5) with the anterior rami of S-1,2. the superior gluteal artery leaves between the lumbosacral trunk and S-1. Below the sciatic nerve lies the inferior gluteal artery above the anterior rami of S-3,4. (From Curtis et al., 1942. By permission of Surg. Gynec. & Obstet.)

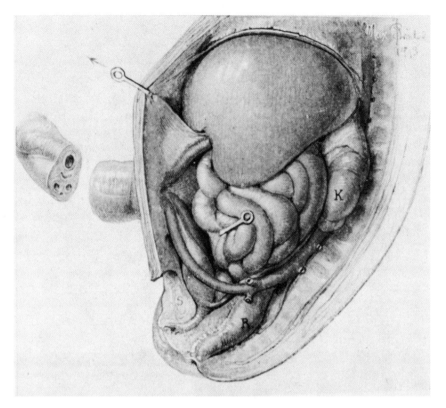

Fig. 4. The umbilical vessels in a fetus. The two umbilical arteries and the single umbilical vein are shown in the abdomen and in the cut section of the cord. The left umbilical artery is seen to be almost as large as the common and internal iliac arteries. (From Cullen, 1916.)

cases, according to Benirschke, a fibrous cord can be seen representing atrophy or closure of the absent artery, leading to the conclusion that the condition does not represent aplasia of one umbilical artery. While the placenta and child may be normal, a somewhat higher than usual incidence of fetal abnormality is observed in these cases.

Postnatally, the umbilical artery maintains a lumen until the origin of the quite constant superior vesical artery. More proximally the uterine artery is a frequent branch, and a corresponding vessel in the male--the vesico-differential artery (Fig. 20) almost always arises from the umbilical (Braithwaite, 1952).

Other visceral branches of the anterior division should be mentioned. The internal pudendal, which may be accounted as the visceral branch to the external genitalia, comes off the anterior division alone, or shares a common stem with the inferior gluteal artery ("gluteopudendal trunk"). A separate dorsal artery of the penis or clitoris ("accessory internal pudendal") may arise from the anterior division of the iliac, or the obturator artery ("normal" or "aberrant" of external iliac origin) to run on the superior aspect of the pelvic diaphragm then beneath the pubic arch (Fig. 20). The inferior vesical artery, usually arises from the pudendal or from a combined gluteopudendal trunk; it gives off five or six branches to the prostate (collectively called the prostatic artery) and to the base of the bladder (Flocks, 1937; Roberts and Krishingner, (1967). Clegg (1955) found that in 32% the prostatic artery arose from the superior rectal at the sides of the rectum. In the female, the uterine artery is most often a branch of the internal pudendal, next often a branch of the umbilical (Roberts and Kirshingner, 1967). The vaginal artery either replaces the inferior vesical, or is combined with it (Fig. 23). Roberts and Krishingner could identify a middle rectal artery in 66% of specimens examined. When present, it most often arose from the internal pudendal, or a gluteopudendal trunk.

D. Effects Of Ligation Of The Internal Iliac Artery

Unilateral or bilateral ligation of the internal iliac has often been performed to control hemorrhage, especially in gynecologic surgery or after obstetrical delivery (Siegel and Mengert, 1961; Reich and Nechtow, 1961) and in fracture of the pelvis; to limit bleeding in operations as for carcinoma of the cervix or abdominoperineal resection of the colon (McGregor, 1959) or in hindquarter amputation; or in the treatment of aneurysm of the internal iliac artery.

The collaterals for the vessel consist of contralateral branches, the inferior mesenteric, lumbar and sacral, and the ovarian arteries--from above; and the medial and lateral femoral circumflex arteries--from below (Edwards and LeMay, 1955). The maneuver has failed to control hemorrhage from fracture of the pelvis (Peltier, 1955; Huittinen and Slätis, 1973) probably by virtue of these collaterals. Rare instances are also reported of ischemic necrosis of the perineum, buttocks, or bladder following the ligation (Finaly, 1925; Tajes, 1956). Failure to restore internal iliac flow in the course of operations on the aorta has also been implicated in the uncommon instance of ischemic damage to the rectum and bladder (Smith and Szilagyi, 1960; Ottinger, et al., 1972).

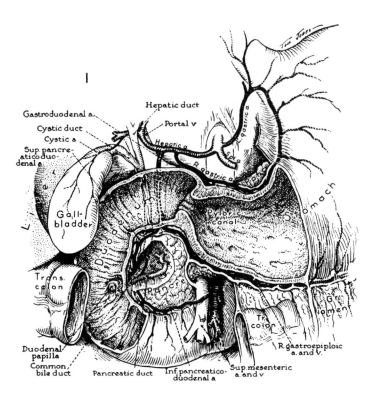

Fig. 5.I. Major branches of the celiac artery. I. Anterior view. (From Jones and Shepard, 1945.)

II. THE SUPRAMESOCOLIC VISCERA

A. Communications Between The Celiac And Superior Mesenteric Arteries

The celiac is the main artery of supply of the supramesocolic viscera, assisted to a variable degree by the superior mesenteric artery. The two arteries communicate quite constantly by the junction of the superior and inferior pancreatico-duodenal arteries, shown in an anterior arcade in Fig. 5, and more accurately as a pair of arcades in Fig. 6 (also Fig. 9). A second set of communications, irregular in their presence and size, exist behind the body of the pancreas and in front of the aorta. These include: (1) the rare common origin of the celiac and superior mesenteric arteries (mentioned above), anastomoses between pancreatic arteries and the middle colic and accessory left colic arteries; (3) the aberrant origin of a pancreatic artery from the superior mesenteric artery, or a colic artery from the celiac (see page 21 of this chapter).

B. The Stomach

The stomach is the most richly vascularized part of the gut. Not only do the gastric and gastroepiploic arteries communicate grossly in so-called inosculations, but communications also exist between small arteries within the wall of the organ. Numerous arteriovenous anastomoses are likewise present (Barlow et al., 1951).

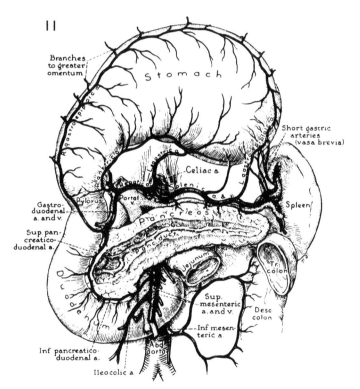

Fig. 5.II Major branches of the celiac artery. II. Retrogastric view. (From Jones and Shepard, 1945.)

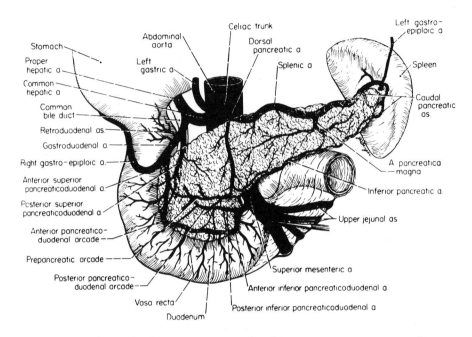

Fig. 6. Arteries of the pancreas and adjacent structures. (From Woodburne, 1973.)

However, the short gastric arteries supplying the fundus may be incapable of supplying distant portions of the stomach. It is generally considered safe in mobilizing the stomach for anastomosis, as to the esophagus, to ligate all but one of the four major arteries. Nakayama (1954) as a result of extensive clinical experience and injection studies, concluded that in order to maintain sufficient circulation in the proximal part of the mobilized stomach, it is safer to leave the arterial arches along the two curvatures "completely untouched," in addition to preserving either the right gastric or the right gastroepiploic artery.

C. The Duodenum and Pancreas

The richness of arterial anastomoses of the stomach abruptly gives way, beyond the pylorus, to a paucity of such connections in the remainder of the gut until the lower rectum. The initial part of the duodenum is supplied by the usually single supraduodenal artery superoanteriorly (Fig. 9), and by one or more retroduodenal arteries posteriorly (Fig. 6). The supraduodenal artery usually originates from the upper gastroduodenal, but may come from the hepatic or the cystic artery. Additional branches to the initial duodenum may stem from the right gastric, right gastroepiploic, or the superior pancreaticoduodenal artery. The remainder of the descending and transverse parts of the duodenum are supplied by vasa recta branching from the anterior and posterior pancreaticoduodenal arcades. The close association of the duodenum and pancreas make it hazardous to separate them, and leads to by-passing operations for benign duodenal disease, or to excision of both in malignant disease such as cancer of the head of the pancreas.

Further detail of the arteries to the pancreas are shown in Fig. 6. The head and neck are supplied mainly by the anterior and posterior pancreaticoduodenal arteries. The anterior and posterior superior arteries usually arise separately from the gastroduodenal, or the right gastroepiploic artery. The two inferior arteries most often possess a common stem (as in this figure). They may arise from a jejunal artery; they then pass posterior to the superior mesenteric vessels and on to their arcades. The remainder of the pancreas is supplied by branches of the splenic artery, two of which are large and quite constant--the dorsal pancreatic and the pancreatica magna. Anomalously, the dorsal pancreatic may arise from the celiac, the common hepatic, the superior mesenteric, or its middle colic or accessory left colic branches (Michels, 1962).

D. <u>The Spleen</u>

The splenic artery instead of arising from the celiac, may stem from the aorta, or the superior mesenteric. It may give off the following arteries: left inferior phrenic, left gastric, right gastroepiploic, left or right hepatic, superior mesenteric, middle colic, dorsal pancreatic (Michels, 1942). The splenic artery may be divided without necrosis of that organ. Collateral flow from gastric vessels is presumably the most significant.

Classically, five or six terminal splenic branches are said to arise at the hilum. Michels described the prehilar branching of the splenic artery as giving rise to two or three "terminal" then several "penultimate," and finally six to 36 (average 17) "ultimate" branches which enter the splenic substance. He observed transverse hilar anastomoses, but made no comment on anastomoses within the organ. Kyber (1870) found that each splenic branch goes to a "lobe" or segment, with no further communications with its neighbors. Gutierrez-Cubillos (1969) has recently confirmed the concept of circumscribed arterial segments in the spleen. "Segmental" resection of the spleen has been performed by Campo-Christo in three cases (1960). Communications between segmental veins were noted by Kyber, and have been shown by injection by Braithwaite and Adams (1957).

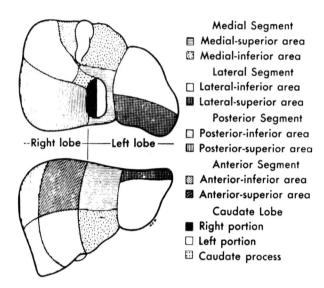

Fig. 7. Hepatic lobes, segments, and subsegments (areas). (From Healey et al., 1953.)

E. The Liver And Biliary Passages

1. Mode of Distribution of the Hepatic Artery

According to standard anatomical description, the liver is divided by the falciform and round ligaments into two lobes of unequal size--a larger right and a smaller left--receiving the right and left hepatic arteries respectively. These, in turn, arise by a bifurcation of the proper hepatic artery, so-called after the common hepatic gives off the right gastric and the gastroduodenal arteries. This description is erroneous as regards the lobulation of the liver, and correct in only 55% of cases as regards the source of the hepatic arteries.

The proper boundary between lobes, indeed of equal size, as first stated by Cantlie in 1897, is a plane extending through the gallbladder and the hepatic segment of the inferior vena cava (Fig. 7). We may call the lobes thus demarcated, the portal lobes, to distinguish them from those defined by current anatomic nomenclature. Numerous workers since Cantlie have established that there exist no communications between the right and left lobar branches of the three elements of the portal pedicle, except for arterioles connecting the two lobar arteries. However, Mays in 1972 and in later papers has demonstrated by angiography that adequate connections between lobar arteries may develop within 15 days. Further details on the branching of the portal elements were contributed by Hjortsjo (1951) and Healey and Schroy (1953), establishing the subdivision of the lobes into segments and subsegments, all with a delimited

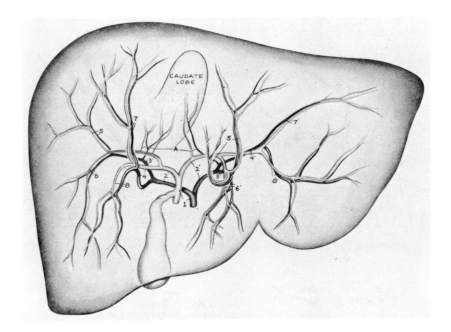

Fig. 8. Intrahepatic arteries and ducts. Numbers refer to artery and duct of each part. (1) Common hepatic duct and artery. (A) Branches to caudate lobe. Branches to right lobe. (2) Lobar. (3) Posterior segment. (4) Anterior segment. (5) Superoposterior subsegment. (6) Interoposterior subsegment. (7) Superoanterior subsegment. (8) Inferoanterior subsegment. Branches to left lobe: 2' Lobar. 3' Medial segment. 4' Lateral segment. 5' Superomedial subsegment. 6' Inferomedial subsegment. 7' Superlateral subsegment. 8' Inferolateral subsegment. (From Healey, Schroy, and Sorensen, 1953.)

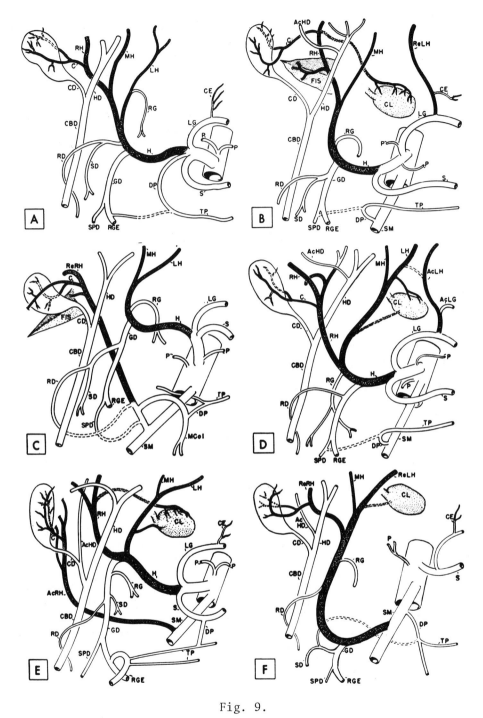

Fig. 9.

←Fig. 9. Sources of the hepatic arteries. A. Entire hepatic artery from the celiac (55%). B. Replaced left hepatic from left gastric (10%). C. Replaced right hepatic from superior mesenteric (11%). D. Accessory left hepatic from left gastric (8%). E. Accessory right hepatic from superior mesenteric (7%). F. Entire hepatic from superior mesenteric (4.5%). Anomalous hepatic arteries arose from both the superior mesenteric and left gastric arteries in 4% and the entire hepatic from the left gastric in .5%. The entire hepatic may stem from the aorta. (Michels uses the term "replaced" for a separate right, left, or middle hepatic artery, and the expression "accessory" for a supernumerary artery supplying other segments or subsegments.) AcHD, accessory hepatic duct; AcLg, accessory left gastric artery; AcLH, accessory left hepatic srtery; AcRH, accessory right hepatic artery; C, cystic artery; CBC, common bile duct; CD, cystic duct; CE, cardio-esophageal branches; CL, caudate lobe; DP, dorsal pancreatic artery (superior pancreatic); Fis, fissured area; GD, gastroduodenal artery; H, hepatic artery; HD, hepatic duct; LG, left gastric artery; LGE, left gastroepiploic artery; LH, left hepatic artery; MCol, middle colic artery; MH, middle hepatic artery; P, inferior phrenic artery; RD, retroduodenal artery (post. sup. pancreaticoduodenal in official nomenclature); ReLH, replaced left hepatic artery; ReRH, replaced right hepatic artery; RG, right gastric artery; RGE, right gastroepiploic artery; RH, right hepatic artery, S, splenic artery; SD, supraduodenal artery of Wilkie; SM, superior mesenteric artery; SPD, superior pancreaticoduodenal artery, (anterior); TP, transverse pancreatic artery (inferior pancreatic). (Figure modified from Michels in Schaeffer (1966); incidences from Michels, 1955.)

territory of portal and hepatic blood supply and duct drainage (Fig. 8). Supernumerary hepatic arteries supply some delimited portion of the liver and are thus not accessory in the sense of acting as collaterals for other arteries. Healey and Schroy have used the term "replaced hepatic artery" to denote a supernumerary artery extending to a hepatic lobe (Fig. 9, B,C.F,) maintaining the expression accessory for a supernumerary artery or duct belonging to a smaller subdivision of the liver (Fig. 9E).

The expressions quadrate and caudate lobes are maintained: the quadrate lobe constitutes the medial segment (of the left portal lobe), while the caudate lobe belongs to both the right and left lobes. Rex (1888) had named the artery to the quadrate lobe (now the medial segment) the middle hepatic artery, for in 25% it could be recognized as a special vessel with an origin from the right or proper hepatic. This appellation is maintained in the newer descriptions.

The hepatic veins are located between lobes and segments receiving venous radicles from each side. The subdivision of the liver into component parts has become an essential principle in the surgery of this organ (Edwards et al., 1975).

a. Blood Supply of the Gallbladder and Bile Duct

The gallbladder receives its own vessel--the cystic artery-- usually derived from the right hepatic, but often a branch of the proper hepatic artery. The bile duct is supplied mainly by the gastroduodenal artery. This vessel, usually a branch of the common hepatic, may originate from the right, middle or left hepatic, or some other nearby artery, including the superior mesenteric (Fig.9).

2. Sources of the Hepatic Artery or its Three Hepatic Branches

In only 55% of individuals does a proper hepatic artery stem from the celiac as usually described (Fig. 9A). In 18% the left hepatic in its entirety, or one of its branches ("accessory artery") comes off the left gastric artery (Fig. 9B,D). In another 18%, the right hepatic entirely, or one of its branches, derives from the superior mesenteric artery (Fig. 9C,E). More rarely, the entire proper hepatic takes origin from the superior mesenteric (Fig. 9F), the aorta, or the left gastric artery. A hepatic artery of superior mesenteric origin usually courses behind the portal vein and the head of the pancreas, less often anterior to the vein and through the pancreas (Edwards et al., 1975).

3. Mortality Following Hepatic Artery Loss, and Available Collaterals

The hepatic artery has been accidentally ligated during cholecystectomy, in pancreatectomy (with anomalous origin from the superior mesenteric), in gastrectomy (with anomalous origin from the left gastric); or occluded during infusion for cancer chemotherapy. Intentional ligation has been performed for cirrhosis, hepatic malignancy, hepatic artery aneurysm and liver injury. The celiac artery at its origin has been intentionally ligated in radical gastrectomy (Appleby, 1953). The older figure for mortality following hepatic artery ligation in cirrhosis was about 30% (Mays, 1967) but 60% after ligation of either the proper hepatic or a lobar artery in other clinical situations (collected series of Graham and Cannel, 1933; Alessandri, 1937; Michels, 1953; Monafo et al., 1965; and Karasewich and Bowden, 1967). However, reviewing their own and reported cases, Brittain and his associates (1964) concluded that the ligation was not ordinarily fatal in the absence of prior hepatic or other disease. Recent reports of intentional ligation are also encouraging, in cases where the indication has been liver trauma (Mays, 1972; Madding and Kennedy, 1972), or carcinoma of the liver (Balasegaram, 1972). Better general support of the patient is undoubtedly responsible for the newer results. As a rule the mortality after hepatic artery obstruction is lessened when the ligature is placed at levels proximal to the portal pedicle. This is attributed to the greater availability of collaterals for the hepatic artery at this level.

Temporary occlusion of the portal pedicle, as by a vascular clamp, is reported to be well tolerated for 15 minutes at normothermic conditions, and for at least 30 minutes with hypothermia (Albo et al., 1969; Yellin et al., 1971).

Michels (1953) divides the collaterals into three categories: (1) Hepatic arteries arising from sources other than the celiac--common hepatic trunk; (2) Pathways outside the hepatic arteries but capable of connecting with them, mainly vessels of the pancreas, stomach, and esophagus; (3) Pathways outside the celiac blood supply. These include arteries of the pancreas, diaphragm, and of the falciform and round ligaments (see also Bengmark and Rosengran, 1970). Child (1954) states that concomitant ligation of the portal vein and the hepatic artery is always fatal.

III. THE BOWEL

A. Distribution Of The Superior Mesenteric Artery

The superior mesenteric shares with the celiac the supply of the duodenum and pancreas (Fig. 6). It is the sole artery of supply to the rest of the small intestine (Fig. 10) as well as to the large intestine from the cecum to a point close to the splenic flexion where it communicates with the left colic branch of the inferior mesenteric (Fig. 12). The jejunal and ileal branches vary from 12 to 15 in number, with variation due to a tendency toward two or more of these vessels to originate from a common stem. There is considerable variation in the territory of the remaining branches of the superior mesenteric artery. Generally the ileocolic supplies the terminal ileum, cecum and appendix, and the ascending colon. The right colic courses to the hepatic flexure; while the middle colic supplies the transverse colon.

The terminal branches of both the small and large intestine--the vasa recta--arise from a series of arcades produced by a primary bifurcation of the named intestinal arteries with junction of adjacent arteries (Figs. 10, 12). The arcades between colic arteries are relatively simple, with only occasional tendency toward further branching and arcade formation prior to the origin of the vasa recta. In the case of the small intestinal arteries, there is a tendency

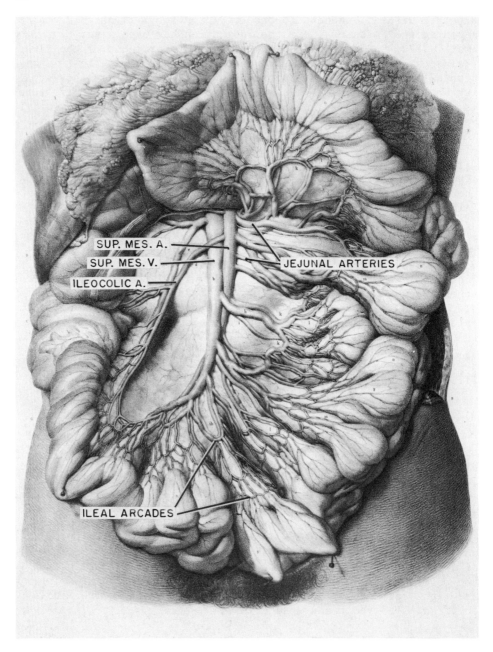

Fig. 10. The superior mesenteric vessels. (From Bourgery and Jacob, 1839.)

toward additional arcade formation before the vasa recta are given off. This tendency increases progressively down the length of the bowel. Thus while a second tier of arcades is rather sporadic in the mesentery of the upper jejunum, two or three tiers are regularly observed lower in the jejunum, increasing to four or five in the terminal ileum. The vasa recta are correspondingly longer on the upper jejunum and progressively shorter on the lower ileum. The vasa recta derived from the first jejunal artery course mainly anterior to the jejunum, a few running posterior. Beyond the first jejunal artery the vasa recta pass to one side or the other of the bowel in a haphazard fashion (Noer, 1943), Noer also noted that the vasa recta may give off some branches to the bowel before reaching that structure, but entered into no anastomoses. The short vasa supply the mesenteric aspect of the bowel, the long extend to the antimesenteric aspect. The vasa recta carry out the principle already noted for the small arteries to the duodenum; namely, that they are effectively end arteries and that the bowel is subject to necrosis if they are injured. Thus tearing of the mesentery close to the bowel for 1 or 2 cm. may result in necrosis.

The right colic artery originates from a common trunk with the middle colic in close to half of all subjects, and from the ileocolic less frequently. The middle colic stems as a single vessel from the superior mesenteric artery in fewer than half of all subjects. The middle colic can originate from the celiac artery or from one of the three branches of that vessel passing behind the pancreas. Michels (1955) found the dorsal pancreatic artery occasionally giving rise to the middle colic, or an accessory left colic artery. Conversely, the accessory middle colic when formed as a branch of the superior mesenteric, often gave off the dorsal pancreatic artery. An accessory left colic artery ("arc of Riolan") springs from the proximal superior mesenteric or the middle colic artery, and joins the left colic artery in the mesocolon close to the duodenojejunal junction. Its incidence is generally given as 10% but Adachi (1928) found it in 20%.

The extent of the colon supplied by the various colic arteries is variable. The left colic artery ascends to reach the splenic flexure in 86% according to Michels et al. (1965). Kahn and Abrams (1964) found by arteriography that the left colic extended to a point proximal to the flexure in 13%. The replacement of one of the colon arteries by an adjacent vessel may be so complete as to justify the statement that the right middle, or rarely the left, colic artery is absent. Contrariwise, two vessels may represent one of these usually single arteries.

Finally, one unusual branch of the superior mesenteric should be noted--a persistent omphalomesenteric artery, most commonly seen extending from a small intestinal artery on one side of a Meckel's diverticulum (Fig. 11). Such a persistent artery has also been observed in extreme rarity in the absence of a diverticulum crossing the peritoneal cavity from mesentery to umbilicus (Hollinshead, 1971).

B. <u>Distribution Of The Inferior Mesenteric Artery</u>: <u>Iliac Branches To The Anorectum</u>

The inferior mesenteric artery supplies the large intestine from about the splenic flexure of the colon to about the middle of the rectum. The initial branch of the inferior mesenteric is the left colic artery, which ascends sharply to supply the descending colon and often the splenic flexure (see above). Next come the sigmoid arteries, generally three in number, the upper of which may be combined with the left colic. The inferior mesenteric continues

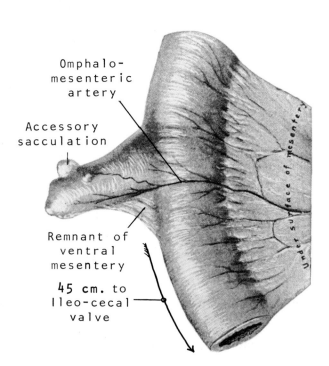

Fig. 11. The omphalomesenteric artery on a Meckel's diverticulum. The distance from the ileocecal valve in this instance is shorter than the average. (From Patten, B.M.: Human Embryology. 3rd ed. Copyright 1968. McGraw-Hill Book Company, New York. Used with permission of McGraw-Hill Book Company.)

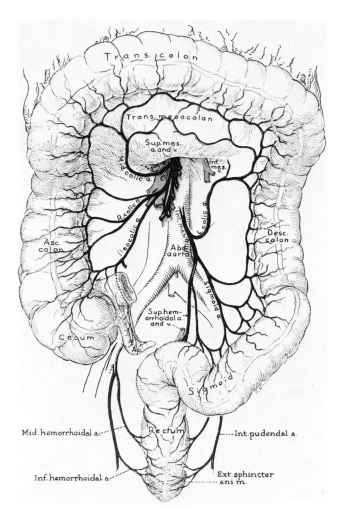

Fig. 12. Arteries of the large intestine. (From Jones and Shepard, 1945.)

downward as the superior rectal (hemorrhoidal) artery (Figs. 2, 12). All the colic and the sigmoidal arteries bifurcate proximal to the bowel. The anastomoses between the bifurcations form a composite vessel termed a marginal artery. A few secondary arcades may also be formed. The marginal artery lies 1 to 8 cm. from the bowel (Steward and Rankin, 1933). Vasa recta are not formed below the distribution of the last sigmoid artery, the anorectum being supplied by interconnecting fine branches of the rectal arteries.

The anorectum is supplied from three sources: the superior rectal from the inferior mesenteric artery, the middle rectal from the internal iliac, and the inferior rectal from the internal pudendal artery. The superior rectal artery bifurcates or sometimes trifurcates unevenly, on the posterior aspect of the rectum, to descend to within a few centimeters above the pelvic floor. The middle rectal is most variable. Angiography shows the rectum to receive many small arteries from the lateral and median sacral, and adjacent visceral arteries. The frequency of the presence of a predominant large middle rectal extending from the internal iliac or one of its branches is variously given from 66 to 98% (Boxall et al., 1963).

C. Collateral Pathways For The Major Intestinal Trunks

The intestinal arteries are generally well supplied with collaterals, except for the vasa recta of the small and large intestine, and the comparable small arteries to the duodenum. Thus in spite of the unusual sensitivity of the intestine to ischemia, slow occlusion as by arteriosclerosis of one of the three major gastrointestinal trunks is well tolerated because of the interconnection of the three trunks. The sites of these connections is clarified on viewing the relative simple alimentary canal of the embryo (Fig. 13). Meager connections exist on the lower esophagus between branches from the thoracic aorta and those from the left gastric artery. However, excellent communications are present between the celiac and superior mesenteric arteries through the superior and inferior pancreaticoduodenal arteries. The occasional anomalous origin of pancreatic or hepatic arteries from the superior mesenteric offer additional pathways between the trunks. In addition, Michels (1955) emphasized that frequent communications exist behind the pancreas between branches of the celiac, and branches of the superior mesenteric arteries, especially the middle colic and, when present, the accessory left colic arteries. The communicating vessels are generally small, but may reach 2 mm. in diameter. The communication between the superior and inferior mesenterics is usually a gross "inosculation" via their adjacent contributions to the marginal artery. When present, the accessory left colic artery mentioned above forms an additional large anastomosis between the superior and inferior mesenteric arteries parallel to the marginal artery, but closer to the base of the mesentery. It is likely that the "meandering mesenteric artery" seen after mesenteric artery obstruction, is made up of an accessory left colic artery plus portions of the colic and marginal arteries (Moskowitz et al., 1964; Gonzales and Jaffe, 1966). It has been mentioned above in considering the aorta, that in chronic occlusion of more than one visceral arterial trunk, the organs will survive through fine collaterals from parietal vessels.

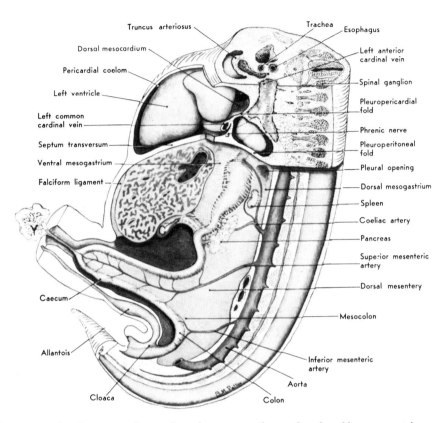

Fig. 13. The human embryo in the seventh week, in diagrammatic longitudinal section. (From Patten, B.M.: Human Embryology, 3rd ed. Copyright 1968, McGraw-Hill Book Company, New York. Used with permission of McGraw-Hill Book Company.)

Sudden occlusion as by division and ligature, of the inferior mesenteric artery is usually well tolerated, with bowel necrosis noted in only a small proportion of cases. However, intestinal necrosis regularly follows acute embolic occlusion of the superior mesenteric artery. The celiac artery has been divided at its origin in operations for carcinoma of the stomach to facilitate a wide removal of the stomach, spleen, tail and body of pancreas and retroperitoneal lymph nodes. The hepatic artery, divided at its origin, is not removed. The remaining head of the pancreas and the duodenum appear to obtain adequate arterial inflow (Appleby, 1953; Grimes and Visalli, 1964).

D. Collaterals Between Intestinal Branches

The collateral system produced by the arcade formation both in the small and large intestine, is in the main good enough to allow the division of one or more arteries contributing to the arcades and the marginal artery, without producing bowel necrosis. The anastomotic system may nevertheless be ineffective because of variations in the size of the major intestinal branches and in their anastomoses, or through the presence of arterial stenosis or thrombosis.

The proximal superior mesenteric artery forms a link between the inferior pancreatico-duodenal and the first jejunal arteries. Lower down, Barlow (1956) found a break in the paraintestinal artery constituted by the convexity of the arcades in 5.7% of individuals, and a "weakness" in 15.7%.

Discontinuity or probable inadequacy of the marginal artery is apt to occur in three locations, with a frequency found in extensive studies by Michels et al. (1963 and 1965) as follows: (1) Between the ileocolic and right colic arteries, 10%; (2) Between the ascending and descending branches of the left colic artery, 39% (Griffiths' point, Griffiths, 1956) although a secondary proximal arcade between these two branches often compensates for the inadequacy of the marginal artery; (3) Between the last sigmoid and the superior rectal arteries, 84% (Sudeck's critical angle, Sudeck, 1907). Basmajian (1954) adds a fourth location, the junction of the right and middle colic arteries, incidence 5%.

Connections between the superior rectal above, with the middle and inferior rectal arteries below, are of importance in connection with rectosigmoid resections for carcinoma. Individual variations in the middle rectal arteries is great. In general, a rectal stump no longer than 8 cm. above the peritoneal reflection of the pelvis will survive on the basis of branches from pelvic vessels.

The presence of a para-intestinal artery in both the small and large intestine allows one to mobilize a pedicled segment of bowel (the Roux loop, Roux, 1897) which may be transposed to re-establish the continuity of the alimentary tract, or to establish a conduit as for a urinary stoma. A mobile jejunal loop may be prepared from the jejunum 30 to 40 cm. beyond the duodenojejunal junction (Rienhoff, 1946; Allison and DaSilva, 1953). A very long intestinal graft may be fashioned, centered on the middle colic artery, consisting either of the terminal ileum with the cecum, ascending, and right half of the transverse colon (Petrov, 1964), or the left colon from the mid transverse part to the sigmoid below the first and second sigmoid arteries (Beck and Baronofsky, 1960).

IV. THE SUPRARENAL GLANDS, KIDNEYS, AND URETERS

A. The Suprarenal Glands

The arteries to the suprarenal gland are termed inferior, middle, and superior. These are in reality sets of fine arteries totaling as many as 60, with an occasional predominant vessel, stemming respectively from the renal artery below, the aorta medially, and the inferior phrenic above (Solotuchin, 1929, Fig. 14). Inferior suprarenal arteries may come off the superior polar renal artery or from other renal branches, the adipose, gonadal, or ureteric. Prior to entering the gland, they ramify within the fat which separates the suprarenal from the kidney. The middle suprarenal artery supplies the celiac ganglion as well as the suprarenal gland. Both structures usually receive additional twigs from the celiac. The celiac may give off the middle suprarenal or the inferior phrenic arteries.

B. The Kidneys

1. Segments of the Kidney: Distribution of the Renal Artery

The division of the kidney into segments, determined by mode of its arterial supply, is important in relation to the development of necrosis secondary to division of a branch of the renal artery or of a supernumerary artery, or hypertension from localized ischemia; and to the possibility of resection of a diseased segment. The lobed appearance of the newborn kidney suggests such localized territories, but no explicit correlation has been made between these lobes and the arterial segments of the adult kidney.

The presence of localized segments was indicated by the observations of Brödel (1901) and older anatomists (see Merklin and Michels, 1958) of a lack of communication between branches of the renal artery except by capsular capillaries and arterioles. The actual segmentation was demonstrated by Graves (1954) and is generally accepted although nomenclature varies (Figs 15, 16). The renal artery divides into anterior and posterior divisions. The anterior division is the larger, supplying the apical and lower segments and the upper and middle segments, the region between the poles anteriorly and along the convex border. The posterior division supplies the posterior segment. Sykes found Graves' pattern of division in 59% of 71 specimens. In half of the remaining 12 kidneys, the renal artery divided into three branches, each subdividing into anterior and posterior arteries. In the other six, two renal arteries of equal size stemmed from the aorta ("dual renal arteries") prior to further division into segmental branches.

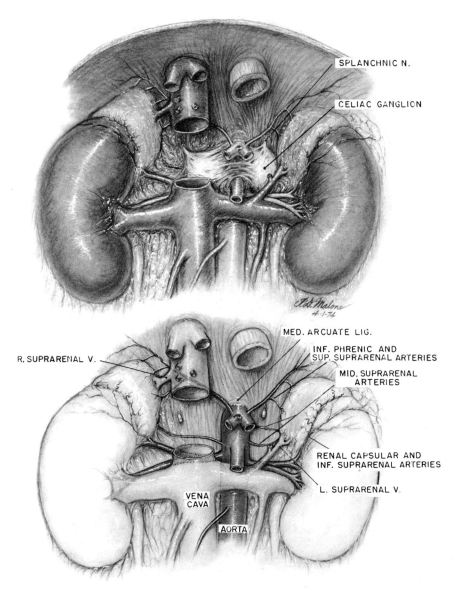

Fig. 14. Arteries to the suprarenal glands and celiac ganglia. (From Edwards et al., 1975.)

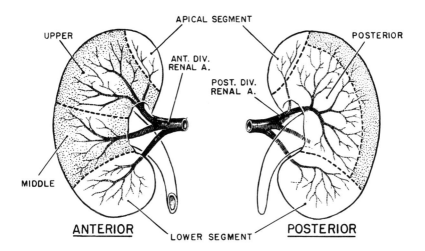

Fig. 15. The segmental arterial supply to the kidney. (From Edwards et al., 1975, after Graves, 1954.)

Further branching, beyond the segmental arteries, as noted by Brödel for the poles, suggests the presence of subsegments; but the "interlobar" branches of the segmental arteries cannot be considered subsegmental since the interlobal arteries supply adjacent pyramids. Löfgren's study (1949) suggests the possibility for a subdivision into subsegments. He found that the kidney medulla of the fetus consists of 14 major pyramids, seven ventral ('V') and seven dorsal ('D'). In the newborn, these are partially fused into a cranial part, comprising the first three ventral and dorsal pyramids (V1-3, D1-3)., an intermediate (V4-5, D4-5), and a caudal (V6-7, D6-7). Boijsen (1959) has at least partially confirmed this subdivision angiographically.

2. Variations in Number and Source of the Renal Arteries

Multiple renal arteries are present in over 20% of all individuals (Adachi, 1928; Pick and Anson, 1940; Merklin and Michels, 1958; Olsson in Abrams, 1971). In usual kidneys, multiple renal arteries occur mainly as the dual renal arteries mentioned above, or as supernumerary arteries (one or more) to either pole of the kidney, those to the lower pole being more frequent. Supernumerary polar arteries usually stem from the aorta, but a superior polar may come off the inferior phrenic artery. As in the liver, such an "accessory" artery cannot act as a collateral for the other renal vessels, for its distribution is limited to some one segment or subsegment. Multiple arteries are the rule in the case of congenitally abnormal kidneys or kidney pelves. There is a wide source for the multiple arteries of renal anomalies, ranging from the sacral, internal, external, and common iliac arteries below--to the aorta, lumbar, inferior and superior mesenteric, hepatic, colic, and inferior phrenic above (Adachi). The final height which the kidney attains determines the probable source of its arteries. A bifid pelvis is often accompanied by dual renal arteries. A horseshoe kidney is often supplied from a single vessel arising from the front of the aorta.

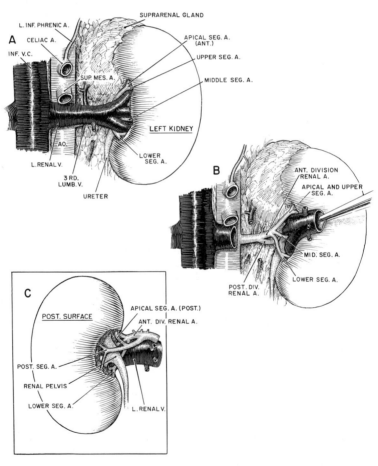

Fig. 16. The pedicle of the left kidney. (From Edwards et al., 1975.)

C. Extrarenal Branches of the Renal Artery and Supply to the Ureter

The proximal part of the renal artery, or its primary divisions prior to their disappearance in the renal sinus quite regularly give off the fine inferior suprarenal, ureteropelvic and capsular branches; anomalously the gonadal vessels may originate here. While small, these arteries are of prime importance to the structures to which they run, and in the case of renal artery obstruction, they constitute significant pathways for collateral flow.

The suprarenal arteries have been described above. The arteries to the kidney, pelvis, and upper ureter are minute vessels lying on the web-like connective tissue about these structures. They are beautifully depicted by Brödel in Kelly and Burnam (1914). These arteries communicate along the ureter with further contributions from the gonadal vessels, the aorta, common and internal iliac arteries, and finally, with arteries of the bladder and internal genitalia (Fig. 17). In spite of this apparent richness of supply, necrosis of the ureter is prone to occur if it is deprived of its blood supply over a distance of 2.5 cm or more. At the upper end this has occurred in connection with kidney transplantation and at its lower end in operations for urinary diversion or extensive lymphatic resection as for carcinoma of the cervix.

The renal capsular arteries supply the perirenal fat. One such vessel is shown in Figure 14, but they are apt to be multiple. They communicate grossly with vessels of the parietes. Occasionally they are large, and I have seen one such artery coursing much like a subcostal artery through the anterolateral part of the abdominal wall.

D. Collaterals for the Renal Artery

Renal damage may be expected after obstruction of the main renal artery under normothermic conditions of more than 20 minutes duration. If prolonged for some hours, gross necrosis is the usual outcome. Collateral circulation is adequate to prevent necrosis when the occlusion develops gradually over a long period of time but the usual outcome is a shrunken kidney and the development of "renovascular" hypertension. The collateral flow extends from the parietes, peritoneum, and organs adjacent to the kidney, reaching the kidney via the extrarenal branches of the renal artery (Abrams and Cornell, 1965).

It should be recalled that in the kidney as in the liver, supernumerary arteries are not accessory in the sense of overcoming ischemia in a specific territory. Segmental necrosis or chronic ischemia with hypertension is thus a likely outcome, when a segmental renal artery is obstructed or divided.

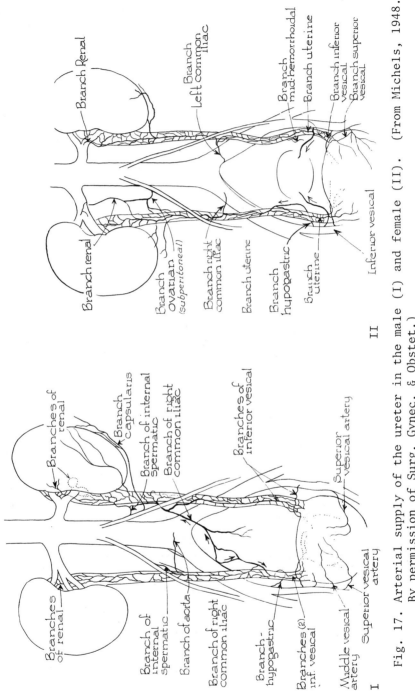

Fig. 17. Arterial supply of the ureter in the male (I) and female (II). (From Michels, 1948. By permission of Surg. Gynec. & Obstet.)

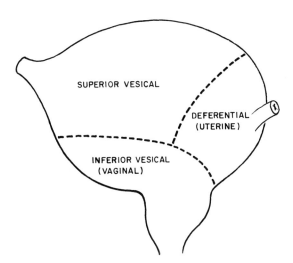

Fig. 18. Arterial territories of the bladder. (From Edwards et al., 1975. After Braithwaite, 1952.)

V. PELVIC AND PERINEAL VISCERA

A. The Bladder

As regards its arterial supply, the bladder of both sexes is divisible into three regions (Figs. 18, 19, 20). The dome and apex are supplied by the branches from the umbilical artery termed superior vesical; the most posterior of branches is sometimes called the middle vesical.

In the male, the lower anterior region is supplied by the inferior vesical; and the lower posterior region by the deferential artery. The inferior vesical artery (Fig. 22) most often stems from the internal pudendal, or from a gluteopudendal trunk (see visceral branches of the internal iliac artery). The deferential, sometimes also called the middle vesical, arises from the umbilical artery (Braithwaite, 1952; Darget et al., 1957) less often from the inferior vesical. It regularly supplies the seminal vesicles, as well as the vas, and plays a role in the blood supply of the testis (Fig. 21). In the female the vaginal artery, when large, represents the inferior vesical, and a "middle vesical" is the counterpart of the deferential. The uterine artery gives a variable contribution to the bladder.

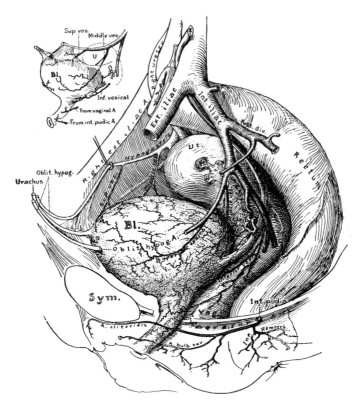

Fig. 19. Arteries to the female bladder and ureter. (From Kelly and Burnam, 1914.)

B. Male Genitalia

1. The Testis

While the testicular (internal spermatic) is the major artery of the testis, it is aided through communications with the deferential and cremasteric arteries (Fig. 21). The terminal branches of the testicular artery are essentially end-arteries (Harrison and Barclay, 1948). The deferential artery, whose origin is shown in Figs. 20 and 22, varies from being threadlike to 1 mm in diameter (Harrison, 1949). Harrison found that the communication of the deferential and testicular arteries takes place via the epididymal branch of the latter, at the lower pole of the testis. Only occasionally was the deferential as large or larger than the testicular, which varied from 0.7 to 1.1 mm in diameter. The cremasteric artery, a branch of the inferior epigastric, is most variable in size, but always the smallest of the three.

Slowly developing obstruction of the testicular artery, quite usual in the arteriosclerotic patient (Edwards and LeMay, 1955), produces no gross effect on the testis. Division of the artery, on the other hand, whether in the young or old, almost always results in atrophy and occasionally in necrosis of the testis. This has been seen in operations for undescended testis, division of the cord in hernia repair, or during kidney transplantation (Edwards et al., 1975).

2. Prostate: Seminal Vesicles, and Penis

The prostate and seminal vesicles are supplied by branches of the inferior vesical and deferential arteries (Figs. 20, 22). Flocks (1937) and Clegg (1956) demonstrated that the two to five prostatic

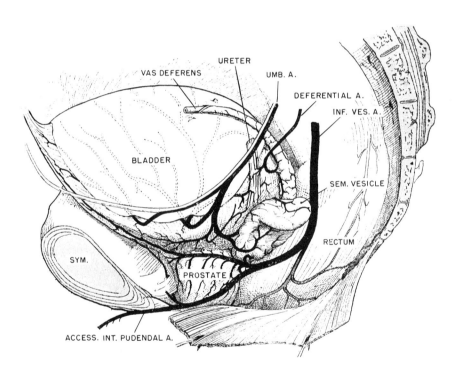

Fig. 20. Arteries to the male bladder and pelvic genital organs. (From Farabeuf, 1905.)

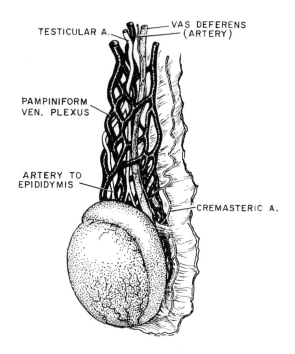

Fig. 21. Arteries to the testis. (From Edwards et al., 1975.)

arteries divide into capsular and urethral groups; the capsular arteries penetrate the prostate to be distributed mainly to the outer zone containing the main prostatic glands. The urethral group of arteries pass through each side of the bladder neck, then turn distally beneath the mucosa to supply the mucosal and submucosal glands. It is the enlargement of these glands which constitutes the mass of benign prostatic hypertrophy.

The penis is supplied by three branches of each internal pudendal artery: the artery to the bulb, the deep, and dorsal arteries of the penis. The deep artery of the penis may arise in the pelvis from the internal iliac or obturator arteries and course on the upper surface of the pelvic diaphragm and beneath the pubic arch, as in Fig. 20.

ABDOMINAL VISCERAL CIRCULATION IN MAN

C. Female Gentalia

1. The Ovaries: Uterus: Vagina: External Genitalia

The ovarian arteries usually take their origin from the aorta but it has been noted that one or both arteries may originate from a renal or inferior phrenic vessel. The ovarian arteries are known to be narrowed or occluded after the menopause. This has been traditionally ascribed to a physiologic shrinkage, but the observation that the testicular artery of old men is often occluded by arteriosclerosis (see above) suggests that the same mechanism may hold for the gonadal vessels of the female.

The ovarian, uterine, and vaginal arteries are in gross communication with each other (Fig. 23) to the extent that no gross effect follows either chronic or sudden occlusion of any one of these vessels. Communications from one side to the other takes place through the small branches and does not appear to be as ample as between vessels of one side. The medial femoral circumflex arteries, through their communications with the lower branches of the internal iliacs, constitute remote collaterals for uterine blood (Reynolds, 1963).

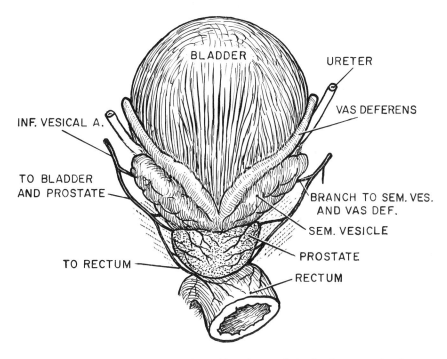

Fig. 22. Posterior view of distribution of inferior vesical arteries in the male. The origin of the deferential artery from the inferior vesical is exceptional. (From Albarran, 1909.)

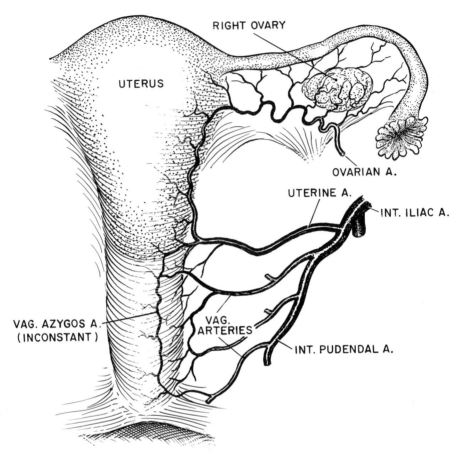

Fig. 23. Arteries supply to the female pelvic organs, viewed from behind. (From Edwards et al., 1975.)

The uterine artery arises most commonly high on the anterior division of the internal iliac artery, near the umbilical or in common with it (Fig. 3). It may come off the proximal internal pudendal or inferior vesical arteries (Roberts and Krishingner, 1967). The vaginal artery may be as large as the uterine. Roberts and Krishingner noted that the vaginal arises somewhat oftener from the internal pudendal within the pelvis than from the uterine. Vaginal branches also arise from adjacent vessels.

The external genitalia are supplied by the internal pudendal arteries, with parts homologous to those of the male, receiving corresponding branches.

a. Adaptation to Pregnancy

The flow to all the female pelvic organs in the nonpregnant state is probably about 75 ml/min. The flow to the pregnant human uterus is maximal at about mid-term, and was estimated by Metcalf and his associates (1955) to average 500 ml/min for a single pregnancy, and was 1150 ml/min to a uterus containing a twin pregnancy. Externally, the uterine arteries are seen to be enlarged, straightened from the spiral form they show in the nonpregnant state, and giving rise to specially large branches to the placental site. The basic change allowing the increased flow in pregnancy is undoubtedly a lowering of resistance in the distribution of the artery, particularly at the placenta, where the branches do not end in a capillary bed, but open directly in the intervillous space in a manner often characterized as an arteriovenous shunt (Spanner, 1936; Burwell, 1938; Reynolds, 1963).

The uterus is not the only structure receiving an increased blood flow during pregnancy. The cardiac output increases from the nonpregnant average of about 4.5 liters/min. to a high of about 6 liters/min in mid-pregnancy (Hytten and Leitch, 1964). The kidney accounts for as much or more of the increased output as does the uterus. Renal blood flow, as measured by Sims and Krantz (1958) averaged 1148 ml/min during pregnancy vs 652 ml/min in the puerperium of the same subjects. Other structures sharing in the increased flow during pregnancy are the skin, the breasts, and the gut.

REFERENCES

Abrams, H.L., 1971 (ed.), "Anigography," 2nd ed., Little Brown, Boston.
Abrams, H.L., and Cornell, S.H., 1965, Patterns of collateral flow in renal ischemia, Radiol., 84:1001-1012.
Adachi, B., 1928 Das Arteriensystem der Japaner, Vol. II, Maruzen, Kyoto.
Albarran, J., 1909, "Médicine Opératoire des Voies Urinaries," Masson, Paris.
Albo, D., Christensen, C., Rasmussen, B.L., and King, T.C., 1969, Massive liver trauma, Am. J. Surg. 118:960-963.
Alessandri, A., 1937, Aneurysm of hepatic artery, in "Nelson Loose-Leaf Surgery" Vol. V, Chapter 10A, Page 608, Nelson, T., New York.
Allison, P.R., and DaSilva, L.T., 1953, The Roux loop, Brit J. Surg. 41:173-180.
Appleby, L.H., 1953, The coeliac axis in the expansion of the operation for gastric carcinoma, Cancer 6:704-707.
Balasegaram, M., 1972, Complete hepatic dearterialization for primary carcinoma of the liver, Am. J. Surg. 124:340-345.

Barlow, T.E., Bentley, F.H., and Walder, D.N., 1951, Arteries, veins, and arteriovenous anastomoses in the human stomach, Surg. Gynec. Obstet. 93:657-671.

Barlow, T.E., 1956, Variations in the blood supply of the upper jejunum, Brit. J. Surg. 43:473-475.

Basmajian, J.V., 1954, The marginal anastomoses of the arteries to the large intestine, Surg. Gynec. Obstet. 99:614-616.

Beck, A.R., and Baronofsky, I.D., 1960, A study of the left colon as a replacement for the resected esophagus, Surg. 48:499-509.

Benirschke, K., 1965, Placental morphogenesis, in "Fetal Homeostasis" (R.M. Wynn, ed), New York Academy of Sciences, New York.

Boijsen, E., 1959, Angiographic studies of the anatomy of single and multiple renal arteries, Acta Radiol. Suppl. 183.

Bourgery, J.M., and Jacob, N.H., 1839, "Traité complet de l'anatomie de l'homme,"Vol 5, Librairie Anatomique, Paris.

Bower, E.G., Smullens, S.N., and Parke, W.W., 1977, Clinical aspects of persistent sciatic artery: Report of two cases and review of the literature, Surg. 81:588-595.

Boxall, T.A., Smart, P.J.G., and Griffiths, J.D., 1963, The blood supply to the distal segment of the rectum in anterior resection, Brit. J. Surg. 50:399-404.

Braithwaite, J.L., 1952, The arterial supply of the male urinary bladder, Brit. J. Urol. 24:64-71.

Braithwaite, J.L., and Adams, D.J., 1957, The venous drainage of the rat spleen, J. Anat. (London) 31:352-356.

Brittain, R.S., Marchioro, T.L., Hermann, G., Waddell, W.R., and Starzl, T.E., 1964, Accidental hepatic artery ligation in humans, Am. J. Surg. 107:822-832.

Brödel, M., 1901, The intrinsic blood vessels of the kidney and their significance in nephrotomy, Bull. J. Hopkins Hosp. 12:10-13.

Burwell, C.S., Strayhorn, W.D., Flickinger, D., Corlette M.B., Bowerman, E.P., and Kennedy, J.A., 1938, Circulation during pregnancy, Arch. Int. Med. 62:979-1001.

Cantlie, J., 1898, On a new arrangement of the right and left lobes of the liver, Proc. Anat. Soc. Gr Brit and Ireland. Pages IV-IX, June 1897, (Bound with J Anat and Physiol 32).

Cauldwell, E.W., and Anson, B.J., 1943, The visceral branches of the abdominal aorta: topographical relationship, Am. J. Anat. 73:27-57.

Chiene, J., 1869, Complete obliteration of the coeliac and their mesenteric arteries: the viscera receiving their blood-supply through the extra-peritoneal system of vessels, J. Anat. Phys. 3:65-72.

Child, C.G., 1954, "The Hepatic Circulation and Portal Hypertension" W. B. Saunders Co., Philadelphia.

Clegg, E.J., 1955, The arterial supply of the human prostate and seminal vesicles, J. Anat. 89:209-216.

Cullen, T.S., 1916, "Embryology, Anatomy and Diseases of the Umbilicus together with Diseases of the Urachus," W.B. Saunders, Philadelphia.

Darget, R., Ballanger, F., and Adano, R., 1957, La vascularization de la prostate, Son interet chirurgical, J. Urol., (Paris) 63:341-349.

Edwards, E.A., and LeMay, M., 1955, Occlusion patterns and collaterals in arteriosclerosis of the lower aorta and iliac arteries, Surg. 38:950-963.

Edwards, E.A., Malone, P.D., and Collins, J.J., Jr., 1972, "Thorax," Lea and Febiger, Philadelphia.

Edwards, E.A., Malone, P.D., and MacArthur, J., 1975, "Operative Anatomy. Abdomen and Pelvis," Lea and Febiger, Philadelphia.

Farabeuf, L.H., 1905, "Les vaisseaux sanguins des organes génitourinaries du perinée et du pelvis," Masson, Paris.

Feller, I., and Woodburne, R.T., 1961, Surgical anatomy of the abdominal aorta, Ann. Surg. 154 Suppl. 239-252.

Finaly, R., 1925, Over stoornissen ten gevolge van onderbiden der beide arteriae hypogastricae, Nederl. Tijdschr. v. Geneseek 69.1:1115-1119.

Flocks, R.H., 1937, The arterial distribution within the prostate gland. Its role in transurethral prostatic resection, J. Urol. 37:524-548.

Gonzalez, L.L., and Jaffe, M.S., 1966, Mesenteric arterial insufficiency following abdominal aortic resection, Arch. Surg. 93:10-20.

Graham, R.R., Cannell, D., 1933, Accidental ligation of the hepatic artery, Brit. J. Surg. 20:566-579.

Graves, F.T., 1954, The anatomy of the intrarenal arteries and its application to segmental resection of the kidney Brit. J. Surg. 42:132-139.

Griffiths, J.D., 1956, Surgical anatomy of the blood supply of the distal colon, Ann. Roy. Coll. Surg. Engl. 19:241-256.

Grimes, O.F., and Visalli, J.A., 1964, The embryologic approach to the surgical management of carcinoma of the upper stomach, Surg. Clin. N. Amer. 44:1227-1237.

Gutierrez-Cubillos, C., 1969, Segmentación espléncia, Rev. Esp. Enferm. Apar. Digest. 29:341-350.

Harrison, R.G., 1949, The distribution of the vasal and cremasteric arteries to the testis and their functional importance, J. Anat. 83:267-282.

Healey, J.E., Jr., Schroy, P.D., 1953, Anatomy of the biliary ducts within the human liver, Arch. Surg. 66:599-616.

Healey, J.E., Jr., Schroy, P.C., and Sorenson, R.J., 1953, The intrahepatic distribution of the hepatic artery in man, J. Internat. Coll. Surg. 20:133-148.

Hjortsjo, C.H., 1951, The topography of the intrahepatic duct systems, Acta Anat. 11:599-615.

Hobbs, J.T., and Edwards, E.A., 1963, Femoral artery flow, limb blood volume and cardiac output through continuously recorded indicator-dilution curves, Ann. Surg. 158:159-171.

Hollinshead, W.A., 1971, "Anatomy for Surgeons, Vol. 2. The Thorax, Abdomen, and Pelvis," 2nd ed. Harper and Row, New York.

Huittinen, V.M., and Slätis, P., 1973, Postmortem angiography and dissection of the hypogastric artery in pelvic fractures, Surg. 73:454-462.

Hyber E., 1870, Ueber die Milz des Menschen und einiger Saügethiere, Arch. Mikr. Anat. 6:540-570.

Hyrtl, J., 1870, "Die Blutgefässe der Menschlichen Nachgeburt in Normalen und Abnormalen Verhältnissen," Braumüller, Wien.

Hytten, F.E., and Leitch, I., 1964, "The physiology of Human Pregnancy," Blackwell, Oxford.

Jones, T., and Shepard, W.C., 1945, "A Manual of Surgical Anatomy," W.B. Saunders, Philadelphia.

Kahn, P., and Abrams, H.L., 1964, Inferior mesenteric arterial patterns. An angiographic study, Radiol. 82:429-441.

Karasewich, E.G., and Bowden, L., 1967, Hepatic artery injury, Surg. Gynec. Obstet. 124:1057-1063.

Kelly, H.A., and Burnam, C.F., 1914, "Diseases of the Kidneys, Ureters and Bladder," Appleton, New York.

Löfgren, F., 1949, "Das Topographische System der Malpighischen Pyramiden der Menschenniere," Ohlsson, Lund.

Madding, G.F., and Kennedy, P.A., 1972, Hepatic artery ligation, Surg, Clin. N. Am. 52:719-728.

Mays, E.T., 1967, Observation and management of hepatic artery ligation, Surg. Gynec. Obstet. 124:801-807.

Mays, E.T., 1972, Lobar dearterialization for exsanguinating wounds of the liver, J. Trauma 12:397-407.

McGregor, R.A., 1959, Ligation of the hypogastric arteries in combined abdominoperineal surgery, Dis. Colon and Rectum 2:166-168.

Merklin, R.J., and Michels, N.A., 1958, The variant renal and suprarenal blood supply with data on the inferior phrenic ureteral, and gonadal arteries, J. Int. Coll. Surg. 29:41-76.

Metcalfe, J., Romney, S.L., Ramsey, L.H., Reid, D.E., and Burwell, C.S., 1955, Estimation of uterine bloodflow in normal human pregnancy at term, J. Clin, Invest. 34:1632-1638.

Michaels, J.P., 1948, Study of ureteral blood supply and its bearing on necrosis of the ureter following the Wertheim operation, Surg. Gynec. Obstet. 86:36-44.

Michels, N.A., 1942, The variational anatomy of the spleen and splenic artery, Amer. J. Anat. 70:21-72.

Michels, N.A., 1953, Collateral arterial pathways to the liver after ligation of the hepatic artery and removal of the celiac axis, Cancer 6:708-724.

Michels, N.A., 1955, "Blood Supply and Anatomy of the Upper Abdominal Organs," Lippincott, Philadelphia.

Michels, N.A., 1962, The anatomic variations of the arterial pancreaticoduodenal arcades, etc., J. Internat. Coll. Surg. 37:13-40.
Michels, N.A., Siddarth, P., Kornblith, P.L., and Parke, W.W., 1963, The variant blood supply to the small and large intestine: its import in regional resections, J. Internat. Coll. Surg. 39:127-170.
Michels, N.A., Siddarth, P., Kornblith, P.L., and Parke, W.W., 1965, The variant blood supply to the descending colon, rectosigmoid and rectum based on 400 dissections. Its importance in regional dissections: a review of medical literature, Dis. Colon and Rectum, 8:251-278.
Michels, N.A.., 1966, in "Morris' Human Anatomy," (J.P. Schaeffer, ed) 12th ed., Blakiston, New York.
Monafo, W.W., Jr., Ternberg, J.L., and Kempson, R., 1965, Accidental ligation of the hepatic artery, Arch. Surg. 921:643-652.
Moskowitz, M., Zimmerman, H., and Felson, B., 1964, The meandering mesenteric artery of the colon, Am. J. Roentg. 92:1088-1099.
Nakayama, K., 1954, Approach to mid thoracic esophageal carcinoma for its radical surgical treatment, Surgery 35:574-589.
Noer, R.J., 1943, The blood vessels of the jejunum and ileum: a comparative study of man and certain laboratory animals, Am. J. Anat. 73:292-334.
Ottinger, L.W., Darling, C., Nathan, M.J., and Linton, R.R., 1972, Left colon ischemia complicating aorto-iliac reconstruction, Arch Surg. 105:841-846.
Patten, B.M., 1968, "Human Embryology," 3rd ed. McGraw-Hill, New York.
Peltier, L.F., 1965, Complications associated with fractures of the pelvis, J.B. and J. Surg. 47-A:1060-1069.
Petrov, B.A., 1964, Retrosternal artificial esophagus created from colon. 100 operations, Surg. 55:520-523.
Pick, J.W., and Anson, B.J., 1940, The renal vascular pedicle: an anatomical study of 430 body-halves, J. Urol. 44:411-434.
Reich, W.J., and Nechtow, M.J., 1961, Ligation of the internal iliac (hypogastric) arteries: a life-saving procedure for uncontrollable gynecologic and obstetric hemorrhage, J. Internat. Coll. Surg. 36:157-168.
Rienhoff, W.J., Jr., 1946, Intrathoracic esophagojejunostomy for lesions of the upper third of the esophagus, South Med. J. 39:928-940.
Rex, H., 1888, Beitrage zur Morphologie der Säugerleber, Morph. Jahr. 14:517-617.
Reynolds, S.R.M., 1963, Maternal blood flow in the uterus and placenta. Chapter 45, in "Handbook of Circulation" (Hamilton, W.F., and Dow, P., eds) Section 2:Vol II, American Physiological Society, Washington.
Rob, C., 1970, Vascular Diseases of the Intestine, in "Modern Trends in Gastro-enterology," (Card, W.I., and Creamer, B., eds), Chapter 12, Vol IV, London, Butterworths.

Roberts, W.H., and Krishingner, G.L., 1967, A comparative study of human internal iliac artery based on Adachi classification, Anat. Rec. 158:191-196.

Roux, C., 1897, De la Gastro-entérostomie, Rev. Gynécol. Chir. Abdom. 1:67-122.

Rudolph, A.M., 1969, Course and distribution of foetal circulation in "Foetal Autonomy," (G.E.W. Wolstenholme, and M. O'Connor, eds), Ciba Symposium, Churchill, London.

Siegel, P., and Mengert, W.F., 1961, Internal iliac artery ligation in obstetrics and gynecology, J. Am. Med. Assn. 178:1059-1062.

Sims, E.A.H., and Krantz, K.E., 1958, Serial studies of renal function during pregnancy and the puerperium in normal women, J. Clin. Investig. 37:1764-1774.

Smith, R.F., and Szilagyi, D.E., 1960, Ischemia of the colon as a complication in the surgery of the abdominal aorta, Arch. Surg. 80:806-821.

Solotuchin, A., 1929, Über die Blutversorgung der Nebennieren, Ztschr. f Anat. 90:288-292.

Spanner, R., 1936, Mutterlicher und kindlicher Kreislauf der menschlicher Placenta und seine Strombahnen, Zeitschr f Anat. u. Entwicklung. 105:163-242.

Steward, J.A., and Rankin, F.W., 1933, Blood supply of the large intestine, Arch. Surg. 26:843-891.

Sudek, P., 1907, Ueber die Gefässversorgung des Mastdarmes in Hinsicht auf die operative Gangrän, Münch. Med. Wochen. 542:1314-1316.

Tajes, R.B., 1956, Ligation of hypogastric arteries and its complications in resection of cancer of the rectum, Am. J. Gastroent. 26:612-618.

Woodburne, R.T., 1973, "Essentials of Human Anatomy," 5th ed. Oxford, New York.

Yellin, A.E., Chaffee, C.B., and Donovan, A.J., 1971, Vascular occlusion in treatment of juxtahepatic venous injuries, Arch. Surg. 102:566-573.

Young, M., 1963, The fetal and neonatal circulation, Chapter 46 in "handbook of Circulation" (Hamilton, W.F., and Dow, P. eds), Section 2:Vol II, American Physiological Society, Washington.

ARTERIAL CIRCULATION OF THE EXTREMITIES

Henry Haimovici, M.D.*

Chief Emeritus, Vascular Surgical Service, Montefiore Hospital and Medical Center, and Clinical Professor Emeritus of Surgery, Albert Einstein College of Medicine, New York, New York

The description of the arterial circulation of the extremities will be divided into:
- I.   MAJOR ARTERIES AND THEIR BRANCHES
- II. MUSCULAR ARTERIES
- III. CUTANEOUS ARTERIES

I. MAJOR ARTERIES AND THEIR BRANCHES

UPPER EXTREMITY

Embryologic Development:

The 10-12 mm long human embryo normally has developed paired ventral and dorsal aortas and six pairs of connecting vascular arches. Early in this development stage, the first and second arches disappear and the fifth arches usually remain rudimentary. The ventral and dorsal aortas cephalad to the third arch, and the ventral aorta between the third and fourth arches form the common, external and internal carotid arteries. The right ventral aorta cephalad to the sixth vascular arch forms the innominate artery. The dorsal aorta disappears between the seventh and eighth dorsal intersegmental arteries.

---

*From the Vascular Surgical Service, Department of Surgery, Montefiore Hospital and Medical Center, and Albert Einstein College of Medicine, New York, New York.

The right subclavian artery is formed from the right fourth vascular arch, and extends as far as the origin of its internal mammary branch. Sometimes the right subclavian artery arises abnormally from the aortic arch distal to the origin of the left subclavian and passes upward and to the right behind the trachea and esophagus. This condition is explained by the persistence of the right dorsal aorta and the obliteration of the fourth right arch.

The left subclavian artery develops from the left seventh dorsal intersegmental artery which arises caudad to the seventh somite and supplies the developing arm bud. By differential growth, the origin of the left subclavian artery migrates cephalad to a position proximal to the ductus arteriosus.

Several segmental arteries later contribute branches to the upper limb-bud and form in it free capillary anastomoses. The subclavian artery is prolonged into the limb, thus providing the arterial stem for the upper arm, forearm, and the interosseous branches to the hand. Later the radial and ulnar arteries are developed as branches of the brachial part of the stem.

Major Arteries and their Branches

1. Subclavian Artery (Fig. 1)

The right subclavian artery is a subdivision of the innominate artery while the left arises directly from the arch of the aorta.

The description of each subclavian artery is divided into three parts. The first extends from the origin of the vessel to the medial border of the scalenus anterior muscle. In the first part of their course, the two subclavian arteries differ in length, direction and relation with neighboring structures. The second lies behind the scalenus anterior muscle. The third extends from the lateral margin of this muscle to the outer border of the first rib where it becomes the axillary artery. For the above mentioned reasons, the first portions of the two subclavian arteries require separate descriptions. The second and third portions are practically alike.

Right Subclavian Artery: First Part. This part arises behind the upper part of the right sternoclavicular joint, and then passes upward and lateralward to the medial margin of the scalenus anterior muscle. It ascends a little above the clavicle, the supraclavicular extent being variable.

Exposure of the first portion of the right subclavian artery can be achieved almost completely, in most instances, through a cervical approach. The artery is covered in front by the skin,

ARTERIAL CIRCULATION OF THE EXTREMITIES 427

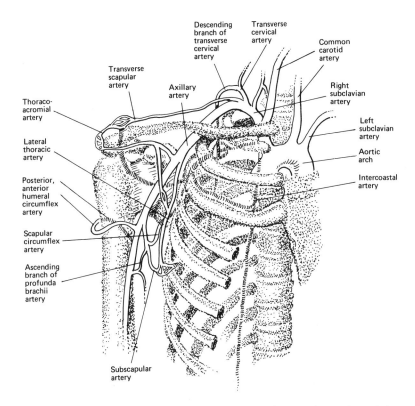

Fig. 1. Diagram of subclavian, axillary and proximal brachial arteries with their major branches. (Redrawn from Quiring, 1949.)

superficial fascia, platysma, deep fascia, the clavicular origin of the sternocleidomastoid, the sternohyoid, and the sternothyroid muscles, and another layer of the deep fascia. It is crossed by the internal jugular and vertebral veins, by the vagus nerve and the cardiac branches of the vagus and sympathetic, and by the subclavian loop of the sympathetic trunk which forms a ring around the vessel. The anterior jugular vein is directed lateralward in front of the artery but is separated from it by the sternohyoid and sternothyroid muscles. Below and behind the artery is the pleura which separates it from the apex of the lung. Behind is the sympathetic trunk, the longus colli muscle and the first thoracic vertebra. The right recurrent nerve winds around the lower and back part of the vessel.

Left Subclavian Artery: First Part. This part arises from the arch of the aorta behind the left common carotid at the level of the fourth thoracic vertebra. Its length is that of the right subclavian artery plus that of the innominate trunk. It ascends in the superior mediastinal cavity to the root of the neck and then arches lateralward to the medial border of the scalenus anterior muscle. In front it is related with the vagus, cardiac and phrenic nerves, the left common carotid and internal jugular and the origin of the left innominate vein. It is covered by the three muscles as mentioned for the right side. Posteriorly it is in relation with the esophagus, thoracic duct, left recurrent nerve, inferior cervical ganglion of the sympathetic trunk, and the longus colli muscle. More proximally the esophagus and the thoracic duct lie to its right, the latter ultimately arching over the vessel to join the angle of the union between the subclavian and internal jugular veins. Medially it is related to the esophagus, trachea, thoracic duct, and left recurrent nerve. Laterally it winds around the pleural dome.

The second part of the subclavian artery is situated behind the scalenus anterior muscle. It is very short and forms the highest part of each of the arches described by the vessel. In front it is covered by the skin, superficial fascia, platysma, deep cervical fascia, sternocleidomastoid and scalenus anterior muscles. On the right side of the neck the phrenic nerve is separated from the second part of the artery by the scalenus anterior while on the left side it crosses the first part of the artery close to the medial edge of the muscle. Posteriorly the vessel is bordered by the pleura and the scalenus medius muscle. Proximally, the brachial plexus of nerves and inferiorly the pleura are the two important relations. The subclavian vein lies below and in front of the artery, separated from it by the scalenus anterior.

The third part of the subclavian artery runs downward and lateralward from the lateral margin of the scalenus anterior muscle to the outer border of the first rib, where it becomes the axillary artery. This is the most superficial portion of the vessel and is contained in the subclavian triangle.

Anteriorly, in addition to the structures mentioned for the second part, the external jugular vein crosses its medial aspect and receives the transverse scapular, transverse cervical and anterior jugular veins which frequently form a plexus in front of the artery. The terminal part of the artery lies behind the clavicle and the subclavius muscle and is crossed by the transverse scapular vessels. The subclavian vein is in front of and at a slightly lower level than the artery. Posteriorly it usually lies on the lowest trunk of the brachial plexus which intervenes between it and the scalenus medius muscle. Above and to its lateral side are the upper trunks of the brachial plexus and the omohyoid muscle. Inferiorly it lies on the upper surface of the first rib.

The portion of the subclavian artery which forms an arch around the cervical pleura is projected above, the rest being below the clavicle. Projection of its origin on the right appears between the second and third thoracic vertebrae, on the left in the middle or above the fourth thoracic vertebra. The average diameter is 0.9 cm on the left and 0.7 to 0.8 cm on the right side (Adachi, 1928).

Variations. The origin of the right subclavian from the innominate takes place in some cases above the sternoclavicular articulation but less frequently below that joint. The artery may arise as a separate trunk from the arch of the aorta either as a first branch or as the second or third. In very rare instances it may arise from the thoracic aorta as low down as the fourth thoracic vertebra. Very rarely this artery may ascend as high as 4 cm above the clavicle.

The left subclavian is occasionally joined at its origin with the left carotid. As a rule it does not reach quite as high a level in the neck as does the right.

Branches of the Subclavian Artery

Four major branches arise from the subclavian artery: the vertebral, internal mammary, thyrocervical and costocervical. On the left side all four branches generally arise from the first portion of the vessel while on the right the costocervical trunk usually springs from the second portion.

The vertebral artery is the first branch of the subclavian and is the most constant of all branches. It has its origin from the posterosuperior aspect, having few, if any, branches until it reaches the upper cervical vertebrae. In 87% of cases, the artery enters through the foramina in the transverse process at the level of the sixth cervical vertebra. In 5% of cases it enters the seventh vertebra, and in approximately 6%, the fifth vertebra (Daseler and Anson, 1959). The artery is surrounded by the plexus of nerve fibers derived from the inferior cervical ganglion of the sympathetic trunk, and ascends around the vessel through the foramina in the transverse processes of the upper six cervical vertebrae.

Two sets of branches arise from the vertebral artery: those given off in the neck and those within the cranium.

The internal mammary artery arises from the inferior aspect of the subclavian and follows a parasternal course. Its origin is opposite the thyrocervical trunk which arises from the anterior surface of the first portion of the artery. It descends behind the cartilages of the upper six ribs at a distance of about 1.25 cm from the margin of the sternum, and at the level of the sixth intercostal space divides into musculophrenic and superior epigastric arteries.

It is directed at first downward, forward and medialward behind the sternal end of the clavicle, the subclavian and internal jugular veins and the first costal cartilage, and passes forward close to the lateral side of the innominate vein. As the artery enters the thorax the phrenic nerve crosses from its lateral to its medial side. Below the first costal cartilage it descends almost vertically to its point of bifurcation. It rests on the pleura, as far as the third costal cartilage. Below this level it is on the transversus thoracis muscle. It is accompanied by a pair of veins. It has anastomoses with the intercostal arteries as well as the costocervical and thoracoscapular vessels. This vessel and its branches then enlarge considerably to form useful collateral pathways. The major branches participating in collateral circulation of the upper and lower extremities are the intercostal, perforating, musculophrenic, and superior epigastric.

The superior epigastric artery continues in the original direction of the internal mammary. It descends through the interval between the costal and sternal attachments of the diaphragm and enters the sheath of the rectus abdominis, at first lying behind the muscle and then perforating and supplying it. It anastomoses with the inferior epigastric artery from the external iliac. Branches perforate the anterior wall of the sheath of the rectus and supply the other muscles of the abdomen and the integument.

The anomalies associated with the internal mammary are rare, usually involving minor variations in its site of origin. Daseler and Anson (1959) found it missing only twice in 800 dissections. In approximately 3.7% of cases it was found to be the origin of the transverse scapular artery which is a major collateral of the scapulohumeral arterial anastomoses. Other rare variations in origin may be from the innominate or brachiocephalic trunk, the aortic arch or axillary artery (Rauber and Kopsch, 1951).

The thyrocervical trunk is short and thick. It arises from the front of the first portion of the subclavian artery, close to the medial border of the scalenus anterior muscle and divides almost immediately into three branches, the inferior thyroid, transverse scapular and transverse cervical. The thyrocervical trunk is the most variable of the potential collateral arteries arising from the subclavian artery. The presumed typical trunk and its three branches are found in only 46% of dissections. Daseler and Anson (1959) found the axis absent in 17% of their cases. The principal variants involved the origin of the three branches.

The inferior thyroid artery passes upward in front of the vertebral artery. Its branches have no anastomotic connections with the arterial circulation to the extremities.

The <u>transverse scapular artery</u> is an important collateral vessel with multiple significant anastomoses arising from the axillary artery. It passes at first downward and lateralward across the scalenus anterior muscle and the phrenic nerve, then crosses the subclavian artery and the brachial plexus and runs behind and parallel with the clavicle and subclavius muscle. It descends behind the neck of the scapula through the great scapular notch and under cover of the inferior transverse ligament to reach the infraspinous fossa where it anastomoses with the scapular circumflex and descending branch of the transverse cervical.

The <u>transverse cervical artery</u> usually passes laterally across the anterior scalene muscle over the phrenic nerve. When it arises from either the second or third portion of the subclavian or passes posterior to the anterior scalene, it may no longer serve as a potential collateral in cases of the scalenus anticus syndrome (Grant, 1962). The artery lies at a higher level than the transverse scapular muscle. It passes transversely beneath the inferior belly of the omohyoid muscle to the anterior margin of the trapezius muscle, beneath which it divides into an ascending and a descending branch. The ascending branch distributes its divisions to the trapezius and to the neighboring muscles. The descending branch supplies the muscles around the scapula. It anastomoses with the transverse scapular and subscapular arteries and with the posterior branches of some of the intercostal arteries. The transverse scapular or the transverse cervical may arise independently from the subclavian artery.

The <u>costocervical trunk</u> is absent in about 10% of cadavers. It arises from the upper and posterior part of the subclavian artery, behind the scalenus anterior on the right side and medial to that muscle on the left side. Passing backward, it gives off the profunda cervicalis, and continuing as the highest intercostal artery, descends behind the pleura in front of the necks of the first and second ribs, and anastomoses with the first aortic intercostal. These arteries because of their small size are probably of minor importance as potential collateral pathways.

2. <u>Axillary Artery</u> (Fig. 1, 2)

The axillary artery, continuing the subclavian, begins below the inferior margin of the clavicle at the outer border of the first rib, then courses through the axillary fossa, and ends at the lower border of the tendon of the teres major, and the lower edge of the pectoralis major muscles, where it takes the name of brachial. In the axillary fossa it lies lateral to and deeper than the axillary vein. At its origin the artery is very deeply situated, but near its termination is more superficial, being covered only by the skin and fascia. Its direction varies with the position of the extremity. When the arm is horizontal the artery is directed at a right angle with the trunk, while it becomes concave upward when the arm is elevated above the shoulder, and convex upward and lateralward when the arm lies by the side. The length of the axillary artery ranges between 8 and 12 cm, its diameter being 0.6 to 0.8 cm (Paturet, 1958).

The description of the axillary artery is divided into three portions: the first part lies proximal, the second behind and the third distal to the pectoralis minor muscle.

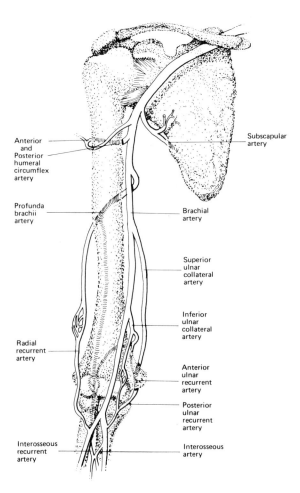

Fig. 2. Arteriogram of axillary and brachial arteries with their major branches.

The <u>first part</u> is covered anteriorly by the clavicular portion of the pectoralis major muscle and the coracoclavicular fascia, and is crossed by the lateral anterior thoracic nerve and the thoracoacromial and cephalic veins. Posteriorly it lies on the first intercostal space, the first and second digitations of the serratus anterior muscle and the long and medial anterior thoracic nerves, and the medial cord of the brachial plexus. Laterally it is bordered by the brachial plexus from which it is separated by a little areolar tissue. On the medial or thoracic side the artery is contiguous to the axillary vein. A fibrous sheath surrounds both the artery and the vein as well as the brachial plexus (the axillary sheath) which is continuous with the deep cervical fascia.

The second part is covered anteriorly by the pectorales major and minor muscles. Posteriorly to it are the posterior cord of the brachial plexus and some areolar tissue which intervenes between it and the subcapularis muscle. On the medial side is the axillary vein separated from the artery by the medial cord of the brachial plexus and the medial anterior thoracic nerve. Laterally to it is the lateral cord of the brachial plexus. Thus the brachial plexus surrounds the artery on three sides and separates it from direct contact with the vein and adjacent muscles.

The <u>third part</u> extends from the lower border of the pectoral minor to the lower border of the tendon of the teres major muscle. Anteriorly it is covered by the lower part of the pectoralis major above, but only by the skin and fascia below. Posteriorly it is in relation with the lower part of the subscapularis and the tendons of the latissimus dorsi and teres major. On the lateral side is the coracobrachialis muscle and on the medial or thoracic side the axillary vein. The brachial plexus and its branches surround the artery as follows: laterally are the lateral head and the trunk of the median; on the medial side the ulnar between the artery and vein; in front, the medial head of the median and the medial antebrachial cutaneous; and behind the radial nerve.

<u>Branches</u> of the axillary artery are: from the first portion the highest thoracic; from the second, thoracoacromial and lateral thoracic; and from the third, subscapular, posterior and anterior humeral circumflex.

The branches are subject to great <u>variations</u>. Indeed the number of arteries that can originate from the axillary varies between six and eleven. Huelke (1959) described seven branches in 86% of 178 cadavers examined. Although in the majority of cases the branches represent a relatively constant configuration, each branch may not always arise directly from the vessel itself. Keen (1961) found a number of variations of the arterial patterns as disclosed by 284 dissections of 142 subjects, variations which appeared to be strikingly asymmetrical.

The origin and distribution of the branches of the axillary artery are as follows:

The highest thoracic artery or superior thoracic artery is a small vessel of variable length supplying the upper two intercostal spaces. It may arise from the thoracoacromial or may be absent. It supplies the pectoralis major and minor muscles as well as the intercostal spaces and anastomoses with the internal mammary and intercostal arteries.

The thoracoacromial artery, an important branch of the axillary, arises at or near the medial border of the pectoralis minor, pierces the coracoclavicular fascia and divides into four branches: pectoral, acromial, clavicular, and deltoid. The pectoral branch is distributed to the two pectoral muscles and to the breast, anastomosing with the intercostal branches of the internal mammary and with the lateral thoracic. The acromial branch participates in the formation of the acromial network. It is situated under the deltoid muscle which it pierces and ends on the acromion in an arterial network formed by branches from the transverse scapular, thoracoacromial and posterior humeral circumflex arteries. The clavicular branch runs upward and supplies the subclavius muscle and the sternoclavicular joint. The deltoid branch, often arising with the acromial, supplies the pectoralis major and the deltoid.

The lateral thoracic artery arises from the middle third of the axillary and runs along the lateral side of the thorax toward the muscles of the chest wall and to the breast. This artery originates from the thoracodorsal branch of the subscapular in 25% of patients (Huelke, 1959). It has collateral connections with the thoracoscapular system of vessels.

The subscapular artery, the largest branch of the axillary, arises at the lower border of the subscapularis muscle, which it follows to the inferior angle of the scapula, where it anastomoses with the lateral thoracic and intercostal arteries and with the descending branch of the transverse cervical and ends in the neighboring muscles. About 4 cm from its origin it gives off a branch, the scapular circumflex artery. The latter is usually larger than the continuation of the subscapular. It curves around the axillary border of the scapula, traversing the space between the subscapularis above and the teres major muscle below and the long head of the triceps laterally. The subscapular artery has an important strategic significance due to its anastomosis with the ascending branch of the profunda brachii artery beneath the lateral head of the triceps muscle. Because of its location, this branch may be injured along with the axillary in abduction-type injuries of the arm. Extrinsic pressure in the axilla, as with a crutch, may also occlude both vessels. Injuries of the upper humerus, particularly dislocations, may damage this important collateral (Mackenzie and Sinclair, 1958).

The posterior humeral circumflex artery arises from the axillary at the lower border of the subscapularis. It winds around the neck of the humerus and is distributed to the deltoid and shoulder joint, anastomosing with the anterior humeral circumflex and profunda brachii.

The anterior humeral circumflex artery, considerably smaller than the posterior, arises nearly opposite it from the lateral side of the axillary artery. Both vessels arise from the third part of the axillary forming an anastomotic loop around the surgical neck of the humerus. The anterior humeral circumflex artery often arises in common with the posterior humeral circumflex or may be represented by three or four very small branches.

Variations of the branches of the axillary artery and those of the brachial artery arise sometimes with the common trunk formed by the union of five vessels. Another variation consists of the radial originating from the axillary artery at a high level as does also the common interosseous artery.

3. Brachial Artery (Fig. 3)

The brachial artery extends from the lower border of the pectoralis major muscle to the cubital fossa where it divides into two branches, the radial and the ulnar arteries. At first the brachial artery lies medial to the humerus then gradually continues in front of the bone and at the bend of the elbow lies midway between its two epicondyles. Its projection lies along the straight line drawn from the middle of the axillary fossa to the center of the line connecting the epicondyles of the humerus. The length of the brachial artery, varying with the site of division, amounts to 15 to 30 cm, its diameter being 0.5 to 0.6 cm (Paturet, 1958).

In contrast to the axillary artery, the brachial is superficial throughout its entire extent, being covered anteriorly by the skin and superficial and deep fasciae. The bicipital fascia lies in front of the artery opposite the elbow and separates it from the medial cubital vein. The median nerve crosses from its lateral to its medial side opposite the insertion of the coracobrachialis. Behind it is separated from the long head of the triceps brachii by the radial nerve and the profunda brachii artery. Laterally, it is in relation above with the median nerve and the coracobrachialis, below with the biceps brachii, the two muscles overlapping the artery to a considerable extent. Medially, its upper half is in relation with

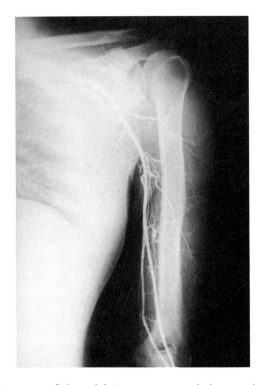

Fig. 3. Diagram of brachial artery and its major branches. (Redrawn from Quiring, 1949.)

the medial antebrachial cutaneous and ulnar nerves, its lower half with the median nerve. The basilic vein lies on its medial side but is separated from it in the lower part of the arm by the deep fascia. The artery is accompanied by two venae comitantes, which lie in close contact with it, and are connected together at intervals by short transverse branches.

At the bend of the elbow the brachial artery descends deeply into the antecubital fossa. This space is triangular in shape and contains in addition to the brachial artery, venae comitantes, the radial and ulnar arteries, the median and radial nerves and the tendon of the biceps brachii. The brachial artery occupies the middle of the space and bifurcates opposite the neck of the radius into its two branches. The artery at this level is also superficial, being covered anteriorly by the skin, the superficial fascia and the medial cubital vein, the latter being separated from the artery by the bicipital fascia. The median nerve lies close to the medial side of the artery proximally but is separated distally by the ulnar head of the pronator teres. The tendon of the biceps lies to the lateral side of the artery, while the radial nerve is situated on the supinator and is concealed by the brachioradialis.

Branches of the Brachial Artery. The profunda brachii or superior profunda artery is a large vessel which arises from the medial and posterior parts of the brachial just below the border of the teres major. It follows closely the radial nerve where it is covered by the lateral head of the biceps brachii to the lateral side of the arm. It pierces then the lateral intermuscular septum and descends between the brachioradialis and the brachialis to the front of the lateral epicondyle of the humerus, and by anastomosing with the radial recurrent artery. It gives branches to the deltoid and to the muscles between which it lies (see below). It provides anastomoses with adjacent arteries through the middle collateral branch, and radial collateral branch.

The nutrient artery to the humerus arises at about the middle of the arm and enters the nutrient canal near the insertion of the coracobrachialis.

The superior and inferior ulnar collateral arteries arise above the elbow on the medial side. Both of these branches provide anastomotic links with recurrent ulnar branches and the radial collateral branch of the profunda. The muscular branches, three or four in number, are distributed to the muscles of the arm.

Variations of the Barchial Artery. The variations relate to the course or the division of the artery. The brachial artery may pass behind the supracondylar process of the humerus then could run beneath, or through the substance of the pronator teres to the bend of the elbow. The major variants of the proximal brachial artery are those involving the first branch, the profunda brachii. Classification of the different patterns of brachial division is complex and has been described by several authors (MacCormack, L.J., et al., 1953). The most common variations seen included different types of a "superficial brachial artery" which is defined as lying superficial to the median nerve. In its course it may bifurcate normally below the elbow or serve as a "high origin" for either the radial or ulnar artery. Frequently the brachial artery divides at a higher level than usual, and the vessels concerned in the high division are three, namely radial, ulnar, and interosseous. Most frequently the radial is given off high up, the other limb of the bifurcation consisting of the ulnar and interosseous. Because of the variability in the course and division of the brachial artery, the location of the latter lying deep to the median nerve is not always the case. Sometimes, in addition, aberrant vessels connect the brachial or the axillary artery with one of the arteries of the forearm or branches from them.

The importance of anatomical variations of the brachial has been recently emphasized both by anatomists and vascular surgeons (Fig. 3).

4. Radial Artery (Fig. 4)

The radial artery, a continuation of the brachial, arises at the bend of the elbow at the level of the neck of the radius. Of the two branches, the radial, smaller in caliber than the ulnar, describes a straight line in the direction of the styloid process of the radius below the brachioradialis muscle. The artery passes directly along the radius in the lower third of the forearm. It then winds backward around the lateral side of the carpus beneath the tendons of the abductor pollicis longus and extensor pollicis longus and brevis to the upper end of the space between the metacarpal bone of the thumb and index fingers. Finally it passes forward between the two heads of the first interosseous dorsalis, into the palm of the hand, where it crosses the metacarpal bone and at the ulnar side of the hand unites with the deep volar branch of the ulnar artery to form the deep volar arch.

The branches of the radial artery are quite numerous but can be divided into the three regions in which the vessel is situated: in the forearm, at the wrist, and in the hand. The major branches are:

The radial recurrent artery which arises immediately below the elbow. It ascends from the beginning of the trunk to the arterial network of the elbow.

The superficial volar branch or palmar branch. It anastomoses occasionally with the terminal portion of the ulnar artery, thus completing the superficial volar arch. It is variable in size, usually very small, and ends in the muscles of the thumb.

The dorsal carpal branch is a small vessel which arises beneath the extensor tendons of the thumb. It crosses the carpus transversely toward the medial border of the hand where it anastomoses with the dorsal carpal branch of the ulnar and with the volar and dorsal interosseous arteries, thus forming the dorsocarpal network. Three slender dorsometacarpal arteries and dorsodigital arteries of the hand are given off from this network.

The principal artery of the thumb is a terminal branch of the radial artery. It originates as it turns medially on gaining the palm. It descends on the palmar aspect of the first metacarpal bone. The artery supplies the thumb and the radial aspect of the forefinger (radial artery of the index finger).

ARTERIAL CIRCULATION OF THE EXTREMITIES

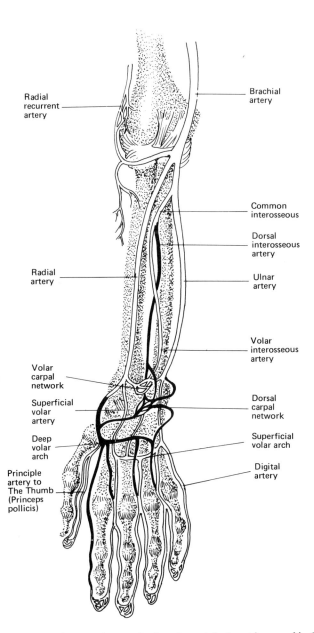

Fig. 4. Diagram of bifurcation of the brachial, the radial, ulnar, interosseous and their terminal branches in the hand and fingers. (Redrawn from Quiring, 1949.)

The deep palmar arch is formed by the anastomosis of the terminal part of the radial artery with the deep volar branch of the ulnar. It lies upon the carpal extremities of the metacarpal bones and on the interossei, being covered by the abductor pollicis obliquus, the flexor tendons of the fingers and the lumbricales. The deep volar arch lies deep to the ulnar nerve in 63% of cases; superficial in 34% of cases. It is occasionally double and will then circle the ulnar nerve (Gray, 1942).

The palmar metacarpal arteries, three or four in number, arise from the convexity of the deep palmar arch. They run along the interosseous muscles toward the fingers. They join the common digital branches of the superficial palmar arch at the clefts of the fingers. They anastomose with the dorsal metacarpal arteries via the perforating branches.

5. Ulnar Artery (Figs. 4, 5, 6)

The ulnar artery, the larger of the two terminal branches of the brachial, begins a little below the bend of the elbow, and passing obliquely downward, reaches the ulnar side of the forearm at a point about midway between the elbow and the wrist. It then runs along the ulnar border of the wrist, crosses the transverse carpal ligament on the radial side of the pisiform bone, and immediately beyond this bone divides into two branches which enter into the formation of the superficial and deep palmar arches.

Proximally the ulnar artery is situated deeply, being covered by the pronator teres and the flexor muscles. The median nerve is in relation with the medial side of the artery for about 2.5 cm and then crosses the vessel being separated from it by the ulnar head of the pronator teres. In the lower half, the artery lies upon the flexor digitorum profundus, being covered by the skin and the superficial and deep fasciae and placed between the flexor carpi ulnaris and flexor digitorum sublimis. It is accompanied by two venae comitantes, and it is overlapped in its middle third by the flexor carpi ulnaris. The ulnar nerve lies on the medial side of the lower two-thirds of the artery and the palmar cutaneous branch of the nerve descends on the lower part of the vessel to the palm of the hand.

At the wrist the ulnar artery is covered by the skin and the volar ligament, and lies upon the transverse carpal ligament. On its medial side is the pisiform bone, and somewhat behind the artery, the ulnar nerve.

The branches of the ulnar artery are:

The ulnar recurrent artery. It is a short and thick vessel, arising immediately below the origin of the main trunk. Its anterior

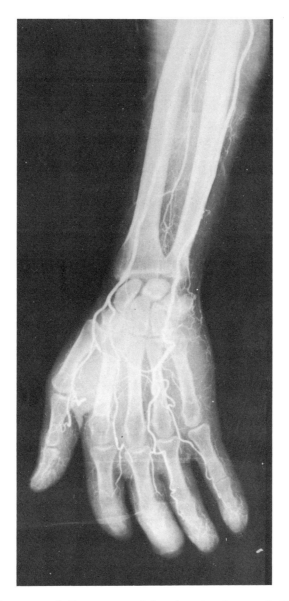

Fig. 5. Arteriogram of forearm and hand arteries and their branches.

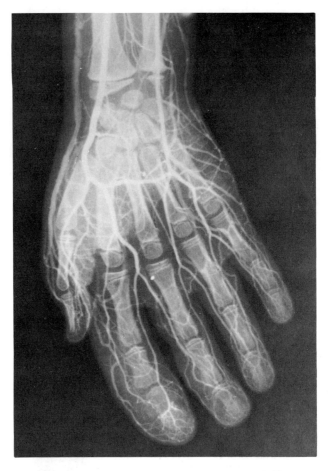

Fig. 6. Arteriogram of the hand arteries indicating arteriovenous and venous phase of the angiogram.

branch anastomoses with the inferior ulnar collateral artery, while the posterior branch ends in the network of the elbow joint.

The common interosseous artery, about 1 cm in length, arises immediately below the tuberosity of the radius, and passing backward to the upper border of the interosseous membrane, divides into two branches, the palmar and the dorsal interosseous arteries. The palmar interosseous artery gives off muscular branches and the nutrient arteries of the radius and ulna. The dorsal interosseous artery gives off, near its origin, the interosseous recurrent artery, which ascends to the interval between the lateral epicondyle and olecranon.

The palmar and dorsal carpal branches are small vessels which cross the carpus and anastomose with the corresponding branches from the radial.

The deep palmar branch, the smaller terminal branch of the main trunk, together with the radial artery, takes part in the formation of the deep palmar arch.

The superficial palmar arch is the larger of the two terminal branches. The three common palmar digital arteries commencing there, bifurcate at the fingers and supply the inner surfaces between the second to fifth fingers (palmar digital arteries).

Blood Supply of the Fingers. The proximal phalanges are supplied by both the dorsal and palmar arteries, the distal phalanges mainly by the palmar vessels. These arteries form superficial arches on the dorsal aspect of the distal phalanges at the height of the superficial base, and form proximal and distal arches beneath the nails.

Variations of the Arteries of the Forearm and the Hand. The radial artery has a highly placed origin in 12% of the cases (Gray, 1954). The superficial palmar branch arises sometimes from the forearm. In such instances the main trunk bends over the dorsal side in its middle third (Kiss, 1963). The radial recurrent artery is occasionally very large or double (Adachi, 1928). A high origin of the ulnar artery is seen in about 8% of dissections. (Gray, 1954). Its course is fairly constant if its origin is normal, and superficial if the origin is placed high.

The arteries of the hand show numerous variations. The three variations of the superficial arterial arch are the ulnar (60%), the radioulnar (31%), and the medioulnar (8%). Insufficient development of the superficial arch is compensated by the deep arterial rch so that in such cases blood is supplied to the fingers by the deep palmar arch alone. Sometimes the radial artery contributes only to the formation of the superficial arch. Edwards (1960) described eleven important variations of the palmar arterial arches. The variations of the vessels of the fingers and particularly those of the forefingers are numerous.

Collateral Circulation of the Upper Extremity. The abundant collateral network of the upper extremities ensures adequate circulation if a major artery becomes occluded.

The main collateral channels after occlusion of the third part of the subclavian artery are the anastomoses between collateral arteries of the shoulder and the upper thorax. The scapular anastomoses connect with each other the vascular systems of the aorta (intercostal arteries), the external carotid (occipital and superior thyroid arteries), the subclavian artery (all branches excepting the vertebral artery), the axillary artery (all branches), and the brachial artery (the deep brachial artery).

The subclavian artery provides the following arteries: the internal mammary, transverse scapular, descending branch of transverse cervical, the subscapular and lateral thoracic. The intercostal branches of the internal mammary anastomose with the subscapular and lateral thoracic arteries. The transverse scapular artery anastomoses with the scapular circumflex and the descending branch of the transverse cervical along the vertebral border of the scapula. An acromial branch anastomoses with the thoracoacromial artery on the acromion.

The descending branch of the transverse cervical artery anastomoses with the subscapular artery, the transverse scapular arteries and some of the intercostal arteries.

The subscapular artery anastomoses with the transverse scapular and the descending branch of the transverse cervical. In the axilla it anastomoses with the ascending branches of the deep brachial artery.

The lateral thoracic artery anastomoses with branches of the internal mammary, the subscapular, some of the intercostal arteries, and with a branch of the thoracoacromial.

Collateral circulation in the arm, in cases of occlusion of the upper third of the brachial, is provided by the humeral circumflex and subscapular arteries with the ascending branches from the profunda brachii. In cases of occlusion of the brachial below the origin of the profunda brachii and superior ulnar collateral, the blood supply is established by the branches of these two arteries anastomosing with the inferior ulnar collateral, the radial and ulnar recurrents, and the dorsal interosseous.

Collateral circulation in the forearm, in cases of radial or ulnar occlusions, is provided by numerous branches of the common interosseous. These anastomotic links extend to the level of the metacarpophalangeal joints.

In the hand the collaterals are provided by either the superficial or the deep arches. In cases of occlusion of both the radial and ulnar arteries, the recurrent branches of the deep arch to the volar carpal network and the interosseous system are probably responsible for the survival of the hand (Quiring, 1949).

The collateral arterial supply of the fingers is provided by the dorsal metacarpals and volar arches.

LOWER EXTREMITY

A. <u>Embryologic Development</u>. The primary arterial trunk or "axis" artery of the embryonic lower limb arises from the dorsal root of the umbilical artery and courses along the dorsal surface of the thigh, knee and leg. The ischiadic, the primary embryonic artery, springs from the umbilical artery and is axial in position. It is a temporary artery, extending originally from the thigh to the foot.

The common iliac arteries represent the roots of the umbilical arteries proximal to the origin of the external and internal iliac arteries. The external iliac, which supplants the ischiadic artery, is also derived from the umbilical artery. As it is prolonged distally to the foot, it is called, depending upon the location: femoral, popliteal, posterior and anterior tibial. Parts of the peroneal, inferior gluteal, and longitudinal anastomatic vessels connecting the perforating branches of the profunda femoris are derived also from the ischiadic artery. An anastomosis between its popliteal portion and the ischiadic artery reduces the size of the lower portion of the latter which becomes the peroneal artery.

Longitudinal anastomoses connect small branches of the ischiadic, directed toward the anterior surface of the leg, to form the anterior tibial artery. This anastomoses with the posterior tibial, and as it is established, the original branches from the ischiadic artery disappear, giving the anterior tibial the appearance of originating from the posterior tibial artery. The original vascular plexus in the foot, like that in the hand, thus has a change in its source of blood. At first supplied by the ischiadic, the trunk regresses and the posterior tibial artery takes over the ischiadic's function. After its establishment, the anterior tibial artery also sends a branch to the foot plexus (Senior, 1919).

B. Major Arteries and their Branches

1. <u>Common Iliac Artery</u> (Figs. 7, 8)

The common iliac arteries, which diverge from the end of the aorta, originate on the left side of the body of the fourth lumbar vertebra. Each common iliac artery passes downward and lateralward and divides opposite the intervertebral fibrocartilage between the last lumbar vertebra and the sacrum into two branches, the external and the internal (hypogastric) iliac arteries. The former supplies the lower extremity, the latter the viscera and the walls of the pelvis.

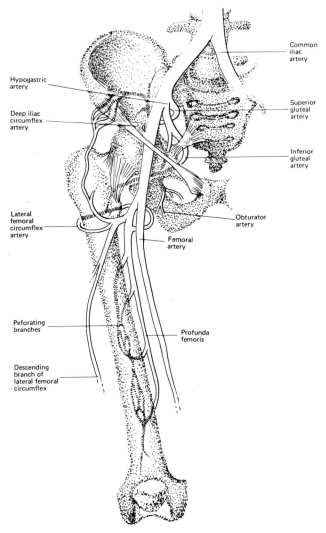

Fig. 7. Diagram of iliac and femoral arteries with their major branches. (Redrawn from Quiring, 1949.)

The right iliac artery is usually somewhat longer than the left and passes more obliquely across the body of the last lumbar vertebra. In front of it are the peritoneum, the small intestines, branches of the sympathetic nerves and, at its point of division, the ureter. Posteriorly, it is separated from the bodies of the fourth and fifth lumbar vertebrae and the intervening fibrocartilage by the terminations of the two common iliac veins and the beginning

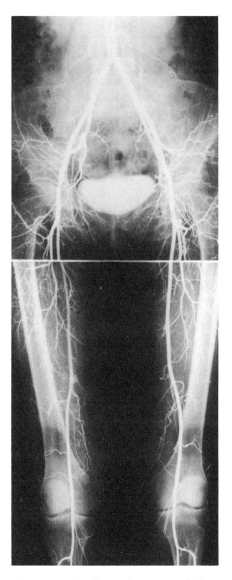

Fig. 8. Aorto-arteriogram showing the aortoiliac and femoropopliteal segments with their major branches.

of the inferior vena cava. Laterally, it is in relation, above, with the inferior vena cava and the right common iliac vein, and below, with the psoas major. Medial to it, above, is the left common iliac vein.

The left common iliac artery is in relation, in front, with the peritoneum, the small intestines, branches of sympathetic nerves, and the superior hemorrhoidal (rectal) artery. The ureter crosses it at this point of bifurcation. The left common iliac vein lies partly medial to, and partly behind it; laterally, the artery is in relation with the pasoas major. The division of the inferior mesenteric artery lies to the left of the common iliac artery.

The angle of division of the aorta on the average is 65°; in males, 75°, in females the range is from 40° to 80°. The right artery has an average length of 6.5 cm, the left 6.3 cm with a range of from 3.5 to 12.0 cm. The diameter measures 0.89 cm on the right side, 0.83 cm on the left. The surface projection of the common iliac arteries beginning 1 cm below the umbilicus, corresponds to a straight line to a point midway between the anterior superior iliac spine and the symphysis pubis (Gray, 1942).

Branches. The common iliac arteries give off small branches to the peritoneum, psoas major, ureters and the surrounding areolar tissue, and occasionally give origin to the iliolumbar, or accessory renal arteries.

Variations. The point of origin of the iliac arteries varies according to the bifurcation of the aorta. Three-fourths of the time, the aorta bifurcates either upon the fourth lumbar vertebra or upon the fibrocartilage between it and the fifth. The bifurcation in one instance out of nine is below and in one out of eleven it is above this point. In about 80% of the cases the aorta bifurcates within 1.25 cm above or below the level of the crest of the ilium; more frequently below than above.

The point of division of the iliacs varies greatly. In two-thirds it is between the last lumbar vertebra and the upper border of the sacrum. The left common iliac artery divides lower down more frequently than the right.

The relative length of the two common iliac arteries varies also. The right common iliac is longer than the left in about 60% of the cases.

2. Internal Iliac Artery (Hypogastric Artery)

The internal iliac artery, arising at the bifurcation of the common iliac, runs to the greater sciatic foramen while dividing into several branches. It provides the blood supply to the pelvic organs, the buttocks and the medial side of the thigh. It is a short, bulky vessel, smaller than the external iliac, and about 4 cm in length. It arises opposite the lumbosacral joint and, passing downward to the upper margin of the greater sciatic foramen, divides into two large trunks, an anterior and a posterior.

In front it is in relation with the ureter; posteriorly, with the internal iliac vein, the lumbosacral trunk, and the pyriformis muscle; laterally near its origin with the external iliac vein, which lies between it and the psoas major muscle; lower down, with the obturator nerve.

As mentioned in the description of embryologic development, the internal iliac artery in the fetus is twice as large as the external iliac and is the direct continuation of the common iliac. With the opposite artery it passes through the umbilical opening where the hypogastric artery is termed umbilical artery and enters the umbilical cord.

Branches. Of the twelve major branches of the internal iliac artery, only a few participate in the circulation of the lower extremity and thus only the latter will be mentioned here. The obturator artery passes forward and downward on the lateral wall of the pelvis to the upper part of the obturator foramen, and escaping from the pelvic cavity to the obturator canal, it divides outside the pelvis into an anterior and a posterior branch.

The anterior branch distributes branches to the obturator externus, pectineus, adductors and gracilis muscles. It anastomoses with the posterior branch and with the medial femoral circumflex artery. The posterior branch gives off smaller branches to the muscles attached to the ischial tuberosity and anastomoses with the inferior gluteal.

The inferior gluteal artery, the larger of the two branches of the anterior trunk of the internal iliac is distributed chiefly to the buttocks and back of the thigh. It passes posteriorly between the first and second or between the second and third sacral nerves, and then descends between the pyriformis and coccygeus muscles through the lower part of the sciatic foramen to the gluteal region. It then descends in the interval between the greater trochanter of the femur and tuberosity of the ischium.

The superior gluteal artery is the largest branch of the posterior division of the internal iliac and appears to be the continuation of the major trunk. It is a short artery thrust backward between the lumbosacral trunk and the first sacral nerve and, passing out of the pelvis, above the upper border of the pyriformis, immediately divides into a superficial and a deep branch. Within the pelvis it gives off a few branches to the iliacus, pyriformis and obturator internus muscles, and before quitting that cavity, a nutrient artery which enters the ilium. Of the two branches of the superior gluteal artery, the deep one lies under the gluteus medius and almost immediately subdivides into two. Of these, the superior division continuing the original course of the vessel passes along the upper border of the gluteus minimus to the anterior superior spine of the ilium, anastomosing with the deep iliac circumflex artery and the ascending branch of the lateral femoral circumflex artery. The inferior division anastomoses with the lateral femoral circumflex artery. The inferior division anastomoses with the lateral femoral circumflex artery.

Variations are rare. If the internal iliac artery is absent, the common iliac artery arches into the pelvis to continue as the external iliac artery. The arteries of the pelvic organs arise from this arch.

3. External Iliac Artery

The external iliac artery supplies the lower extremity. It is the larger branch of the common iliac artery which descends from the sacroiliac joint forward and laterally to the inguinal ligament and, passing below it, midway between the anterior superior spine of the ilium and the symphysis pubis, where it enters the thigh and becomes the femoral artery.

In front and medially the artery is in relation with the peritoneum, subperitoneal areolar tissue, the termination of the ileum on the right side and, on the left, the sigmoid colon and a thin layer of fascia, which surrounds the artery and vein. In the female it is crossed at its origin by the ovarian vessels, and occasionally by the ureter. The internal spermatic vessels lie for some distance upon it or near its termination, and it is crossed in this situation by the external spermatic branch of the genitofemoral nerve and the deep iliac circumflex vein. Behind and laterally it is in relation with the psoas major, from which it is separated by the iliac fascia. Numerous lymphatic vessels and lymph glands lie in front and on the medial side of the vessel.

Branches of the External Iliac Artery

The inferior epigastric artery arises immediately above the inguinal ligament. The vessel follows a medial and then an ascending course. It runs on the dorsal surface of the rectus abdominis muscle and communicates with the terminal branches of the internal thoracic artery. After dividing into numerous branches, it anastomoses above the umbilicus with the superior epigastric branch of the internal mammary and with the lower intercostal arteries. The ductus deferens, as it leaves the spermatic cord in the male, or the round ligament of the uterus in the female, winds around the lateral and posterior aspect of the artery. The inferior epigastric artery gives off the external spermatic artery, a pubic branch and muscular branches, some of which are distributed to the abdominal muscles and peritoneum, anastomosing with the iliac circumflex and lumbar arteries. This branch is an important potential collateral between the upper and lower extremities.

The deep iliac circumflex artery arises from the lateral aspect of the external iliac opposite the inferior epigastric. It anastomoses with a branch of the deep femoral and with the ilio-lumbar and superior gluteal arteries. This branch represents an important bridge for establishing the collateral circulation between external iliac, abdominal aorta and femoral artery.

4. Femoral Artery

The femoral artery is the direct continuation of the external iliac and provides the greater part of the arterial supply of the lower extremity. It begins immediately behind the inguinal ligament, midway between the anterior superior spine of the ilium and the symphysis pubis, and passes down the front and medial side of the thigh. It ends at the junction of the middle with the lower one-third of the thigh where it passes through an opening in the adductor magnus to become the popliteal artery. The vessel, at the upper part of the thigh, lies in front of the hip joint. In the lower thigh it lies to the medial side of the shaft of the femur.

The common femoral artery is contained in the femoral or Scarpa's triangle, which corresponds to the depression seen immediately below the fold of the groin. The apex of the triangle is directed downward. It is bordered laterally by the sartorius and medially by the adductor longus muscles and above by the inguinal ligament. The floor of the space is formed by the iliacus, psoas major, pectineus and in some cases a small part of the adductor brevis and longus muscles. Besides the vessels and nerves, the space contains some fat and lymphatics.

The superficial femoral is contained in the adductor canal (Hunter's canal) which is an aponeurotic tunnel in the middle third of the thigh extending from the apex of the femoral triangle to the foramen of the adductor magnus. It is bounded anteriorly and laterally by the vastus medialis, posteriorly by the adductors longus and magnus and is covered by a strong aponeurosis which extends from the vastus medialis across the femoral vessels to the adductors. Lying on the aponeurosis is the sartorius muscle. In addition to the uperficial femoral artery, the canal contains the superficial femoral vein, the saphenous nerve and the nerve to the vastus medialis.

Variations. The femoral artery is rarely absent. When it is, it is replaced by the accompanying artery of the sciatic nerve (Luzsa, 1974). A doubling of the artery may be due to islet formation below the origin of the deep femoral artery, while the lower segment of the vessel consists again of a single trunk. Complete doubling also occurs, but very rarely.

The branches of the femoral artery are:

The superficial epigastric artery arises from the front of the femoral about 1 cm below the inguinal ligament and passes through the femoral sheath and the fascia, turns upward in front of the inguinal ligament and anastomoses with branches of the inferior epigastric and with its fellow opposite. The superficial circumflex iliac artery runs to the anterior superior iliac spine. It often arises together with the superficial epigastric artery. The external pudendal arteries consist of two or three trunks distributed to the groin, the scrotum or the labium.

The profunda femoris artery is the largest branch. It usually arises 3-4 cm below the inguinal ligament. It then passes below the adductor longus muscle and lies against the bone on theterior side of the thigh. The diameter of the deep femoral artery ranges between 3-7 mm. The takeoff of the profunda femoris from the common femoral is variable. In 50% of the cases it is found 3.5 to 5.0 cm below the ligament; in 24%, 5.0 to 8.5 cm below the ligament, and in 25% under the ligament or above. The profunda femoris, whose caliber at its origin is somewhat less than that of the superficial femoral, arises from the lateral and posterior aspect of the common femoral. At first it lies lateral to it and then runs behind it and the femoral vein, coursing toward the medial side of the femur. It then passes downward behind the adductor longus, ending at the lower third of the thigh in a branch which pierces the adductor magnus muscle.

The profunda gives off the following branches: medial femoral circumflex, lateral femoral circumflex, perforating, and muscular.

The medial femoral circumflex arises most often from the origin of the profunda, although in 25% of the cases it may arise from the posterior aspect of the common femoral proximal to the origin of the profunda femoral. It usually winds around the medial side of the femur passing between the pectineus and psoas major.

The lateral femoral circumflex arises from the lateral side of the profunda, passes horizontally between the divisions of the femoral nerve and behind the sartorius and the rectus femoris and divides into ascending, transverse and descending branches. The ascending branch passes upward to the lateral aspect of the hip, the transverse passes lateralward over the vastus intermedius and thence around the femur, and the descending branch runs downward behind the rectus femoris. One long branch of the latter descends in the muscle as far as the knee and anastomoses with the superior lateral genicular branch of the popliteal artery. This descending branch, often called the <u>artery of the quadriceps</u>, plays a significant role in the anastomotic network of the thigh.

The <u>perforating arteries</u>, usually three in number, pass backward close to the linea aspera of the femur by perforating the tendon of the adductor magnus, thus reaching the back of the thigh. The three perforating arteries give off branches joining with each other both in front of and behind the linea aspera. The termination of the profunda, as already mentioned, is sometimes called the fourth perforating artery.

The profunda gives off numerous muscular branches for the adductors and the hamstrings. These muscular branches participate actively in the collateral networks of the profunda with the popliteal collaterals. It should be pointed out that when the origin of the profunda is under the inguinal ligament, it is the artery that gives off all the branches which normally arise from the common femoral.

<u>Collaterals</u>. The branches of the profunda represent the pivotal anastomotic pathways with those of the iliac above and popliteal below. The first perforating artery which gives off branches to the adductors, biceps femoris and gluteus maximus, provides anastomoses with the inferior gluteal, medial and lateral femoral circumflex and second perforating artery. The latter provides anastomoses with the terminal branches of the profunda and the muscular branches of the popliteal. Thus the various anastomoses represent an important chain which extends from the gluteal arteries above, to the popliteal and tibial muscular branches below. When the superficial femoral or proximal popliteal artery is occluded, the profunda and its collateral branches enlarge and may transmit distally a pulsatile flow. Indeed this collateral circulation may be so effective that intermittent claudication may sometimes go unnoticed, especially in the presence of short segmental occlusions.

The <u>highest genicular artery</u>. This branch also called <u>anastomotica magna</u>, has important collateral potential. It arises from the femoral just before it passes through the foramen of the adductor magnus, and immediately divides into the saphenous and musculoarticular branches. The saphenous branch pierces the aponeurotic covering of the adductor canal and accompanies the saphenous nerve to the medial side of the knee. The musculoarticular branch descends in the substance of the vastus medialis and in front of the tendon of the adductor magnus to the medial side of the knee where it anastomoses with the medial superior genicular and anterior recurrent tibial artery. As shown by us in a series of femoral arteriograms, the highest genicular artery is an important stem for a reentry channel in cases of popliteal artery occlusion (Haimovici, 1967). This branch also provides an increased skin temperature of the knee in those occlusive patterns. This sign, which we call the "hyperemic knee sign" (Haimovici, 1976), is indicative of the increased collateral supply provided by the highest genicular artery and the anastomotic branches from the popliteal in the presence of an occlusion of the latter artery.

5. Popliteal Artery (Figs. 9, 10)

The popliteal artery, the continuation of the femoral, extends from the adductor magnus foramen to the lower border of the popliteus muscle where it divides into the anterior and posterior tibial. Its entire course is through the popliteal fossa which is a lozenge-shaped base behind the knee joint. Anteriorly the popliteal artery from above downward is situated behind the floor of the popliteal surface of the femur, the oblique popliteal ligament of the knee joint, the upper end of the tibia and the fascia covering the popliteus. Posteriorly it is overlapped by the semimembranosus above and is covered by the gastrocnemius and plantaris muscles below. In the middle part of its course the artery is separated from the skin and fascia by a quantity of fat and is crossed from the lateral to the medial side by the tibial nerve and the popliteal vein. On its lateral side, above, are the biceps femoris, the tibiial nerve, the popliteal vein and the lateral condyle of the femur. Below are the plantaris and the lateral head of the gastrocnemius. On its medial side, above, are the semimembranosus and the medial condyle of the femur, while below are the tibial nerve, the popliteal vein and the medial head of the gastrocnemius. The popliteal lymph nodes, six or seven in number, are imbedded in adipose tissue. One of the lymph nodes is beneath the popliteal fasia near the ending of the external saphenous vein, another is between the popliteal artery and the back of the knee joint, or alongside the popliteal vessel.

Variations of Popliteal Bifurcation. Anatomic variations relative to the division of the popliteal artery, although infrequent, may be encountered in the infragenual portion. Recognition of such variations may assume important significance in the techniques used for anastomosing a graft into the popliteal artery or its branches. The localization of these branches by a good arteriogram is essential to preoperative planning. Morris et al. (1960) and Bardsley and Staple (1970) evaluated the variations in branching of the popliteal artery, based on several hundred arteriograms. From these studies it is possible to describe four main patterns of branching of the popliteal artery. In approximately 90% of the cases the

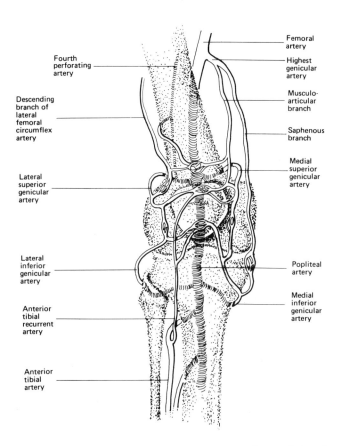

Fig. 9. Diagram depicting the popliteal arteries with the collateral circulation around the knee. (Redrawn from Quiring, 1949.)

most common pattern consists of the popliteal artery division more than 1 cm below the joint space into the anterior tibial and the posterior tibial trunk, the peroneal artery arising from the latter as a branch. A second pattern consists of a high division of the popliteal artery, usually behind the knee joint space. A third pattern consists also of a high division of the popliteal as in pattern two, but the peroneal arises from the anterior tibial. A fourth pattern consists of a trifurcation at a normal level but with the peroneal arising from the anterior tibial. Although other minor variations are also described, from a surgical point of view it is well to remember that the normal pattern is present in almost 90% of the extremities. In patterns two and three, instead of the infragenual popliteal, two tibial branches are present.

Branches of the Popliteal Artery. The superior genicular arteries, two in number, arise in either side of the popliteal, and wind around the femur immediately above its condyles to the front of the knee joint. They leave the popliteal at nearly right angles. The medial superior genicular divides into two branches and anastomoses with the highest genicular and the medial inferior genicular arteries. The lateral superior genicular also divides into a superficial and a deep branch, and anastomoses with the descending branch of the lateral femoral circumflex and the lateral inferior genicular artery.

The middle genicular artery is a small branch arising opposite the back of the knee joint. It supplies the ligaments and synovial membrane of the knee articulation.

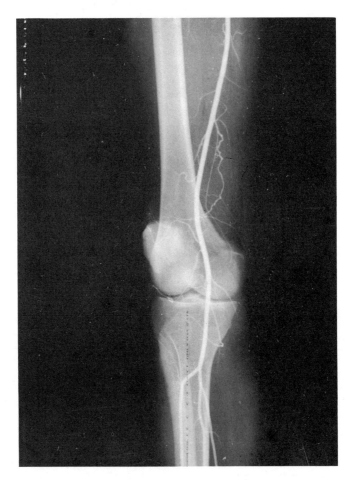

Fig. 10. Arteriogram of the popliteal artery and its bifurcation.

The inferior genicular arteries, two in number, arise from the popliteal beneath the gastrocnemius. The medial and lateral inferior genicular surround the condyles of the tibia and contribute to the formation of the arterial network around the knee. They are in communication with the anterior tibial recurrent artery.

The sural arteries are two large branches which are distributed to the gastrocnemius, soleus and plantaris. They arise from the popliteal artery opposite the knee joint. A few other arteries, two or three in number, arise from the upper part of the artery and are distributed to the lower part of the adductor magnus and hamstring muscles and anastomose with the endings of the profunda femoris. The sural arteries provide little potential for collateral pathways. This is in contrast to the better capability as collaterals of the genicular arteries. They form two plexuses, one superficial and the other deep, around the knee joint. The former is situated in the superficial fascia around the patella and the latter lies on the lower end of the femur and upper end of the tibia around the articular surfaces.

The Terminal Branches of the Popliteal (Figs. 11, 12)

6. Anterior tibial artery is one of the terminal branches of the popliteal and arises at the lower border of the popliteus muscle. After crossing the upper edge of the interosseous membrane, it enters the anterior compartment and courses upon the interosseous membrane. In the upper part it runs between the extensor muscles of the leg and then descends steeply. The course of its projection extends from the midpoint of the line between the head of the fibula and the tuberosity of the tibia to the line midway between the two malleoli. The vessel is continued on the dorsum of the foot as the dorsalis pedis artery. The diameter of the anterior tibial artery is 1.19 to 3.5 mm (Luzsa, 1974). The artery is accompanied by a pair of venae comitantes which lie one on either side of it. The deep peroneal nerve, coursing around the lateral side of the neck of the fibula, comes into relation with the lateral side of the artery shortly after it has reached the front of the leg.

Variations. The anterior tibial may be rudimentary in size or may be absent, in which case its place is taken over by perforating branches from the posterior tibial or by the perforating branches of the peroneal artery. In rare instances the artery may be superficial from the midleg distally, being covered merely by the skin and fascia below that point.

Branches. The anterior tibial recurrent artery arises from the anterior tibial as soon as the vessel has passed through the interosseous space. It ascends in the tibialis anterior and participates in the formation of the patellar plexus by anastomosing with

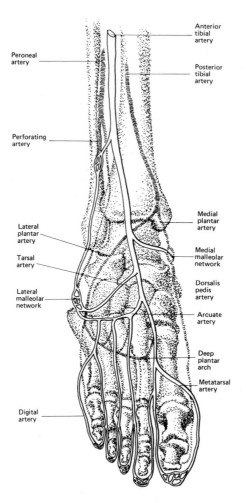

Fig. 11. Diagram of arterial tree of the lower third of the leg and foot with its major branches and collaterals. (Redrawn from Quiring, 1949.)

the genicular branches of the popliteal, and with the highest genicular artery.

The anterior medial and lateral malleolar arteries arise above the ankle joint and participate in the formation of the arterial network around the ankle.

The muscular branches are numerous. They are distributed to the muscles which lie on either side of the tibial artery, some piercing the deep fascia to supply the skin, others passing through the interosseous membrane and anastomosing with branches of the posterior tibial and peroneal arteries.

7. <u>Posterior tibial artery</u>, the other terminal branch of the popliteal, begins at the lower border of the popliteus, opposite the interval between the tibia and fibula. It passes then between the superficial and deep muscles in a straight downward line from the middle of the popliteal space to midway between the medial malleolus and the medial tubercle of the calcaneus to divide into end branches on the sole of the foot.

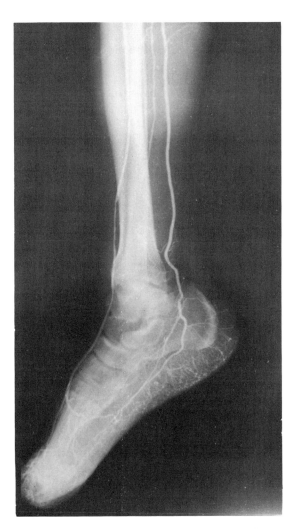

Fig. 12. Arteriogram showing the leg and foot arteries.

The posterior tibial artery lies upon the tibialis posterior the flexor digitorum longus, the tibia and the back of the ankle joint. It is covered by the deep transverse fascia of the leg which separates it above from the gastrocnemius and soleus and, at its termination it is covered by the abductor hallucis. In the lower third of the leg it becomes more ficial and it is covered only by the skin and fascia and runs parallel with the medial border of the tendon Achilles. It is accompanied by two satellite veins and by the tibial nerve which lies at first to the medial side of the artery, but soon crosses it posteriorly and in the greater part of its course on its lateral side. The diameter of the posterior tibial artery is 2.23 to 4.15 mm (Luzsa, 1974).

The posterior tibial is not infrequently smaller than usual or absent. It is then supplanted by a large peroneal artery which either joins the small posterior tibial artery or continues alone to the sole of the foot.

The branches of the posterior tibial artery are:

The circumflex fibular branch passes around the neck of the fibula. The muscular branches of the posterior tibial are distributed to the soleus and the deep muscles along the back of the leg. The posterior medial malleolar artery is a small branch which winds around the tibial malleolus and ends in the medial malleolar network. The medial calcaneal arteries arise from the posterior tibial just before its division. They anastomose with the peroneal and medial malleolar and end on the back of the heel with the lateral calcaneal arteries.

8. Peroneal artery, a branch of the posterior tibial trunk, is deeply seated on the back of the fibular side of the leg. It arises about 2.5 cm from the posterior tibial below the lower border of the popliteus, passes obliquely toward the fibula, and then descends along the medial side of that bone contained in a fibrous canal between the tibialis posterior and the flexor hallucis longus, or is in the substance of the latter muscle. Distally it divides into lateral calcaneal branches which ramify on the lateral and posterior surfaces of the calcaneus.

As already mentioned above, the peroneal artery may arise 7 or 8 cm below the popliteus or from the posterior tibial high up or even from the popliteal. Rarely, it also may arise from the anterior tibial. Its size is usually increased rather than diminished. In some instances it takes altogether the place of the posterior tibial in the lower part of the leg and foot.

The peroneal branches provide supply to the muscles of the soleus, tibialis posterior, flexor hallucis longus and the peronei.

The perforating branch pierces the interosseous membrane about 5 cm above the lateral malleolus to reach the front of the leg where it anastomoses with the anterior lateral malleolar. This branch is sometimes enlarged and takes the place of the dorsalis pedis artery. In half of the cases it anastomoses with the lateral branch of the anterior tibial, proximal to the anterior lateral malleolar. The communicating branch is given off from the peroneal about 2.5 cm from its lower end and joins the communicating branch of the posterior tibial. The lateral calcaneal arteries are terminal branches of the peroneal artery. They pass to the lateral side of the heel and communicate with the lateral malleolar and with the medial calcaneal arteries.

9. <u>Dorsalis Pedis Artery</u>. This artery is a continuation of the anterior tibial, passes forward from the ankle joint along the tibial side on the dorsum of the foot to the proximal part of the first intermetatarsal space, where it divides into two branches, the first dorsal metatarsal and the deep plantar.

In its course the vessel rests upon the articular capsule of the ankle joint, the talus, navicular and second cuneiform bones. Near its termination it is crossed by the first tendon of the extensor digitorum brevis. On the tibial side is the tendon of the extensor hallucis longus, on its fibular side, the first tendon of the extensor digitorum longus and the termination of the deep peroneal nerve. The artery is accompanied by two veins.

This vessel is rarely absent. Occasionally it may be larger than usual to compensate for a deficient plantar artery. If the branches to the dorsum of the metatarsus and toes are absent, the arterial supply is provided by the medial plantar artery.

<u>Branches</u>. The lateral tarsal artery passes in an arched direction lateralward and supplies the extensor digitorum brevis and the joints of the tarsus. The medial tarsal arteries are two or three small branches which ramify on the medial border of the foot and join the medial malleolar network.

The arcuate artery arises a little anterior to the lateral tarsal artery and passes lateralward to anastomose with the lateral tarsal and lateral plantar arteries. This vessel gives off the second, third and fourth dorsal metatarsal arteries.

The first dorsal metatarsal artery runs forward on the first interosseous dorsalis, and at the cleft between the first and second toes divides into two branches, one of which passes beneath the tendon of the extensor hallucis longus and is distributed to the medial border of the great toe. The other bifurcates to supply the adjoining side of the great and second toes.

The deep plantar artery descends into the sole of the foot between the two heads of the first interosseous dorsalis and unites with the termination of the lateral plantar artery to complete the plantar arch.

10. Plantar Arteries. The posterior tibial artery divides beneath the origin of the abductor hallucis into the medial and lateral plantar arteries.

The medial plantar artery, much smaller than the lateral, passes forward along the medial side of the foot; at the base of the first metatarsal bone it passes along the medial border of the first toe anastomosing with the first dorsal metatarsal artery. Small superficial digital branches accompany the digital branches of the medial plantar nerve and join the plantar metatarsal arteries of the first three spaces.

The lateral plantar artery, much larger than the medial, passes laterally and forward to the base of the fifth metatarsal bone. It unites with the deep plantar branch of the dorsalis pedis artery thus completing the plantar arch. The arch is deeply situated and extends from the base of the fifth metatarsal bone to the proximal part of the first interosseous space and forms the plantar arch. It is convexed forward, lies below the base of the second, third and fourth metatarsal bones and the corresponding interossei. The plantar arch, besides distributing numerous branches to the muscles, skin and fasciae in the sole, gives off the perforating and plantar metatarsal branches.

C. Collateral Circulation of the Lower Extremity

The collateral circulation of the lower extremity is insured by three groups of anastomoses.

1. The system of iliac anastomoses is not as good as that of the scapular region. Connections between the aorta (lumbar, medial sacral, testicular, superior and inferior mesenteric arteries), the internal iliac artery (superior and inferior gluteal, internal pudendal, obturator, iliolumbar arteries), external iliac artery (inferior epigastric, deep circumflex iliac arteries), and the femoral artery (superficial epigastric, superficial circumflex iliac, external pudendal, deep femoral, medial and lateral circumflex femoral arteries) are able to compensate for the insufficiency of the iliac arteries. No adequate circulation in the lower extremity can develop if the proximal part of the femoral artery is obstructed because of the impaired function of the deep femoral in such cases, whereas sufficient collateral circulation in the lower third can be maintained to the anastomoses between the femoral and popliteal arteries, if the deep femoral is fully patent.

2. <u>Genicular anastomoses</u> are formed between the common femoral, the deep femoral (third perforating artery, descending branch of the lateral circumflex femoral artery), the popliteal (superior, middle and inferior genicular arteries), and the anterior tibial (anterior and posterior tibial recurrent arteries). The communication is not always adequate. Collateral circulation is functionally sufficient only if the arterial network of the knee can compensate satisfactorily.

3. <u>Anastomoses in the legs and feet</u> between the three arteries of the leg are formed in the region of the ankles (malleolar), and in that of the heels through the connections of the peroneal artery and by way of the vessels which compose the plantar arch. The initial segment of the posterior tibial artery is of primary importance in this connection. Its deficiency can be compensated only by the sural arteries. The anterior tibial may be replaced by its communication with the peroneal artery and the arterial network of the knee. The peroneal artery has so many anastomoses that its elmination elicits no clinical manifestation (Quiring, 1949).

II <u>MUSCULAR ARTERIES</u>

1. <u>General Considerations</u>

The muscular arteries have inspired little interest or investigation, although the muscles as an anatomic entity constitute about 42% of the body weight. Recently, however, the significance of the muscular arteries from a physiopathologic point of view has been receiving greater attention. Thus their role in reestablishing collateral circulation in the presence of occlusive arteriopathies has been the object of several anatomic and arteriographic studies.

In the past the description in the anatomical textbooks of the vascular patterns of human muscles was usually confined mostly to the larger branches and thus offered little information concerning the intramuscular terminal arteriolar networks. One of the reasons for this incompleteness of their description is due to the method of investigation. Indeed, the earlier studies of the blood supply to the muscles in man relied essentially upon dissection alone. In more recent years, use of arteriography in cadavers and in vivo has added a new dimension to their study. Although radiographic investigations of the blood supply of the muscles was carried out as early as 1915 and 1919, it was not until the work of Salmon and Dor (1933) that such a systematic study was applied to its evaluation. By means of the radiographic technique, these authors have been able to carefully assess the origin of these arteries, their course, number, caliber, relations, and mode of distribution and anastomoses. They divided the study of these arteries into: extramuscular, intramuscular and their patterns of anastomoses. More recently Saunders

(1957), using more refined arteriographic techniques, has studied the vascular patterns in the muscle in various ways: (1) by a standard x-ray unit he performed straight and stereoarteriograms of the intact limbs, as well as of the excised and isolated muscle groups; (2) by microarteriography he evaluated the finer detail of the intramuscular vessels; and (3) by stereomicroarteriography he studied the smallest intramuscular vessels in the third dimension, thus removing the difficulties of interpretation occasioned by the superimposition of shadows inherent in flat film radiography.

Knowledge of the arterial supply to individual or group of muscles should provide the basis for the mechanism of localized pathologic changes often noted in occlusive arterial disease. A classic example is that of intermittent claudication or even necrosis of the anterior tibial group of muscles alone. Recent advances in vascular surgery have been instrumental in emphasizing the role not only of the major arteries, but also that of the branches supplying the muscles. The study that follows is based primarily on arteriographic data.

a. Extramuscular Arteries

The muscular arteries originate either from a proximal major trunk or from one close to the muscle. While the origin of the smaller arterioles may be variable, that of the major arteries is relatively constant. The number of arteries received by each muscle is also constant. However, their number may increase proportionately to the mass of the muscle. The caliber or diameter of the muscular arteries is probably the most variable element. It may vary not only with the mass of the muscle, but especially with age, sex, muscular activity, and obviously with the presence or absence of any occlusive disease of the major trunks. The course of the muscular arteries, although it varies with each muscle, is always the same for a given muscle. Thus the arteries to the brachial biceps are always directed transversely while those of the vastus lateralis of the quadriceps are always oblique (Fig. 13). The length of the extramuscular course of the arteries is almost always the same for a given muscle. Likewise the point of entry of these arteries into the muscular mass is always constant.

In brief, these radiologic studies have revealed that the muscular arteries display a regularity of their origin and distribution. These findings are not only true for the major named muscular arteries, but are applicable to the arrangement of many of the minor and unnamed arteries.

Arterial distribution to skeletal muscle is in general the same on both sides of the human body, and exhibits a characteristic pattern of being bilateral and symmetrical. These studies have further

ARTERIAL CIRCULATION OF THE EXTREMITIES 465

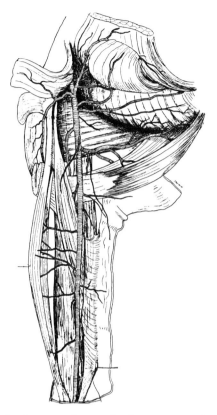

Fig. 13. Arteries of the muscles of the pectoral and brachial areas arising from the axillary and brachial arteries. Note their perpendicular origin and direction irrigating the various muscles. (From M. Salmon and J. Dor, 1933.)

revealed that the arterial supply to the skeletal muscles is provided by segmental arteries which arise either from the main arterial axes or their major named or unnamed muscular branches. The length of the segmental branches varies, but it seems to be related primarily to function. These segmental arteries end in various ways. They may end in a muscle belly or may bifurcate and supply two or three or more muscle bellies, or may traverse a muscle in order to terminate in another muscle. Often the muscular branches are interconnected with those of the skin, subcutaneous tissue and bone (Salmon and Dor, 1933).

In stereoarteriograms it is often possible to trace one limb of a bifurcating segmental branch to the subcutaneous vascular network, and thus be able to observe the interconnections between the intramuscular and subcutaneous vascular networks. The blood supply of a muscle belly appears to be related to its volume. In some instances a given muscle may derive its segmental supply from multiple axes that do not necessarily lie on the same side of the limb. The clinical significance of this pattern is related to ischemic changes of a major arterial trunk usually supplying the branches to a given muscle (Saunders 1957).

Based on both the anatomical and functional features of the muscles, the following groupings of human skeletal muscles were proposed by Saunders (1957). They should prove of practical value to both the clinician and investigator.

Class I: Free muscle belly with grouped supply. This group consists of those muscles that have attachments of limited area and derive their major blood supply from a short or localized segment of a main arterial trunk. Such muscles are particularly vulnerable to interruption of their blood supply since they are dependent upon a principal source. An example of this type is represented by the supply to the rectus femoris and gastrocnemius. Their arterial supply is derived from the lateral femoral circumflex and the popliteal arteries respectively. Other outstanding examples of this group of muscles are the biceps brachii, and biceps femoris. From a clinical viewpoint it is fortunate that relatively few muscles belong to this class.

Class II: Fixed muscle belly and dispersed supply. This group consists of muscles having an extensive fleshy attachment and hence a relatively thick muscle belly. Their blood supply consists of a series of separate or segmental arteries which arise from a long segment of a main arterial trunk.

Most of the deeply placed muscles belong to this class with few exceptions of superficial muscles. Examples of such muscles in the upper limb are the brachialis, triceps brachii, superficial and deep forearm flexors and extensors; in the lower limb, the vasti, sartorius, soleus, peronei, all extensors and deep flexors of the leg, and the hamstring muscles.

Class III: Mixed type. This group comprises those muscles which exhibit features of both the preceding classes in that they possess a freely moving part as well as an extensive and relatively fixed part. Their blood supply is provided by one or more large muscular arteries and also by numerous small arteries from multiple sources. Most of the large trunk muscles fall within this class. The deltoid and spinati are similarly supplied, receiving their blood vessels from the anastomotic circles about the humeral and glenoid neck. Likewise the gluteal muscles also fall within this class, the distribution of the vessels being not unlike those of the deltoid.

b. Intramuscular Arteries

Intramuscular vascular patterns in man have been described by several investigators (Campbell and Pennefather (1919), Salmon and Dor (1933), Blomfield (1945), Saunders (1957)). Although flat film arteriography was first instrumental in demonstrating the distribution of the intramuscular arterial patterns, however, it became apparent subsequently that reliance of their evaluation cannot be placed on two dimensional radiography alone. Suitable contrast media using both stereoradiographic and stereomicroradiographic techniques are the methods of choice to avoid misinterpretation due to superimposition of shadows inherent in flat film radiography. The larger muscular arteries usually enter perpendicular to the direction of the muscular fibers. In the muscular belly the arteries are surrounded by two satellite veins and sometimes by a nerve often accompanying the artery as far as its distribution, as part of a neurovascular bundle. The artery, vein and nerve are surrounded by a thin sheath in miniature which separates them from the muscle fibers and facilitates their dissection. The artery divides into two or several branches. Usually one of the branches (recurrent branch) ascends toward the superior insertion of the muscle. Others are directed toward the opposite end of the muscle. The anastomosing channels between the different muscular arteries most often form a mesh-type of vessel pattern. Although intramuscular networks of anastomotic vessels have been previously described, Saunders has pointed out the existence of two types of such vascular networks composed of a series of anastomosing arterioarterial arcades. This latter author introduced the two terms, "macromesh" and "micromesh" to designate two types of vascular networks, one readily seen in gross arteriograms and the other requiring enlargements of such muscle arteriograms permitting greater detailed studies. Using microradiographic methods, Saunders could determine the relationship between the macro- and micromesh and the capillary bed. Thus arising at intervals from arterioles forming the micromesh, and passing off to one side or the other, one can determine the presence of small tortuous branches. They are best designated as precapillary arterioles. By means of stereomi croarteriograms the capillaries can be delineated which pass off and run parallel to one another between and about the muscle fibers and thereby contribute to a capillary mesh, which by means of steroscopically observed vessels indicate a longitudinal disposition of the network both about and within each bundle of fibers.

Arteriovenous Shunts. Arteriovenous anastomoses (AVA) or shunts are well known to exist in various structures of the human body (Haimovici et al., 1966). Their existence within muscle was also demonstrated recently by Saunders using the techniques as described above. Experiments conducted by the latter author provided anatomical evidence of the existence of two pathways whereby blood might

flow in its passage from the arterial to the venous components of
the micromesh, not only in fresh fetal and adult material, but also
in the living animal. These arteriovenous shunts do not resemble
the specialized structures described in the human finger in that
they do not show the so-called epithelial cells in their wall. This
may be of great significance from a physiologic and pathologic point
of view. Peripheral resistance to blood flow is confined to the arterioles and capillaries. They represent the chief areas of blood
flow regulation in the periphery. The newly discovered available
pathways in human muscle, as illustrated by certain features of the
peripheral mesh patterns, appear to call for revision of traditional
ideas regarding the systemic circuit. *"Thus, the dampening of intermittent flow, maintenance of local blood flow conditions, and accommodatory or release mechanisms which augment or reduce the rate of peripheral blood flow in relation to changing cardiac output, must be considered against the existence of such mesh patterns in the major tissues as muscle and skin and alternate pathways for the passage of blood as indicated by two types of arteriovenous communications,"* (Saunders, 1957).

2. Muscular Arteries of the Upper Extremity

a. Muscular Arteries of the Shoulder. The arterial supply to
the muscles of the shoulder arise primarily from the axillary and
brachial. A few branches are supplied from the subclavian artery.
The deltoid receives its arterial supply from the posterior circumflex and its branches. The acromial branch of the thoracoacromial
and the posterior humeral circumflex provide the majority of the arterial supply. The supraspinatus is supplied by one or two branches
of the transverse cervical. The infraspinatus receives its main
supply from the scapular circumflex and some of the branches from
the termination of the subscapular artery. The teres major receives
its principal supply from the inferior scapular artery in 65% and
from the axillary in about 12% of the cases. The teres minor muscle
receives its supply from the posterior circumflex branch of the
brachial artery. The various arteries providing the muscles of the
shoulder display frequent and constant anastomotic networks in the
extra- as well as intramuscular compartments.

b. Muscular Arteries of the Arm. The arterial supply of the
biceps is variable. Two main types have been isolated by dissection
and arteriographic studies. One type consists of two major branches.
A second type consists of four to eight branches arising from the
brachial artery at various levels. Between these two types a great
variety of patterns of intermediate vascularization of the muscle
have been described. As a rule all these branches arise from the
brachial although a number of variations exist. The coracobrachial
receives its major artery from the axillary. Accessory branches are
provided by the acromiothoracic and the anterior humeral circumflex.

The brachialis or anterior brachial muscle receives its supply from the brachial artery by means of two or three small branches. On arteriograms it is possible to note common branches to the biceps and brachialis. There is a long intramuscular anastomosis between these two branches. The triceps is irrigated principally by the deep brachial artery and the superior collateral and the posterior circumflex. Arteriograms have disclosed the triceps to be the site of a rich anastomotic network linking the anterior and posterior group of muscles and thus playing a significant role in the occlusion of the brachial artery.

  c. Muscular Arteries of the Forearm. The arterial supply of the muscles of the forearm has been the focus of somewhat greater attention from both the anatomists and surgeons because of the relatively more frequent vascular involvement of this area (Volkman's ischemic contracture). The arterial supply to the forearm is considered separately for the two groups: the volar antebrachial and the dorsal antebrachial muscles.

  The volar antebrachial muscles are divided in turn into two groups, the superficial and deep. The superficial group on the lateral side receives its arterial supply from the radial recurrent, superficial volar and muscular branches arising from the radial artery. On the medial side of the anterior ulnar, recurrent artery and muscular branches provide the balance of the arterial supply. The arteries anastomose in the depth of the muscles, rarely outside of them. The deep group of muscles receive their arterial supply either directly from the ulnar artery or from the origin of the anterior interosseous branch, more rarely from the trunk of the interosseous artery itself.

  Dorsal Antebrachial Muscles. The arterial supply to the muscles of the posterior or dorsal antebrachial group separately penetrate the superficial and deep muscles. Most are irrigated by the interosseous artery and its branches. Each of these branches of the interosseous artery provide twigs to the individual muscles. It should be pointed out again that these vessels anatomose quite frequently intramuscularly. The intramuscular arterial network is about equal in the extensor and flexor muscles. As mentioned before, variations in the origin and distribution of the arteries to these muscles occur with a moderate frequency.

  Muscles of the Hand. Anatomical variations of the arteries supplying the muscles of the hand are so frequent that it is difficult to describe a typical pattern. In general terms the muscles of the hand are surrounded by a true arterial plexus constituted by the arterioles which arise from the superficial arch, the deep arch, and from the trunk of the radial artery. The arterioles anastomose very frequently and irregularly either on the lateral side near the first metacarpal or on the inner border of the superficial flexor brevis. On the medial side of the hand, the arterial pattern is similar to that of the lateral side.

3. Muscular Arteries of the Lower Extremity

The muscles of the lower extremity are subdivided into groups corresponding with the different regions of the limb; (1) iliac; (2) thigh; (3) leg; and (4) foot.

   a. The muscles of the iliac region include the psoas major, psoas minor and iliacus. These muscles belong to the iliolumbar region and only their lower portion, especially that of the psoas major, is part of the lower extremity. However, some of the arterial branches providing the arterial supply to these muscles anastomose with some of the collaterals from the lower extremity. The psoas major and the iliacus (often regarded as a single muscle under the name of iliopsoas) receive their arterial supply from the lumbar, iliolumbar, obturator, external iliac, and common femoral, thus providing them with a rich arterial network. Of these arteries, the external iliac, the obturator, the common and deep femoral are the major source of their blood supply.

   b. The Muscles of the thigh are described in four groups: anterior, medial, gluteal, and posterior.

   The anterior group includes the sartorius and the quadriceps femoris. The saratorius in its proximal portion receives a branch from the superficial circumflex iliac often supplemented by a branch from the external circumflex. These branches are small and inconstant. The major arterial supply is provided by the femoral artery, in its proximal portion arising from the common, and in its medial and inferior portion, from the superficial femoral.

   The quadriceps femoris which is composed of four muscles, receives almost its entire arterial supply from a branch of the deep femoral artery, the artery of the quadriceps (Fig. 14.) While in 80% it originates from the profunda, in 20% it arises from the common femoral or from the superficial, at variable levels. The diameter of this artery is significant since often it is equal to that of the deep femoral. The artery of the quadriceps provides branches to the individual muscles, the rectus femoris, the vastus lateralis, the vastus intermedius and also the vastus medialis. The quadriceps femoris is thus richly irrigated. Except for the vastus medialis which is mainly supplied directly from the superficial femoral, all the other three muscles receive their arterial branches from a common trunk originating from the deep femoral.

ARTERIAL CIRCULATION OF THE EXTREMITIES 471

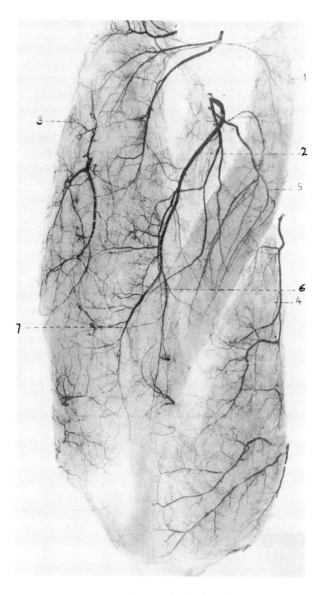

Fig. 14. Intramuscular arteries and their branches of the quadriceps femoris: (1) rectus femoris, (2) vastus intermedius, (3) vastus lateralis, (4) vastus medialis, (5) superior artery of the rectus femoris, (6) inferior artery of the rectus femoris, (7) major branch of the vastus lateralis. (From M. Salmon and J. Dor, 1933.)

The medial femoral muscles include the adductor longus, brevis and magnus as well as gracilis and pectineus. The arterial supply to these muscles is provided by the major arterial trunks of the anterior region of the thigh and a few from the posterior: the medial femoral circumflex artery, the deep femoral, the superficial femoral and the obturator. The medial circumflex arises from the medial aspect of the profunda and winds around the medial side of the femur passing first between the pectineus and the psoas major and then between the obturator externus and the adductor brevis. It then divides into superficial and deep branches. The superficial branch appears between the quadratus femoris and upper border of the adductor magnus and anastomoses with the inferior gluteal, lateral femoral circumflex and the first perforating arteries. The deep branch runs obliquely upward upon the tendon of the obturator externus and in front of the quadratus femoris toward the trochanteric fossa where it anastomoses with twigs from the gluteal arteries. The deep femoral artery provides a major branch called the artery of the adductors which is often 3 mm in diameter. Its branches are distributed to the three adductors as well as to the vastus medialis. The adductor brevis receives in addition to the branches just mentioned, some branches from the obturator artery.

The muscles of the gluteal region include the three gluteal muscles, two obturator, the quadratus femoris and two small muscular fasciculae accessories to the tendon of the obturator internus. The arterial supply to these gluteal muscles is provided by the superior and inferior gluteal branches of the internal iliac. The superior gluteal artery is the largest. In turn, it is divided into a superficial and a deep branch. Before quitting the pelvic cavity, it provides branches to the iliacus, pyriformis and obturator internus. The superficial branch enters the deep surface of the gluteus maximus and divides into numerous branches, some of which supply the muscle and anastomose with the inferior gluteal while others perforate its tendinous origin and supply the skin covering the posterior surface of the sacrum, anastomosing with the posterior branches of the lateral sacral arteries. The deep branch lies on the gluteus medius and almost immediately subdivides into two: the superior and inferior divisions. These, in turn, distribute branches to the glutei and anastomose with the lateral femoral circumflex artery.

The inferior gluteal artery is distributed chiefly to the buttocks and back of the thigh. Outside the pelvis the inferior gluteal provides a number of branches that supply the gluteus maximus, anastomosing with the superior gluteal within the substance of the muscle, the external rotators anastomosing with the internal pudendal and the muscles attached to the tuberosity of the ischium, anastomosing further with the posterior branch of the obturator and the medial femoral circumflex arteries.

The posterior group of muscles includes the biceps femoris the semitendinosus, and the semimembranosus. The biceps femoris has two heads of origin: (1) the long head, arises from the lower and inner impression of the back of the tuberosity of the ischium by a tendon common to it and the semitendinosus, and (2) the short head, arises from the lateral lip of the linea aspera, extending up almost as high as the insertion of the gluteus maximus. The fibers of the long head end in an aponeurosis which covers the posterior surface of the muscle and receives the fibers of the short head. Their common tendon is inserted on the head of the fibula and lateral condyle of the tibia. The arterial supply of the long head is provided by the first perforating branch of the deep femoral. The distal portion of the long head is also supplied by branches from the second perforating artery. Besides these two major sources of arterial supply, there are a few smaller accessory arteries arising from the inferior gluteal and the descending branch of the medial circumflex artery. The short head of the biceps receives its main supply from the third perforating artery. Occasionally a few branches, or twigs, arising from the second perforating provide some of the arterial irrigation of the short head.

The semitendinosus receives two principal arteries, one arising from the descending branch of the circumflex or from the first perforating. The semimembranosus' arterial supply is variable. In general, the perforating arteries provide most of the blood supply.

c. The Muscles of the leg are divided into three groups: anterior, posterior and lateral. The anterior group includes the anterior tibial, the extensor hallucis longus and the extensor digitorum longus with the peroneus tertius as part of the extensor digitorum longus. These muscles are perfused by the anterior tibial artery which gives off 25 to 30 small-caliber muscular branches of a transverse direction. The anterior tibial muscle in its proximal portion also receives branches from the anterior tibial recurrent and the peroneal arteries. Arteriographic study of the supply to the lateral crural muscles has disclosed a great number of anastomoses between these arteries. This is in contrast to the lesser number of arterial branches supplying the anterior group of muscles.

The posterior crural muscles situated on the back of the leg are subdivided into two groups: superficial and deep. The superficial group includes the gastrocnemius, the soleus, and the plantaris. The gastrocnemius, the most superficial muscle, arises by two heads which are connected to the condyles of the femur. The medial and larger head takes its origin from a depression at the upper and back part of the medial condyle. The lateral head arises from an impression on the side of the lateral condyle. The fibers unite at an angle in the middle line of the muscle in a tendinous structure which expands into a broad aponeurosis on the anterior

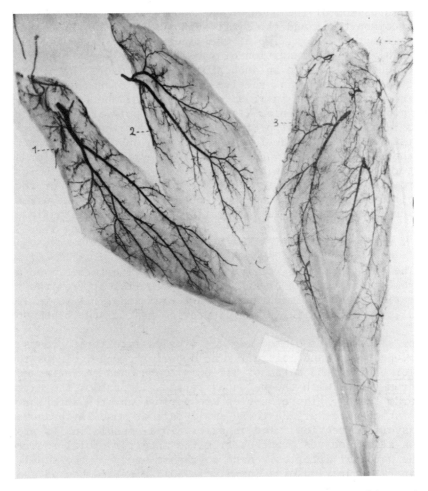

Fig. 15. Arteries of the gastrocnemius (1, 2), soleus (3), and plantaris (4). (From M. Salmon and J. Dor, 1933.)

surface of the muscle. The aponeurosis, gradually contracting, unites with a tendon of the soleus, and forms with it the tendon Achilles. Each head of the gastrocnemius receives a separate branch arising from the popliteal artery, most often opposite the knee joint space. Each head possesses an arterial branch which penetrates into the substance of the muscle close to the axis of the muscle at the level of the medial portion of the popliteal space. The arteries branch off in the substance of the muscle near its anterior surface.

The arteriographic study provides a clear illustration of this point (Fig. 15). The irrigation of the soleus is variable. It receives its supply through two principal arteries, one superior and one inferior, and several small accessory arterioles. They originate from the posterior tibial trunk at the level of the arcade of the soleus. The arteriographic study demonstrates that the main arteries bifurcate in the substance of the muscle and extend into the major part of the body of the muscle. The posterior tibial artery provides the accessory branches to the distal portion of the soleus. Very few anastomotic branches exist between these various arterioles. It appears, therefore, that the arteries perfusing these muscles are of the terminal type. The plantaris muscle is irrigated by a branch arising from that of the lateral head of the gastrocnemius and through a branch arising directly from the popliteal artery.

The deep crural group of muscles (popliteus, flexor hallucis longus, flexor digitorum longus, and tibialis posterior) are supplied normally by a great number of branches of small caliber. The flexor digitorum is penetrated by a group of arterioles arising from the posterior tibial, while the flexor hallucis is irrigated by branches arising from the peroneal. The latter provides 10 to 12 small anterioles penetrating the body of the muscle. The posterior tibial muscle receives two small arterial branches. Some of the twigs arising from these branches are distributed also to the posterior tibial nerve. The popliteus receives a number of branches forming a plexus around the muscle. The arteriograms disclose that these muscles are well vascularized although no anastomoses are seen. The arteries of the deep group of muscles are also providing blood to the periosteum of the adjacent bones and to the integument surrounding these muscles.

The muscles of the foot receive their arterial supply from the dorsalis pedis and the two plantar arteries and their branches. It is difficult to describe a standard type of vascularization of the pedal muscles, because of the inconstancy and great variability of the origin of the arterioles supplying the various muscles of the foot (Salmon and Dor, 1933.)

III. CUTANEOUS ARTERIES

The arterial supply of the skin like that of the muscles has been the object of little investigation. Yet from a purely morphologic point of view, the importance of origin, distribution and patterns of arterial networks of the skin and especially their significance from a surgical point of view, are undeniable. For example, the general direction of the arterioles supplying the skin should indicate the direction of skin incisions in a given area so as to avoid cutting off critical blood supply to the integument. The comprehensive work by Salmon on the arteries of the skin, although dating back to 1936, is still the most important source of

our information to date on this subject. The description to follow
will be largely based on Salmon's richly documented and illustrated
publications.

The arteries supplying the skin form a network in the subcutaneous tissue. From this network branches are given off to supply the corium and appendages of the skin. The description of the arteries of the skin include: (1) gross anatomy dealing with the origin, course, macroscopic distribution; and (2) microcirculation (fine anatomy) dealing with the fine details of the distribution of the arteries and anastomoses as disclosed by means of arteriography (mostly in cadavers).

1. General Considerations. The origin of the main arteries of the skin, contrary to data secured from dissection alone, is constant in the majority of cases. They may originate either from the major trunks from a given region, or from a common trunk of a muscular artery or nerve, or that of an osteoarticular branch. In general, the arteries of the skin may be direct or indirect. The former are more frequently seen in the region of the head, scalp, neck, and the anterior aspect of the trunk. By contrast the type of muscular-cutaneous arteries are relatively more frequently encountered at the level of the extremities. Three common patterns are noted: (1) the cutaneous collateral appears as a simple branch of the muscular arterial trunk, or (2) the cutaneous artery is larger, the muscular branches obviously being collaterals from the latter, and (3) the cutaneous and muscular branches display no particular pattern and appear to be of equal length and diameter.

The number of cutaneous arteries in a given region usually varies very little. However, they vary comparatively from one region to another, their number being inversely proportional to their diameter.

The diameter is constant for the same reason in most subjects. The amount of subcutaneous tissue may determine the size of the arteries.

The course of the cutaneous arteries is both subfascial and extra-fascial. The subfacial course can be long and tortuous, depending on the region. It is short in the region of the hand and much longer in the region of the thigh. In the extrafascial course the cutaneous arteries are always tortuous, thereby adapting to the mobility of the skin. The direction of the arteries in the extremities is very irregular: sometimes oblique, or horizontal; occasionally vertical without any apparent reason. The cutaneous arteries have anatomical connections with the muscles as mentioned above and also with the fascial layers through which they perforate to enter the extrafascial planes. Most of the cutaneous arteries are accompanied by satellite veins which provide anastomotic links with the superficial and deep venous networks. In addition, the cutaneous arteries are accompanied by satellite nerves. Some of the arterioles are paraneural and some are perineural. The inflammatory process or the occlusive lesions of the smaller arterioles are often the origin of painful syndromes.

The distribution of the cutaneous arteries includes collateral and terminal branches. The collateral branches include muscular, fascial, periosteal and capsular and cutaneous collaterals. The terminal branches are seen as a bifurcation or trifurcation. They usually form numerous networks, superficial and deep, which communicate with each other and with those of the subcutaneous tissue by oblique branches. The problem of distribution is that of the cutaneous arterial territories. Unlike the distribution of the nerves, that of the cutaneous arteries is much less well defined for two reasons: (1) the majority of cutaneous arteries being largely composed of anastomoses, there is no precise line of demarcation between the various arterial systems of the skin, and (2) the various cutaneous arteries, again unlike the constancy of the nerve distribution, may have a reciprocal supply of adjacent territories.

2. Arterial Supply of the Skin of the Upper extremity (Fig. 16)

a. Shoulder. The shoulder includes several sources of arteries for the skin: in the deltoid region the transverse cervical or the subscapular (subclavian origin) anteriorly to the deltopectoral and its perforating branches of the posterior circumflex are provided by the axillary system.

b. Arm. Three territories are described for the arterial supply: medial, middle, and lateral. The medial territory receives its supply from direct branches of the brachial. The middle or anterior territory receives its arterial supply from indirect branches of the brachial, namely those of the biceps or from the anterior brachial. The lateral territory occupies the outer surface of the arm and extends into the posterior aspect. It is irrigated by the deep brachial, including direct branches and some from the branches of the deltoid.

c. Elbow. Because of the great number of anastomoses around the elbow, identification of the branches providing the skin is somewhat more difficult. Anteriorly the ulnar artery and its epicondyle branches, the radial with its epicondyle branches, and the brachial with its branches in the fold of the elbow constitute the major supply of the skin.

d. Forearm. The ulnar and radial arterial systems share the major part of the irrigation of the skin in the volar surface. In

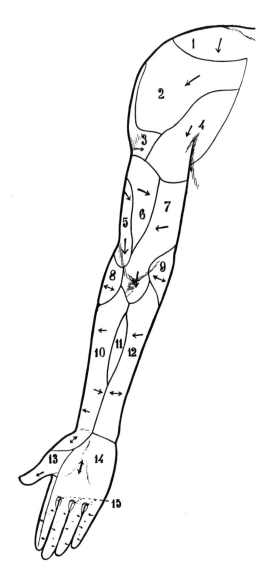

Fig. 16. Cutaneous arterial territories of the upper extremity of the volar surface: (1) transverse cervical, and transverse scapular arteries, (2) deltopectoral artery, (3) posterior circumflex artery, (4) thoracic branch of the axillary artery, (5) profunda brachii artery, (6) and (7) brachial, (8) and (9) inferior ulnar and radial collateral branches, (10) radial artery, (11) interosseous artery, (12) ulnar artery, (13) deep palmar arch, (14) superficial palmar arch, (15) digital arterioles. (From M. Salmon, 1936.)

the posterior aspect of the forearm, three sources of the skin supply are identifiable: the ulnar on the medial side, the posterior interosseous artery on the lateral side, and the anterior interosseous for the inferior portion of the forearm through the perforating branches. The region of the wrist does not possess a well defined arterial supply.

3. <u>Hand</u>. (Fig. 17) The palmar region receives its cutaneous supply from the radial through the radial palmar branch for the thenar region. The thumb and the lateral aspect of the index are supplied by the deep palmar arch.

The ulnar irrigates the rest of the thenar skin, either directly through the superficial arch or indirectly through the digital arteries (Fig. 15). The line of division between the radial and ulnar systems is represented by a line passing through the middle of the index finger. The line of division, however, is not constant since variations of the arteries of the palmar region of the hand are not infrequent.

On the dorsal region of the hand, the medial side is provided by the ulnar while the lateral is provided by the radial system, including the interosseous branches. The skin of the posterior surface of the last four fingers is irrigated by the digital arteries which originate from the ulnar except for the interdigital spaces (palmar webs) which is irrigated by the anterior interosseous branches. Therefore the line of separation between the two sources on the dorsal surface of the fingers is roughly the same as for the palmar area.

3. <u>Arterial Supply of the Skin of the Lower Extremity</u> (Fig. 18)

a. <u>Gluteal Region</u>. The skin of this region is supplied by the lumbar arteries, the superficial circumflex iliac and the gluteal arteries. The inferior gluteal artery provides the most distal portion of the skin of the gluteal region.

b. <u>Thigh</u>. On the lateral aspect, the skin is irrigated by the superficial circumflex iliac, the common branches from the common femoral and the pudendal arteries. Anteriorly the thigh skin is perfused by the superficial femoral by direct and indirect branches. The medial aspect is dependent on the deep femoral through the branches of the adductors. The deep femoral also provides some blood flow to the skin on the lateral aspect. The posterior aspect of the thigh is supplied by the perforating branches of the profunda and branches of the sciatic artery as well as the branches from the adductor vessels which, in turn, are branches of the deep femoral. Briefly, the profunda femoris through its muscular collaterals, or either through the perforating branches, provides the greatest part

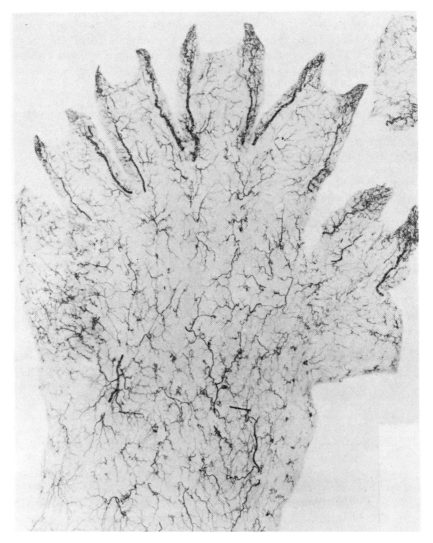

Fig. 17. Cutaneous arteries of the distal third of the forearm, hand, and digits. (From M. Salmon, 1936.)

of the arterial supply of the skin of the thigh. The common as well as the superficial femoral and their collaterals provide only a limited zone of skin irrigation.

   c. <u>Knee</u>. (Fig 19) Due to the great number of anastomotic branches surrounding the patella, it is difficult to identify with precision the various arterial branches in a given area of the skin. Anteriorly the lateral and middle portions of the knee are irrigated

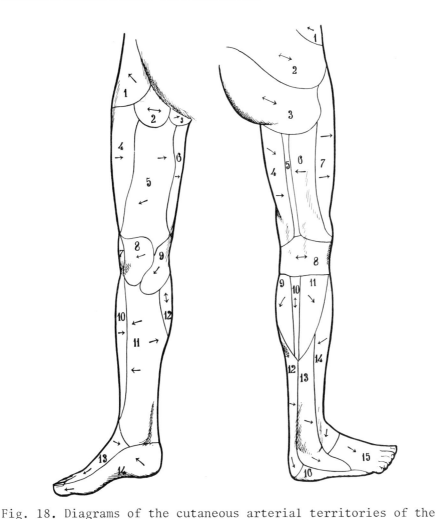

Fig. 18. Diagrams of the cutaneous arterial territories of the lower extremity, left anteromedial aspect, right postero-lateral aspect.
Left diagram: (1), (2), and (3) common femoral artery, (4) and (6) profunda femoris artery, (5) superficial femoral artery, (7) and (8) popliteal artery, (9) superior genicular artery, (10) anterior tibial artery, (11) posterior tibial artery, (12) branch of the gastrocnemius artery, (13) dorsalis pedis artery, (14) medial plantar artery.
Right diagram: (1) superior iliac artery, (2) superior gluteal artery, (3) inferior gluteal artery, (4), (6), and (7) profunda femoris artery, (5) branch of the sciatic nerve, (8), (9), (10), and (11) popliteal artery, (12) posterior tibial artery, (13) peroneal artery, (14) anterior tibial artery, (15) dorsalis pedis artery, (16) lateral plantar artery. (From M. Salmon, 1936.)

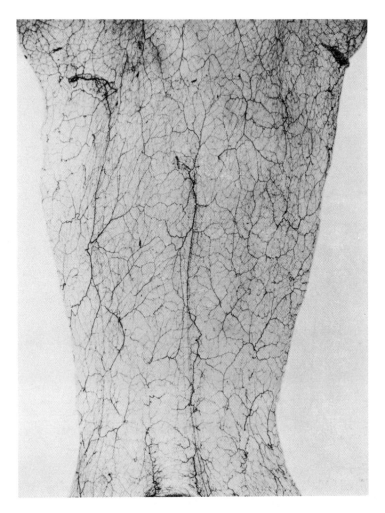

Fig. 19. Cutaneous arteries of the knee and leg. (From M. Salmon, 1936.)

by branches supplied by the popliteal, while the medial side is supplied by the highest genicular originating in the superficial femoral. Posteriorly all the branches originate in the popliteal artery.

d. <u>Leg</u>. The anterolateral aspect of the leg is irrigated by the anterior tibial branches, the peroneal provides the branches for the skin extending between the anterolateral and posterior surface of the calf. The posterior tibial artery provides blood to the skin between the crest of the tibia and the tendon Achilles.

   e. <u>Ankle</u>. The anterior tibial provides the irrigation of the skin of the anterior and lateral aspects of the ankle, the peroneal perfuses its lateral and posterior aspects and the posterior tibial

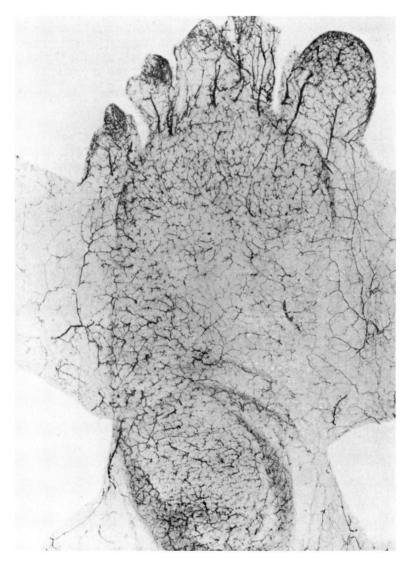

Fig. 20. Cutaneous arteries of the plantar skin without the underlying adipose plantar tissue. (From M. Salmon, 1936.)

artery cares for its posterior and medial aspect. All these territories are only an extension of the same territories of the leg. Their limits are not very precise.

f. Foot. The dorsum of the foot includes the branches supplied by the dorsalis pedis. Only a small area of the lateral aspect of the foot is supplied by the peroneal branches.

The plantar surface (Fig. 20) is divided between the medial and lateral plantar arteries which share the entire surface of the skin. The posterior tibial and the dorsalis pedis per se provide only a very limited irrigation of the skin, the former provides some circulation to the posterior surface of the heel and the latter irrigates only a space between the first and second toes.

REFERENCES

Adachi, B., 1928, "Das Arteriensystem der Japaner, "Kaiserlich Japanischen Universitat zu Kyoto, Kyoto.
Bardsley, J. L., and Staple, T. W., 1970, Variations in branching of the popliteal artery, Radiology, 94:581.
Barry, A., 1951, The aortic arch derivatives in the human adult, Anat. Rec., 111:221.
Blomfield, L. B., 1945, Intramuscular vascular patterns in man, Proc. R. Soc. M. Lond., 38:617.
Campbell, J., and Pennefather, C.M., 1919, An investigationto the blood supply of muscles with special reference to war surgery, Lancet, I:294.
Coleman, S., and Anson, B., 1961, Arterial patterns in the hand based upon a study of 650 specimens, Surg., Gynec. & Obstet., 113:4.
Daseler, E.H., and Anson, B.J., 1959, Surgical anatomy of the subclavian artery and its branches, Surg., Gynec. & Obstet., 108:149.
Edwards, E.A., 1953, The anatomic basis for ischemia localized to certain muscles of the lower limb, Surg., Gynec., & Obstet., 97:87.
Edwards, E.A., 1960, Organization of the small arteries of hand and digits, Am. J. Surg., 99:837.
Grant, J.C.B., 1962, "An Atlas of Anatomy," Ed. 5., The Williams & Wilkins, Co., Baltimore.
Gray, H., 1942, "Anatomy of the Human Body" (W. H. Lewis, ed.), Lea & Febiger, Philadelphia.
Haimovici, H., Steinman, C., and Caplan, L.H., 1966, Role of arteriovenous anastomoses in vascular diseases of the lower extremity, Ann. Surg., 164:990.
Haimovici, H., 1967, Patterns of arteriosclerotic lesions of the lower extremity, Arch. Surg., 95:918.
Haimovici, H., Hyperemic knee sign, in press, 1978.

Huelke, D.F., 1959, Variation in the origins of the branches of the axillary artery, Anat. Rec., 135:33.

Keen, J.A., 1961, A study of the arterial variations in the limbs, with special reference to symmetry of vascular patterns, Amer. J. Anat., 108:245.

Kiss, F. in G. Luzsa.

Luzsa, G., 1974, "X-ray Anatomy of the Vascular System," Butterworth and Co., (Publishers) ltd., London, and J. B. Lippincott Co., Philadelphia.

McCormack, L.J., Cauldwell, E.W., and Anson, B.J., 1953, Brachial and antebrachial arterial patterns: a study of 750 extremities, Surg., Gynec. & Obstet., 96:43.

Morris. G.C., Jr., Beall, A.C., Jr., Berry, W.B., Feste, J. and DeBakey, M.E., 1960, Anatomical studies of the distal popliteal artery and its branches. Surg. Forum, 10:498.

Paturet, G., 1958, "Traite d'Anatomie Humaine," Vols. 2-3, Masson et Cie., Paris.

Quiring, D.P., 1949, "Collateral Circulation-Anatomical Aspects," Lea dn Febriger, Philadelphia.

Rauber, A., and Kopsch, F., 1951, "Lehrbuch und Atlas der Anatomie des Menschen," (F. Kopsch, ed.) Vol. 2, Thieme, Leipzig.

Ricciardi, L., and Compostella, A., 1962, The arterial cutaneous circulation of the foot, Panminerva Med., 4:280.

Salmon, M., and Dor, J., 1933, "Les Artères des Muscles des Membres et du Tronc," Masson et Cie., Paris.

Salmon, M., 1936, "Artères de la Peau," Masson et Cie., Paris.

Saunders, R.L. de C.H., Lawrence, J., Maciver, D.A., and Nemethy, N., 1957, The anatomic basis of peripheral circulation in man, in "Peripheral Circulation in Health and Disease," (W. Redisch, F. F. Tangco, and R. L. de C. H. Saunders, eds.) Part V, Grune & Stratton, Inc., New York.

Senior. H.D., 1919, The development of the arteries of the human lower extremity, Am. J. Anat., 25:55.

BIOLOGY OF THE COLLATERAL CIRCULATION

D. E. Strandness, Jr.,

Professor of Surgery, University of Washington

School of Medicine, Seattle, Washington 98195

INTRODUCTION

Collateral vascular channels are those arteries and veins which are interposed as connecting links capable of bypassing areas of vascular occlusion. Their basic function is to carry blood in sufficient quantities to maintain the circulatory equilibrium of the part they serve. In the arterial system, most diseases which lead to the development of collateral circulation interfere with delivery of blood to the tissues, often resulting in either acute or chronic ischemia. The extent to which the collateral arteries can assume the role of the parent, diseased artery is dependent upon many factors which include the location of the obstruction, the rapidity with which the occlusion occurs, the size and distribution of the preformed collateral channels, and the metabolic requirements of the organ system.

On the venous side of the circulation, the collateral channels function to prevent the development of venous hypertension and its sequelae. Tissue viability is rarely a problem with venous occlusion. Thus the time it takes to reestablish venous function is usually not as critical a factor as it often is in the arterial circulation. However, many aspects of collateral development appear to be common to both circulations in terms of their ultimate function. One major difference which is of great importance on the venous side of the circulation and must be taken into account is the presence of valves. While there appears to be a great capacity for collateral development in the venous system, the end result is commonly valvular incompetence, a point of great importance in terms of the functional disability which commonly occurs.

ANATOMY OF COLLATERAL SYSTEMS

Those vessels which are called upon to assume the transport role in response to acute and chronic occlusion are nearly always preformed (Bellman et al., 1959). While there is no doubt that neogenic collaterals can develop in specific circumstances, the clinically important channels are present and able to function to varying degrees when the need arises.

The simplest and most efficient set of collaterals develops in response to a single area of occlusion where the bridging collaterals follow a short course, returning blood flow to the parent distributing channel immediately distal to the site of involvement (Fig. 1). It is useful for discussion purposes to classify the collaterals in terms of their three major components (Longland, 1953). The first is the stem vessel(s) which with its origin proximal to the diseased segment, distributes blood to the second major subdivision, the mid-zone. The mid-zone vessels are those myriads of tiny channels which are most critical in terms of both immediate and long-term function. When sequential arteriograms have been performed in experimental animals to follow the time-course of collateral development in response to acute arterial occlusion, these arteries are often not seen initially, indicating their very small size--probably less than 200µ in diameter (Cresti and Steger, 1962; Winblad et al., 1959). However, with time, the mid-zone vessels enlarge, occasionally reaching the size of 1-3 mm.

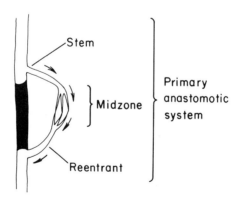

Fig. 1. Schematic diagram indicating the terminology applied to a collateral bypassing circuit.

BIOLOGY OF THE COLLATERAL CIRCULATION   489

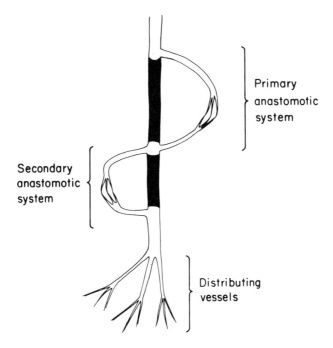

Fig. 2. When two obstructions are in series, more than one collateral bed must develop. The resistances of these two anastomic systems are additive.

The third subdivision is that vessel(s) which serves as the re-entry point back into the low resistance parent arterial trunk. Flow in the re-entrant channel is the reverse of that normally present. In most circumstances, particularly with a short bridging system of collaterals, the stem and re-entrant arteries are of the same approximate diameter. In physiologic terms, the mid-zone vessels offer the greatest resistance to flow as compared to the other two major components of the collateral system.

The situation is complicated when more than one segment of the vascular system becomes occluded (Fig. 2). Under these circumstances, a secondary anastomotic system is required. This adds greatly to the problem in terms of satisfying tissue requirements. With collateral systems in series, the resistances to flow for each anastomotic network are additive, leading to a further reduction in blood flow to the capillary bed (Learnmonth, 1950).

REGIONAL ANATOMICAL DIFFERENCES

Extensive experience with arterial occlusion, both acute and chronic, in the human has clearly indicated the anatomic differences which exist from one organ system to another. Indeed, this is observed from one muscle group to another in terms of the capability of the collateral circulation. Before considering some of the specific regional differences, there are some general concepts concerning regional collateral beds which are of great importance. Situations in which major branch vessels run a relatively parallel course to the parent trunk which is occluded provide the most efficient collateral pathways (Fig. 3). However, when the stem vessels arise and run at right angles to the involved major trunk, the anastomotic connections which ultimately develop in response to the occlusion are often very small, offering a much higher resistance to flow. Examples of parallel channels are best shown by the brachial and superficial femoral artery where acute occlusions result in a low incidence of limb loss. This is in contrast to the popliteal artery whose branches are small and arise at right angles to the vessel. Here the incidence of limb loss is high.

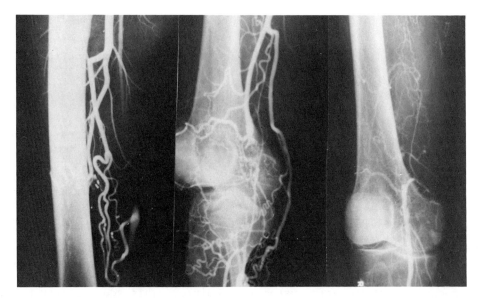

Fig. 3. Examples where the major collateral pathways follow a relatively parallel course to the parent trunk. The mid-zone vessels in the collaterals in the right panel are much smaller and less efficient in providing blood flow to the limb.

Individual muscles vary greatly in the nature of their arterial blood supply and their ability to withstand acute arterial occlusion and ischemia. The arterial blood supply to the gastrocnemius is largely from a single blood vessel, the sural artery which takes its origin from the popliteal (Fig. 4). The soleus muscle, on the other hand, is usually supplied from five vessels entering the muscle successively and contributing to a longitudinal anastomotic chain within the muscle itself. Blomfield (1945) has discussed this matter, showing that there are at least five basic patterns represented which are of importance in appreciating the basis for the variable responses which can develop when the blood supply to a muscle group is acutely occluded: (1) a longitudinal anastomotic chain formed from a succession of vessels entering it throughout the length of the muscle; (2) a longitudinal distribution derived from a single artery entering one end of the muscle; (3) a radiating pattern arising from a single vessel entering the mid-portion of the muscle; (4) a pattern formed by a series of anastomotic loops throughout the muscle originating from several entering vessels; and (5) an open quadrilateral pattern with sparse anastomoses.

It is also important that the collateral circulation supplying an ischemic area can come from connecting branches of the artery which is diseased, or from anastomotic links with other arteries in the area (Fig. 4). An example of the first case is the profunda femoris artery which connects with the geniculate arteries to bypass disease of the superficial femoral artery (Fig. 3). The second example is exemplified by one internal carotid artery supplying the contralateral cerebral hemisphere via the circle of Willis when the ipsilateral carotid becomes occluded.

Structural Changes

As collateral arteries develop in response to disease, there are definite structural changes in the vessel wall (Schaper, 1967). The most important adaptation to occur is an increase in the cross-sectional area which decreases the resistance to flow. The magnitude of the change is, of course, variable but vessels less than 100μ may enlarge up to a few millimeters in size. The time required for this to occur depends upon many factors which are not yet fully understood.

The major changes occur in those segments of the collateral bed in the mid-zone. As the vessels progressively increase their cross-sectional area, there is initially a thinning of the media with gaps appearing in the internal elastic lamina (Liebow, et al., 1959; Longland, 1953). During the latter stages of maturation, the media again increases in thickness and there may appear aggregates of smooth muscle that run longitudinally, dissecting beneath or replacing the internal elastic lamina. These latter changes have been seen in the heart but apparently occur most extensively in the lung when

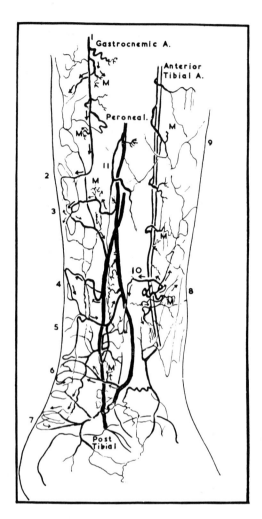

Fig. 4. Diagram made from a stereoarteriogram of an intact leg. The segmental arteries which arise at nearly right angles from the major trunks, divide into a muscular branch (M) or end in the subcutaneous tissue. The segmental arteries branch extensively to intercommunicate with other segmental branches at proximal and distal levels. The arrows indicate potential directions of blood flow. (From Saunders, R.L. de C.H., Lawrence, J., Macirer, D.A., and Nemethy, H. in Redisch, W., Tangco, F.F., and Saunders, R.L. de C.H.: Peripheral Circulation in Health and Disease. New York, Grune and Stratton, Inc., 1957.)

collateral circulation via the bronchial arteries develops (Loring and Liebow, 1954). Another interesting facet of the structural changes is that there may be some lengthening of the collaterals so that they appear to follow a spiral course. This phenomenon also begins in the mid-zone, progressively extending distally until the entire length of the collateral efferents is involved.

PHYSIOLOGY OF COLLATERAL DEVELOPMENT

It is clear that collateral arteries, while largely derived from pre-existing channels, do not begin to change unless there appears to be a need for them to do so. The key and yet unsolved questions are: how do they get the message and what are the factors responsible for bringing about the well-known anatomic changes that take place (Liebow, 1962). In the case of acute arterial occlusion, the ability of the part to survive is dependent nearly entirely upon the cross-sectional area of the bridging collaterals as well as the availability of blood from other uninvolved arteries in the vicinity which have pre-existing intercommunications. The ischemia produced is variable and dependent upon the location and extent of the occlusion and the collateral capacity. If the part survives the initial anoxic insult, the collaterals will undergo progressive changes which may require weeks to months to reach a final stable level.

If a major artery is gradually narrowed, the collateral arteries have time to develop in response to the need. At this point, it is important to consider in some detail the hemodynamics of arterial narrowing and occlusion because this has relevance to our understanding of the physiology of collateral development. It has been recognized for a long period of time that if an artery is progressively narrowed, there is a point at which there is a drop in blood flow in the artery across the stenotic segment. Corresponding to the decrease in flow, there is a fall in pressure distal to the narrowed segment resulting in an abnormal pressure gradient (ΔP). This pressure gradient (ΔP) and the factors which produce it can be expressed in the following formula (May et al., 1963):

$$\Delta P = \frac{8\eta L}{r_1^2} V_1 \cdot \left(\frac{A_1}{A_2}\right)^2 + \frac{4 \cdot 8\eta}{r_1} V_1 \left(\frac{A_1}{A_2}\right)^{1\ 1/2} + \rho V_1^2 \left(\frac{A_1}{A_2}\right)^2$$

A_1 and R_1 = cross-sectional area and radius of the pre-stenotic segment

A_2 = area of the stenotic segment η = viscosity

L = length of stenosis ρ = density of blood

V_1 = velocity in the pre-stenotic segment

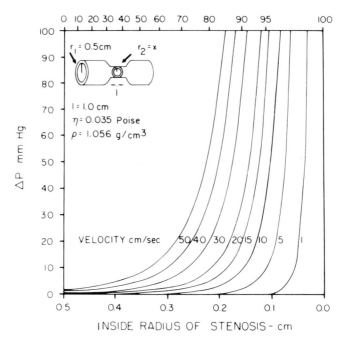

Fig. 5. Relationship of pressure drop across a stenosis to the radius of the narrowed segment. Curves are based upon the formula of May et al., 1963. See text for explanation. (From Strandness, D.E., Jr., and Sumner, D.S.: Hemodynamics for Surgeons. Grune and Stratton, Inc., New York, 1975.)

From this formula, it is possible to generate curves representing these factors. If, as shown in Fig. 5, the prestenotic segment is assumed to have an inside diameter of 1.0 cm, the vessels' cross-sectional area must be reduced by 90-95% at a flow rate of 3.9 cm^3/sec before the pressure gradient (ΔP) increases. However, if the flow rate is increased to 39 cm^3/sec (a ten-fold increase), a stenosis of 60% will produce an appreciable pressure drop. Thus, the need for collateral artery function is determined, not simply by the presence of a stenosis, but upon those factors such as resting flow rate and the changes brought about by such situations as exercise. For example, a narrowed segment in an iliac artery may not be significant under resting conditions where limb blood flow is low, but when the exercising muscle requires more blood, the stenosis may become "hemodynamically significant" at the higher flow rates. The above simple formula is, of course, incomplete and refers largely to a model system. It does not take into account the effects of changes in sympathetic tone, i.e., resistance to flow distal to the site of narrowing. It is true that in the slowly developing arterial stenosis seen with atherosclerosis, the resistance of both the distal arteries and arterioles, as well as the resistance of the collateral arteries, tends to decrease, ensuring that the total resting blood flow will remain relatively constant.

Once the pressure and flow, either at rest or with exercise, begin to be affected, what are the factors that begin to act on the collateral channels to increase their cross-sectional area, thus reducing their resistance to flow? Those factors which have been considered as important in influencing collateral artery function can be subdivided into mechanical, chemical, and neural.

Mechanical Theory

It is well established that arterial obstruction leads to the development of an abnormal pressure gradient across the involved segment which is a direct result of the high resistance offered by the collateral arteries. As blood to the limb increases via the collaterial arteries, the pressure gradient diminishes until it ultimately stabilizes at some unpredictable level. It is also true that the major changes which occur involve the collateral arteries themselves and not in the parent distributing arteries which are the site of the problem.

Because of the very good correlation between the sequential anatomic changes in the collateral vessels and the rise in pressure distal to the involved segment, some authors have concluded that the pressure gradient itself is the major factor which stimulates the growth of the collateral arteries (Volpel, 1959). John and Warren (1961) concluded that proof of a causal relationship would be fulfilled if the following conditions were met:

(1) The rate of return of blood flow to a limb paralleled the return in pressure;

(2) The return in distal pressure was related to the arteriographic appearance of the collateral channels;

(3) Other factors are shown to be unimportant.

The evidence that (1) and (2) are true is indeed very strong. Winblad et al., (1959) measured simultaneous pressures from the anterior tibial and common femoral arteries of dogs before, during and after acute occlusion of the superficial femoral artery. Tissue PO_2 was also measured from the calf muscles during the same course of events. The above physiologic variables were then correlated with serial arteriograms performed to document the anatomic changes in the collateral arteries. With acute occlusion, the distal arterial pressure fell to between 35 and 60 mm Hg with a marked decrease in the

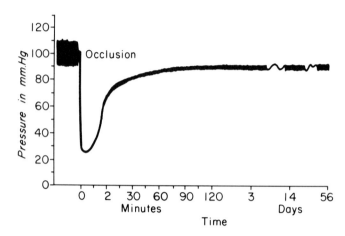

Fig. 6. Time-course of blood pressure return in the anterior tibial artery of the dog distal to a femoral artery occlusion. (Redrawn from John, H.T., and Warren, R.: The stimulus to collateral circulation. Surgery 49:14, 1961.)

pulse pressure. The tissue PO_2 dropped to between 84 and 95% of normal but rapidly increased to the preocclusion level within five to eight minutes. The opening of collateral channels, particularly in the mid-zone as documented arteriographically, correlated very well with the rise in pressure recorded from the anterior tibial artery.

John and Warren (1961, in a similar set of experiments, found that the increase in size as well as the number of collaterals bypassing an obstruction of the superficial femoral artery correlated well with the rise in pressure in the anterior tibial artery (Fig. 6). Calf blood flow was also measured plethysmographically, examining both resting and post-hyperemic flow levels (Fig. 7). Resting flow levels were restored to normal between three and nine days, but even up to four months, the hyperemic response never returned to normal.

Perfusion of the limb with the occlusion with both oxygenated and hypoxic blood at the preocclusion pressure level resulted in either a diminution or disappearance of the collaterals which appeared during the acute phase of the obstruction. When the limb was amputated at the knee level, the pressure recovery distal to the block was unaffected during the subacute or chronic recovery phase.

Keenan and Rodbard (1973) have presented another theory of collateral development based upon hydrodynamic drag on the endothelium. The hypothesis was suggested by the observation that when two arteries supply the same tissue segment, one vessel often tends to regress.

BIOLOGY OF THE COLLATERAL CIRCULATION

It is suggested that the endothelial cells have mechanoreceptors that are sensitive to drag. The rate of change in radius may be expressed as a first approximation as follows:

$$\frac{dr}{dt} = K(\tau - \tau_c)$$

r = vessel radius
K = endothelial sensitivity
τ = existing endothelial shear stress
τ_c = critical or preferred shear stress

As the shear stress increases, there is relaxation in the smooth muscle which thus reduces the velocity and the shear stress.

There is no direct evidence available to support this hypothesis, but there is indirect supporting evidence that it may play a role. In the case of long-standing arteriovenous fistulas, the flow through the proximal feeding arteries remains higher than normal. This is in contrast to the situation with arterial obstruction where the flow

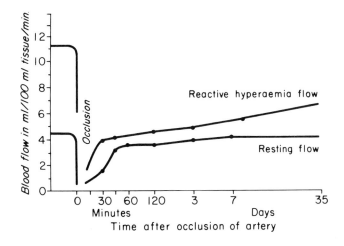

Fig. 7. Pattern of return of mean blood flow measured plethysmographically in six dogs with a femoral artery occlusion. (Redrawn from John, H.T., and Warren, R.: The stimulus to collateral circulation. Surgery 49:14, 1961.)

rate, i.e., shear rate, in the artery feeding the stem vessel is usually within a normal range, particularly at rest. If the arteriovenous fistula is not repaired, there may be progressive dilatation of all the arterial elements down to and including the fistula and stem arteries (Holman, 1937; Ingebrigtsen et al., 1960; Reid, 1925). The pressure gradient theory does not, of course, explain this interesting phenomenon.

A complicating feature which remains to be explored, however, in the case of arteriovenous fistulas is the effect of turbulence on the arterial wall of those vessels feeding the fistula. It is well known that bruits are present in the arteries leading into the fistula which represent arterial wall vibration. While it is not known for certain, these arterial wall vibrations may, with time, result in damage to the major structural elements in the arterial wall leading to dilatation.

Schaper (1967), in a careful study of the development of canine coronary collaterals, has related the morphologic features of changes in the collaterals to the effect of tangential wall stress. With arterial occlusion, there is hypoxic arteriolar dilatation resulting in an increase in the tangential wall stress on that vessel according to the Law of Laplace: $T = P \cdot \bar{R}$ where T is tension, P is pressure and $\bar{R}$ equals the mean radius. This law is applicable only to those structures where there is a very small wall thickness to radius ratio. Since the wall thickness cannot be ignored in the determination of tension, it is appropriate to modify the equation to the following:

$$T = \frac{\bar{R} \cdot P}{d_c \cdot n}$$

here, d_c is the diameter of the smooth muscle cells

n = number of smooth muscle layers

This formulation appears to fit well with the morphologic changes that are known to occur. During the early phase after arterial occlusion, the diameter change is much greater than the vessel wall volume, i.e., the tangential wall stress is high. During the second phase of collateral development, mitotic figures appear in the smooth muscle, adding more layers to the media which would tend to reduce the stress on the wall. During this period of time, the cross-sectional area of the collateral remains stable.

The effect of the increase in tangential wall stress during the early phase of collateral dilatation could also explain the fragmentation which occurs in the internal elastic membrane. However, the appearance of the longitudinal muscle layer that often dissects or replaces the internal elastic lamina cannot be explained by this theory. In the dog heart, this phenomenon is marked by the eighth week, but surprisingly after this time, the longitudinal smooth muscle tends to become much less prominent and may even disappear.

Chemical Theory

It is an attractive hypothesis that as a result of arterial occlusion some substance(s) is released in greater than normal amounts which may participate in the progressive enlargement of the collateral channels (Lewis, 1939). This could occur by regulation of vasodilation, controlling the proliferation of new vessels, stimulating the growth of vessels and guiding the collaterals to their needed location.

It is important to remember that collateral development can be described in two phases--acute and chronic. Metabolically, the greatest changes take place when the arterial supply to a part is acutely interrupted. There is no doubt during this time that the PO_2 goes down, the PCO_2 increases and the ph decreases. These changes occur most dramatically in the tissues farthest removed from the site of the occlusion. This is clearly evident when an arm or leg is seen immediately after an occlusion. The most distal tissues are always the most ischemic, with the tissues in the immediate vicinity of the collaterals appearing normal.

It is known that an anoxic insult will result in vasodilatation (Carrier et al., 1964) either due to the reduced PO_2, the appearance of vasodilator substances or both. It is also possible that the development of collaterals, particularly to newly-formed tissues, may be, in part at least, due to chemical factors. This is most evident in organs with a dual blood supply, such as the lung (Weibel, 1960). Here the blood supply of primary malignant tumors is from systemic vessels and not the pulmonary arteries. Also, in cases where granulation tissue is formed in response to a disease process such as bronchiectasis (Liebow et al., 1949); Marchand et al., 1950), the bronchial arteries, not the pulmonary arteries, in large part are the primary supplier of blood to the area.

Hormones may also play a role in the development of collaterals, particularly in newborn animals, such as puppies with pulmonary artery ligations (Meffert and Liebow, 1966). In the rat, there is evidence that with pulmonary artery ligation, the initial dilatation of the collaterals is followed by a phase of active proliferation of new vessels which are attracted to the tissues distal to the ligation. Whether or not this represents a hormonal influence or other chemical factors remains unsettled at the moment.

Neural Factors

Neural factors, particularly as related to lysis of sympathetic tone, could have marked influences on limb blood flow by decreasing further arteriolar resistance distal to the occlusion as well as promoting vasodilatation in the collateral arteries themselves (Ferris and Harvey, 1924; Longland, 1953). Evidence for support of this theory is also conflicting.

Shepherd (1950) studied calf blood flow before and after infusion of tetraethylammonium bromide in normal subjects while flow was interrupted by external occlusion of the femoral artery in the groin. Calf blood flow was measured after exercise in the normal state, with the femoral artery occluded and before and after drug infusion. Normally, exercise increased blood flow to 35 ml/100 ml/min. After femoral artery occlusion, the same exercise resulted in flows in the five to seven ml/100 ml/min range. With tetraethylammonium, flows were measured at 13 to 20 ml/100 ml/min. The supposition was that the increase in flow occurring with the drug was attributable to a direct action on the collateral arteries.

Dornhorst and Sharpey-Schafer (1951) measured limb blood pressure and flow in human lower extremities distal to sites of chronic arterial obstruction. These authors found that collateral artery resistance decreased following sympathectomy in the ten patients studied, but it was very transient in seven, rapidly returning to the preoperative level.

Thulesius (1962) studied the effect of sympathectomy on skin and muscle blood flow and blood pressure in cats with acute occlusion of the femoral artery. With sectioning of the sympathetic chain, skin blood flow increased markedly, with muscle flow going up only slightly. The distal blood pressure rose slightly, reflecting the dilatation of the collateral arteries. The collateral resistance was reduced to 70% of the presympathectomy level.

These results are at variance with the studies of Rosenthal and Guyton (1968) who evaluated the hemodynamics of collateral artery vasodilatation in the dog. These authors studied the effects of acute occlusion of the femoral artery by measuring the pressure and flow changes in the anterior tibial artery and pressure in the femoral artery proximal to the occlusion. The following formula was used to calculate collateral conductance:

$$C_c = \frac{F_{at}}{P_b - P_{at}}$$

C_c = collateral conductance
F_{at} = flow in the anterior tibial artery
P_{at} = pressure in the anterior tibial artery
P_b = pressure in the femoral artery

Collateral conductance reached a minimum within eight to ten seconds following occlusion, but rose 277 ± 67%* from this low value during the first 70 seconds and 178 ± 66% during the ensuing hour. To study the effects of sympathetic denervation, the animals were rendered totally areflexic by total spinal anesthesia. This did not alter the course of vasodilatation of the collateral channels. It is of interest, however, that when the systemic arterial blood pressure was lowered to 45 mm Hg for eight minutes prior to the femoral artery occlusion, the collaterals did dilate before the obstruction was instituted. This certainly suggests that anoxia is, in some way, related to collateral vasodilatation.

Comment

It is not feasible to draw any certain conclusion regarding the major factor(s) that control collateral artery development. This is difficult for many reasons, some of which are as follows: (1) the mechanisms may be different in various vascular beds; (2) some factors may be operating at different times during collateral artery development. For example, the time period immediately following acute occlusion may have different factors operating than during the ensuing days to months--a period in which changes are still known to occur; and (3) it is clearly possible that there are multiple factors operating, each of which contributes to varying degrees in collateral artery dilatation.

REACTIVITY OF COLLATERAL CHANNELS

Another method of evaluating the functional capability of collateral arteries is to study their response to vasoactive agents. Coffman (1966), in one of the most extensive studies of this type, evaluated the following in dogs with both acute and chronic arterial ligations of hindlimb arteries: (1) the effect of exercise simulated by electrical stimulation; (2) the changes that occurred with infusions of bradykinin, norepinephrine bitartrate or epinephrine given intravenously; (3) the changes occurring with elevation or depression of systemic blood pressure; and (4) the effect of reactive hyperemia

---

*Standard error of the mean.

produced by 60 seconds of aortic occlusion. Collateral blood flow to the limb was considered as equal to the venous outflow. Collateral artery resistance was calculated as the pressure difference across the collateral channels divided by the venous outflow which was measured with a rotameter.

Coffman (1966) found in these studies that exercise, intra-aortic and intra-arterial bradykinin, increasing systemic blood pressure and transient aortic occlusion, all produced large increases in collateral blood flow. The norepinephrine given intravenously produced vasodilatation probably as a reflex sympathetic response to the rise in systemic blood pressure. Epinephrine had an effect by possibly two modes of action--increasing systemic blood pressure and direct vasodilation of skeletal muscle blood vessels. Isoproterenol produced similar increases in collateral flow but these were generally not as great. When epinephrine and norepinephrine were administered distal to the arterial occlusion, collateral blood flow decreased. Reducing systemic blood pressure produced an increase in collateral artery resistance. The demonstration that transient aortic occlusion producing reactive hyperemia resulted in an increase in collateral blood flow shows that the collaterals are sensitive to an ischemic procedure. The reactive hyperemia showed a lower peak flow and required a longer time to return to the baseline levels. Coffman (1966) interpreted the data as demonstrating that collateral arteries react to vasoactive stimuli in a manner similar to other blood vessels in the limb.

Thulesius (1962) studied the effects of various vasodilator drugs on collateral arteries with obstruction of the femoral artery in the cat. It was found that intravenous administration had little, if any, effect on collateral blood flow and, indeed, if the drug led to a decrease in systemic blood pressure, this often led to a fall in limb blood flow via the collateral channels.

NEOGENIC COLLATERALS

While it is true that most collateral channels which become functionally important are pre-existing vessels, there are situations in which both new arteries and veins must develop. The most obvious situation in which this occurs is with transplanted tissue, such as pedicle flaps where after an appropriate time interval the parent source of arterial flow is severed, yet the flap survives by virtue of the new vascular connections which have taken place.

Presumably when tissue is transferred to a new location with its blood supply intact, the initial communications take place at the capillary level. The process occurs as the capillary beds from the transplant grow to join those of the recipient tissue. This provides the initial neogenic mid-zone (Dollery et al., 1967; Dollery 1968;

North et al., 1960; Virkkula et al., 1971). From this capillary network small arteries and veins form which will enlarge to reach sizes of approximately 100μ on the arterial side and 150μ on the venous side (North et al., 1960).

One of the most interesting and intensive efforts to develop new collateral channels has been in the heart. Dr. Claude Beck was one of the earliest pioneers in this field, demonstrating that it was indeed feasible to promote the formation of neo-collaterals between the coronary arteries and extra cardiac sources such as the internal mammary artery which was implanted into the myocardium (Beck et al., 1935; Mautz, 1957). In fact, the observations that such collateral connections can develop led to the widespread application of internal mammary implantation into the myocardium for the treatment of coronary artery insufficiency (Vineberg et al., 1955). This operation has been largely superseded by the more physiological and direct bypass grafting techniques which are used today.

The factors responsible for the development of neo-collaterals are even less certain than the situation which exists with the preformed bypassing channels.

IMPORTANCE OF COLLATERALS IN CLINICAL MEDICINE

The preceding discussion has dealt primarily with factors thought to be of importance in the development of collateral arterial channels in response to both an acute and chronic occlusion. Since vascular obstruction, both on the arterial and venous side of the circulation, constitutes the single most important end-point of most vascular disorders, interest in the function of the collateral circulation is of more than academic interest. Two examples are sufficient to emphasize the importance of the problem. First, atherosclerosis is the most common arterial disorder of the Western World, affecting to varying degrees nearly every major artery in the body. While the etiology and pathogenesis of this disorder remains obscure, the end-point or end-stage of the disease is represented by total occlusion of the affected arterial segment. Second, acute venous thrombosis (thrombophlebitis) is the most serious, life-threatening, disabling disorder on the venous side of the circulation which also has as its end-point complete venous occlusion.

In both of the situations mentioned, the survival and function of the part depends to a great extent upon the ability of the preexisting collateral channels to meet the newly imposed demands upon them. It is now recognized that there is a collateral potential that exists for each organ and tissue in the body. Thus, it is not reasonable to attempt to generalize concerning the collateral circulation without taking into account both the anatomy and function of the organ system which may be the site of the disease. While it is not feasible to cover every organ system in the body, a brief review of those which are of greatest clinical importance will be presented.

THE HEART

Atherosclerosis is the most common vascular disorder affecting the coronary arteries. Stenosis or occlusion of one or more of the coronary arteries may lead either to sudden death or damage to the myocardium which produces severe chronic disability in the form of angina pectoris, congestive heart failure, or both. Thus, it would appear that the circulation to the heart would produce a good model both experimentally and clinically to evaluate the role of the collateral circulation in providing protection to an area of myocardium supplied by a diseased artery. While there is no doubt that collaterals do develop in the heart in response to arterial narrowing and occlusion, there is ongoing debate as to their functional effectiveness in providing protection to the myocardium.

Before entering into a discussion of existing data, it is important to state where the problems appear to be in the debate over coronary collaterals and their significance.

Accepted Facts

(1) Areas of myocardium supplied by a narrowed or occluded coronary artery may be supplied by three potential collateral sources (Baroldi et al., 1956):

   (A) Homocollaterals - connecting or bridging vessels of the same artery

   (B) Intercoronary collaterals - connecting branches from other major coronaries or their tributaries

   (C) Extracoronary collaterals - vascular connections can develop between extracoronary sources and the coronary arteries. However, these vessels are not considered to be important in the protection of the myocardium.

(2) The collateral channels which do develop in response to ischemia are pre-existing channels (Baroldi and Somazzon, 1967).

(3) Those collaterals which ultimately appear in response to arterial narrowing are not normally visible angiographically, and functionally are considered to be of little or no importance (James, 1961; Gensini and da Costa, 1969).

(4) When a coronary artery becomes significantly narrowed (greater than 75% of luminal area), collaterals will begin to be seen angiographically (Harris et al., 1972).

(5) As the number of coronary arteries involved increases, there is a progressive increase in the number of collateral arteries visualized (Helfant et al., 1970).

(6) The appearance of collaterals does not prevent the development of exercise produced myocardial ischemia as measured by electrocardiographic ST changes (Helfant et al., 1971).

(7) There is as yet no convincing evidence that collaterals are able to improve or prevent the development of localized myocardial asynergy (Helfant et al., 1971).

(8) There is as yet no data to support the observation that the hemodynamic performance of the left ventricle is related to the presence or absence of collateral channels.

The above facts would appear to warrant the conclusion that collaterals can and do develop in response to arterial narrowing, but probably play a minimal role in either protecting the myocardium from acute occlusion or permitting better myocardial performance in the area involved. How can this be the case? If collateral channels have no influence on the course of events, then every area of myocardium served by an occluded coronary artery would become infarcted, which we also know is not the case. Obviously, the myocardial segments which remain viable must derive their life support from somewhere and that has to be from either homo- or intercoronary collateral channels.

Factors not yet considered but of obvious importance in assessing the functional role of coronary collaterals include the following:

(1) the size and location of the coronary artery which becomes diseased, i.e., the amount of myocardium supplied by the arterial segment;

(2) the rapidity with which the artery becomes narrowed and occluded;

(3) the physiologic circumstances under which collateral artery function is measured. It may well be that collateral flow is adequate to meet the needs of the myocardium under resting conditions but not in response to the stress of exercise.

While final answers to the problems posed by the coronary circulation are not yet in, there is considerable experimental and clinical data which provides at least partial answers to some of these questions.

Experimental Studies

It is possible to demonstrate by arteriography the presence of new, alternate channels which bring blood to segments of the myocardium distal to areas of occlusion. These studies are primarily anatomic and do not provide physiologic data with regard to the effect of this new blood supply on myocardial performance. How, then, can collateral function be described physiologically in the heart?

The extent to which collaterals develop in response to acute, subacute or chronic coronary artery occlusion can be assessed by a variety of physiologic studies which include: (1) pressure in the coronary system distal to the occlusion (Schaper, 1967); (2) coronary blood flow available to the ischemic area as estimated by measured retrograde flow when the artery is exposed to atmospheric pressure (Gundel et al., 1970); (3) myocardial blood flow as estimated by isotope clearance techniques or distribution of microspheres injected into the coronary circulation (Becker and Pitt, 1971; Johansson et al., 1964; Rees and Redding, 1967); (4) quantitative assessment of ventricular performance as related to the degree of collateralization (Helfant et al., 1970); and (5) quantitative examination of infarct size as a function of the collateral circulation (Schaper et al., 1969).

One of the most complete studies performed with regard to the use of back pressure in the coronary arteries distal to areas of occlusion was by Schaper (1967). This investigator has convincingly demonstrated that coronary artery pressure distal to an obstructed artery is a reflection of the total resistance to flow, i.e., collateral artery resistance bringing blood to an ischemic area of myocardium. It is clear that flow through the collaterals results in the development of a pressure gradient which is a reflection of the collateral artery diameter and length. It is also true, of course, that pressure in an arterial conduit can be influenced by the level of the arteriolar resistance. The extent to which the late diastolic back pressure is related to time after gradual occlusion of a coronary artery in the dog is shown in Fig. 8.

Schaper (1967) applied an ameroid-constrictor to the circumflex branch of the left coronary artery which resulted in an occlusion between the second and third week. As a result, 34% of the dogs died suddenly, with 17% demonstrating a myocardial infarction but surviving the insult. Forty percent of the animals survived without myocardial infarction. Collateral artery enlargement became apparent just prior to the total occlusion. Coronary collaterals, as estimated by the measurement of back pressure, proceeded at about 10 mm Hg/day beginning to level off at the beginning of the third week. During the period of collateral transformation, vessels the size of arterioles increase their radius by about ten times and the vessel wall volume by 50 which can come about only by cellular hyperplasia in the media.

BIOLOGY OF THE COLLATERAL CIRCULATION 507

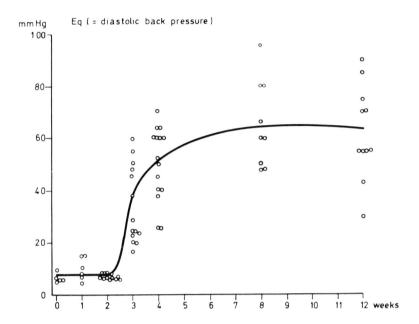

Fig. 8. Time-course of the rise in diastolic back pressure in dogs
with arterial stenoses and occlusions. (From Schaper, W.:
The Collateral Circulation in the Canine Coronary System.
ⓒ Janssen Pharmaceutica NU, Holland, 1957.)

This growth process was substantiated by the demonstration of mitosis
in the smooth muscle cells. This excellent study by Schaper (1967)
demonstrates clearly that collateral artery function may not be suf-
ficent to prevent death or myocardial injury with a constrictor on
the circumflex artery in slightly over 50% of the dogs studied. However, other animals may be protected from myocardial damage presumably by the collaterals which developed in response to the occlusion.

While measurements of pressure in the coronary arteries distal
to the area of occlusion would appear to be a valid indicator of the
amount of resistance offered by the collateral channels, other methods
have been used experimentally to evaluate the functional characteristics of collaterals in the heart. Rees and Redding (1967) tied the
anterior descending artery in dogs and injected Xe^{133} into the ischemic area via a fine catheter over a period of ten days. With
acute occlusion, the clearance rate averaged 25% (10-60%) of the normal level. By the tenth day, the flow reached the preocclusion level.

Eliot et al., 1971, in a similar study, also found that the clearance of Xenon also rose to the control levels when the circumflex artery was progressively constricted over a period of three to four days. The pressure distal to the constriction also rose to four times the maximum decrease.

Becker and Pitt (1971) estimated collateral blood flow in conscious dogs with chronic coronary artery occlusion using radioactive microspheres. In eleven dogs with complete ameroid occlusion of the left circumflex coronary artery, six were found to have infarcts, with the remaining five being grossly normal. The ratio of radioactivity in the left circumflex (LC) distribution to the left anterior descending (LAD) was the same in both groups, suggesting equal blood flow in spite of complete chronic occlusion. Their conclusion was that extensive intercoronary anastomoses developed. The failure to find a low LC/LAD ratio may be explained by the small size of the apparent infarcts.

In the human, the problems in evaluating the presence and, more importantly, the functional states of coronary collaterals are formidable. The advent and widespread application of coronary arteriography has permitted localization of areas of coronary artery narrowing and occlusion as well as the visualization of those collateral channels which are of sufficient size to be seen with the technique. Harris et al., 1972, examined 181 patients angiographically who had coronary artery disease and found that 39 (21.5%) had demonstrable collaterals. It was also noted that collaterals did not become apparent until the stenosis exceeded 75% of the diameter. Also, as the extent of the disease (from one to four vessels) increased, the collaterals became more evident. However, they also concluded that the appearance of collaterals per se did not appear to protect the patient against myocardial ischemia.

Helfant et al., (1970, 1971) attempted to relate the effect of intercoronary artery collaterals on some of the clinical features of the disease and included (1) physical activity; (2) duration of angina; (3) frequency of prior myocardial infarcts; (4) electrocardiographic changes; and (5) hemodynamic and ventriculographic abnormalities. There were no apparent differences found in the 61 patients with visible collaterals as compared to the 58 without obvious collateral circulation. Their data suggested, but did not prove, that the presence of collaterals may decrease the mortality rate.

Goldstein et al., (1974) estimated the immediately available potential collateral that existed in seven patients without coronary artery disease at the time of aortic valve placement. These investigators measured peripheral coronary pressure (PCP) and retrograde flow (RF) from the right and left coronary arteries during periods of brief occlusion of each artery. PCP was expressed as a fraction of perfusion pressure (PP) and collateral resistance (CR) calculated as PP/RF. The results were as follows:

Occluded Perfuser	RF (ml/min)	CR	PCP	PCP/PP
LCA	2.4	48	15	0.19
RCA	1.7	64	16	0.24

These data show that collaterals in the absence of disease have an extremely limited capacity to transmit flow or pressure. It is also of interest that administration of nitroglycerin showed no improvement in collateral flow.

Goldstein et al., (1974) also measured RCF and PCP at operation in patients undergoing aortocoronary bypass grafting. A total of 29 patients were evaluated with regard to the above parameters before and after nitroglycerin. The results were as follows:

Baseline		After Nitroglycerin
Mean aortic pressure	79 mm Hg	↓ 18%
PCP	30 mm Hg	↑ 9.9% ($P<0.02$)
RF	2.7 cc/min	Not significant
CR	28.5 mm Hg/ml/min	↓ 28% ($P<0.05$)

It appeared that nitroglycerin diminished collateral artery resistance despite severe multivessel involvement. They also related the functional performance of the collaterals with the angiographic results and found a good correlation. Those patients with no visible collaterals on angiography had the least evidence of collateral function (CR 94 mm Hg/ml/min). In contrast, those patients with maximal coronary collaterals had the lowest collateral resistance (5.1 mm Hg/ml/min).

THE BRAIN

The brain is a unique organ which has four major sources of arterial input, all of which intercommunicate by the Circle of Willis if it is intact. Furthermore, the resistance vessels within the central nervous system are not subject to the same neurohumoral influences as the peripheral arteries. From a theoretical standpoint, the anatomic arrangement of the distributing arteries at least would appear to be ideal in terms of compensating for disease of one or more of the major arterial inputs. The carotid and vertebral arteries are relatively large vessels which can accommodate high flow rates when called upon to do so.

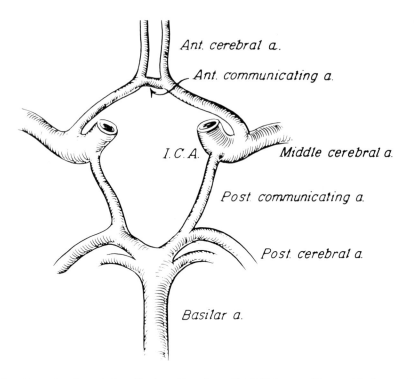

Fig. 9. Standard model of the Circle of Willis. (From Alskne, J.F.: Normal arterial system in Strandness, D.E., Jr. (Ed) Collateral Circulation in Clinical Surgery. W.B. Saunders Co., Philadelphia, 1969.)

Anatomically, it is important to consider the potential sources of supply when either the extracranial or intracranial arteries become diseased. The distribution of blood flow via alternate channels can be considered in terms of their origin and distribution.

The first and most readily available collateral is by vessels within the cranial cavity itself. The most obvious and theoretically important connecting system is the Circle of Willis. The standard model of the Circle is shown in Fig. 9. If the Circle is intact and if the vessels are of sufficient size, it should provide the best potential for collateral flow between the two internal carotid arteries and between the basilar and either the right or left carotid.

In a study of 700 autopsy specimens, Riggs and Rupp (1963) found that the components of the Circle were rarely absent but the

BIOLOGY OF THE COLLATERAL CIRCULATION 511

system was frequently incomplete functionally secondary to hypoplastic vessels. Some of the more common variants are shown in Fig. 10. In 62%, the Circle contained hypoplastic segments with the following distribution: (1) anterior communicating - 9.0%; (2) anterior cerebral - 11%; (3) posterior communicating - 21%; (4) combined - 21%. In 17% of cases the posterior cerebral artery took its origin from the internal carotid artery. In only 21% of the cases studied did the Circle appear to be functionally complete.

DeSousa Pereira (1974), in a very thorough anatomic study of the Circle of Willis in 50 Portugese individuals, noted eight different patterns. In Group I (44%), the Circle was complete and appeared functional. In the second Group (8%), there was right carotid dominance, while in the third Group (20%), there was left carotid dominance. In Group IV (8%), the posterior communicating arteries were

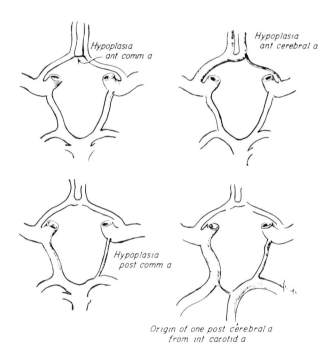

Fig. 10. Common variants in the Circle of Willis (From Alaskne, J.F.: Collateral circulation in Strandness, D.E., Jr. (Ed) Collateral Circulation in Clinical Surgery. W.B. Saunders Co., Philadelphia, 1969.)

inadequate. In Group V (8%), there was right vertebro-basilar dominance, while in Group VI (2%), left vertebro-basilar dominance was noted. In Group VII (6%), there was bilateral carotid dominance over the vertebrobasilar system, and in Group VIII (4%), there was independence of the afferent arterial system as a result of inadequate communicating vessels.

While there are differences in the pattern and incidence of variations in the Circle of Willis in the studies by Riggs and Rupp (1963) and DeSousa Pereria (1974), it is clear that the functionally complete Circle is frequently absent. This, of course, may have a profound effect on the patient's response to an obstructing lesion, depending upon, of course, its exact location in reference to the deficient Circle of Willis. These variations undoubtedly explain as well why one patient may tolerate a major vessel occlusion very well, while in another, it may have a devastating effect. This would appear to be supported by the study of Battarcharji et al., (1967), who studied the Circle of Willis in normal and infarcted brains. There were 49 cases of cerebral infarction to be compared with 88 controls. The incidence of anomalies was higher in the infarct group. The most striking finding was the very high incidence of unilateral string-like posterior communicating vessels combined with narrowing of one or more of the extracranial arteries.

While it is clear that the anatomic arrangement available to the patient is possibly a key factor in terms of the location and extent of the cerebral ischemia, it is also necessary to review the physiologic changes that can and do occur in response to arterial narrowing and occlusion at all levels.

Experimental Studies

In describing the effect of acute arterial occlusion on the function of the cerebral cortex, there are known changes which occur to varying degrees, depending upon the immediately available collateral circulation: (1) the blood pressure distal to the occlusion falls; (2) the cortical oxygen level falls in the distribution of the occluded vessel; (3) the EEG will demonstrate slowing; and (4) an injury potential will appear if there is neuronal damage (Meyer and Denny-Brown, 1957).

In the distribution of the occluded artery, three distinct zones have been described in relation to cortical oxygen availability and temperature. In the center, i.e., the most ischemic area, the cortical oxygen shows the maximum depression. There is next to this a borderline area where the tissue oxygen is unstable. Lastly, there is the outer zone where the tissue oxygen is increased as well as the cortical temperature which both reflect the increase in collateral blood flow. If the systemic systolic pressure falls below 50 mm Hg, the collateral circulation can no longer function and infarction ensues.

Meyer and Denny-Brown (1957) studied cerebral collateral function in monkeys whose middle cerebral and common carotid arteries had been ligated. With occlusion, there was a rapid fall in cortical oxygen tension which was less compensated for when the middle cerebral was occluded as compared to the common carotid artery. The most prompt and thought to be the most important compensatory mechanism for increasing collateral flow was the decrease in intraluminal pressure. This resulted in a prompt (within 30 sec) dilatation of arteries of 50 to 250 mµ with changes of greater than 50% commonly occurring. This was similar to the Bayliss effect described in limb vessels.

Meyer and Denny-Brown (1957) also evaluated the effects of arterial oxygen tension and pH as local vasodilators. As the arterial oxygen tension was reduced, vasodilatation did occur with a corresponding increase in blood flow. Also, application of a solution of ammonium chloride with a pH of 0.6 to the cortex led to vasodilatation, not only in the area of application, but in neighboring anastomotic vessels as well. These were not deemed as important as the Bayliss effect in promotion of blood flow after acute arterial occlusion.

It is well known that cerebral blood flow is controlled by two primary methods--chemical regulation in response to changes in arterial carbon dioxide tension and autoregulation where cerebrovascular resistance changes in response to alterations in perfusion pressure. In the normally supplied cortex, cerebral blood flow increases as the arterial carbon dioxide goes up and vice versa. With regard to pressure changes, resistance increases with a rise in pressure and vice versa.

However, when an area of cortex becomes ischemic, these normal regulatory mechanisms no longer may result in the same changes in flow when either the PCO_2 level or intra-arterial pressure changes. While it is somewhat controversial, current evidence would suggest that since there is maximal or nearly maximal vasodilatation in the ischemic area, these vessels can no longer respond to changes in PCO_2. The bordering normal cortex vessels can, however, dilate, resulting in a shunting of blood away from the ischemic area; termed the "intracerebral steal" by Lassen (1968). The reverse appears to occur with with hypocapnic hyperventilation. It also appears that ischemic brain may exhibit dysregulation with regard to blood flow changes with induced hypertension, i.e., blood flow in the ischemic area increases.

When occlusions of the extracranial arteries are produced, there appears to be an attempt immediately to increase flow to the potentially ischemic zone by corresponding increases in flow via the contralateral artery. For example, Iriuchijima and Koihe (1970) showed in dogs that a unilateral common carotid occlusion resulted in a 30% increase in flow through the contralateral carotid artery. The effectiveness of such collateral flow is, of course, dependent upon intact communicating vessels within the cranial cavity.

Clinical Problems

The blood supply to a segment or a hemisphere of the brain may be interrupted by a variety of mechanisms, both acute and chronic. By far the most important disease which leads to the development of transient ischemic attacks, completed strokes, or the so-called "Steal" syndromes is atherosclerosis. The mechanisms responsible for the ischemic episodes appear to be largely the result of one or more of the following: (1) a reduction in blood flow by arterial occlusion at one or more levels of the arterial supply with inadequate collateral circulation; (2) embolic occlusion by particulate matter released from the base of the ulcerated, complicated atherosclerotic plaque often within the carotid bulb; and (3) diversion of blood flow away from an area of brain and extracranial artery as a source of collateral circulation for the arm. This latter is most often observed in the subclavian steal syndrome.

Since the ischemia must be reversible for survival of the affected cortex, the available collaterals must be able to respond time-wise for this to occur. Obviously, an acute arterial occlusion stresses the system maximally, since time is of the essence and is measured in terms of minutes before irreversible damage is to occur. If the occlusion is outside the confines of the skull, the Circle of Willis will be the key since, if intact functionally, blood from the opposite vessels may be effectively conveyed via the communicating arteries. However, as already noted, up to 44% may have one or more communicating vessel inadequate to meet the need.

With occlusions of the intracerebral arteries, such as the middle cerebral, collateral flow must come via the leptomeningeal vessels which anastomose on the surface of the cortex. Van der Eecken (1959) studied these vessels and found there were numerous interconnecting arteries, 200 to 700 microns in diameter between the anterior, middle, and posterior cerebral arteries. The communications were either by direct end-to-end anastomoses between adjacent arteries, or what he termed "candelabra" anastomoses in which the arteries ramified into smaller twigs which then joined together. Similar anastomoses were described for the cerebellum. However, there are no significant anastomoses between the perforating arteries which are branches of the superficial vessels and supply the deeper layers of the brain.

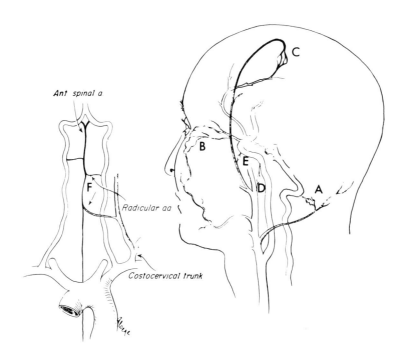

Fig. 11. Examples of the potential extra-intracranial collateral connections. A. Anastomoses between muscular branches of occipital artery and the vertebral artery. B. Communications between orbital branches of the external carotid artery and the opthalmic. C. Rete mirabile between dural branches of middle meningeal artery and pial branches of the anterior cerebral artery. D. Anastomoses between terminal branches of the carotico tympanic artery and tympanic branches of external carotid artery. E. Anastomoses between dural branches of carotid siphon and dural branches of external carotid artery. F. Connections between anterior spinal artery, radicular branches of costocervical trunk and radicular branches of vertebral artery. (From Alskne, J.F.: Collateral circulation in Strandness, D.E., Jr. (Ed) Collarteral Circulation in Clinical Surgery, W.B. Saunders Co., Philadelphia, 1969.)

In addition to the communicating branches of the Circle of Willis and the leptomeningeal anastomoses, there are other sources of blood supply to the brain from arteries outside the skull. These are shown diagramatically in Fig. 11. While there is little doubt that these collateral channels can and do function in response to chronic arterial occlusion, the magnitude of the flow contribution and its importance in maintaining cerebral blood flow has not been demonstrated.

THE VISCERA

The extensive anatomical studies done by Michels (1951) on the blood supply to the abdominal organs stands as the classic, anatomical definition of the normal and potential collateral of this region. It is not feasible to cover in detail the nature of the collateral circulation to each organ in this chapter. Rather, the author has chosen to review the salient features of the available supply to the liver, kidney, and intestine-organs which commonly require collateral input for function and survival of the tissues. As in many other areas of the arterial circulation, atherosclerosis is the major disease which affects the nutritional supply to viscera requiring the development of collateral circulation.

The liver is a unique organ which has two major vascular inputs--the hepatic artery and the portal vein. Normally, the portal vein flow accounts for about two-thirds of the total hepatic supply (Schenk et al., 1962). It is possible that following arterial occlusion, the involved liver parenchyma may meet its oxygen requirements, in part at least, by increasing the oxygen extracted from the portal vein. Studies by Tygstrup et al., (1962) did show that transient clamping of the hepatic artery distal to the gastroduodenal did result in an increase in the portal-hepatic venous oxygen difference. These studies suggest that portal venous flow may play a part at least in sustaining portions of the liver deprived of normal arterial flow.

The normal blood supply and potential extrahepatic pathways are shown in Figs. 12 and 13. A few general comments with regard to the blood supply to the liver are in order:

(1) The finding of anomalous patterns in terms of the arterial blood supply occurs so commonly that Michels in 1951 suggested that this "variation is constant.". Indeed, the conventional textbook picture shown in Fig. 12 is found in only 55% of the population;
(2) If an artery supplying the liver arises from a source other than the normal terminal branches of the celiac artery, it is termed an aberrant vessel;
(3) Aberrant vessels are of two types - accessory and replaced. Accessory vessels are those which supply a portion of the

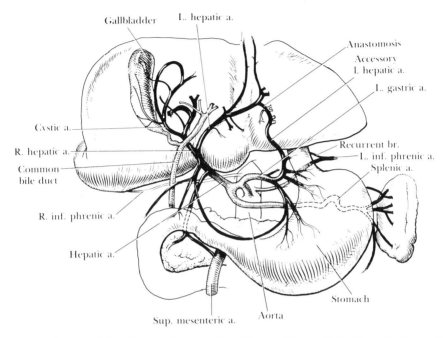

Fig. 12. Common blood supply to the liver (From Schultz, R.D.: Collateral circulation to the liver. In Strandness, D.E., Jr. (Ed) Collateral Circulation in Clinical Surgery, W.B. Saunders Co., Philadelphia, 1969.)

liver when the normal right or left celiac hepatic is present. When either the normal right or the normal left celiac hepatic is missing and the supply comes from another source, the term replaced is used;

(4) Since the aberrant and accessory arteries are end-arteries, they cannot nourish segments of liver supplied from other sources and, hence, are not truly collateral arteries;

(5) The important collateral pathways are those channels from extrahepatic sources that can provide blood to those hepatic arteries interrupted for whatever cause (Fig. 13).

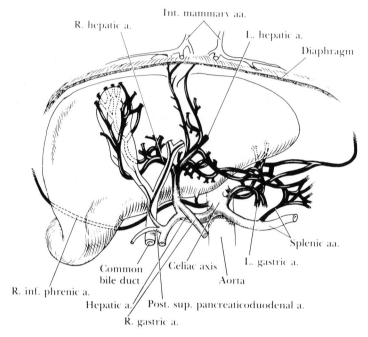

Fig. 13. Potential extrahepatic collateral pathways to the liver. (From Schultz, R.D.: Collateral circulation to the liver. In Strandness, D.E., Jr. (Ed) Collateral Circulation in Clinical Surgery, W.B. Saunders Co., Philadelphia, 1969.)

Experimental Studies

Animal studies have largely dealt with the effects of hepatic artery ligation on survival of the liver. The results vary from species to species, probably because of differing collateral pathways and the normal presence of anaerobic bacteria within the liver (the dog, for example) which materially influences the outcome of arterial interruption. For example, Horvath et al., (1957) studied this problem in dogs and showed clearly that unless antibiotics were given, none of the animals survived hepatic artery ligation. In contrast, Child et al., (1952) removed the hepatic artery as completely as possible in 19 Macaca mullata monkeys. Five animals received antibiotics. Liver biopsies or autopsies were performed at intervals of from two hours to 128 days. Eighteen survived without complications. No infarcts were found in the antibiotic group but were present to varying degrees in five of the 13 not receiving antimicrobial therapy.

Clinical Problems

Most of the data with regard to the collateral circulation of the liver in the human has come from clinical observations when the arterial supply has been interrupted from a variety of causes.

Most of the information has come from deliberate ligation of the hepatic arteries in patients with cirrhosis and portal hypertension (Berman and Hull, 1952; McFadzean and Cook, 1953; Rienhoff and Woods, 1953). The results were somewhat variable, but it appears that in those patients who died, massive liver necrosis was not invariably present but microscopic areas of necrosis were observed. These clinical experiences cannot necessarily be transferred to the outlook for the normal liver, since the prior existence of vascular abnormalties known to occur in cirrhosis may have played a role in protecting the liver.

Other reports in which inadvertent hepatic artery ligation had occurred suggest that the involved liver segment may survive even though there will be transient abnormalities in liver function tests (Brittain et al., 1964; Andreassen et al., 1962). The role of antibiotics in promoting survival is difficult to state with certainty at this time.

THE KIDNEY

The normal blood supply to the kidney is by a single branch from the lateral aspect of the aorta. Multiple renal arteries occur in approximately 32% of subjects. They occur as commonly on one side as on the other, and usually do not exceed more than two or three vessels per kidney, but as many as five arteries to one kidney have been reported (Pick and Anson, 1940). These nearly always arise from the aorta and may be quite large. Anson et al., (1936) showed that approximately half of these accessory arteries enter the hilum, with the remaining entering directly either the upper or lower pole of the kidney. Accessory arteries from other sources are infrequent and quite small.

The collateral circulation to the kidney largely comes from the lumbar, internal iliac, testicular or ovarian, inferior adrenal, and intercostal arteries (Fig. 14). As Abrams and Cornell (1965) have shown, the third lumbar artery is the most common source of significant arterial supply when the renal artery is significantly narrowed or occluded.

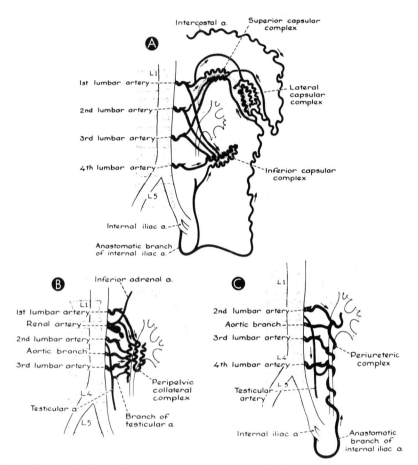

Fig. 14. Potential collateral sources to the kidney. A. First four lumbar arteries along with the intercostal and internal iliac arteries. B. The aorta, first three lumbar arteries, testicular or ovarian arteries contribute to the region of the renal pelvis. C. Collateral source to the ureter via the internal iliac artery. (From Abrams, H.D., and Cornell, S.H.: Patterns of collateral flow in renal ischemia. Radiology 84:1001, 1965.)

Experimental Studies

The most common clinical situation in which renal artery narrowing becomes of importance is in patients with hypertension. If the patient is hypertensive secondary to the ischemia, it is postulated that excess renin is produced. Indeed, measuring the ratios of renal venous plasma renin activity between kidneys is commonly used to identify not only the basis for the hypertension, but also indicate which kidney is indeed at fault.

A point of interest is to what extent does collateral development affect the generation of renin by the involved kidney. Ernst et al., (1976) studied this problem in nine dogs, equating the lateralizing renal venous renin activity with the degree of arterial stenosis and the arteriographic appearance of collateral channels. These studies showed that as collaterals develop, the ratios of the constricted to the nonconstricted side do, in fact, decrease. This indicates that renin production does, in fact, decrease as the collateral circulation improves.

Clinical Problems

Acute arterial occlusion of the main renal artery will result in prompt and total loss of renal function. The available collaterals as shown in Fig. 14 do not have time, nor do they have the potential, for maintaining sufficient pressure and flow to preserve renal function. If accessory renal arteries are present, they may provide varying degrees of protection, depending entirely upon the intrarenal distribution of the artery(s).

If an accessory renal artery is lost by surgical ligation, the ischemic portion of the kidney may produce renin and lead to the development of hypertension. Geyskes et al., (1972) reported a case in which occlusion of a branch of a renal artery resulted in partial infarction only and led to the development of malignant hypertension.

THE INTESTINE

Atherosclerosis involving the blood supply to the gut can lead to both acute and chronic ischemia if the collateral blood supply is inadequate. The normal blood supply is from three large vessels: the celiac axis, the superior mesenteric artery, and the inferior mesenteric artery. If one or more of these arteries become narrowed or occluded, there are four major anastomotic pathways which become of critical importance: (1) the left inferior phrenic artery which communicates with the left gastric artery and, in turn; (2) the pancreaticoduodenal arcade. This constant collateral network joins the celiac axis with the superior mesenteric artery; (3) the left colic-middle colic arterial anastomoses which forms the key anastomotic network joining the superior and inferior mesenteric arteries as well as the branches of the internal iliac artery; and (4) the internal iliac artery which has anastomoses with the superior rectal artery, a branch of the inferior mesenteric artery.

This collateral interchange provides an extremely effective method of supplying the small and large bowel in cases of chronic arterial input.

Clinical Problems

The two major problems to be dealt with are embolism, usually to the superior mesenteric artery, and chronic occlusion secondary to atherosclerosis.

When emboli lodge in the superior mesenteric artery, it is usually distal to the origin of the middle colic artery. If the artery is occluded at this point, it is virtually an end vessel and acute necrosis of the small bowel results.

The remarkable potential for collateral circulation to the gut is evident by the extent of the chronic occlusive process required to produce intestinal angina. Fry (1969) and Morris et al., (1962) have noted that at least two of the main vessels must be involved before the symptoms of visceral ischemia develop.

THE EXTREMITIES

There are multiple collateral pathways available in the arms and legs which have the remarkable capacity of meeting resting flow requirements even with extensive, multilevel chronic arterial occlusions. Anastomically these can best be described in terms of the potentially involved artery and what pathways are available (Radke, 1969).

The Arm

The subclavian artery may be effectively bypassed by the extensive cervical arterial collateral network shown in Fig. 15. Functionally, the vertebral arteries form an important input by the reversal of flow that can take place in this vessel.

With respect to the brachial artery, all of its branches, with the exception of the supply to the biceps muscle, have significant collateral function. These arteries, in large part, all accompany the major nerves (Fig. 16).

Below the elbow, there are communications between the radial and ulnar, and the volar and dorsal intraosseous branches. Likewise,

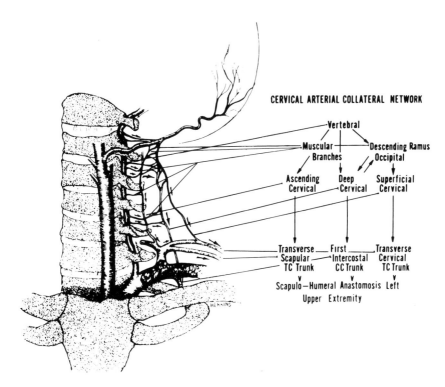

Fig. 15. Cervical collateral network available for occlusions of the subclavian artery. (From Levowitz, B.S., Young, S., and Kantrowitz, A.: Distal occlusion of the subclavian artery in the steal syndrome. Arch Surg 93:980, 1966.)

if the palmar arch is intact, this communication between the terminal branches of the radial and ulnar artery may serve a useful collateral function for occlusions involving the more proximal arteries.

In the hand, the superficial palmar arch is incomplete in about 20% of cases, whereas the deep palmar arch has a more constant pattern (97% of cases) (Coleman and Anson, 1961). This regular, nearly predictable intercommunication of the deep palmar arch accounts for the nearly perfect collateral network in the hand.

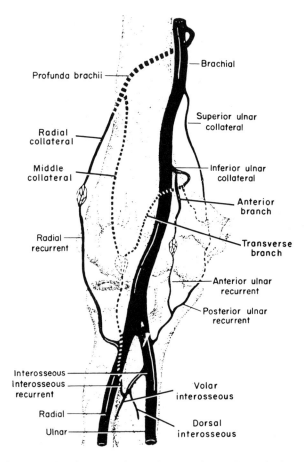

Fig. 16. Collateral circulation about the elbow joint (adapted from Quiring, D.P.: Collateral Circulation, Anatomical Aspects. Lea and Febiger, Philadelphia, 1949.)

The Leg

When there is disease involving the abdominal aorta and iliac arteries, the major collateral pathways available can be subdivided into the major parietal pathways (Fig. 17) and the visceral anastomotic interconnections (Fig. 18). The femoropopliteal segment is bypassed by an unusually effective system of collaterals which are composed of the profunda femoris artery and its connections with the geniculate arteries (Fig. 3).

Below the knee, there is extensive intercommunication between the anterior tibial, posterior tibial, and peroneal arteries (Fig. 4). This provides the area of the leg with one of the most effective collateral networks in the body.

Experimental Studies

As mentioned earlier, Winblad et al., (1959) studied intra-arterial pressure, muscle PO_2, and performed serial arteriograms to evaluate the time-course of events that followed acute arterial occlusion in the dog. Immediately after femoral artery occlusion, the distal arterial pressure fell to between 35 mm Hg and 60 mm Hg with a marked reduction in pulse pressure. Tissue PO_2 dropped to 84 to 95% of normal but rapidly increased to reach preocclusion levels within five to eight minutes. These experimental studies were among the first to stress the importance of the relationship between a rise in pressure in the distal artery and the enlargement of the collateral mid-zone.

Another interesting experimental study which has important clinical implications is that carried out by Jacobson and McAllister (1957). These investigators studied the extent to which collateral arteries regress once the need for them is no longer present and to which extent they may respond when the need is again imposed. Six dogs had the abdominal aorta, inferior mesenteric artery, and lowermost lumbar arteries interrupted. The animals were continually retested until their preligation exercise tolerance was reached. At this point, aortic continuity was restored with a pneumatic cuff left in place that would permit acute occlusion at will. With repeat aortic occlusion, a shorter time was required for the animals to reach maximal walking tolerance, suggesting that the collateral arteries had not reached their original preligation state.

Clinical Problems

Considerable experience has been gained in the understanding of how collateral arteries respond to a variety of situations. While the exact mechanisms involved are still a mystery, the time-course of response of these vessels and how it is manifested clinically is now well understood.

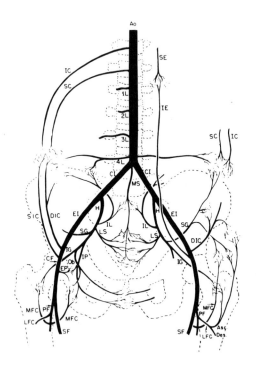

Ao	Aorta	IP	Internal pudendal
Asc	Ascending branch	L	Lumbar
CF	Common femoral	LFC	Lateral femoral circumflex
CI	Common iliac	LS	Lateral sacral
Des	Descending branch	MFC	Medial femoral circumflex
DIC	Deep iliac circumflex	MS	Middle sacral
EI	External iliac	Ob	Obturator
EP	External pudendal	PF	Profunda femoris
H	Hypogastric	SC	Subcostal
IC	Intercostal	SE	Superior epigastric
IE	Inferior epigastric	SF	Superficial femoral
IG	Inferior gluteal	SG	Superior gluteal
H	Hiolumbar	SIC	Superficial iliac circumflex

Fig. 17. Major parietal pathways bypassing areas of aorto-ilio-femoral occlusive disease. (From Friedenberg, M.J., and Perez, C.A.: Collateral circulation in aorto-ilio-femoral occlusive disease. Amer J Roentgen 94:145, 1965.)

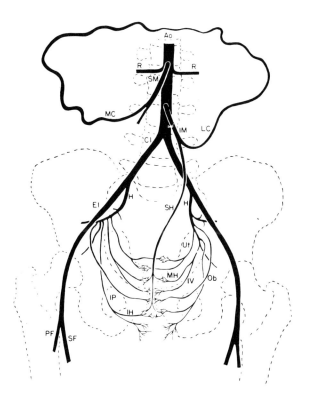

Ao	Aorta	MC	Middle colic
CI	Common iliac	MH	Middle hemorrhoidal
EI	External iliac	Ob	Obturator
H	Hypogastric	PF	Profunda femoris
IH	Inferior hemorrhoidal	R	Renal
IM	Inferior mesenteric	SF	Superficial femoral
IP	Internal pudendal	SH	Superior hemorrhoidal
IV	Inferior vesical	SM	Superior mesenteric
LC	Left colic	Ut	Uterine

Fig. 18. Major visceral collateral pathways in aorto-ilio-femoral occlusive disease. From Friedenberg, M.J., and Perez, C.A.: Collateral circulation in aorto-ilio-femoral occlusive disease. Amer J. Roentgen 94:145, 1965.)

Arteriovenous Fistula

The classic experimental model for studying the development of the collateral circulation is the side-side (H-type) arteriovenous fistula. In the experimental animal, the degree of collateralization was always less impressive in limbs with a ligated artery than it was in legs with an arteriovenous fistula (Holman, 1949; Reid, 1925). Holman (1949) was the first to establish the importance of retrograde flow in this process by a series of ingenious experiments. If the artery was ligated just distal to the fistula, collateral circulation was markedly reduced. In contrast, when the artery was ligated distal to a single branch still permitting retrograde flow, collateral development was again prodigious.

Holman (1949) postulated a purely mechanistic theory for the collateral development. Sir Thomas Lewis (1940) objected, suggesting that there was some "chemical stimulant" produced locally by the tissues which were presumed to be ischemic. To counter this argument, Holman was able to show that collaterals continued to develop even if the limb distal to the fistula had been amputated.

The importance of the collaterals which develop in response to a side-side fistula permitted the application of quadruple ligation as the method of treatment prior to the development of modern vascular surgery. This form of therapy was accompanied by a very low rate of limb loss, presumably because of the prodigious collateral available to supply the distal limb.

Acute Arterial Occlusion

When a major artery in the limb is suddenly occluded, there is an immediate drop in distal pressure and flow. Peripheral ischemia results, the extent of which is dependent entirely upon the magnitude of the flow reduction. Clinical observations of collateral function have led to the following conclusions which are of importance in patient evaluation (Strandness, 1969):

(1) There is considerable variation in the magnitude of the ischemia which is dependent upon the site of the occlusion. For example, acute aortic obstruction produces profound ischemia which must be corrected for tissue survival. In contrast, with occlusion of the superficial femoral artery (between the profunda femoris and re-entry geniculate arteries), the distal limb will be transiently ischemic but will usually survive even if the occlusion is not removed. Below the knee, acute occlusion of the tibial arteries is usually of minimal consequence because of the excellent, immediately available collateral vessels;

(2) If the limb remains viable over the first several hours following occlusion, the collaterals can and will continue to improve over the ensuing days to weeks. This has been documented by following the changes in limb blood pressure distal to the occlusion. As the cross-sectional area of the collateral circulation increases, i.e., the resistance falls, distal pressure will rise;

(3) The absolute level of the pressure distal to the occlusion has some predictive value both immediately after the event and later, in terms of estimating its effect upon function. For example, limb systolic pressures below the level of 40 mm Hg are commonly associated with ischemic rest pain. Conversely, pressures above this level are rarely associated with severe rest pain;

(4) If the limb survives the initial insult without surgical correction of the occlusion, the ultimate level of functional recovery cannot be predicted with certainty until several months have passed. Improvement in collateral function with acute occlusion has a rapid phase which occurs during the first few days and then a much slower time-course which may take months to plateau.

Chronic Arterial Occlusion

An important question which may bear relevance to other vascular beds, particularly the heart, is "Can collateral function be improved after a period of stability?" The one mode of therapy which has been used to test this is exercise. Extensive studies in patients with intermittent claudication have shown that graded exercise can improve walking distance significantly (Larsen and Lassen, 1966; Schlüssel, 1965; Schoop, 1964). Skinner and Strandness (1967) examined the distal limb blood pressure response to a rigorously controlled exercise protocol and were able to show that ankle blood pressure and its response to exercise improved without a corresponding change in systemic pressure. This can occur only because of a decrease in resistance to flow at the level of the collateral arteries.

VENOUS COLLATERALS

There is no doubt that the capacity to develop collaterals on the venous side of the circulation is equal to, if not greater than, those on the arterial side. As with arteries, venous collaterals always respond to obstruction by enlarging, thus decreasing their resistance to flow. A major anatomic difference between arteries and veins is the presence of valves. These structures are absolutely essential for proper venous function in the lower extremities. Normally the valves permit only unidirectional flow as follows: (1) from the superficial to the deep veins via the communicating or perforating veins; and (2) distal to proximal flow in the deep venous system even with the most forceful muscular contraction.

With acute venous thrombosis, ischemia is rarely a problem unless the process is so extensive as to essentially block all venous return. Under these circumstances "venous gangrene" may occur, but this is a rare event. A secondary complication of the development of the venous collateral is that the valves often lose their functional capabilities either secondary to the inflammatory process associated with the thrombophlebitis, or because of an inability for the valves to coapt properly as the vein dilates. This loss of valvular function essentially nullifies the unidirectional flow pattern and results in edema, stasis pigmentation, and ulceration--the so-called postphlebitic syndrome.

The mechanisms promoting venous collateral development are not understood at all, and virtually no work in this area has been done. However, there are excellent anatomic studies which detail the pathways available and the time-course of events that follow acute venous obstruction (Dale, 1958; Edwards and Robuck, 1947; Mavor and Galloway, 1967).

REFERENCES

Abrams, H.L., and Cornell, S.H.: Patterns of collateral flow in renal ischemia. Radiology 84:1001, 1965.
Andreassen, M., Lindenberg, J., and Winkler, K.: Peripheral ligation of the hepatic artery during surgery in non-cirrhotic patients. Gut 3:167, 1962.
Anson, B.J., Richardson, G.A., and Minear, W.L.: Variations in number and arrangement of renal vessels: study of blood supply of 400 kidneys. J Urol 36:211, 1936.
Baroldi, G., and Scomazzoni, G.: Coronary circulation in normal and pathologic heart. Office of Surgeon General, Department of Army, Washington, D.C., 1967.
Baroldi, G., Mantero, O., and Scomazzoni, G.: The collaterals of the coronary arteries in normal and pathologic hearts. Circ. Res. 4:223, 1956.
Battacharji, S.K., Hutchinson, E.C., and McCall, A.J.: The Circle of Willis--the incidence of developmental abnormalities in normal and infarcted brains. Brain 90:747, 1967.
Beck, C.S., Tichy, V.L., and Moritz, A.R.: Production of a collateral circulation to the heart. Proc Soc Exper Biol and Med 32:759, 1935.
Becker, L.C., and Pitt, B.: Collateral blood flow in conscious dogs with chronic coronary artery occlusion. Amer J Physiol 221: 1507, 1971.
Bellman, S., Frank, H.A., Lambert, P.B., and Roy, A.: Studies of collateral vascular responses. I. Effects of selective occlusions of major trunks within an extensively anastomosing arterial system. Angiology 10:214, 1959.

Berman, J.K., and Hull, J.E.: Hepatic, splenic and left gastric arterial ligations in advanced portal cirrhosis. Arch Surg 65:37, 1952.

Blomfield, L.B.: Intramuscular vascular patterns in man. Proc Roy Soc Med 38:617, 1945.

Brittain, R.S., Marchioro, T.L., Hermann, G., Waddell, W.R., and Starzl, T.E.: Accidental hepatic artery ligations in humans. Amer J. Surg 107:822, 1964.

Carrier, O., Jr., Walker, J.R., and Guyton, A.C.: Role of oxygen in the autoregulation of blood flow in isolated vessels. Amer J. Physiol 206:951, 1964.

Child, C.G., III, Hays, D.M., and McClure, R.D., Jr.: Studies of the hepatic circulation in the Macaca mulatta monkey and in man. Surg Forum 2:140, 1952.

Coffman, J.D.: Peripheral collateral blood flow and vascular reactivity in the dog. J Clin Invest 45:923, 1966.

Coleman, S.S., and Anson, B.J.: Arterial patterns in the hand based upon a study of 650 specimens. Surg Gynecol Obstet 113:409, 1961.

Cresti, M., and Steger, C.: The correlation between the effects of morphology and function on the collateral circulation in the obliterative arteriopathy of the inferior limbs. Angiology 13:271, 1962.

Dale, W.A.: Ligation of the inferior vena cava for thromboembolism. Surgery 43:24, 1958.

DeSousa Pereria, J.M.M.: Circulacão colateral do cerebro. Inova-Artes Gráficas, Porto, 1974.

Dollery, C.T.: The formation of collaterals. Postgrad Med J 44:28, 1968.

Dollery, C.T., Hill, D.W., Patterson, J.W., Ramalho, P.S., and Kohner, E.M.: Collateral blood flow after branch arteriolar occlusion in the human retina. Brit J Opthal 51:249, 1967.

Dornhorst, A.C., and Sharpey-Schafer, E.P.: Collateral resistance in limbs with arterial obstruction: spontaneous changes and the effects of sympathectomy. Clin Sci 10:371, 1951.

Edwards, E.A., and Robuck, J.D., Jr.: Applied anatomy of femoral vein and its tributaries. Surg Gynec Obstet 85:547, 1947.

Eliot, E.C., Bloor, C.M., Jones, E.L., Mitchell, W.J., and Gregg, D.E.: Effect of controlled coronary occlusion on collateral circulation in conscious dogs. Amer J Physiol 220:857, 1971.

Ernst, C.E., Daugherty, M.E., and Kotchen, T.A.: Relationship between collateral development and renin in experimental renal arterial stenosis. Surgery (in press).

Ferris, H.W., and Harvey, S.C.: A physiologic study of the development of the collateral circulation in the leg of the dog. Proc Soc Exptl Biol Med 22:383, 1924-1925.

Fry, W.L.: Arterial Circulation of the Small and Large Instestine. Ch 24 in Collateral Circulation in Clinical Surgery, Strandness, D.E., Jr. (Ed), W.B. Saunders Co, Philadelphia, 1969.

Gensini, G.G., and da Costa, B.C.B.: Coronary collateral circulation. Amer J Cardiol 24:393, 1969.

Geyskes, G.G., Misage, J.R., Bron, K., Haas, J.E., Berg, G., and Shapiro, A.P.: Malignant hypertension following renal artery branch obstruction. JAMA 222:457, 1972.

Goldstein, R.E., Stinson, E.B., Scherer, J.L., Seningen, R.P., Grehl, T.M., and Epstein, S.E.: Intraoperative coronary collateral function in patients with coronary occlusive disease: Nitroglycerin responsiveness and angiographic correlations. Circulation 49:298, 1974.

Gundel, W.D., Brown, B.G., and Gott, V.L.: Coronary collateral flow studies during variable aortic pressure waveforms. J Appl Physiol 29:579, 1970.

Harris, C.N., Kaplan, M.A., Parker, D.P., Aronow, W.S., and Ellestad, M.H.: Anatomical and functional correlates of intercoronary collateral vessels. Amer J Cardiol 30:611, 1972.

Helfant, R.H., Kemp, H.G., and Gorlin, R.: Coronary atherosclerosis, coronary collaterals and their relation to cardiac function. Ann Intern Med 73:189, 1970.

Helfant, R.H., Vokonas, P.S., and Gorlin, R.: Functional importance of the human collateral circulation. New Engl J Med 284:1277, 1971.

Hilton, S.M.: A peripheral arterial conducting mechanism underlying dilatation of the femoral artery and concerned in functional vasodilatation in skeletal muscle. J Physiol (London) 149:93, 1959.

Holman, E.: Arteriovenous Aneurysm: Abnormal Communications Between the Arterial and Venous Circulations. New York, the Macmillan Co., 1937.

Holman, E.: Problems in the dynamics of blood flow. I. Conditions controlling collateral circulation in the presence of an arteriovenous fistula, following ligation of an artery. Surgery 26:880, 1949.

Horvath, S.M., Farrand, E.A., and Larsen, R.: Effect of hepatic arterial ligation on hepatic blood flow and related metabolic function. Arch Surg 74:565, 1957.

Ingebrigtsen, R., Johansen, K., Müller, O., and Wehn, P.: Blood pressure of the proximal artery in experimental arteriovenous fistulas of long standing. Acta Chir Scand 253 (Suppl):131, 1960.

Iriuchijima, J., and Koike, H.: Carotid blood flow, intrasinusal pressure and collateral flow during carotid occlusion. Amer J Physiol 218:876, 1970.

Jacobson, J.H., and McAllister, F.F.: The harmful effect of arterial grafting on existing collateral circulation. Surgery 42:148, 1957.

James, T.N.: The Anatomy of the Coronary Arteries. New York, Paul B. Hoeber Inc., 1961, p. 145.

Johansson, B., Linder, E., and Seeman, T.: Collateral blood flow in the myocardium of dogs measured with Krypton^{85}. Acta Physiol Scand 62:263, 1964.
John, H.T., and Warren, R.: The stimulus to collateral circulation. Surgery 49:14, 1961.
Keenan, R.L., and Rodbard, S.: Competition between collateral vessels. Cardiovasc Res 7:670, 1973.
Larsen, O.A., and Lassen, N.A.: Effect of daily muscular exercise in patients with intermittent claudication. Lancet 2:1093, 1966.
Lassen, N.A.: The luxury-perfusion syndrome and its possible relation to acute metabolic acidosis localized within the brain. Lancet 2:1113, 1968.
Learnmonth, J.: Collateral circulation: natural and artificial. Surg Gynecol Obstet 90:385, 1950.
Lewis, T.: The adjustment of blood flow to affected limb in arteriovenous fistula. Clin Sci 4:277, 1940.
Liebow, A.A.: Situations which lead to changes in vascular patterns. In Handbook of Physiology, Sec 2, Vol 2, Circulation. Ed by W.F. Hamilton and P. Dow. Washington, D.C., American Physiological Society, 1962, p. 1251.
Liebow, A.A., Hales, M.R., and Lindskog, G.E.: Enlargement of bronchial arteries and their anastomoses with the pulmonary arteries in bronchiectasis. Amer J. Pathol 25:211, 1949.
Liebow, A.A., Hales, M.R., Bloomer, W.E., Harrison, W., and Linskog, G.E.: Studies on the lung after ligation of the pulmonary artery. II. Anatomical changes. Amer J. Pathol 26:177, 1950.
Longland, C.J.: The collateral circulation of the limb. Ann Roy Coll Surg Engl 13:161, 1953.
Loring, W.E., and Liebow, A.A.: Effects of bronchial collateral circulation on heart and blood volume. Lab Invest 3:175, 1954.
Marchand, P., Gilroy, J.C., and Wilson, V.H.: An anatomical study of the bronchial vascular system and its variation in disease. Thorax 5:207, 1950.
Mautz, F.R.: Anatomical and physiological considerations in the development of a collateral circulation to the myocardium. Dis Chest 31:265, 1957.
Mavor, G.E., and Galloway, J.M.D.: Collaterals of the deep venous circulation of the lower limb. Surg Gynec Obstet 125:561, 1967.
May, A.G., DeWeese, J.A., and Rob, C.O.: Hemodynamic effects of arterial stenosis. Surgery 53:513, 1963.
McFadzean, A.J.S., and Cook, J.: Ligation of the splenic and hepatic arteries in portal hypertension. Lancet 1:615, 1953.
Meffert, W., and Liebow, A.A.: Hormonal control of collateral circulation. Cir Res 18:228, 1966.
Meyer, J.S., and Denny-Brown, D.: The cerebral collateral circulation. 1. Factors influencing collateral blood flow. Neurology 7:447, 1957.
Michels, N.A.: The hepatic, cystic and retroduodenal arteries and their relations to the biliary ducts. Ann Surg 133:503, 1951.

Morris, G.C., Crawford, E.S., Coolery, D.A., and DeBakey, M.E.: Revascularization of the celiac and superior mesenteric arteries. Arch Surg 84:95, 1962.

North, K.A.K., Sanders, A.G., and Florey, H.W.: The development of an anastomotic circulation to transplanted tissue. Brit J Exptl Pathol 41:520, 1960.

Pick, J.W., and Anson, B.J.: The renal vascular pedicle: an anatomical study of 430 body halves. J Urol 44:411, 1940.

Radke, H.M.: Arterial Circulation of the Upper and Lower Extremity. Chapters 11, 12 and 13 in Collateral Circulation in Clinical Surgery. Strandness, D.E., Jr. (Ed), W.B. Saunders Co., Philadelphia, 1969.

Rees, J.R., and Redding, V.J.: Anastomotic blood flow in experimental myocardial infarction. Cardiovasc Res 1:169, 1967.

Reid, M.R.: Abnormal arteriovenous communications, acquired and congenital. III. The effects of abnormal arteriovenous communications on the heart, blood vessels and other structures. Arch Surg 11:25, 1925.

Rienhoff, W.F., Jr., and Woods, A.C.: Ligation of splenic and hepatic arteries in treatment of cirrhosis with ascites. JAMA 152:687, 1953.

Riggs, H.E., and Rupp, C.: Variation in the form of the Circle of Willis: The relation of the variations to collateral circulation: anatomic analysis. Arch Neurol 8:8, 1963.

Rosenthal, S.L., and Guyton, A.C.: Hemodynamics of collateral vasodilatation following femoral artery occlusion in anesthetized dogs. Circ Res 23:239, 1968.

Schaper, W.: The collateral circulation in the canine coronary system. Ⓒ Janssen Pharmaceutica NU, Holland, 1967.

Schaper, W., Remijsen, P., and Xhonneux, R.: Size of myocardial infarction after experimental coronary artery ligation. Arch Kreislaufforsch 58:904, 1969.

Schenk, W.G., Jr., McDonald, J.C., McDonald, K., and Drapanas, T.: Direct measurement of hepatic blood flow in surgical patients with related observations in experimental animals. Ann Surg 156:463, 1962.

Schlüssel, H.: Das gehtraining bei älteren potienten mit chronisihe Beinarleieverschluss. Med Wett 3:145, 1965.

Schoop, W.: Bewegungstherapie bei peripheren durch blutungsstörungen. Med Wett Part 1:503, 1964.

Shepherd, J.T.: The effect of acute occlusion of the femoral artery on the blood supply to the calf of the leg before and after release of sympathetic vasomotor tone. Clin Sci 9:355, 1950.

Skinner, J.S., and Strandness, D.E., Jr.: Exercise and intermittent claudication. II. Effect of physical training. Circulation 36:23, 1967.

Strandness, D.E., Jr.: Peripheral Arterial Disease: A Physiologic Approach. Little, Brown & Co., Boston, 1969.

Thulesuis, O.: Hemodynamic studies on experimental obstruction of the femoral artery in the cat. Acta Physiol Scand 57 (Suppl 199), 1962.

Tygstrup, N., Winkler, K., Mellemgaard, K., and Andreassen, M.: Determination of the hepatic arterial blood flow and oxygen supply in man by clamping the hepatic artery during surgery. J Clin Invest 41:447, 1962.

Van der Eecken, H.M.: The Anastomoses Between the Leptomeningeal Arteries of the Brain. Springfield, Ill, Charles C. Thomas Co., 1959.

Vineberg, A., Munro, D.D., Cohen, H., and Buller, W.: Four years' clinical experience with internal mammary artery implantation in the treatment of human coronary artery insufficiency including additional experimental studies. J Thorac Surg 29:1, 1955.

Virkkula, L., Sittnikow, K., Ervasti, O., and Perttala, Y.: Study of the neogenic collateral arteries. Arch Chir Scand 137:766, 1971.

Volpel, W.: Über die entstehungsbedingungen des arteriellen kollaterals kreislaufes. Acta Biol et Med Ger 3:557, 1959.

Weibel, E.R.: Early stages in the development of collateral circulation to the lung in the rat. Circ Res 8:353, 1960.

Winblad, J.N., Reemstma, K., Vernhet, J.L., Laville, L.P.. and Greech, O., Jr.: Etiologic mechanisms in the development of collateral circulation. Surgery 45:105, 1959.

MEASUREMENT OF BLOOD PRESSURE, BLOOD FLOW AND RESISTANCE TO BLOOD FLOW IN THE SYSTEMIC CIRCULATION

John Ludbrook, M.D. (Otago), Ch.M., B. Med. Sci., (N.Z.),
F.R.C.S., F.R.A.C.S.
Mortlock Professor of Surgery, University of Adelaide,
and Visiting Surgeon, Royal Adelaide Hospital, Adelaide,
South Australia.

INTRODUCTION

The ramifications of blood pressure, blood flow, and resistance are vast subjects to discuss, and some means of containing them has had to be devised. The account that follows deals with blood pressure, blood flow, and their interrelationship in a manner that is intended to provide by text and illustration both a clear description of principles, and an indication of aspects that are especially important or contentious. Where hard data are provided, they are concentrated into tabular form. The method of citing references has followed a rather similar pattern. There are a number of recent, fully referenced, monographs and review articles that deal with these subjects. Where a source of information is not specifically cited, it can be located by reference to these reviews which are separately listed. The sources of information that are especially important, contentious, or new are individually cited and listed.

DEFINITIONS OF BLOOD PRESSURE

While the term blood pressure may seem to be one that is self-evident, its precise meaning depends very much on the method that is used to measure it. The definitions that follow will appear simplistic to all (or most) who actually make measurements of blood pressure, but they represent information that is essential for those who merely intend to do so.

1. Reference points

By any method of measurement the term blood pressure is a relative one: it is the pressure difference between the point of measurement and some external reference point. The most widely used convention is to relate blood pressure to atmospheric pressure. However, particular anatomic and physiologic considerations may determine a preference for other reference points; this is particularly so in the low pressure components of the cardiovascular system, Thus, pressure in intrathoracic vessels may be related to intrathoracic pressure; and portal pressures, especially venous, may be appropriately related to inferior vena caval pressure. Systemic venous pressure are not uncommonly related to the level of the heart (or phlebostatic axis). One purpose of using these reference points as alternatives to the atmosphere is to dissect out the hydraulic from the hydrostatic components of measured pressure, and to better express the driving force of perfusion pressure (q.v.).

2. Hydrostatic vs hydraulic pressure (Fig. 1)

As implied from the etymology, hydrostatic pressure is that exerted by a static column of water (or in this context, blood). Its physiologic importance rests in the fact that the cardiovascular system is in a sense--or at least in some circumstances--a freely communicating system of tubes, and that most animals with cardiovascular systems have a vertical dimension. Thus pressure differences between points in the vasculature are in part hydrostatic or gravitational, the hydrostatic component being calculable knowing the specific gravity of blood (approximately 1.055), and the vertical distance between the points of measurement.

Hydraulic pressure is that transmitted from the appropriate pressure-generator. In most circumstances this is the heart, but there are other generators such as those resulting from the action of skeletal or smooth muscle contraction on valved veins. The transmission of hydraulic pressure from its source in the cardiovascular system is imperfect and complex, in that energy loss occurs from the elasticity of blood vessel walls, from friction within the bloodstream, and from friction between the bloodstream and the walls of blood vessels. There results a general decay of hydraulic pressure within the bloodstream in a centrifugal fashion away from the pressure-generator (e.g., heart).

Thus the measured pressure within any blood vessel is a sum of the hydrostatic pressure, determined by the height of the continuous column of blood above that point, and the hydraulic pressure.

3. Internal vs transmural pressure (Fig. 1)

One normally conceives of blood pressure as if the point of measurement were within the blood vessel, and as if its wall and the surrounding tissues were at atmospheric pressure. This is appropriate when a parameter such as the flow of blood along the axis of

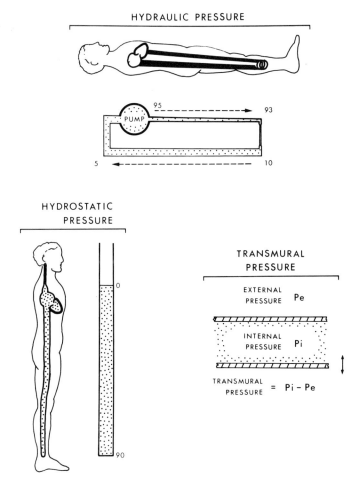

Fig. 1. Definitions of pressure. Hydraulic pressure gradients generated by the interaction of the cardiac pump with the resistance offered by arterial, microvascular, and venous elements of the vascular tree. Hydrostatic pressure in the venous system in the upright posture, resulting from the weight of a column of blood in a system that is rendered "open-ended" by collapse of the veins above heart level. Transmural (distending) pressure: the difference between intravascular and tissue pressure (from Ludbrook, J. and Walsh, J.A., 1973, Physiology, in "Peripheral vascular surgery," (M. Birnstingl, ed.), p. 4, Heinemann, London, by kind permission of editor and publishers).

the vessel is under consideration. There are at least two circumstances in which the pressure difference between the blood and the surrounding tissues--the transmural pressure--is important. These are when one is considering the flow or transport of substances across the wall of the vessel (as in the microvasculature); or when one wishes to consider the effects of fluctuating external tissue pressure on the cross-section of, or resistance to flow within the vessel (as in the thorax, abdomen, and within or between muscle compartments).

4. Static vs dynamic pressure (Fig. 2)

Under most resting conditions the internal pressure in a vein is almost constant with time. However in arteries, in the smaller pre-resistance vessels, and even in some circumstances in the post-resistance vessels, blood pressure is pulsatile as a consequence of the reciprocating-pump action of the pressure-generator (the heart). While this phenomenon is self-evident, the consequences to the student of pressure and flow are not immediately so. Quite unsophisticated devices, such as mercury or water manometers, will faithfully and accurately record mean or static pressures. In contrast, highly sophisticated instrumentation is essential if the rapidly fluctuating pressure changes that occur within the heart and within arteries are to be precisely measured. With the over-damped instruments of a generation ago, some broad descriptions of the behavior of blood pressure, and of the interrelations of pressure and flow, were able to be evolved. However, for many present-day purposes--and especially for study of the properties of the arterial wall, or of the more complex interrelations of blood pressure and blood flow--accurate reproduction of the shape, magnitude, and timing of the pressure pulse is essential.

5. Directional pressures

"True" pressure is an abstract concept, that could be realized only by a measuring device that was infinitely small, and which absorbed no energy. In practice, intravascular (direct) blood pressure measuring devices are finite in size, do absorb energy, and in addition, are directionally sensitive to pressure changes.

Thus a further dimension of pressure--or at any rate of pressure measurement--is the angle between the pressure-sensing device and the axis of the blood stream. "Upstream" pressure is higher than "downstream" by a factor derived from the conversion of kinetic energy into pressure. That is, the greater the velocity of flow of the blood stream, the greater the pressure difference between the upstream and downstream directions (to the point that this phenomenon has been used to measure blood flow). "Lateral" pressure is intermediate between these two.

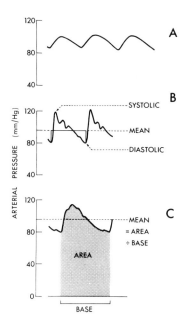

Fig. 2. Comparisons between damped arterial pressure record as obtained with a mercury manometer (A); dynamic arterial pressure as recorded with a high-fidelity electromanometer (B); and the mathematical basis for calculating mean arterial pressure (C) (After Geddes, L.A., 1970, by kind permission of author and publishers).

6. Units of measurement

In biology the most widely used unit has been the millimeter of mercury (mm Hg), based historically on the use of the mercury manometer as a pressure measuring device. For measurements in low-pressure capacity vessels, centimeters of water (cm H_2O), cm saline, or cm blood are quite widely used. The derivation is the saline (water) manometer, and persistence of the units into the electronic age has been assured by clinical practice and by the concept of a static column of blood. It is likely that the units mm Hg and cm H_2O will continue to be used in laboratory and clinical practice for some time yet. However, those concerned with blood pressure measurement should be acquainted with the SI (Système Internationale) unit of pressure, the pascal (Pa = Nm^{-2}), and its derivation most applicable in biology, the hectopascal (hPa). Approximate equivalence of the units is:

$$1 \text{ hPa} = 0.75 \text{ mm Hg} = 1.02 \text{ cm } H_2O$$

TECHNIQUES OF BLOOD PRESSURE MEASUREMENT

1. Milestones

There are a number of excellent reference sources in which more or less detailed accounts of the early history of blood pressure measurement can be found (Franklin, 1933; Fishman and Richards, 1964; Geddes, 1970).

It is usually accepted that the first deliberate attempt to measure arterial pressure was made by the Reverend Stephen Hales in 1733, who by direct cannulation of the femoral artery of a horse, estimated its mean arterial pressure as 8 ft. 3 in. of blood. The more convenient mercury U-tube manometer was not used systematically for this purpose until Poiseuille did so as a medical student in 1828. Graphic recording of the pressure pulse-form by way of a mercury manometer, float, and smoked-drum kymograph was established by Ludwig in 1847, and this technique is still to be found in use in some student laboratories today. There was then a long gap until methods of exact reproduction of the pressure pulse wave were evolved by Wiggers in the early 1900's, and especially by Hamilton in the 1930's. However, while their instruments were capable of remarkably faithful reproduction of the pressure pulse, they were intricate and not suited to routine clinical--or even routine laboratory--use. It is only in the past 30 years that various forms of electromanometer have been developed in a reliable, commercially-available form.

The other notable milestone was the description by Korotkoff in 1905 of the method of indirect arterial pressure measurement in man by way of inflatable cuff and stethoscope that has been a mainstay of clinical practice for the past 50 years.

2. Direct methods

These include all circumstances in which a pressure measuring system is introduced into the bloodstream. The direct method has advanced in sophistication from the 9 ft. long glass tube that Stephen Hales connected by way of a brass pipe to the femoral artery of a horse in 1733, to the minimanometers of today in which the pressure-sensing element itself is introduced into the lumen of the blood vessel. In the account that follows, all (or at any rate, most) of the devices currently used in laboratory or clinical practice will be described, together with their advantages and limitations.

Static methods

These involve the use of a simple fluid manometer, and the variations on this theme are quite limited. Despite their short-comings in reproducing blood pressure, they retain a place in both clinical and laboratory practice because of their simplicity.

Saline manometers, whether of the U-tube or vertical-tube type, are still useful for making measurements in the low-pressure sections of the vasculature. The tube itself may be of autoclavable glass, or of pre-sterilized pre-packaged transparent plastic. The fluid content may be of any convenient isomolar, sterile, physiologic electrolyte, solution. Thus there is safe, direct contact of the fluid of the manometer with the bloodstream, and the only consequent technical difficulty is clotting or sludging of blood in the cannula. In order to maintain a frequency-response that at least permits patency of the cannula to be ascertained, its bore should be not less than 2 mm; to maintain patency of the system over hours or days, intermittent infusion by way of a T-tap is essential; and a low concentration of heparin (e.g., 1000 i.u./l) in the infusion fluid is desirable.

The advantages of this technique are its extreme simplicity, ready availability, and low cost. Its defects are the spatial limitations to measurement of pressures greater than 50 cm H_2O (34 mm Hg), such a low frequency-response that it will do little more than, for instance, indicate the presence of respiratory waves in the venous system, and that it is not capable of providing graphic recording of pressure. Apart from such obvious sources of gross error as malplacement or blockage of the cannula, a major practical difficulty is in determining the exact anatomic location and horizontal plane of the tip of the intravascular catheter. On this latter determination depends the absolute accuracy of static pressure measurement, and the accuracy with which changes in pressure can be observed in face of variations in posture.

Nevertheless it has gained a firm place in clinical practice for monitoring right atrial pressure in intensive-care situations or in the operating room, by way of a catheter introduced through an antecubital or subclavian vein; and for measuring portal venous pressure by pre-operative cannulation of mesenteric veins, or by transcutaneous splenic puncture at the time of splenic portography. The office practice of estimating right atrial pressure by observing the height of filling of the jugular veins is still commonplace. While in general it correlates remarkably well with direct pressure measurement (Borst and Molhuysen, 1952), there are errors in individual patients that are in many circumstances inacceptable (Davison and Cannon, 1974).

Mercury manometers of the U-tube type have been traditional in student laboratories for well over a century as a means for measuring arterial pressure. However, the only real basis for their continued use is on grounds of low cost. This type of manometer scarcely requires description, and the same requirements for maintaining patency of the intravascular cannula obtain as for the saline manometer.

The defects of the mercury manometer as a means for measuring arterial pressure are almost too numerous to be listed. The transduction of arterial pressure waveform is so damped as to render interpretation impossible except as a means of counting heart rate. The apparatus is not readily cleaned nor sterilized, and an open channel between mercury and the bloodstream is not tolerable in clinical practice. While graphical recording of the pressure trace is possible, either by an adaptation of the Ludwig float-to-smoked drum technique, or electrically by way of variable-capacitance-detecting circuitry, at best the methods are clumsy ones for registering changes in mean arterial pressure.

Anaeroid manometers have from time to time been used to measure arterial pressure, in order to overcome some of the problems associated with the mercury manometer. Some form of flaccid diaphragm is employed at the air-fluid interface. The method has little to commend it. These manometers do, however, have a role in clinical or laboratory practice as a less cumbersome means than the mercury manometer for regular calibration of the response of electromanometers.

Dynamic methods

These depend on the transduction of pressure energy into an alternative form which is compatible with a graphical recording system. Their history has been one of a search for methods that will achieve exact reproduction of the pressure fluctuations in the cardiovascular system, and which are at the same time practicable in the laboratory or clinical setting.

The characteristics of an ideal dynamic intravascular pressure measuring system have been fairly clearly defined, usually by Fourier analysis of the harmonic contributions to the arterial wave form. These data are summarized in Table 1. It seems that for good reproduction of the waveform there should be faithful reproduction of the 5th to 10th harmonics (approximately from 0 Hz to between 5 and 10 Hz); while for some special purposes, or if the heart rate is rapid, it may be desirable to extend the range to 20 Hz or more.

It is no mean task to devise a test-system to determine the frequency-response characteristics of a manometer system. It must be appreciated that it is the entire system from the pressure-sampling point in the bloodstream to the ultimate visual record that must be tested. Some mathematical predictions of the behaviors of a system can be made (McDonald, 1968; Geddes, 1970), but the method usually accepted as most effective involves the construction of a

TABLE 1

Approximate percentage amplitude of harmonic contributions to the arterial pressure waveform, and corresponding frequency-responses necessary for instrumental reproduction, at a heart rate of 100/min which is taken to be the 1st harmonic (cf., McDonald, 1968; Geddes, 1970; Taylor, 1973).

	NUMBER OF HARMONIC				
	1st	5th	10th	15th	20th
Contribution of harmonic to waveform	100%	10%	1%	0.1%	0.01%
Frequency response of manometer necessary for reproduction	1.7 Hz	8.4 Hz	17 Hz	25 Hz	34 Hz

sine-wave generator (e.g., Noble, 1959; Yanof et al., 1963; Stegall, 1967), and direct measurement of the frequency-response of the system. There is still the limiting factor of the incapability of the generator to maintain constant amplitude. A less direct, but elegant, method is to analyze the response of the manometer system to a sudden pressure decrement produced by "popping" a rubber membrane enclosing a pressure chamber (McDonald, 1968; Gabe, 1972). Given that there are inevitable imperfections in the frequency-response of a system, and provided they are predictable, it is possible to correct them by rather tedious Fourier analysis, or by way of an analog computer (Melbin and Spohr, 1969).

A great many of the imperfections of frequency-response in direct blood pressure measuring systems derive from the needle, cannulae, catheters and stopcocks placed between the bloodstream and the pressure-sensing device. The viscosity of the contained fluid results in damping, while the characteristics of the wall and lumen of a catheter and its connections and stopcocks can result in resonant amplification of some frequencies. Gas bubbles in the system cause a dramatic reduction in frequency-response. While attention to the bore- and wall-characteristics of the tubing and connections can minimize the resonant peaks (or extend the band-width within which they are absent), and great care can be taken to eliminate bubbles, the resultant system is a compromise. Nor does this eliminate other mechanical distortions that are not fully predictable, such as result from longitudinal acceleration of the catheter ("catheter whip"). For these reasons there has been increasing interest in catheter-tip mounted pressure sensing devices. Before selecting a manometer for direct blood pressure measurements, and especially before setting out to use it, the novice at blood pressure measurement (and even the habitué) would be well advised to consult a good account of the theoretical and practical sources of error (e.g., Geddes, 1970; Gabe, 1972).

Optical manometers operate by way of a light beam reflected from a diaphragm, which is variably deformed in response to variations of pressure across it. The original versions used the reflected light beam itself for recording pressure changes on moving photographic film. These reached their most sophisticated form in that used by Hamilton et al., (1934), in which the diaphragm was of brass to which was cemented a mirror. The improvement in natural frequency to a very acceptable level (over 200 Hz) from the use of a stiff metallic diaphragm was offset by the 5 meter light-path necessary for amplification, the great sensitivity of the apparatus to extraneous vibration, and the inconveniences of photographic recording.

Recently a light beam has been used in a somewhat different fashion as part of a pressure measuring device (Clarc et al., 1965; Ramirez et al., 1969; Lindström, 1970). There are some variations of design, but in most of these systems a reflective diaphragm is mounted at the tip of a hollow catheter containing optical fiberglass bundles. These are divided into two sets: one for transmitted light, the other for reflected. Change in the difference in intensity between incident and reflected light is converted into an electrical signal by what is in effect an optical bridge, and is a function of pressure (Fig. 3). A very satisfactory frequency-response is attainable (Table 2). There are not great technical difficulties with calibration of the static response, which is done by applying known pressure decrements to the back of the diaphragm by way of an

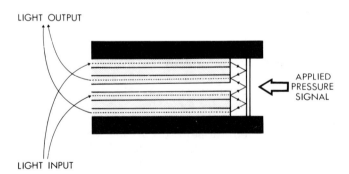

Fig. 3. Principle of fiberoptic catheter manometer, in which light input-output relationship varies according to the movement of a reflective diaphragm (after Ramirez, A. et al., 1969, by kind permission of authors and publishers).

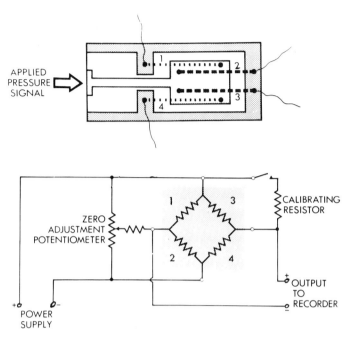

Fig. 4. Principle of an unbonded wire strain gage manometer. Above, tension and relaxation of wire resistances in response to an applied pressure signal. Below, simplified diagram of disposition of wire resistances in bridge circuit (courtesy of Statham Medical Instruments, Oxnard, California).

air-inlet port (which also contributes to thermal stability). The zero stability with time is good, but special measures have to be taken to maintain a check on the pressure zero-point; for instance, the provision of a second portal to the outside of the catheter-tip. A fringe-benefit of these manometers is their capacity to record sound, and there are exciting possibilities of using a fiberoptic system to also transmit information relating to blood flow and oxygen saturation (e.g., Taylor et al., 1972).

Strain gage manometers are more widely used than any other single sort. The principle of their operation is remarkably simple, and depends on the increase in electrical resistance of a wire as it is stretched. Most commonly a 4-element wire strain gage is used, forming the arms of a bridge. Increase in pressure stretches two opposite arms, and relaxes the other two (Fig. 4). Either an AC or DC activating voltage can be used, usually of 5-10V, which gives an electrical output within the 2-30μV/V/mm Hg. A natural frequency of over 150 Hz is attainable, and Lambert and Wood (1947) showed that with a suitable catheter and recording system, a flat frequency-response up to 27 Hz is possible. The other feature is a notably time- and temperature-stable zero baseline and responsivity. Thus this type of instrument fulfills many of the prerequisites of the "ideal" manometer.

As an alternative to the wire gage, a silicon bar bridge can be used (Angelakos, 1964), silicon having a 60-fold greater $\Delta R/\Delta L$ characteristic, though with the penalty of a higher temperature coefficient.

Both wire (Warnick and Drake, 1958) and silicon (Angelakos, 1964) strain gages have been miniaturized into catheter-tip-mountable forms, with acceptable external diameter and frequency-response characteristics (Table 2).

Capacitance manometers represent another manner in which the movements of a stiff diaphragm can be converted into an electrical signal. They have been used in much Scandinavian work on blood pressure from the impetus of the work of Hansen (1949). The movement of a flexible metallic diaphragm in relation to a fixed plate, from which it is separated by air as a dielectric, creates the variable electrical capacity that reflects changes in pressure. The advantages are a high natural frequency and small displacement volume. Against these are set the serious defects of marked thermal sensitivity and susceptibility to random interference from stray capacitance in the cables.

The inductive manometer uses a magnetic core fixed to the back of a flexible diaphragm, with transformer windings that create an inductance bridge. The system has an acceptable natural frequency, good temperature stability, and good signal/output ratio. The main use to which this type of transducer has been put is as a catheter-tip-mounted instrument. This was first done by Wetterer in 1943, but has been more recently developed and sophisticated for human use by Gauer and Gienapp (1950) and Laurens et al. (1959).

Catheter tip manometers (minimanometers) have been referred to earlier, but deserve separate and special consideration. Their great advantages are that the inherent natural frequency of the transducer can be used, without the damping or resonance caused by transmission of the pressure signal through a fluid-filled catheter; without the problems of establishing a hydrostatic zero-level and with minimal effects from longitudinal acceleration. They are seen to best advantage as pressure-measuring devices by comparison with external manometers when the rate of ventricular pressure rise is to be measured (Gould et al., 1973). A fringe benefit in diagnosis is that their natural frequencies reach the audio range, so that by suitable filtering, both pressure and heart sounds can be recorded simultaneously. Potential or actual problems with these devices

TABLE 2

Some characteristics of catheter-tip manometers. Sources of data: general (Geddes, 1970); wire strain gage (Warnick and Drake, 1958); silicon strain gage (Angelakos, 1964); inductive (Gauer and Gienapp, 1950; Laurens et al., 1959; Allard, 1962); fiberoptic (Ramirez et al., 1969; Lindström, 1970).

Type of transducer	Diameter of catheter (mm)	Upper limit of flat frequency-response (Hz)	Working pressure range (mm Hg)	Methods of sterilization
Wire strain-gage	1.7 - 2.2	1000-2000	-50 to +300	Chemical
Silicon strain-gage	1.0 - 2.3	>2000	0 to +250	Chemical
Inductive	2.7	>1000	-50 to +250	Chemical
Fiberoptic	1.5 - 2.7	1600-15000	-50 to +250	Ethylene oxide or chemical
Stegall et al., 1968	(K) + 1.9 (Dw) - 0.1	0.97 0.99	- 0.2 + 0.3	0.96 (U) 0.98
Kazamias et al., 1971	(Df) + 2 to -3	0.99	+ 2	0.95

stem from three sources. Perfect electrical insulation is essential if they are to be used close to the heart. Miniaturization heightens mechanical problems, such as fibrin-deposition on the pressure diaphragm and contact with the wall of the vessel or heart. Perhaps the greatest practical problem is inaccessibility of the transducer, so that the true zero cannot readily be checked while it is in use (the input/output characteristics can usually be checked electrically or by way of a tube leading from the back of the diaphragm to the atmosphere). The ultimate problem is the technical one of miniaturizing the system to the point that its size is small in relation to the diameter of the vessel, so that there is no artefactual distortion of blood flow and blood pressure.

There are available catheter-tip manometers that employ a strain gage (Warnick and Drake, 1958; Angelakos, 1964), an inductive or differential transformer system (Gauer and Gienapp, 1950; Laurens et al., 1959; Allard, 1962), an optical system (Clarc et al., 1965; Ramirez et al., 1969, Lindström 1970) and an ultrasonic system (Noble, 1957). Some of their characteristics are listed in Table 2. Recently the electrolytic resistor has been revived as a transducer for catheter-tip use (Hök, 1974).

Micromanometers are in increasing demand as attempts are made to quantitate the behavior of the microvasculature (Baez, 1977). Null-point micromanometers, using a bubble or dye as indicator, together with almost any form of manometer, have been in use since Landis (1934) measured capillary pressure in animals and in man. Some approximtion to mean pressure is measured with this type of instrument. Methods of reproducing with reasonable accuracy the pulsatile pressure changes in microscopic vessels have not yet been fully developed. With some modifications and great attention to detail, a conventional transducer can be used (Bloch, 1966; Raper and Lavasseur, 1971). Wiederhielm and Rushmer (1964) have described a servo-null method, using a 0.5-1.0 μm pipette and resistance bridge circuit to maintain a blood-saline interface in constant position. Intaglietta et al. (1970) have reported satisfactory measurements with a similar instrument.

The arterial loop technique is one which has been used more and more frequently to overcome the problem of repeated blood pressure measurements in animals. Indirect ("non-invasive") measuring techniques are commonly of dubious accuracy in animals, while to gain repeated access to an artery of sufficient size for direct measurements often requires exceptional skill, or at least an abnormal posture (for the animal). Construction of a subcutaneous carotid arterial loop using conventional surgical techniques has successfully overcome these problems in animals ranging in size from rabbits to cattle. The use of arterial or vein grafts, and/or microsurgical techniques, should in future provide repeated access to the arteries of smaller animals.

3. Indirect methods

As might be expected, these provide only approximate values for intravascular pressure. Their great merit is that to a greater or lesser degree they are "non-invasive," and such an approach may be necessary for a variety of practical reasons. Arterial pressure measurement is so important as a clinical tool in human medicine that a method must be available to every physician. Long term changes in arterial pressure must also sometimes be measured in laboratory animals, and an indirect method may therefore be appropriate. Finally, most pressure measurements in the microvasculature have been made by indirect means, because of the technical difficulties of direct methods.

Sphygmomanometry

The arrest of flow down an axial artery by the application of an external pneumatic cuff within which the pressure is manometrically registered, and the detection of the level of applied pressure at which cessation or recommencement of flow occurs, has been extensively used in man and laboratory animals as an indicator of arterial pressure.

At the outset, it is essential that geometric considerations are satisfied, so that the external pressure exerted by the pneumatic cuff is indeed transmitted to the entire cross-section of the limb. It is not unfair to say that the isobaric geometry has been but little studied by those using sphygmomanometry (but see Ludbrook and Collins, 1967), and that the accuracy of results published using this general method is thus often in doubt. The extremity to which the pressure cuff is applied should be approximately cylindrical, and should be well-covered with soft tissue; while the length of the extremity over which external pressure is applied must be such as to leave no room for doubt that in the center of the limb the tissue pressure is equal to the applied pressure. So far as the human arm is concerned, standards for cuff size have been set (American Heart Association, 1967); but these should indeed be regarded merely as standards and not necessarily as based on generally-applicable, scientifically-established, data (Geddes, 1970). Other sites at which the external cuff technique has been used with some expectation of adequate pressure transmission to the axial artery are the human lower leg, the human finger, the sub-human primate upper limb, the rat or mouse tail, and the legs of birds.

The second element to this technique is the means for detecting cessation or onset of blood flow past the cuff as it is inflated or deflated. Digital palpation of a distal pulse is one such method, though a relatively insensitive and subjective one. A plethora of supposedly more sensitive and objective methods have been described:

direct auscultation; microphone-amplifier recording; visible capillary refilling or pulsation; oscillometric, plethysmographic, or mercury-in-rubber strain gauge detection of pulsation; and most recently, transcutaneous ultrasonic detection of flow or of arterial wall movement.

Geddes (1970) has reviewed very thoroughly and critically the variants of this technique that have been employed in animals (and man). However the sphygmomanometric method of measuring arterial pressure in man deserves special attention from three points of view: the ability to detect diastolic arterial pressure, the accuracy with which systolic and diastolic pressure can be determined, and the capability of the method to be automated. The discussion of these points can be used to appraise old (or new) methods of sphygmomanometric blood pressure determination in animals.

The enormous success of sphygmomanometry in man stems from Korotkoff's description of stethoscopically audible sounds over the brachial artery at the lower border of the cuff, which are related to systolic and diastolic blood pressure. The basis for the sounds is complex (and probably even yet not fully understood), but their empirical relation to blood pressure is well-established. The main source of debate is whether they arise from movement of the arterial wall, or from movement of blood (or more likely both). Geddes (1970) reviews the experimental work that bears on this. In ordinary clinical practice, using a stethoscope to detect the Korotkoff sounds, there are three notable episodes as the cuff pressure is lowered steadily from a suprasystolic level to zero: a tapping sound that is related to systolic pressure; a sudden muffling of sound; and finally disappearance of sound. Both the latter events bear a relation to diastolic pressure.

The tapping sound is detected at a cuff pressure that is invariably below true systolic pressure, but the relation is a reasonably consistent one. There is much less consistency in determining diastolic pressure, and the point of muffling has generally been found to be higher than, and the disappearance of sound lower than, true diastolic pressure. While the relation between true arterial pressure and that determined by sphygmomanometry is documented for normal human subjects, there are less data on humans that are hypotensive, hypertensive, or tachycardic, though the effect of obesity has been studied (Nielsen and Janniche, 1974). Such data as there are suggest that considerable inaccuracies may occur, especially in the estimation of diastolic pressure. Another source of error that is important in epidemiologic surveys of blood pressure is bias on the part of the observer, and random-scale manometers have been designed to overcome this (Yarrow, 1963; Rose et al., 1964; Wright and Dore, 1970).

Other methods of detecting the systolic and diastolic end-points have been devised, originally in response to a need for more remote monitoring than permitted by a stethoscope, and latterly with respect to automated systems for use in intensive-care and operating-room conditions. All the automated methods involve automatic, intermittent, rapid inflation and constant-rate deflation of the pneumatic arterial occlusion cuff. The pressure in the cuff is displayed, usually by a transducer and recorder. Parallel with, or superimposed on, this trace is the record of some parameter designed to indicate systolic (and sometimes diastolic) pressure. Those parameters that have been most useful are described below. Doppler-shift of ultrasound has been used in two ways. Stegall et al. (1968) used a beam directed across the brachial arterial wall to register changes in wall movement, while Kazamias et al., (1971) used a beam directed to register changes in blood flow. Both groups reported excellent correlation with directly measured pressure in small groups of normal subjects. However it is likely that, as with Korotkoff sounds, major errors will occur if there is vasoconstriction and a low vascular capacity distal to the cuff. In these circumstances arterial inflow may fill this capacity before the cuff pressure falls to the diastolic level. Other methods have employed a microphone with frequency-filtration of the output (Roman et al., 1965), or some form of oscillometry, to indicate systolic and diastolic pressures with varying success. There is still an unfortunate lack of data to indicate in just what circumstances automated methods are reliable or otherwise. However, for many clinical purposes the frequency of sampling of blood pressure, and the relative absence of observer bias, may more than make up for a deficiency of accuracy.

In Table 3 are listed data obtained from comparison of direct intraarterial pressure measurements, with indirectly measured upper limb arterial pressures by conventional and Doppler techniques. These data should not be accepted uncritically, for with one exception (Raftery and Ward, 1968) the studies are notable for the lack of information on the accuracy of the electromanometer systems employed, lack of precise definitions of the end-points for diastolic pressure by indirect measurement, and lack of control of observer bias.

One other application of sphygmomanometry should receive brief mention. While subcutaneous arterial loops are most suited to direct pressure measuring methods, they can be used to estimate systolic and diastolic pressure indirectly by the use of Korotkoff sounds.

Counterpressure techniques

The principle is one of applying direct counterpressure, and determining systolic pressure by direct, magnified, observation of

TABLE 3

Some reported comparisons in the human upper limb of directly versus indirectly measured arterial pressure. Indirect methods used were based on Korotkoff sounds (K), Doppler-recorded arterial wall movement (Dw), or Doppler-recorded blood flow (Df). The end-points for diastolic pressure by the Korotkoff sound method were either muffling (M), disappearance (D), or unspecified (U).

Investigators and indirect method used		Systolic pressure		Diastolic pressure	
		mean direct-indirect (mm Hg)	correlation coefficient	mean direct-indirect (mm Hg)	correlation coefficient
Roberts et al., 1953	(K) (K)	+ 12.0	0.89	+ 3.4 + 7.4	0.89 (M) 0.60 (D)
Moss and Adams, 1963	(K) (K)			- 13 + 6	0.36 (M) 0.32 (D)
Holland and Humerfelt, 1964	(K) (K)	+ 24.6	0.95	+ 5.3 +13.1	0.83 (M) 0.93 (D)
Stegall et al., 1968	(K) (Dw)	+ 1.9 - 0.1	0.97 0.99	- 0.2 + 0.3	0.96 (U) 0.98

TABLE 3 (continued)

Investigators and indirect method used		Systolic pressure		Diastolic pressure	
		mean direct-indirect (mm Hg)	correlation coefficient	mean direct-indirect (mm Hg)	correlation coefficient
Raftery and Ward, 1968	(K)	+ 5.4	0.86	- 11.1 - 6.6	0.87 (M) 0.83 (D)
Kazamias et al., 1971	(Df)	+ 2 to - 3	0.99	+ 2	0.95
Nielsen and Janniche, 1974	(K)	+ 11.2	0.95	- 7.2	0.92 (D)

the compressed vessels. This method has been used in man to compare retinal artery pressures, though with dubious accuracy (Lowe and Stephens, 1961; Russell and Cranston, 1961), and to measure pressure in nailfold capillaries (Landis, 1934). In animals, a similar technique has been used for mesenteric and ear vessels. It is doubtful whether any of these techniques are now of more than historic interest.

Wedge pressure

This indirect (but invasive) technique has been used to measure pressure in inaccessible vessels. If flow through a resistance is arrested distally, the mean pressure distal to the resistance becomes equal to that proximal, the column of blood acting as a pressure-transmitting device. Thus a catheter passed retrogradely in an hepatic vein to the point that the vein is occluded or "wedged" will reflect portal venous pressure. Similarly a catheter passed antegradely to wedge a pulmonary artery will reflect pulmonary venous pressure.

DEFINITIONS OF BLOOD FLOW

As in the case of blood pressure, we have thought it necessary to clearly define the several meanings of the term blood flow.

1. Single vessel vs tissue flow

Most times the investigator is concerned to measure some aspect of flow at a single point in a single, anatomically-defined, blood vessel. Conceptually, the measurement of cardiac output falls also into this category. However, even in this circumstance the real aim of the investigator may be to measure the flow through (i.e., to or from) a discrete organ or volume of tissue. There is often some uncertainty as to the precise volume of tissue that an artery is supplying (or a vein draining), and even more uncertainty that with change in cardiovascular function this tissue-volume will remain constant. This matter will be taken up in detail later.

There are also methods designed to measure flow through a tissue in a more direct fashion: that is, the total through-flow of blood in a defined volume of tissue in unit time. Very often the same problems as referred to above exist, in that there may be uncertainty about the volume of tissue being studied. Another approach is to measure the partitioning of blood flow among the various tissues or organs in a direct fashion: for instance, the distribution of cardiac output.

2. Velocity of flow vs volume-flow

These are important distinctions, often not fully comprehended by investigators. In the case of single vessels, it is most often the velocity of flow that is actually measured, volume-flow in unit time being calculated from the internal cross-sectional area of the vessel. In the above sense, velocity is a measure of the mean linear velocity of a notional wave-front of blood flowing through the vessel. However, velocity of flow can also be conceived of (and measured) as that existing at a single point in the lumen of the vessel. The shape of the notional wave-front, and its mean velocity, is thus some integral of the point-velocities.

On the other hand, it is usually the volume-flow that the investigator actually seeks to measure. In the case of a single vessel, it is a matter of the volume of blood passing the point of measurement in unit time. In the case of tissue flow, it is the volume through-flow in unit time per unit volume of tissue.

3. Constant flow vs pulsatile flow

It is rare in the vasculature for blood to flow at a steady velocity. As the pressure-generators responsible for flow (most importantly the heart) have reciprocating actions, flow in most segments of the vasculature is correspondingly pulsatile.

Steady flow of blood, or indeed of any other fluid, can be measured fairly easily and fairly accurately. Conversely, techniques for measuring pulse-by-pulse changes in flow have only quite recently been devised, and most of the older measurements of blood flow that was actually pulsatile have been capable of measuring only the mean net forward flow in a blood vessel. Some rather important phenomena in the vascular system were consequently obscured. For instance, because the heart ejects its pulsatile output into a branching elastic-compliant system of tubes, at the end of which there is considerable resistance to flow offered by their diminution in size, it does not follow that the direction of flow in arteries is always cardiofugal. These matters will be discussed in more detail, but it is sufficient for the moment to recognize that in such a pulsatile flow system there is opportunity for momentary reversal of direction of flow.

4. Laminar vs turbulent flow

In day-to-day experience of fluid flow (most often from domestic faucets), there is no occasion to believe that the advancing wave front is other than perpendicular to the axis of the pipe. In actual fact--and it is no new discovery--the wave front of any fluid flowing through a tube is approximately parabolic. The conceptual explanation of this is not simple to follow, but in essence is that there is frictional resistance or "drag" at the interface of fluid and wall, as well as frictional or viscous resistance to flow within the fluid. The net result is an axial core of fluid that moves at a greater velocity than the peripheral "boundary zone." Given a uniform bore to the tube, and uniform properties of its lining, the characteristics of this laminar flow, and the shape of the wave front, are constant and can be predicted with some confidence.

However, when the above desiderata are not met, a variety of disturbances to laminar flow may occur. The simplest to comprehend is turbulent flow, in which there is a notable motion of units of the fluid (laminae, or more simply, individual red cells) in a direction perpendicular to the axis of the vessel. Transient episodes of turbulence can be produced in a uniform bore tube when flow is pulsatile. Turbulence more readily occurs when the viscosity of the fluid is low (i.e., there is anemia). Finally, a variety of non-uniformities of the wall of the tube may cause transient turbulence, whether these be cardiac valves and arterial branches, or pathological stenoses, irregularities, and arteriovenous fistulae. The biophysics of these phenomena will be considered elsewhere. Suffice to say that the major consequence in terms of flow is that at a site of turbulence energy losses are greater, and the hydraulic pressure gradient is steeper, so that net flow through the system is less for a given energy input.

5. Distribution of blood flow

This term has a great many meanings, and indeed in any particular circumstance the most useful one is strictly operational and based on methodology. It is nevertheless worth outlining what investigators would like it to mean in various circumstances.

There is interest in defining the distribution of cardiac output to all of the various tissues and organs of the body with changes in body function (rather than just one or two). That is, to study the borrowing/lending phenomena of blood flow, given a maximum limit to cardiac output, or even merely a need for economy of cardiac work.

For other reasons, there is growing interest in discerning the changing pattern of blood flow distribution within a single organ or region, whether under varying physiologic conditions or as a result of disease process. Examples are the distribution of cerebral blood flow with normal or disturbed mentation, and the exaggerated borrowing/lending phenomena that occur with arterial stenosis or occlusion.

Finally, there is intense interest in making qualitative and quantitative observations at a microcirculatory level on the redistribution of blood flow among the various parallel elements connecting arterioles with venules that occurs in response to local, neural, or humoral stimuli. In this context the real problem to be examined is the effects of such redistributions of blood flow on blood-tissue exchange of water, solutes, and even cells. Hence the origin of such terminology as "nutritive" flow (to indicate flow through microvessels across the walls of which such exchanges can take place), as against "shunt" flow. As will become apparent, the tools for quantitating such functions of the microvasculature are as yet crude and often indirect or even speculative, but interest runs high and will continue to do so.

6. Short-term (acute) vs long-term (chronic) flow measurements

This is not so much a matter of definition as one of introduction and explanation. The methodology of making blood flow measurements over a matter of minutes or hours presents far fewer difficulties than estimations of the behavior of flow over longer periods of time. In the short term, such factors as the geometric location of the flow measuring device, blood clotting, or tissue reaction that may interfere with measurement, and the calibration of the instrument, are usually controllable (though nonetheless important). However, physiologists--and clinical scientists--are becoming more concerned with variations in blood flow over a matter of days, weeks, or months; and in free-ranging animals or men rather than those tied to the laboratory bench. This latter approach presents formidable problems of methodology in all the matters referred to above (most notably that of maintaining a check on continuing accuracy), as well as presenting the new problem of information-transmission from the fully or partly free-ranging subject of the investigation.

7. Units of measurement of flow

There is a great deal of variety in the conventions used in biology to quantitate blood flow. The simplest is linear velocity, readily expressed in regular SI units as meters per second (ms^{-1}). Linear volume flow in a single blood vessel should be expressed as cubic volume per unit of time (e.g., $m^3 s^{-1}$). However, it is still common practice to describe flow in liters (or submultiples of liters) per unit time. In describing volume-flow in a tissue, there is a multiplicity of conventions used by physiologists, depending on the measuring technique. Clearly, the proper SI unit should be cubic volume of blood per cubic volume of tissue per unit of time ($m^3 m^{-3} s^{-1}$ or s^{-1}). In practice such units as ml/100ml/min or ml/100g/min are used. In the case of cardiac output, the cardiac index $1/m^2/min$) is often used to make intercomparisons.

TECHNIQUES OF BLOOD FLOW MEASUREMENT

1. Milestones

It has been a very much more difficult matter to measure blood flow than blood pressure. It has been especially difficult to devise methods that do not of themselves alter blood flow, and that are applicable to man. This is clear from the immense number of blood flow measuring techniques that have been devised, usually either to be subsequently discarded or to find only very limited application. In this introduction only the principles and method that have resulted in major discovery or widespread application have been regarded as milestones. More details of these--and of many methods that might better be described as tombstones--can be found in various refference works (Franklin, 1933; Potter, 1948; Hyman and Winsor, 1961; Fishman and Richards, 1964; Rushmer, 1966).

In this field, Stephen Hales (1733) again seems to have been a pioneer. On a basis of post-mortem left ventricular volume, and of heart rate in life, he calculated the cardiac output of a horse to be 360 cubic inches per minute (in the event, a gross under-estimate!), and extended these calculations to man. The name of Ludwig also reappears (Ludwig and Dogiel, 1867) with the invention of the stromuhr for measuring flow in a cannulated vessel. Other early methods were a paddle-wheel in an interposed cannula (Chauveau, 1860), and circulation time as indicated by the injected marker potassium ferrocyanide (Hering, 1829).

In 1870 Fick published his now famous principle, used later as an "indirect" (Plesch, 1909) and a "direct" (Cournand, 1945) method for measuring cardiac output in man. The Fick principle has also been the basis of indicator-dilution methods, the indicators at first being gases like nitrous oxide (Krogh and Lindhard, 1912) or chemicals (Stewart, 1897). In the modern era have come dye dilution (Moore et al., 1929) and thermal dilution (Bennett et al., 1944; Fegler, 1954).

If measurement of cardiac output has presented a problem, no less so has measurement of flow in (other) single vessels. Some of the many devices that have been employed for this purpose will be discussed briefly later in this account, but few can be described as milestones. It is only in the past 40 years that relatively non-invasive, non-artefact-producing, reasonably reliable, methods have been developed, notably the electromagnetic flowmeter (Kolin, 1936), and the Doppler ultrasonic flowmeter (Franklin et al., 1961).

Finally, among the milestones should be considered methods for measuring tissue flow--that is, volume-flow with respect to time in some known or estimated volume of tissue. Two techniques are notable, not because of their versatility and accuracy, but because of the valuable data that have been derived from their application, especially to man. The first is venous occlusion plethysmography, originally used to measure renal blood flow (Brodie and Russell, 1905), and then applied by two generations of physiologists to human limbs. The second is the tissue clearance of injected, diffusible, gamma-emitting radionuclides. Valuable data have been obtained on skin, muscle, renal, and especially cerebral blood flow by this means. The clearance methods derive from the pioneer work of Kety and his colleagues, who evolved the mathematics and methodology for measuring cerebral blood flow by nitrous oxide inhalation (Kety and Schmidt, 1948), and tissue blood flow by radiosodium clearance (Kety, 1949).

2. Methods for single vessels

Included in this account will be methods for estimating cardiac output (regarding the aorta or pulmonary artery as a single vessel), methods applicable to other macroscopic vessels, and the few methods that can be used in microscopic vessels. All these methods except those for microscopic vessels can be used both in animals and in man, the only limiting factor being one of vessel size. Only those methods that carry with them proven high standards of accuracy will be considered in any detail here. Some of the multitudinous methods that have been employed in the past, but which cannot now be recommended either on grounds of inaccuracy or because of the production of artefactual changes in flow, will be separately listed.

Methods employing the Fick principle:

In its simplest form the Fick principle can be described as:

$$F = \frac{\Delta I/\Delta t}{[I]_d - [I]_u}$$

that is, volume-flow (F) in a given period of time (t) is equal to the amount of indicator substance (I) entering the stream of flow in that period of time, divided by the difference in concentration of the indicator upstream $[I]_u$ from that downstream $[I]_d$ to the points of entry (or removal) of the indicator (Fig. 5).

The Fick principle forms the basis of more methods of measuring blood flow than any other. An enormous variety of indicators have been used, and there has been a great deal of ingenuity in devising methods of introducing or removing the indicator from the bloodstream. It is beyond the scope of this account to describe the details of these numerous permutations, but reference sources will be indicated.

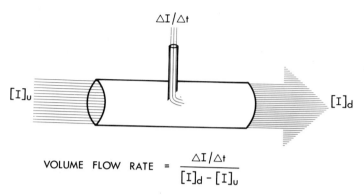

Fig. 5. Diagrammatic illustration of application of Fick principle to the measurement of single vessel, steady-state, volume blood flow by continuous infusion of an indicator. ' $\Delta I/\Delta t$ = rate of infusion of indicator; $[I]_u$ = upstream indicator concentration; $[I]_d$ = downstream indicator concentration.

For the Fick principle to be validly applied, some desiderata must be met. An indicator must be innocuous, and not vasoactive. It must not leave the vasculature between the points of injection and sampling. There must be complete and uniform mixing of the indicator with the blood stream shortly beyond the injection point; indeed, in the branching vascular system there must have been complete mixing between the point of injection and the first branch of the blood vessel. Because the vasculature is a circular and closed (rather than longitudinal, open-ended) set of tubes, there will usually be recirculation of the indicator past the point of injection; this must be allowed for.

Perhaps the most notable application of the Fick principle, at least in the historic sense, has been to measure cardiac output using oxygen as the indicator. In the direct method the rate of O_2 consumption by the man or animal is measured, the right heart being used as the "downstream" sampling point, and any peripheral artery used for "upstream" sampling. Since Cournand and his associates (Cournand, 1945) demonstrated the feasibility of this method in man, the methods of measuring O_2 consumption and blood O_2 content have been sophisticated (e.g., Guyton et al., 1973), but otherwise the method is unchanged. The method is believed to be accurate to within a few percent, and because of its simplicity of concept, has been usually used as the reference against which to compare newly-devised techniques. The main problems with the method are that the O_2 consumption of the subject must be in a steady state, that the method measures "mean" cardiac output over the period of time necessary to estimate O_2 consumption reliably (1 to 5 minutes), and that both right heart catheterization and systemic arterial cannulation are necessary. A variety of <u>indirect</u> methods have been devised to avoid the need for right heart catheterization, using CO_2, or inhaled foreign gases such as acetylene or nitrous oxide as indicators (Guyton et al., 1973). It is doubtful whether their accuracy entitles them to continued use in man despite their non-invasiveness.

The Fick principle has also been put to use in measuring renal, hepatic, and cerebral blood flow, but these will be considered under the heading of tissue blood flow techniques. It also forms the basis for the indicator dilution techniques described in the next section.

Indicator dilution techniques

If the Fick principle can be used to measure cardiac output, it can clearly be used to measure flow in other single vessels by appropriate adaptation and provided the necessary desiderata are met.

In theory, a continuous measure of flow in a vessel such as the femoral artery is possible by constant-rate infusion of a suitable indicator into the femoral artery, and continuous sampling from the femoral vein. There are, however, practical difficulties. Complete mixing must be attained before the first downstream branch, and mere infusion through a needle or catheter will not accomplish this: streaming occurs, and can be prevented only by some form of jet injection (Overbeck et al., 1969). Another major practical difficulty is the build-up of recirculating indicator, which sets limits to the duration of flow measurement--especially if the indicator is a dye or radionuclide, and if the subject of investigation is man. There is also the problem of the cumulative loss of blood volume due to sampling. Nevertheless, such continuous infusion methods have been described and have found limited application in man (Overbeck et al., 1969; Jorfeldt and Wahren, 1971).

The practical breakthrough derives from the theoretical and experimental work of Stewart, and of Hamilton and his colleagues (Guyton et al., 1973): the single injection methods. The notion behind these is that if a single dose of indicator is injected over a brief time, then provided complete mixing is obtained at the site of injection, a record of the concentration-time course of indicator as it passes a downstream point will allow calculation of flow past the point of injection. That is, the Fick equation is modified to:

$$\dot{Q} = \frac{D_I}{\overline{C}_I \times t}$$

where $\dot{Q}$ is the volume flow-past in unit time, D_I the injected dose of indicator, t is the time it takes to pass the sampling point, and $\bar{C}$ its mean concentration over that time (see Fig. 6).

The greatest difficulty with the single-injection method is that recirculation of the indicator is likely to occur before the "tail" of the concentration-time curve has passed the sampling point. This is overcome by replotting the curve on a basis that its initial exponential decay continues in this form (Fig. 6), or by using an on-line computer to achieve the same end. There is not such technical difficulty in ensuring that the duration of the injection is brief by comparison with the time taken to pass the sampling point, and the necessary high-pressure injection aids mixing. Guyton et al. (1973) discuss these points in more detail, and the mathematical bases for dilution curves have been analyzed by Harris and Newman (1970); it seems that either a compartmental or a distributional model will

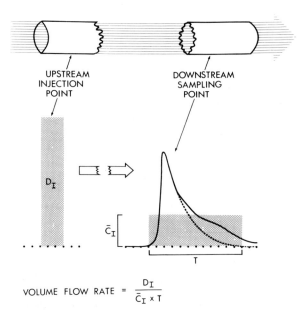

Fig. 6. Diagrammatic illustration of <u>Stewart-Hamilton method</u> of single vessel, steady-state, <u>volume blood flow measurement</u>. D_I = dose of indicator introduced upstream by brief injection; $\bar{C}_I$ mean downstream concentration of indicator, determined by continuous downstream sampling (continuous line) during time (T) of passage of indicator and corrected for recirculation (broken line).

fit observed data equally well. A comprehensive theoretical and experimental review of 25 methods of indicator dilution curve analysis has been published by Spieckerman and Bretschneider (1968).

A great variety of indicators has been used. Non-diffusable dyes have been traditional, first T-1824 (Evans Blue), and more recently indocyanine green. The latter has an advantage in terms of spectral discrimination from hemoglobin; and because it is very rapidly taken up by the liver and excreted in bile, one of the difficulties with repeated flow measurements is minimized: the build-up of a background concentration of indicator.

The other indicator that has been receiving increasing attention is heat (or cold). This is most successfully delivered by injecting room-temperature saline, the concentration-time curve being measured by a thermistor. The enormous potential advantages are that the indicator is cheap and innocuous, and that there is no background build-up so that unlimited repetition of measurements can be made.

Some of the many other indicators that have come and gone are: saline, plasma or dextran with sensing of electrical conductivity or optical transmission; gamma-emitting radionuclides, with the promise of non-invasive detection of the concentration-time curve; and hydrogen, with a hydrogen electrode sensor.

A great variety of sampling devices has also been used (Guyton et al., 1973). For dyes, these have evolved from tedious external spectrophotometric analysis of sequential blood samples, to a cuvette through which the blood is drawn, to a transmitted light/filter/photocell device fitted to the ear, to intravascular sensing by way of a fiberoptic catheter (e.g., Singh et al., 1970). For thermal dilution, the low heat-capacity thermocouple has had to be discarded in favor of the thermistor, with its much greater electrical output, in order that the dose of "heat" or "cold" can be kept low.

Indicator-dilution techniques have found their greatest place in cardiac output measurement in human clinical and physiologic practice and in the animal laboratory. Dye dilution is more firmly established, and has been repeatedly validated against the direct Fick method (Guyton et al., 1973), though a recent study in dogs of indocyanine green dilution versus both electromagnetic flowmetry and timed volume collection suggests a consistent overestimate of flow by the dye method (Donald and Yipintsai, 1973). The most usual technique is still to make the injection into the right atrium, and to sample by withdrawal through a peripheral arterial cannula and cuvette densitometer, with or without on-line bench computer analysis. Thermal dilution, using room temperature saline injection into the right atrium through a catheter (which may be thermistor-tipped), and temperature-sensing by way of a thermistor-tipped catheter in the pulmonary artery (Fig. 7), has been carefully tested (Fegler, 1957; Singh et al., 1970; Ganz et al., 1971; Wilson et al., 1972; Forrester et al., 1972) and found to correlate well with single-injection dye dilution (r = 0.90), and with direct Fick measurements (Branthwaite and Bradley, 1968), though sometimes with systematic differences. The alternative sampling site of the aorta carries with it the hazard of heat exchange in the pulmonary vascular bed. Though the thermal dilution method with aortic sampling compares well with dye dilution and Fick determinations (e.g., Goodyer et al., 1959; Evonuk et al., 1961), the uncertainties of pulmonary heat exchange in changing circumstances limits its reliability when the injection is made into the right heart (Singh et al., 1970: Warren and Ledingham, 1974). Thus either right atrial injection with pulmonary artery sampling, or left atrial injection with aortic sampling, appear to be reliable methods. Thermal dilution has been used to measure cardiac output in rats (Cooper et al., 1963), rabbits (Korner, 1965; Warren and Ledingham, 1974) and dogs (Khalil et al., 1966) as well as in man. In conscious rabbits thermodilution overestimates flow

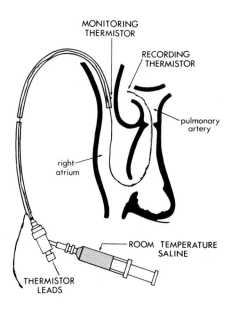

Fig. 7. Room-temperature saline thermal dilution technique for cardiac output measurement. Dose of "cold" calculated from volume injected, and temperature of injectate (monitoring thermistor) at point of entry into right atrium. Dilution curve of indicator obtained from thermistor in pulmonary artery. (After Ganz, W. et al., 1971, by kind permission of authors and publishers.)

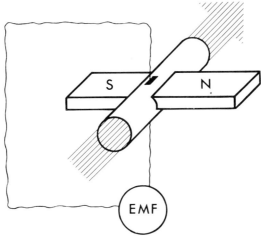

Fig. 8. Diagrammatic illustration of principle of blood flow measurement by <u>electromagnetic flowmeter</u> applied to the outside of a vessel. The magnetic field applied across the vessels sets up an e.m.f. which is at right angles to the axis of flow and to the magnetic field, and which is a function of the flow rate of blood.

at high cardiac output, compared with the Doppler or electromagnetic flowmeters (White et al., 1974). There are other methods for using thermal dilution that will be considered later.

Bolus injection of dye has also found some applications in measuring volume-flow in peripheral vessels, especially in intact humans in whom externally applied flowmeters are inapplicable (e.g., Folse, 1965). Recently a non-invasive method applicable to coronary arteries has been described, using the transit-time of radiocontrast medium and estimating vessel cross-section by biplane orthogonal radiography; good correlation (r = 0.95) with electromagnetic flowmetry in dogs is claimed (Smith et al., 1973).

Electromagnetic flowmeters (Fig. 8)

The principle of these is simple: that a magnetic field operating on a flowing conductor sets up an e.m.f. in a direction perpendicular to both the magnetic field and the axis of flow, and this e.m.f. is a function of the velocity of flow of the conductor. It is only the practice that is difficult, but since Kolin's original (1936) proposals there has been enormous development of the principle into practical and reliable flow-measuring instruments (Berger and Gessner, 1966; Cappelen, 1968; Mills, 1972).

Most attention has been given to probes that can be applied to the outside of a vessel, and these are available for an approximate range of vessel diameters of 1 to 40 mm. The initial problem of polarization potential has been overcome by using a magnetic field of alternating polarity (square wave, or "chopped" sine wave). Some sources of artefact remain, in the forms of transformer effects on electrical leads, variation in contact of the electrodes with the vessel wall, galvanic coupling between the magnet winding and the electrodes, imperfect axial symmetry of the probe with the flowing blood, interference with the magnetic field by nearby metallic objects, the consequences of imperfect grounding, and variations in hematocrit (Bergel and Gessner, 1966: Mills, 1972). The greatest practical difficulty is that the zero-flow baseline must be repeatedly determined during long-term measurements, and that this can be done with accuracy only by arresting blood flow; the signal obtaining when the magnet is switched off cannot be used as a reliable substitute. This source of error is least when the flowmeter is chronically implanted around the ascending aorta in order to measure cardiac output, when end-diastolic flow can be considered to be zero.

The enormously advantageous properties of electromagnetic flowmeters are that they measure instantaneous flow; that they are direction-sensitive; and that they can be chronically implanted for long-term measurement. A flat amplitude-frequency response is possible up to at least 10 Hz and probably beyond, but with a phase-lag that increases with frequency (Bergel and Gessner, 1966; Hainsworth et al., 1968; Mills, 1972). Silicon rubber embedded probes (Folts et al., 1971) can be readily sterilized, and are relatively non-reactive to tissues and non-erosive, making long-term implantation with plug-in leads feasible. On the other hand, measurement of flow in the free-ranging animal has been better accomplished by telemetered Doppler flowmeters (Vatner et al., 1970).

There have been limited applications of the extravascular probe to man, mainly as a means for determining in the operating-room that blood flow is adequate in a surgically created arterial graft or prosthesis. As a pure research tool, these probes have found their greatest application in examining oscillatory pressure-flow relationships (Taylor, 1973), as a conventional (if expensive) method of measuring flow in acute animal experiments, and for measuring long-term aortic or peripheral arterial flow responses by chronic implantation.

Electromagnetic flowmeters can also be miniaturized (down to 0.5 mm diameter) for catheter-tip mounting (Mills, 1972; Taylor, 1973) and have been used in the human heart, aorta and venae cavae as a clinical research tool (Mason et al., 1970). When the instrument is used in an intravascular location it merely measures velocity

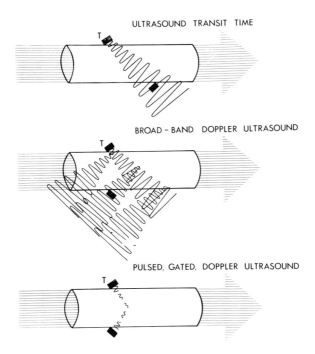

Fig. 9. Illustration of three ways in which ultrasound can be used to measure single-vessel blood flow. Above: transit time of ultrasound. Middle: Doppler-shift of broad band ultrasound, sampling the entire wave front of flow. Below: Doppler-shift of pulsed, gated, ultrasound, sampling from a small portion of wave front of flow.

of flow, and only when there is an assurance that the velocity profile is flat can assumptions be made about volume-flow with any confidence.

Ultrasonic flowmeters (Fig. 9)

These were originally designed to measure differences in upstream and downstream transit-time of an ultrasound beam (Franklin et al., 1962), but the Doppler-effect flowmeter (Franklin et al., 1961) has found more general use. In its simplest form the latter consists of separate transmitting and receiving crystals, so angled that maximum reception is from the center of the lumen of the vessel. When applied directly to the vessel, the two crystals are on opposed sides, or "ring" the vessel; when applied transcutaneously, the crystals are closely adjacent. It is usual to "gate" the received ultrasound to between 0.1 and 100 k Hz in order to exclude low frequencies deriving from vascular wall movements, and high frequency noise. These sorts of arrangements have given excellent qualitative estimations of velocity of flow, but their ability to quantitate flow is debatable (Bergel, 1972). There is doubt that an adequate zero-setting can be obtained, especially with the wide range of ultrasound frequencies beamed. The earlier instruments were insensitive to direction of flow, but this has to an extent been overcome. The difficulties that remain are in part technical (or electronic), and in part conceptual: both forward and reverse flow can sometimes be registered simultaneously with these instruments. This phenomenon is readily explicable in terms of wave-front theory, because with rapid pressure change a complex wave-front may include prograde and retrograde elements. However it does not give confidence that the integrated signal fairly represents net movements of blood volume in the axis of the vessel. Nevertheless, direct comparisons made by Reneman et al., (1973) in dogs between implanted Doppler flow-probes connected to a high threshold zero-crossing meter, and electromagnetic flow-probes, showed that the instantaneous flow tracings were closely similar.

In an attempt to overcome some of these difficulties, the pulsed ultrasonic flowmeter has been devised, using a single crystal with brief pulses of transmission and time- and frequency-gated reception (Peronneau et al., 1969). The theoretical advantages of this technique are a reliable electronic zero, non-invasive measurement of velocity profile, and even an ability to estimate the internal diameter of the vessel. The pulsed method also allows sampling from very small volumes of the bloodstream and Histand et al. (1973) have used this property to measure velocity profiles across major arteries transcutaneously.

Despite the technical and conceptual difficulties with the Doppler ultrasound method, there have been certain notable gains. The signal is capable of being telemetered, and chronically implanted probes with telemetry have performed well by direct comparison with the electromagnetic flowmeter (Vatner et al., 1970). Blood flow has been telemetered by Franklin and his colleagues in such diverse circumstances as free-roving baboons, horses and dogs; diving seals; sharks, alligators; and even man (Guyton, 1973). A variety of clinical applications for transcutaneous probes has been found, and the low price of commercially available instruments has encouraged their use as semi-quantitative clinical tools.

Ultrasonic instruments, both of the transit-time type, and of the Doppler-effect type, have been used in a catheter-tip mounted form (Mills, 1972).

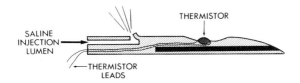

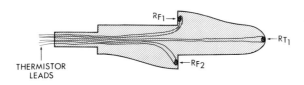

Fig. 10. <u>Single-vessel blood flow</u> measurement by thermal means. <u>Above</u>: intermittent measurement by thermal dilution probe, with room temperature saline injection about 10 mm above thermistor sampling point (after White et al., 1967, by kind permission of authors and publishers). <u>Below</u>: isothermal flowmeter, in which flow is a function of electric current necessary to maintain thermistors RF_1 and RF_2 at constant temperature, thermistor RT_1 acting as a reference (after Grahn et al., 1969, by kind permission of authors and publishers).

Thermal dilution probes (Fig. 10)

While these have been described in a variety of shapes and forms, they can be divided into two major categories.

In the first, a pulse of thermal indicator is released at an upstream point on the probe, and an indicator-dilution curve recorded by way of a probe-mounted thermistor. Such an instrument using cold saline was originally described by Fronek and Ganz (1960), and used extensively for long-term experiments in small animals by Korner and his colleagues (White et al., 1967). A known volume dose of thermal injectate is delivered, and the concentration-time curve registered by a thermistor located approximately 1 cm downstream, the catheter diameter being approximately 1.0 mm. The attractiveness of the device is the simplicity of design of the probe and its electronics.

Its defects are that while in vivo comparison with alternative flow measuring methods has been made for specific applications of the technique, geometric considerations make it unlikely that complete mixing of indicator with the bloodstream will occur in all circumstances. Nevertheless this type of instrument is capable of measuring mean volume flow, while most other forms of catheter-mounted flowmeter measure only velocity.

There has also been interest in thermal flowmeters designed to continuously monitor flow, dating back to the original thermostromuhr of Rein (1928) which was applied to the outside of the vessel, there being a central heating element and upstream and downstream thermocouples. Despite some development of this instrument (Guyton, 1973), conduction of heat along the vessel wall destroyed its accuracy. There has been a revival of interest in the principle, applied to intravascular probes. In general, constant current flow to the heating elements, with recording of downstream, or upstream-downstream temperature changes (or the use of a single thermistor as the heating and temperature sensing element), has not been satisfactory, if only because overheating is possible. More fundamentally, the instrument measures flow-velocity rather than volume-flow, and its frequency-response is poor because it is determined by a thermal time-constant.

More promising have been isothermal flowmeters, in which high frequency-response electronic circuitry maintains catheter-tip temperature constant in face of varying flow. Grahn et al. (1969) have devised an instrument incorporating three thermistors, approximately 2 mm in diameter, and capable of direction-sensitive flow velocity measurements with a frequency response comparable to the electromagnetic flowmeter.

The frequency response of thermal flowmeters is still limited by the heat capacity of the probe, including the thermistors. Thus the newest development in this field has been hot-film anemometry using a constant-temperature, heated, noble metal, film coating a dielectric substrate. These have reached a point of being able to be hypodermic-needle-mounted and direction-sensitive (Mills, 1972; Taylor, 1973), with a frequency response of up to 8 Hz, and capable of mapping velocity distribution across vascular lumina (Seed and Wood, 1970; Nerem et al., 1974).

Electrical impedance cardiac output measurement

In man the changes of electrical impedance across the chest wall during the cardiac cycle, measured by applying a constant simusoidal current, have been found to reflect stroke volume (Kubicek et al., 1966). The method has considerable promise as a non-invasive method for use in man, especially for within-subject measurement of the changes in cardiac output which occur in such circumstances as exercise (Denniston et al., 1976).

Microvascular flow techniques

Obviously a quintessence of technical ingenuity is necessary to estimate flow in single vessels of the microcirculation, and the art is scarcely beginning to be evolved (Baez, 1977).

Several methods for estimating erythrocyte velocity have been evolved, using purely optical methods by high-speed cinematography, or by employing electronic counting through an optical or flying-spot television system (Rushmer, 1966; Zweifach, 1973; Bollinger et al., 1974).

3. Methods of measuring tissue flow

Included in this section are all methods in which the blood flow measured is not in a single vessel, but is rather the volume-flow of blood through some circumscribed volume of tissue.

Venous occlusion plethysmography

This is deserving of first place, because of its simplicity of concept and execution, its antiquity, and the information that has been gathered by this means about blood flow in human limbs. A good description is given by Greenfield (1960). The principle scarcely requires stating: complete, temporary occlusion of the venous outflow from a limb or organ results in a rate of increase of volume distal to the venous occlusion cuff which is equal to arterial inflow. It has been well established that provided the external pressure of venous occlusion is less than arterial diastolic, there is no hindrance to arterial inflow.

There are a variety of sources of artefact: incomplete venous occlusion, or overcomplete; inadequate size of distal capacity vessels to accommodate the arterial inflow without the generation of back pressure; and imperfections in the volume-sensing device. The same considerations apply to choice of cuff width and inflation pressure as do for the indirect measurement of arterial pressure. The most useful volume-sensing device is the doublejacketed, temperature-controlled, stirred, water plethysmograph, which controls skin temperature and which readily allows recording by means ranging from a float and smoked drum, to an electric volumeter-amplifier-penwriter system. A variety of air-filled plethysmographs have been used that have little advantage except that the subject is less firmly anchored. The mercury-in-rubber strain gage is less cumbersome still, and is in itself an electrical transducer. While the tissue volume within which changes are measured is less certain than is the case with the water plethysmograph, studies have shown excellent correlation of calf blood flow by strain gage plethysmography and by water plethysmography (Dahn and Hallböök, 1970, and excellent correlation of forearm blood flow measured by strain gage plethysmography with brachial artery flow measured by electromagnetic flowmetry (Longhurst et al., 1974). A variety of other volume-change sensing devices have been used, including the capacitance plethysmograph.

There can be little doubt of the accuracy of venous occlusion plethysmography--provided the plethysmographer is discriminating in his application of the technique. Favorable loci are the human hand and forearm, to a lesser extent the human lower leg and foot, and perhaps the animal spleen. Most other loci are unfavorable because of their geometry, or because of inadequate capacity vessels. Favorable circumstances are the selected normal human subject at rest in a warm, comfortable environment; in most other circumstances there is hazard of error. Human limb plethysmography measures flow in a mixture of skin and muscle: mostly muscle in the forearm and calf, and mostly skin in the hand and foot. These are, however, static anatomic proportions, and the exact proportioning of changes in limb blood flow between skin and muscle is not easy.

Thermal methods

Calorimetry has been used to measure flow in certain specific situations, usually in an extremity. If the Fick principle were able to be applied in practice, and the temperature of arterial and venous blood entering and leaving the part were measured as well as the rate of heat gain or loss to the calorimeter, blood through-flow in conventional units could be calculated. In reality, measurements of blood temperature are not made, and the method is used only to observe the direction of changes in blood flow, and estimate semiquantitatively their magnitude. It has found its greatest application in the case of the human finger (Greenfield, 1960). An alternative to the rather cumbersome traditional calorimeter has been the copper-tellurium heat flow disc, applicable to any part of the skin surface (Catchpole and Jepson, 1955).

As an alternative to calorimetry Hensel (Rushmer, 1966) has devised a variety of instruments to measure skin or tissue "conductivity": or rather the changes in apparent thermal conductivity caused by variations in tissue blood flow, as estimated by heat conduction from a source of known magnitude to neighboring temperature-sensing elements. However, despite their sophisticated mathematical basis, the accuracy of these instruments is dubious, and their response-time extremely slow.

It is, in fact, doubtful whether calorimetry or the Hensel thermal conductivity meter have any real advantage in estimating skin blood flow over simple skin temperature measurement, or the more graphic method of thermography (e.g., Freundlich, 1972). All in reality provide no more than qualitative estimates.

Tissue clearance techniques

These have found a real place in vascular physiology and clinical investigation for estimation of tissue through-flow, as well as for examining some of the blood-tissue fluid exchange functions.

The most successful methods have employed lipophilic, gaseous, gamma-emitting nucleotides such as Xenon-133 or Krypton-85 that have the property of being very rapidly excreted by the lungs so that recirculation does not occur and the radiation dose is thus kept low. These have been employed in two different ways. The most used method has been to inject a small depot of the gas dissolved in saline into skin, subcutaneous tissue, or skeletal muscle, and to measure by an externally applied gamma-detector the logarithmic decay curve of removal. Particularly in the case of human skeletal muscle, good correlation has been found with other flow-measuring techniques such as venous occlusion plethysmography (e.g., Lassen et al., 1965). Excellent correlation of ^{133}Xe clearance and directly recorded flow has been reported recently for adipose tissue (Nielson, 1972).

An alternative approach has been to "saturate" a region with the indicator, either by prolonged preliminary infusion or inhalation, or by bolus injection, and measure the rate of its removal. This approach derives from the original Kety and Schmidt (1948) nitrous oxide method for measuring cerebral blood flow, but has been able to be extended and elaborated by the availability of appropriate radionuclides, and by further mathematical analysis (e.g., Zierler, 1965). The method has reached its greatest elaboration in measuring regional cerebral blood flow by bolus injection of Xenon-133 and simultaneous multiregional detection and analysis of transit curves by way of a gamma camera and analogue-to-digital computer (Lassen and Ingvar, 1972). By compartmental analysis it has been possible to identify multiple in-parallel regions of flow, and in some instances to give these anatomical reality by autoradiography (Thorburn et al., 1963). By using combinations of lipid- and water-soluble radionuclides, capillary exchange functions in various vascular beds and in varying circumstances, have been estimated (Mellander, 1970).

Methods of estimating proportional distribution among tissues and organs

These have developed mainly from the work of Sapirstein and his colleagues (1957). The original concept was of a technique that could be used to quantitate the distribution of left ventricular output in small animals by the injections of ^{42}K, ^{86}Rb or ^{131}I-antipyrine, rapid killing of the animal, and gamma-counting the individual organs and tissues. The method had two major drawbacks: cerebral perfusion was not able to be estimated because of failure of the former two radionuclides to pass the blood-brain barrier; and the inevitable death of the animal (not such a problem when populations of small animals can be studied).

The most recent version of this approach, now being extensively (and sometimes uncritically) used, is to inject gamma-emitting radionuclide-labeled carbonized microspheres into the left atrium, and by post-mortem analysis of organs and tissues, to quantitate the distribution of systemic blood flow (Rudolph and Heymann, 1967; Kaihara et al., 1968). Any one of 27 different radionuclides can be used to label the microspheres. By gamma spectroscopy it is possible to differentiate between at least four separate labels, so that retrospective flow measurements can be made in the same animal at four different points in time. It is not necessary to make a direct measurement of cardiac output, for a reference flow rate can be determined by constant-speed withdrawal of an arterial blood sample over the period of injection and circulatory distribution of the microspheres. These matters have been reviewed (Hales, 1974; Warren and Ledingham, 1974), as well as the desiderata that must be met if errors of observation and interpretation are to be avoided. To ensure adequate mixing, the injection must be made into the left atrium, rather than into the left ventricle or aorta. The dose injected must fall within prescribed upper and lower limits. The size of microsphere selected is critical, for if it is too small, then microspheres may pass through the systemic resistance vessels to the lungs. There is, furthermore, the conceptual and practical problem of whether "total" flow, "nutritive" flow, or "shunt" flow is being measured, and whether the same type of flow is being measured in all tissues and in all circumstances (Hales, 1974); Warren and Ledingham, 1974). The method is applicable to conscious animals and has already been used to measure the distribution of cardiac output, and flow in single organs and tissues, in animals ranging from rats to sub-human primates.

Selective-uptake techniques

Special properties of specific organs can be used to measure organ blood flow, and such methods have been made use of in man because of their relative non-invasiveness. The nitrous oxide method of measuring cerebral blood flow (Kety and Schmidt, 1948) is one example already discussed. Others are the bromsulphthalein (Bradley et al., 1945) or indocyanine green (Banaszak et al., 1960) methods for measuring hepatic blood flow (Bradley, 1974), and the para-amino hippurate or diodrast methods of measuring renal blood flow (Levinsky and Levy, 1973). All rely on the Fick principle, and on a high extraction ratio by the specific organ for the specific indicator.

Historial methods of measuring blood flow

Brief mention is made of these for completeness. They have been largely discarded from use in physiologic or clinical work, because of their inaccuracy or clumsiness compared with the methods described earlier in this account. The justification for including them under this heading is that the naive investigator may consider using one of these methods: hopefully this will make him aware that there are limitations, so that he will seek more details (Fry and Ross, 1966).

Open outflow recorders, ranging from the venous effluent bucket-and-stopwatch technique, to more continuous monitoring methods--drop counters, manometers or volumeters--may still be useful in animal preparations as references to more sophisticated techniques, but have little other merit.

The differential-pressure instruments: orifice meters, venturi meters, elbow meters, or pitot tubes all demand not merely sophisticated measurement of small pressure differences, but geometric and hydrodynamic circumstances that are rarely attainable in practice.

Drag measuring techniques are fine for many fluids--but not blood, because of the destruction caused by the insertion of a bristle, ball or vane into a coagulable, mechanically-destructable fluid.

Transit-time techniques may have limited application when an indicator such as a bubble can be timed through an in-line geometrically uniform artificial tube; but the transit times of injected indicators such as radionuclides or dyes give little reliable information because of the series-parallel variable geometry of so much of the vasculature.

Estimation of resistance to blood flow

Resistance to flow is not directly measurable. In its simplest form it is defined as pressure ÷ flow; it is analogous to the ohm in electrical circuitry. As P/F it has some meaning in a model of the circulatory system in which pressure and flow are steady--a circumstance that in reality is non-existent. Nevertheless mean perfusion pressure ÷ mean flow is used in circulatory physiology to provide a crude measure of the resistance to cardiac output, or to flow through a single vessel, organ, or defined volume of tissue. The simplest unit--the peripheral resistance unit or p.r.u.--may have as much meaning as any other and is defined as perfusion pressure (e.g., mm Hg) divided by flow (e.g., liters/min). There has been some enthusiasm for an apparently more sophisticated unit in describing whole body resistance to cardiac output: $dyne\ sec\ cm^{-5}$. The unit has little to commend it either in conceptual meaning or in terms of SI units. The appropriate SI unit would be Pa m^{-3} or J m^{-6}s, and Coulter (1966) has suggested this be entitled the hemodynamic ohm or hohm.

The more fundamental lack of meaning in these units is that they are strictly applicable only to steady pressure and flow states, and that just as a more sophisticated method of calculating energy loss is necessary for alternating electric current, so is one necessary in the case of pulsatile pressure and flow. Thus there has been developed the mathematical concept of input impedance, broadly defined as the ratio of pulsatile pressure to pulsatile flow, and derived by Fourier analysis of measured pressure and flow. Whereas resistance (mean pressure/mean flow) gives some indication of the state of constriction of resistance vessels, impedance also includes information about the stiffness of the vasculature and wave reflection. The interpretation of impedance changes in physiologic terms is, however, still in its infancy. A largely non-mathematical description of the phenomena that go to make up vascular impedance, with special reference to deduction and speculation applicable to man, is given by Milnor (1972). More mathematically critical description and comment is found in McDonald (1968, 1974), Gessner, (1972) and Taylor (1973). Coulter (1966) has suggested that the impedance unit should again be the hohm, defined in this instance as:

$$\frac{\text{Amplitude of pressure sinusoid in J cm}^{-3}}{\text{Amplitude of flow sinusoid in cm}^3\text{s}^{-1}}$$

or J cm^{-6}s for the particular harmonic.

GENERAL REVIEWS

Baez, S., 1977, Microcirculation, Ann. Rev. Physiol. 39:391.
Bergel, D.H. (ed.), 1972, "Cardiovascular fluid dynamics," Vol. 1, Academic, New York.
Fishman, A.P., and Richards, D.W., (eds.), 1964, "Circulation of the blood: men and ideas," Oxford Univ. Press, New York.
Franklin, K.J., 1933, "A short history of physiology," Bale, London.
Geddes, L.A., 1970, "The direct and indirect measurement of blood pressure," Year Book, Chicago.
Guyton, A.C., Jones, C.E., and Coleman, T.G., 1973, "Circulatory physiology: cardiac output and its regulation," Saunders, Philadelphia.
McDonald, D.A., 1974, "Blood flow in arteries," 2nd edition, Arnold, London.
McDonald, D.A., 1968, Hemodynamics, Ann. Rev. Physiol. 30:525.
Potter, V.R. (ed.), 1948, "Methods in Medical Research," Vol. 1, Year Book, Chicago.
Rushmer, R.F. (ed.), 1966, "Methods in Medical Research," Vol. 11, Year Book, Chicago.

Taylor, M.G., 1973, Hemodynamics, Ann. Rev. Physiol. 35:87.
Woodcock, J.P., 1975, "Theory and practice of blood flow measurement," Butterworths, London.

REFERENCES

Allard, E.M., 1962, Sound and pressure signals obtained from a single intracardiac transducer, IRE Trans. Biomed. Electron. BME 9:74.
American Heart Association (Kirkendall, W.M., Burton, A.C., Epstein, F.H., and Freis, E.D.), 1967, Recommendations for human blood pressure determination by sphygmomanometers, Circulation 36:980.
Angelakos, E.T., 1964, Semiconductor pressure microtransducers for measuring velocity and acceleration of intraventricular pressures, Am. J. Med. Electron. 3:266.
Banazak, E.F., Stekiel, W.J., Grace, R.A., and Smith, J.J., 1960. Estimation of hepatic blood flow using a single injection dye clearance method, Am. J. Physiol. 198:877.
Bennett, H.S., Sweet, W.H., and Bassett, D.L., 1944, Heated thermocouple, flowmeter, J. Clin. Invest. 23:200.
Bergel, D.H., and Gessner, U., 1966, The electromagnetic flowmeter, in "Methods in medical research" (R.F. Rushmer, ed.), Vol. 11, pp. 70-82, Year Book, Chicago.
Bloch, E.H., 1966, Low-compliance pressure gauge, in "Methods in medical research" (R.F. Rushmer, ed.), Vol. 11, pp. 190-194, Year Book, Chicago.
Bollinger, A., Butti, P., Barras, J.-P., Trachsler, H., and Siegenthaler, W., 1974, Red blood cell velocity in nailfold capillaries of man measured by a television microscopy technique, Microvasc. Res. 7:61.
Borst, J.G.G., and Molhuysen, J.A., 1952, Exact determination of the central venous pressure by a simple clinical method, Lancet 2:304.
Bradley, E.L., 1974, Measurement of helatic blood flow in man, Surgery, 75:783.
Bradley, S.E., Ingelfinger, F.J., Bradley, G.D., and Curry, J.J., 1945, Estimation of hepatic blood flow in man, J. Clin. Invest. 24:890.
Branthwaite, M.A., and Bradley, R.D., 1968, Measurement of cardiac output by thermal dilution in man, J. Appl. Physiol. 24:434.
Brodie, T.G., and Russell, A.E., 1905, On the determination of the rate of blood flow through an organ, J. Physiol. (London) 32:47.
Cappelen, C. Jr., (ed.). 1968, "New findings in blood flowmetry," Universitetsforlaget, Oslo.
Catchpole, B.N., and Jepson, R.P., 1955, Hand and finger blood flow, Clin. Sci. 14:109.
Chauveau, A., Bertolus, G., and Laroyenne, L., 1860, Vitesse de la circulation dans lest artères du cheval d'après les indications d'un nouvel hémodromètre, J. Physiol. (Brown-Sequard) 3:695.

Clarc, F.H., Schmidt, E.M., and de la Croiz, R.F., 1965, Fiber optic blood pressure catheter with frequency response from dc into the audio range, Proc. Nat. Electron Conf. 21:213.

Cooper, T., Pinakatt, T., and Richardson, A.W., 1963, The use of the thermal dilution principle for measurement of cardiac output in the rat, Med. Electron. Biol. Eng. 1:61.

Coulter, N.A., Jr., 1966, Toward a rational system of units in hemodynamics. IEEE Trans. Biomed. Eng. BME-13:207.

Cournand, A., 1945, Measurement of the cardiac output in man using the right heart catheterization. Description of technique, discussion of validity and of place in the study of the circulation, Federation Proc. 4:207.

Dahn, I., and Hallböök, T., 1970, Simultaneous blood flow measurements by water and strain gauge plethysmography, Scand. J. Clin. Lab. Invest. 25:419.

Davison, R., and Cannon, R., 1974, Estimation of central venous pressure by examination of jugular veins, Am. Heart J. 87:279.

Denniston, J.C., Mahr, J.T., Reeves, J.T., Cruz, J.C., Cymerman, A., and Grover, R.F., 1976, Measurement of cardiac output by electrical impedance at rest and during exercise, J. Appl. Physiol. 40:91.

Donald, D.E., and Yipintsoi, T., 1973, Comparison of measured and indocyanine green blood flows in various organs and systems, Proc. Staff Meetings Mayo Clinic 48:492.

Evonuk, E., Imig, C.J., Greenfield, W., and Eckstein, J.W., 1961, Cardiac output measured by thermal dilution of room temperature injectate, J. Appl. Physiol. 16:271.

Fegler, G., 1954, Measurement of cardiac output in anaesthetised animals by a thermodilution method, Quart. J. Exp. Physiol. 39:153.

Fegler, G., 1957, The reliability of the thermodilution method for determination of the cardiac output and the blood flow in central veins, Quart. J. Exp. Physiol. 42:254.

Fick, A., 1870, Ueber die Messung des Blutquantums in den Herzbentrikeln, S.B. Phys. Med. ges. Würzburg, July 9.

Folse, R., 1965, Application of the sudden injection dye dilution principle to the study of the femoral circulation, Surg. Gynecol. Obstet. 120:1194.

Folts, J.D., and Rowe, G.G., 1971, A nonerosive electromagnetic flowmeter probe for chronic aortic implantation, J. Appl. Physiol. 31:782.

Forrester, J.S., Ganz, W., Diamond, G., McHugh, T., Chonette, D.W., and Swan, H.J.C., 1972, Thermodilution cardiac output determination with a single flow-directed catheter, Amer. Heart J. 83:306.

Forsyth, R.P., 1972, Sympathetic nervous system control of distribution of cardiac output in unanaesthetized monkeys. Federation Proc. 31:1240.

Franklin, D.J., Baker, D. W., and Rushmer, R.F., 1962, Pulsed ultrasonic transit time flowmeter, IRE Trans. Biomed. Electron. 9:44.

Franklin, D.L., Schlegel, W.A., and Rushmer, R.F., 1961, Blood flow measured by Doppler frequency shift of backscattered ultrasound, Science 134:564.

Freundlich, I.M., 1972, Thermography, New Engl. J. Med. 287:880.

Fronek, A., and Ganz, V., 1960, Measurement of flow in single blood vessels including cardiac output by local thermodilution, Circulation Res. 8:175.

Fry, D.L. and Ross, J. Jr., 1966, Survey of flow detection technics, in "Methods in medical research", Vol. 11, pp.50-69, Year Book, Chicago.

Gabe, I.T., 1972, Pressure measurement in experimental physiology, in "Cardiovascular fluid dynamics" (D.H. Bergel, ed.) Vol. 1, pp.11-50, Academic, New York.

Ganz, W., Donoso, R., Marcus, H.S., Forrester, J.S., and Swan, H.J.C., 1971, A new technique for measurement of cardiac output by thermal dilution in man, Am. J. Cardiol. 27:392.

Garrow, J.S., 1963, Zero-muddler for unprejudiced sphygmomanometry, Lancet 2:1205.

Gauer, O.H., and Gienapp, E., 1950, A miniature pressure recording device, Science 112:404.

Gessner, U., 1972, Vascular input impedance in "Cardiovascular fluid dynamics" (D.H. Bergel, ed.), Vol. 1, pp.315-349, Academic, New York.

Goodyer, A.V.N., Huros, A., Eckhardt, W.F., and Ostberg, R., 1959, Thermal dilution curves in the intact animal, Circulation Res. 7:432.

Gould, K.L., Trenholme, S., and Kennedy, J.W., 1973, In vivo comparison of catheter manometer systems with the catheter-tip micromanometer, J. Appl. Physiol. 34:263.

Grahn, A.R., Paul, M.H., and Wessel, H.U., 1969, A new direction-sensitive probe for catheter-tip thermal velocity measurement, J. Appl. Physiol. 27:407.

Greenfield, A.D.M., 1960, II, Electromechanical methods. Venous occlusion plethysmography. Peripheral blood flow by calorimetry, in "Methods in Medical Research" (H.D. Bruner, ed.), Vol. 8, pp.293-307, Year Book, Chicago.

Hainsworth, R., Ledsome, J.R., and Snow, H.M., 1968, Dynamic testing of electromagnetic flowmeters by mechanical and electronic methods, J. Appl. Physiol. 25:469.

Hales, J.R.S., 1974, Radioactive microsphere techniques for studies of the circulation, Clin. Exper. Pharmacol. Physiol., 1 Suppl 1:31.

Hales, S., 1733, "Statistical essays, Vol. 2, Haemostaticks," Innys and Manby, London.

Hamilton, W.F., Brewer, G., and Brotman, I., 1934, Pressure contours in the intact animal, Am. J. Physiol. 107:427.

Hansen, A.T., 1949, Pressure measurements in the human organism, Acta Physiol. Scand. 19 Suppl. 68:1.

Harris, T.R., and Newman, E.V., 1970, An analysis of mathematical models of circulatory indicator-dilution curves, J. Appl. Physiol. 28:840.

Hering, E., 1829, Versuche, die Schnelligkeit des Blutlaufs und der Absonderung zu bestimmen, Z. Physiol. 3:85.

Histand, M.B., Miller, C.W., and McLeod, F.D., 1973, Transcutaneous measurement of blood velocity profiles and flow, Cardiovasc. Res. 7:703.

Hök, B., 1974, Electrolytic catheter-tip pressure transducer, Med. Biol. Eng. 12:355.

Holland, W.W., and Humerfelt, S., 1964, Measurement of blood-pressure: comparison of intra-arterial and cuff values, Brit. Med. J. 2:1241.

Hyman, C., and Winsor, T., 1961, History of plethysmography, J. Cardiovasc. Surg. 2:506.

Intaglietta, M., Pawula, R.F., and Tompkins, W.R., 1970, Pressure measurements in the mammalian microvasculature, Microvasc. Res. 2:212.

Jorfeldt, L., and Wahren, J., 1971, Leg blood flow during exercise in man, Clin. Sci. 41:459.

Kaihara, S., Van Heerden, P.D., Migita, T., and Wagner, H.N., 1968, Measurement of the distribution of cardiac output, J. Appl. Physiol. 25:696.

Kazamias, T.M., Grander, M.P., Franklin, D.L., and Ross, J. Jr., 1971, Blood pressure measurment with Doppler ultrasonic flowmeter, J. Appl. Physiol. 30:585.

Kety, S.S., 1949, Measurement of regional circulation by the local clearance of radioactive sodium, Am. Heart J. 38:321.

Kety, S.S. and Schmidt, C.F., 1948, Nitrous oxide method for the quantitative determinations of cerebral blood flow in man: theory, procedure and normal values, J. Clin. Invest. 27:476.

Khalil, H.H., Richardson, T.O., and Guyton, A.C., 1966, Measurement of cardiac output by thermal-dilution and direct Fick methods in dogs, J. Appl. Physiol. 21:1131.

Kolin, A., 1936, An electromagnetic flowmeter. Principle of the method and its application to blood flow measurement, Proc. Soc. Exp. Biol. Med. 35:53.

Korner, P.I., 1965, The effect of section of the carotid sinus and aortic nerves on the cardiac output of the rabbit, J. Physiol. (London), 180:226.

Korotkoff, N.S., 1905, On methods of studying blood pressure, Izv. Voennomed. Akad. St. Petersburg 11:365.

Krogh, A., and Lindhard, J., 1912, Measurements of the blood flow through the lungs of man, Skand. Arch. Physiol. 27:100.

Kubicek, W.G., Karnegis, J.N., Patterson, R.P., Witsoe, D.A., and Mattson, R.H., 1966, Development and evaluation of an impedance cardiac output system, Aerospace Med. 37:1208.

Lambert, E.H., and Wood, E.H., 1947, The use of a resistance wire strain gauge manometer to measure intraarterial pressure, Proc. Soc. Exp. Biol. Med. 64:186.

Landis, E.M., 1934, Capillary pressure and capillary permeability, Physiol. Rev. 14:404.

Lassen, N.A., and Ingvar, D.H., 1972, Radioisotopic assessment of regional blood flow, in "Progress in nuclear medicine" (E.J. Potchen and V.R. McCready, eds.) Vol. 1, pp.376-409, Karger, Basel.

Lassen, N.A., Lindbjerg, I.F., and Dahn, I., 1965, Validity of the Xenon-133 method for measurement of muscle blood flow evaluated by simultaneous venous occlusion plethysmography; observations in the calf of normal man and in patients with occlusive vascular disease, Circulation Res. 16:287.

Laurens, P., Bouchard, F., Brial, E., Cornu, C., Baculard, P., and Soulie, P., 1959, Bruits et pressions cardiovasculaires enregistrés in situ à l'aide d'un micromanomètre, Arch Maladies Coeur Valisseaux 52:121.

Levinsky, N.G., and Levy, M., 1973, Clearance technics, in "Handbook of physiology, Section 8: Renal physiology" pp.103-117, American Physiological Society, Washington.

Lindström, L.H., 1970, Miniaturized pressure transducer intended for intravascular use, IEEE Trans. Bio-Med. Eng. BME 17:207.

Longhurst, J., Capone, R.J., Mason, D.T., and Zelis, R., 1974, Comparison of blood flow measured by plethysmograph and flowmeter during steady state forearm exercise, Circulation 44:535.

Lowe, R.D., and Stephens, N.L., 1961, Carotid occlusion; Diagnosis by opthalmodynamometry during carotid compression, Lancet 1:1241.

Ludbrook, J., and Collins, G.M., 1967, Venous occlusion plethysmography in the human upper limb, Circulat. Res. 2:139.

Ludwig, C., 1847, Loc. cit. "A short history of physiology" (K.J. Franklin), 1933, Bale, London.

Ludwig, C., and Dogiel, J., 1867, "A short history of physiology," (K.J. Franklin), 1933, Bale, London.

Mason, D.T., Gabe, I.T., Mills, C.J., Gault, J.H., Ross, J. Jr., Braunwald, E., and Shillingford, J.P., 1970, Applications of the catheter-tip electromagnetic velocity probe in the study of the central circulation in man, Am. J. Med. 49:465.

Melbin, J., and Spohr, M., 1969, Evaluation and correction of manometer systems with two degrees of freedom, J. Appl. Physiol. 27:749.

Mellander, S., 1970, Systemic circulation: local control, Ann. Rev. Physiol. 32.313.

Milnor, W.R., 1972, Pulsatile blood flow, N. Engl. J. Med. 287:27.

Mills, C.J., 1972, Measurement of pulsatile flow and flow velocity, in "Cardiovascular fluid dynamics" (D.H. Bergel, ed.), Vol. 1, pp.51-90, Academic, New York.

Moore, J.W., Kinsman, J.M., Hamilton, W.F., and Spurling, R.G., 1929, Studies on the circulation. II. Cardiac output determinations: comparison of the injection method with the direct Fick procedure, Am. J. Physiol. 89:331.

Moss, A.J., and Adams, F.H., 1963, Index of indirect estimation of diastolic blood pressure, Am. J. Diseases Children 106:364.

Nerem, R.M., Rumberger, J.A., Gross, D.R., Hamlin, R.L., and Geiger, G.L., 1974, Hot-film anemometer velocity measurements of arterial blood flow in horses, Circulation Res. 34:193.

Nielsen, P.E., and Jannicke, H., 1974, The accuracy of auscultatory measurement of arm blood pressure in very obese subjects, Acta Med. Scand. 195:403.

Nielsen, S.L., 1972, Measurement of blood flow in adipose tissue from the washout of Xenon-133 after atraumatic labelling, Acta Physiol. Scand. 84:187.

Noble, F.W., 1957, The sonic valve pressure gauge, IRE Trans. Med. Electron. PGME 8:38.

Noble, F.W., 1959, A hydraulic pressure generator for testing the dynamic characteristics of blood pressure manometers, J. Lab. Clin. Med. 54:897.

Overbeck, H.W., Daugherty, R.M. Jr., and Haddy, F.J., 1969, Continuous infusion indicator dilution measurement of limb blood flow and vascular response to magnesium sulfate in normotensive and hypertensive men, J. Clin. Invest. 48:1944.

Peronneau, P., Deloche, A., Bui-Mong-Hung, and Hinglais, J., 1969, Débitmétrie ultrasonore: développements et applications expérimentales, Europ. Surg. Res. 1:147.

Plesch, J., 1909, Hämodynamische Studien, Z. Exp. Path. Ther. 6:380.

Poiseuille, J.L.M., 1828, "Recherches sur la force du coeur aortique" Didot, Paris.

Raftery, E.B., and Ward, A.P., 1968, The indirect method of recording blood pressure, Cardiovasc. Res. 2:210.

Ramirez, A., Hood, W.B. Jr., Polany, M., Wagner, R., Yankopoulos, N.A., and Abelmann, W.H., 1969, Registration of intravascular pressure and sound by a fiberoptic catheter, J. Appl. Physiol. 26:679.

Raper, A.J., and Levasseur, J.E., 1971, Accurate sustained measurement of intra-luminal pressure from the microvasculature, Cardiovasc. Res. 5:589.

Rein, H., 1928, Die Thermo-Stromuhr, Z. Biol. 87:394.

Reneman, R.S., Clarke, H.F., Simmons, N., and Spencer, M.P., 1973, In vivo comparison of electromagnetic and Doppler flowmeters: with special attention to the processing of the analogue Doppler flow signal, Cardiovasc. Res. 7:557.

Roberts, L.N., Smiley, J.R., and Manning, G.W., 1953, A comparison of direct and indirect blood-pressure determination, Circulation 8:232.

Roman, J., Henry, J.P., and Meehan, J.P., 1965, Validity of flight blood pressure data, Aerospace Med. 36:436.

Rose, G.A., Holland, W.W., and Crowley, E.A., 1964, Sphygmomanometer for epidemiologists, Lancet 1:296.

Rudolph, A.M., and Heymann, M.A., 1967, The circulation of the fetus in utero: methods for studying distribution of blood flow, cardiac output and organ blood flow, Circulation Res. 21:163.

Russell, R.W.R., and Cranston, W.I., 1961, Ophthalmodynamometry in carotid artery disease, J. Neurol. Neurosurg. Psych. 24:281.

Sapirstein, L.A. 1957, Regional blood flow by fractional distribution of indicators, Am. J. Physiol. 193:161.

Seed, W.A., and Wood, N.B., 1970, Development and evaluation of a hot-film velocity probe for cardiovascular studies, Cardiovasc. Res. 4:253.

Singh, R., Ranieri, A.J. Jr., Vest, H.R. Jr., Bowers, D.L., and Dammann, J.F. Jr., 1970, Simultaneous determinations of cardiac output by thermal dilution, fiberoptic and dye-dilution methods, Am. J. Cardiol. 25:579.

Smith, H.C., Sturm, R.E., and Wood, E.H., 1973, Videodensitometric system for measurement of vessel blood flow, particularly in the coronoary arteries, in man, Am. J. Cardiol. 32:144.

Spieckermann, P.G., and Bretschneider, H.J., 1968, Vereinsfachte quantitative Answertung von Indikatorverdünnungskurven, Arch. Kreislaufforsch 55:211.

Stegall, H.F., 1967, A simple inexpensive sinusoidal pressure generator, J. Appl. Physiol. 22:591.

Stegall, H.F., Kardon, M.B., and Kemmerer, W.T., 1968, Indirect measurement of arterial blood pressure by Doppler ultrasonic sphygmomanometry, J. Appl. Physiol. 25:793.

Stewart, G.N., 1897, Researches on the circulation time and on the influences which affect it. IV. The output of the heart. J. Physiol. (London) 22:159.

Taylor, J.B., Lowen, B., and Polyani, M., 1972, In vivo monitoring with a fiber optic catheter, J. Am. Med. Assoc. 221:667.

Thorburn, G.D., Kopald, H.H., Herd, J.A., Hollenberg, M., O'Morchoe, C.C.C., and Barger, A.C., 1963, Intrarenal distribution of nutrient blood flow determined with Krypton 85 in the unanesthetised dog. Circulation Res. 13:290.

Vatner, S.F., Franklin, D., and Van Citters, R.L., 1970, Simultaneous comparison and calibration of the Doppler and electromagnetic flowmeters, J. Appl. Physiol. 29:907.

Warnick, A., and Drake, E.H., 1958, A new intracardiac pressure measuring system for infants and adults, IRE Nat. Conv. Rec. 9:68.

Warren, D.J., and Ledingham, J.G.G., 1974, Measurement of cardiac output distribution using microspheres: some practical and theoretical considerations, Cardiovasc. Res. 8:570.

Warren, D.J. and Ledingham, J.G.G., 1974, Cardiac output in the conscious rabbit: an analysis of the thermodilution technique, J. Appl. Physiol. 36:246.

Wetterer, E., 1943, Eine neuer manometrische Sonde mit elekitrischer Transmission, Z. Biol. 101:332.

White, S.W., McRitchie, R.J., and Porges, W.L., 1974, A comparison between thermodilution, electromagnetic and Doppler methods for cardiac output measurement in the rabbit, Clin. Exper. Pharmacol. Physiol. 1:175.

White, S.W., Chalmers, J.P., Hilder, R., and Korner, P.I., 1967, Local thermodilution method for measuring blood flow in the portal and renal veins of the unanaesthetized rabbit, Aust. J. Exp. Biol. Med. Sci. 45:453.

Wiederhielm, C.A., and Rushmer, R.F., 1964, Pre- and post-arteriolar resistance changes in the blood vessels of the frog's mesentery, Bibliotheca Anat. 4:234.

Wiggers, C.J., 1928, "The pressure pulses in the cardiovascular system," Longmans, London.

Wilson, E.M., Ranieri, A.J. Jr., Updike, O.L., and Dammann, J.F. Jr., 1972, An evaluation of thermal dilution for obtaining measurements of cardiac output, Med. Biol. Eng. 10:179.

Wright, B.M., and Dore, C.F., 1970, A random zero sphygmomanometer, Lancet 1:337.

Yanof, H.M., Rosen, A.L., McDonald, N.M., and McDonald, D.A., 1963, A critical study of the responses of manometers to forced oscillations, Phys. Med. Biol. 8:407.

Zierler, K.L., 1965, Equations for measuring blood flow by external monitoring of radioisotopes, Circulation Res. 16:309.

Zweifach, B.W., 1973, Microcirculation, Ann. Rev. Physiol. 35:117.

REGULATION OF ARTERIAL BLOOD FLOW, PRESSURE AND RESISTANCE IN THE SYSTEMIC CIRCULATION

John Ludbrook, M.D., Ch.M., B. Med. Sci., F.R.C.S., F.R.A.C.S.
Mortlock Professor of Surgery, University of Adelaide, and Visiting Surgeon, Royal Adelaide Hospital, Adelaide, South Australia.

INTRODUCTION

The first milestone in regard to the systemic circulation was William Harvey's recognition, and publication in 1628, of the fact that blood does indeed circulate. The events leading up to, and following on, this revolutionary announcement are brilliantly discussed in a publication that followed the tercentenary of his death (Fishman and Richards, 1964). The second milestone was the visible demonstration that there are in fact invisible connections between arteries and veins, by Malpighi in 1661 in the lung, and by Leeuwenhoeck in 1674 in the systemic circulation.

From that time to the present day the milestones may be smaller, but they have come thick and fast as a result of observations of increasing sophistication, from those in the early 18th century by Stephen Hales on blood pressure and flow, to those of the twentieth century. The latter are perhaps most easily summarized by citing the appropriate Nobel laureates (Table 1).

The anatomic disposition and composition of blood vessels have been presented earlier in this volume, as have the structure and functions of the cardiovascular receptor and effector apparatuses. The biophysical and rheological phenomena that attend blood pressure and flow are to follow.

To describe comprehensively the functions of the systemic circulation and their control is, of course, an impossible brief. But having accepted it, the most appropriate course would seem to be to describe some of the effector mechanisms that are known to be able

TABLE 1

NOBEL PRIZEWINNERS IN MEDICINE AND PHYSIOLOGY
1901-1973 WHOSE WORK ILLUMINATED THE FUNCTIONING
OF THE CARDIOVASCULAR SYSTEM

Date and Name	Citation
1920 A. Krogh	For his discovery of the regulation of the motor mechanism of capillaries.
1924 W. Einthoven	For his discovery of the regulation of of the electrocardiogram.
1946 H. Dale, O. Loewi	For their discoveries relating to the chemical transmission of nerve impulses.
1938 C. Heymans	For his discovery of the role played by the sinus and aortic mechanisms in the regulation of respiration.
1956 A.F. Cournard, W. Forssmann, D.W. Richards	For their discoveries concerning heart catheterization and pathological changes in the circulatory system.
1957 D. Bovet	For his discoveries relating to synthetic compounds that inhibit the action of certain body substances, and especially their action on the vascular system and the skeletal muscles.
1970 U. von Euler, J. Axelrod, B. Katz	For their discoveries concerning the humoral transmitters in the nerve terminals and the mechanism for their storage, release and inactivation.

to produce variations in pressure and flow in the systemic circulation, something of the hemodynamic changes that take place under certain real-life circumstances, and some of the control systems that regulate cardiovascular functions. Companion chapters on arterial receptor mechanisms, the innervation of arteries, and the coronary, pulmonary circulation, and renal circulation will reinforce many facets of the present chapter, and together should provide a comprehensive overview of the nature of the circulation as a whole.

I. Mechanisms that produce changes in systemic blood pressure and flow

These mechanisms can be neatly--undoubtedly too neatly--categorized into pumps, pipes, resistances, and reservoirs, with the properties of the blood itself as a fifth parameter. All except pumps and resistances are considered in detail elsewhere in this volume, and will be referred to only as necessary in this present account.

PUMPS.

The chief generator of pressure and flow in the systemic circulation is the heart, and in particular its left ventricle.

It is a commonplace observation in biology that provided the cardiac musculature is adequately perfused, the heart will continue to function autonomously as a reciprocating pump, as in the isolated frog and mammalian heart preparations of Starling. This phenomenon stems from fundamental properties of the cardiac muscle cells themselves, which develop rhythmic electrical activity and contractility early in embryonic life (Hoff et al., 1939; Goss, 1940), and can maintain these properties even in tissue culture (Mark and Strasser, 1966), although as the heart develops these properties become restricted to the specialized pacemaker and conducting cardiac cells. Like skeletal muscle (but unlike smooth muscle), contraction of cardiac muscle is always preceded by a transmitted depolarizing wave moving along the cell membrane (sarcolemma). Unlike skeletal muscle (but like smooth muscle), the force-velocity characteristics of contraction are greatly influenced by external neural, humoral, and electrolytic factors.

The ultrastructure and function of the contractile apparatus of cardiac muscle is well described in several reviews (Blinks and Jewell, 1972; Ross and Sobel, 1972; Langer, 1973; Fuch, 1974; Korner, 1974; Braunwald, et al., 1976). It fits the Huxley sliding filament hypothesis, but with some important differences from skeletal and smooth muscle. There are deep invaginations of the sarcolemma in cardiac muscle ("transverse tubules"). There is a complicated sarcoplasmic reticulum enveloping the myofibrils ("longitudinal component"), with specialized cisternae adjacent to, but not in communication with, the transverse tubules. The transverse tubules are thought to play an important part in excitation-contraction coupling by carrying an electrotonic depolarization, or possibly a Na^+ dependent form of action potential (Fuchs, 1974), to the cisternae. These latter release the Ca^{++} ions (Reuter, 1974) necessary for the sequence of myosin activation, ATP hydrolysis, energy release that complexes actin to myosin, and consequent contraction. The sarcoplasmic reticulum then takes up Ca^{++} again, resulting in relaxation.

Membrane functions of cardiac muscle have also been elucidated, chiefly by way of intracellular microelectrode recording and voltage clamp techniques (Noble, 1966; Weidmann, 1967; De Mello, 1972; Trautwein, 1973; Fozzard and Gibbons, 1973). The electrophysiological details need not concern us in the present chapter, and can be best summarized in diagrammatic form (Fig. 1). Ordinary cardiac muscle fibers maintain a steady resting membrane potential in between action potentials. However in certain specialized conducting fibers there is a steady depolarizing current, which when it reaches threshold sets off an action potential (Fig. 1). The slope of this depolarizing current is greatest in the fibers of the sino-atrial node, less in those of the atrio-ventricular node, and least in the ventricular Purkinje fibers. From this property stems the hierarchy of cardiac pacemakers, headed up by the sino-atrial node.

The heart has three fundamental properties, consistently demonstrable within the limits of experimental possibility in studies of single fibers (Meijler and Brutsaert, 1971; De Mello, 1972), muscle strips (Sonnenblick, 1962), in isolated heart preparations (Sarnoff, 1955), and in the intact animal (Rushmer, 1970; Folkow and Neil, 1971; Guyton et al., 1973; Braunwald et al., 1976). One is a capacity to vary its rate of contraction. This stems from variations in the rate of diastolic steady depolarization in the conducting tissue (normally the fibers of the SA node). Thus an increased depolarization gradient is caused by increased temperature, hyperthyroidism, and catecholamines (whether circulating or released by sympathetic nerve activity), and results in cardio-acceleration. A diminution of the depolarizing gradient is caused by cold, hypothyroidism, or by acetyl choline released by parasympathetic activity, and results in bradycardia.

Increase in circulating catecholamines (and in Ca^{++}) increases the velocity of shortening of cardiac muscle fibers (Sonnenblick, 1962), or causes an increased rate of ventricular pressure generation with time (Sarnoff and Mitchell, 1962), as well as a greater maximal tension. The precise mechanism is not yet entirely clarified, but is presumed to be due to facilitation of Ca^{++} transfer from the endoplasmic reticulum to the contractile system.

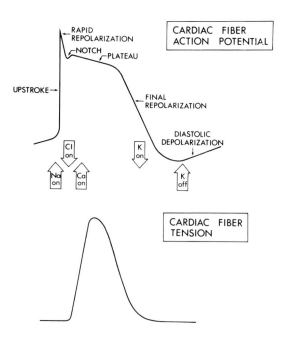

Fig. 1. Diagrammatic display of action potential and contractile response in a cardiac muscle fiber. Above: single fiber action potential, with type and direction of ionic movements indicated. Below: development of tension in single cardiac fiber on same time scale. (Redrawn from Fozzard and Gibbons, 1973, Figs. 1 and 2, by kind permission of the publishers.)

The most fundamental functional characteristic of cardiac muscle and of the heart was of course revealed by Starling (1918), and has been further elucidated in subsequent studies of ventricular function (Sarnoff and Mitchell, 1962) and of the properties of cardiac muscle strips (Sonnenblick, 1962; 1965). It is that the rate of development of isometric contractile tension, or the rate of isotonic shortening, is a function of initial muscle fiber length. As Starling (1918) himself put it: "*The law of the heart is therefore the same as that of skeletal muscle, namely that the mechanical energy set free on passage from the resting to the contracted state depends on the area of 'chemically active surfaces,' i.e., on the length of the muscle fiber. This simple formula serves to 'explain' the whole behaviour of the isolated mammalian heart.*"

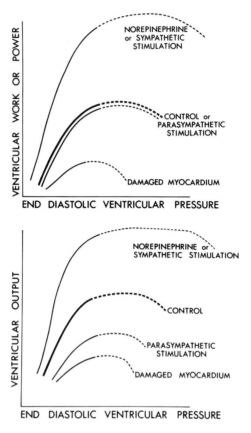

Fig. 2. Diagrammatic illustration of ventricular function curves. Above: ventricular work vs end-diastolic pressure. Below: ventricular (cardiac) output vs end-diastolic pressure. In each case the effect of neural and humoral influences, and of myocardial damage, are shown (cf. Sarnoff and Mitchell, 1962).

In summary, then, there are three mechanisms by which the heart considered as an isolated entity makes a contribution to the control of the systemic circulation. There is an autoregulatory mechanism, in which speed and strength of contraction are functions of initial stretch (Fig. 2). When this mechanism is considered in relation to the intact organism, it is implied that the greater the filling (end-diastolic) volume of the left ventricle, the more rapidly and powerfully it empties. It is indeed still an implication rather than an attested fact, because of the great difficulty of accurately measuring ventricular volume in the unanesthetized animal (Ross and

Sobel, 1972; Braunwald, 1974; Sandler and Alderman, 1974). Superimposed on this autoregulatory mechanism are two direct and specific ways in which cardiac function can be modulated from without (Fig. 2). The frequency of contraction is determined by the interplay of neuro-cholinergic, and neural and humoral adrenergic activity on the sino-atrial node. The velocity and force of ventricular muscular contraction are increased by neural and humoral adrenergic action on the cardiac muscle fibers themselves.

The capacity of the heart to autoregulate its functions in the absence of a nerve supply has gained practical expression in the remarkable--if transitory--success of cardiac allotransplantation in man. More precise observations of cardiac function have been made in laboratory animals whose hearts are denervated, and in whom adrenalectomy or adrenergic blockade has removed the major external sources of modification of cardiac behavior. Some of the most interesting work has been done on the performance of trained racing greyhounds: either denervation or adrenalectomy cause only a modest diminution in performance, and while the combination of beta-blockade and denervation cause a marked reduction in performance, the animals can nevertheless still run (Donald, 1974). Within the obvious limitations it has been possible to apply a similar experimental approach to man (Rushmer, 1970; Braunwald, 1974).

Apart from the heart, there are some sources of energy input into the circulation concerned with special circulatory functions. These are the various musculo-venous pumps, in which the driving force is skeletal muscle contraction, and in which the reciprocating action is derived from the presence of venous valves. These are described in more detail elsewhere in this volume, but one of their functions is to increase flow through an organ or region without calling on the heart to perform additional work. That is, they act as local boosters to the driving force of the heart (Ludbrook, 1966; Stegall, 1966). Most developed, and most studied are the musculo-venous pumps of the lower limb. Others that have a possible or actual effect are created by the action of the abdominal musculature on the portal venous system, and by the action of respiratory movements on high compliance intrathoracic and intra-abdominal vessels.

RESISTANCES

The magnitude of the total resistance of the systemic vasculature to the output of the left ventricle can be perceived in various ways. The mean blood pressure in the root of the aorta is about 95 mm Hg in man, and in the right atrium only about 3 mm Hg, though the volume blood-flow at the two points is virtually identical. That is, most of the component of left ventricular power that is concerned with generating pressure is dissipated in the vasculature.

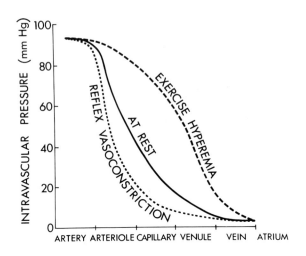

Fig. 3. Diagrammatic representation of pressure gradients in the systemic circulation, as through an upper-limb skeletal muscle vascular bed. Gradients shown at rest (solid line), during maximal exercise hyperemia (interrupted line), and with reflex vasoconstriction, as in hypovolemia (dotted line).

The elements of the vasculature in which this energy loss occurs are most simply indicated by a diagrammatic plot of the mean blood pressure gradient through the vasculature (Fig. 3); it is clear that under conditions of physical inactivity the arterioles are the major site of resistance to flow. However, this may not be true in all tissues at all times: in exercising muscle or in erectile tissue it appears that the location of maximum resistance is shifted distally (Fig. 3). These sorts of phenomena can be studied in a quite direct fashion by micropressure studies (Zweifach, 1973; 1974) with similar general conclusions.

While arterioles contribute the major resistance to blood flow in the systemic circulation, individually they also have the greatest capability for varying their resistance. This might be predicted from considering La Place's law in relation to the thickness of smooth muscle in the walls of blood vessels, and deducing that the ability of a vessel to actively constrict will be an inverse function of its diameter, an inverse function of the transmural pressure, and a direct function of the thickness of smooth muscle in its wall. This is a gross mathematical oversimplification (Somlyo and Somlyo, 1968), but points up the ability of arterioles to vary their luminal diameter, and thus the resistance they offer to flow.

Certain mechanisms that control the contractile state of the smooth muscle in resistance vessels have been able to be more or less clearly identified, and have been elegantly summarized in diagrammatic form by Folkow and Neil (Fig. 4).

It is clear that the resistance vessels in some vascular beds possess a resting or basal tone that is independent of all identifiable external influences (Folkow and Neil, 1971). Such for instance is the case in skeletal muscle and salivary glands, for autonomic denervation and exclusion of circulating vasoactive substances has very little effect on blood flow. On the other hand, resistance vessels in other tissues have virtually no basal tone, and tone is determined almost exclusively by the intensity of autonomic nervous activity: such is the case in the skin. What accounts for these very great differences in behavior has not been able to be directly

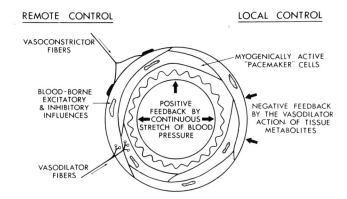

Fig. 4. Diagram of the mechanisms that determine resistance vessel tone (after Folkow and Neil, 1971, Fig. 16-1, by kind permission of authors and publishers).

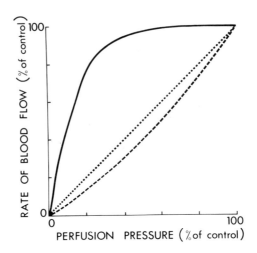

Fig. 5. Diagrammatic expression of the relation of perfusion pressure to flow. Dotted line: the situation for a Newtonian fluid flowing through rigid vessels. Interrupted line: blood flow through a non-autoregulating vascular bed (cf. skin). Solid line: blood flow through an autoregulating vascular bed (cf. brain, muscle).

determined, but the suggestion that automatic smooth muscle cell pacemakers account for basal tone is an attractive one (Folkow and Neil, 1971).

Then there are the various factors that modulate basal tone. The first is the so-called myogenic response, propounded originally by Bayliss (1902) under circumstances that would not nowadays be thought to offer rigorous proof that the phenomenon was indeed myogenic. However the phenomenon has been confirmed in vessels such as the umbilical artery or isolated perfused rabbit ear artery (Mellander and Johansson, 1968) in which the vascular smooth muscle appears to be completely isolated from external metabolic, nervous, or humoral influences. Under these circumstances the vascular smooth muscle responds to stretch by active contraction, and to release of stretch by relaxation. That is, flow through the vessel tends to be maintained constant in face of variation of perfusion (or distending) blood pressure--one example of the phenomenon of autoregulation. However, if this is a fundamental property of vascular smooth muscle, it is not at all clear what mechanism underlies it: whether it resides in the contractile mechanism itself, or whether it stems from variations in membrane potential. There is good evidence of membrane depolarization and increased spontaneous firing rates in response to stretch of nonvascular visceral (single unit) smooth muscle (Bülbring, 1964), and there is no reason to suppose that the same phenomenon may not occur in some vascular smooth muscle (Somlyo and Somlyo, 1968). In other words, there is at least a mechanism available for vascular smooth muscle to act as both mechano-receptor and effector. This point of view has been put cogently by Folkow (1964).

The phenomenon of autoregulation (Fig. 5) of blood flow is one that has attracted a good deal of attention (Mellander and Johansson, 1968; Stainsby, 1973). While myogenic responses may in part contribute to autoregulation, it is clear that other factors are equally important, and in many tissues more important. In certain tissues or organs, when they are denervated and isolated from the effect of vasoactive hormones, blood flow remains remarkably constant over a very wide range of perfusion pressures. The relative abilities of various tissues and organs in this regard are expressed semi-quantitatively in Table 2. It can be seen that (with the exception of the kidney) there is an inverse association between density of autonomic innervation and propensity for autoregulation. The mechanisms responsible for autoregulation are ill-understood, except within the framework of general statements that (again with the exception of the kidney) they appear to be concerned with maintaining a blood flow that will satisfy the metabolic demands of the tissue, and that the same mechanisms are likely to be responsible for reactive hyperemia: the sharp drop in vascular resistance that follows tissue ischemia.

TABLE 2

The capacity for autoregulation of blood flow in regional vascular beds, in relation to the potency of constrictor innervation.

Vascular bed	Potency of vasoconstrictor innervation	Capacity for autoregulation
Skin	++++	±
Kidney	++++	++++
Liver (arterial)	+++	++
Gut	+++	++
Skeletal muscle	++	+++
Heart	±	+++
Brain	−	++++

In the case of the brain there is good evidence of a close link between tissue pCO_2-pH coupled changes and vascular smooth muscle tone (Betz, 1972; Lassen, 1974). In the case of other tissues--and skeletal and cardiac muscle have received most attention--an enormous variety of metabolically-evolved local transmitters have been examined: pCO_2, pO_2, pH, K^+, lactate, phosphate, magnesium, adenosine, osmalality, and a variety of kinins, to name but some. The problem has been recently reviewed by Stainsby (1973), and there are several earlier reviews (Spencer, 1966; Haddy and Scott, 1968).

However inadequate the explanations, the phenomenon itself is unchallenged, and has been able to be detected in such a range of preparations that it is now virtually immune from challenge. Thus to the self-regulatory cardiac phenomena must be added a self-regulation, if not of peripheral resistance, at least of tissue blood flow. Taken together these properties of heart and of resistance vessels point up the intrinsic stability and automaticity of the cardiovascular system.

Nearly all sections of the vascular tree that contain smooth muscle have an autonomic effector innervation. The probable exceptions are the umbilical and many placental vessels; the ductus arteriosus; much of the cerebral vasculature; and of rather more general importance, precapillary sphincters (Somlyo and Somlyo, 1968; Burnstock, 1970; Speden, 1970; Folkow and Neil, 1971).

In general, the nerve endings do not extend nearer the lumen than the outermost layers of the media. While there is specialization of the fine terminal nerve fibers, there are rather large distances between these presumptive transmitter stores and the smooth muscle cell membranes (Somlyo and Somlyo, 1968; Burnstock, 1970). These two features support the concept of a visceral, single-unit type, arrangement of vascular smooth muscle, with a rather diffuse response to nervous activity that spreads longitudinally by way of smooth muscle intercellular impulse propagation. The supposition must be that the inner layers of smooth muscle in a blood vessel wall are accessible only to circulating vasoactive hormones that gain access by diffusion through the intima and through the endothelium of the vasa vasorum. However, in the case of many arterioles, the ratio of fine nerve terminals to smooth muscle fibers is high, and the opportunity for multi-unit activity correspondingly greater.

There has, of course, been intense interest in the forms of innervation of vascular smooth muscle, and in its pharmacological responsiveness. These matters are discussed in detail in a number of recent reviews (Somlyo and Somlyo, 1968, 1970; Burnstock, 1970; Speden, 1970; Folkow and Neil, 1971; Bevan and Su, 1973; Bohr, 1973). Some summary generalizations are made in the following paragraphs.

Sympathetic noradrenergic fibers are ubiquitous in vascular smooth muscle, with the exceptions referred to earlier. They are associated with alpha-receptors, except in the coronary vasculature where beta-receptors predominate. The details of norepinephrine synthesis, release, receptor-activation, and removal need not concern us here, but are extensively reviewed (Somlyo and Somlyo, 1968, 1970; Speden, 1970; Bevan and Su, 1973; Krnjevíc, 1974).

Non-innervated beta-receptors are found in relation to resistance vessels in some tissues, notably in skeletal and cardiac muscle and possibly in the umbilical artery. What is not quite clear is whether these beta-receptors are located in the vascular smooth muscle itself. Beta-receptors have been described in the cells of skeletal and cardiac muscle, in adipocytes, in hepatocytes and may exist in other body cells. When stimulated they cause "acceleration" of cellular metabolism. Because it is incontrovertibly established that the tone of resistance vessels of skeletal and cardiac muscle is exquisitely sensitive to changes in metabolic activity in the surrounding tissue, it is difficult to dissect out direct effects of catecholamines on vascular smooth muscle from indirect effects mediated by way of metabolic changes in surrounding striated muscle.

In the sympathetic outflow there are cholinergic, as well as adrenergic, fibers (Campbell, 1970). The feature of cholinergic transmission in general have been recently reviewed (Koelle, 1972; Krnjevíc, 1974). These fibers have been shown to be distributed to skeletal muscle resistance vessels in at least some non-primate mammalian species, notably the cat and dog (Uvnas, 1970). There is inferential evidence that they are present in man (Shepherd, 1963; Whelan, 1967), but this evidence has been strongly challenged by Uvnas (1970). It was at one time suggested that the cholinergic action of sympathetic nerves might be merely due to incidental release of ACh which was acting as a step in the chain of events leading to release of norepinephrine, but it now seems firmly established that cholinergic fibers are present in sympathetic nerves as a separate and specific system.

Parasympathetic cholinergic effectors have vasodilatory effects in some vascular beds. In the case of genital erectile tissue, the action seems to be direct and specific. The great increase in blood flow in salivary glands and gastric mucosa that follows appropriate parasympathetic stimulation probably has a more complcated explanation. In the case of salivary glands there is suggestive evidence that the major vascular effect derives from a kinin released from the secretory cells (Somlyo and Somlyo, 1970; Folkow and Neil, 1971). The situation may well be similar in the gastric mucosa, though with a different intermediary (histamine has been suggested but the case is far from proven). There is still a good deal of debate whether thermally-induced cutaneous vasodilatation is in part mediated by a kinin released by cholinergically-innervated sweat glands, as suggested by Fox and Hilton (1958).

Many other neuro-humoral transmitter systems to vascular smooth muscle have been proposed, and one that is currently being strongly argued is that purinergic fibers exist within the autonomic outflow (Burnstock, 1972). Another is a vasodilator, possibly histaminergic, system distributed to the blood vessels of the limbs (Rolewicz and Zimmerman, 1972; Bell et al., 1973; Lioy and White, 1973).

There is one other information-transfer system that is not strictly neural (though presumptively membrane-based), and which is not certainly real. There is evidence for a distal-to-proximal information-transfer system that resides in the vasculature itself. Hilton (1962) observed a phenomenon of centripetal conduction of a wave of vascular smooth muscle relaxation from the periphery, independent of nerve supply, and perhaps due to a conducted membrane potential change in the smooth muscle itself.

The other known extrinsic source of variation in vascular smooth muscle tone is humoral, by way of vasoactive substances carried by the bloodstream. The adrenal medullary hormones, epinephrine and norepinephrine, in varying proportions according to the animal species, are of uncontestable importance. That is, they can be detected in the bloodstream in concentrations that can be demonstrated to have a peripheral effect on vascular smooth muscle (Folkow and Neil, 1970; Somlyo and Somlyo, 1970; Speden, 1970). The direction of the effect --whether contraction or relaxation--depends on the presence and relative proportions of alpha- and beta-receptors in particular sections of the vasculature, and in particular tissues or organs.

A great number of other vasoactive hormones have been identified: for instance, posterior pituitary antidiuretic hormone (ADH), angiotensin II, 5-hydroxytryptamine (serotonin), kinins such as bradykinin, histamine, and prostaglandins (Somlyo and Somlyo, 1970). While the vasoactivity of these agents is not in doubt, nor that they have physiological roles in the body, it is not at all clear whether in real life they reach a sufficient concentration in the systemic bloodstream to have remote effects on vascular smooth muscle (Folkow and Neil, 1971). There is a suggestion that the skin vasoconstriction attendant on fainting or acute hypovolemia may be induced by ADH, but there is little evidence to suggest a direct role for the remainder in cardiovascular regulation under physiologic conditions.

II. Cardiovascular sensors

By this term is meant biological sensing devices that are responsive to some change in cardiovascular function, and which transmit this information centripetally by way of the afferent limb of a reflex arc (whether the mechanism of the information-transfer in the arc be neural, humoral, or both). Excluded from consideration under this heading are those sensorimotor systems in which information transfer is purely local: these have already been considered. This is not to say that the latter are unimportant, on the contrary, they are responsible for a great deal of the basic automaticity of cardiovascular functions.

HIGH-PRESSURE MECHANORECEPTORS (Folkow and Neil, 1971; Korner, 1971; Neil, 1972; Paintal, 1973; Pelletier and Shepherd, 1973).

These are located in the carotid sinus, and in the aortic arch and right subclavian arteries of at least some mammals. Their modes of action appear similar, the differences being mainly quantitative rather than qualitative (Hainsworth et al., 1970). However, it has been suggested that the range of arterial pressures over which the aortic arch receptors respond is higher than that for those in the carotid sinus (Pelletier and Shepherd, 1973). The microstructure of these mechanoreceptors (Rees, 1967) has received remarkably little recent attention. That they exist is most times inferred by gross and single-fiber recording from their afferent nerves, or by stimulation of their central cut ends. Their transducer action is presumed to be analogous to that of other mechanoreceptors. Afferent fibers from them enter the CNS by way of the parasympathetic system: in the carotid sinus nerves, or in the case of aortic-subclavian receptors by way of the vagus nerves. Both medullated and non-medullated fibers are present in the carotid sinus and aortic nerves, but there is no current evidence that these subserve different functions (Douglas et al., 1956).

These sensors are strictly stretch mechanoreceptors. Neurophysiologic studies show that single fibers discharge in time with systolic pressure peaks, and that an increase in mean arterial pressure results both in increased discharge-frequency and in fiber-recruitment (Fig. 6). There are sparsely distributed, sympathetically-innervated, smooth muscle cells in the adventiti of the carotid sinus. It is suggested that these may form the effector mechanism of a peripheral feedback loop that damps or exaggerates baroreceptor discharge (Rees, 1967; Kirchheim, 1976).

LOW-PRESSURE MECHANORECEPTORS (Folkow and Neil, 1971; Korner, 1971; Neil, 1972; Paintal, 1973; Pelletier and Shepherd, 1973; Linden, 1973; Thorén et al., 1976).

These are stretch mechanoreceptors located in the central, low-pressure, compartment of the circulation. Unencapsulated terminals of myelinated fibers, and "end-nets" have been described (Pelletier and Shepherd, 1973). Neurophysiological mapping (Paintal, 1972) has revealed concentrations of the receptors in the right atrium and its

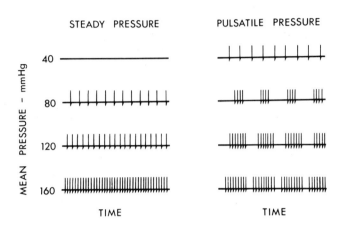

Fig. 6. Diagram to illustrate the ability of a single high pressure baroreceptor to respond to step increases in steady arterial pressure (left), and in pulsatile arterial pressure (right). (After Green, 1954 loc. cit. Folkow and Neil, 1971, Fig. 18-1, by kind permission of authors and publishers.)

adjacent great veins, and in the left atrium--pulmonary vein area (the veno-atrial junctions). It is further inferred that they are more properly considered as volume receptors, if only because of the high wall-compliance of the structures in which they are located. They too have a phasic discharge dependent either on the cyclical filling and emptying of the venous structures in which they lie (B receptors), or an atrial contraction (A receptors). This distinction has, however, been disputed (Kappagoda et al., 1977). Their central connection has been regarded as being by way of medullated and non-medullated fibers in the vagus nerve, but sympathetically-innervated receptors have also been described (Malliani et al., 1973; Mancia et al., 1976). While the behavior of right and left atrial receptors is similar, it is likely that the central connections of their afferents are different.

VENTRICULAR MECHANORECEPTORS (Korner, 1971; Paintal, 1973; Pelletier and Shepherd, 1973; Lindon, 1973; Thorén et al., 1976; Mancia et al., 1976).

These are located in the ventricular epicardium and myocardium, and their microscopic structure appears similar to that of the atrial receptors. They are vagally innervated, and fire at the time of isometric ventricular contraction and possibly during rapid ventricular emptying. Sympathetic afferents from the ventricles have also been described (Malliani et al., 1973).

Other cardiac sensors

There has recently been adduced from several sources tantalizing evidence that there may be coronary vascular receptors, sensitive to changes in coronary pressure or flow, and operating reflexes through the medulla and spinal cord (Smith, 1974).

CHEMORECEPTORS (Folkow and Neil, 1971; Korner, 1971, 1978; Neil and Howe, 1972; Mitchell and Wildenthal, 1974; Vatner and Pagani, 1976).

So-called ischemic receptors, presumably responsive to metabolic products of ischemia, are by inference found in skeletal muscle and brain. There is excellent physiologic evidence of neural afferent pathways from skeletal muscle, activated by ischemic exercise, and capable of eliciting a pressor response by way of the CNS (Folkow and Neil, 1971; Shepherd, 1971). Cerebral ischemia, whether resulting from raised intracranial pressure or from profound arterial hypotension has also been recognized to elicit a pressor response. Perhaps the receptors are those same $pH-pCO_2$ sensors in the medulla that are normally associated with control of respiration. In neither case is the exact nature of the receptors, of the effective stimulus, nor of the central reflex pathways, known.

The carotid bodies, and in some animals aortic arch bodies, are the most discrete chemoreceptors. The former especially have been extensively studied by classical physiologic techniques (Korner, 1971, 1978; Paintal, 1973). The most powerful effective stimuli are a lowered arterial pO_2, or histotoxic anoxia induced by cyanide. The afferent pathway is along the parasympathetic carotid sinus and vagus nerves.

PERIPHERAL RECEPTORS

Activation of almost every sort of peripheral receptor produces some sort of cardiovascular response. This is true of pain and thermal receptors in the skin, and of pain receptors in musculo-skeletal and visceral structures. That the reflex arcs can operate at a level below that of consciousness has become clear from studies in spinal man (Corbett et al., 1971). There has been special interest recently in facio-naso-pharyngeal receptors in relation to profound reflex cardiovascular changes when regularly-diving mammals--and even occasionally-diving mammals such as rabbits and man--submerse (Andersen, 1966; Korner, 1971; Daly, 1972; White and McRitchie, 1973). Pulmonary stretch receptors have long been recognized to have central connections with central cardiovascular control centers as well as with respiratory centers (Korner, 1971, 1978; Daly, 1972). Activation of bladder stretch receptors, and to a lesser extent activation of stretch receptors in the gastrointestinal tract (Whitteridge, 1960; Corbett et al., 1971) produce quite profound cardiovascular effects in intact and in spinal man. It is likely, but not absolutely certain, that the mechanoreceptors of muscles, bones, and joints have central connections with the cardiovascular reflex centers.

Auditory and visual stimuli may result in reflex changes in cardiovascular functions, especially when the input is emotive in content. To postulate corticothalamic "psychic" receptors is perhaps carrying the concept of receptors too far, but it is clear that simple or complex psychic stimuli, and a great variety of psychoactive drugs, can produce profound short- and long-term effects on cardiovascular functions (Bartorelli and Zanchetti, 1971; Zanchetti, 1972; Smith, 1974). Indeed there is growing evidence that feedback learning techniques (operant conditioning) from which somatic responses are excluded can modify cardiovascular functions in lower animals, subhuman primates, and man (Smith, 1974).

III Cardiovascular control centers (Folkow and Neil, 1971; Korner, 1971, 1978; Smith, 1974; Calaresu et al., 1975).

Study of these has gone through a number of phases. In the mid 19th century it was first recognized that stimulation of the central ends of autonomic nerves produces reflex changes in cardiovascular functions such as heart rate and arterial pressure. By the 1870's the importance of the medulla oblongata as a control center had been recognized. During the next 50 years neuro-anatomical techniques were used to map the locations of clusters of neurones concerned with autonomic effector functions, particularly those in the ponto-medullary region; and the number of discrete, eponymous, cardiovascular reflexes multiplied. Since the 1930's, and particularly since the 1950's, there have been two major approaches. One has been the development of control theory and its application to cardiovascular control. The other has been the execution of progressively more sophisticated experiments on cardiovascular input-output relationships with reference to the location and functioning of CNS control centers, in animals whose state has progressively approached the free-ranging, and whose status in the anthropocentric mammalian hierarchy has progressively approached that of man. Currently the subhuman primate is the subject of intense investigation in this regard, as in others. In this account it is impossible to do more than indicate the anatomic and functional complexity of CNS cardiovascular control centers, while at the same time giving a simplistic description of their location.

In Fig. 7 are illustrated the major cardiovascular control centers: discrete volumes of the CNS within which are located synaptic

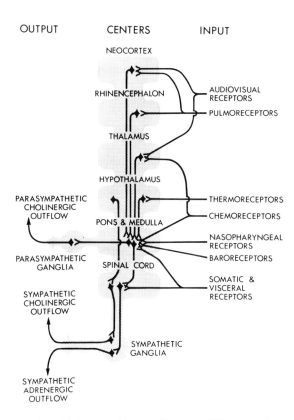

Fig. 7. Diagrammatic illustration of some CNS cardiovascular control centers, and the anatomy of the input to, and output from, them (after Korner, 1971).

connections upon which converge information from receptors, and from which diverge an integrated output that results in change, or potential change, in cardiovascular functions. Figure 7 is a greatly simplified and condensed version of the enormous accumulation of data on the location and connections of control centers, but does emphasize that integration occurs at all levels of the CNS from neocortex to autonomic ganglion. It also highlights the artificiality of attempts to locate discrete cardiovascular "centers in the CNS, often under experimental conditions that render the results at least partly invalid (such as the use of general anesthesia). This point is made strongly by Korner (1971), whose account of the subject is authoritative.

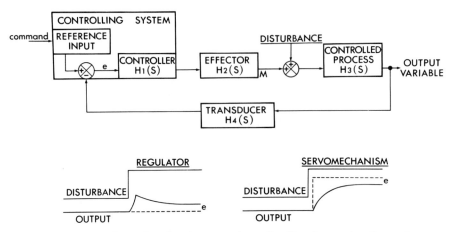

Fig. 8. Schema of a physical negative-feedback control system. $H(s)$ = the transfer action of each block. e = a function of the difference between output variable and reference input, and which determines the magnitude of the effector output M. (After Korner, 1971, Fig. 1., by kind permission of author and publisher.)

Anatomic accounts of the organization of cardiovascular control--even though based on studies of function--also fail to reveal the conception of cardiovascular control (or at least of some of its components such as control of arterial pressure, cardiac output, etc.) as a sophisticated servo-control mechanism, closely analogous to those used in control engineering. This concept is illustrated in Figs. 8 and 9. Even this is inadequate for the consideration of more than one controlled variable, whereas the totality of cardiovascular function is made up of a great number of controllable variables that are to a varying extent independent of each other except through the unbelievably complex interactions of the totality of cardiovascular control loops.

IV Specific open-loop cardiovascular reflexes

Considered here will be the responses to enhanced or reduced activation of single, specific sets of receptors: open-loop reflex responses. The difficulties of designing experiments in the field of cardiovascular function are nowhere better illustrated than in these conceptually-simple studies. For a start, it is extremely difficult to ensure that only the desired receptors are stimulated, for any cardiovascular response tends to activate other sensors. It is also difficult to ensure that the mode of stimulation bears a close relation to the mode that occurs in real life. General anesthesia is likely to alter, sometimes quite radically, the action and interaction of the integrating mechanisms. These considerations, as well as others such as posture, make it essential that studies be carried out in unanesthetized animals. If the ultimate goal is to understand patterns of behavior in man, there is merit in the now-popular approach of commencing with initial rigorous studies of sub-human primates and then extending these to man in a less-rigorously controlled fashion. Another problem that receives less attention than it perhaps deserves is that of identifying the "pure" output side of reflex arcs, for if output is measured by observing change in cardiovascular (rather than neural) functions, it may be blurred by the highly-developed autoregulatory mechanisms that modify both cardiac and peripheral vascular functions. A promising means for quantitating autonomic neural output in man has only recently been described (Delius et al., 1972; Hagbarth et al., 1972).

After these cautionary words, there follow some summary descriptions of changes in cardiovascular functions after change in input from single receptor-groups, emphasis having been placed on data gathered by methods that meet most closely the above desiderata.

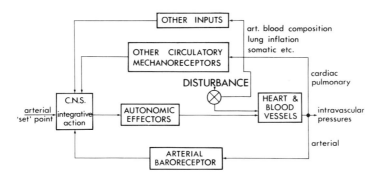

Fig. 9. Schema of a neural circulatory control system, for comparison with Figure 8. (After Korner, 1971, Fig. 2, by kind permission of author and publisher.)

LOW-PRESSURE BARORECEPTOR STIMULATION (Korner, 1971, 1978; Pelletier and Shepherd, 1973; Linden, 1973; Thorén et al., 1976; Mancia et al., 1976; Table 3).

TABLE 3

Effect of change in atrial pressure (after Korner, 1971; Pelletier and Shepherd, 1973). + = increased smooth muscle tone, or neural discharge rate. - = reduced smooth muscle tone, or neural discharge rate. ± = no change. ? = magnitude and direction of change uncertain. An idea of the gain is given by the number of + or -.

Autonomic effector	Atrial pressure increased	Atrial pressure reduced
Cardiac vagus	? -	? -
Cardiac sympathetic	? +	? ++
Splanchnic vascular tone		
• resistance	±	? ++
• capacitance	?	? +
Renal vascular tone	- - -	? ++
Muscle vascular tone		
• resistance	-	? +
• capacitance	?	?
Skin vascular tone		
• resistance	±	? ±
• capacitance	?	?
Adrenal catecholamine output	?	? +
ADH secretion	- -	?

The usually applied stimuli have been sudden increase or decrease in atrial volume, by fluid infusion or blood withdrawal, or by the use of inflatable atrial balloons. The reflex consequences are not yet quite clear-cut, because of variety in experimental design and experimental animal, and because of greater or lesser control of other consequent reflexes such as from the high pressure baroreceptors. In other words, a truly open-loop situation is difficult to contrive.

In general, atrial distension has caused cardioacceleration from increased sympathetic discharge to the sinatrial node, though with a considerable variability in the effect depending on the initial control heart rate. There has been a suggestion that activation of type A receptors cause increased sympathetic output to the heart, while type B receptor stimulation has a converse effect. It is doubtful whether there is alteration in sympathetic discharge to the myocardium more generally, or change in adrenal catecholamine output. There is reflex reduction in sympathetic discharge-rate to the renal vasculature, and probably also to the limb vasculature, including both resistance and capacity vessels. ADH secretion is reduced. There is convincing evidence that in man a fall of central venous pressure causes reflex vasoconstriction in the splanchnic area and in skeletal muscle, before there is a fall in arterial pressure (Johnson et al., 1974). There is less direct evidence of reflex renal vasoconstriction with activation of the renin-angiotensin system, and of reflex increase in ADH secretion.

All in all, decisive experiments that measure cardiovascular responses to isolated changes in low-pressure baroreceptor stimulation in intact animals are lacking. This should be of particular concern to those who seek to elucidate the basis for the cardiovascular changes that result from acute changes in blood volume, or acute changes in posture.

HIGH-PRESSURE BAROCEPTOR STIMULATION (Korner, 1971, 1978; Pelletier and Shepherd, 1973; Oberg, 1976; Kirchheim, 1976; Abboud et al., 1976; Ludbrook, 1978; Table 4).

Of all the sensors of cardiovascular function, the arterial baroreceptors are most amenable to stimulation and destimulation. The techniques of study in anesthetized animals have included alteration of pressure in the isolated, perfused, carotid sinus or aortic arch, and electrical stimulation of the central end of the carotid sinus or aortic nerves. In conscious animals (and man) electrical "pacemaking" of the carotid sinus nerve has been possible, temporary occlusion of the common carotid artery, or observation before and after baroreceptor denervation. Graded alteration of systemic arterial pressure has allowed the heart rate response to be quantified: in conscious man this has been accomplished by injecting vasoactive drugs, and in conscious animals either by the same means or by inflating cuffs around the descending aorta or inferior vena cava. There are limitations to the above techniques: they are either inapplicable to the conscious state, do not provide the physiologically appropriate stimulus, offer only a unidirectional stimulus, or allow only the response of heart rate to be measured. In many ways preferable is the technique used of altering carotid transmural pressure in conscious man in a graded fashion by means of a variable-pressure neck chamber, and in conscious animals by the chronic implantation of a variable-pressure capsule around the carotid sinus.

TABLE 4

Effect of changes in carotid sinus pressure (after Korner, 1971; Pelletier and Shepherd, 1973; Smith, 1974). + = increased smooth muscle tone, or neural discharge rate. - = reduced smooth muscle tone, or neural discharge rate. ± = no change. ? = magnitude and direction of change uncertain. An idea of the gain is given by the number of + or -.

Autonomic effector	Sinus pressure increased	Sinus pressure decreased
Cardiac vagus	++++	-
Cardiac sympathetic	-	+++
Splanchnic vascular tone		
• resistance	--	++
• capacitance	--	++
Renal vascular tone	-	+
Muscle vascular tone		
• resistance	---	++++
• capacitance	-	+
Skin vascular tone		
• resistance	-	++
• capacitance	?	±
Adrenal catecholamine output	±	++
ADH secretion	?	++

This emphasis on methodology is occasioned by a desire to be able to detect changes in the operating characteristics of these reflexes, resulting from age, hypertension, and alterations in CNS activity caused by differences in physiologic state or by drug action. Nevertheless the cardiovascular responses to alteration in arterial baroreceptor activity are qualitatively well-established. There is relatively little variation among species except as regards relative contribution of the various effector mechanisms.

VENTRICULAR MECHANORECEPTOR STIMULATION (Korner 1971, 1978; Pelletier and Shepherd, 1973; Linden, 1973; Thorén et al., 1976; Mancia et al., 1976).

The consequences of this are quite speculative, the best suggestion being that the contraction of an empty ventricle may elicit profound vagal and sympathetic cholinergic discharge, and sympathetic adrenergic inhibition: that is, this may be a mechanism operating in the vaso-vagal attack or common faint, especially when it is induced by mild hypovolemia. It is nevertheless clear that so much information is transmitted to the CNS about ventricular performance that it is unreasonable not to believe that these receptors have an inportant, if yet undiscovered, role in cardiovascular control.

ARTERIAL CHEMORECEPTOR STIMULATION (Korner, 1971; Daly, 1972; Table 5).

TABLE 5

Effect of arterial chemoreceptor activation (after Korner, 1971; Daly, 1972). + = increased smooth muscle tone, or neural discharge rate. - = reduced smooth muscle tone, or neural discharge rate. ± = no change. ? = magnitude and direction of change uncertain. An idea of the gain is given by the number of + or -.

Autonomic effector	Chemoreceptor stimulation
Cardiac vagus	++
Cardiac sympathetic	--
Splanchnic vascular tone	
• resistance	+
• capacitance	?
Renal vascular tone	+
Muscle vascular tone	
• resistance	+
• capacitance	?
Skin vascular tone	
• resistance	+
• capacitance	?
Adrenal catecholamine output	+
ADH secretion	?

TABLE 6

Cardiovascular responses to hypovolemia (after Cournand et al., 1943; Skillman et al., 1967; Baue, 1968; Falkow and Neil, 1971; Johnson et al., 1974). + = increase in listed function. - = decrease in listed function. ± = no change. Number of + or - indicate magnitude of change on a 4 point scale.

	Mild, acute, hypovolemia - early - vasovagal faint		Profound, acute, hypovolemia - early - late	
CENTRAL CHANGES				
Right atrial pressure	-	-	- - -	- - - -
Heart rate	±	- - - -	++	+++
Cardiac output	-	- - -	- -	- - -
Arterial pressure	±	- - - -	- -	- - - -
Overall peripheral resistance	+	- - -	+	- -
PERIPHERAL CHANGES				
Splanchnic				
• flow	-	-	- - - -	- -
• resistance	+	±	++++	++
• capacitance	-	-	- - -	-
Renal				
• flow	-	- -	- -	- - -
• resistance	+	±	++++	+++

Muscle				
• flow	−	+	− − −	− − −
• resistance	+	− − −	+ + + +	− −
• capacitance	−	+	−	±
Skin				
• flow	±	− −	− − −	− − − −
• resistance	±	±	+ + + +	+ + + +
• capacitance	±	+	−	− − −
Cerebral				
• flow	±	−	−	− −
• resistance	±	− − −	− −	− − − −
Adrenal catecholamine output	?	?	+ + + +	+ + + +
ADH secretion	?	?	+ +	+ +

The chemoreceptors are powerful afferents in respiratory reflex arcs, but in an appropriate setting hypoxia also induces a sequence of cardiovascular changes. As a generalization it can be said that it is only with a rather profound hypoxic stimulus that the reflex cardiovascular changes are remarkable. Moreover, the cardiovascular changes are only in part the direct result of afferent input from the chemoreceptors, and in part are the consequence of alterations in pulmonary ventilatory mechanics.

V. Effects of some complex stimuli on cardiovascular function

The cardiovascular responses to certain more or less complex forms of behavior have been studied in unanesthetized, more or less free-ranging, animals and especially in man. The patterns of cardiovascular response are fairly well established, even if (as has been made clear already) the mechanisms responsible for these changes are not. Thus the following description of the changes observed in response to certain changes of circumstance is reasonably accurate, though the postulated mechanisms are more or less speculative. The descriptions will be primarily tabular, with some supporting discussion. The stimuli to change considered will be sometimes complex, sometimes simple: the latter are included here because these reflexes have not been studied in a fully open-loop fashion.

BLOOD VOLUME CHANGE AND SYNCOPE (Table 6)

The normal diurnal changes in blood volume are quite modest, and amount to little more than ±5%. Situations in everyday life in which blood volume tends to fall are during prolonged standing (hydrostatic loss into tissue spaces), after heavy meals (loss into the transcellular space), or during exposure to excessive heat (evaporative water loss and sodium depletion). There is a tendency for blood volume to rise at night (resorption from tissue spaces).

However the human organism may sustain a more or less profound and acute reduction of blood volume in abnormal circumstances (external blood or plasma loss, gastrointestinal fluid and electrolyte loss). Conversely, the human may at times sustain great increases in blood volume, whether chronically (as in congestive heart failure or renal failure), or acutely and iatrogenically (over-enthusiastic intravenous infusion). It is rather doubtful whether man has built-in, specific mechanisms for coping with these situations, if only because when acute blood loss is sustained by a lower animal in its natural state, the sequel is usually that the animal is ingested by a predator. Nevertheless, there are recognizable cardiovascular consequences to acute loss of blood volume in man, and these are to an extent homeostatic if not always self-protective. Man's exposure to such environmental hazards as the motor vehicle and warfare for a few more millenia may result in the evolution of more specifically self-protective mechanisms.

Acute loss of 10% or so of blood volume may result in no easily detectable changes in cardiovascular functions: a small reduction in central venous pressure and cardiac output, with a corresponding small rise in peripheral resistance. However in some subjects there may result a sequence of responses closely similar to those sometimes also found in emotional stress, or prolonged standing: the so-called vasovagal or syncopal attack. A preliminary mild tachycardia is followed by a dramatic, vagally-induced, bradycardia that may result in cardiac standstill for as long as 20 seconds, accompanied by profound muscle resistance vessel dilation (thought to be mediated by sympathetic cholinergic fibers) so that arterial blood pressure falls below the level (40-50 mm Hg) at which cerebral blood flow is sufficient to maintain consciousness. Profuse sympathetic cholinergic sweating is a concomitant feature. The usual consequent fall of the person to the horizontal posture is associated with a gradually evolving tachycardia, recovery of arterial pressure, and extreme skin pallor (possibly due to massive ADH release). It is not clear what receptor mechanisms are responsible for triggering off this reflex sequence. Audio-visual-olfactory-pain (psychic) stimuli alone may do so. A similar form of syncope may also occur in emotionally neutral circumstances involving modest blood volume reduction: it has been suggested that ventricular mechano-receptors may constitute the receptor mechanism. Nor is it clear what the outcome of a vaso-vagal episode would be if the horizontal posture were not able to be assumed: the rarity of crucifixions nowadays makes it difficult for the physiologist to clarify this particular problem.

More profound, acute loss of blood volume activates the sympathoadrenal effector system, presumably by way of the low- and high-pressure baroreceptors. There then arises a conflict between the widespread vasoconstriction induced by these mechanisms, and the local self-protective autoregulatory dilator mechanisms that reside at tissue level. There is evidence that in the end the latter will win out, but there is also a suggestion that the powerful invocation of sympatho-adrenal vasoconstrictor mechanisms may be ultimately destructive rather than protective to the organism, (Nickerson, 1970), and that rather more emphasis on tissue perfusion and less on arterial pressure maintenance might be more beneficial.

CHANGE IN POSTURE (Table 7)

This activity is most remarkable in bipeds, and most highly developed in humans. The ability to change rapidly from a horizontal to a standing posture (and as rapidly back again) demands a specialized geometry of the cardiovascular system, or specifically-developed cardiovascular reflexes.

In man, standing up results in a substantial hydrostatic redistribution of the blood volume within the capacity vessels. Pulmonary

TABLE 7

Cardiovascular responses to assumption of the upright posture (after Rushmer, 1970; Folkow and Neil, 1971; Shepherd, 1971; Ludbrook, 1972). + = increase in listed fundtion. - = decrease in listed function. ± = no change. Number of + or - indicates magnitude of change on a 4 point scale.

	Immediate	Late
CENTRAL CHANGES		
Right atrial pressure	-	--
Heart rate	+	+++
Cardiac output	-	---
Arterial pressure	±	-
Overall peripheral resistance	+	++
PERIPHERAL CHANGES		
Splanchnic		
• flow	-	--
• resistance	+	++
• capacitance	-	±
Renal		
• flow	-	--
• resistance	+	++
Muscle		
• flow	±	-
• resistance	±	+
• capacitance (in lower limbs)	++	+++
Skin		
• flow	±	-
• resistance	±	+
• capacitance	±	-
Cerebral		
• flow	±	-
• resistance	-	--
Adrenal catecholamine output	+	++
ADH secretion	+	++

blood volume falls, splanchnic blood volume remains much the same, and lower limb blood volume increases. Despite this, right atrial pressure remains constant, or falls only slightly. This relative constancy of cardiac filling pressure seems to be maintained in part by the geometric form of the low-pressure circulation, within which the heart is effectively centrally placed (in a functional if not a structural sense); in part it is maintained because of reflex capacity-vessel constriction; in part from reduced compliance of the chambers of the heart; and in part from activity of musculovenous pumps. By these means rapid and automatic cardiovascular compensation is achieved.

Long-continued, motionless standing brings in another element: a gradual loss of blood volume into the tissue spaces of the dependent parts of the body. In these circumstances the reflexes described under hyopovolemia may be activated, more (or less, as in "guardsman's syncope") effectively.

MUSCULAR EXERCISE (Table 8)

During the emotive period preceding anticipated exercise, a number of quite profound psychically induced changes may occur ("flight reflex"). There is a variable increase in rate and strength of cardiac contraction, together with reduced flow and capacitance and increased resistance to flow in a great many vascular beds. These changes come about from increased sympathetic adrenergic discharge, and increased adrenal catecholamine output. There is also a marked and generalized increase in muscle blood flow, occasioned by increased sympathetic cholinergic discharge to muscle vessels (in sub-primates) and perhaps contributed to by the action of circulating catecholamines on beta receptors. These changes are usually regarded teleologically as a preparation for flight. However one of the mysteries of physiology is the emotion-induced muscle vasodilatation, for the increased flow seems to be through shunt rather than nutritive vessels, and it is in any event difficult to perceive how anticipatory hyperperfusion can be of more than very transitory value to the fleeing organism.

Once dynamic exercise has commenced, there is rapid and profound resistance-vessel dilatation and flow-increase in the exercising muscles, induced by local and presumably metabolic effects on the vascular smooth muscle. At the same time there is continued resistance-vessel and capacity-vessel reflex constriction in virtually all other vascular beds, including non-exercising muscles. Cardiac output and heart-rate are greatly increased, though the relative contribution to this increase of the several identifiable mechanisms is not yet clear: the initial increase in venous return from lower-limb musculo-venous pumps, continued sympatho-adrenal discharge, increasing quantities of circulating products of muscle metabolism, and a marked

TABLE 8

Cardiovascular responses to muscular exercise (after Rushmer, 1970; Folkow and Neil, 1971; Shepherd, 1971; Braunwald, 1974; Rowell, 1974; Clement and Shepherd, 1976; Mitchell and Wildenthal, 1976; Smith et al., 1976; Vatner and Pagani, 1976).
+ = increase in listed function. - = decrease in listed function. ± = no change. Number of + or - indicates magnitude of change on a 4 point scale.

	Immediately before exercise	During exercise isometric	During exercise dynamic	Immediately after dynamic exercise
CENTRAL CHANGES				
Right atrial pressure	±	±	±	-
Heart rate	+	+++	++++	++++
Cardiac output	+	+++	++++	+++
Arterial pressure	+	+++	+	-
Overall peripheral resistance	+	±	- - -	- - - -
PERIPHERAL CHANGES				
Splanchnic				
• flow	-	±	- -	-
• resistance	+	±	+++	++
• capacitance	±	±	- -	±
Renal				
• flow	±	±	- -	-
• resistance	±	±	+++	++

	1	2	3	4
Muscle *				
• flow	+	----	++++	++++
• resistance	-	++++	----	----
• capacitance	±	----	-	++
Skin				
• flow	-	±	++	+++
• resistance	+	±	-	---
• capacitance	-	±	++	+++
Cerebral				
• flow	±	±	±	-
• resistance	±	±	±	-
Adrenal catechlolamine output	+	?	++	+
ADH secretion	+	?	-	±

* exercising muscle

drop in overall peripheral resistance from vasodilatation in the exercising muscles. Arterial pressure usually rises somewhat, and in exhausting exercise may rise greatly. Right atrial pressure remains normal or is slightly raised: the tendency for blood volume in exercising muscles to increase because of capillary opening-up is offset by the action of musculovenous pumps, and by reflex capacity vessel constriction in some other vascular beds. Later in exercise (and afterward) thermo-regulatory reflexes are activated by the greatly increased rate of heat production, so that an initial cutaneous vasoconstriction is replaced by vasodilatation, venodilatation, and sweating.

Immediately after the cessation of dynamic exercise there may be some problems of cardiovascular homeostasis, especially if the exercise has been exhausting and if the upright posture is maintained. Muscle vasodilatation persists until the metabolic debt is repaid, cutaneous vasodilatation may be profound, and cessation of action of the lower limb musculovenous pumps allows hydrostatic pooling of blood in capacity vessels. There may thus result a marked fall in right atrial pressure and in aterial pressure, occasionally to the point of syncope.

Isometric (static) muscle contraction causes quite profound changes in cardiovascular functions (Lind et al., 1964), which are rather different from those which occur during dynamic exercise. The most notable feature is the very great increase in arterial pressure, accompanied by a parallel rise in heart rate and cardiac output and with little change in overall peripheral resistance. The acute hypertension implies either the consent of, or assistance by, the arterial baroreceptor reflex control system. Which of the two explanations is the correct one is not clear: one group has found the sensitivity of this reflex greatly depressed (Petersen et al., 1972), while another has reported it to be unchanged (Ludbrook et al., 1978). There is rather more information about the stimulus responsible for the cardiovascular changes: this is a combination of volition, and of afferent input from metabolic receptors in the contracting muscle (in which blood flow may be arrested because of the sustained pressure rise within the muscle).

THERMAL CHANGE (Table 9)

The stimulus to change in body temperature may come from variation of heat loss due to change in the thermal environment; by change in body heat production consequent on change of muscular activity; or as a result of infection by pyrogenic microorganisms. The latter is beyond the scope of this account, and involves resetting of the central thermal control mechanism itself. At least in the case of bacterial infections this resetting is caused by endogenous leucocytic pyrogens released by contact with bacteria or bacterial pyrogens (Wolstenholme and Birch, 1971).

There are two main sensing devices that actuate thermoregulatory reflexes. The more important are thermally-sensitive receptors located principally in the hypothalamic control center itself (single cell neurophysiological studies have revealed some with extremely temperature-sensitive firing rates). The less powerful are the thermal receptors of the skin, responsive to changes in heat flux across the skin. The functioning of skin thermoreceptors as part of the body's temperature regulating mechanism may, however, be confused by vasomotor changes induced in the skin by other than thermoregulatory reflexes: for instance, the cutaneous vasoconstriction accompanying a hypovolemic state tends to produce the same thermoreceptor response as exposure to environmental cold.

The main effector mechanisms for combating a threatened fall in body temperature are heat conservation by constriction of cutaneous resistance and capacity vessels, and increased metabolic heat production from shivering. There is a suggestion that in newborn mammals an adrenergic beta-stimulated increase in metabolic heat production may also occur. The effector mechanisms that combat a threat of raised body temperature are neurally-mediated dilatation of cutaneous resistance and capacity vessels and sweating (with the possiblity of further vasodilatation resulting from kinin release by sweat glands). It has recently become apparent that the variations in caliber of the thickly-muscled, densely sympathetically-innervated, cutaneous and subcutaneous veins are more intimately concerned with regulation of heat exchange than the regulation of vascular capacity. In fur-coated animals, in which heat exchange between skin and the environment is less direct, increased heat production by shivering and increased heat loss by hyperventilation are more important than in man.

VI. Overall cardiovascular control

It will be clear from the foregoing that it is simply not possible to give an all-embracing yet coherent account of the behavior of the cardiovascular system by conventional verbal or diagrammatic means. This is not merely a matter of the limitations of language or of conventional static visual display. The problem is a more fundamental one that derives from the data-sources. Even in the anesthetized laboratory animal the effects of only a limited permutation of receptor inputs can be measured in terms of only a limited number of effector functions. Even if with considerable ingenuity the range of inputs and outputs tested is increased, there remains a fundamental flaw of experimental design in that CNS function, metabolism, and even posture of the anesthetized labora bear little relation to those that obtain in its normal free-ranging state--and still less to those of man. This last consideration is being vigorously overcome by the chronic implantation of stimulating and function-detecting devices into free-ranging animals (including sub-human

TABLE 9

Cardiovascular responses to body temperature changes (after Webb-Peploe and Shepherd, 1968; Folkow and Neil, 1971; Hensel, 1973; Rowell, 1974; Cabanac, 1975; Shepherd and Vanhoutle, 1975). + = increase in listed function. - = decrease in listed function. ± = no change. Number of + or - indicates magnitude of change on a 4 point scale.

	Threat of body temperature rise	Threat of body temperature fall
CENTRAL CHANGES		
Right atrial pressure	-	±
Heart rate	+	+
Cardiac output	+	±
Arterial pressure	-	+
Overall peripheral resistance	-	+
PERIPHERAL CHANGES		
Splanchnic		
• flow	-	±
• resistance	+	±
• capacitance	-	±
Renal		
• flow	-	+
• resistance	+	-

REGULATION OF BLOOD FLOW AND PRESSURE

Muscle		
• flow	+	−
• resistance	−	+
• capacitance	±	±
Skin		
• flow	− − −	+ + + +
• resistance	+ + + +	− − − −
• capacitance	− − −	+ + + +
Cerebral		
• flow	±	±
• resistance	+	−
Adrenal catecholamine output	+ +	−
ADH secretion	+	−

primates), the data being retrieved by leads from trained animals, or by telemetry. Man has always been in many ways an ideal experimental animal, and the stimulus of space medical research has led to the development or refinement of a number of relatively non-invasive ways of studying cardiovascular functions in man, if not in a fully-free ranging state, at least in circumstances closely analogous to those of real life. Yet an ultimate difficulty remains: the impossibility of varying in an exactly controlled fashion multiple inputs while simultaneously making exact measurements of changes in multiple output functions.

The future of much cardiovascular research, especially in man, clearly lies in methods of multivariate analysis. In clinical situations, multiple vector techniques have been used to analyze complex data gathered from large groups of patients, so that clusters of patients with similar cardiovascular disturbances can be recognized and perhaps rationally treated (Siegel et al., 1970). Computational techniques can be used in a somewhat different fashion in intensive-care situations to interpret data, and prescribe or even execute therapy, in a way that may be more reliable than even the above-average M.D. (Sheppard et al., 1968).

More exciting in prospect is the use of analogue computational techniques to predict the "complete" behavior of the cardiovascular system in man (Guyton et al., 1972; Guyton et al., 1973). The prospect of using the analogue model to predict situational changes in cardiovascular dynamics, or to predict causes of such changes, and then to test hypotheses by realizable and limited sets of observations is an exciting one.

GENERAL REVIEWS

Braunwald, E., 1974, Regulation of the Circulation, New Engl. J. Med. 290:1124 and 1420.
Braunwald, E., Ross, J., and Sonnenblick, E.H., 1976, "Mechanisms of contraction of the normal and failing heart," 2nd edition, Little Brown, Boston.
Calaresu, F.R., Faiers, A.A., and Mogenson, G.J., 1975, Central neural regulation of heart and blood vessels in mammals, Progr. Neurobiol. 5:1.
Fishman, A.P., and Richards, D.W., (eds.), 1964, "Circulation of the blood; men and ideas," Oxford Univ. Press, Oxford.
Folkow, B., and Neil, E., 1971, "Circulation," Oxford Univ. Press, New York.
Guyton, A.C., Jones, C.E., and Coleman, T.G., 1973, "Circulatory physiology: cardiac output and its regulation," Saunders, Philadelphia.
Kirchheim, H., 1976, Systemic arterial baroreceptor revlexes, Physiol. Rev. 56:100.

Korner, P.I., 1971, Integrative neurovascular control, Physiol. Rev. 51:312.

Korner, P.I., 1978, Central nervous control of the heart and circulation, in "Handbook of Physiology: Circulation" 2nd edition (in press), American Physiological Society, Washington.

Mancia, G., Lorenz, R.R., and Shepherd, J.T., 1976, Reflex control of circulation by heart and lungs, Internat. Rev. Physiol.: Cardiovasc. Physiol. II 9:111.

Meijler, F.L., Brutsaert, D.L., (eds.), 1971, Contractile behaviour of heart muscle, Cardiovasc. Res. 5 Suppl. 1.

Mellander, S., and Johansson, B., 1968, Control of resistance, exchange, and capacitance function in the peripheral circulation, Pharmacol. Rev. 20:117.

Neil, E., (ed.), 1972, "Handbook of sensory physiology: enteroceptors," III/I, Springer, Berlin.

Oberg, B., 1976, Overall cardiovascular regulation, Ann. Rev. Physiol. 38:537.

Rushmer, R.F., 1970, "Cardiovascular dynamics," Saunders, Philadelphia.

Stainsby, W.N., 1973, Local control of regional blood flow, Ann. Rev. Physiol. 35:151

REFERENCES

Abboud, F.M., Heistad, D.D., Mark, A.L., and Schmid, P.G., 1976, Reflex control of the peripheral circulation, Progr. Cardiovasc. Dis. 18:371.

Andersen, H.T., 1966, Physiological adaptations in diving vertebrates, Physiol. Rev. 46:212.

Bartorelli, C., and Zanchetti, A., (eds.), 1971, "Cardiovascular regulation in health and disease," Instituto de Ricerche Cardiovascolari, Milano.

Baue, A.E., 1968, Recent developments in the study and treatment of shock, Surg. Gynecol. Obstet. 127:849.

Bayliss, W.M., 1902, On the local reactions of the arterial wall to changes in internal pressure, J. Physiol. (London), 28:220.

Bell, C., Lang, W.J., and Tsilemanis, C., 1973, Non-cholinergic vasodilatation in the canine hind limb evoked by hypothalamic stimulation, Brain Res. 56:392.

Betz, E., 1972, Cerebral blood flow: its measurement and regulation, Physiol. Rev. 52:595.

Bevan, J.A., and Su, C., 1973, Sympathetic mechanisms in blood vessels: nerve and muscle relationships, Ann. Rev. Pharmacol. 13:269.

Blinks, J.R., and Jewell, B.R., 1972, The meaning and measurement of myocardial contractility, in "Cardiovascular fluid dynamics," (Bergel, D.H., ed.), pp.225-260, Academic, New York.

Bohr, D.F., 1973, Vascular smooth muscle updated, Circulation Res. 32:665.

Bülbring, E., 1964, "Pharmacology of smooth muscle," Pergamon, Oxford.
Burnstock, G., 1970, Structure of smooth muscle and its innervation, in "Smooth muscle," (Bülbring, E., Brading, A.F., Jones, A.W., Tadao, T., eds.) pp.1-69, Arnold, London.
Burnstock, G., 1972, Purinergic nerves, Pharmacol. Rev. 24:509.
Cabanac, M., 1975, Temperature regulation, Ann. Rev. Physiol. 37:415.
Campbell, G., 1970, Autonomic nervous supply to effector tissues, in "Smooth muscle," (Bülbring, E., Brading, A.F., Jones, A.W., Tadao, T., eds.), pp.451-495, Arnold, London.
Clement, D.L., and Shepherd, J.T., 1976, Regulation of peripheral circulation during muscular exercise, Progr. Cardiovasc. Dis. 19:23.
Corbett, J.L., Frankel, H.L., and Harris, P.J., 1971, Cardiovascular reflex responses to cutaneous and visceral stimuli in spinal man, J. Physiol. (London) 215:395.
Cournand, A., Riley, R.L., Bradley, S.E., Breed, E.S., Noble, R.P., Lauson, H.D., Gregerson, M.I., and Richards, D.W., 1943, Studies of the circulation in clinical shock, Surgery 13:964.
Cunningham, D.J.C., Petersen, E.S., Peto, R., Pickering, T.G., and Sleight, P., 1972, Comparison of the effect of different types of exercise on the baroreflex regulation of heart rate, Acta. Physiol. Scand. 86:444.
Daly, M. de B., 1972, Interaction of cardiovascular reflexes, Sci. Basis Med. 20:307.
Delius, W., Hagbarth, K-E., Hongell, A., and Wallin, B.G., 1972, General characteristics of sympathetic activity in human muscle nerves, Acta. Physiol. Scand. 84:65.
De Mello, W.D., (ed.) 1972, "Electrical phenomena in the heart," Academic, New York.
Donald, D.E., 1974, Myocardial performance after excision of the extrinsic cardiac nerves in the dog, Circulation Res. 34:417.
Douglas, W.W., Ritchie, J.M., and Schaumann, W., 1956, Depressor reflexes from medullated and non-medullated fibres in the rabbit's aortic nerve, J. Physiol. (London) 132:187.
Folkow, B., 1964, Description of the myogenic hypothesis, Circulation Res. 15 Suppl. 1:279.
Fox, R.H., and Hilton, S.M., 1958, Bradykinin formation in human skin as a factor in heat vasodilatation, J. Physiol. (London), 142:219.
Fozzard, H.A., and Gibbons, W.R., 1973, Action potential and contraction of heart muscle, Am. J. Cardiol. 31:182.
Fuch, F., 1974, Striated muscle, Ann. Rev. Physiol. 36:461.
Goss, C.M., 1940, First contractions of the heart without cytological differentiation, Anat. Rec. 76:19.
Guyton, A.C., Coleman, T.G., and Granger, H.J., 1972, Circulation: overall regulation, Ann. Rev. Physiol. 34:13.

Haddy, F.J., and Scott, J.B., 1968, Metabolically linked vasoactive chemicals in local regulation of blood flow, Physiol. Rev. 48:688.

Hagbarth, K-E., Hallin, R.G., Hongell, A., Torbjörk, H.E., and Wallin, B.G., 1972, General characteristics of sympathetic activity in human skin nerves, Acta. Physiol. Scand. 84:164.

Hainsworth, R., Ledsome, J.R., and Carswell, F., 1970, Reflex responses from aortic baroreceptors, Am. J. Physiol. 218:423.

Hensel, H., 1973, Neural processes in thermoregulation, Physiol. Rev. 53:948.

Hilton, S.M., 1962, Local mechanisms regulating peripheral blood flow, Physiol. Rev. 42, Suppl. 5:265.

Hoff, E.C., Kramer, T.C., DuBois, D., and Patten, B.M., 1939, The development of the electrocardiogram of the embryonic heart, Am. Heart J. 17:470.

Johnson, J.M., Rowell, L.B., Niederberger, M., and Elsman, M.M., 1974, Human splanchnic and forearm vasoconstrictor responses to reductions of right atrial and aortic pressures, Circulation Res. 34:515.

Koelle, G.B., 1972, Acetyl choline - current status in physiology, pharmacology and medicine, New Engl. J. Med. 286:1086.

Korner, P.I., 1974, Present concepts about the myocardium, in "Advances in cardiology" (Reader, R., ed.), Vol. 12, pp.1-14, Karger, Basel.

Krnjevíc, K., 1974, Chemical nature of synaptic transmission in vertebrates, Physiol. Rev. 54:418.

Langer, G.A., 1973, Heart: excitation - contraction coupling, Ann. Rev. Physiol. 35:55.

Lassen, N.A., 1974, Control of cerebral circulation in health and disease, Circulation Res. 34:749.

Lind, A.R., Taylor, S.H., Hymphreys, P.S., Kennelly, B.M., and Donald, K.W., 1964, The circulatory effects of sustained voluntary muscle contraction, Clin. Sci. 27:229.

Linden, R.J., 1973, Function of cardiac receptors, Circulation 48:463.

Lioy F., and White, K.P., 1973 ^{14}C-histamine release during vasodilatation induced by lumbar ventral root stimulation, Pflügers Archiv. 342:319.

Ludbrook, J., 1966, "Aspects of venous function in the lower limbs," Thomas, Springfield.

Ludbrook, J., 1972, "The analysis of the venous system," Huber, Bern.

Ludbrook, J., 1978, Techniques for examining the arterial baroreceptor reflexes in the conscious state in "Studies in neurophysiology," Ed. R. Porter (in press), Cambridge University Press, Cambridge.

Ludbrook, J., Faris, I.B., Iannos, J., Jamieson, G.G., and Russell, W.J., 1978, Lack of effect of isometric handgrip exercise on the responses of the carotid sinus baroreceptor reflex in man, Clin. Sci. (in press).

Malliani, A., Recordati, G., and Schwartz, P.G., 1973, Nervous activity of afferent cardiac sympathetic fibres with atrial and ventricular endings, J. Physiol. (London) 229:457.

Mark, G.E., and Stasser, F.F., 1966, Pacemaker activity and mitosis in cultures of newborn rat heart ventricle cells, Exp. Cell. Res. 44:217.

Mitchell, J.H., and Wildenthal, K., 1974, Static (isometric) exercise and the heart: physiological and clinical considerations, Ann. Rev. Med. 25:369.

Neil, E., and Howe, A., 1972, Arterial chemoreceptors, in "Handbook of sensory physiology: enteroreceptors" (Neil, E., ed.) Vol. III/1, pp. 47-80, Springer, Berlin.

Nickerson, M., 1970, Vascular adjustments during the development of shock, Can. Med. Assoc. J. 103:853.

Noble, D., 1966, Applications of Hodgkin-Huxley equations to excitable tissue, Physiol. Rev. 46:1.

Paintal, A.S., 1972, Cardiovascular receptors, in "Handbook of sensory physiology: enteroreceptors" (Neil, E., ed.) Vol. III/1, pp.1-45, Springer, Berlin.

Paintal, A.S., 1973, Vagal sensory receptors and their reflex effects, Physiol. Rev. 53:159.

Pelletier, C.L., and Shepherd, J.T., 1973, Circulatory reflexes from mechanoreceptors in the cardio-aortic area, Circulation Res. 33:131.

Rees, P.M., 1967, Observations on fine structure and distribution of presumptive baroreceptor nerves at the carotid sinus, J. Comp. Neurol. 131:517.

Rolewicz, T.F., and Zimmerman, B.D., 1972, Peripheral distribution of cutaneous sympathetic system, Am. J. Physiol. 223:939.

Rowell, L.B., 1974, Human cardiovascular adjustments to exercise and thermal stress, Physiol. Rev. 51:75.

Ross, J., Jr., and Sobel, B.E., 1972, Regulation of cardiac contraction, Ann. Rev. Physiol. 34:47.

Reuter, H., 1974, Exchange of calcium ions in the mammalian myocardium, Circulation Res. 34:599.

Sandler, H., and Alderman, E., 1974, Determination of left ventricular size and shape, Circulation Res. 34:1.

Sarnoff, S.J., 1955, Myocardial contractility as described by ventricular function curves: observations on Starling's law of the heart, Physiol. Rev. 35:107.

Sarnoff, S.J., and Mitchell, J.H., 1962, Control of function of the heart, in "Handbook of Physiology: 2; Circulation:1" (Hamilton, W.F., ed.) pp. 489-532, American Physiological Society, Washington.

Shepherd, J.T., 1963, "Physiology in the circulation in human limbs in health and disease," Saunders, Philadelphia.

Shepherd, J.T., 1971, Cardiovascular reflexes during exercise, in "Cardiovascular regulation in health and disease" (Bartorelli, C., and Zanchetti, A., eds.), pp. 63-70, Istituto de Ricerche Cardiovascolari, Milano.

Shepherd, J.T. and Vanhoutte, P.M., 1975, "Veins and their control," Saunders, Philadelphia.

Sheppard, L.C., Koschoukos, N.T., Kurtts, M.A., and Kirklin, J.W., 1968, Automated treatment of critically ill patients following operation, Ann. Surg. 168:596.

Siegel, J.H., Strom, B.L., Miller, M., Lewin, I., and Goldwyn, R.M., 1970, A computer-based clinical assessment, research and education system to facilitate continuing education in the care of the critically ill patient, Surgery 68:238.

Skillman, J.J., Olson, J.E., Lyons, J.H., and Moore, F.D., 1967, The hemodynamic effect of acute blood loss in normal man, with observations on the effect of the Valsalva maneuver and breath holding, Ann. Surg. 166:713.

Smith, E.E., Guyton, A.C. Manning, R.D., and White, R.J., 1976, Integrated mechanisms of cardiovascular response and control during exercise in the normal human, Progr. Cardiovasc. Dis. 18:421.

Smith, O.A., 1974, Reflex and central mechanisms involved in the control of the heart and circulation, Ann. Rev. Physiol. 36:93.

Somlyo, A.P., and Somlyo, A.V., 1968, Vascular smooth muscle, I. Normal structure, pathology, biochemistry and biophysics, Pharmacol. Rev. 20:197.

Somlyo, A.P., and Somlyo, A.V., 1970, Vascular smooth muscle, II. Pharmacology of normal and hypertensive vessels, Pharmacol Rev. 22:249.

Sonnenblick, E.H., 1962, Force-velocity relations in mammalian heart muscle, Am. J. Physiol. 202:931.

Sonnenblick, E.H., 1965, Determinants of active state in heart muscle: Force, velocity, instantaneous muscle length, time, Federation Proc. 24:1396.

Speden, R.N., 1970, Excitation of vascular smooth muscle, in "Smooth muscle" (Bülbring, E., Brading, A.F., Jones, A.W., Tadao, T., eds.) pp. 558-588, Arnold, London.

Spencer, M.P., 1966, Systemic circulation, Ann. Rev. Physiol. 28:311.

Starling, E.H., 1918, "The Linacre lecture on the law of the heart given at Cambridge, 1915," Longmans, London.

Stegall, H.F., 1966, Muscle pumping in the dependent leg, Circulation Res. 19:180.

Thorén, P.M., Donald, D.E., and Shepherd, J.T., 1976, Role of heart and lung receptors with non-medullated vagal afferents in circulatory control, Circulation Res. 38 Suppl. II:2.

Trautwein, W., 1973, Membrane currents in cardial muscle fibers, Physiol. Rev. 53:793.

Uvnäs, B., 1970, Cholinergic muscle vasodilatation, in "Cardiovascular regulation in health and disease" (Bartorelli, C., Zanchetti, A., eds.) pp. 7-16 Istituto di Ricerche Cardiovascolari, Milano.

Vatner, S.F., and Pagani, M., 1976, Cardiovascular adjustments to exercise: hemodynamics and mechanisms, Progr. Cardiovasc. Dis. 19:91.

Webb-Peploe, M.M., and Shepherd, J.T., 1968, Veins and their control, New Engl. J. Med. 278:317.

Weidmann, S., 1965-66, Cardiac electrophysiology in the light of recent morphological findings, in "The Harvey Lectures Series" Vol. 61, pp. 1-16, Academic, New York.

Whelan, R.F., 1967, "Control of the peripheral circulation in man," Thomas, Springfield.

White, S.W., and McRitchie, R.J., 1973, Nasopharyngeal reflexes: integrative analysis of evoked respiratory and cardiovascular effects, Australian J. Exp. Biol. Med. Sci. 51:17.

Whitteridge, D., 1960, Cardiovascular reflexes initiated from afferent sites other than the cardiovascular system itself, Physiol. Rev. 40 Suppl. 4:198.

Wolstenholme, G.E.W., and Birch, J., (eds.), 1971, "Pyrogens and fever", Livingstone, Edinburgh.

Zanchetti, A., (ed.), 1972, "Neural and psychological mechanisms in cardiovascular disease", Casa Editrice "Il Ponte," Milano.

Zweifach, B.W., 1973, Microcirculation, Ann. Rev. Physiol. 35:117.

Zweifach, B.W., 1974, Quantitative studies of microcirculatory structure and function. 1. Analysis of pressure distribution in the terminal vascular bed in cat mesentery, Circulation Res. 34:843.

THE ANATOMY OF THE RENAL CIRCULATION

Kim Solez,[1] M.D., Robert H. Heptinstall,[2] M.D.

[1]Assistant Professor of Pathology and Medicine, Johns
Hopkins University and Hospital, Baltimore, MD 21205
[2]Baxley Professor and Director, Department of Pathology,
Johns Hopkins University and Hospital, Baltimore, MD
21205

There is no organ with a blood vascular system more complex than that of the mammalian kidney. To draw an analogy with electrical circuits, the renal vasculature represents an intricate system of constantly varying resistances, arranged in series and parallel, which control the flow of energy required for the many homeostatic roles the kidney must play. Morphologic studies demonstrate the many different routes which blood may take in its passages through the kidney. Such studies cannot capture the dynamic nature of the renal circulation but are important nonetheless in delineating the anatomic substrate upon which functional and disease-related alterations in vessel structure and caliber take place.

I. THE RENAL VASCULATURE IN MAN

A. SEGMENTAL DISTRIBUTION OF ARTERIAL BLOOD

Before the branchings of the renal artery are described it should be understood that from a vascular point of view the kidney is divided up into a number of segments (Graves, 1954). These are described as apical, upper, middle, lower, and posterior (Fig. 1). Each has its own arterial supply--the branches taking the name of the segment supplied (Fig. 2)--and there is no appreciable arterial communication between the segments. This concept of Graves (1954) has been widely accepted although modifications have been made by some authors (Boijsen, 1959; Faller and Ungvary, 1962; Sykes, 1963; Sykes, 1964).

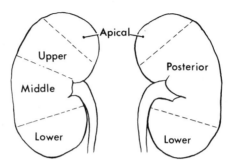

Fig. 1. Different segments of kidney.

The idea of a paucity of cross circulation between segments is of importance in providing relatively bloodless lines which are of value to the surgeon. Brödel (1901) pointed out that because of lack of communication between anterior and posterior arterial fields, there was a relatively bloodless line along the lateral border of the kidney. However, by studying casts of renal arteries, Sykes (1963) has demonstrated that Brödel's line is not completely avascular because the terminal branches of the anterior and posterior divisions overlap to a considerable degree.

B. MAIN RENAL ARTERY AND EXTRARENAL BRANCHES

Each kidney is supplied by a main renal artery which originates from the lateral aspects of the abdominal aorta just below the suerior mesenteric artery. Each proceeds laterally and slightly downward, crossing the crus of the diaphragm. The right renal artery passes behind the inferior vena cava, the right renal vein, the head of the pancreas, and the second part of the duodenum. The left renal artery is shorter in length because the abdominal aorta is placed to the left of the median plane. It is situated slightly higher than the right and lies behind the left renal vein, the body of the pancreas, and the splenic vein.

As the main renal artery approaches the hilus of the kidney, it assumes one of several different patterns (Brödel, 1901; Kuprijanoff, 1924; Graves, 1954; Greenblatt, 1963; Hegedus, 1972). It usually divides into an anterior and posterior division (Fig. 2), the one passing in front of the renal pelvis and the other behind. In most cases the anterior division divides into 3 segmental branches;

the upper passes anterior to the upper major calyx to supply the
upper segment; the middle courses laterally to supply the middle
segment; and the lower passes in front of the lower major calyx to
supply the lower segment. The posterior main division, after passing behind the renal pelvis, gives rise to several branches which
supply the posterior segment. The apical segmental artery is subject to great variation; it usually originates from the upper segmental artery but may arise directly from the main renal artery
before division occurs into the anterior and posterior divisions
(accessory polar artery). Many other variants of the apical segmental artery have been described, as have other variations in the
extrarenal branches (Boijsen, 1959; Graves, 1956a; Merklin and
Michels, 1958; Poisel and Spängler, 1969). One of these is a branch
supplying the lower segment directly from the aorta; such branches
may be associated with hydronephrosis and it is often assumed, perhaps erroneously, that the aberrant vessel is the cause of the hydronephrosis.

A more detailed account of the arteries is given by Sykes (1964)
who elaborates on the arrangement described by Graves (1954). To
understand his description it is necessary to recall that Löfgren

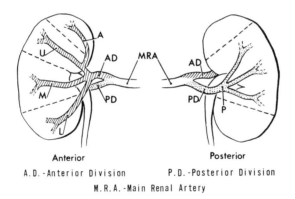

Anterior   Posterior
A.D.-Anterior Division   P.D.-Posterior Division
M.R.A.-Main Renal Artery

Fig. 2. Main arterial divisions of kidney. MRA, main renal artery;
PD, posterior division; AD, anterior division. Segmental
arteries indicated by A (apical), U (upper), M (middle),
L (lower), P (posterior).

(1949) considered there were 3 lobes on each surface of the kidney and 7 colliculi majores* on each surface. Further, the upper lobe consisted of colliculi 1, 2 and 3, the hilar lobe of colliculi 4 and 5, and the lower lobe of colliculi 6 and 7. Sykes (1964) considered the distribution of the segmental arteries as follows:

1. Apical artery supplies colliculi A1, A2, P1, and P2.
2. Upper segmental artery supplies A3.
3. Middle segmental artery supplies A4 and A5.
4. Lower segmental artery supplies A6, A7, P6, and P7.
5. Posterior segmental artery supplies P3, P4, and P5.

While the arteries are still in an extrarenal position, small branches are given off to supply the ureter and the suprarenal gland; the latter are known as the inferior suprarenal arteries. Branches are given off to the pelvic mucosa by arteries as they gain access to the parenchyma and by the arcuate arteries. The pelvic arteries are characteristically tortuous and this may be related to the fact that the pelvis is a distensible structure.

C. INTRARENAL BRANCHES

Once the segmental arteries have gained access to the renal parenchyma, they divide into the arcuate arteries which proceed in a gently curving way (Fig. 3), giving rise at regular intervals to the interlobular arteries which proceed in a straight line to the outermost part of the cortex (Fig. 4). Because of different concepts of the lobule, some prefer the name intralobular artery to interlobular artery. Whatever name is employed--and we prefer interlobular--these arteries give off the afferent arterioles to individual glomeruli. The distal part of the afferent arteriole contains characteristic granules making up part of the juxtaglomerular apparatus. These granules are considered to be concerned in the production of renin.

The afferent arterioles give way to a specialized collection of capillaries called the glomerular tuft, and here the process of filtration takes place. Blood leaves the glomerular tuft through the efferent arterioles which is more delicate than the afferent branch and which breaks up into the intertubular capillary plexus.

---

*The terminology needs explaining because the word lobule is used in two different senses. On the one hand, it is used as a synonym for colliculus major and is, therefore, a large unit of renal tissue. On the other hand, it is used to describe the medullary ray of the cortex with its adjacent glomeruli and tubules; here it is a microscopic unit. In this chapter the term colliculus major will be used to describe the gross structure, with lobule being reserved for the microscopic units in the cortex.

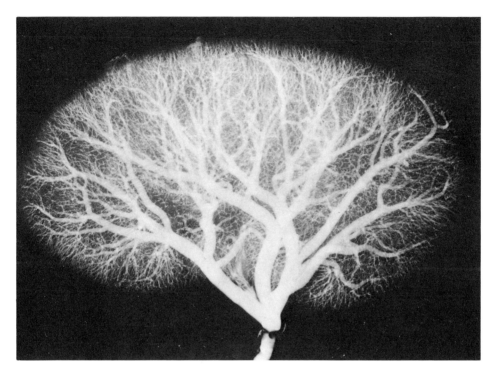

Fig. 3. Angiogram of kidney prepared postmortem. The segmental arteries give rise to the gently curving arcuate arteries which in turn give rise to the interlobular arteries.

The tubules are supplied with blood by this complex intertubular capillary plexus. Blood drains from these capillaries into a system of thin-walled veins whose names are similar to the vessels on the arterial side situated in close proximity to the arteries. There is no segmental arrangement of the veins (Graves, 1956b), and blood leaves the kidney through 3 main lobar veins (upper, hilar, and lower) which unite to form the main renal vein which discharges into the vena cava (Fourman and Moffat, 1971). In addition there is a superficial system of veins referred to as stellate which drain into the interlobular veins.

D. BLOOD SUPPLY TO THE MEDULLA

Blood is carried to the medulla by the vasa recta which comprise the arteriolae rectae spuriae and the arteriolae rectae verae. The former are the more plentiful and are a series of branches which

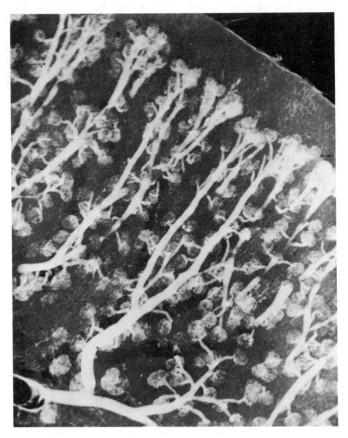

Fig. 4. Microangiogram of rabbit kidney showing interlobular arteries pursuing straight course to periphery. From these are given off the afferent arterioles, each of which supplies a glomerulus. Glomeruli are well filled, and the fine intercapillary plexus can just be discerned. X30.

originate from the efferent arterioles of the juxtamedullary glomeruli. The arteriolae rectae verae take their origin directly from interlobular arteries (Fig. 5). Some, at least, are formed when a deep glomerulus becomes sclerosed with the result that the afferent and efferent arterioles become continuous and continue down into the medulla. The vasa recta travel down through the medulla in vascular bundles which are composed of vasa recta (both ascending and descending), limbs of Henle, and collecting ducts (Kriz and

Lever, 1969; Röllhauser et al., 1964). The arrangement in the vascular bundle is depicted in Figs. 6 and 7. At intervals the descending vasa recta break out from the bundles to form a capillary network which is more complex in the outer than in the inner medulla. Capillaries join together to form the ascending vasa recta which gain access to a vascular bundle and travel outward to join the venous drainage. At one time it was believed that the ascending and descending vasa recta were a continuous tube with a hairpin bend occurring at different levels in the medulla; it is now appreciated that this arrangment is uncommon.

The number of arteriolae recta verae increases with age but these aglomerular channels are usually vastly outnumbered by arteriolae rectae spuriae and are unlikely to be hemodynamically significant (Maxwell et al., 1950; More and Duff, 1951; Ljungqvist and Lagergren, 1962). It has been suggested that these vessels allow a

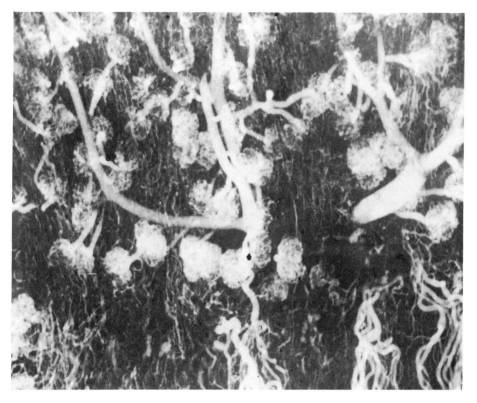

Fig. 5. Microangiogram of rabbit kidney to show corticomedullary junction and beginning of vasa recta. An arteriola recta spuria can be clearly seen in the lower center. X50.

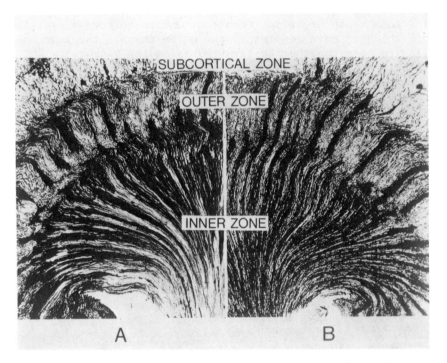

Fig. 6. Two vertical sections of rat kidney showing the vascular bundles of outer and inner zones. A. Following venous injection. B. Following arterial injection of India ink. X16.

diversion of blood away from the cortex and into the medulla during shock. This concept of a bypassing of the cortical glomeruli has been called the "Trueta shunt." However, it is clear from Trueta's book (Trueta et al., 1947) that the shunt proposed involved the juxtamedullary glomeruli and arteriolae rectae spuriae rather than arteriolae recta verae:

"*.... the kidney has two potential circulations, a greater and a lesser, and in extreme conditions the blood may pass either almost exclusively through one or the other of two pathways, or in less abnormal circumstances, to a varying degree through both. The vessels making up the pathway of the greater circulation are those associated with the cortical glomeruli; the channels of the lesser circulation are those associated with the juxtamedullary glomeruli.*"

The origins of the confusion about Trueta's beliefs have been well described by More and Duff (1951).

THE ANATOMY OF THE RENAL CIRCULATION

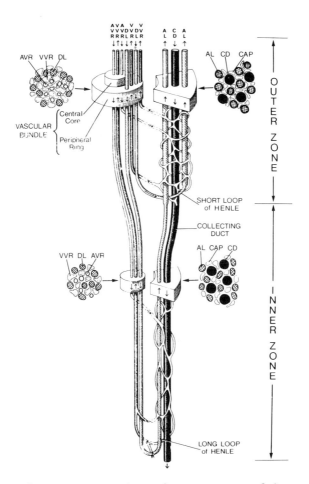

Fig. 7. Schematic representation of structures of inner and outer zones of renal medulla in rat. A short loop of Henle is shown crossing between the bundle and the collecting duct in the outer zone. A long loop of Henle crosses in a similar way in the inner zone. In addition, arterial and venous vasa recta pass between the bundle and the capillary plexus which surrounds the collecting duct and the ascending limb of Henle's loop. AVR, arterial vas rectum; VVR, venous vas rectum; DL, descending limb of Henle's loop; AL, ascending limb of Henle's loop; CD, collecting duct; CAP, capillaries.

E. HISTOLOGY

The extrarenal parts of the renal arteries resemble other muscular arteries derived from the aorta. There is a well-formed adventitia separated from the media by the external elastic lamella. The media is relatively thick and accounts for the main thickness of the wall. It is made up of smooth muscle fibers, collagen, and fine elastic fibers. The internal elastic lamella is well developed and may even show two or three layers in apparently normal kidneys with advancing years. It separates the media from the thin intima. The extrarenal branches may develop arteriosclerotic changes, but large collections of lipid in the wall are relatively uncommon except in the very proximal part of the main artery. When renovascular hypertension occurs on the basis of arteriosclerosis, it is usually a plaque at the ostium of the main artery that is responsible. The extrarenal arteries may be the seat of fibromusclar dysplasia, a poorly understood condition which affects young women in particular; it is a potent cause of hypertension.

The segmental and arcuate arteries also show the same structure as the main renal artery, and with increasing age and in the presence of hypertension, show fibrous intimal thickening. In addition, these arteries may show intimal "cushions" which are situated in particular at the origin of collateral branches (Rotter, 1952). Medial defects have also been described (Rotter, 1952) as have collections of smooth muscle fibers forming a bundle along one side of the arcuate vessels (Barrie et al., 1950).

A similar histologic structure is possessed by the interlobular arteries, and these vessels are particularly susceptible to the effects of hypertension. With modest levels of hypertension, they develop medial hypertrophy and intimal thickening containing much new elastic tissue. In the more severe forms of hypertension there is a profound intimal thickening of these branches; this is frequently very cellular and presents a rather mucinous appearance.

The afferent arterioles are made up almost entirely of muscle and have only a very delicate internal elastic lamella which disappears in the terminal part of the arteriole. The terminal part contains granules and constitutes part of the juxtaglomerular apparatus. These arterioles show a diffuse bland eosinophilic thickening with increasing age, hypertension and diabetes. Actual necroses are found in the more severe forms of hypertension.

The intrarenal veins are extremely thin walled and contain very little muscle (Fig. 8). They are situated in close proximity to the corresponding arteries and often appear as endothelial lined spaces. The extrarenal veins have the picture of veins in other parts of the body with plentiful muscularis.

THE ANATOMY OF THE RENAL CIRCULATION   641

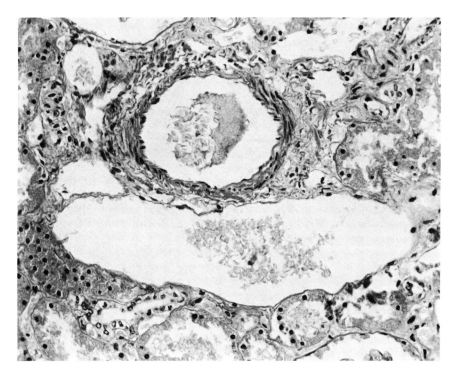

Fig. 8. Thin walled vein in close proximity to artery. From a young adult. H & E. X250.

In the outer medulla the vasa recta have muscular walls, but in the inner medulla the muscle is absent and the vessels consist of endothelium with perivascular cells (pericytes) on the outside. The control of the medullary circulation is poorly understood and is discussed in a later chapter.

F. JUXTAGLOMERULAR APPARATUS

The juxtaglomerular apparatus (Goormaghtigh, 1932; Goormaghtigh, 1940) is situated at the hilus of the glomerulus. It is made up of 3 components (Fig. 9), first, the granular cells found in the terminal part of the afferent arteriole; second, the macula densa, a specialized area of the distal convoluted tubule; and third, the lacis cells, a group of nongranulated cells lying between the macula densa and the glomerular tuft. The granules, which are regarded as containing renin (Edelman and Hartroft, 1961; Cook, 1964), can be

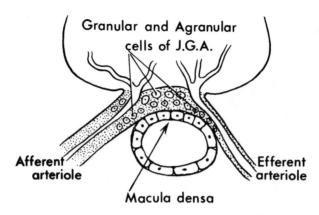

Fig. 9. Location and constituents of juxtaglomerular apparatus (JGA). Most of the granular cells are situated in the terminal part of the afferent arteriole and to a lesser degree in the early part of the efferent. The cells at the root of the glomerulus are mainly lacis cells.

demonstrated by such stains as Bowie's, using light microscopy, although in man the results are not so satisfactory as in the rat. There have been relatively few detailed anatomic studies of the juxtaglomerular apparatus in man (Rouiller and Orci, 1971; Christensen et al., 1975). The functional aspects of the juxtaglomerular apparatus are described in another chapter.

G. GLOMERULUS

The glomerulus is a highly complex system of capillary loops (Fig. 10) to which blood is delivered by the afferent arteriole and from which blood is drained by the efferent arteriole. This is described in detail elsewhere (Heptinstall, 1974). At the hilus is the juxtaglomerular apparatus. Filtration occurs across these capillaries into Bowman's capsule from which the fluid flows into the proximal convoluted tubule. The collection of capillaries is known as the glomerular or capillary tuft and it is made up of some 8 lobules, each of which has a central connective tissue core. Anastomoses exist between the capillaries in a lobule but there is a difference of opinion on the matter of anastomoses between lobules (Boyer, 1956; Hall, 1954).

The central supportive part of a lobule is referred to as the mesangium (Fig. 11) and consists of mesangial cells and material with the same staining characteristics as basement membrane; this is referred to as mesangial matrix or basement membrane-like material. Little is known about the mesangium apart from its supportive and phagocytic roles. Around the mesangium are situated capillary loops whose lumens are separated from the mesangial cells by endothelial cells. The capillary walls consist of a basement membrane--resolvable into 3 layers by electron microscopy--which is lined on the inside by endothelium and covered on the outside by foot processes which join up to form the epithelial cells of the tuft (Fig. 12). The foot processes are separated by epithelial slits, which at their basement membrane connection are bridged by a thin line referred to as the slit membrane or diaphragm. It is considered that the slit membrane plays a role in regulating the retention of macromolecules.

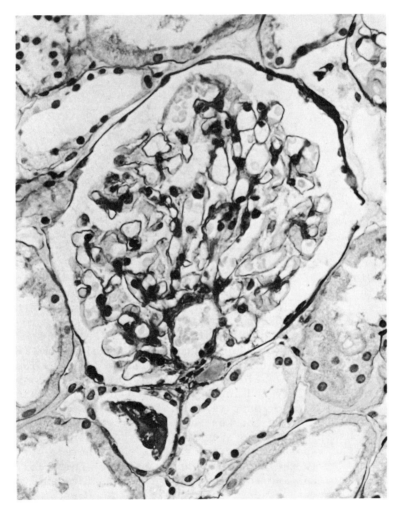

Fig. 10. Normal glomerulus from renal biopsy on young man. Stained with silver to show delicate basement membrane. X400.

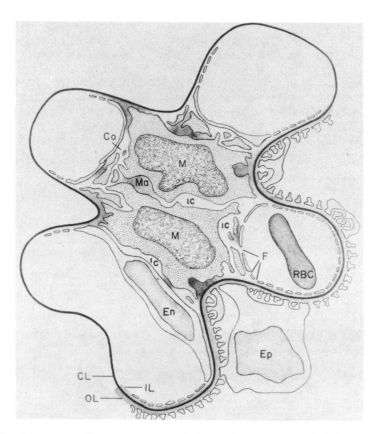

Fig. 11. Diagram of a rat mesangial region surrounded by capillaries. The central layer (CL) of the basement membrane invests the capillaries like a sheet and also passes over the mesangial region. Another boundary of the mesangium is formed by the central portions of endothelial cells (En). Mesangial cells (M) are partially surrounded by mesangial matrix (MA), in which bundles of collagen (Co) can sometimes be found. Fenestrations (F) allow passage of particles and plasma into intercapillary and intercellular channels (IC). The outer layer (OL) and inner layer (IL) of the basement membrane have been drawn thicker than they are to show their relationships to the adjacent cells. The central layer is thin in the rabbit, nouse, and rat (as shown here). It is much thicker in the monkey and human beings. Epithelial cell bodies (Ep) and their primary processes frequently cover the foot processes. The capillary diameter is usually somewhat greater than the diameter of a red blood cell (RBC).

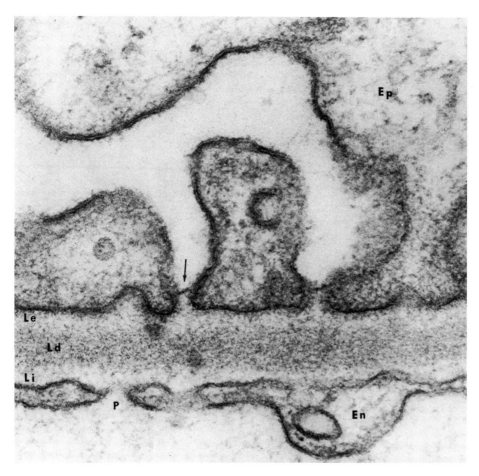

Fig. 12. High magnification electron micrograph of capillary basement membrane from rat glomerulus. The basement membrane has three zones: the lamina rara externa (Le), the lamina densa (Ld), and the lamina rara interna (Li). Also shown are epithelium (Ep) and endothelium (En). There is a prominent plasma membrane showing the "unit-membrane" structure over the epithelial cell with a covering of "fuzz." The filtration slit membrane (arrow) is clearly seen and appears to be continuous with the outer portion of the "unit membrane." The basement membrane is composed of small fibrillar structures that are tightly packed in the lamina densa and further apart in the laminae rarae. The endothelial cells exhibit pores (P) or fenestrations. Specimen stained in 1% OsO_4 in Palade's buffer, section double stained with uranyl acetate and lead hydroxide. X120,000.

Although the glomeruli, which Malpighi compared to apples on an arterial tree, make up only 3% of the kidney by volume, the total length of all the glomerular capillaries in one adult human kidney is more than 10 kilometers (Elias and Henning, 1965; Jorgensen, 1966). The total area available for filtration in the same kidney is about $0.78m^2$ (Vimtrup, 1928, Elias and Henning, 1965), slightly greater than the area of the bottom of an ordinary bathtub. While considerable, this surface area is much smaller than that of the alveolar capillaries in one human lung (35 m^2, Weibel, 1965). In man, unlike many other animals, few new glomeruli are formed after birth and the number of glomeruli declines slightly from infancy to adulthood (Vimtrup, 1928; Smith, 1951; Elias and Henning, 1965). However, glomerular size doubles in this time so that there is a net increase in the total volume of glomeruli (Vimtrup, 1928; Fahr, 1925; Elias and Henning, 1965; Jørgensen, 1966).

II. THE RENAL VASCULATURE IN OTHER MAMMALIAN SPECIES

The comparative anatomy of the renal circulation in mammals has been well studied by others (Bremer, 1915; Kazzar and Shanklin, 1951; Nissen, 1969; Fourman and Moffat, 1971; Fuller and Huelke, 1973; Sokabe and Ogawa, 1974). The present review will restrict itself to significant anatomic differences between the renal vasculature of various experimental animals which may influence either the functional capacities of the kidney or its susceptibility to disease, or the appropriateness of a given animal as a model for the study of human renal disease.

A. SEGMENTAL DISTRIBUTION OF ARTERIAL BLOOD

With the exception of the pig and the squirrel monkey, all mammals commonly used in the laboratory are unlike man in that they have an unipapillary rather than a multipapillary kidney (Strauss, 1934; Christensen, 1964; Walker, 1967; Hodsen et al., 1969; Moffat, 1975). It might be expected that the blood supply to the various cortical segments might be less independent in the more simplified unipapillary kidney as compared with the human kidney, but there is no evidence that this is the case. In the rat there are anastomoses between the efferent arterioles of the different cortical segments but, as in man, these anastomoses are insufficient to prevent segmental infarction following obstruction of one of the renal artery branches (Loomis and Jett-Jackson, 1942).

B. MAIN RENAL ARTERY AND EXTRARENAL BRANCHES

The caliber of the renal artery in mammalian species is roughly proportional to body size, while the caliber of the afferent arteriole and glomerular capillaries (8-15μ) is nearly the same in animals of widely differing sizes (data derived from Shonyo and Mann, 1944;

THE ANATOMY OF THE RENAL CIRCULATION

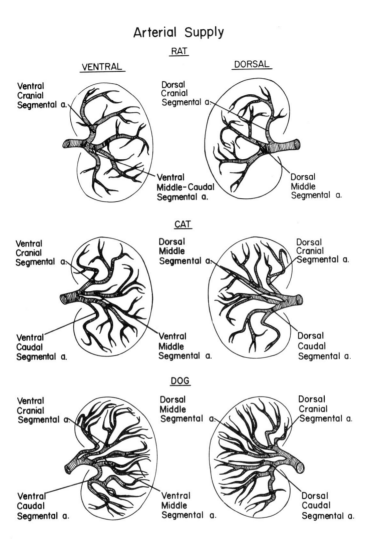

Fig. 13. The arterial supply of the rat, cat, and dog kidney. In the rat, which has the most simplified arrangement of segmental arteries, the caudal pole is supplied only by a branch from the ventral-caudal segmental artery.

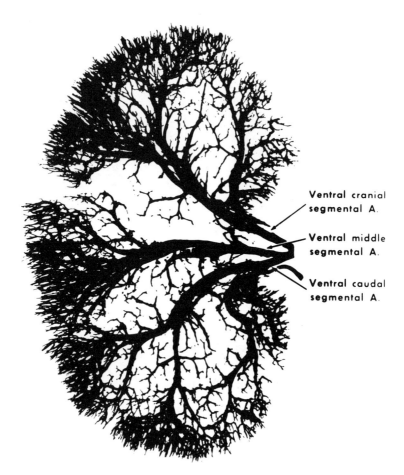

Fig. 14. Vinyl acetate injection specimen of the ventral half of the renal vascular tree of a dog kidney. There is little overlap in the distribution of the three segmental arteries.

Jørgensen, 1966; Rouiller, 1969; Rhodin, 1971). In man the diameter of the renal artery is more than 500 times the diameter of the glomerular capillaries, whereas in the rat and mouse this ratio is less than 50. Changes in caliber of the renal artery in the mouse or rat may considerably alter the resistance to blood flow through the kidney. It is unlikely that changes in caliber of the much larger renal artery in man and other larger animals commonly exert a significant effect on vascular resistance. In man and the dog, the larger arteries and smaller arterioles of the kidney respond quite differently to vasoactive drugs such as angiotensin. If this differential responsiveness is primarily a function of vessel size, then it would be expected to be less prominent in smaller animals.

The arrangement of the extrarenal vessels in most mammalian species is quite similar to that in man. Near the hilus of the kidney the renal artery divides into dorsal and ventral branches. Each of these branches then divides into two or three segmental arteries (Fig. 13). In the rat the caudal pole of the kidney is supplied solely by vessels which have their origin in the ventral branch of the main renal artery (Fuller and Huelke, 1973). In the dog and cat the caudal pole is supplied by both dorsal and ventral branches. There is little overlapping of the regions supplied by the individual segmental arteries (Fig. 14).

C. INTRARENAL BRANCHES

The segmental arteries may undergo further branching as they course along the renal pelvis before entering the renal parenchyma at the junction between cortex and medulla. At the base of the medullary pyramids they change their direction as they subdivide into arcuate arteries which run parallel to the kidney surface at the corticomedullary junction. In the dog, cat, and hamster secondary branches arise from the arcuate arteries and then subdivide in various ways to form the interlobular arteries (Fig. 15) (Fourman and Moffat, 1971). In the rat and the rabbit the interlobular arteries arise directly from the arcuate arteries. Afferent arterioles supplying individual glomeruli are given off at right angles as the interlobular arteries course toward the surface of the kidney. In the dog and cat (and in man) some of the interlobular arteries continue to the surface of the kidney where they give off branches to the capsular plexus. In the cat kidney, which has a very thick capsule, there is a prominent subcapsular network of arteries and veins surrounding the kidney (Fig. 16). In this species, the subcapsular veins, which can be easily cannulated for blood flow studies, drain the outer cortex, while the arcuate veins, which parallel the arcuate arteries, drain the inner cortex (Nissen, 1969).

In experimental animals, as in man, the capsular circulation plays an important role in providing collateral circulation when the main renal artery is obstructed. When the occlusion of the main renal artery occurs slowly, surprisingly good renal function may be retained owing to the hypertrophy of collateral vessels (Johansson, 1926; Cook and Pearson, 1946). Experimentally, the collateral circulation of the kidney has been most extensively studied in the rabbit (Moses and Schlegel, 1952; O'Morchoe, 1961) and dog (Belt and Joelsen, 1925; Abrams and Cornell, 1965; and Eliska, 1968). In the rabbit (Fig. 17) the ureteric arteries are the most important collateral vessels. Perforating capsular vessels anastomosing with

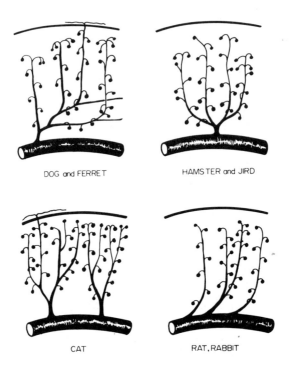

Fig. 15. The arrangement of the interlobular arteries in various species.

arteries of the perirenal region are an inconstant finding (O'Morchoe, 1961). In the dog, in contrast, there are six relatively large perforating arteries per kidney which pass directly from the arcuate arteries to the capsular plexus (Eliska, 1968). These aglomerular vessels may be capable in some cases of maintaining adequate renal blood flow in the face of renal artery occlusion (Fourman and Moffat, 1971).

D. BLOOD SUPPLY TO THE MEDULLA

In most mammals, the blood supply to the medulla is essentially similar to that described in man. In the mouse, the vascular bundles which contain the ascending and descending vasa recta come together to form large secondary bundles as they pass from the outer stripe into the inner stripe of the medulla. These giant vascular bundles then dissolve into the original number of individual bundles in what Kriz and Koepsell (1974) have dubbed the "innermost stripe" of the outer medulla before entering the inner medulla. These authors have suggested that the giant bundles and the "innermost stripe" may represent an adaptation which improves the kidney's concentrating function.

The blood supply of the renal pelvis comes from vessels which branch directly off the segmental or interlobar arteries. The pelvic vascular plexus thus formed (Fig. 18) contains ectopic glomeruli. The efferent arterioles of these glomeruli give off vasa recta which

Fig. 16. Injection preparation showing the superficial veins of the cat kidney.

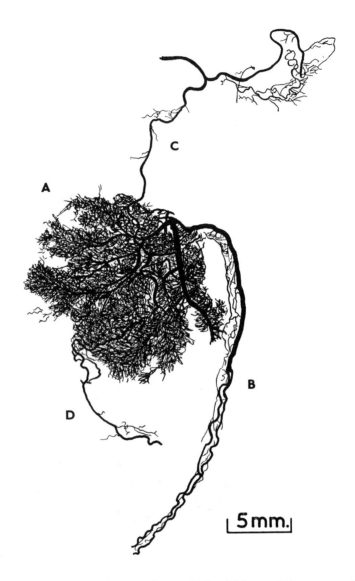

Fig. 17. Latex injection specimen of a rabbit kidney (A) one week after ligation of the renal artery. Collateral circulation is provided by the ureteric vessels (B), the perihilar vessels (C), and the perforating capsular artery (D).

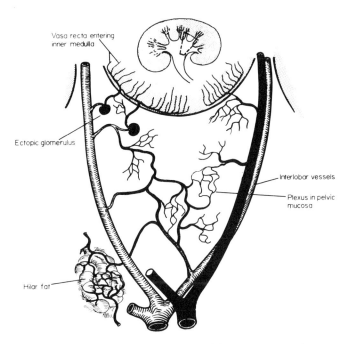

Fig. 18. Diagram showing the relationship between the vessels of the renal pelvis and the inner medulla of the rat. Ectopic glomeruli lie deep to the pelvic mucosa. Both the afferent and efferent arterioles of these glomeruli give contributions to the pelvic capillary plexus. The efferent arterioles anastomose to form a marginal artery, from which vasa recta supplying the inner medulla are given off. A glomerular nutrient branches from the interlobular arteries also supplies the pelvic plexus as well as the hilar fat.

contribute to the blood supply of the inner medulla in most species excepting the rabbit (Fig. 18 and 19) (Fouman and Moffat, 1971). Hill (1966) observed that rats with experimental pyelonephritis seldom develop papillary necrosis, whereas pyelonephritic rabbits do. He related this finding to the absence of a pelvic collateral circulation in the rabbit papilla. In man and the dog, the pelvic ectopic glomeruli usually degenerate early in life leaving a direct aglomerular communication between the interlobar arteries and the superficial vasa recta of the inner medulla (Fig. 19) (Fourman and Moffat, 1971).

E. HISTOLOGY

Except for the size differences commented upon earlier, the histology of the renal vessels in other mammalian species differs

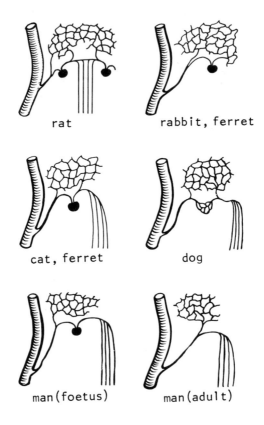

Fig. 19. Relationship between the pelvic vessels and the vasa recta in different species. In the rabbit, no vasa recta are formed from the efferent arterioles of the ectopic pelvic glomeruli. In adult man and the dog, the ectopic glomeruli degenerate, leaving only aglomerular pathways.

little from that described in man. Most of the interesting renal structural differences between species seem related to tubular rather than vascular elements (Spargo, 1968; Kriz et al., 1972; Schwartz and Venkatachalam, 1974).

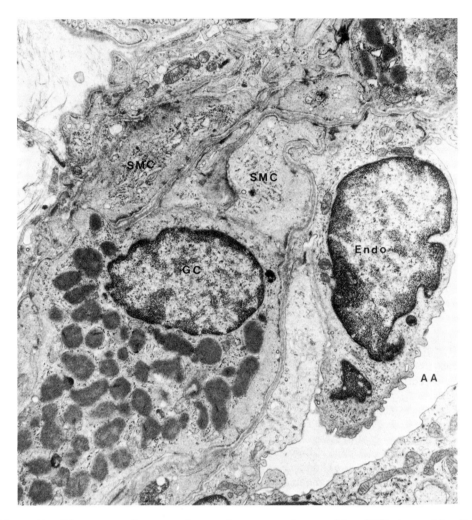

Fig. 20. Electron micrograph to show granular cells (GC) in wall of afferent arteriole (AA) of a rat. Granule shapes in other species differ slightly. SMC, smooth muscle cell; Endo, endothelial cell. X9,750.

F. JUXTAGLOMERULAR APPARATUS

The juxtaglomerular apparatus is more prominent in the rat and rabbit kidney than it is in man, but its structure is basically similar in all mammalian species. Minor species differences have been noted, such as the shape of the granules in the granular cells of the afferent arteriole (Fig. 20), and the relationship of the macula densa to the efferent arteriole (Barajas and Latta, 1963; Barajas, 1966; Faarup, 1965; Biava and West, 1965, 1966). Sokabe (1974) and Sokabe and Ogawa (1974) have done extensive comparative studies of the juxtaglomerular apparatus in the different vertebrate classes; a well-defined macula densa and lacis area apparently only found in mammals.

G. GLOMERULUS

The human kidney has its full complement of glomeruli at birth, while in the kidney of the rat and several other species the number of glomeruli approximately doubles between birth and maturity (Smith, 1951). In most animals glomerular filtration rate increases with age. Within a given species total glomerular volume tends to correlate with body weight (Moffat, 1975). Desert animals tend to have fewer glomeruli than related animals living in damper environments (Moffat, 1975).

REFERENCES

Abrams, H.L., and Cornell, S.H., 1965, Patterns of collateral flow in renal ischemia, Radiology 84:1001.
Barajas, L., 1966, The development and ultrastructure of the juxtaglomerular cell granule, J. Ultrastruct. Res. 15:400.
Barajas, L., and Latta, H., 1963, A three dimensional study of the juxtaglomerular cell granule, J. Ultrastruct. Res. 15:400.
Barrie, H.J., Klebanoff, S.J., and Cates, G.W., 1950, Direct medullary arterioles and arteriovenous anastomoses in the arcuate sponges of the kidney; preliminary communication, Lancet 1:23.
Belt, A.E., and Joelson, J.J., 1925, The effect of ligation of branches of the renal artery, Arch. Surg. 10:117.
Biava, C.G., and West, M., 1965, Lipofuscin-like granules in vascular smooth muscle and juxtaglomerular cells of human kidneys, Am. J. Pathol. 47:287.
Biava, C.G., and West., M., 1966, Fine structure of normal human juxtaglomerular cells. I. General structure and intercellular relationship. II. Specific and non-specific cytoplasmic granules, Am. J. Pathol. 49:679, 955.
Boijsen, E., 1959, Angiographic studies of the anatomy of single and multiple renal arteries, Acta. Radiol. Suppl. 183.

Boyer, C.C., 1956, The vascular pattern of the renal glomerulus as revealed by plastic reconstruction from serial sections, Anat. Rec. 125:433.
Bremen, J.L., 1915, The origin of the renal arteries in mammals and the anomalies, Am. J. Anat. 18:119.
Brödel, M., 1901, The intrinsic blood-vessels of the kidney and their significance in nephrotomy. Bull. Johns Hopkins Hosp. 12:10.
Christensen, G.C., 1964, The urogenital system and mammary glands, in "Anatomy of the Dog" (M.E. Miller, author), pp. 741-747, W.B. Saunders, Philadelphia.
Christensen, J.A., Meyer, D.S., and Bohle, A., 1975, The structure of the human juxtaglomerular apparatus, Virchows Arch. A 367:83.
Cook. W.F., 1964, The detection of renin in juxtaglomerular cells, J. Physiol. 194:73P.
Cook, G.T., and Pearson, R.S.B., 1946, Hyperpiesis with atheromatous obstruction of the renal arteries, J. Pathol. Bacteriol. 58:564.
Edelman, R., and Hartroft, P.M., 1961, Localization of renin in juxtaglomerular cells of the rabbit and dog through the use of the fluorescent-antibody technique, Circ. Res. 9:1069.
Elias, H., and Hennig, A., 1965, Stereology of the human renal glomerulus, in "Quantitative Methods in Morphology" (E.R. Weibel and H. Elias, eds.) pp. 130-166, Springer-Verlag, New York.
Eliska, O., 1968, The perforating arteries and their role in the collateral circulation of the kidneys, Acta Anat. 70:184.
Faarup, P., 1965, On the morphology of the JG apparatus, Acta Anat. 60:20.
Fahr, T., 1925, Pathologische Anatomie des Morbus Brightii, in "Handbuck der speziellen Pathologischen Anatomie und Histologie" (F. Henke and O. Lubarsch, eds.) Vol. 6, p. 158, Springer-Verlag, Berlin.
Faller, J., and Ungvary, G., 1962, Die arterielle Segmentation der Niere, Zbl. Chir. 87:972.
Fourman, J., and Moffat, D.B., 1971, "The Blood Vessels of the Kidney," Blackwell Scientific Publications, Oxford.
Fuller, P.M., and Huelke, D.F., 1973, Kidney vascular supply in the rat, cat and dog, Acta Anat. 84:516.
Goormaghtigh, N., 1932, Les segments neuro-myo-artériels juxtaglomérulaires du rein, Arch. Biol. (Liege) 43:575.
Goormaghtigh, N., 1940, Histologic changes in the ischemic kidney: With special reference to the juxtaglomerular apparatus, Am. J. Pathol. 16:409.
Graves, F.T., 1954, The anatomy of the intrarenal arteries and its application to segmental resection of the kidney, Br. J. Surg. 42:132.
Graves, F.T., 1956a, The aberrant renal artery, J. Anat. 90:553.
Graves, F.T., 1956b, The anatomy of the intrarenal arteries in health and disease, Br. J. Surg. 43:605.

Greenblatt, M., 1963, Primary renal arteriosclerosis: A comparative angiographic study of hypertensive and normotensive individuals, Lab. Invest. 12:1270.

Hall, B.V., 1954, Further studies of the normal structure of the renal glomerulus. Proceedings of the Sixth Annual Conference on the Nephrotic Syndrome, National Nephrosis Foundation, New York, pg. 1.

Hegedüs, V., 1972, Arterial anatomy of the kidney: A three-dimensional angiographic investigation, Acta. Radiol. Diag. 12:604.

Heptinstall, R.H., 1974, "Pathology of the Kidney," 2nd edition, Little, Brown and Co., Boston, p. 12.

Hill, G.S., 1966, Experimental pyelonephritis: Compensatory vascular alterations and their relation to the development of papillary necrosis, Bull. Johns Hopkins Hosp. 119:100.

Hodson, C.J., Craven, J.D., Lewis, D.G., Matz, L.R., Clarke, R.J., and Ross, E.J., 1969, Experimental Obstructive Nephropathy in the Pig (Suppl to Br. J. Urol. vol. 41, no. 6) Livingstone, Edinburgh.

Johanssen, S., 1926, Question of collateral circulation in the renal capsule, Acta Chir. Scand. 61:181.

Jørgensen, F., 1966, "The Ultrastructure of the Normal Human Glomerulus," Munksgaard, Copenhagen.

Kazzar, D., and Shanklin, W.M., 1951, Comparative anatomy of the superficial vessels of the mammalian kidney demonstrated by plastic (vinyl acetate) injections and corrosion, J. Anat. 85:163.

Kimmelstiel, P., Kim, O.J., and Beres, J., 1962, Studies on renal biopsy specimens, with the aid of the electron microscope: I. Glomeruli in diabetes, Am. J. Clin. Pathol. 38:270.

Kriz, W., and Koepsell, H., 1974, The structural organization of the mouse kidney, Z. Anat. Entwickl.-Gesch. 144:137.

Kriz, W., and Lever, A.F., 1969, Renal countercurrent mechanisms: Structure and function, Am. Heart J. 78:101.

Kriz, W., Schnermann, J., and Dieterich, H.J., 1972, Differences in the morphology of descending limbs of short and long loops of Henle in the rat kidney, in "International Symposium of Renal Handling of Sodium," Brestenberg, 1971, pp. 140-144, S. Karger, Basel.

Kuprijanoff, P.A., 1924, Das intrarenale arterielle System gesunder und pathologischer Nieren, Dtsch. Z. Chir. 188:206.

Lewis, O.J., 1959, The vascular architecture of the developing human renal glomerulus, Anat. Rec. 135:93.

Ljungvist, A., and Lagergren, C., 1962, Normal intrarenal arterial pattern in adult and aging human kidney, J. Anat. Lond. 96:285.

Lofgren, F., 1949, Das Topographische System der Malpighischen Pyramiden der Menschenniere, A.B. Gleerupska Univ.-Bokhandeln, Lund.

Loomis, D., and Jett-Jackson, C.E., 1942, Plastic studies in abnormal renal architecture, VI. An investigation of the circulation in infarcts of the kidney, Arch. Pathol. 33:735.

Maxwell, M.H., Breed, E.S., and Smith, H.W., 1950, Significance of the renal juxtamedullary circulation in man, Am. J. Med. 9:216.

Merklin, R.J., and Michels, N.A., 1958, The variant renal and suprarenal blood supply with data on the inferior phrenic, ureteral and gonadal arteries. A statistical analysis based on 185 dissections and review of the literature, J. Int. Coll. Surg. 29:41.

Moffat, D.B., 1975, "The Mammalian Kidney," Cambridge University Press, New York.

More, R.H., and Duff, G.L., 1951, The renal arterial vasculature in man, Am. J. Pathol. 27:95.

Moses, J.B., and Schlegel, J.U., 1952, Preservation of the juxtamedullary circulation following ligation of the renal artery in the rabbit, Anat. Rec. 114:149.

Nissen, O.I., 1969, "The Function of Superficial and Deep Areas of the Cat Kidney," Virum Costers Bogtrykkeri, Copenhagen.

O'Morchoe, C.C.C., 1961, Collateral blood supply to the rabbit kidney after ligation of the renal artery, Br. J. Urol. 33:278.

Poisel, S., and Spängler, H.P., 1969, Über aberrante und akzessorische Nierenarterien bei Nieren in typischer Lage, Anat. Anz. 124:244.

Rhodin, J.A.G., 1971, Structure of the kidney in "Diseases of the Kidney" (M.B. Strauss and L.G. Welt, eds.). pp. 1-30, Little, Brown and Co., Boston.

Rollhäuser, H., Kriz, W., and Heinke, W., 1964, Das Gefäss-System der Rattenniere, Z. Zellforsch. Mikrosk. Anat. 64:381.

Rotter, W., 1952, Die Sperr-(Polster-BZW Drossel) Arterien der Nieren des Menschen, Z. Zellforsch. 37:101.

Rouiller, C., 1969, General anatomy and histology of the kidney, in "The Kidney," (C. Rouiller and A.F. Muller, eds.), Vol. I, p. 61-156, Academic Press, New York.

Rouiller, C., and Orci, L., 1971, The Structure of the Juxtaglomerular Complex, in "The Kidney" (C. Rouiller, and A.F. Muller, eds.), Vol. 4, p. 1, Academic Press, New York.

Schwartz, M.M., and Venkatachalam, M.A., 1974, Structural differences in thin limbs of Henle: Physiological implications, Kidney Int. 6:193.

Shonyo, E.S., and Mann, F.C., 1944, An experimental investigation of renal circulation, Arch. Pathol. 38:287.

Smith, H.W., 1951, "The Kidney" Structure and Function in Health and Disease," Oxford University Press, New York.

Sokabe, H., 1974, Phylogeny of the renal effects of angiotensin, Kidney Int. 6:263.

Sokabe, H., and Ogawa, M., 1974, Comparative studies of the juxtaglomerular apparatus, Int. Rev. Cytol. 37:271.

Spargo, B.H., 1968, Renal changes with potassium depletion, in "Structural Basis of Renal Disease," (E.L. Baker, ed.), pp. 565, Harper and Row, New York.
Strauss, W.L., 1934, The structure of the primate kidney, J. Anat. 69:93.
Sykes, D., 1963, The arterial supply of the human kidney with special reference to accessory renal arteries, Br. J. Surg. 50:368.
Sykes, D., 1964, The correlation between renal vascularisation and lobulation of the kidney, Br. J. Urol. 36:549.
Trueta, J., Barclay, A.E., Daniel, P.M., Franklin, K.J., and Richard, M.M.L., 1947, "Studies of the Renal Circulation," Charles C. Thomas, Springfield, Ill.
Vimtrup, B.J., 1928, On number, shape, structure, and surface area of the glomeruli in kidneys of man and animals, Am. J. Anat. 41:123.
Walker, W.F., 1967, "A Study of the Cat," W.B. Saunders, Philadelphia.
Weibel, E.R., 1965, Morphometry and lung models, in "Quantitative Methods in Morphology," (E.R. Weibel and H. Elias, eds.), pp. 253-267, Springer-Verlag, New York.

ACKNOWLEDGMENTS

This work was supported in part by U.S. Public Health Service Grants HL-07835 and GM-00415. The authors wish to express their appreciation to Nancy H. Lambert for her excellent secretarial assistance.

The figures were borrowed from other publications with the permission of the authors and publishers. Figures 1, 2, 3, 4, 5, 8, 9, 10, 12, and 20 are from R. H. Heptinstall, Pathology of the Kidney, Little, Brown and Co., Boston 1974. Figures 6 and 7 are from Kriz, W. and Lever, A.F., Am. Heart J. 78:101, 1969. Figure 11 is from Latta, H. in S. R. Geiger, Handbook of Physiology, Williams & Wilkins, Baltimore, 1973. Figurs 13 and 14 are from Fuller, P.M., and Huelke, D.F., Acta Anat. 84:516, 1973. Figures 15, 16, 18 and 19 are from Fourman J. and Moffat, D.B., The Blood Vessels of the Kidney, Blackwell Scientific Publications, Oxford, 1971. Figure 17 is from O'Morchoe, C.C.C., Br. J. Urol. 33:278, 1961.

THE RENAL CIRCULATION: PHYSIOLOGY AND HORMONAL CONTROL

Kim Solez,[1] M.D., Robert H. Heptinstall,[2] M.D.

[1]Assistant Professor of Pathology and Medicine, Johns
 Hopkins University and Hospital, Baltimore MD 21205
[2]Baxley Professor and Director, Department of Pathology,
 Johns Hopkins University and Hospital, Baltimore, MD
 21205

Although much attention has been paid to the immunologic aspects of kidney disease (Germuth and Rodriguez, 1973; McCluskey, 1974; Wilson and Dixon, 1974), it is clear that many of the more common afflictions have a predominantly circulatory rather than an immunologic basis. Vascular abnormalities and altered renal perfusion play a key role in the renal disorders associated with arteriosclerosis, hypertension, shock, liver disease, sepsis, trauma, diabetes, and the toxic effects of certain drugs (Barger, 1966; Siperstein et al., 1968; Hollenberg, 1973; Hollenberg and Adams, 1974). Disturbed renal hemodynamics are pathogenetically important even in disorders of undisputed immunologic origin, such as renal transplant rejection (Hollenberg et al., 1972).

In order to understand circulatory disorders of the kidney it is necessary first to understand the physiology of the renal circulation. Two excellent reviews of this subject have been published covering the period up to 1971 (Thurau and Levine, 1971; Barger and Herd, 1973). The present chapter does not attempt to duplicate the coverage of these comprehensive reviews but rather endeavors to provide complementary information.

Part I of this chapter briefly discusses some of the more important aspects of the physiology of the renal circulation, emphasizing work published between 1971 and June of 1975. Part II deals with the influence of hormones on the circulation of the kidney, a subject not extensively discussed in the review articles mentioned above. (The anatomy of the renal circulation has been described in an earlier chapter.)

I. PHYSIOLOGY OF THE RENAL CIRCULATION

A. GENERAL CONSIDERATIONS

Renal blood flow in man and other animals accounts for approximately 25% of the cardiac output at rest despite the fact that the kidneys constitute less than 0.5% of body mass (Kiil, 1971; Valtin, 1973). Average renal blood flow per gram is approximately 4 ml/min, about four times the level of perfusion in such metabolically active organs as the liver, brain, gut, heart, and exercising skeletal muscle (Fig. 1). Renal blood flow clearly exceeds that which is necessary to provide the kidney with oxygen and metabolic substrates. Renal oxygen consumption does not exceed 10% of the oxygen consumption of the whole body and the a.v. oxygen difference (1.7 vol %) is less than half the difference in oxygen content between arterial and mixed venous blood (Kiil, 1971). Renal oxygen consumption is proportional to renal blood flow at levels above 3 ml/min/100g and thus the renal arteriovenous difference remains relatively constant.
There is a linear relationship between sodium reabsorption and oxygen consumption. This relationship appears to hold even when sodium reabsorption is changed without a similar change in glomerular filtration rate (Valtin, 1973). Approximately 80% of renal oxygen consumption is thought to support tubular transport functions. The remaining 20% (100 μmoles/100g/min) represents basal oxygen consumption of the kidney tissue itself and is comparable to that of other epithelial tissues (Valtin, 1973).

Alterations in total renal blood flow in man brought about by disease are often masked by the extremely wide normal range (Hollenberg et al., 1975). Studies of total renal blood flow in dogs using the chronically implanted electromagnetic flow probe have demonstrated considerable day-to-day variability in resting blood flow in individual animals (Reinhardt et al., 1975).

Large changes in medullary blood flow are not detectable as changes in total blood flow, since 93% of total renal blood flow perfuses the cortex alone (Thurau and Levine, 1971). In the rat, the inner medulla makes up only about 3% of kidney mass and has a plasma flow of 32 ml/min/100g as opposed to 300 ml/min/100g for the kidney as a whole (Solez et al., 1974b; Arendshorst et al., 1975). Nevertheless, alterations in inner medullary perfusion which may have no significant effect on total renal blood flow can have a profound influence on renal concentrating ability (see Section D) (Thurau, 1964).

B. INTRARENAL DISTRIBUTION OF BLOOD FLOW

Trueta et al., (1947) generated considerable interest in the distribution of blood flow within the kidney by suggesting that there was a "diversion" of blood flow from the cortex to the medulla in

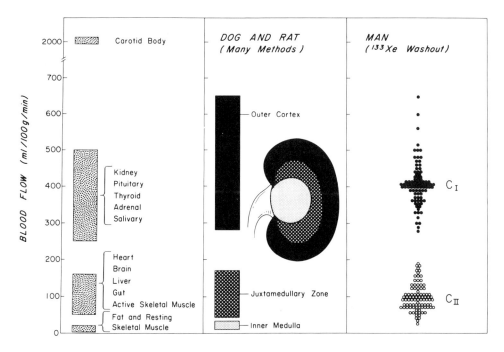

Fig. 1. A comparison of renal blood flow to other parenchymal organs. Outer cortical blood flow per 100 gm of tissue is four to five times that of such active organs as the heart, brain, liver, and the gastrointestinal tract. Inner medullary blood flow (approximately 40 ml/100g/min in the rat and therefore slightly misrepresented in the figure) is slightly lower than blood flow to these active organs. Components one and two of the xenon washout curve correspond to blood flow to the outer and juxtamedullary cortex. Difficulties encountered using the xenon washout method to measure intrarenal distribution of blood flow are discussed in the text. (From Hollenberg, 1973, used by permission of the author and publisher.)

certain pathologic states. The presence or absence of vascular pathways that would permit an aglomerular shunting of blood had been extensively debated (Moffat and Fourman, 1963; Plakke and Pfeiffer, 1964; Kriz, 1967). Vascular channels between the afferent and efferent arterioles bypassing the glomeruli have been described by Ljungqvist (1964) and Ljungqvist and Wagermark (1970) but their presence is denied by Spinelli et al., (1972). Lameire et al., (1972) concluded on the basis of studies using radioactive microspheres (see below) that there was a 5% aglomerular shunting of blood through the medulla in the normal rat. Using similar techniques, O'Dorisio et al., (1973) and Stein et al., (1974) were unable to demonstrate such a shunt pathway in the dog. It seems likely that aglomerular vascular pathways leading to the medulla do develop with age as a result of glomerular obsolescence but are of little functional importance.

More recently interest in the intrarenal distribution of blood flow has centered around the proposal by Barger (1966) that important functional differences existed between the superficial and deep cortical nephrons. He postulated that the superficial nephrons tended to lose relatively more sodium than the deep glomeruli. Thus a redistribution of blood flow toward the deep cortex would cause sodium retention, while a redistribution toward the superficial cortex would lead to sodium loss. As will be discussed more fully below, recent evidence is in conflict with this proposal (Stein et al., 1973; Boonjarern et al., 1974; Kleinman and Reuter, 1974).

1. Measurement of the Intrarenal Distribution of Blood Flow

Many methods for determining renal blood flow distribution have been described. Each method has inherent advantages and limitations. For a detailed account of methods published before 1971, the reader is referred to the chapter by Ladefoged and Munck (1971). The more recent review by Stein et al. (1973) is recommended for its assessment of the methods widely used at present. The present section concerns itself primarily with two methods of determining the distribution of cortical blood flow: the inert gas washout method and the radioactive microsphere method. Methods for determining medullary blood flow are dealt with in a later section.

a. Inert gas washout methods

Curves which describe the disappearance of radioactivity from the kidney after an intra-arterial injection of a radioactive gas such as ^{85}Kr or ^{133}Xe have been widely used to study renal blood flow distribution. Curves relating radioactivity and time are plotted on semilogarithmic paper and then resolved into several straight lines, each representing one "compartment" of renal blood flow. Each of these lines is described by the equation: $F/V = \dfrac{K \times \lambda}{\rho}$ where

F = blood flow in ml/min, V = volume of distribution of the radioactive gas in grams, K = the rate constant for disappearance of the gas (or $0.693/T_{1/2}$ where $T_{1/2}$ is the half time of each line), λ = the partition coefficient of the radioactive gas between tissue and blood, and ρ = specific gravity of the tissue (Thorburn et al., 1963; Ladefoged and Munck, 1971; Stein et al., 1973).

This method allows repeated determinations in the same animal and does not require anesthesia. It is the only method suitable for determining renal blood flow distribution in man. The use of this method in man has been limited because of the necessity of selective renal artery catheterization, a relatively major procedure. However, data on intrarenal hemodynamics in normal man have been obtained from potential kidney donors undergoing selective renal arteriography as part of their assessment (Hollenberg, 1973).

Using autoradiography, the components of the washout curve have been shown to correspond to different anatomic areas of the kidney in the normal dog (Thorburn et al., 1963). However, after experimental manipulation, a given component of the washout curve may not correspond to the same area of the kidney as it does in the control situation, since all circulations having the same volume-flow relationship merge into the same component of the washout curve (Slotkoff et al., 1971; Hollenberg et al., 1974). An attempt has been made to circumvent this problem by implanting a detector in the kidney itself and measuring the passage of the tracer substance through a well-defined area of the kidney (Aukland et al., 1967; Wolgast, 1968). This procedure would seem to have little to recommend it since the damage inflicted on the kidney parenchyma by implantation of the detector almost certainly alters intrarenal hemodynamics.

There are theoretical and practical objections to the inert gas washout methods. Mowat et al., (1972) have reported that the injection of a bolus of ^{133}Xe into the renal artery causes an abrupt fall in renal blood flow as measured by an electromagnetic flowmeter. Renal blood flow remained depressed for about a minute and thereafter sometimes increased above baseline levels. This response was not specific for ^{133}Xe, but was also seen following the bolus injections of normal saline. Mowat et al., (1972) found that the washout estimate of blood flow did not correlate well with the flowmeter blood flow reading recorded before ^{133}Xe injection, but did correlate with the minimum flowmeter reading after injection. The results reported by Mowat et al. suggest that measurements of total renal blood flow and blood flow distribution using inert gas washout techniques may be highly influenced by the vascular effects of injections into the renal artery and may not reflect the true hemodynamic state of the kidney prior to injection. On the other hand, other investigators have found an excellent correlation between blood flow measurements determined using PAH clearance or the electromagnetic flow probe

(Mangel et al., 1970; Ladefoged and Munck, 1971). It is likely that the phenomenon reported by Mowat et al. occurs only in response to extremely brisk injections made through a fine catheter or injections of very hypertonic solutions (Hollenberg, 1975, personal communication; Solez and Altman, unpublished).

The inert gas washout methods do not measure blood flow in the strict sense but rather flow per unit volume. Thus these methods may not give accurate results in saline diuresis and ureteral obstruction, situations in which renal volume is markedly increased (Stein et al., 1973).

Inert gas methods also assume that the partition coefficient λ is the same throughout the kidney and remains constant, and that complete equilibration of the gas occurs. As Stein et al. (1973) have emphasized, these assumptions have yet to be validated.

b. Radioactive microsphere methods

If a bolus of radioisotope-labeled plastic microspheres 15μ in diameter suspended in dextran is injected into the left ventricle or root of the aorta, these particles will lodge in capillary beds throughout the body, theoretically in proportion to tissue blood flow. By measuring tissue radioactivity after miscrosphere injection, one may determine not only the distribution of cardiac output to various organs, but also the distribution of blood flow within a given organ. In the past five years this method has been used extensively to study the distribution of intrarenal blood flow (Slotkoff et al., 1971; McNay and Abe, 1970; Stein et al., 1973). The spheres lodge in glomerular capillaries before reaching the medullary circulation. Thus, this method is capable only of measuring the distribution of cortical blood flow.

The method is based on four assumptions (Slotkoff et al., 1971; Stein et al., 1973):

1. The microspheres must be uniformly mixed with the blood during injection and must follow the distribution of red cells within the kidney.

2. All the microspheres entering the renal circulation must be trapped within the kidney, i.e., extraction must be complete.

3. The injection of microspheres must not alter systemic hemodynamics.

4. The injection of microspheres must not alter renal hemodynamics or renal function.

In the dog, the last three assumptions have been shown to be valid (McNay and Abe, 1970; Slotkoff et al., 1971). The first assumption is more doubtful. Assuming that good mixing of the spheres is obtained, there remains no assurance that the microspheres will distribute themselves in the same fashion as the smaller, more deformable red cells. Baehler et al. (1973) compared the distribution of microspheres and glutaraldehyde-fixed frog red blood cells (which have a different size, shape, and density from microspheres) in the dog kidney and could demonstrate no differences. They concluded that "*differences in size, shape, or density of a given particle may have only a minor effect on determining its distribution in comparison with the effect of its interaction with the surrounding (normal) red cells.*"

The microsphere method as it is employed in the dog cannot be used without modification in other animals. The rat, for instance, has an idiosyncratic anaphylactoid response to dextran, in which the microspheres are usually suspended, and thus glucose must be used instead (Selye, 1968; Jaenike, 1972).

The microsphere method has recently been combined with microdisection to study blood flow to individual glomeruli (Kallskog et al., 1972; Bankir et al., 1973). These studies have shown that regional differences in blood flow per glomerulus are much smaller than regional differences in blood flow per gram of cortex. The lower blood flow in the inner cortex as compared to the outer cortex is, in part, a reflection of the relative sparsity of glomeruli in the juxtamedullary cortex.

One drawback of the radioactive microsphere method as it is currently employed is that it is usually used in anesthetized animals. In the kidney of the anesthetized animal there is likely to be increased sympathetic tone, and increased renin and prostaglandin release, all of which may affect blood flow (Zins, 1975). However, the microsphere method can be used in unanesthetized animals (Warren and Ledingham, 1974), so this is not a serious objection.

A comparison of blood flow distribution measurements made using the inert gas washout and microsphere methods has clarified the interpretation of the first component of inert gas washout curve and has raised questions about the interpretation of the other components (Stein et al. 1973). The first (most rapid) component of the washout curve seems invariably to change in the same direction as total renal blood flow and frequently in a direction opposite to that of outer cortical blood flow as measured by the microsphere method, suggesting that it is a measure of flow to a majority of cortical nephrons rather than just those in the outer cortex. Flow in the second component of the washout curve does not correlate with flow to the inner cortex as measured by the microsphere method. Slotkoff et al. (1971)

have been unable to correlate this second component with the flow to any specific area of the cortex.

2. Blood Flow Distribution and Renal Function

The notion that superficial and deep cortical nephrons are characterized by different rates of sodium excretion (Barger, 1966) has recently been questioned by several groups. In contrast to Barger's proposal that sodium-losing states are associated with a redistribution of blood flow toward the outer cortex, Stein et al. (1973) observed a redistribution of cortical blood flow toward the inner cortex during the natriuresis associated with saline loading, or administration of acetylcholine or furosemide. Kleinman and Reuter (1974) observed an increase in the ratio of inner cortical flow to outer cortical flow following saline loading in the adult dog. In the puppy, which shows a blunted natriuretic response to saline loading as compared with the adult dog, this ratio decreased, indicating a redistribution toward the outer cortex. There was no correlation between the magnitude of change in sodium excretion and the change in cortical blood flow distribution. In another study of saline loaded adult dogs, Bruns et al. (1974) have reported a significant increase in glomerular plasma flow (determined using microspheres) and a significant decrease in filtration fraction in juxtamedullary nephrons but not in superficial nephrons. Since medullary blood flow comes from the juxtamedullary nephrons, these results suggest that there is a marked increase in medullary blood flow following saline loading, a fact that has been demonstrated by Solez et al. (1974b). Boonjaren et al. (1974) recently demonstrated that renal sodium retention may exist in experimental congestive heart failure without a change in cortical blood flow distribution, a finding which is in direct opposition to Barger's earlier work (Barger, 1966). Most of the studies cited above study situations in which total renal blood flow changes. It remains possible that redistribution of cortical blood flow plays a role in altering sodium excretion in situations in which total renal blood flow does not change. The relationship between renal hemodynamics and sodium excretion is a very confused issue, and it can only be stated that at the present time there is no firm evidence that changes in renal sodium excretion are brought about by redistribution of renal blood flow.

C. MEDULLARY BLOOD FLOW

The intracortical distribution of blood flow probably has an important, albeit indirect, influence on renal water excretion by producing alterations in medullary blood flow. Elevations of medullary blood flow would be expected to interfere with the countercurrent concentration mechanism by "washing out" the medullary osmotic gradient (Thurau, 1964), while a reduction in medullary blood flow sufficient to produce ischemic damage to medullary tubules,

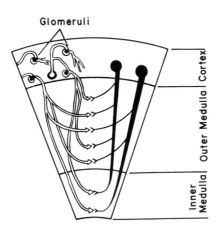

Fig. 2. The vascular architecture of the kidney. The medullary circulation as a whole is "in series" with the juxtamedullary cortical circulation, and thus blood flow in these two regions of the kidney tends to change in the same direction. However, the circulation of the inner and outer medulla are "in parallel" and it is theoretically possible that inner and outer medullary blood flow may change in opposite directions under certain circumstances.

such as occurs in ureteral obstruction of 18 hours duration (Solez et al., 1976a), would also be expected to interfere with the concentration mechanism.

The juxtamedullary glomeruli, from which the vessels supplying the medulla emerge, are larger than other glomeruli and have larger efferent arterioles (Fourman and Moffat, 1971). Recent studies of blood flow to individual glomeruli using microsphere techniques (Bankir et al., 1973; Bruns et al., 1974) have shown that the blood flow to deep glomeruli is significantly greater than the blood flow to more superficial glomeruli, even though blood flow per gram cortex is significantly lower in the juxtamedullary cortex than it is in the superficial cortex. Since branches of the efferent arterioles of juxtamedullary glomeruli supply the deep cortex as well as the medulla, the flow to these glomeruli is somewhat greater than total medullary blood flow.

Since the circulations of the inner and outer medulla are "in parallel" (using an analogy with electrical circuits) (Fig. 2), it is theoretically possible for inner medullary blood flow to change in a direction opposite to that of outer medullary or inner cortical blood flow. The only circumstance in which such an alteration is at all likely is during elevation of ureteral pressure where mechanical compression of the inner medulla reduces its blood flow despite a probable elevation in inner cortical blood flow (Solez, 1976a). In most other circumstances, it seems likely that medullary blood flow changes in the same direction as inner cortical blood flow.

1. Methods for Determining Medullary Blood Flow

Recent review articles have emphasized the severe technical and interpretive difficulties inherent in the measurement of medullary blood flow using available methods (inert gas or hydrogen washout, methods utilizing implanted detectors, etc.) (Thurau and Levine, 1971; Ladefoged and Munck, 1971; Stein et al., 1973; Hollenberg, 1973). We have developed a simple method of determining medullary plasma flow* in the rat which overcomes many of these difficulties (Solez et al., 1974b).

The method is based on the fact that the papillary transit time for circulating albumin (time required for albumin-containing plasma to pass through the papilla) in the normal rat is about 40 seconds (Fig. 3). If radioactive albumin is allowed to circulate through the kidney for 30 seconds, the accumulation rate of papillary radioactivity during this period is proportional to plasma flow rate. By comparing papillary radioactivity with the radioactivity of the arterial plasma during the same 30 second period, an accurate measure of medullary plasma flow is obtained. This determination of medullary plasma flow does not take into account glomerular filtration. However, since the glomerular filtrate is largely reabsorbed before the papillary portion of the descending loop of Henle is reached, it contributes directly to medullary plasma flow. If glomerular filtration is not taken into account, less than a 5% error is introduced in the determination of medullary plasma flow.

In the normal animal there is extensive leakage of albumin into the interstitial space where it equilibrates with the large extravascular albumin pool (Slotkoff and Lilienfield, 1967; Wilde and Vorburger, 1967; Venkatchalem and Karnovsky, 1972). The fenestrated ascending vasa recta are quite permeable to albumin and other small molecular weight proteins. Shimamura and Morrison (1973) have recently demonstrated extensive leakage of ferritin and horseradish peroxidase out of the ascending vasa recta 30 seconds after intravenous injection in normal rats. There is no evidence that the conduit vessels bringing blood to the medulla are similarly permeable to albumin. It deserves to be emphasized that extensive leakage of

---

*The method described measures inner medullary (papillary) plasma flow, referred to in this chapter as "medullary plasma flow."

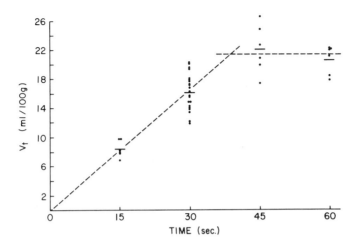

Fig. 3. Volume of distribution of radioactive albumin containing plasma in the papillae of 20 control rats at 15, 30, 45, and 60 sec after the infused ^{125}I-albumin first entered the renal arteries. The apparent circulation time for albumin is 40 sec. Volume of distribution increases linearly for the first 30 sec, suggesting that little radioactive albumin leaves the papilla during this time. (From Solez et al., 1974b. Used by permission of the publisher.)

albumin into the papillary interstitium from the ascending vasa recta interferes in no way with the determination of medullary plasma flow described by the albumin infusion method. It makes no difference whether the radioactive albumin brought to the medulla via the vasa recta is intravascular or extravascular at the time papillary radioactivity is measured.

The method used to determine medullary plasma flow is based on two major assumptions: (1) that radioactive albumin reaches the papilla only via medullary blood flow, and (2) that little radioactive albumin leaves the papilla during the infusion period. There is no evidence that substantial amounts of radioactive albumin are brought to the papilla via the glomerular filtrate or other routes and, thus, it seems likely that the first assumption is valid (Solez et al., 1974b). The fact that the volume of distribution of radioactive albumin increases linearly for the first 30 seconds in control animals (and, thus that medullary plasma flow determined at 15 seconds is the same as that determined at 30 seconds; see Fig. 3) suggests that the second assumption is also valid. If substantial amounts of radioactive albumin left the papilla during the infusion period, then the plot of the volume of distribution during the first 30 seconds would be expected to be nonlinear, with a decreasing slope.

This method of measuring medullary plasma flow seems to be both more reliable and subject to fewer interpretive difficulties than the other methods which have been described in the literature (Kramer et al., 1960; Lilienfeld et al., 1961; Thurau, 1964; Harsing et al., 1969; Grunfeld et al., 1971; Ladefoged and Munck, 1971; Thurau and Levine, 1971; Ganguli and Tobian, 1972; Hollenberg, 1973).

To convert medullary plasma flow to medullary blood flow, it is necessary to know the hematocrit of the blood perfusing the medulla. Pappenheimer and Kinter (1956) postulated the existence of "plasma skimming" in the renal cortex whereby the hematocrit of the blood flowing through the interlobular arteries would increase toward the superficial cortex due to the tendency for red cells to congregate in the center of the vessels while the plasma was concentrated along the wall of the vessel. Direct measurements of superficial nephron capillary hematocrits in the dog have not supported the existence of plasma skimming in the renal cortex (Stein et al., 1973). However, it seems likely that a phenomenon akin to plasma skimming occurs in the medulla (Lundgren and Jodal, 1975). Studies by Rasmussen (1973) and Wolgast (1973) using plasma and red cell markers have demonstrated that the hematocrit in the inner medulla is approximately 40-50% of arterial hematocrit. If these values are correct, medullary blood flow is approximately 20% greater than medullary plasma flow. Rasmussen (1972) and Schmid-Schönbein et al. (1973) have suggested that the low hematocrit in vasa recta blood is necessary to prevent a marked increase in viscosity of the blood caused by the hyperoncotic environment of the medulla.

D. GLOMERULAR BLOOD FLOW

1. Intrarenal Resistances

Recent investigations by Brenner et al. (1974) using the mutant Wistar rat with superficial glomeruli accessible to micropuncture have suggested that the resistance to blood flow in the afferent arteriole is much higher than had previously been believed, with a corresponding reduction in glomerular filtration pressure (Table 1). Afferent arteriolar resistance is approximately twice that in the efferent arteriole. Possibly because of viscosity effects, efferent arteriolar resistance (and hence filtration fraction) increases with increasing hematocrit (Table 1) (Byers, et al., 1975). Measurements of these parameters of glomerular hemodynamics have been carried out only for superficial glomeruli. However, if the pressure relationships in juxtamedullary glomeruli are similar, this would mean that the pressure in the vasa recta must be much lower than pressures

TABLE 1

Pressures in Vessels of the Rat Kidney (from Meyers et al., 1975)

Hematocrit	Renal Artery	Glomerular Capillaries	Peritubular Capillaries	Resistance ($\times 10^{10}$ dynes-sec-cm^{-5})		
				Afferent	Efferent	Total
50.8% (normal)	117 mm Hg	44.4 mm Hg	7.7 mm Hg	3.6	2.2	5.8
21.3% (lowered by isovolemic exchange)	100 mm Hg	43.1 mm Hg	9.4 mm Hg	2.3	1.7	4.0
62.5% (raised by isovolemic exchange)	115 mm Hg	56.3 mm Hg	9.9 mm Hg	3.8	4.0	7.8

which occur in the renal pelvis during ureteral obstruction (up to 80 mmHg), a fact which probably accounts for the marked decrease in inner medullary plasma flow in obstructed kidney (Solez et al., 1976a).

2. Dynamics of Glomerular Ultrafiltration

Direct sampling of blood from intrarenal vessels has demonstrated that afferent and efferent arteriolar oncotic pressures in the rat are 20 and 35 mmHg, respectively (Brenner et al., 1974). Since the pressure in glomerular capillaries averages 45 mmHg and the pressure in Bowman's space averages 10 mmHg, this means that at the afferent end of the glomerular capillaries there is a net filtration pressure of 15 mmHg and that filtration equilibrium (zero net filtration pressure) is reached at the efferent end of the glomerular capillaries. As Lassiter (1975) has emphasized, the finding of filtration equilibrium (and of a strong dependence of glomerular filtration rate on glomerular plasma flow) in the surface glomeruli of a mutant strain of Wistar rats cannot necessarily be generalized. It remains possible that in other strains or species (or in other glomeruli of the mutant Wistar rat) the situation might be quite different.

The permeability of the glomerular capillaries represented by the ultrafiltration coefficient K_f is approximately 50 times greater than capillary permeability in skeletal muscle (Lassiter, 1975). It is possible that changes in filtration fraction which occur in certain disease states or after treatment with vasoactive drugs in the absence of changes in mean arterial pressure may prove to be caused in part by a change in the permeability of the glomerular capillaries. In the past such changes in filtration fraction have been attributed to other factors such as resistance changes in the afferent and efferent arterioles.

E. THE RENAL NERVES AND THE RENAL CIRCULATION

The elegant fluorescence studies of Ljungqvist and Wagermark (1970) have demonstrated extensive adrenergic innervation of intrarenal vessels. Stimulation of the renal nerves in the intact dog decreases renal blood flow without changing the cortical distribution of blood flow (Katz and Shear, 1975a). Similar effects are produced by infusions of catecholamines (p. 42). Denervation and transplantation experiments suggest that there is little resting neurogenic sympathetic constrictor tone in the kidney of the unanesthetized dog or human (Thurau and Levine, 1971). Sympathetic tone appears to be greatly increased by anesthesia, although renal vasoconstrictor effects of anesthetics are probably partly caused by activation of the renin-angiotensin system. A full evaluation of the influence of the renal nerves on renal function is difficult to obtain because complete denervation involves extensive surgical trauma to the kidney (Katz and Shear, 1975b). The finding of relatively normal renal hemodynamics and function in the denervated or transplanted kidney must not be accepted uncritically as evidence that the renal nerves play only a minor role in the regulation of renal hemodynamics. It must be remembered that the denervated kidney becomes hypersensitive to circulating neurohumoral agents and that nerves begin to regenerate within weeks (Barger and Herd, 1973). The very rapid onset of changes in renal blood flow associated with emotional stimuli in dogs (Schramm et al., 1975) suggest that such changes are mediated via the renal nerves. The effects of the renal nerves on renin release are dealt with in a later section of this chapter.

F. AUTOREGULATION OF RENAL BLOOD FLOW

Relative constancy of renal blood flow when mean perfusion pressure varies between 80 and 200 mmHg is a well-established phenomenon. The older work on autoregulation of renal blood flow, beginning with the observations of Burton-Opitz and Lucas (1911), has been well summarized by Thurau and Levine (1971). The finding that glomerular filtration rate is also autoregulated (Forster and Meas, 1947) constituted the first persuasive evidence that autoregulation was mediated by preglomerular resistance changes. Autoregulatory responses to changes in perfusion pressure are almost instantaneous, suggesting that the control mechanisms involved are highly efficient (Arendhorst et al., 1975). The mechanism of autoregulation is not well understood. One theory of renal autoregulation is that the stimulus for constriction of preglomerular vessels in response to increased perfusion pressure is a rise in the transmural pressure (Sample and de Wardoner, 1959). This theory does not adequately account for autoregulatory phenomena in the kidney, since vessel caliber changes which would maintain constant wall tension in the face of changing perfusion pressures would not be expected to keep blood flow constant, according to the equations of Poiseville and LaPlace (Thurau and Levine, 1971). Autoregulation is not observed below a perfusion pressure of 70 mmHg, where glomerular filtration ceases. This fact has been used to support the idea that autoregulation is mediated via some humoral feedback mechanism, with the initial stimulus being an alteration in the composition or flow of the glomerular filtrate. The renin-angiotensin and prostaglandin systems have been implicated (Thurau and Levine, 1971; Herbaczynska-Cedro and Vane, 1973; Thurau and Mason, 1974). However, recent work seems to argue against involvement of these systems (Schmid, 1972; Eide et al., 1973; Gagnon et al., 1974; Bell et al., 1975; Zins, 1975). This issue is further discussed in appropriate sections of the second part of this chapter.

Using an indicator dilution technique with labeled red cells and small needle-shaped detectors implanted in the renal medulla,

Grängsjö and Wolgast (1972) demonstrated that the medullary circulation is autoregulated just as the cortical circulation is. The method of blood flow measurement used by these authors seems less than optimal, since it involves impaling the kidney. Nonetheless, the results obtained are probably valid. Autoregulation of the medullary circulation is also suggested by the finding that inner medullary plasma flow (Fig. 4) is only slightly reduced in moderate hemorrhagic hypotension in the rat (mean blood pressure 50-70 mmHg) (Solez et al., 1974b).

G. THE RENAL LYMPHATICS AND THE RENAL CIRCULATION

Virtually no work has been done on the relationship between the renal lymphatics and renal hemodynamics, despite the fact that renal lymph flow may have important interactions with other parameters of

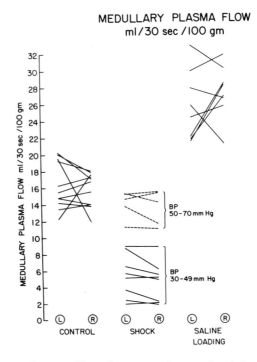

Fig. 4. Medullary plasma flow in rats determined by using a 30-second infusion of radioactive albumin. Lines connect the values for the left and right kidneys of individual animals. (From Solez et al., 1974b. Used by permission of the publisher.)

renal function. Renal lymph flow equals urine flow in most circumstances, and the renal lymph transports considerable quantities of protein and other substances out of the kidney (Yoffey and Courtice, 1970; Stolarczyk and Carone, 1975).

Recent evidence indicates that the renal lymph has its origin predominantly in the renal cortex (Keyl et al., 1972; Bell et al., 1973; Szabo and Magyar, 1974). The renal lymphatics are distributed primarily adjacent to arcuate blood vessels, in the interlobular spaces, and around Bowman's capsule (Bell et al., 1968). Acute occlusion of the lymphatics draining the rat kidney leads to a marked increase in urine flow and solute excretion without any change in urine osmolarity or glomerular filtration rate (Stolarczyk and Carone, 1975). Urea concentration in the medulla is significantly elevated. These intriguing changes may be brought about by reduced medullary blood flow secondary to elevated interstitial pressure.

H. THE RENAL CIRCULATION AND THE AGING KIDNEY

In man, renal mass decreases by 10-20% in old age (Bell, 1950; Rao and Wagner, 1972). This decrease in mass is accompanied by a 20% decrease in blood flow per unit volume as measured by inert gas washout studies (Hollenberg et al., 1974) suggesting that the decrease in kidney mass is caused by decreased perfusion rather than by primary parenchymal atrophy, which would be expected to result in comparable reductions in blood flow and mass. Vasodilating responses to acetylcholine and salt loading are increasingly blunted with age (Hollenberg et al., 1974), and it seems likely that the decreased blood flow in the aging kidney is caused by vascular damage rather than by a functional abnormality. Increased sympathetic tone would be expected to result in a potentiated response to vasodilators (Hollenberg et al., 1975), and is thus a much less likely explanation for the age-related reduction in renal perfusion. A progressive, age-related decrease in catecholamine content of vascular tissue from the brain and oral cavity has been documented (Robinson et al., 1972; Waterson et al., 1974). No similar studies have been carried out in the kidney, but if a similar phenomenon occurs, one would expect a decrease in sympathetic tone with age.

In an autopsy study of 105 patients who died suddenly and had no history of hypertension or renal disease, Darmady et al. (1973) demonstrated a progressive reduplication of elastic tissue and intimal thickening in renal vessels with age. Similar changes had been demonstrated previously by Moritz and Oldt (1937) and Smith (1955). These lesions undoubtedly reduce renal blood flow. Kilo et al. (1972) have shown that the thickness of capillary basement membranes increases with age in normal subjects. The possible influence of this capillary alteration on renal hemodynamics is unknown.

It is of interest that the male rat, which continues to grow during adulthood, shows an increase in kidney weight with age (Altman and Dittmer, 1962; Leon and Bloor, 1974) and apparently has no striking age-associated renal vascular alterations (Bras, 1969).

Aging changes in the human kidney may be of importance in the field of renal transplantation. In an analysis of 6883 human kidney transplants, Darmady (1974) found that the duration of survival of the recipient seemed to be inversely correlated with donor age. The shorter survival of recipients of kidneys from older donors suggests that the vascular alterations in the aged kidney may make the kidney less able to withstand transplantation.

II. HORMONAL CONTROL OF THE RENAL CIRCULATION

A. GENERAL CONSIDERATIONS

There are four groups of hormones which are believed to influence the renal vasculature directly: renin-angiotensin, the prostaglandins, the catecholamines, and kallikrein-kinin. Although these hormonal systems are often considered as separate entities, there are many interactions between them and it is likely that all four, with the possible exception of the kinin-kallikrein system, simultaneously exert an important influence on the renal circulation. The vasoactive compounds in each system have important noncirculatory effects. Angiotension II and its heptapeptide metabolite des-asp-angiotensin II stimulate the adrenal gland to secrete aldosterone (Mulrow et al., 1962; Peach and Chiu, 1974). Prostaglandin E influences sodium transport, gastrointestinal tract motility, the inflammatory response, and body temperature (Von Euler, et al., 1937; Johnston et al., 1967; Misiewicz, 1969; Kaley and Weiner, 1971a and b; Milton and Wendlandt, 1971; Lee, 1972). Catecholamines are important neural transmitter substances (von Euler, 1956), while kinins cause pain, increase capillary permeability, and influence the cellular inflammatory response (Colman, 1974).

B. THE RENIN-ANGIOTENSIN SYSTEM

1. Renin

Renin is an enzyme of approximately 40,000 molecular weight which initiates the formation of angiotensin (Kemp and Rubin, 1964). It was first described by Tigerstedt and Bergman (1898) and later studied by Braun-Menendez et al. (1940), Page and Helmer (1940), and Hill and Pickering (1940). Renin cleaves renin substrate to form the decapeptide angiotensin I. Angiotensin I is subsequently converted to the octapeptide angiotensin II, a potent vasoconstrictor substance (Peart, 1956; Skeggs et al., 1956). Renin is produced

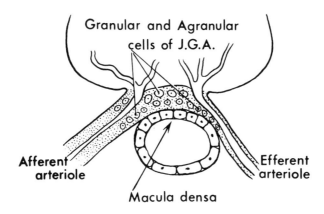

Fig. 5. Location and constituents of the juxtaglomerular apparatus (JGA). (From Heptinstall, 1974, Fig. 1-30, p. 26. Used by permission of the publisher.)

and stored in the granular myoepithelial cells of the juxtaglomerular apparatus in the wall of the glomerular afferent arteriole (Goormaghtigh, 1939; Hartroft, 1963; Barajas and Latta, 1967; Faarup, 1971) (Fig. 5). Recent data obtained, using micropuncture techniques and a specific ultramicroradioimmunoassay in the rat, suggest that the granular cells do not release renin directly into the lumen of arterioles but rather into the interstitium from whence it enters the circulation via the peritubular capillaries (Davis and Morgan, 1975; Johnston et al., 1975). Renin content per gram of kidney tissue varies in different species and correlates roughly with the prominence of the juxtaglomerular apparatus in histologic sections. Renin content is relatively great in the rabbit (12 Goldblatt units per gram), intermediate in the rat (1.2 Goldblatt units per gram), and low in man (0.013 Goldblatt units per gram) (Schaffenburg et al., 1960). In the dog and the rabbit, the renin activity of the juxtaglomerular apparatus from superficial glomeruli is greater than that of the juxtaglomerular apparatus from deeper glomeruli (Cook et al., 1956; Brown et al., 1965; Moriuchi et al., 1971). Rouffignac et al. (1974) and Flamenbaum and Hamburger (1974) have observed a similar difference in the rat, but Granger et al. (1972) found renin activity to be the same in superficial and deep glomeruli in this species.

Using calculations based on renin activity of the isolated juxtaglomerular apparatus, it has been claimed that juxtaglomerular renin activity is 10^{16} times higher than that of a similar volume (10^{-9} ml) of peripheral venous plasma (Thurau and Mason, 1974). Renin activity in the lymph surrounding the juxtaglomerular apparatus has been estimated to be 10^9 times higher than that in peripheral plasma (Thurau and Mason, 1974). These calculations may exaggerate the true capacity of juxtaglomerular apparatus to produce renin in vivo. Nevertheless, they suggest that peripheral renin levels may not always accurately reflect intrarenal renin activity.

Renin or renin-like enzymes (Peart, 1975) are found in other organs beside the kidney (Ferris et al., 1967; Eskildsen, 1973), principally in the uterus where they play an important role in regulating uterine blood flow (Ferris et al., 1972). Human endometrial and myometrial cells have been shown to produce renin in tissue culture (Symonds et al., 1968). It appears that the submaxillary gland also produces renin, and in some species submaxillary renin production may exert an influence on blood pressure (Takeda et al., 1969; Menzie and Michelakis, 1972). The walls of arteries and veins, as well as the vagina, adrenals, liver, and ureter, all contain a higher concentration of renin than plasma, but it has not been demonstrated that renin is actually produced at these sites (Gould et al., 1964; Eskildsen, 1973). Plasma renin decreases substantially with age in normal man and in laboratory animals, a fact which must be taken into consideration in population studies and in the assessment of "low-renin" hypertension (Hayduk et al., 1973; Sen et al., 1972; Sassard et al., 1975).

2. The Control of Renin Release

This topic has recently been the subject of several comprehensive review articles (Vander, 1967; Davis, 1973; Oparil and Haber, 1974), and therefore only a brief account of it will be given here. Renin release may be triggered by various humoral agents (see Section IV.G), by the renal sympathetic nerves, and by intrarenal receptors in the renal afferent arterioles and the macula densa. It appears that the afferent arteriole may act as a baroreceptor, responding to changes in wall tension and playing a primary role in the control of renin release (Davis, 1973). The macula densa is thought to act as a sensor of distal tubule sodium delivery, although the exact signal perceived is unknown. In the intact animal, decreasing distal tubule sodium delivery by occluding the ureter in the presence of a mannitol diuresis results in renin release (DiBona, 1971). However, increasing sodium concentration in distal tubular fluid by treatment with ethacrynic acid also stimulates renin release (Cooke et al., 1970). In the isolated juxtaglomerular apparatus, perfusion of the macula densa with isotonic sodium chloride increases renin activity and angiotensin II concentration (Thurau et al., 1972). Thus, it is not clear whether the stimulus to renin release is an increase or a decrease in macula densa sodium delivery or concentration.

In man and in experimental animals, potassium administration inhibits renin release (Brunner et al., 1970; Shade et al., 1972).

This effect is thought to be mediated through the macula densa receptor, since it is not observed in the nonfiltering kidney (Shade et al., 1972).

In spite of these studies on the role of the macula densa, there is good evidence to show that changes in perfusion pressure in the afferent arteriole can play an important part in the control of renin production and release (Tobian, Tomboulian, and Janecek, 1959; Davis, 1973). Tribe and Heptinstall (1965) demonstrated that hyperplasia of the juxtaglomerular apparatus--indicative of increased renin content--could be induced in rats in which the macula densa had either been destroyed or removed from contact with glomerular filtrate.

Stimulation of the renal nerves (see Section III.G.) results in renin release accompanied by renal vasoconstriction (Vander, 1965; Johnson et al., 1971). Neural stimulation apparently also can inhibit renin release.* Electrical stimulation of widespread areas of the dog hypothalamus has been alleged to result in a 50% reduction in plasma renin activity without an alteration in renal blood flow (Zehr and Feigl, 1973). This response was not observed following renal denervation. Hall and Buckalew (1975) observed a significant increase in renal blood flow in the dog after stimulation of hypothalamic nucleii and concluded from the time course of the response that a humoral mechanism was involved. Renin levels were not determined.

Renin secretion may be regulated by three parallel feedback loops (Fig. 6). (Oparil and Haber, 1974). In the first of these (Loop 1 in Fig. 6), renin release, acting through angiotensin II,

---

*There is recent evidence (Kurz et al., 1975; Yun et al., 1974, 1975) that left atrial volume receptors influence renin secretion. Increasing left atrial transmural pressure in the dog results in suppression of renin secretion. This response which apparently is not accompanied by any change in renal blood flow (Kurz et al., 1975) is not observed after sectioning of either the vagi or the renal nerves. Sectioning of the vagi or denervating the atrioventricular node in dogs maintained on a high salt diet causes an increase in plasma renin activity (Yun et al., 1974, 1975). These results suggest that changes in intravascular volume influence renin secretion through a CNS neurosomal pathway that involves a vagal efferent circuit and a renal nerve afferent circuit. Mancia et al. (1975) have demonstrated that interruption of afferent vagal nerve traffic by bilateral vagal cooling causes an increase in renin release, the magnitude of which appears to be related to pressure in the carotid sinus. These authors have concluded that *"vagal afferents from the cardiopulmonary region exert a tonic restraint on the release of renin; this restraint occurs in circumstances in which these afferents cause little change in total renal blood flow."*

stimulates the adrenal to release aldosterone. This increases extracellular fluid volume and renal perfusion, tending to turn off renin secretion. The second feedback loop (Loop IIa, b, and c in Fig. 6) operates through angiotensin II directly. Angiotensin II has a direct inhibitory effect on renin release (Loop IIa), and also increases renal perfusion pressure (and therefore reduces renin release) via its indirect (Loop IIb) and direct (Loop IIc) pressor effect. The third feedback loop (Loop III, Fig. 6) is the most controversial. Proponents of the existence of this loop postulate that renin release leads to the local formation of angiotensin II which causes constriction of the afferent arteriole, decreasing glomerular filtration rate and sodium delivery to the macula densa and turning off renin secretion (Guyton et al., 1964; Thurau and Mason, 1974). As will be discussed below, there is good evidence for the intrarenal generation of angiotensin II, but it is not established that reduced sodium delivery to the macula densa inhibits renin release.

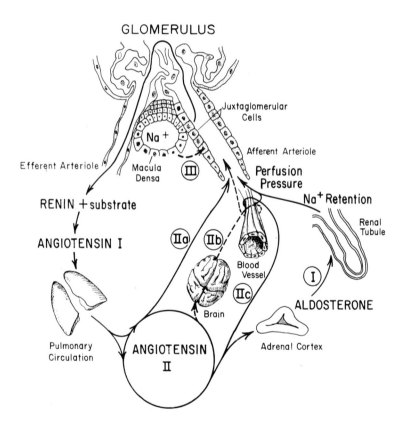

Fig. 6. Feedback control of renin release. (From Oparil and Haber, 1974, Fig. 4. Reprinted, by permission, from <u>New Engl. J. Med.</u> 291:389, 1974.)

There is evidence that the stimulus for renin synthesis by the kidney and renin release may not be the same. Silverman and Barajas (1974) have recently demonstrated that reserpine treatment in rats reduces plasma renin to half of normal while increasing renal renin and juxtaglomerular index two and one half times (see Section IV.D, Catecholamines). Similarly, Krieger (1970) found a 70% increase in renal renin content, but a marked reduction in plasma renin activity in rats 72-96 hours after a sublethal dose of $HgCl_2$.

3. Renin Substrate

Renin substrate consists of a group of glycoproteins of about 60,000 molecular weight, which are formed in the liver and which release angiotensin I when acted upon by renin (Page et al., 1941; Skeggs et al., 1963). There appear to be several different forms of renin substrate, all of which have similar molecular weight and amino acid composition, differing mainly in their carbohydrate composition (Skeggs et al., 1964) but possibly also in their reactivity. Skeggs et al. (1957) isolated an amino terminal tetradecapeptide by tryptic digestion of hog renin substrate. This tetradecapeptide releases angiotensin I at the same rate as native hog renin substrate when acted upon by renin in aqueous solution, but reacts feebly with renin in serum. This suggests that a portion of the larger native renin substrate molecule prevents inhibition by serum, but is not directly involved in the angiotensin I releasing reaction (Skeggs et al., 1963).

The reaction between renin and renin substrate is only moderately species specific. Human renin substrate reacts only with primate renins (Braun-Menendez, et al., 1946). Human renin reacts poorly with rat renin substrate (Shipley and Helmer, 1948), and mouse renin substrate appears to be cleaved to angiotensin I only by mouse renin (Oliver and Gross, 1966). On the other hand, human renin does not react with renin substrate from pigs, dogs, horses, cows, and goats (Fasciolo et al., 1940). Human renin reacts with sheep renin substrate five times faster than it does with human renin substrate (Skinner, 1967). With cat renin substrate, the reaction is twice as fast as with human renin substrate (Poulson, 1968).

Altered renin substrate levels are thought to play an important role in the so-called "hepatorenal syndrome," in which reduced renin substrate levels are alleged to contribute to hemodynamic instability and redistribution of blood flow in the kidney (Berkowitz et al., 1974). Elevated renin substrate levels may contribute to the hypertension sometimes observed in women using estrogen-containing oral contraceptives (Skinner et al., 1969).

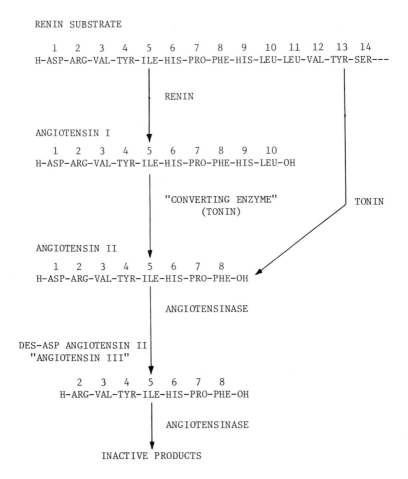

Fig. 7. Biochemistry of the renin-angiotensin system.

4. Angiotensin I

The decapeptide angiotensin I is produced when renin splits the leucine-leucine bond of renin substrate (Fig. 7) (Skeggs et al., 1974). It is unlikely that angiotensin I has any direct circulatory action. Infusion of angiotensin I into the renal artery of the intact dog has a significant vasoconstrictor effect which, however, is blocked by inhibition of converting enzyme (DiSalvo et al., 1971), (see B5 below), or by antibody to angiotensin II (Merrill et al., 1973). Itskovitz and McGiff (1974) have demonstrated a vasoconstrictor effect of angiotensin I in the isolated perfused dog kidney even after treatment with converting enzyme inhibitor. However, since the enzyme tonin is present in the kidney and can convert angiotensin I to angiotensin II in the presence of converting enzyme inhibitor (Boucher et al., 1974; Genest, 1974), there remains no proof that angiotensin I itself is a vasoactive compound.

5. Converting Enzyme

The isolation of an enzyme from horse plasma, which converted angiotensin I to angiotensin II by removing the dipeptide histidyl-leucine from the C-terminal end, was first carried out by Skeggs et al. (1956). Subsequently a similar enzyme has been found in all mammalian species investigated, including man (Vane, 1974). Even when highly purified, preparations of converting enzyme hydrolyze and inactivate bradykinin (Eliseiva et al., 1971; Yang et al., 1971), and it is likely that converting enzyme is identical with bradykininase. Plasma-converting enzyme is not present in sufficient quantities to account for the rapid pressor effect of angiotensin I following an intravenous injection (Ng and Vane, 1967; Poulsen and Poulsen, 1971). With the demonstration of much higher levels of converting enzyme activity in the lung and other organs (Ng and Vane, 1967, 1968), it has recently been suggested that the plasma enzyme merely represents "leakage" from tissue bound sites of enzyme activity (Vane, 1974). However, it has not been demonstrated that the tissue and plasma enzymes have identical properties, and the site of production of the plasma enzyme is unclear. Converting enzyme activity has been found in the isolated juxtaglomerular apparatus of the rat (Thurau, 1974), raising the possibility that angiotensin II is formed locally within the kidney. Most converting enzymes isolated from animals require the chloride ion for their activity, although this is apparently not the case for converting enzyme prepared from human lung (Fitz et al., 1974). The relationship between the human lung converting enzyme and tonin, which also does not require chloride, remains unclear (see below).

6. Tonin

Boucher et al., (1972, 1974) have described a new enzyme, termed "tonin," which not only converts angiotensin I to angiotensin II, but also can cleave the phenylalanine-histidine bond of tetradecapeptide renin substrate to form angiotensin II directly. They have found this enzyme in all tissues examined so far, and in particularly high concentrations in rat submaxillary gland. In the kidney, it is found to be present in higher concentrations in the medulla and inner cortex than in the outer cortex. The cortical distribution of tonin is thus the reverse of that for renin (see Section B1). Tonin does not appear to be a bradykinase and does not require chloride ions for its action. Its action is completely inhibited by the presence of serum, and thus it is not active in the circulating blood. The molecular weight of tonin is about 31,000 (Boucher et al., 1974) as compared with 200,000 to 600,000 for converting enzyme isolated from human lung (Erdös, 1974; Fitz et al., 1974). It has been suggested that this enzyme may play a fundamental role in the local formation of angiotensin II and the control of the distribution of intrarenal blood flow (Boucher et al., 1974).

7. Angiotensin II and Renal Blood Flow

The octopeptide angiotensin II is the most potent naturally occurring pressor agent known. Its effects on the renal vasculature seem paradoxical. In the intact kidney it has a marked vasoconstrictor effect even in subpressor doses (Herrick et al., 1942; DeBono et al., 1963; Navar and Langford, 1974). Large arterial doses cause a transient reduction of renal blood flow almost to zero (DiSalvo and Fell, 1970), although the greatest sustained reduction that can be obtained with an intra-arterial infusion is 50% (Furuayama et al., 1967; Fourcade et al., 1971). These findings are in marked contrast to the unresponsiveness of kidney vascular smooth muscle to angiotensin II in vitro. Bohr and Uchida (1967) showed that smooth muscle from segmental and arcuate arteries of dog kidney is completely unresponsive to angiotensin II. Somlyo and Somlyo (1971) showed that the maximal response of smooth muscle from the main renal artery to angiotensin II is only 6% of the maximum response to epinephrine or norepinephrine. Selective renal arteriograms in normal human subjects (Hollenberg and Adams, 1974) show no constriction of the segmental and arcuate arteries in response to an angiotensin II infusion, whereas there is marked constriction of those vessels after a comparable infusion of norepinephrine. These data suggest that the renal vasoconstrictor effects of angiotensin II are confined primarily to intrarenal vessels smaller than the arcuate arteries. Kincaid-Smith et al. (1974) have observed alternating areas of dilation and constriction in interlobular arteries and afferent arterioles of fetal kidney allografts in a rabbit ear chamber after injection of angiotensin II (1 µg/kg) into the opposite ear vein. Glomerular blood flow decreased progressively over 3 to 4 minutes, then appeared to cease altogether for 30 seconds before normal blood flow again commenced.

Angiotensin II increases filtration fraction. The reduction it causes in glomerular filtration rate is consistently less than the reduction in renal blood flow (Bock et al., 1958; Laragh et al., 1963; Earley and Friedler, 1966; Navar and Langford, 1974). This increase in filtration fraction suggests that angiotensin II causes constriction of the efferent arterioles, thus preventing the marked decrease in glomerular filtration pressure which usually accompanies renal vasoconstriction. However, recent micropuncture studies in mutant Wistar rats with surface glomeruli (Andreucci et al., 1975) raise questions about the occurrence of selective vasoconstriction of the efferent arteriole.

Unlike other hormones which affect the renal circulation, angiotensin II does not appear to alter the intracortical distribution of renal blood flow (Navar and Langford, 1974).

Angiotensin II has little effect on renal autoregulation. Doses sufficient to cause a 30-50% reduction in renal blood flow do not impair the normal intrarenal vasodilator response to acute decreases in perfusion pressure within the range of 80-160 mmHg (Belleau and Earley, 1967; Kiil et al., 1969; Gagnon et al., 1970). Under conditions of reduced perfusion pressure, the renal vasculature is hyporesponsive to angiotensin II, a fact which suggests that angiotensin II does not play an important role in renal autoregulatory responses (McNay and Kishimoto, 1969; Navar and Langford 1974).

The doses of angiotensin II used in many of the studies cited above result in plasma levels many fold higher than levels which occur physiologically (0-24 pg/ml in normal subjects, 4-135 pg/ml in patients with malignant hypertension), and therefore the physiologic relevance of these studies is subject to question (Peart, 1971; Ruiz-Maza et al., 1974).

Repeated or continued administration of angiotensin II has a diminishing effect on blood pressure and renal blood flow (Stewart, 1974), a phenomenon known as tachyphylaxis. This means that relatively large doses must be infused to produce a sustained reduction of renal blood flow.

Caution must be exercised in extrapolating from data obtained using exogenous angiotensin II to the intrarenal role of endogenous angiotensin II. Angiotensin II may be formed within the juxtaglomerular apparatus (Thurau, 1974), and act directly on the afferent and efferent arterioles from the adventitial rather than the luminal side, producing a situation quite different from that which exists during an intravenous or intraarterial infusion of angiotensin.

8. Angiotensinases

Angiotensin II is not degraded by the lungs (Biron et al., 1968, 1969), but is inactivated in the arterial circulation very rapidly. After intravenous injection of very large amounts, blood pressure rises promptly and then returns to normal levels within 3 minutes (Laragh and Sealey, 1973). While a small portion of infused angiotensin II is probably removed from the circulation by binding at sites of action, most is degraded by plasma and tissue enzymes, including enzymes in the kidney (Ryan, 1974). The half life of infused angiotensin measured by radioimmunoassay is about one minute (Boyd et al., 1969; Cain et al., 1970). It has been suggested that the half life for endogenously produced angiotensin II may be closer to 15-20 seconds since it has been demonstrated that most vascular beds are capable of destroying a major portion of infused angiotensin in a single circulation (Hodge et al., 1967; Doyle et al., 1968). The enzymes responsible for degrading angiotensin II are not specific for angiotensin and include a variety of amino-, endo-, and carboxypeptidises, chymotrypsin, trypsin and pepsin (Ledingham and Leary, 1974).

9. Angiotensin III and Aldosterone

Most of the metabolites of angiotensin II seem to be inactive and are currently regarded as of interest only in that they may be confused with angiotensin II in a radioimmunoassay. One metabolite, des-asp-angiotensin II is a potent stimulus for aldosterone secretion (Peach and Chiu, 1974; Spielman et al., 1974) and has been named "angiotensin III." The physiologic importance of this compound has not yet been established, and its effects on the kidney are unknown.

Aldosterone has no direct effect on the renal circulation, although it may influence renal perfusion by altering intravascular volume through its effect on sodium secretion.

10. Inhibitors of the renin-angiotensin system

Anephric animals are hyperresponsive to renin. This hyperresponsiveness cannot be accounted for simply on the basis of increased renin substrate levels, and suggests that the kidney normally produces a renin inhibitor. Smeby and Bumpus (1971) have prepared a lysophospholipid from dog kidney tissue which inhibits renin in vitro and in vivo. This compound contains a large amount of arachidonic acid, raising the possibility that it is a prostaglandin precursor (see Section C.1). Poulsen (1971) could find no evidence for the presence of renin inhibitors in rat plasma, and it is possible that the increased responsiveness which follows bilateral nephrectomy is due to an increased responsiveness of angiotensin vascular receptors (Laragh and Sealey, 1973) rather than to circulating inhibitors.

Renin is inhibited by heparin (Sealey et al., 1967), bile (Hiwada et al., 1969), and pepstatin, a proteinase inhibitor derived from bacteria (Peart, 1975). Prostaglandin A inhibits renin, but only in excessive amounts (Kotchen et al., 1974).

Angiotensin I converting enzyme is inhibited by chelating agents such as EDTA, BAL, and O-phenanthroline and also by specifi peptide inhibitors isolated from snake venom (Erdös, 1975). The nonopeptide Pca-Trp-Pro-Arg-Pro-Gly-Ile-Pro-Pro (SQ 20,881) from Bothrops jararaca and the undecapeptide Pca-Gly-Leu-Pro-Arg-Pro-Lys-Ile-Pro-Pro (potentiator B) from Aghistrodon halys blomhoffii are potent competitive inhibitors of converting enzyme. The pentapeptide Pca-Lys-Trp-Ala-Pro (SQ 20,475) from Bothrops jararaca inhibits converting enzyme but is also degraded by it (Bakhle, 1974).

The biologic activity of angiotensin II is dependent on the structure of the C-terminal end around phenylalanine (Peart, 1975). Replacement of this phenylalanine by another amino acid results in a compound without substantial biologic activity which is a competitive inhibitor of angiotensin II.

A number of such angiotensin analogs have been synthesized. The most potent of these are [Ile^8]-angiotensin II and [Sar^1, Ile^8]-angiotensin II (Türker et al., 1974). [Sar^1, Ile^8]-angiotensin II appears to be more stable and is degraded more slowly in vivo than [Ile^8]-angiotensin II. [Sar^1, Ala^8]-angiotensin II (Pals et al., 1971) is only slightly less active than the [Sar^1, Ile^8] analog and is more widely used at the present time. Depending on experimental conditions and the species in which these compounds are used, angiotensin analogs may have considerable agonistic (angiotensin-like) effects. Thus in the conscious normotensive dog [Sar^1, Ala^8]-angiotensin II causes a decrease in renal blood flow (Satoh et al., 1975), an agonistic effect. In animals with increased endogenous angiotensin levels secondary to anesthesia or some other stimulus, the same dose of analog increases renal blood flow, presumably by inhibiting the vasoconstrictor effect of angiotensin II.

In addition to the direct pharmacological antagonists, the response to angiotensin II is diminished by concurrent administration of various vasodilator drugs (Bianchi et al., 1960; Barer, 1963). Prostaglandin E_1 decreases the pressor response to angiotensin in rats without changing responsiveness to norepinephrine (Türker et al., 1968).

Renal denervation reduces renal vascular reactivity to angiotensin II in the dog; agents such as guanethidine, bretylium, and hydralazine, which decrease sympathetic nervous activity by inhibiting the release of neural transmitter, have a similar effect (McGiff and Fasy, 1965).

C. PROSTAGLANDINS

1. Structure and Synthesis

Prostaglandins are 20-carbon unsaturated fatty acids containing cyclopentane ring with two adjacent side chains, one of which bears a carboxyl group at the terminal position (Bergstrom and Sjovall, 1960a, b; Bergstrom and Samuelsson, 1965). Prostaglandins are ubiquitous vasoactive compounds and are synthesized in a wide variety of tissues, most notably the prostate and the kidney (Hickler et al., 1964; Lee et al., 1964). Depending on the substitutions on the cyclopentane ring, the compounds are designated as prostaglandins (PG) A,B,C,D,E, or F with a subscript indicating the number of side chain double bonds (Caton, 1973; Jones, 1972; Flower, 1974).

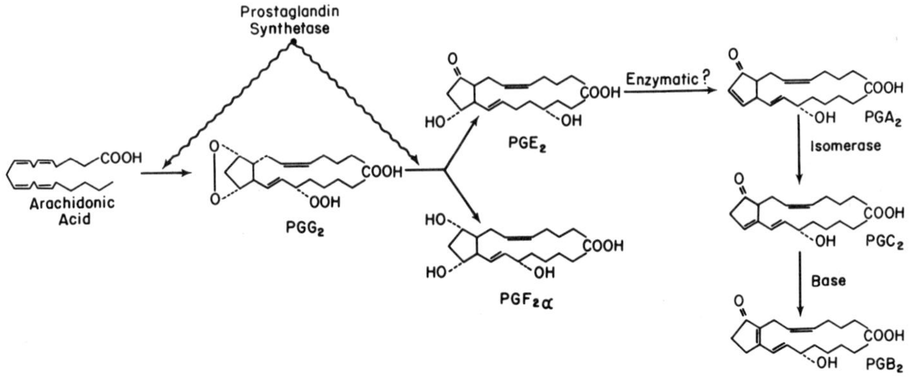

Fig. 8. The biochemistry of prostaglandin synthesis. Besides the pathways shown, there is recent evidence that PGE_2 may be converted to $PGF_{2\alpha}$ by an enzyme present in sheep blood (Hensby et al., 1974).

The terms PGG_2 and PGH_2 have been used to designate the biologically active endoperoxide precursors of prostaglandins (Samuelsson and Hamberg, 1974).

The naturally occurring prostaglandins which have an important influence on the kidney are PGE_2, $PGF_{2\alpha}$ and PGA_2 (McGiff and Itskovitz, 1973). PGE_2, and $PGF_{2\alpha}$ are rapidly degraded by the lungs, whereas PGA_2 is not and may act as a systemic hormone. A scheme for the biosynthesis of these compounds is shown in Fig. 8. Deformation of the cell membrane, or other abnormalities activates phospholipase in the membrane, which in turn cleaves phospholipids to form arachidonic acid. Arachidonic acid is then acted upon by "prostaglandin synthetase," a fatty acid cyclooxygenase, to form the endoperoxide PGG_2 (Samuelsson and Hamberg, 1974). This highly unstable endoperoxide, which is a vasoactive compound in its own right, is then transformed into PGE_2 or $PGF_{2\alpha}$. The factors determining which of these compounds will be formed from the endoperoxide pool are unknown (Wlodawer and Samuelsson, 1973), although it has been suggested that other vasoactive drugs such as bradykinin and angiotensin II may influence the way in which the endoperoxide breaks up into PGE_2 and $PGF_{2\alpha}$. PGE_2 is relatively unstable and readily undergoes dehydration within the cyclopentane ring to form PGA_2. This fact makes it difficult to establish that any organ actually produces PGA_2 (Hamburg,

1969; Lee et al., 1967; McGiff and Itskovitz, 1973), although there is little doubt that PGA_2 occurs naturally in plasma (Zusman et al., 1973; Jaffee et al., 1973) and in seminal vesicular fluid (Hamberg and Samuelsson, 1969). PGA isomerase, an enzyme which transforms PGA_2 to PGC_2, is present in the plasma of the cat, dog, rabbit, pig and rat, but not of man or the guinea pig (Jones and Cammock, 1973). PGC_2 is a more potent vasodilator than PGA_2. Like PGA_2, it is not degraded by the lung as are PGE_2 and $PGF_{2\alpha}$.

Under mild alkaline conditions PGC_2 readily undergoes isomerization to PGB_2 a potent vasoconstrictor (Greenberg et al., 1974).

Renal prostaglandin synthesis takes place predominantly in the medulla (Janszen and Nugteren, 1971; Crowshaw, 1971, 1973), although recent studies suggest that prostaglandin synthesis may occur in the renal cortex under certain circumstances (Larsson and Änggård, 1973; McGiff and Itskovitz, 1973; Zins, 1975). Histochemical studies allegedly demonstrate prostaglandin production in the epithelium of the medullary collecting ducts (Janszen and Nugteren, 1971). Renal medullary interstitial cells grown in tissue culture have been shown to produce prostaglandins (Muirhead et al., 1972).

Prostaglandins produced in the renal medulla are thought to be transported to the cortex via both the vasa recta and the ascending limbs of the loop of Henle (Zins, 1975). Prostaglandins of the E and the F series have been found in human urine (Frolich et al., 1973). Stop flow studies have suggested that prostaglandins enter the tubular filtrate at the ascending limb of the loop of Henle (Williams et al., 1974), raising questions about the collecting duct origin of prostaglandins.

Prostaglandins are rapidly metabolized in the renal cortex. The major degrading enzyme, 15-hydroxy prostaglandin dehydrogenase, is present in high concentration in the cytoplasm of distal tube epithelial cells, and is found in lesser amounts in the media of cortical arteries and arterioles and in the epithelium of Bowman's capsule (Nissen and Andersen, 1968).

2. Prostaglandin E_2

PGE_2 is a potent vasodilator which increases renal blood flow and decreases renal vascular resistance on intra-arterial administration (Vander, 1968; Nakano, 1973). However, since PGE_2 is not normally detectable in arterial blood, owing to metabolism by the lungs, these observations shed little light on the renal circulatory effects of endogenous PGE_2.

Some information about the possible role of prostaglandins in the regulation of renal blood flow has come from recent studies by

Lonigro et al. (1973); these have demonstrated that inhibition of prostaglandin synthesis by indomethacin results in a 45% decrease in renal blood flow which coincides with a marked decrease in renal vein PGE efflux. This effect is seen both in the anesthetized dog and in the isolated perfused dog kidney, and has been used to support the notion that PGE_2 controls renal vascular tone at rest, a function which previously had been attributed to "intrinsic arteriolar activity" (McGiff and Itskovitz, 1973). The idea that prostaglandins control resting renal blood flow has been challenged by studies in the unanesthetized dog in which inhibition of prostaglandin synthesis had no significant effect on renal blood flow (Satoh and Zimmerman, 1974; Zins, 1975). It may be that prostaglandins are an important determinant of "resting" blood flow only in kidneys in which prostaglandin production by the renal medulla is increased secondary to anesthesia or surgery (Swain et al., 1974; Zins, 1975).

Herbaczynska-Cedro and Vane (1973) have implicated PGE release in renal autoregulation. They found that reduction of renal perfusion pressure resulted in release of prostaglandins into the renal vein coincident with compensatory renal vasodilatation. Inhibition of prostaglandin synthesis by indomethacin treatment abolished the renal autoregulatory response to decreased perfusion pressure. Indomethacin also abolished the reactive hyperemia which follows relief of renal artery occlusion (Herbaczynska-Cedro and Vane, 1974). However, others have not been able to confirm that indomethacin impairs renal autoregulation in the dog (Owen et al., 1974; Bell et al., 1975). The role of PGE_2 in autoregulation and in the control of renal vascular tone remains uncertain at this time.

A redistribution of renal blood flow toward the superficial cortex following indomethacin treatment has been noted by several investigators (Itskovitz et al., 1973, 1974; Kirschenbaum et al., 1974), while an increase in juxtamedullary blood flow has been observed after stimulation of intrarenal PGE_2 synthesis (Larsson and Änggård, 1974; Chang et al., 1975). Based on this indirect evidence, McGiff and Itskovitz (1973) have postulated that PGE_2 functions as a local hormone controlling medullary blood flow. Using the I^{125}-albumin infusion technique to measure inner medullary plasma flow in rats, we have observed a 33% reduction in medullary plasma flow after inhibition of prostaglandin synthesis by indomethacin (5 µg/kg i.p.). The inhibition of prostaglandin synthesis was short-lived and was followed by a "rebound" 64% elevation in prostaglandin E and A levels associated with 57% elevation in medullary plasma flow (Figs. 9a, b) (Solez et al., 1974a and b). These data strongly suggest that renal prostaglandin release controls medullary blood flow.

Angiotensin II, norepinephrine, and bradykinin all increase renal PGE_2 synthesis (McGiff and Itskovitz, 1973). These interactions are discussed in Section IV.G.

THE RENAL CIRCULATION: PHYSIOLOGY AND HORMONAL CONTROL 693

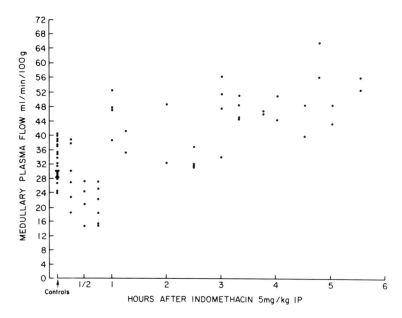

Fig. 9 a. Renal medullary plasma flow after indomethacin treatment.

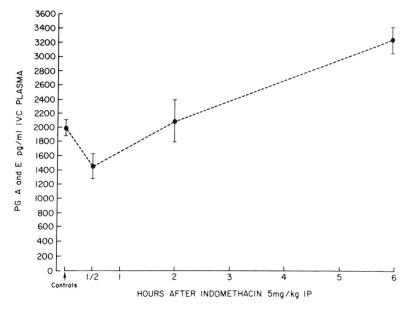

Fig. 9 b. Levels of prostaglandins A and E in vena cava plasma after
indomethacin treatment (± standard error of the mean).
(Both figures from Solez et al., 1974b, Figs. 1 and 2.
used by permission of the publisher.)

3. Prostaglandin A_2

Opinions are divided on the question of whether PGA_2 is produced by the kidney or is formed from PGE_2 in the circulating blood (Attallah et al., 1974). PGA_2 has biologic properties similar to PGE_2, but it is more likely to function as a circulating hormone than is PGE_2 since it escapes degradation by the lung. Its potency as a vasodilator is one-fifth of that of PGE_2.

Attallah and Lee (1973) demonstrated that a soluble fraction of a homogenate of rabbit papilla had the capacity to bind tritiated PGA_1 in an apparently specific manner. Kuehl et al. (1974) demonstrated that these preparations would also bind tritiated PGA_2 in the same fashion. Unlabeled PGA decreased the binding of labeled PGA. PGE_2 bound approximately one-tenth as well as PGA_2. These observations suggest that there is a specific PGA receptor in the medulla.

Both PGA and PGE increase urinary sodium excretion when infused into the renal artery, and it has been suggested that PGA functions as a natriuretic hormone. However, circulating levels of PGA in man decrease on a high sodium intake and increase on a low sodium intake, whereas PGE concentrations do not change under these circumstances (Zusman et al., 1973), suggesting that PGE and PGA have independent actions, and that endogenous PGA may play an important role in sodium homeostasis as an "antinatriuretic hormone" (Tobian, 1975).

PGA_2 (0.07 to 0.2 µg/min infused into the dog renal artery) has been alleged to cause an increase in renal cortical blood flow and a decrease in medullary blood flow based on studies of blood flow distribution using inert gas washout techniques (Birtch and Zakheim, 1967, quoted in Lee, 1968). If PGA_2 truly decreased medullary blood flow, this would make its renal hemodynamic effects quite unlike those of PGE_2 which is thought to increase medullary blood flow. However, we have observed a marked increase in inner medullary plasma flow determined with the I^{125}-albumin infusion technique in the rat after intravenous injection of PGA_2 (4-16 µg/kg) Solez, Buono, and Vernon, unpublished) and it therefore seems likely that PGA_2 and PGE_2 have similar effects on renal blood flow.

4. Prostaglandin $F_{2\alpha}$

$PGF_{2\alpha}$ decreases blood pressure in cats and rabbits (Änggård and Bergström, 1963; Horton and Main, 1963) and increases blood pressure in rats and dogs (DuCharme et al., 1968; Nakano and McCurdy, 1968; Nakano and Cole, 1969). The renal hemodynamic effects of this prostaglandin have not been well studied. Nakano and McCurdy (1967, 1968) observed that intra-arterial injection of $PGF_{2\alpha}$ (0.1 µg/kg) increased peripheral vascular resistance and decreased renal blood flow in dogs. In a study of bradykinin-prostaglandin interactions by McGiff et al., (1972), brief mention is made of the fact that intra-arterial infusion of equivalent amounts of $PGF_{2\alpha}$ had no effect on renal blood flow, renal resistance, or systemic blood pressure in dogs. Fülgraff et al., (1974) and Tannenbaum et al., (1975) also observed no alteration in these parameters after $PGF_{2\alpha}$ administration.

$PGF_{2\alpha}$ increases cardiac output, an effect which may be related to enhanced venomotor tone and increased venous return (DuCharme et al., 1968). Hinman (1967) has suggested that $PGF_{2\alpha}$ protects renal function by increasing cardiac output in response to decreased renal perfusion.

The half-life of an intravenous dose of $PGF_{2\alpha}$ in man and other primates is only a few seconds (Granström, 1975). Recently the synthetic $PGF_{2\alpha}$ analog, 17-phenyl-18, 19, 20-tris $PGF_{2\alpha}$ has been reported to have a longer half life, five times the presso potency, and three times the uterus-contracting potency of $PGF_{2\alpha}$ (Granström, 1975; Miller et al., 1975).

5. Control of Prostaglandin Synthesis

Prostaglandins are not stored in the kidney (Änggård et al., 1972), but are synthesized on demand by cells of the renal medulla and immediately released into the extracellular compartment (McGiff and Itskovitz, 1973). A number of stimuli cause increased synthesis of prostaglandins by the kidney. These include clamping or narrowing of the renal artery, stimulation of the renal nerves, and administration of norepinephrine, epinephrine, angiotensin, vasopressin, and bradykinin (Zins, 1975). With the exception of bradykinin, all of these stimuli produce renal ischemia or renal vasoconstriction. Prostaglandins thus seem to protect the kidney against ischemia and modulate the action of vasoconstrictor substances.

The availability of arachidonic acid (Fig. 8) appears to be the rate-limiting step in the biosynthesis of prostaglandins (Samuelsson, 1970). The administration of an excess of arachidonic acid, which in itself is pharmacologically inactive (Dakhill and Vogt, 1962), increases intrarenal prostaglandin synthesis by approximately 50% (Larsson and Änggård, 1974).

Recent studies by Schramm and Carlson (1975) have demonstrated that an acute elevation in ureteral pressure lowers renal vascular resistance and inhibits renal vasoconstriction elicited by catecholamines or by stimulation of either the renal nerves or the "defense regions" of the central nervous system in cats. This effect of increased ureteral pressure is diminished following inhibition of prostaglandin synthesis by indomethacin and is therefore felt to be dependent, in part, on intrarenal prostaglandin synthesis (Fig. 10).

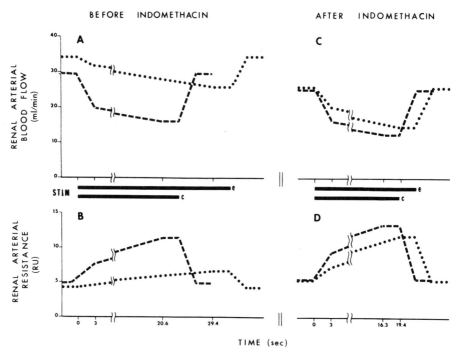

Fig. 10. Effect of elevated ureteral pressure on renal vasoconstrictions in cats, elicited by renal stimulation before and after indomethacin. Dashes represent responses during normal ureteral pressure; dots represent resonses during elevated ureteral pressure. STIM trace represents the period of stimulation during (c) control and (e) after elevation of ureteral pressure. Indomethacin treatment greatly reduces the renal vasodilatation (and refractoriness to vasoconstriction) caused by acutely elevated ureteral pressure. (From Schramm and Carlson, 1975, Fig. 3. Used by permission of the authors and the publisher.)

Prostaglandins are such potent substances that it is tempting to suggest that certain pathologic conditions are caused by a prostaglandin deficiency. However, most pathologic conditions which affect the kidney are attended by elevated prostaglandin levels. Increased renal vein PGE levels have been found in patients with renal artery stenosis (Edwards et al., 1969). Spontaneously hypertensive rats have higher concentrations of PGA in their plasma and kidneys than do normal rats (Zusman et al., 1973). Elevated PGE levels have been observed in experimental nephrotoxic acute renal failure (Torres et al., 1973). Prostaglandin levels are also increased in endotoxic and hemorrhagic shock (Collier et al., 1973; Selkurt, 1974).

The uterus and placenta produce substantial amounts of PGE which appears to be an important regulator of uteroplacental blood flow (Venuto et al., 1975). Speroff (1973) has suggested that toxemia of pregnancy results from defective prostaglandin production by the uteroplacental unit, but there is no evidence that the renal manifestations of toxemia are due to defective prostaglandin synthesis.

There is evidence that the increase in renal blood flow brought about by the diuretic ethacrynic acid is dependent on prostaglandin synthesis (Williamson et al., 1974).

6. Inhibitors of Prostaglandin Synthesis

The subject of prostaglandin synthesis inhibitors has been well covered in a recent symposium (Robinson and Vane, 1974). Five inhibitors in common use are meclofenamate, indomethacin, paracetamol, phenylbutazone, and aspirin. It is not clear whether inhibitors of prostaglandin synthesis prevent the formation of $PGG_{2\alpha}$ endoperoxide (Kuehl et al., 1974) or its subsequent conversion to PGE_2 and $PGF_{2\alpha}$ (Fig. 8). There is evidence that phenylbutazone interferes with the breakdown of the endoperoxide and may favor the formation of some prostaglandins over others (Flower et al., 1973). Paracetamol is only 6% as effective as aspirin in inhibiting prostaglandin synthesis in dog spleen (Flower et al., 1972) but is 4 times as effective as aspirin in inhibiting prostaglandin synthesis in rabbit kidney (Flower and Vane, 1974) suggesting that these two prostaglandin synthetase systems have different inhibitor specificities.

Beside inhibiting prostaglandin synthesis, indomethacin also inhibits prostaglandin-15-dehydrogenase, the major prostaglandin degrading enzyme (Hansen, 1974; Flower and Vane, 1974). This effect of indomethacin may provide one explanation for the increased plasma levels of prostaglandins which are found in the indomethacin-treated rat after the initial inhibitory effect of indomethaci on prostaglandin synthesis wears off (Fig. 5). It may also provide an explanation for the observation of Dr. J. B. Lee (1975, personal communication) that an alternate day regimen of indomethacin in human patients does not result in a decrease in plasma PGA levels.

Recent work by Rome and Lands (1975) suggests that indomethacin and aspirin are not competitive inhibitors of prostaglandin synthetase, but rather act by destroying (in a concentration-dependent, time-dependent fashion) the dioxygenase component of "prostaglandin synthetase" which is responsible for production of the PGG_2 endoperoxide intermediate (Fig. 8). It is possible that a compensatory increase in the rate of dioxygenase production after indomethacin treatment accounts for the "overshoot" effects described in the preceding paragraph.

D. THE KALLIKREIN-KININ SYSTEM

1. Biosynthesis of Kinins

The polypeptides bradykinin, lysyl-bradykinin (killidin), and methionyl-lysyl-bradykinin represent a group of compounds known collectively as kinins. The kinins represent the most potent mammalian vasodilator substances known. The complicated subject of kinin formation has been well discussed in a recent article by Colman (1974) and will be only briefly outlined here.

Kinin formation is initiated when factor XII (Hageman factor) of the coagulation system is converted to activated factor XII, a proteolytic enzyme. The fibrinolytic enzyme plasmin then acts on activated factor XIII, splitting it into proteolytic fragments of 30,000 to 70,000 molecular weight known as prekallikrein activators (Kaplan and Austen, 1971; Bagdasarian et al., 1973). These prekallikrein activators then attack prekallikrein, an inactive plasma proenzyme, converting it to the proteolytic and esterolytic enzyme kallikrein. Kallikrein then acts on kininogens, a family of α-globulin precursors, to form kinins. The kinins may then be degraded to inactive peptides by plasma peptidases, the kininases.

2. The Kallikrein-Kinin System and the Kidney

When the kinin-forming enzyme kallikrein was first detected in urine, it was assumed to originate from the pancreas (Frey, 1929; Kraut et al., 1930). However, it has subsequently been found that the saline-perfused kidney contains a kallikrein indistinguishable from that present in urine (Nustad, 1970a) and it seems likely that the kidney itself produces kallikrein. Urinary kallikrein releases lysyl-bradykinin from kininogen, while plasma kallikrein releases bradykinin (Webster and Pierce, 1963). In addition, urinary, plasma, and pancreatic kallikrein differ in electrophoretic mobility, susceptibility to proteolytic inhibitors, and absorption to DEAE-cellulose (Kellermeyer and Graham, 1968). It remains possible, however, that urinary kallikrein is formed from plasma kallikrein by the kidney and not synthesized by the kidney de novo.

Renal kallikrein activity is located primarily in the cortex (Nustad, 1970b). Using stop-flow technique, Scicli et al. (1975) have localized the site of kallikrein secretion to the distal tubule or a nearby structure such as the juxtaglomerular apparatus.

It has been suggested that renin and kallikrein may be located in the same subcellular particles (granules) in the mouse salivary gland (Chiang et al., 1968) and this may also hold true in the kidney (Carvalho and Diniz, 1963, 1964).

Infusion of bradykinin into the renal artery increases renal blood flow (Barac, 1957; Goldberg et al., 1965; Freed et al., 1968; McGiff et al., 1972). The effects of lysyl-bradykinin and methionyl-lysyl-bradykinin on the renal circulation have not been well studied, but it is likely that these compounds have similar effects since they are rapidly converted to bradykinin by tissue aminopeptidases (Prado et al., 1975).

McGiff et al. (1972) noted that the renal vasodilation brought about by infusion of bradykinin is accompanied by an 8-fold increase in renal vein PGE efflux. These investigators suggested that PGE participates in the renal vasodilator response to bradykinin. Vane and Ferreira (1975) have reported that bradykinin also causes prostaglandin release by the lung.

Attempts to delineate the physiologic role in kinins in the kidney have been frustrated by the lack of a specific kinin inhibitor. Grez (1974) has recently studied the influence of antibradykinin antisera on renal function. He observed that antibodies against bradykinin resulted in a decreased natriuretic and diuretic response to saline loading but had no effect on renal blood flow as determined by PAH clearance. These results suggest that bradykinin is not a determinant of resting renal blood flow.

3. Kininases

There are two plasma peptidases that degrade the kinins by removing amino acids from the carboxy terminal end (Erdös and Yang, 1970). The activity of these kinases is responsible for the extremely short half life of injected bradykinin in man (0.3-0.4 minutes). Kininase I is a carboxypeptidase which splits off the C-terminal arginine from bradykinin. Kininase II is a dipeptidyl carboxypeptidase, probably identical to angiotensin I converting enzyme, which cleaves the Pro^7-Phe^8 bond of bradykinin. Besides acting on bradykinin and angiotensin I, this enzyme also removes the two C-terminal peptides from the B chain of insulin and from a number of other polypeptides (Erdös, 1975). Kininase I and II constitute the major routes of inactivation of bradykinin. However, since splitting any bond in bradykinin inactivates the peptide, there are a wide range of enzymes which may participate in bradykinin inactivation.

E. CATECHOLAMINES

Catecholamines have been previously discussed in the section on the influence of the renal nerves. The present section will concern itself with the renal hemodynamic influences of the catecholamines acting as circulating hormones.

Epinephrine and norepinephrine, the two endogenous catecholamines which are present in detectable amounts in human plasma and urine, each exert a combination of alpha- and beta-adrenergic effects. Epinephrine is the major hormone produced by the adrenal medulla; norepinephrine accounts for only 10-18% of the catecholamine content of the human adrenal medulla (Innes and Nickerson, 1970).

Systemic administration of catecholamines brings about effects on blood pressure and cardiac output which tend to obscure any direct effects on the renal vasculature (Schrier, 1974). Thus, infusion of beta- and alpha-adrenergic agents directly into the renal artery has been necessary to delineate the renal circulatory effects of these compounds. The kidney is an effective site of catecholamine inactivation (Gryglewski and Vane, 1970), and thus intrarenal arterial infusions of catecholamine may not have systemic effects.

Norepinephrine, primarily an alpha adrenergic agonist, appears to produce vasoconstriction of several different sites in the kidney, including the afferent and efferent arterioles as well as venules (Schrier, 1974). Modest doses of norepinephrine reduce renal blood flow without reducing glomerular filtration rate; this phenomenon has been attributed to a simultaneous constriction of the afferent and efferent arterioles (Schrier and Berl, 1973).

Epinephrine, which is both an alpha- and a beta-adrenergic agonist, reduces renal blood flow (PAH clearance) in man less than norepinephrine, when given in comparable doses (Werko et al., 1951). This is consistent with the observation that the beta-adrenergic agonist, isoproterenol, increases renal blood flow (Mark et al., 1969).

In contrast to other vasoactive compounds such as prostaglandin E, acetylcholine, and bradykinin (Stein et al., 1971; McGiff and Itskovitz, 1973), norepinephrine does not appear to cause a redistribution of renal blood flow (Rector et al., 1972).

F. OTHER HORMONAL SUBSTANCES

1. Antidiuretic Hormone (ADH)

Total renal blood flow, especially the cortical component, increases after sectioning of the hypothalamic-hypophysial tract in dogs. This effect has been attributed to the lack of ADH (Fisher et al., 1970). Infusion of ADH into these dogs reduces renal blood flow to normal levels. These observations suggest that ADH may play an important role in the control of renal blood flow and in the renal circulatory responses to contraction and expansion of intravascular volume (Grantham, 1974).

2. Adrenal Steroids

In physiological doses, mineralocorticoids (e.g., aldosterone) do not appear to have a direct effect on renal hemodynamics (Hierholzer and Lange, 1974). Glucocorticoids, on the other hand, increase glomerular filtration rate and renal blood flow in man (Levitt and Bader, 1951). In rats, the increased blood flow associated with methylprednisolone treatment is associated with a redistribution of blood flow toward the juxtamedullary cortex (de Bermudez and Hayslett, 1972).

3. Estrogens

The estrogens seem to have little effect on renal hemodynamics. Neither bilateral ovariectomy nor administration of estradiol alters renal plasma flow in man (Smith, 1951; Dignam et al., 1956). In dogs, stilbesterol and, to a lesser extent, estradiol appear to increase renal plasma flow (Dance et al., 1959; Johnson et al., 1972). Estrogens also increase renin substrate levels, although the influence of this effect on the renal circulation is unknown (Christy and Shaver, 1974).

4. Thyroid Hormones

Renal blood flow is increased in patients with hyperthyroidism and reduced in patients with hypothyroidism (Bradley et al., 1974). It has not been demonstrated that these changes in renal hemodynamics represent a direct effect of thyroid hormones on the renal vasculature.

G. MISCELLANEOUS HORMONAL INTERACTIONS

The hormonal systems discussed in the previous sections are interrelated in many ways. This section surveys some of the hormonal interactions which may play a role in the control of the renal circulation. Most of these interactions have not been studied in sufficient detail to evaluate their possible physiologic significance. In almost all of the studies the hormones have been used in concentrations much higher than those which might conceivably occur naturally.

1. Pressor Hormones and Prostaglandin Release

A prostaglandin E-like substance has been shown to be released by the dog kidney in response to an intrarenal infusion of angiotensin II or norepinephrine (McGiff et al., 1970, 1972). While it is possible that prostaglandin release is caused by the hemodynamic changes, these pressor hormones may exert a direct influence on cells which produce prostaglandins. Limas (1974) demonstrated that angiotensin caused a marked stimulation of prostaglandin synthesis in rat heart slices. Norepinephrine had a similar effect. Omission of tyrosine, a catecholamine precursor, from the medium abolished the prostaglandin stimulating effect of angiotensin. Depletion of catecholamines using 6-hydroxydopamine reduced angiotensin-induced stimulation of prostaglandin synthesis. Limas (1974) interprets these results as an indication that angiotensin stimulates prostaglandin biosynthesis by enhancing the release of newly synthesized norepinephrine. Gimbrone and Alexander (1975) have recently demonstrated that angiotensin II stimulates prostaglandin E production by human vascular endothelium grown in tissue culture. It is likely that similar interactions take place in renal medullary cells which produce prostaglandins (Danon and Chang, 1973).

2. Angiotensin, Bradykinin, and Catecholamines

The release of adrenal catecholamines by angiotensin was first demonstrated by Braun-Menendez et al., (1940). More recently bradykinin, as well as angiotensin, has been shown to cause release of epinephrine by the adrenal medulla (Feldberg and Lewis, 1964).

3. Prostaglandins and Catecholamine Release

Prostaglandins have been shown to potentiate the stimulatory effects of epinephrine, norepinephrine, and serotonin on isolated seminal vesicles (Eliasson and Risley, 1967). On the other hand, PGE inhibits norepinephrine release by adrenergic nerve endings (Stjärne, 1973) and it has been suggested that prostaglandins may act as a "brake" to the release of catecholamines in a variety of tissues (Hedquist et al., 1971; Samuelsson and Wennmalm, 1971).

4. Prostaglandins and the Adrenergic Nervous System

Brody and Kadowitz (1974) have suggested that prostaglandins act as modulators of the autonomic nervous system with PGE inhibiting autonomic transmission and PGF facilitating transmission. Hedquist (1970) has shown that stimulation of the adrenergic nerves of the heart and the spleen causes release of PGE by these organs. However, McGiff and Itskovitz (1973) have reported that stimulation of the renal nerves to an extent which causes a moderate reduction in renal blood flow does not cause the release of PGE by the kidney. Under conditions of increased renal perfusion pressure, stimulation of the renal nerves does cause prostaglandin release (Dunham and Zimmerman, 1970).

5. Prostaglandins and Inhibitions of the Renin-Angiotensin System

PGA in very high concentrations acts as a competitive inhibitor of renin (Kotchen, et al., 1974). The renin inhibitor described by Smeby and Bumpus (1971) contains substantial amounts of arachidonic acid, suggesting that it may be a prostaglandin precursor. In the rabbit, arachidonic acid has been shown to increase plasma renin activity, whereas indomethacin, an inhibitor of prostaglandin synthetase, decreases plasma renin activity (Larsson et al., 1974).

6. Catecholamines and Renin Release

Michelakis et al. (1969) have demonstrated that epinephrine, norepinephrine, and cyclic AMP stimulate renin production by dog renal cortical cell suspension. Beck et al. (1975) have recently shown that isoproterenol increases plasma renin activity even after inhibition of catecholamine dependent cyclic AMP generation.

7. Insulin and Angiotensin I Converting Enzyme

Bing et al. (1974) have shown that insulin and its peptide fragments are effective inhibitors of angiotensin I converting enzyme at concentrations of 10^{-4} to 10^{-5} molar. None of the insulin fragments approach the potency of the nonpeptide inhibitor SQ 20, 881 which is active at a concentration of 10^{-9} molar.

8. Bradykinin and Prostaglandins

Bradykinin and PGB_2, $PGF_{2\alpha}$, PGE_2, and PGE have synergistic effects on the isolated rat uterus (Denko et al., 1972). As mentioned previously, bradykinin infusion appears to stimulate production of PGE in the dog kidney (McGiff et al., 1972).

9. Bradykinin, PGE_2, and ADH Release

Intracarotid injections of bradykinin and PGE_2 have been shown to cause ADH release in the dog (Rocha e Silva, 1970; Stokes and Bartter, 1974).

H. CONCLUDING COMMENTS

In this chapter we have discussed the major known hormones which are believed to influence the renal circulation. In so doing we have probably omitted many substances which will ultimately be shown to have important effects on the vasculature of the kidney. Such ill-defined compounds as "antihypertensive neutral renomedullary lipids" (Muirhead et al., 1971), "rabbit aorta contracting substance" (Piper and Vane, 1969), "angionecrosis producing substance" (Nakamura et al., 1975), "nephrotensin" (Grollman and Krishnamurty, 1971), and "substance R" (Nishiyama et al., 1972), when better characterized, may prove to be of great importance. The recent finding by Burzynski et al. (1974) of 90 different vasoactive peptides in human urine emphasizes the fact that the hormones discussed in this chapter may represent only a small fraction of the hormonal substances which effect the renal circulation.

REFERENCES

Altman, P.L., and Dittmer, D.S. (eds.), 1962, "Growth Including Reproduction and Morphological Development," Federation of American Societies Experimental Biology, Washington, D.C., p. 365.

Andreucci, V.E., Dal Canton, A., and Corradi, A., 1975, The role of the efferent arteriole in cortical renal hemodynamics: A micropuncture study, Kidney Int. 7:187.

Änggård, E., and Bergström, S., 1963, Biologic effects of an unsaturated tryhydroxy acid ($PGF_{2\alpha}$) from normal swine lung, Acta Physiol. Scand. 58:1.

Änggård, E., Bohman, S.O., Griffin, J.E. III, Larsson, C., and Maunsbach, A.B., 1972, Subcellular localization of the prostaglandin system in the rabbit renal papilla, Acta Physiol. Scand. 84:231.

Arendshorst, W.J., Finn, W.F., and Gottschalk, C.W., 1975, Autoregulation of blood flow in the rat kidney, Am. J. Physiol. 228:127.

Attallah, A.A., and Lee, J.B., 1973, Specific binding sites in the rabbit kidney for prostaglandin A, Prostaglandins 4:703.

Attallah, A.A., Payakkapan, W., Lee, J.B., Carr, A., and Brazelton, E., 1974, PGA: Fact, not artifact, Prostaglandins 5:69.

Aukland, K., 1967, Study of renal circulation with inert gas: Measurements in tissue, in "Proceedings of the Third International Congress of Nephrology" (Washington, D.C., 1966) Vol. I: Physiology (G.E. Schreiner, ed.) pp. 188-200, S. Karger, Basel.

Baehler, R.W., Catanzaro, A.J., Stein, J.H., and Hunter, W., 1973, The radiolabeled frog red blood cell: A new marker of cortical blood flow distribution in the kidney of the dog, Circ. Res. 32:718.

Bagdasarian, A., Lahiri, B., and Colman, R.W., 1973, Origin of the high molecular weight activator of prekallikrein, J. Biol. Chem. 248:7742.

Bakhle, Y.S., 1974, Converting enzyme in vitro measurement and properties, in "Handbook of Experimental Pharmacology, Vol. 37, Angiotensin" (I.H. Page and F.M. Bumpus, eds.), pp. 41-80, Springer-Verlag, New York.

Bankir, L., Farman, N., Grünfeld, J.-P., Huet de la Tour, E., and Funck-Brentano, J.-L., 1973, Radioactive microsphere distribution and single glomerular blood flow in the normal rabbit kidney, Pflug. Arch. 342:111.

Barac, G., 1957, Effet renal de la bradykinine chez le Chien, C.R. Soc. Biol. (Paris) 151:1771.
Barajas, L., and Latta, H., 1967, Structure of the juxtaglomerular apparatus, Circ. Res. 20/21 (Suppl. II):15.
Barer, G.R., 1963, The action of vasopressin, a vasopressin analog (PLV_2), oxytocin, angiotensin, brakykinin and theophyline ethylene diamine on renal blood flood in the anesthetized cat, J. Physiol. 169:62.
Barger, A.C., 1966, Renal hemodynamic factors in congestive heart failure, Ann. N.Y. Acad. Sci. 139:276.
Barger, A.C., and Herd, J.A., 1973, Renal vascular anatomy and distribution of blood flow, in "Handbook of Physiology, Section 8, Renal Physiology," (J. Orloff and R.W. Berliner, eds.) pp. 249-313, Amer. Physiol. Soc., Washington.
Beck, N., Kim, K.S., and Davis, B.B., 1975, Catecholamine-dependent cyclic adenosine monophosphate and renin in the dog kidney, Circ. Res. 36:401.
Bell, E.T., 1950, "Renal Diseases," 2nd Edition, Lea and Febiger, Philadelphia, p. 337.
Bell, R.D., Keyl, M.J., Shrader, F.R., Jones, E., and Henry, L.P., 1968, Renal lymphatics: The internal distribution, Nephron 5:454.
Bell, R.D., Parry, W.L., and Grundy, W.G., 1973, Renal lymph sodium and potassium concentrations following renal vasodilation, Proc. Soc. Exp. Biol. Med. 143:499.
Bell, R.D., Sinclair, R.J., and Parry, W.L., 1975, The effects of indomethacin on autoregulation and the renal response to hemorrhage, Circ. Shock 2:57.
Belleau, L.J., and Earley, L.E., 1967, Autoregulation of renal blood flow in the presence of angiotensin infusion, Amer. J. Physiol. 213:1590.
Bergström, S., and Samuelsson, B., 1965, Prostaglandins, Ann. Rev. Biochem. 34:101.
Bergström, S., and Sjövall, J., 1960a, The isolation of prostaglandin F from sheep prostate glands, Acta Chem. Scand. 14:1693.
Bergström, S., and Sjövall, J., 1960b, The isolation of prostaglandin E from sheep prostate glands, Acta. Chem. Scand. 14:1701.
Berkowitz, H.D., Galvin, C.C., and Miller, L.D., 1974, The control of renal cortical perfusion by the renin angiotensin system, Ann. Surg. 179:238.
Bianchi, A., DeSchaepdryver, A.F., DeVleeschhouver, G.R., and Preziosi, P., 1960, On the pharmacology of synthetic hypertensine, Arch. Int. Pharmacodyn. 124:21.
Bing, J., Poulson, K., and Markussen, J., 1974, The ability of various insulins and insulin fragments to inhibit the angiotensin I converting enzyme, Acta Path. Microbiol. Scand., Sect. A. 82:777.

Biron, P., Campeau, L., and David, P., 1969, Fate of angiotensin I and II in the human pulmonary circulation, Amer. J. Cardiol. 24:544.

Biron, P., Meyer, P., and Panisset, V.C., 1968, Removal of angiotensins from the systemic circulation, Canad. J. Physiol. Pharmacol. 46:175.

Bock, K.D., Dengler, H., Krecke, H.J., and Reichel, G., 1958, Untersuchurgen über die Wirkung von synthetischem Hypertensin II auf Elektrolythaushalt, Nierenfunktion und Kreislauf beim Menschen, Klin. Woch. 36:808.

Bohr, D.F., and Uchida, E., 1967, Individualities of vascular smooth muscle in response to angiotensin, Circ. Res. 21 (Suppl. II):135.

Boonjarern, S., Wertenfelder, C., Arruda, J.A.L., Lockwood, R., and Kurtzman, N.A., 1974, Distribution of renal cortical blood flow (CBF) in congestive heart failure (CHF) with radioactive microspheres, (MS), Kidney Int. 6:26A.

Boucher, R., Asselin, J., and Genest, J., 1974, A new enzyme leading to the direct formation of angiotensin II, Circ. Res. 34 (Suppl. I):203.

Boucher, R., Saidi, M., and Genest, J., 1972, A new "angiotensin I converting enzyme" system in "Hypertension 72" (J. Genest and E. Koiw, eds.) pp. 512-523, Springer-Verlag, New York.

Boyd, G.W., Landon, J., and Peart, W.S., 1969, The radioimmunoassay of angiotensin II, Proc. Roy. Soc. B. 173:327.

Bradley, S.E., Stéphan, F., Coelho, J.B., and Réville, P., 1974, The thyroid and the kidney, Kidney Int. 6:346.

Bras, G., 1969, Age-associated kidney lesions in the rat, J. Infect. Dis. 120:131.

Braun-Menendez, E., Fasciolo, J.C., Leloir, L.F., Munoz, V.M., and Taquini, A.C., 1946, "Renal Hypertension," Charles C. Thomas, Springfield, pp. 134-135.

Braun-Menendez, E., Fasciolo, J.C., Leloir, L.F., and Munoz, J.M., 1940, The substance causing renal hypertension, J. Physiol. 98:283.

Brenner, B.M., Deen, W.M., and Robertson, C.R., 1974, The physiological basis of glomerular ultrafiltration, in "MTP International Review of Science, Physiology Series 1, Vol. 6, Kidney and Urinary Tract Physiology," (K. Thurau, ed.), pp. 335-356, Butterworths, London.

Brody, M.J., and Kadowitz, P.J., 1974, Prostaglandins as modulators of the autonomic nervous system, Fed. Proc. 33:48.

Brown, J.J., Davies, D.L., Lever, A.F., Parker, R., and Robertson, J.I.S., 1965, The assay of renin in single glomeruli in the normal rabbit and the appearance of the juxtaglomerular apparatus, J. Physiol. 176:418.

Brunner, H.R., Baer, L., and Sealey, J.E., Ledingham, J.G.G., and Laragh, J.H., 1970, The influence of potassium administration and of potassium deprivation on plasma renin in normal and hypertensive subjects, J. Clin. Invest. 49:2128.

Bruns, F.J., Alexander, E.A., Riley, A.L., and Levinsky, N.G., 1974, Superficial and juxtamedullary nephron function during saline loading in the dog, J. Clin. Invest. 53:971.

Burton-Optiz, R., and Lucas, D.R., 1911, The blood supply of the kidney. V. The influence of the vagus nerve on the vascularity of the left organ, J. Exp. Med. 13:308.

Burzynski, S.R., Ungar, A.L., and Lubanski, E., 1974, Biologically active peptides in human urine. II. Effect on intestinal smooth muscle and heart, Physiol. Chem. Physics 6:457.

Byers, B.D., Deen, W.M., Robertson, C.R., and Brenner, B.M., 1975, Dynamics of glomerular ultrafiltration in the rat. VIII. Effects of hematocrit, Circ. Res. 36:425.

Cain, M.D., Catt, K.J., Coghlan, J.P., and Blair-West, J.R., 1970, Evaluation of angiotensin II metabolism in sheep by radio-immunoassay, Endocrinology 86:955.

Carvalho, I.F., and Diniz, C.R., 1963, Kininogenin in rat kidney tissue, Ciencia Cult. (S. Paulo) 15:286.

Carvalho, I.F., and Diniz, C.R., 1964, Cellular localization of renin and kininogenin, Ciencia Cult. (S. Paulo) 16:263.

Caton, M.P.L., 1973, Chemical structure and availability, in "The Prostaglandins" (M.F. Cuthbert, ed.) pp. 1-22, Heinemann, London.

Chang, L.C.T., Splawinski, J.A., Oates, J.A., and Nies, A.S., 1975, Enhanced renal prostaglandin production in the dog. II. Effects on intrarenal hemodynamics, Circ. Res. 36:204.

Chiang, T.S., Erdös, E.G., Miwa, I., Tague, L.L., and Coalson, J.J., 1968, Isolation from a salivary gland of granules containing renin and kallikrein, Circ. Res. 23:507.

Christy, N.P., and Shaver, J.C., 1974, Estrogens and the kidney, Kidney Int. 6:366.

Collier, J.G., Herman, A.G., and Vane, J.R., 1973, Appearance of prostaglandins in the renal venous blood of dogs in response to acute systemic hypotension produced by bleeding or endotoxin, J. Physiol. 230:19P.

Colman, R.W., 1974, Formation of human plasma kinin, New Engl. J. Med. 291:509.

Cook, W.F., 1971, Cellular localization of renins in "Kidney Hormones" (J.W. Fisher, ed.), Academic Press, New York.

Cook, W.F., Gordon, D.B., and Peart, W.S., 1956, The location of renin in the rabbit kidney, J. Physiol. 135:46P.

Cooke, C.R., Brown, T.C., Zacherle, B.J., and Walker, W.G., 1970, The effect of altered sodium concentration in the distal nephron segments on renin release, J. Clin. Invest. 49:1630.

Crowshaw, K., 1971, Prostaglandin biosynthesis from endogenous precursors in rabbit kidney, Nature, New Biol. 231:240.

Crowshaw, K., 1973, The incorporation of $[1-^{14}C]$ arachidonic acid into the lipids of rabbit renal slices and conversion to prostaglandins E_2 and $F_{2\alpha}$, Prostaglandins 3:607.

Dakhill, T., and Vogt, W., 1962, Hydroperoxyde als Thäges der darmerregenden Wirkung hochungesättigen Fettsäuren, Arch. Exp. Pathol. Pharmakol. 243:174.

Dance, P., Lloyd, S., and Pickford, M., 1959, The effects of stilbestrol on the renal activity of conscious dogs, J. Physiol. 145:225.

Danon, A., and Chang, L.C.T., 1973, Release of prostaglandins from rat renal papilla in vitro: Effects of arachidonic acid and angiotensin II, Fed. Proc. 32:788.

Darmady, E.M., 1974, Transplantation and the aging kidney, Lancet 2:1046.

Darmady, E.M., Offer, J., and Woodhouse, M.A., 1973, The parameters of the aging kidney, J. Pathol. 109:195.

Davis, J.M., and Morgan, T.O., 1975, Renin release from the juxtaglomerular apparatus of the rat, Abstracts of Free Communications, VIth International Congress of Nephrology, June 8-12, 1975, Florence, Italy (Abstract 550).

Davis, J.O., 1973, The control of renin release, Am. J. Med. 55:333.

deBermudez, L., and Hayslett, J.P., 1972, Effect of methylprednisolone on renal function and zonal distribution of blood flow in the rat, Circ. Res. 31:44.

De Bono, E., Lee, G. deJ., Mottram, F.R., Pickering, G.W., Brown, J.J., Keen, H., Peart, W.S., and Sanderson, P.H., 1963, The action of angiotensin in man, Clin. Sci. 25:123.

Denko, C.W., Moskowitz, R.W., and Heinrich, G., 1972, Interrelated pharmacologic effects of prostaglandins and bradykinin, Pharmacology 8:353.

DiBona, G.F., 1971, Effect of mannitol diuresis and ureteral occlusion on distal tubular reabsorption, Amer. J. Physiol. 221:511.

Dignam, W.S., Voskian, J., and Assali, N.S., 1956, Effects of estrogens on renal hemodynamics and excretion of electrolytes in human subjects, J. Clin. Endocrinol. 16:1032.

DiSalvo, J., and Fell, C., 1970, Effects of angiotensin on canine renal blood flow, Proc. Soc. Exp. Biol. Med. 133:1432.

DiSalvo, J., Peterson, A., Montefusco, C., and Menta, M., 1971, Intrarenal conversion of angiotensin I to angiotensin II in the dog, Circ. Res. 29:398.

Doyle, A.E., Louis, W.J., Jerums, G., and Osborn, E.C., 1968, Metabolism and blood levels following infusion of a radioactive analog of angiotensin, Amer. J. Physiol. 215:164.

DuCharme, D.W., Weeks, J.R., and Montgomery, R.G., 1968, Studies on the mechanism of the hypertensive effect of prostaglandin $F_{2\alpha}$, J. Pharmacol. Exp. Ther. 160:1

Dunham, E.W., and Zimmerman, B.G., 1970, Release of prostaglandin-like material from dog kidney during nerve stimulation, Amer. J. Physiol. 219:1279.

Earley, L.E., and Friedler, R.M., 1966, The effects of combined renal vasodilation and pressor agents on renal hemodynamics and the tubular reabsorption of sodium, J. Clin. Invest. 45:542.

Edwards, W.G., Strong, C.G., and Hunt, J.C., 1969, A vasodepressor lipid resembling prostaglandin E_2 (PGE_2) in the renal venous blood of hypertensive patients, J. Lab. Clin. Med. 74:389.

Eide, I., Löyning, E., Aars, H., and Ikre, S., 1973, Autoregulation of renal blood flow in rabbits immunized with angiotensin II, Scand. J. Clin. Lab. Invest. 31:123.

Eliasson, R., and Risley, P.S., 1967, Potentiated response of isolated seminal vesicles to catecholamines and acetylcholine in the presence of prostaglandins, in "Prostaglandins, Nobel Symposium, Vol. 2, (S. Bergstrom and B. Samuelsson, eds.) pp. 85-90, Interscience, New York.

Eliseiva, Y.E., Orekhovich, V.N., Pavlikhina, L.V., and Aleseenko, L.P., 1971, Carboxycathepsin - a key regulatory component of two physiological systems involved in the regulation of blood pressure, Clin. Chim. Acta 31:413.

Erdös, E.G., 1974, Discussion, Circ. Res. 34 (Suppl. I):210.

Erdös, E.G., 1975, Angiotensin I converting enzyme, Circ. Res. 36:247.

Erdös, E.G., and Yang, H.Y.T., 1970, Kinases, in "Handbook of Experimental Pharmacology, Vol. 25, Bradykinin, kallidin, and kallikrein" (E.G. Erdös, ed). pp. 289-323, Springer-Verlag, New York.

Eskildsen, P.C., 1973, Renin in different tissues, amniotic fluid and plasma of pregnant and non-pregnant rabbits, Acta. Pathol. Microbiol. Scand. Section A, 81:263.

Faarup, P., 1971, Morphological aspects of the renin-angiotensin system, Acta Pathol. Microbiol. Scand. (A) Suppl. 222:1.

Fasciolo, J.C., Leloir, L.F., Moñoz, J.M., and Braun-Menendez, E., 1940, On the specificity of renin, Science 92:554.

Feldberg, M., and Lewis, G.P., 1964, The action of peptides on the adrenal medulla. Release of adrenaline by bradykinin and angiotensin, J. Physiol. 171:98.

Ferris, T.F., Gorden, P., and Mulrow, P.J., 1967, Rabbit uterus as a source of renin, Am. J. Physiol. 212:698.

Ferris, T.F., Stein, J.H., and Kauffman, J., 1972, Uterine blood flow and uterine renin secretion, J. Clin. Invest. 51:2827.

Fichman, M.P., Littenberg, G., Brooker, G., Brooker, G., and Horton, R., 1972, Effect of prostaglandin A_1 on renal and adrenal function in man, Circ. Res. 31 (Suppl. II):19.

Fisher, R.D., Grünfeld, J.-P., and Barger, A.C., 1970, Intrarenal distribution of blood flow in diabetes insipidus: role of ADH, Am. J. Physiol. 219:1348.

Fitz, A., Boaz, D., and Wyatt, S., 1974, Studies of human lung angiotensin I converting enzyme, Circulation, 50 (Suppl. III):30.

Flamenbaum, W., and Hamburger, R.J., 1974, Juxtaglomerular apparatus renin activity: Role of the renin-angiotensin system in acute renal failure, Circulation, 50, (Suppl. III):134.

Flower, R.J., 1974, Discussion, in "Prostaglandin Synthetase Inhibitors," (H.J. Robinson and J.R. Vane, eds.) p. 376, Raven Press, New York.

Flower, R.J., Cheung, H.S., and Cushman, D.W., 1973, Quantitative determination of prostaglandins and malondialdehyde formed by the arachidonate oxygenase (prostaglandin synthetase) system of bovine seminal vesicles, Prostaglandins 4:325.

Flower, R.J., Gryglewski, R., Herbaczynska-Cedro, K., and Vane, J.R., 1972, Effects of anti-inflammatory drugs on prostaglandin biosynthesis, Nature, New Biol. 238:104.

Flower, R.J., and Vane, J.R., 1974, Commentary - Inhibition of prostaglandin biosynthesis, Biochem. Pharmacol. 23:1439.

Forster, R.P., and Mess, J.P., 1947, Effects of experimental neurogenic hypertension on renal blood flow and glomerular filtration rates in intact denervated kidneys of unanesthetized rabbits with adrenal glands demedullated, Am. J. Physiol. 150:534.

Fourcade, J.C., Navar, L.G., and Guyton, A.C., 1971, Possibility that angiotensin resulting from unilateral kidney disease affects contralateral renal function, Nephron 8:11.

Fourman, J., and Moffat, D.B., 1971, "The Blood Vessels of the kidney," Blackwell Scientific Publications, Oxford.

Freed, T.A., Neal, M.P. Jr., and Vinik, M., 1968, The effect of bradykinin on renal arteriography, Amer. J. Roentgenol. 102:776.

Frey, E.K., 1929, Kreislaufhormon und innere Sekretion, Munch. med. Wschr. 76:1951.

Frolich, V.C., Sweetman, B.J., Carr, K., Splawinski, J., Watson, V.T., Änggård, E., Oates, V.A., 1973, Occurrence of prostaglandins in human urine, Adv. Biosci. 9:321.

Fülgraff, G., Brandenbusch, G., and Heintze, K., 1974, Dose response relation of the renal effects of PGA_1 PGE_2, and $PGF_{2\alpha}$ in dogs, Prostaglandins 8:21.

Furuayama, T., Suzuki, C., Shioji, R., Saito, H., Maebashi, M., and Aida, M., 1967, Effect of angiotensin on renal function of the dog, Tohoku J. Exp. Med. 92:73.

Gagnon, J.A., Keller, H.I., Kokotis, W., and Schrier, R.W., 1970, Analysis of role of renin-angiotensin system in autoregulation of glomerular filtration, Amer. J. Physiol. 219:491.

Gagnon, J.A., Rice, M.K., and Flamenbaum, W., 1974, Effect of angiotensin converting enzyme inhibition on renal autoregulation, Proc. Soc. Exp. Biol. Med. 146:414.

Ganguli, M.D., Tobian, L., 1972, Renal medullary plasma flow in hypertension, Fed. Proc. 31:394, 1972.

Genest, J., 1974, Renin and aldosterone in benign essential hypertension, Circulation, 50 (Suppl. III):1.

Germuth, F.G., and Rodriquez, E., 1973, "Immunopathology of the Renal Glomerulus," Little, Brown and Co., Boston.

Gimbrone, M.A., and Alexander, R.W., 1975, Angiotensin II stimulation of prostaglandin production in cultured human vascular endothelium, Science, 189:219, 1975.

Goldberg, L.I., Dollery, C.T., and Pentecost, B.L., 1965, Effects of intrarenal infusions of bradykinin and acetylcholine on renal blood flow in man, J. Clin. Invest. 44:1052.

Goormaghtigh, N., 1932, Les segments neuro-myarteriels juxtaglomérulaires du rein, Arch. Biol. (Liege) 43:575.

Goormaghtigh, N., 1939, Existence of an endocrine gland in the media of the renal arterioles, Proc. Soc. Exp. Biol. Med. 42:688.

Gould, A.B., Skeggs, L.T. Jr., and Kahn, J.R., 1964, The presence of renin activity in blood vessel walls, J. Exp. Med. 119:389.

Granger, P., Dahlheim, H., and Thurau, K., 1972, Enzyme activities of the single juxtaglomerular apparatus in the rat kidney, Kidney Int. 1:78.

Grängsjö, G., and Wolgast, M., 1972, The pressure-flow relationship in renal cortical and medullary circulation, Acta Physiol. Scand. 82:228.

Granström, E., 1975, Metabolism of 17-phenyl-18, 19, 20-trinor-prostaglandin $F_{2\alpha}$ in the cynomolgus monkey and the human female, Prostaglandins 9:19.

Grantham, J.J., 1974, Action of antidiuretic hormones in the mammalian kidney, in "MTP International Review of Science, Physiology Series One, Vol. 6, Kidney and Urinary Tract Physiology," (A.C. Guyton and K. Thurau, eds) pp. 247-272.

Greenberg, S., Wilson, W.R., and Howard, L., 1974, Mechanism of the vasosconstrictor action of prostaglandin B, J. Pharmacol. Exp. Ther. 190:59.

Grez, M.M., 1974, The influence of antibodies against bradykinin on isotonic saline diuresis in the rat, Pflug. Arch. 350:231.

Grollman, A., and Krishnamusty, V.S.R., 1971, A new pressor agent of renal origin: Its differentiation from renin and angiotensin, Amer. J. Physiol. 221:1499.

Gross, F., 1971, Renin stores in the kidney and plasma renin activity, in "Kidney Hormones" (J.W. Fisher, ed.) pp. 93-116, Academic Press, New York.

Grünfeld, J.P., Raphael, J.E., Bunkir, L., Kleinknecht, D., and Barger, A.C., 1971, Intrarenal distribution of blood flow, in "Advances in Nephrology," Vol. 1, (J. Hamburger, J. Crosnier, and M.H. Maxwell, eds.) Yearbook Medical Publishers, Chicago, pp. 125-143.

Gryglewski, R., and Vane, J.R., 1970, The inactivation of nonadrenaline and isoprenaline in dogs, Br. J. Pharmacol. 39:573.

Guyton, A.C., Langston, J.B., and Navar, B., 1964, Theory for renal autoregulation by feedback at the juxtaglomerular apparatus, Circ. Res. 15 (Suppl. I):187.

Hall, R.E., and Buckalew, V.M., 1975, Effect of hypothalamic stimulation on renal blood flow and sodium excretion, Clin. Res. 23:363A.

Hamberg, M., 1969, Biosynthesis of prostaglandin in the renal medulla of the rabbit, FEBS Lett. 5:127.

Hamberg, M., and Samuelsson, B., 1966, Prostoglandins in human seminal plasma, J. Biol. Chem., 241:257.

Hansen, H.S., 1974, Inhibition by indomethacin and aspirin of 15-hydroxy-prostaglandin dehydrogenase in vitro, Prostaglandins 8:95.

Hársing, L., Kállay, K., Bartha, J., and Debreczeni, L., 1969, Effect of hyperoncotic albumin and isotonic saline infusion on renal medullary blood flow, Acta Physiol. Acad. Sci. Hung. 36:227.

Hartroft, P.M., 1963, Juxtaglomerular cells, Circ. Res. 12:525.

Hayduk, K., Krause, D.K., Kaufmann, W., Huenges, R., Schillmoller, U., and Unbehaun, V., 1973, Age dependent changes of plasma renin concentration in humans, Clin. Sci. Mol. Med. 45:273s.

Hays, R.M., and Levine, S.D., 1974, Vasopressin, Kidney Int. 6:307.

Hedquist, P., 1970, Studies on the effect of prostaglandin E_1 and E_2 on the sympathetic neuromuscular transmission in some animal tissues, Acta Physiol. Scand. (Suppl.) 345:1.

Hedquist, P., Stjärne, L., and Wemmalm, A., 1971, Facilitation of sympathetic neurotransmission in the cat spleen after inhibition of prostaglandin synthesis, Acta Physiol. Scand. 83:430.

Hensby, C.N., 1974, Reduction of prostaglandin E_2 to prostaglandin $F_{2\alpha}$ by an enzyme in sheep blood, Biochim. Biophys. Acta 348:1

Herbaczynska-Cedro, K., and Vane, J.R., 1973, Contribution of intrarenal generation of prostaglandin in autoregulation of renal blood flow in the dog, Circ. Res. 33:428.

Herbaczynska-Cedro, K., and Vane, J.R., 1974, Prostaglandins as mediators or reactive hyperemia in kidney, Nature 247:492.

Herrick, J.F., Corcoran, A.C., and Essex, H.E., 1942, The effects of renin and of angiotensin on the renal blood flow and blood pressure of the dog, Am. J. Physiol. 135:88.

Hickler, R.B., Lauler, D.P., Saravis, C.A., Vugnucci, A.I., Steiner, G., and Thorn, G.W., 1964, Vasodepressor lipid from the renal medulla, Can. Med. Assoc. J. 90:280.

Hierholzer, K., and Lange, S., 1974, The effects of adrenal steroids on renal function, in "MTP International Review of Science, Physiology Series One, Vol. 6, Kidney and Urinary Tract Physiology" (A.C. Guyton and K. Thurau, eds.) pp. 273-333, Butterworths, London.

Hill, J.R., and Pickering, G.W., 1940, Hypertension produced in the rabbit by prolonged renin infusion, Clin. Sci. 4:207.

Hinman, J.W., 1967, The prostaglandins, Bioscience 17:779.

Hiwada, K., Kokubu, T., Yamamura, Y., 1969, Inhibitory effect of rabbit bile on renin angiotensin reaction system, Jap. Circ. J. 33:1231.

Hodge, R.L., Ng, K.K.F., and Vane, J.R., 1967, Disappearance of angiotensin from the circulation of the dog, Nature 215:138.

Hollenberg, N.K., Birtch, A., Rashid, A., Mangel, R., Briggs, W., Epstein, M., Murray, J.E., and Merrill, J.P., 1970, "Relationships between intrarenal perfusion and function: Serial hemodynamic studies in the transplanted human kidney," Medicine 51:95.

Hollenberg, N.K., 1973, Renal disease, in "The Microcirculation in Clinical Medicine" (R. Wells, ed.) pp. 61-80, Academic Press, New York.

Hollenberg, N.K., and Adams, D.F., 1974, Vascular factors in the pathogenesis of acute renal failure in man, in "Proceedings: Renal Failure Conference, 1973" (E.A. Friedman and H.E. Eliahou, eds.) DHEW Publ. No. (NIH) 74-603, pp. 209-229.

Hollenberg, N.K., Adams, D.F., Solomon, H., Clenitz, W.R., Burger, B.M., Abrams, H.L., and Merrill, J.P., 1975 "Renal vascular tone in essential and secondary hypertension: Hemodynamic and angiographic responses to vasodilators, Medicine 54:29.

Horton, E.W., and Main, I.H.M., 1963, A comparison of the biological activities of four prostaglandins, Brit. J. Pharmacol. 21:182.

Innes, I.R., and Nickerson, M., 1970, Drugs acting on postganglionic adrenergic nerve ending and structures innervated by them (sympathomimetic drugs), in "The Pharmacological Basis of Therapeutics" (L.S. Goodman and A. Gilman, eds.) pp. 478-523, MacMillan Co., New York.

Itskovitz, H.D., and McGiff, J.C., 1974, Hormonal regulation of the renal circulation, Circ. Res. 34 (Suppl. I):65.

Itskovitz, H.D., Stempler, J., Pacholczyk, D., and McGiff, J.C., 1973, Renal prostaglandins: Determinants of intrarenal distribution of blood flow in the dog, Clin. Sci. Mol. Med. 45 (Suppl. I):321.

Itskovitz, H.D., Terragno, N.A., and McGiff, J.C., 1974, Effect of a renal prostaglandin on distribution of blood flow in the isolated canine kidney, Circ. Res. 34:770.

Jaenike, J.R., 1972, The renal functional defect of postobstructive nephropathy: The effects of bilateral ureteral obstruction in the rat, J. Clin. Invest. 51:2999.

Jaffe, B.M., Behrman, H.R., and Parker, C.W., 1973, Radioimmunoassay measurement of prostaglandins E, A, and F in human plasma, J. Clin. Invest. 52:398.

Janszen, F.H.A., and Nugteren, D.H., 1971, Histochemical localization of prostaglandin synthetase, Histochemie 27:159.

Johnson, J.A., Davis, J.O., and Witty, R.T., 1971, Effects of catecholamines and renal nerve stimulation on renin release in the nonfiltering kidney, Circ. Res. 29:646.

Johnson, J.A., Davis, J.O., Brown, P.R., Wheeler, P.D., and Witty, R.T., 1972, Effects of estradiol on sodium and potassium balances in adrenalectomized dogs, Am. J. Physiol. 223:194.

Johnston, C.I., Matthews, P.G., Davis, J.M., and Morgan, T., 1975, Renin measurement in blood collected from the efferent arteriole of the kidney of the rat, Pflug. Arch. 356:277.

Johnston, H.H., Herzog, J.P., and Lauler, D.P., 1967, Effect of prostaglandin E_1 on renal hemodynamics, sodium and water excretion, Am. J. Physiol. 213:939.

Jones, R.L., 1972, Properties of a new prostaglandin, Brit. J. Pharmacol. 45:144.

Jones, R.L., and Cammock, S., 1973, Purification, properties, and biological significance of prostaglandin A isomerase, Adv. Biosci. 9:61.

Kaley, G., and Weiner, R., 1971(a), Effect of prostaglandin E_1 on leucocyte migration, Nature (New Biol.) 234:114.

Kaley, G., and Weiner, R., 1971, Prostaglandin E_1: A potential mediator of the inflammatory response, Ann. N.Y. Acad. Sci. 180:338.

Källskog, Ö., Ulfendahl, H.R., and Wolgast, M., 1972, Single glomerular blood flow as measured with carbonized 141-Ce labeled microspheres, Acta Physiol. Scand. 85:408.

Kaplan, A.P., Austen, K.F., 1971, A prealbumin activator of prekallikrein. II. Derivation of activators of prekallikrein from active Hageman factor by digestion with plasmin, J. Exp. Med. 133:696.

Katz, M.A., and Shear, L., 1975a, Effects of renal nerves on renal hemodynamics: I. Direct stimulation and carotid occlusion Nephron 14:246.

Katz, M.A., and Shear, L., 1975b, Effects of renal nerves on renal hemodynamics: II. Renal denervation models, Nephron 14:390.

Kellermeyer, R.W., and Graham, R.C., 1968, Kinins - possible physiologic and pathologic roles in man, New Engl. J. Med. 279:754.

Kemp, E., and Rubin, I., 1964, Molecular weight of renin determined by Sephedex gel filtration, Acta. Chem. Scand. 18:2403.

Keyl, M.J., Bell, R.D., and Parry, W.L., 1972, Summary of renal lymphatic studies, Trans. Am. Assoc. Genito-Urinary Surg. 64:140.

Kiil, F., 1971, Blood flow and oxygen utilization by the kidney, in "Kidney Hormones" (J.W. Fisher, ed.), pp. 1-30, Academic Press, New York.

Kiil, F., Kjekshus, J., and Löyning, E., 1969, Renal autoregulation during infusion of noradrenaline, angiotensin, and acetylcholine, Acta Physiol. Scand. 76:10.

Kilo, C., Vogler, N., and Williamson, J.R., 1972, Muscle Capillary basement membrane changes related to aging and to diabetes mellitus, Diabetes 21:881.

Kincaid-Smith, P., Friedman, A., and Hobbs, J.B., 1974, Morphological effects of angiotensin in arteries, in "Handbook of Experimental Pharmacology, Vol. 37, Angiotensin" (I.H. Page and F.M. Bumpus, eds.) pp. 490-499, Springer-Verlag, New York.

Kirschenbaum, M.A., White, N., Stein, J.H., and Ferris, T.F., 1974, Redistribution of renal cortical blood flow during inhibition of prostaglandin synthesis, Am. J. Physiol. 227:801.

Kleinman, L.I., and Reuter, V.H., 1974, Renal response of the newborn dog to a saline load: The role of intrarenal blood flow distribution, J. Physiol. 239:225.

Kotchen, T.A., Hedrick, J.L., Miller, M.C., and Talwalker, R.T., 1974, Effect of prostaglandins on the velocity of the reaction between human renin and homologous renin substrate, J. Clin. Endocrinol. Metab. 39:530.

Kramer, K., Thurau, K., and Deetjen, P., 1960, Haemodynamik des Neirenmarks: I. Capillarie Passagezet, Durchblutung, Gewebshaematokrit, und O_2-Verbrauch des Nierenmarks in situ, Arch. Ges. Physiol. 270:251.

Kraut, H., Frey, E.K., and Werle, E., 1930, Der Nachweis Eines Kreislaufhormons in der Pankreasdrüse, Hoppe-Seyler's Z. Physiol. Chem. 189:97.

Kreiger, G., 1970, Thesis, University of Heidelberg, quoted in Gross (1971), p. 112.

Kriz, W., 1967, Der architektonische und funktionelle Aufbau der Rattenniere, Z. Zellforsch, Mikrosk. Anat. 82:495.

Keuhl, F.A., Oien, H.G., and Ham, E.A., 1974, Prostaglandins and prostaglandin synthetase inhibitors: Actions on cell function, in "Prostaglandin Synthetase Inhibitors," (H.J. Robinson and J.R. Vane, eds.) pp. 53-65, Raven Press, New York.

Kurz, K.D., Hasbargen, J.A., and Zehr, J.E., 1975, Effects of increased left atrial transmural pressure on renin secretion, Ped. Proc. 34:367.

Ladefoged, L., and Munck, O., 1971, Distribution of blood flow in the kidney, in "Kidney Hormones" (J.W. Fisher, ed.) pp. 31-58, Academic Press, New York.

Lameire, N., Schaper, W., and Ringoir, S., 1972, Circulation intrarénale dans l'intoxication mercurielle du rat étudié par microsphères marquées, J. Urol. Nephrol. 78:771.

Laragh, J.H., Cannon, P.J., Bentzel, C.J., Sicinski, A.M., and Meltzer, J.I., 1963, Angiotensin II, norepinephrine, and renal transport of electrolytes and water in normal man and in cirrhosis with ascites, J. Clin. Invest. 42:1179.

Laragh, J.H., and Sealey, J.E., 1973, The renin-angiotensin-aldosterone hormonal system and regulation of sodium, postassium, and blood pressure homeostasis, in "Handbook of Physiology, Section 8: Renal Physiology," (J. Orloff and R.W. Berliner, eds.) pp. 831-908, Williams and Wilkins, Baltimore.

Larsson, C., and Änggård, E., 1973, Regional differences in the formation and metabolism of prostaglandins in the rabbit kidney, Eur. J. Pharmacol. 21:30.

Larsson, C., Weber, P., and Änggård, E., 1974, Arachidonic acid increases and indomethacin decreases plasma renin activity in the rabbit, Eur. J. Pharmacol. 28:391.

Lassiter, W.E., 1975, Kidney, Ann. Rev. Physiol. 37:371.

Ledingham, J.G., and Leary, W.P., 1974, Catabolism of Angiotensin II, in "Handbook of Experimental Pharmacology, Vol. 37, Angiotensin," (I.H. Page and F.M. Bumpus, eds.) pp. 111-125, Springer-Verlag, New York.

Lee, J.B., 1968, Cardiovascular implications of the renal prostaglandins in "Prostaglandin Symposium of the Worcester Foundation," (P.W. Ramwell and J.E. Shaw, eds.) pp. 131-146, Interscience, New York.

Lee, J.B., 1972, Natriuretic "hormone" and the renal prostaglandins, Prostaglandins 1:55.

Lee, J.B., Mazzeo, M.A., and Takman, B.H., 1964, The acidic lipid characteristics of sustained renomedullary depressor activity, Clin. Res. 12:254.

Lee, J.B., Crowshaw, K., Takman, B.H., Attrep, K.A., and Gougoutas, J.Z., 1967, The identification of prostaglandins E_2, $F_{2\alpha}$, and A_2 from rabbit kindney medulla, Biochem. J. 105: 1251.

Lee, J.B., Crowshaw, K., Takman, B.H., Attrep, K.A., and Gougoutas, J.Z., 1967, Identification of prostaglandins E_2, $F_{2\alpha}$, and A_2 from rabbit renal medulla, Biochem. J. 105:1251.

Lee, S.J., Johnson, J.G., Smith, C.J., and Hatch, F.E., 1972, Renal effects of prostaglandin A_1 in patients with essential hypertension, Kidney Int. 1:254.

Leon, A.S., and Bloor, C.M., 1974, Effects of chronic exercise on the kidneys at different ages, Life Sci. 15:29.

Levitt, M.F., and Bader, M.E., 1951, Effect of cortisone and ACTH on fluid and electrolyte distribution in man, Am. J. Med. 11:715.

Lilienfield, L.S., Maganzini, H.C., and Bauer, M.H., 1961, Blood flow in the renal medulla, Circ. Res. 9:614.

Ljungqvist, A., 1964, Structure of the arteriole-glomerular units in different zones of the kidney, Nephron 1:329.

Ljungqvist, A., Wagermark, J., 1970, The adrenergic innervation of intrarenal glomerular and extra-glomerular circulatory routes, Nephron 7:218.

Limas, C.J., 1974, Stimulation by angiotensin of myocardial prostaglandin synthesis, Biochim. Biophys. Acta 337:417.

Lonigro, A.J., Itskovitz, H.D., Crowshaw, K., and McGiff, J.C., 1973, Dependency of renal blood flow on prostaglandin synthesis in the dog, Circ. Res. 32:712.

Lundgren, O., and Jodal, M., 1975, Regional blood flow, Ann. Rev. Physiol. 37:395.

Mancia, G., Romero, J.C., and Shepherd, J.T., 1975, Control of renin release by cardiopulmonary receptors, Circulation 50 (Suppl. III):63.

Mangel, R., Hollenberg, N.K., and Merrill, J.P., 1970, Radioisotope washout for measurement of renal blood flow, Fed. Proc. 29:397A.

Mark, A.L., Eckstein, J.W., Abboud, F.M., and Wendling, M.G., 1969, Renal vascular responses to isoproterenol, Am. J. Physiol. 217:764.

McCluskey, R.T., 1974, Immunologic mechanisms in renal disease, in R. H. Heptinstall "Pathology of the Kidney," pp. 273-317, Little, Brown and Co., Boston.

McGiff, J.C., Crowshaw, K., Terragno, N.A., and Lonigro, A.J., 1970, Renal prostaglandins: Possible regulators of the renal actions of pressor hormones, Nature 227:1255.

McGiff, J.C., Crowshaw, K., Terragno, N.A., Malik, K.U., and Lonigro, A.J., 1972, Differential effect of noradrenalin and renal nerve stimulation on vascular resistance in the dog kidney and the release of prostaglandin E-like substance, Clin. Sci. 42:223.

McGiff, J.C., and Fasy, T.M., 1965, The relationship of the renal vascular activity of angiotensin II to the autonomic nervous system, J. Clin. Invest. 44:1911.

McGiff, J.C., and Itskovitz, H.D., 1973, Prostaglandins and the kidney, Circ. Res. 33:479.

McGiff, J.C., Terragno, N.A., Malik, K.D., and Lonigro, A.J., 1972, Release of a prostaglandin E-like substance from canine kidney by bradykinin, Circ. Res. 31:36.

McGiff, J.C., Terragno, D.A., Terragno, N.A., and Mulik, K.U., 1975, Vascular reactivity and prostaglandins, (remarks at Symposium on "The role of prostaglandins in vascular homeostasis," 59th Annual Meeting Federation of American Societies for Experimental Biology, Atlantic City, N.J., April 17, 1975).

McNay, J.L., and Kishimoto, T., 1969, Association between autoregulation and pressure dependency of renal vascular responsiveness in dogs, Circ. Res. 24:599.

McNay, J.L., and Abe., Y., 1970, Pressure-dependent heterogeneity of renal cortical blood flow in dogs, Circ. Res. 25:571.

Menzie, J.W., and Michelakis, A.M., 1972, Renin release from the submaxillary gland by alpha-adrenergic agonists, Fed. Proc. 31:511.

Merrill, J.E., Peach, M.J., and Gilmore, J.P., 1973, Angiotensin I conversion in the kidney and its modulation by sodium balance, Am. J. Physiol. 224:1104.

Michelakis, A.M., Caudle, J., and Liddle, G.W., 1969, In vitro stimulation of renin production by epinephrine, norepinephrine, and cyclic AMP, Proc. Soc. Exp. Biol. Med. 130:748.

Miller, W.L., Weeks, J.R., Lauderdale, J.W., and Kirton, K.T., 1975, Biological activites of 17-phenyl-18, 19, 20-trinorprostaglandins, Prostaglandins 9:9.

Milton, A.S., and Wendlandt, S., 1971, Effects on body temperature of prostaglandins of the A, E, and F series on injection into the third ventricle of unanesthetized cats and rabbits, J. Physiol. (London) 218:325.

Misiewicz, J.J., Waller, S.L., and Kiley, N., 1969, Effects of prostaglandin E_1 on intestinal transit in man, Lancet 1:648.

Moffat, D.B., and Fourman, J., 1963, The vascular pattern of the rat kidney, J. Anat. 97:543.

Moritz, A.R., and Oldt, M.R., 1937, Arteriolar sclerosis in hypertensive and non-hypertensive individuals, Am. J. Pathol. 13:679.

Moriuchi, K., Tanaka, H., Yamamoto, K., and Ueda, J., 1971, Distribution of renin in the dog kidney, Life Sci. 10:727.

Mowat, P., Lupu, A.N., and Maxwell, M.H., 1972, Limitations of ^{133}Xe washout technique in estimation of renal blood flow, Am. J. Physiol. 223:682.

Muirhead, E.E., Leach, B.E., Byers, L.W., Brooks, B., Daniels, E.G., and Hinman, J.W., 1971, Antihypertensive neutral renomedullary lipids (ANRL), in "Kidney Hormones" (J.W. Fisher, ed.) pp. 405-506, Academic Press, New York.

Muirhead, E.E., Germain, G., Leach, B.E., Pitcock, V.A., Stephenson, P., Brooks, B., Brosius, W.L., Daniels, E.G., and Hinman, J.W., 1972, Production of renomedullary prostaglandins by renomedullary interstitial cells grown in tissue culture, Circ. Res. 31 (Suppl. II):161.

Mulrow, P.J., Ganong, W.F., Cera, G., and Kuljian, A., 1962, The nature of the aldosterone-stimulating factor in dog kidneys, J. Clin. Invest. 41:505.

Murakami, K., and Inagami, T., 1975, Isolation of pure and stable renin from hog kidney, Biochem. Biophys. Res. Comm. 62:757.

Myers, B.D., Deen, W.M., Robertson, C.R., and Brenner, B.M., 1975, Dynamics of glomerular ultrafiltration in the rat, VIII. Effects of hematocrit, Circ. Res. 36:425.

Nakamura, M., Ezaki, I., Sumiyashi, A., Kai, M., Kanaide, H., Naito, D., and Kato, K., 1975, Renal subcellular fractions producing angionecrosis and increased vascular permeability, Br. J. Exp. Pathol. 56:62.

Nakano, J., 1973, General pharmacology of prostaglandins, in "The Prostaglandins," (M.F. Cuthbert, ed.) pp. 23-124, Lippincott, Philadelphia.

Nakano, J., and Cole, B., 1969, Effects of prostaglandin E_1 and $F_{2\alpha}$ on systemic, pulmonary and splanchnic circulations in dogs, Am. J. Physiol. 217:222.

Nakano, J., and McCurdy, J.R., 1967, Cardiovascular effects of prostaglandin E_1, J. Pharmacol. Exp. Ther. 156:538.

Nakano, J., and McCurdy, J.R., 1968, Hemodynamic effects of prostaglandins E_1, A_1, and $F_{2\alpha}$ in dogs, Proc. Soc. Exp. Biol, Med. 128:39.

Navar, L.G., and Langford, H.G., 1974, Effects of angiotensin on the renal circulation, in "Handbook of Experimental Pharmacology, Vol. 37, Angiotensin" (I.H. Page and F.M. Bumpus, eds.) pp. 455-474, Springer-Verlag, New York.

Nissen, H.M., and Andersen, H., 1968, On the localization of a prostaglandin dehydrogenase activity in the kidney, Histochemie 14:189.

Ng, K.K.F., and Vane, J.R., 1967, Conversion of angiotensin I to angiotensin II, Nature 216:762.

Ng, K.K.F., and Vane, J.R., 1968, Fate of angiotensin I in the circulation, Nature 218:144.
Nishiyama, A., Rikimaru, A., Fukushi, Y., and Suzuki, T., 1972, Effect of "substance R" on the cat blood pressure, Tohoku J. Exp. Med. 106:207.
Nustad, K., 1970a, The relationship between kidney and urinary kininogenase, Br. J. Pharmacol. 39:73.
Nustad, K., 1970b, Localization of kininogenase in the rat kidney, Br. J. Pharmacol. 39:87.
O'Dorisio, T.M., Stein, J.H., Osgood, R.W., and Ferris, T.F., 1973, Absence of aglomerular blood flow during renal vasodilatation and hemorrhage in the dog, Proc. Soc. Exp. Biol. Med. 143:612.
Oliver, W.J., and Gross, F., 1966, Unique specificity of mouse angiotensin to homologous renin, Proc. Soc. Exp. Biol. Med. 122:923.
Oparil, S., and Haber, E., 1974, The renin-angiotensin system, New Engl. J. Med. 291:389.
Owen T., Ehrhart, I., Weider, J., Haddy, F., and Scott, J., 1974, Effects of indomethacin on blood flow, reactive hyperemia and autoregulation in the dog kidney, Fed. Proc. 33:348.
Page, I.H., and Helmer, O.M., 1940, A crystalline pressor substance (angiotonin) resulting from the reaction between renin and renin-activator, J. Exp. Med. 71:29.
Page, I.H., McSwain, B., Knapp, G.M., and Andrus, W.D., 1941, The origin of renin-activator, Am. J. Physiol. 135:214.
Pals, D.T., Masucci, F.D., Sipos, F., and Denning, G.S. Jr., 1971, A specific competitive antagonist of the vascular action of angiotensin II, Circ. Res. 29:664.
Pappenheimer, J.R., and Kinter, W.B., 1956, Hematocrit ratio of blood within mammalian kidney and its significance for renal hemodynamics, Am. J. Physiol. 185:377.
Peach, M.J., and Chiu, A.T., 1974, Stimulation and inhibition of aldosterone biosynthesis in vitro by angiotensin II and analogs, Circ. Res. 34 (Suppl. I):7.
Peart, W.S., 1956, The isolation of a hypertensin, Biochem. J. 62:520.
Peart, W.S., 1971, Renin-angiotensin system in hypertensive disease, in "Kidney Hormones" (J.W. Fisher, ed.), pp. 217-242, Academic Press, New York.
Peart, W.S., 1975, Renin-angiotensin system, New Engl. J. Med. 292:302.
Piper, P.J., and Vane, J.R., 1969, Release of additional factors in anaphylaxis and its antagonism by anti-inflammatory drugs, Nature 223:29.
Plakke, R.K., Pfeiffer, E.W., 1964, Blood vessels of the mammalian renal medulla, Science 146:1683.
Poulsen, K., 1968, Measurement of renin in human plasma, Scand. J. Clin. Lab. Invest. 21:49.

Poulsen, K., 1971, No evidence of active renin-inhibition in plasma. The kinetics of the reaction between renin and substrate in non-pretreated plasma, Scand. J. Clin. Lab. Invest. 27:37.

Poulsen, K., and Poulsen, L.L., 1971, Simultaneous determination of plasma converting enzyme and angiotensinase activity by radioimmunoassay, Clin. Sci. 40:443.

Prado, J.L., Limãos, E.A., Roblero, J., Nuñez, S.T., and Prado, E.S., 1975, In vivo conversion of kinins by exsanguinated rat preparations, Fed. Proc. 34:819.

Rao, U.V.G., and Wagner, H.N. Jr., 1972, Normal weights of human organs, Radiology 102:337.

Rasmussen, S.N., 1972, Influence of plasma hypertonicity on blood viscosity studied in vitro and in an isolated vascular bed, Acta Physiol. Scand 84:472.

Rasmussen, S.N., 1973, Intrarenal red cell and plasma volumes in the nondiuretic rat, Pflug. Arch. 342:61.

Rector, J.B., Stein, J.H., Bay, W.H., Osgood, R.W., and Ferris, T.F., 1972, Effect of hemorrhage and vasopressor agents on distribution of renal blood flow, Am. J. Physiol. 222:1125.

Reinhardt, H.W., Kaczmarczyk, G., Fahrenhorst, K., Bledinger, I., Gatzka, M., Kuhl, U., and Riedel, J., 1975, Post-prandial changes of renal blood flow, Pflug. Arch. 354:287.

Robinson, D.S., Nies, A., Davis, J.N., Bunney, W.E., Davis, J.M., Colburn, R.W., Bourne, H.R., Shaw, D.M., and Coppen, A.J., 1972, Ageing, monoamines and monoamine-oxidase levels, Lancet 1:290.

Robinson, H.J., and Vane, J.R., (eds.), 1974, "Prostaglandin Synthetase Inhibitors," Raven Press, New York.

Rocha e Silva, M., 1970, "Kinin Hormones," p. 164, Charles C. Thomas, Springfield.

Rome, L.H., and Lands, W.E.M., 1975, Inhibition of prostaglandin synthesis, Fed. Proc. 34:790.

Rouffignac, C. de, Bonvalet, J.P., and Menard, J., 1974, Renin content in superficial and deep glomeruli of normal and salt-loaded rat, Am. J. Physiol. 226:150.

Ruiz-Maza, F., Tiller, D.J., and Walker, W.G., 1974, Measurement of angiotensin II in an ultrafiltrate of plasma, Johns Hopkins Med. J. 135:211.

Ryan, J.W., 1974, The fate of angiotensin II, in "Handbook of Experimental Pharmacology, Vol. 37, Angiotensin" (I.H. Page and F.M. Bumpus, eds.) pp. 81-110, Springer-Verlag, New York.

Ryan, J.W., Smith, U., and Niemeyer, R.S., 1972, Angiotensin I: Metabolism by plasma membrane of lung, Science 176:64.

Sample, S.J.G., and de Wardener, H.E., 1959, Effect of increased renal venous pressure on circulatory autoregulation of isolated dog kidneys, Circ. Res. 7:643.

Samuelsson, B., 1970, in "Pharmacology of the Prostaglandins," Proc. 4th Inter. Congr. Pharmacol., p. 12, Schwabe and Co., Basel.

Samuelsson, B., and Hamberg, M., 1974, Role of endoperoxides in the biosynthesis and action of prostaglandin, in "Prostaglandin Synthetase Inhibitors" (H.J. Robinson and J.R. Vane, eds.) pp. 107-119, Raven Press, New York.

Samuelsson, B., and Wennmalm, Å., 1971, Increased nerve stimulation induced release of noradrenalin from the rabbit heart after inhibition of prostaglandin synthesis, Acta. Physiol. Scand. 83:163.

Sassard, J., Sann, L., Vincent, M., Francois, R., and Cier, J.F., 1975, Plasma renin activity in normal subjects from infancy to puberty, J. Clin. Endocrinol. Metab. 40:524.

Satoh, S., and Zimmerman, B.G., 1974, Participation of intrarenal prostaglandin in renal ischemia, Fed. Proc. 33:576.

Satoh, S., Kraft, E., and Zimmerman, B., 1975, Effect of 1-Sar-8-Ala angiotensin II (P-113) on blood pressure, renal blood flow and plasma renin in conscious dog, Fed. Proc. 34:770.

Schaffenburg, C.A., Haas, E., and Goldblatt, H., 1960, Concentration of renin in kidneys and angiotensinogen in serum of various species, Am. J. Physiol. 199:788.

Schmid, H.E., 1972, Renal autoregulation and renin release during changes in renal perfusion pressure, Am. J. Physiol. 222:1132.

Schmid-Schönbein, H., Wells, R.E., Goldstone, J., 1973, Effect of ultrafiltration and plasma osmolarity upon the flow properties of blood: A possible mechanism for control of blood flow in the renal medullary vasa recta, Pflug. Arch. 338:93.

Schramm, L.P., and Carlson, D.E., 1975, Inhibition of renal vasoconstriction by elevated ureteral pressure, Am. J. Physiol. 228:1126.

Schramm, L.P., Anderson, D.E., and Randall, D.C., 1975, Renal blood flow changes during aversive conditioning in the dog. Experientia 31:71.

Schrier, R.W., 1974, Effects of adrenergic nerous system and catecholamines on systemic and renal hemodynamics, sodium and water excretion, and renin secretion, Kidney Int. 6:291.

Schrier, R.W., and Berl, T., 1973, Mechanism of effect of alpha adrenergic stimulation with norepinephrine on renal water excretion, J. Clin. Invest. 52:502.

Scicli, A.G., Carretero, O.A., Hampton, A., and Oza, N.B., 1975, Kallikrein localization in the nephron by the stop-flow technique, Fed. Proc. 34:378.

Sealey, J.E., Gerten, J.N., Ledingham, J.G.G., and Laragh, J.H., 1967, Inhibition of renin by heparin, J. Clin. Endocrin. 27:699.

Selkurt, E.E., 1974, Concise review: Current status of renal circulation and related nephron function in hemorrhage and experimental hemorrhagic shock. II. Neurohumoral and tubular mechanisms, Circ. Shock 1:89.

Selye, H., 1968, "Anaphylactoid Edema," p. 21, Warren H. Green, Inc., St. Louis.

Sen, S., Smeby, R.R., and Bumpus, F.M., 1972, Renin in rats with spontaneous hypertension, Circ. Res. 31:876.

Shade, R.E., Davis, J.O., Johnson, J.A., Witty, R.T., 1972, Effects of renal arterial infusion of sodium and potassium on renin secretion in the dog, Circ. Res. 31:719.

Shanser, J.D., Korobkin, M., Seidlitz, L., Carlson, E.L., and Shames, D.M., 1974, Hazards in interpretation of xenon washout studies of the canine kidney, Radiology 111:461.

Shimamura, T., and Morrison, A.B., 1973, Vascular permeability of the renal medullary vessels in the mouse and rat, Am. J. Pathol. 71:155.

Shipley, R.E., and Helmer, O.M., 1948, Observation on "the sustained pressor principle" in different animal species, Am. J. Physiol. 153:341.

Silverman, A.J., and Barajas, L., 1974, Effect of reserpine on the juxtaglomerular granular cells and renal nerves, Lab. Invest. 30:723.

Siperstein, M.D., Unger, R.H., and Madison, L.L., 1968, Studies of muscle capillary basement membranes in normal subjects, diabetic, and prediabetic patients, J. Clin. Invest. 47:1973.

Skeggs, L.T., Dorer, F.E., Kahn, J.E., Lentz, K.E., and Levine, M., 1974, The biological production of angiotensin, in "Handbook of Experimental Pharmacology, Vol. 37, Angiotensin" (I.H. Page and F.M. Bumpus, eds.) pp. 1-16, Springer-Verlag, New York.

Skeggs, L.T., Kahn, J.R., Lentz, K.E., and Shumway, N.P., 1957, The preparation, purification, and amino acid sequence of a polypeptide renin substrate, J. Exp. Med. 106:439.

Skeggs, L.T., Kahn, J.R., and Shumway, N.P., 1956, The preparation and function of the hypertensin-converting enzyme, J. Exp. Med. 103:295.

Skeggs, L.T., Lentz, K.E., and Hochstrasser, H., and Kahn, J.R., 1963, The purification and partial characterization of several forms of hog renin substrate, J. Exp. Med. 118:73.

Skeggs, L.T., Lentz, K.E., and Hochstrasser, H., and Kahn, J.R., 1964, The chemistry of renin substrate, Canad. Med. Assoc. J. 90:185.

Skeggs, L.T., Lentz, K.E., Kahn, J.R., Shumway, N.P., and Woods, K.R., 1956, The amino acid sequence of hypertensin II, J. Exp. Med. 104:193.

Skinner, S.L., 1967, Improved assay methods for renin "concentration" and "activity" in human plasma, Circ. Res. 20:391.

Skinner, S.L., Lumbers, E.R., and Symonds, E.M., 1969, Alteration by oral contraceptives of normal menstrual changes in plasma renin activity, concentration, and substrate, Clin. Sci. 36:67.

Slotkoff, L.M., and Lilienfield, L.S., 1967, Extravascular renal albumin, Am. J. Physiol. 212:400.

Slotkoff, L.M., Logan, A., Jose, P., D'Avella, J., and Eisner, G.M., 1971, Microsphere measurement of intrarenal circulation of the dog, Circ. Res. 28:158.

Smeby, R.R., and Bumpus, F.M., 1971, Renin inhibitors, in "Kidney Hormones" (J.W. Fisher, ed.) pp. 207-216, Academic Press, New York.

Smith, H.W., 1951, "The Kidney: Structure and Function in Health and Disease," p. 467, Oxford, New York.

Smith, J.P., 1955, Hyaline arteriolosclerosis in the kidney, J. Pathol. Bacteriol. 69:147.

Solez, K., Fox, J.A., Miller, M., and Heptinstall, R.H., 1974a, Effects of indomethacin on renal inner medullary plasma flow, Prostaglandins 7:91.

Solez, K., Kramer, E.C., Fox, J.A., and Heptinstall, R.H., 1974b, Medullary plasma flow and intravascular leukocyte accumulation in acute renal failure, Kidney Int. 6:24.

Solez, K., Ponchak, S., Buono, R.A., Vernon, N., Finer, P.M., Miller, M., and Heptinstall, R.H., 1976a, Inner medullary plasma flow in the kidney with ureteral obstruction, Am. J. Physiol. 231:1315.

Solez, K., D'Agostini, R.J., Buono, R.A., Vernon, N., Wang, A.L., and Heptinstall, R.H., 1976b, The renal medulla and mechanisms of hypertension inthe spontaneously hypertensive rat, Am. J. Path. 85:555.

Somlyo, A.P., and Somlyo, A.V., 1971, Electrophysiological correlates of the inequality of maximal vascular smooth muscle contraction elicited by drugs, Proc. Symp. Physiol. Pharmacol. Vascular Neuroeffector Systems, p. 216, S. Karger, Basel.

Speroff, L., 1973, An essay: Prostaglandins and toxemia in pregnancy, Prostaglandins 3:721.

Spielman, W.S., Davis, J.O., Freeman, R.H., and Johnson, J.A., 1974, Stimulation of aldosterone by a heptapeptide fragment of angiotensin II in the rat, Fed. Proc. 32:254.

Spinelli, F.R., Wirz, H., Brucher, C., and Pehling, G., 1972, Non-existence of shunts between afferent and efferent arterioles of juxtamedullary glomeruli in dog and rat kidneys, Nephron 9:123.

Stein, J.H., Ferris, T.F., Huprich, J.E., Smith, T.C., and Osgood, R.W., 1971, Effect of renal vasodilatation on the distribution of cortical blood flow in the kidney of the dog, J. Clin. Invest. 50:1429.

Stein, J.H., Boonjarern, S., Wilson, C.B., and Ferris, R.F., 1973, Alterations in intrarenal blood flow distribution: Methods of measurement and relationship to sodium balance, Circ. Res. 32 (Suppl. I):61.

Stein, J.H., Baehler, R.W., Cox, J.W., and Ferris, T.F., 1974, Renal hemodynamic alterations during drug-induced renal vasoconstriction and acute hemorrhagic hypotension, in "Conference on Acute Renal Failure," (E.A. Friedman, and H.E. Eliohou, eds.) Washington, D.C., Government Printing Office, DHEW (NIH) Pub. No. 74-608.

Stewart, J., 1974, Tachyphylaxis to angiotensin, in "Handbook of Experimental Pharmacology, Vol. 37, Angiotensin" (I.H. Page and F.M. Bumpus, eds.) pp. 170-184, Springer-Verlag, New York.

Stjarne, L., 1973, Alpha-adrenoceptor mediated feedback control of sympathetic neurotransmitter secretion in the guinea pig vas deferens, Nature (New Biol.) 241:190.

Stokes, J.B. III, and Bartter, F.C., 1974, Antidiuretic effect of intracarotid infusion of prostaglandin E_2 in dogs, Circulation 50 (Suppl. III):136.

Stolarczyk, J., and Carone, F.A., 1975, Effects of renal lymphatic occlusion and venous constriction on renal function, Am. J. Pathol. 78:285.

Swain, J. Heyndricks, G., Boettcher, D., and Vatner, S., 1974, Prostaglandin regulation of renal blood flow in unanesthetized dogs and baboons, Circulation 50 (Suppl. III):134.

Symonds, E.M., Stanley, M.A., and Skinner, S.L., 1968, Production of renin by in vitro cultures of human chorion and uterine muscle, Nature 217:1152.

Szabo, G., and Magyar, Z., 1974, The origin of hilar and capsular renal lymph, Res. Exp. Med. 163:171.

Takeda, T., DeBusk, J., and Grollman, A., 1969, Physiologic role of renin-like constituent of submaxillary gland of the mouse, Am. J. Physiol. 216:1194.

Tannenbaum, J., Splawinski, J.A., Oates, J.A., and Nies, A.S., 1975, Enhanced renal prostaglandin production in the dog. I. Effects on renal function, Circ. Res. 36:197.

Thorburn, G.D., Kopald, H.H., Herd, J.A., Hollenberg, N.K., O'Marchoe, C.C.C., and Barger, A.C., 1963, Intrarenal distribution of nutrient blood flow determined with Krypton[85] in the unanesthetized dog, Circ. Res. 13:290.

Thurau, K., 1964, Renal hemodynamics, Am. J. Med. 36:698.

Thurau, K., 1974, Intrarenal action of angiotensin, in "Handbook of Experimental Pharmacology, Vol. 37, Angiotensin" (I.H. Page and F.M. Bumpus, eds.) pp. 475-489, Springer-Verlag, New York.

Thurau, K., Dahlheim, H., Grüner, A., Mason, J., Granger, P., 1972, Activation of renin in the single juxtaglomerular apparatus by sodium chloride in the tubular fluid at the macula densa, Circ. Res. 30 (Suppl. II):182.

Thurau, K., and Mason, J., 1974, The intrarenal function of the juxtaglomerular apparatus, in "MTP International Review of Science, Physiology Series One, Vol. 6, Kidney and Urinary Tract Physiology" (K. Thurau, ed.) pp. 357-389, Butterworths, London.

Thurau, K., and Levine, D.Z., 1971, The renal circulation, in "The Kidney" (C.L. Rouiller and A.F. Muller, eds.) Vol. III, pp. 1-70, Academic Press, New York.

Tigerstedt, R., and Bergman, P.G., 1898, Niere und Kreislauf, Scand. Arch. Physiol. 8:223.

Tobian, L., 1975, Prostaglandins, hypertension, and sodium, Remarks at Symposium on the Role of Prostaglandins in Vascular Homeostasis, 59th Annual Meeting of Delegation of American Societies for Experimental Biology, Atlantic City, N.J., April 17, 1975.

Tobian, L., Tomboulian, A., and Janecek, J., 1959, The effect of high perfusion pressures on the granulation of juxtaglomerular cells in an isolated kidney, J. Clin. Invest. 38:605.

Torres, V.E., Strong, C.G., Romero, J.C., and Wilson, D.M., 1973, Prostaglandins and renin in glycerol induced rabbit acute renal failure, Abstracts 1973 American Society of Nephrology, 6th Annual Meeting, p. 106.

Tribe, C.R., and Heptinstall, R.H., 1965, The juxtaglomerular apparatus in scarred kidneys, Brit. J. Exp. Pathol. 46:339.

Trueta, J., Barclay, A.E., Daniel, P.M., Franklin, K.J., Pritchard, M.M.L., 1947, "Studies of the Renal Circulation," Oxford, Blackwell.

Türker, R.K., Kaymakcalan, S., and Ayhan, I.H., 1968, Effect of prostaglandin E_1 (PGE_1) on the vascular responsiveness to norepinephrine and angiotensin in the anesthetized rat, Arzheimittel-Forsch. 18:1310.

Türker, R.K., Page, I.H., and Bumpus, F.M., 1974, Antagonists of angiotensin II, in "Handbook of Experimental Pharmacology, Vol. 37, Angiotensin" (I.H. Page and F.M. Bumpus, eds.) pp. 490-499, Springer-Verlag, New York.

Valtin, H., 1973, "Renal Function," Little, Brown and Company, Boston.

Vander, A.J., 1965, Effect of catecholamines and the renal nerves on renin secretion in anesthetized dogs, Am. J. Physiol. 209:659.

Vander, A.J., 1967, Control of renin release, Physiol. Rev. 47:359.

Vander, A.J., 1968, Direct effects of prostaglandin on renal function and renin release in anesthetized dog, Am. J. Physiol. 214:218.

Vane, J.R., 1974, The fate of angiotensin I, in "Handbook of Experimental Pharmacology, Vol. 37, Angiotensin," (I.H. Page and F.M. Bumpus, eds.) pp. 17-40.

Vane, J.R, and Ferreira, S.H., 1975, Interactions between bradykinin and prostaglandins, Life Sci. 16:804.

Venkatchalam, M.A., Karnovsky, M.S., 1972, Extravascular protein in the kidney, Lab. Invest. 27:435.

Venuto, R.C., O'Dorisio, T., Stein, J.H., and Ferris, T.F., 1975, Uterine prostaglandin E secretion and uterine blood flow in the pregnant rabbit, J. Clin. Invest. 55:193.

von Euler, U.S., 1937, On the specific vasodilating and plain muscle stimulating substances from accessory genital glands in man and certain animals (prostaglandins and vesiglandin), J. Physiol. 88:213.

von Euler, U.S., 1956, "Noradrenaline," Charles C. Thomas, Springfield.

Warren, D.J., and Ledingham, J.G.G., 1974, Renal blood flow distribution in the conscious rabbit using microspheres: The effects of hemorrhage, renal denervation and adrenalectomy, in European Colloquium on Nephron Physiology, May 16-18, 1974, p. 199, INSERM Publ. #2199, Paris.

Waterson, J.G., Frewin, D.B., and Soltys, J.S., 1974, Age-related differences in catecholamine fluorescence of human vascular tissue, Blood Vessels 11:79.

Webster, M.E., and Pierce, J.V., 1963, The nature of the kallidins released from human plasma by kallikreins and other enzymes, Ann. N.Y. Acad. Sci. 104:91.

Werkö, L., Bucht, H., Josephson, B., and Ek, J., 1951, The effect of noradrenaline and adrenaline on renal hemodynamics and renal function in man, Scand. J. Clin. Lab. Invest. 3:255.

Wilde, W.S., and Vorburger, C., 1967, Albumin multiplier in kidney vasa recta analyzed by microspectrophotometry of T-1824, Am. J. Physiol. 213:1233.

Williams, M.W., Wilson, T.W., Oates, J.A., Nies, A.S., Frolich, J.C., 1974, Origin of urinary prostaglandins, J. Clin. Invest. 53:85a.

Williamson, H.E, Bourland, W.A., and Marchand, G.R., 1974, Inhibition of ethacrynic acid induced increase in renal blood flow by indomethacin, Prostaglandins 8:297.

Wilson, C.B., and Dixon, F.J., 1974, Diagnosis of immunopathologic renal disease, Kidney Int. 5:389.

Wlodawer, P., and Samuelsson, B., 1973, On the organization and mechanism of prostaglandin synthetase, J. Biol. Chem. 248:5673.

Wolgast, M., 1973, Renal medullary red cell and plasma flow as studied with labeled indicators and internal detection, Acta Physiol. Scand. 88:215.

Wolgast, M., 1968, Studies in regional renal blood flow with P^{32}-labeled red cells and small beta-sensitive semiconductor detectors, Acta. Physiol. Scand. Suppl. 313:1.

Yang, H.Y.T., Erdös, E.G., and Levin, Y., 1971, Characterization of a dipeptide hydrolase (Kininase II: angiotensin I converting enzyme), J. Pharmacol. Exp. Ther. 177:291.

Yoffey, J.M., and Courtice, F.C., 1970, "Lymphatics, Lymph, and the Lymphomyeloid Complex," pp. 236-249, Academic Press, London.

Yun, J.C.H., Delea, C.S., Bartter, F.C., and Kelly, G., 1974, Vagal control of renin secretion, Fed. Proc. 33:253.

Yun, J.C.H., Delea, C.S., Bartter, F.C., and Kelly, G., 1975, Vagal control of renin secretion II, Fed. Proc. 34:311.

Zehr, J.E, and Feigl, E.O. 1973, Suppression of renin activity by hypothalamic stimulation, Circ. Res. 32 (Suppl. I):17.

Zins, G.R., 1975, Renal prostaglandins, Am. J. Med. 58:14.

Zusman, R.M., Spector, D., Caldwell, B.V., Speroff, L., Forman, B., Schneider, G., and Mulrow, P., 1973, Prostaglandin A concentrations in plasma of normal and hypertensive humans, J. Clin. Invest. 52:93a.

Zusman, R.M., Spector, D., Caldwell, B.V., Speroff, L., Schneider, G., and Mulrow, P.J., 1973, The effect of chronic sodium loading and restriction on plasma prostaglandin A, E, and F concentrations in normal humans, J. Clin. Invest. 52:1093.

THE INNERVATION OF ARTERIES

G. Burnstock,[1] D.Sc.,F.A.A., J. H. Chamley,[2] Ph.D., and
G. R. Campbell,[2] Ph.D. *

[1]Department of Anatomy and Embryology, University College
London, Gower Street, London, WCIE 6BT
[2]Baker Medical Research Institute, Commercial Road,
Prahran, Victoria, Australia, 3181

I. INTRODUCTION

Knowledge of the innervation of arteries was limited for many years to light microscopic examination of silver and methylene blue stained preparations (Hillarp, 1959; Grigor'eva, 1962). These methods cannot distinguish between different autonomic nerve types or reveal the relationship of individual nerve fibers to smooth muscle cells. There are still many gaps in our knowledge of the pattern, density, and mechanism of innervation of the wide variety of arteries that are present in the vascular system, but considerable advances have been made in this field in the past decade following the application of histochemical and electron-microscopical methods (see reviews Ehinger et al., 1967; Somlyo and Somlyo, 1968; Holman, 1969; Burnstock, 1970, 1974, 1975a, 1975d, 1978; Burnstock et al., 1970; Speden, 1970; Bell and Burnstock, 1971; Bevan et al., 1972; Bevan and Su, 1973, 1974; Bukinich, 1973; Hudlická, 1973; Rosenblueth, 1976; Su and Lee, 1976; Hung and Loosli, 1977).

Before considering the details of artery innervation, it is essential to consider the broad definition of the autonomic neuromuscular junction since it differs in several important respects from the better known skeletal neuromuscular junction. In the general

---

*Much of the work reported in this article was carried out while the authors were working in the Department of Zoology, University of Melbourne, where it was supported by grants from the National Heart Foundation of Australia, Life Insurance Medical Research Fund of Australia and New Zealand, and the National Health and Medical Research Council of Australia.

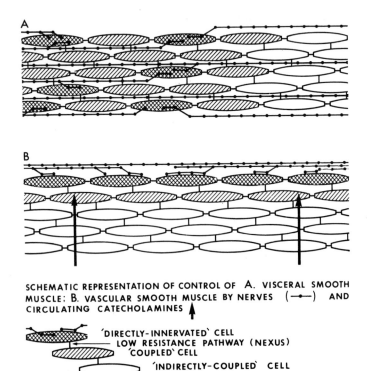

Fig. 1. Models of innervation of:
  A. Visceral smooth muscle
     (modified from Burnstock, in 'Smooth muscle,' Edward Arnold, London, p. 1-69, 1970).
  B. Vascular smooth muscle
     (modified from Burnstock, Clin. exp. Pharmacol. Physiol. (Suppl. 2, 7-20, 1975).

model of the autonomic neuromuscular junction first proposed by Burnstock (1970), the essential features (Fig. 1A) are that the terminal portions of autonomic nerve fibers are varicose (Fig. 2A), transmitter being released "en passage" from varicosities during conduction of an impulse, and that the effector is a muscle bundle rather than a single smooth muscle cell. Individual muscle cells are connected by low resistance pathways which allow electrotonic spread of activity within the effector bundle. The sites of these low resistance pathways appear to be the "gap junctions" or "nexuses" (Fig. 2B, C, and Section III). In most organs, some, but not all, muscle cells are directly-innervated, i.e., in close (20-120nm) apposition with nerve varicosities and are directly affected by transmitter released from them. "Coupled cells" adjoin "directly-innervated cells" via low resistance pathways so that excitatory junction potentials can be recorded. When the muscle cells in an area of an effector bundle become depolarized in this way, an all-or-none action potential is initiated which propagates through the tissue, activating many muscle cells. Thus, in some tissues, many cells (termed "indirectly-coupled cells") are neither "directly-innervated" nor "coupled" and yet respond on stimulation of the nerves supplying the organ.

There is a wide variation in the pattern and density of innervation of smooth muscle in different systems (Burnstock, 1970; Burnstock and Bell, 1974). For example, all the muscle cells in tissues such as vas deferens appear to be "directly-innervated," but only a small proportion of the muscle cells in tissues such as the uterus, ureter, and most blood vessels are "directly innervated." The minimum separation of nerve and muscle also varies considerably in different tissues. In general, it is closest (15-20nm) in densely innervated tissues such as the iris and vas deferens, but may be as wide as 2,000nm in some large elastic arteries.

The pattern of innervation of smooth muscle strongly influences the response of the muscle to activation of its nerve supply, the susceptibility of transmission to modification by drugs, and the sensitivity of the muscle to agonists (Trendelenburg, 1972).

II. GENERAL MODEL OF ARTERY INNERVATION

Adrenergic nerves are the dominant autonomic nerve component supplying most arteries, although cholinergic, purinergic, and possibly peptidergic nerves also play a role (Burnstock, 1978, 1979). An important aspect of sympathetic neuroeffector transmission is that norepinephrine released from sympathetic nerves is inactivated largely by re-uptake into the terminals following its action on the effector cell membrane. In densely-innervated smooth muscle tissues such as the vas deferens and iris, circulating catecholamine plays little part in the control of activity, since nerves are distributed throughout the tissue (Figs. 1A and 3A) and exogenous catecholamine

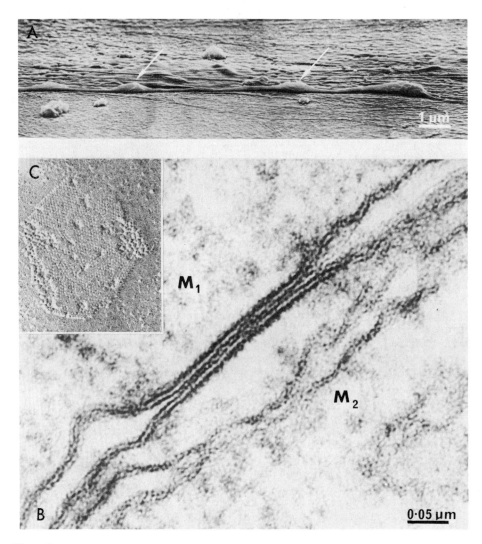

Fig. 2.
- A. Scanning electronmicrograph of varicosities (arrows), in a single nerve fiber growing in a culture of newborn guinea-pig sympathetic ganglia. Photographed at an angle of 70° (from Burnstock, in 'Methods in Pharmacology,' Vol. 3, Plenum Press, New York, pp. 113-137, 1975).
- B. "Gap junction" between two cultured smooth muscle cells (M_1 and M_2) from embryo chicken gizzard. A gap of up to about 30Å can be seen between the outer leaflets of the

unit membrane; there are a few short areas of fusion. The inner leaflets of the membranes are lined by an accumulation of electron-opaque material. x 224,000 (from Campbell, Uehara, Mark, and Burnstock, J. Cell Biol., 49, 21-34, 1971).

C. High magnification view of a freeze-fracture preparation of a gap junction in the guinea-pig sphincter pupillae showing 80-90Å particles on the A face and 30-40Å pits on the B face. The hexagonal arrangement of the pits on the B face also shows some dislocations of the lattice. x 100,000 (from Fry, Devine, and Burnstock, submitted for publication).

---

is rapidly taken up into the dense nerve plexus. Thus, in these tissues, adrenergic fibers have a dual role; not only do they provide nervous control of smooth muscle, but they "protect" the organ from unwanted effects of circulating catecholamines. These organs perform functions that are unrelated to general homeostatic functions of the sympatho-adrenal system, while the sparsely-innervated parts of the intestine, for example, are very sensitive to circulating adrenal-medullary catecholamines (Furness and Burnstock, 1975).

Blood vessels are also concerned in homeostasis responses. Their pattern of adrenergic innervation, with nerves confined to the adventitial side of the media (Norberg and Hamberger, 1964; Ehinger et al., 1967; Burnstock et al., 1970; Bevan et al., 1972; Burnstock, 1975a), appears to be suitable for control by both circulatory catecholamines and adrenergic nerves (Figs. 1B and 3B). Stimulation of the adrenergic nerves results in a prompt vasoconstriction. Excitatory junction potentials have been recorded from cells of the inner media remote from the nerve plexus, indicating that the muscle cells are electrically coupled (Holman, 1969; Speden, 1970). However, the asymmetric disposition of the nerves leaves most of the medial smooth muscle devoid of adrenergic fibers and therefore of sites of norepinephrine inactivation, so that the muscles are more responsive to circulatory catecholamine. The non-innervated media behaves as a supersensitive visceral smooth muscle following denervation or blockade of neural uptake by cocaine (Folkow and Neil, 1971), and application of norepinephrine intraluminally in isolated rabbit ear artery is two to fifteen times more active than when applied extraluminally (De la Lande et al., 1966). The absence of adrenergic nerves in the media of most blood vessels also facilitates the diffusion of catecholamine from the adrenals to their peripheral targets.

While the importance of sympathetic nerves in control of different vessels is becoming clearer, the relative importance of circulating catecholamines is less well understood. It may be that

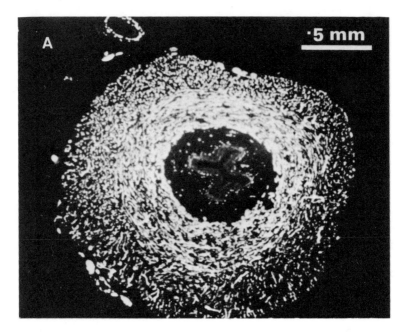

Fig. 3. Fluorescence histochemical demonstration of adrenergic nerve fibers.

    A. Rat vas deferens T.S. Note dense innervation of muscle coats (from Burnstock and Bell, in 'The Peripheral Nervous System,' Plenum Press, New York, pp. 277-327, 1974).

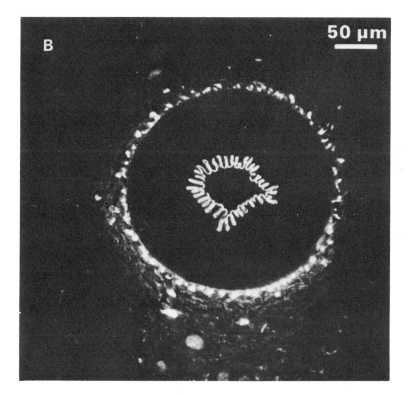

Fig. 3. Fluorescence histochemical demonstration of adrenergic nerve fibers.

  B. Rabbit ear artery T.S. Note restriction of adrenergic fibers to adventitial-medial border (from Burnstock, McCulloch, Storey, and Wright, Brit. J. Pharmacol., 46, 243-253, 1972).

circulating amines have little effect on the activity of most vessels in inactive animals, but during situations such as fright or heavy exercise, the resulting high levels of circulatory catecholamines have profound effects on vessel tone, although probably there are differential effects on different types of vessel and in different species. In most situations both circulating catecholamines and nerves produce vasoconstriction of vessels (Powis, 1974). However, in some vessels, such as those in skeletal muscle, the effect of circulating adrenaline has an antagonistic (vasodilatation) action to that of adrenergic nerves (Hudlická, 1973). It is of interest, too, that non-innervated vessels such as mesenteric precapillary arterioles are sensitive to concentrations of norepinephrine as low as 10^{-10} g/ml (Furness and Marshall, 1974), which compares with a figure of 7.45×10^{-9} g/ml for levels of circulating norepinephrine reported in rats (Chin and Evanuk, 1971). Non-innervated umbilical arteries are also affected by catecholamines (Von Euler, 1938; Somlyo, Woo, and Somlyo, 1965).

Norepinephrine uptake into nerves appears to be an important mechanism for terminating sympathetic vasomotor action in muscular arteries and arterioles, but it appears to be insignificant in elastic arteries (Bell, 1974). Extraneuronal uptake of catecholamines into smooth muscle cells is likely to be more important for termination of the action of circulating catecholamines (Nedergaard and Bevan, 1971; Wyse, 1973; De la Lande et al., 1974).

Presynaptic inhibition of norepinephrine release from sympathetic nerves by acetylcholine, adenosine triphosphate, and adenosine is an important peripheral mechanism of vasomotor control, as well as negative feedback by norepinephrine itself (Westfall, 1977).

III. GAP JUNCTIONS OR "NEXUSES" AND ELECTRICAL COUPLING OF ARTERIAL SMOOTH MUSCLE CELLS

Areas of close apposition or "bridges" between apposing membranes of smooth muscle cells were first recognized by Bergman (1958) and Prosser et al., (1960). This type of junction was later termed the "nexus" by Dewey and Barr (1962) and high resolution transmission electronmicroscopy has revealed that most, if not all, of these nexuses are "gap junctions" (Figs. 2B, C; Revel et al., 1967; Uehara and Burnstock, 1970). The freeze-fracture technique has considerable advantages for studies of the detailed ultrastructure of intercellular junctions (McNutt and Weinstein, 1973), and a study of the size, distribution, and organization of nexuses in a variety of smooth muscles has been carried out (Fry et al., 1976). These authors reported that the largest nexus seen covered a surface area of 0_26 μm^2, while the smallest recognizable nexus had an area of 0.001 μm^2. Regardless of size, nexuses are oriented in parallel rows, with their long axis parallel to the long axis of the cell.

All the nexuses seen have the characteristic hexagonal particle array on the A face and corresponding pits on the B face (Fig. 2C). A figure of approximately 100 nexuses per smooth muscle cell was calculated for the guinea-pig sphincter papillae.

The nexus has been considered by most workers to constitute the morphological basis of the low resistance pathways that allow electronic coupling of activity between adjacent smooth muscle cells within muscle effector bundles (Dewey and Barr, 1968; Bennett and Burnstock, 1968). However, doubts were raised about this conclusion following the discovery that very few, if any, nexuses are present in the longitudinal muscle of the intestine of dog (Henderson et al., 1971) and guinea-pig (Gabella, 1972), both of which preparations exhibit propagated action potentials. The punctate nexuses revealed in freeze-fracture preparation (Fry et al., 1976) may partially explain this apparent anomaly. Nexuses have been described in rat and mouse vas deferens, where every cell appears to be innervated (Burnstock, 1970). Many nexuses show a close relationship with sarcoplasmic reticulum (Gabella, 1974; Fry et al., 1976), and it has been suggested that this might facilitate the process of excitation-contraction coupling following current flow across this specialized junction by analogy with the release of bound calcium from the sarcotubular system of skeletal muscle following current flow from the transverse tubules (Fry et al., 1976). It should be pointed out that while experiments have not yet been carried out on smooth muscle, nexuses in other tissues have been shown to allow the passage of molecules as large as 10^3 daltons in molecular weight from one cell to another (Furshpan and Potter, 1968; Bennett, 1973; Lowenstein, 1973).

Nexuses have been recognized in a variety of visceral smooth muscle preparations, including chicken and pigeon gizzard, guinea-pig taenia coli, mouse and guinea-pig vas deferens (Cobb and Bennett, 1969; Barr et al., 1968), dog duodenum (Henderson et al., 1971), rat duodenum (Friend and Gilula, 1972) and guinea-pig ileum and sphincter pupillae (Gabella, 1972, 1974). They have also been described in cultured smooth muscle (Campbell et al., 1971).

There are a growing number of reports of nexuses in vascular smooth muscle. Nexuses have been identified in the aorta of rats (Cliff, 1967) and man (Fig. 4; Burnstock et al., 1970), although most experimental evidence suggests that electrotonic coupling between muscle cells is not a strongly represented mechanism in large elastic vessels (Burnstock and Prosser, 1960; Somlyo and Somlyo, 1968; Bevan and Su, 1973; Mekata, 1971; Bevan and Ljung, 1974; but see Mekata, 1974). They have been described in rabbit pulmonary artery (Devine et al., 1972), guinea-pig pulmonary artery (Fry et al., 1976), chick anterior mesenteric artery (Bolton, 1974; Fry et al., 1976), rabbit carotid artery and aorta (Henderson, 1975), and umbilical arteries

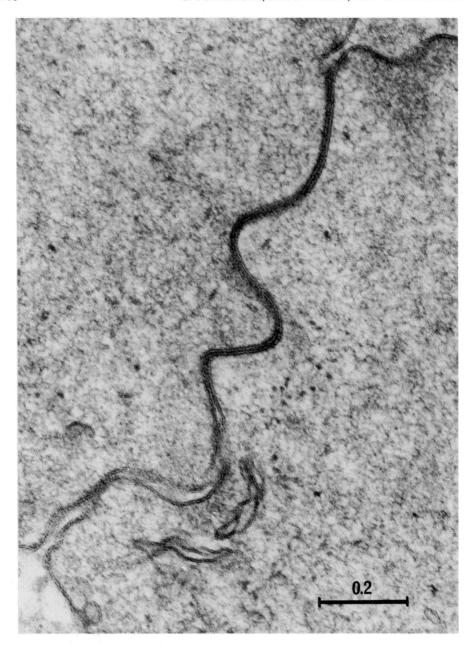

Fig. 4. Electron micrograph of a gap junction between two smooth muscle cells in the human aorta which probably represents the site of a low-resistance pathway allowing electronic coupling. Calibration 0.2μ., (from Burnstock, Gannon, and Iwayama, Circulation Res., 27 (Suppl. II), 5-24, 1970).

(Spiteri et al., 1966). Regions of close contact comparable to nexuses were observed between smooth muscle cells of arterioles and precapillary sphincters (Rhodin, 1967; Dahl, 1973). Nexuses have also been observed between muscle cells of the rabbit (Holman et al., 1968) and guinea-pig (Fry et al., 1976) portal vein. In addition, areas of close intercellular contact have been described in rabbit aorta (Seifert, 1963; Bierring and Kobayasi, 1963) and carotid artery (Henderson, 1975). Desmosomes have been observed in rabbit carotid artery (Henderson, 1975) and bulbous projections of one cell into another have been observed in a variety of vessels (Zelander et al., 1962; Mathews and Gardner, 1966; Tapp, 1969).

IV. DISTRIBUTION AND IDENTIFICATION OF DIFFERENT TYPES OF AUTONOMIC NERVES TO ARTERIES

The most important motor components in autonomic control of the vasculature are the sympathetic nerves, which usually exert vasoconstrictor control via β-adrenoceptor sites. The adrenergic nerve fibers that supply most vessels in the body have their origin in the pre- or para-vertebral ganglia of the sympathetic nervous system, but it is possible that some blood vessels in the brain may be innervated by nerve fibers originating in central catecholamine-containing neurons (Edvinsson et al., 1973). Some non-adrenergic vasomotor nerves to various vessels have also been described (Su and Lee, 1976).

Inhibitory (vasodilator) nervous control is present in many blood vessels, but the nature of this control is less well understood. Cholinergic nerves have been identified in some vessels (Uvnäs, 1966; Schenk and El Badawi, 1968; Bell, 1969; Iwayama et al., 1970; Borodulya and Pletchkova, 1973; Motavkin and Palaschenko, 1973; Hudlická, 1973). In addition, a variety of other substances have been suggested to exert physiological vasodilator control, including histamine, prostaglandin, adenosine triphosphate, serotonin, dopamine, angiotensin, and bradykinin (Burnstock, 1972, 1975b; Su and Lee, 1976). Evidence has been presented that adenosine triphosphate is the transmitter released from non-adrenergic, non-cholinergic ("purinergic") nerves supplying the gastrointestinal tract, lung, bladder, trachea, esophagus, seminal vesicles, and since many vessels are known to be highly sensitive to adenine nucleotides and nucleosides, the possibility that purinergic nerves provide vasodilator control of parts of the vascular system has also been considered (Burnstock, 1972, 1975b). There is evidence for purinergic innervation of the rabbit portal vein (Hughes and Vane, 1967, 1970; Su and Sum, 1974; Su, 1975; Su and Lee, 1976) and for the physiological regulation of coronary, renal, skin, and skeletal muscle blood flow by adenine nucleosides or nucleotides (Berne, 1964; Forrester, 1972; Kiernan, 1972a,b; Paddle and Burnstock, 1974).

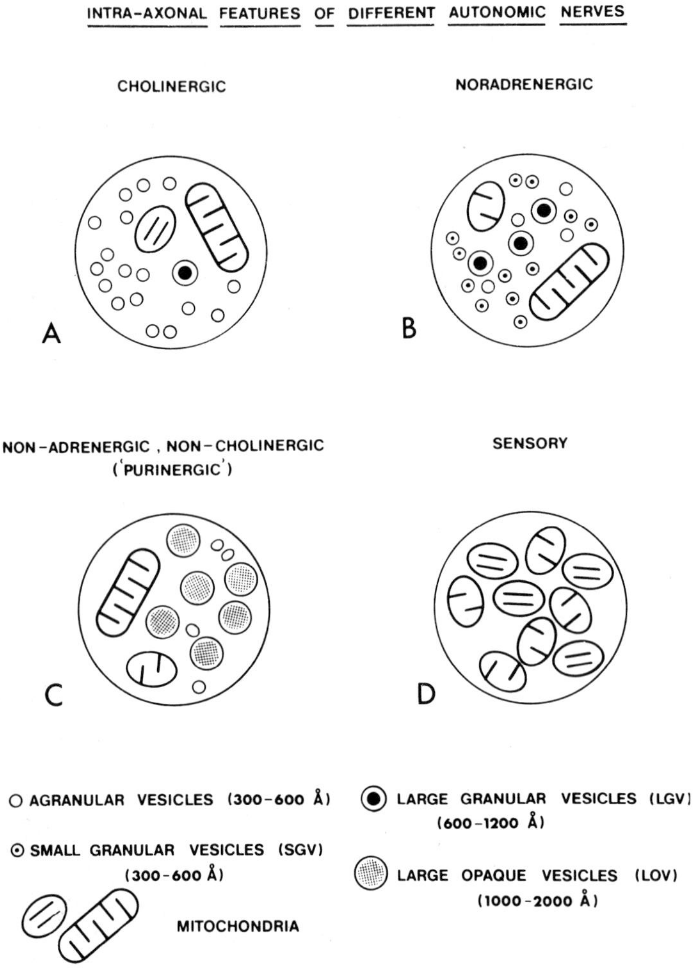

Fig. 5. Diagrammatic representation of sections through the terminal varicosities of autonomic nerves. For explanation see text (modified from Burnstock and Iwayama, Progr. in Brain Res., 34, 389-404, 1971).

Sensory neurons, many of which are myelinated, are commonly found in many vessels, particularly those supplying carotid, pial, coronary, and cerebral arteries (Hagen and Wittkowski, 1969; Burnstock et al., 1970) and vessels in skeletal muscle (Abraham, 1964; Hudlická, 1973).

The types of vesicles found within autonomic axon profiles (Fig. 5) have been used to identify the different types of autonomic nerves found in relation to blood vessels (Lever and Esterhuizen, 1961; Rhodin, 1967; Devine and Simpson, 1968; Lever et al., 1968; Iwayama et al., 1970; Nielsen et al, 1971; Lorez et al., 1973; Burnstock and Bell, 1974; Cervós-Navarro and Matakas, 1974; Burnstock, 1975c).

1. Adrenergic Nerves

Considerable evidence has accumulated that nerves containing norepinephrine are characterized by the predominance of small granular vesicles (30-60nm) with a dense core (Fig. 5B; Burnstock and Costa, 1975). Some large granular vesicles (60-120nm) are also present; there is evidence that these are also capable of taking up and storing catecholamines (Tranzer and Thoenen, 1968a, b; Hökfelt, 1968, 1971; Bennett et al., 1970; Furness et al., 1970a). The low percentage of small agranular vesicles present in varicosities of adrenergic nerves represent "empty" small granular vesicles, since drug treatment leading to increase in levels of norepinephrine in the nerves is associated with a significant increase in the percentage of small granular vesicles (Thoenen et al., 1966; Van Orden et al., 1966; Tranzer et al., 1969). Profiles of this kind have been described in a wide variety of vessels. Great care must be taken in relating norepinephrine storage levels to granular vesicles, because of the problems of granule preservation which varies widely with different fixatives in different tissues and species (Tranzer et al., 1969; Bloom, 1970; Hökfelt, 1971; Iwayama and Furness, 1971; Burnstock, 1975c). In lower vertebrates, many adrenergic profiles contain predominantly large granular vesicles.

2. Cholinergic Nerves

It is generally accepted that profiles containing predominantly small agranular vesicles (35-60nm) represent cholinergic nerves (Fig. 5A; De Robertis and Bennett, 1955; Whittaker et al., 1964; Grillo, 1966; Burnstock and Robinson, 1967; Burnstock and Iwayama, 1971). A few large granular vesicles (60-120nm) are usually also present; they do not take up catecholamines, 5- or 6-hydroxydopamine (Tranzer et al., 1969; Bennett et al., 1970) and their function is not known. Profiles with these features have been described in a number of blood vessels, including pial arteries (Fig. 10, and Iwayama et al., 1970; Edvinsson et al., 1972; Denn and Stone, 1976; Burnstock, 1979) and uterine arteries (Bell, 1969).

3. Purinergic Nerves

Purinergic nerve profiles appear to be characterized by a predominance of large vesicles, termed "large opaque vesicles," in order to distinguish them from the "large granular vesicles" found in small numbers (3-5%) in most adrenergic and cholinergic nerves (Fig. 5C; Robinson et al., 1971; Burnstock and Iwayama, 1971; Burnstock, 1972, 1975b). Large opaque vesicles differ from large granular vesicles in the following ways: large opaque vesicles are larger (100-200nm) than large granular vesicles (60-120nm); while large granular vesicles are characterized by a well-defined electron-transparent halo between the granular core and the vesicle membrane, large opaque vesicles do not have a prominent halo, although there appears to be some species variation in this respect; the granular matrix in large opaque vesicles is usually less dense than the granular core of large granular vesicles; large granular vesicles are at first "loaded," then depleted and destroyed by 6-hydroxydopamine, while large opaque vesicles are unaffected, as is the response to purinergic nerve stimulation. Large opaque vesicles are also unaffected by the catecholamine-depleting drugs reserpine, metaraminol, and guanethidine. Evidence for purinergic innervation of some blood vessels has been reviewed by Burnstock, 1978, 1979.

4. Sensory Nerves

Axon profiles which are quite different from those characteristic of cholinergic and adrenergic nerves have been observed in a variety of tissues, including vas deferens (Merrillees, 1968), pial artery (Hagen and Wittkowski, 1969), rat anterior cerebral artery (Burnstock et al., 1970) and avian ureter (Burnstock and Iwayama, 1971). They contain few, if any, vesicles and are packed with small, oval mitochondria (Fig. 5D). In the rat cerebral artery, these profiles have been traced back by serial sampling to myelinated nerve fibers, suggesting that they may represent sensory nerves (Burnstock et al., 1970), a feature which has also been demonstrated in longitudinal section by Hagen and Wittkowski, 1969.

5. Extra-adrenal Chromaffin Cells

Finally, the contribution of processes of extra-adrenal chromaffin cells to the nerve plexuses around blood vessels must be considered (McLean and Burnstock 1966; Costa and Furness, 1973; Lorez et al., 1973; Bennett et al., 1974). These processes contain very large irregular granules up to 300nm in diameter (Burnstock et al., 1970). Drugs used to deplete or denervate adrenergic nerves do not necessarily affect the monoamines contained in chromaffin cells, so care must be exercised concerning the conclusions reached from pharmacological experiments of this kind on control of vessel function.

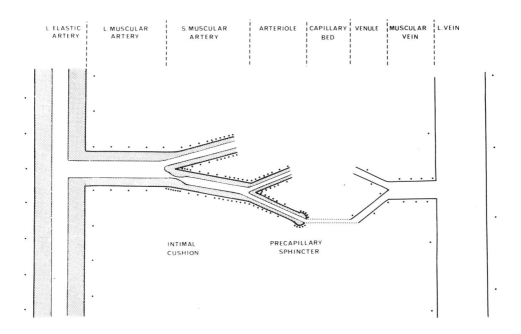

Fig. 6. Diagrammatic representation of innervation density in different regions of the vascular system (from Burnstock, Clin. exp. Pharmacol. physiol., Suppl. 2, 7-20, 1975).

V. VARIATIONS IN INNERVATION OF DIFFERENT ARTERIES

There is wide variation in the pattern and density of innervation of different vessels (Burnstock et al., 1970; Gillespie and Rae, 1972; Burnstock, 1975a; Su and Lee, 1976), depending to a large extent on their physiological role in relation to the particular organ system they supply. Nevertheless, some broad generalizations can be made (Fig. 6). Most large elastic arteries are sparsely innervated. As the arteries get smaller, the density of innervation increases, with a peak in small arteries and large arterioles. "Intimal cushions" at branching sites of arteries are also densely innervated. Most precapillary arterioles are not innervated, but some (although not all) precapillary sphincters are richly supplied by nerves. It is usually assumed that capillaries are not innervated, and most collecting venules and small veins have an extremely sparse nerve supply. Most large veins are not well supplied by adrenergic nerves, but some medium-size muscular veins have rich innervation. A more detailed survey of the variations in innervation pattern of different vessels follows.

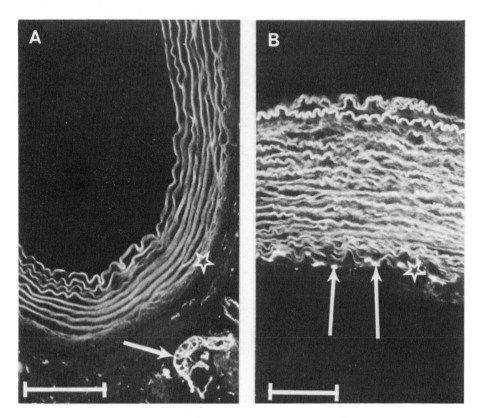

Fig. 7. Fluorescence histochemical localization of adrenergic nerves in various arteries.

    A. Rat aorta, T.S. Note that no fluorescent nerves are seen in the adventitia (asterisk) of the aorta, but some (arrow) are associated with small blood vessels in the surrounding tissue.

    B. Adrenergic innervation of the rabbit thoracic aorta. Transverse section of aorta showing a plexus of adrenergic nerves (arrows) at the border of the adventitia (asterisk) and the media. Calibration 100 mm. Freeze-dried tissues incubated in formaldehyde vapor for 1 hour at 80°C. (from Wright, MSc thesis, Melbourne University, (1972).

1. Large Elastic Arteries

While it is generally true that large elastic arteries are not well supplied by sympathetic nerves, there is considerable species variation in density of innervation (Figs. 7A, B; and Čech and Doležel, 1967; Burnstock, et al., 1970; Bevan et al., 1972; Doležel et al., 1973). For example, few, if any, adrenergic nerves can be seen in the rat aorta; in the cat, dog, calf, pig, and rhesus monkey there is moderate innervation; while the guinea-pig aorta has a relatively dense nerve supply. The density of innervation of rabbit thoracic aorta was calculated to be about 25% of that of ear artery in terms of the number of varicosities/mm^2 surface area in the perivascular plexus (Bevan et al., 1972). Nerve varicosities are widely separated from muscle, with a minimum of about 2,000nm reported for rabbit pulmonary artery (Bevan and Su, 1974) and rabbit thoracic aorta (Bell and Vogt, 1971; Bevan et al., 1972). The response of elastic arteries to nerve stimulation is slow in onset, and weak compared to that of smaller arteries; a higher frequency of stimulation is needed to elicit a response, and no discrete junction potentials have been recorded. These large blood vessels therefore appear to be mainly under humoral control by circulating catecholamines (at least under certain physiological conditions), while adrenergic nerve control is weak and sluggish. Preparations of smooth muscle from the inner part of the media of sheep carotid artery gave 50% maximal contractions in response to approximately 1/15 the concentration of norepinephrine needed to produce similar responses from outer smooth muscle; this difference persisted after sympathetic denervation, so

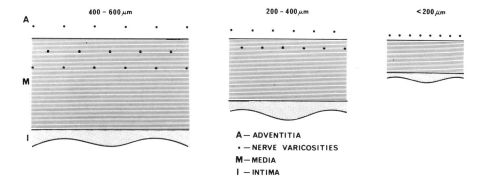

Fig. 8. Arrangement of nerves in arteries of different wall thickness. A, adventitia; •, nerve varicosities; M, media; I, intima. (From Burnstock, Clin. exp. Pharmacol. Physiol. Suppl. 2, 7-20, 1975.)

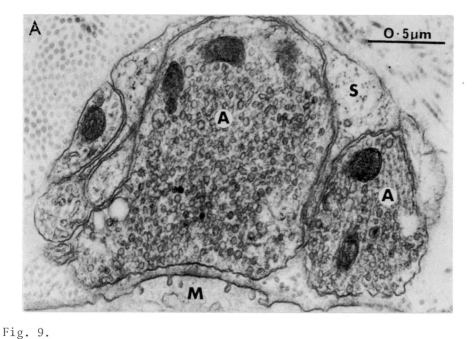

Fig. 9.
    A. Relation of axons (A) and smooth muscle (M) at the adventitial-medial border of the anterior cerebral artery of the rat. Axons are devoid of Schwann cytoplasm (S) on the side facing the muscle and approach the muscle surface as near as 800Å. Basement membrane material is interposed between axonal and smooth muscle membranes. There are many synaptic vesicles and mitochondria in the terminal varicosities. Some small and large granular vesicles are also present.

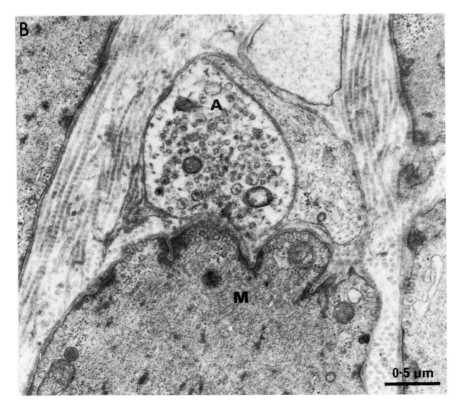

Fig. 9.
B. Adrenergic nerve fibers within the media of the sheep carotid artery. Electron micrograph of an axon profile (A) with Schwann cell (S) approaching to within 1,000Å of the surface of a smooth muscle cell (M). Basement membrane lies between the axon and the muscle cell, and many synaptic vesicles, some of which are granular, are seen in the axon profile.

cannot be attributed to uptake of norepinephrine by nerve fibers in the outer smooth muscle preparations (Graham and Keatinge, 1972; Keatinge and Torrie, 1975).

While the adrenergic nerves supplying elastic arteries in some species appear to be confined to the adventitial-medial border, there have been a number of reports of adrenergic nerves penetrating into the media of large arteries in others (Fig. 9B; Keatinge, 1966; Ehinger et al., 1967; Fillenz, 1967; Tsunekawa et al., 1967; Mohri et al., 1969; Burnstock et al., 1970; Bevan and Purdy, 1973; White, Ikeda, and Elsner, 1973). There is some evidence to suggest that the degree of penetration is largely related to wall thickness (Fig. 8; Ehinger et al., 1967; Doležel, 1972; Bevan and Purdy, 1973; Costa and Furness, 1974). Thus, in large animals such as dog, pig, sheep, cow, seal, and man, nerves commonly penetrate the media of large elastic and muscular arteries, whereas penetration of even the largest vessels in mouse, rat, guinea pig, and even rabbit is rare. However, there are some reports of adrenergic nerve penetration into the media of relatively small vessels, e.g., rabbit saphenous artery (Bevan and Purdy, 1973) and the aorta of fish and amphibians (Kirby and Burnstock, 1969), so that other factors may also need to be taken into account. It must also be pointed out that many of the nerves found inside the media of very large vessels accompany the vasa vasorum (Ehinger et al., 1967; Norberg, 1967; Doležel, 1972; Ellison, 1974).

2. Muscular Arteries

In general, these arteries are more heavily innervated than elastic arteries (De la Lande and Waterson, 1967; Ehinger et al., 1967; Fillenz, 1967). Minimum nerve-muscle separation is less (200-500nm) in large and medium-size muscular arteries such as rabbit ear, guinea-pig uterine, rabbit saphenous and popliteal, and rat mesenteric arteries (Fig. 9A; Bell and Vogt, 1971; Bevan and Purdy, 1973; Bevan and Su, 1974), but in smaller muscular arteries can be as little as 80nm (Bevan and Su, 1974). Evidence for myogenic propagation of excitation through the thickness of the vessel wall from the adventitia has been presented (Bevan and Waterson, 1971), and excitatory junction potentials have been recorded from cells of the inner media remote from the nerve plexus, indicating that the muscle cells are electrically coupled (Holman, 1969; Speden, 1970). Stimulation of the adrenergic nerves results in a prompt vasoconstriction.

In many arteries there is a fine plexus around the media itself ("perivascular" nerves) that should be distinguished from larger nerves ("paravascular" nerves) which are on their way farther down to supply both vascular and non-vascular tissue (Costa and Furness, 1974). Arterial anastomoses such as those in the wall of the stomach are innervated to about the same extent as the bordering arteries.

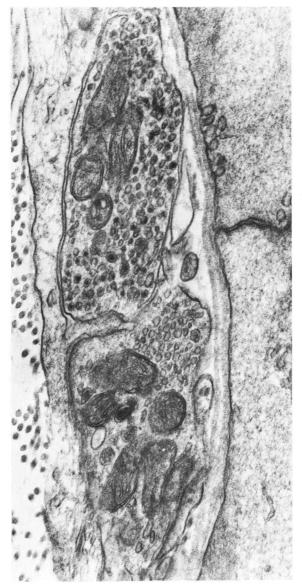

Fig. 10. Axon profiles near the anterior cerebral artery of a rat injected with 6-hydroxydopamine (250 mg/kg) 1 hour before decapitation. One axon has many small vesicles with distinct electron-opaque cores and also large vesicles with dense granulation. The other axon contains no small granular vesicles but many small agranular vesicles and a few large vesicles with moderately granular cores. Osmium fixation. x 48,000 (from Iwayama, Furness, and Burnstock, Circulation Res., 26, 635-646, 1970).

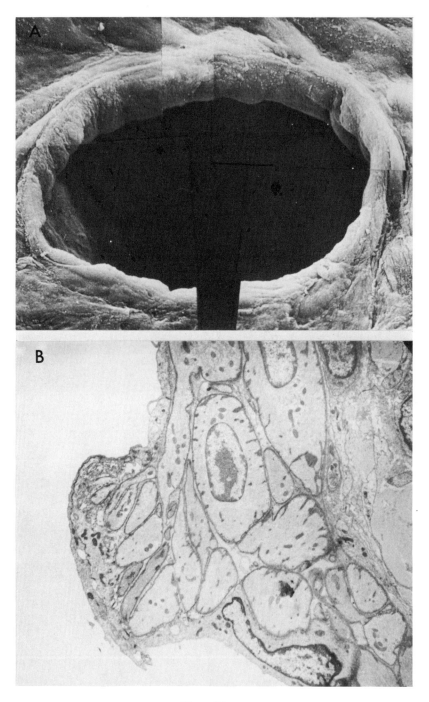

Fig. 11.

The innervation of the anterior mesenteric artery of the chicken is unusual, since in addition to the usual asymmetrical localization of nerves supplying a circular layer of smooth muscle, there is an outer longitudinal layer of smooth muscle which has an innervation pattern comparable to that seen in visceral muscle (Bolton, 1968; Campbell, Johnston, and Bell, personal communication).

It is now well established that extracerebral (pial) arteries have a rich supply of both adrenergic and cholinergic nerves (Fig. 10; Neilsen and Owman, 1967: Iwayama et al., 1970; Edvinsson et al., 1972). The existence of close apposition (about 25nm) of adrenergic and cholinergic nerve profiles in the wall of these vessels (Neilsen et al., 1971) raises the interesting possibility of physiological interaction between these two autonomic nerve types. The innervation of intracerebral arteries and arterioles is more controversial. Until recently, most workers considered that innervation of intracerebral vessels was largely absent (Rhodin, 1962). However, it now seems likely that many of these vessels are supplied by a sparse plexus of adrenergic nerves (Owman et al., 1974). The point of contention is whether some intracerebral arteries are normally supplied by adrenergic fibers of non-sympathetic origin, originating from an intracerebral source, probably the locus coeruleus (Hartman et al., 1972; Owman et al., 1974). There is little doubt that reinnervation of denervated vessels by central adrenergic neurons can occur (Katzman et al., 1971; Björklund et al., 1975).

3. "Intimal Cushions"

Thickenings have been described, known as intimal cushions, at the branching sites of a number of vessels, including the coronary (Yohro and Burnstock, 1973a), cerebral (Hassler, 1962; Takayanagi et al., 1972), renal (Moffat and Creasey, 1971), thyroid (Reale and Luziano, 1966), and ciliary (Jellinger, 1974) arteries. They consist of an elaborate arrangement of processes of smooth muscle cells and are often heavily innervated (Fig. 11).

←Fig. 11.
- A. A scanning electronmicrograph of an intimal cushion at the orifice of the septal (coronary) artery of rat. The ridge is well-developed around the orifice. x 1100.
- B. A transmission electronmicrograph of an intimal cushion at a branching site of a smaller coronary artery (rat). The ridge contains only the processes of smooth muscles. The top of the ridge is nearly flat. Compare with the scanning micrograph shown in Fig. A. Note the nerve profiles at the inner border of the muscle cells. x 5000. (From Yohro and Burnstock, Z. Anat. Entwickl. Gesch., 140, 187-202, 1973.)

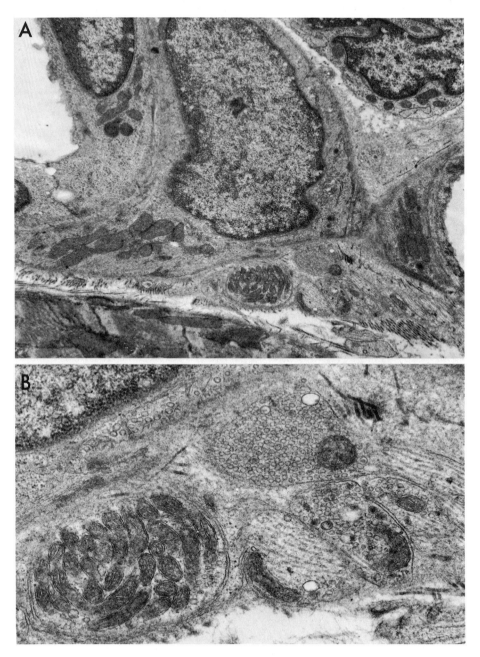

Fig. 12.

← Fig. 12.
  A. Precapillary sphincter region in rat coronary vascular bed. Note several nerve profiles at the branching site. Glutaraldehyde-osmium fixation. x 12,000.
  B. A higher magnification of part of Fig. 12A, showing axon profiles of at least three types: one is packed with mitochondria and may be sensory; another contains a predominance of small agranular vesicles and is probably cholinergic; another contains granular vesicles and is probably adrenergic. Glutaraldehyde-osmium fixation x 36,000. (Courtesy of Yohro and Burnstock, unpublished data.)

4. Arterioles and Precapillary Sphincters

The border between small arteries and large arterioles is not well defined, but both are very important in the regulation of blood pressure and are heavily supplied by adrenergic nerves (Norberg and Hamberger, 1964; Dahlström and Fuxe, 1965; Lever et al., 1967; Devine and Simpson, 1968). There is wide variation, however, in the density of innervation of smaller arterioles, which have a sparse nerve supply in the coronary or cerebral circulation (Samarasinghe, 1965; Dahl, 1973; Cervós-Navarro and Matakas, 1974), or are relatively well supplied with nerves in the mesenteric circulation (Furness, 1973). There is a similar variation in innervation density of precapillary sphincters from dense innervation by sensory, adrenergic, and cholinergic nerves in the coronary circulation (Fig. 12; Yohro and Burnstock, unpublished results) to absence of any nerves in the mesenteric vasculature of the rat (Furness, 1973). However, since these latter vessels are very sensitive to exogenously applied catecholamines, they may be controlled to some extent by circulating hormones (Furness and Marshall, 1974; Altura, 1971; Baez, 1977).

5. Capillaries

It is still not clear whether capillaries are innervated or not (Hudlická, 1973; Burnstock, 1975). Few nerves have been seen in relation to capillaries in the heart. On the other hand, capillaries in the cat hypothalamus appear to be innervated, possibly by fibers derived from central neurons (Rennels and Nelson, 1975). There are also many nerve fibers closely aligned to capillaries in the intestine, but it may be that the nerves and capillaries are merely running in a common space between the muscle cells. The separation of nerves from these capillaries is less than their separation from smooth muscle cells in the intestine which are considered to be affected by released transmitter, and contractile filaments have been demonstrated in vascular endothelial cells (Yohro and Burnstock, 1973b). Nevertheless, functional transmission cannot be assumed on the basis of morphology alone.

6. Veins

In general, veins have a less dense perivascular nerve plexus than arteries. For example, in the mesentery, the plexus about veins is considerably less dense than that surrounding the arteries that run alongside them. However, there is wide variation in density and pattern of innervation of different veins (Bevan et al., 1974). For example, the density of innervation of mesenteric veins (as well as of cephalic, pulmonary, and small saphenous veins of rabbit) is greater than that of renal, jugular, central ear veins, and inferior vena cava, while the femoral, brachial, deep circumflex, and superficial cervical veins are not innervated at all (Pegram et al., 1973). Minimum separation of nerve varicosities and muscle cells is about 100-150nm, for example, in rat portal vein (Holman et al., 1968; Booz, 1971). Collecting venules and small veins appear to be sparsely supplied, if at all, by adrenergic nerves (Samarasinghe, 1965; Rhodin, 1967; Furness, 1973). However, some muscular veins are heavily innervated by adrenergic nerves. For example, portal-mesenteric or renal veins have, in addition to a plexus at the adventitial-medial border, a longitudinal muscle coat that resembles visceral muscle. This coat lies outside the perivascular plexus and is richly supplied throughout by varicose adrenergic nerves (Holman and McLean, 1967; Holman et al., 1968; Burnstock et al., 1970; Bennett and Malmfors, 1970; Johansson et al., 1970). Larger muscular veins, including the saphenous, facial, and penile veins are also quite heavily innervated, but the large conducting veins have a sparse supply (Ehinger et al., 1967). The wide veins of the erectile tissue in the inferior concha have an unusually rich adrenergic innervation (Dahlström and Fuxe, 1965). The subendothelial "cushions" found in veins in nasal mucosa also appear to be richly supplied by nerves (Cauna and Cauna, 1975).

IV. DEVELOPMENT OF SYMPATHETIC INNERVATION OF BLOOD VESSELS

The development of innervation of most blood vessels consists of a gradual increase during pre- and postnatal stages with some decline in old age. However, the innervation of some vessels such as femoral and brachial arteries of dog, aorta of rabbit, and temporal arteries of man, declines in early postnatal life, perhaps reflecting a progressive reduction in the role of these arteries in regulation of peripheral resistance (Shibata et al., 1971; Doležel, 1972, 1973; Gerová et al., 1974; Burnstock, 1975; Rickenbacher and Ruflin, 1974).

The ductus arteriosus is one of the first effector organs to receive adrenergic fibers during ontogenesis. During fetus, it becomes fully innervated, but it is obliterated by constriction just after birth, possibly via adrenergic and/or prostaglandin mechanisms

THE INNERVATION OF ARTERIES

(Folkow and Neil, 1971; Olley et al., 1975). Adrenergic vasoconstrictor fibers to the vasculature of the hindlimb of the dog only become effective two weeks after birth (Boatman et al., 1965), and it has been suggested that the increase of blood pressure during postnatal development is due to increasing activity of the autonomic nervous system with the establishment of a sympathetic vasoconstrictor tone in the systemic circulation (Mott, 1961). Extra-adrenal chromaffin tissue is in prominent relation to many blood vessels during early stages of development (Coupland, 1965). It seems likely that nerve growth factors is involved in the programming of sequential development of innervation of different vascular smooth muscles (Burnstock and Costa, 1975).

SUMMARY

1. Attention is drawn to differences in the pattern of adrenergic innervation of visceral smooth muscles such as the vas deferens and iris (where there is a dense varicose nerve plexus throughout and close, 20nm neuromuscular separations) compared to that of most blood vessels (where the adrenergic nerves are confined to the adventitial-medial border and neuromuscular separations are greater than 80nm. In both types of tissue there are gap junctions ("nexuses") representing low resistance pathways between smooth muscle cells.

2. A general model of artery innervation is proposed in which it is suggested that the asymmetric geometry of innervation of most vascular smooth muscles favors dual control by both nerves (responses of "directly-innervated" muscle cells at the adventitial side being propagated through to the intimal side) and by circulating catecholamines (which diffuse from the intimal to adventitial side, their action unimpaired by uptake into adrenergic nerves). Furthermore, the absence of adrenergic nerves in the media facilitates the passage of catecholamines from the adrenal glands to their peripheral target sites. In contrast, it is suggested that the geometry of innervation of most visceral smooth muscles not only provides efficient control by nerves, but also "protects" the organ from the unwanted rapid uptake into the abundant nerve varicosities.

3. Most vasomotor nerves supplying arteries are adrenergic, and some cholinergic vasodilator nerves have been demonstrated. However, there is growing evidence for non-adrenergic non-cholinergic innervation of many vessels. Many of these appear to be purinergic, but other substances claimed to be involved in different vessels include histamine, prostaglandin, serotonin, dopamine, angiotensin, bradykinin, vasoactive intestinal peptide, and adenosine triphosphate.

4. There are differences in the relative importance of nerves in control of different parts of the vascular tree. Most large

elastic arteries are sparsely innervated and nerve muscle separations are wide (1,000-2,000nm). As arteries get smaller, the density of innervation increases with a peak in small arteries and large arterioles. "Intimal cushions" at branching sites of arteries are also densely innervated. Most precapillary arterioles are not innervated, but some precapillary sphincters are richly supplied by nerves. With the exception of some muscular veins, the venous system is less densely innervated than the arterial. The relative importance of circulating catecholamines on different vessels is not well understood. Some precapillary arterioles are sensitive to the low concentrations of catecholamine normally circulating in the blood. Other vessels may only be affected by the higher concentrations of circulating catecholamines that occur during such situations as fright or heavy exercise.

5. While the development of innervation of most blood vessels consists of a gradual increase during pre- and postnatal stages, with some decline in old age, the innervation of some vessels such as femoral, brachial, and temporal arteries declines in early postnatal life.

REFERENCES

Abraham, A. (1964) Die mikroskopische innervation des herzens und der blutgefässe von vertebraten. Akadĕmiai Kiradó, Budapest.
Altura, B.M. (1971) Chemical and humoral regulation of blood flow through the precapillary sphincter. Microvasc. Res. 3, 361-384.
Baez, S. (1977) Central neural influence on precapillary microvessels and sphincter. Amer. J. Physiol. 233, H141-H147.
Barr, L., Berger, W., and Dewey, M.M. (1968) Electrical transmission at the nexus between smooth muscle cells. J. Gen. Physiol., 51, 347-368.
Bell, C. (1969) Fine structural localization of acetylcholinesterase at a cholinergic vasodilator nerve-arterial smooth muscle synapse. Circ. Res. 24, 61-70.
Bell, C. (1974) Release of endogenous noradrenaline from an isolated elastic artery. J. Physiol. (Lond.), 236, 473-482.
Bell, C., and Burnstock, G. (1971) Cholinergic vasomotor neuroeffector junctions. In: Symposium on the Physiology and Pharmacology of Vascular Neuroeffector Systems, Interlaken, 1969. (Ed. J.A. Bevan, R.F. Furchgott, R.A. Maxwell, and A.P. Somlyo), pp. 37-46, S. Karger, Basel.
Bell, C., and Vogt, M. (1971) Release of endogenous noradrenaline from an isolated muscular artery. J. Physiol. (Lond.). 215, 509-520.
Bennett, M.R. (1973) Structure and electrical properties of the autonomic neuromuscular junction. Philos. Trans. R. Soc. Lond. (B), 265, 25-34.

Bennett, M.R., and Burnstock, G. (1968) Electrophysiology of the innervation of intestinal smooth muscle. In: Handbook of Physiology, Section 6, Alimentary Canal, IV Motility. AM Physiol. Sec. (Washington) 1709-1732.

Bennett, T., Burnstock, G., Cobb, J.L.S., and Malmfors, T. (1970) An ultrastructural and histochemical study of the short term effects of 6-hydroxydopamine on adrenergic nerves in the domestic fowl. Brit. J. Pharmacol. Chemother. 38, 802-809.

Bennett, T., Cobb, J.L., and Malmfors, T. (1974) The vasomotor innervation of the inferior vena cava of the domestic fowl (Gallus gallus domesticus L.). Cell & Tiss. Res. 148, 521-533.

Bennett, T., and Malmfors, T. (1970) The adrenergic nervous system of the domestic fowl (Gallus domesticus). Z. Zellforsch., 106, 22-50.

Bergman, R.A. (1958) Intercellular bridges in ureteral smooth muscle Johns Hopkins Hosp. Bull. 102, 195-202.

Berne, R.M. (1964) Regulation of coronary flow. Physiol. Rev. 44, 1-29.

Bevan, J.A., Bevan, R.D., Purdy, R.E., Robinson, C.P., Su, C., and Waterson, J.G. (1972) Comparison of adrenergic mechanisms in an elastic and a muscular artery of the rabbit. Circ. Res. 30, 541-548.

Bevan, J.A., Hosmer, D.W., Ljung, B., Pegram, B.L., and Su, C. (1974) Norepinephrine uptake, smooth muscle sensitivity and metabolizing enzyme activity in rabbit veins. Circ. Res. 34, 541-548.

Bevan, J.A., and Ljung, B. (1974) Longitudinal propagation of myogenic activity in rabbit arteries and in the rat portal vein. Acta. Physiol. Scand., 90, 703-715.

Bevan, J.A., and Purdy, R.E. (1973) Variations in the adrenergic innervation and contractile responses of the rabbit saphenous artery. Circ. Res. 32, 746-751.

Bevan, J.A., and Su, C. (1973) Sympathetic mechanisms in blood vessels: nerve and muscle relationships. Ann. Rev. Pharmacol., 13, 269-285.

Bevan, J.A., and Su, C. (1974) Variation of intra- and perisynaptic adrenergic transmitter concentrations with width of synaptic cleft in vascular tissue. J. Pharmacol., 190, 30-38.

Bevan, J.A., and Waterson, J.G. (1971) Biphasic constrictor response of the rabbit ear artery. Circ. Res., 28, 655-661.

Bierring, F., and Kobayasi, T. (1963) Electron microscopy of the normal rabbit aorta. Acta. Path. Microbiol. Scand., 57, 154-168.

Björklund, A., Johansson, B., Stenevi, U., and Svendgaard, N.A. (1975) Re-establishment of functional connections by regenerating central adrenergic and cholinergic axons. Nature 253, 446-448.

Bloom, F.E. (1970) The fine structural localization of biogenic monoamines in nervous tissue. Int. Rev. Neurobiol., 13, 27-66.

Boatman, D.L., Shaffer, R.A., Dixon, R.L., and Brody, M.J. (1965) Function of vascular smooth muscle and its sympathetic innervation in the newborn dog. J. Clin. Invest., 44, 241-246.

Bolton, T.B. (1968) Studies on the longitudinal muscle of the anterior mesenteric artery of the domestic fowl. J. Physiol. (Lond.), 196, 273-292.

Bolton, T.B. (1974) Electrical properties and constants of longitudinamuscle from the avian anterior mesenteric artery. Blood Vessels, 11, 65-78.

Booz, K.H. (1971) Zur Innervation der autonom pulsierenden Vena portae der weissen Ratte. Eine histochemische und elektronenmikroskopische Untersuchung. Z. Zellforsch, 117, 394-418.

Borodulya, A.V., and Pletchkova, E.K. (1973) Distribution of cholinergic and adrenergic nerves in the internal carotid artery; histochemical study. Acta anat., 86, 376-393.

Bukinich, A.D. (1973) Structure of adrenergic apparatus of mammalian and avian blood vessels. J. Evol. Biochem. Physiol. 9, 185-187.

Burnstock G. (1970) Structure of smooth muscle and its innervation in: Smooth Muscle (Eds. E. Bülbring, A. Brading, A. Jones, and T. Tomita), pp. 1-69. Edward Arnold, London.

Burnstock, G. (1972) Purinergic nerves. Pharmacol. Rev. 24, 509-581.

Burnstock, G. (1974) Innervation of blood vessels. In: The Smooth Muscle of the Arterial Wall. Plenum Press, New York (In press).

Burnstock, G. (1975a) Innervation of vascular smooth muscle: histochemistry and electron microscopy. Clin. exp. Pharmacol. Physiol., Suppl. 2, 7-20.

Burnstock, G. (1975b) Purinergic transmission. In: Handbook of Psycho-pharmacology (Eds. L. Iversen, S. Iversen and S. Snyder) Vol. 5, pp. 131-194. Plenum Press, New York.

Burnstock, G. (1975c) Ultrastructure of autonomic nerves and neuroeffector junctions; analysis of drug action. In: Methods in Pharmacology, Vol. 3, Smooth Muscle (Eds. E.E. Daniel and D.M. Paton), Chapter 5, pp. 113-137. Plenum Press, New York.

Burnstock, G. (1975d) Innervation of vascular smooth muscle. In: The Smooth Muscle of the Artery, (Eds. S. Wolf and N. T. Werthessen), Advances in Exp. Medicind and Biology, Vol. 57, pp. 20-34. Plenum Press, New York.

Burnstock, G. (1978) An assessment of purinergic nerve involvement in vasodilatation. In: Mechanisms of Vasodilatation, pp. 278-284, (Eds. P.M. Vanhoutte and I. Leusen), Karger, Basel.

Burnstock, G. (1979) Cholinergic and purinergic regulation of blood vessels. In: Handbook of Physiology, (Vascular Smooth Muscle), (Eds. D. Bohr, A.D. Somlyo, and H.V. Sparks), Amer. Physiol. Soc., The Williams & Wilkins Co., Baltimore, (in press).

Burnstock, G., and Bell, C. (1974) Peripheral autonomic transmission. In: The Peripheral Nervous System. (Ed. J.I. Hubbard), pp. 277-327. Plenum Press, New York.

Burnstock, G., and Costa, M. (1975) Adrenergic Neurons, Chapman and Hall Ltd., London.

Burnstock, G., Gannon, B., and Iwayama, T. (1970) Sympathetic innervation of vascular smooth muscle in normal and hypertensive animals. Circ. Res., 26 and 27 (Suppl. 11). 5-24.

Burnstock, G., and Iwayama, T. (1971) Fine structural identification of autonomic nerves and their relation to smooth muscle. Prog. Brain Res. 34, 389-404.

Burnstock, G., and Prosser, C.L. (1960) Conduction in smooth muscles; comparative electrical properties. Am. J. Physiol., 199, 553-559.

Burnstock, G., and Robinson, P.M. (1967) Localization of catecholamines and acetylcholinesterase in autonomic nerves. Circ. Res. 21, (Suppl. 3), 43-55.

Campbell, G.R., Uehara, Y., Mark, G., and Burnstock, G. (1971) Fine structure of smooth muscle cells grown in tissue culture. J. Cell Biol., 49, 21-34.

Cauna, N., and Cauna, D. (1975) The fine structure and innervation of the cushion veins of the human nasal respiratory mucosa. Anat. Rec., 181, 1-16.

Cech, S., and Doležel, S. (1967) Monoaminergic innervation of the pulmonary vessels in various laboratory animals (rat, rabbit, cat). Experientia, 23, 113-114.

Cervós-Navarro, J., and Matakas, F. (1974) Electron microscopic evidence for innervation of intracerebral arterioles in the cat. Neurol. (Minneapolis), 24, 282-286.

Chin, A.K., and Evanuk, E. (1971) Changes in plasma catecholamine and corticosterone levels after muscular exercise. J. Appl. Physiol., 30 205-207.

Cliff, W.J. (1967) The aortic tunica media in growing rats studied with the electron microscope. Lab. Invest., 17, 599-615.

Cobb, J.L.S., and Bennett, T. (1969) A study of nexuses in visceral smooth muscle. J. Cell. Biol., 41, 287-297.

Coupland, R.E. (1965) The Natural History of the Chromaffin Cell, Longmans Green, London.

Costa, M., and Furness, J.B. (1973) Observations on the anatomy and amine histochemistry of the nerves and ganglia which supply the pelvic viscera and on the associated chromaffin tissues in the guinea pig. Z. Anat. Entw-Gesch., 140, 85-105.

Costa, M., and Furness, J.B. (1974) (Personal communication).

Dahl, E. (1973) The innervation of the cerebral arteries. J. Anat. 115, 53-63.

Dahlström, A., Fuxe, K. (1965) The adrenergic innervation of the nasal mucosa of certain mammals. Acta oto-laryng., Stockh., 59, 65-72.

De la Lande, I.S., Cannell, V.A., and Waterson, J.G. (1966) The interaction of serotonin and noradrenaline on the perfused artery. Brit. J. Pharmacol. 28, 255-272.

De la Lande, I.S., Jellett, L.B., Lazner, M.A., Parker, A.S., and Waterson, J.G. (1974) Histochemical analysis of the diffusion of noradrenaline across the artery wall. Aust. J. exp. Biol. med. Sci., 52, 193-200.

De la Lande, I.S., and Waterson, J.G. (1967). Site of action of cocaine on the perfused artery. Nature, 214, 313.

Denn, M.J., and Stone, H.L. (1976) Cholinergic innervation of monkey cerebral vessels. Brain Res., 113, 394-399.

De Robertis, E., and Bennett, H.S. (1955) Some features of the submicroscopic morphology of synapses in frog and earthworm. J. biophys. biochem. Cytol., 1, 47-65.

Devine, C.E., and Simpson, F.O. (1968) Localization of tritiated norepinephrine in vascular sympathetic axons of the rat intestine and mesentery by electron microscope radioautography. J. Cell Biol., 38, 184-192.

Devine, C.E., Somlyo, A.V., and Somlyo, A.P. (1972) Sarcoplasmic reticulum and excitation-contraction coupling in mammalian smooth muscles. J. Cell Biol., 52, 690-718.

Dewey, M.M., and Barr, L. (1962) Intercellular connection between smooth muscle cells: The Nexus. Science (N.Y.) 137, 670-671.

Dewey, M.M., and Barr, L. (1968) Structure of vertebrate intestinal smooth muscle. In: Handbook of Physiology, Section 6. Alimentary canal, Vol. IV, Motility, Am. Physiol. Soc. (Washington), 1629-1654.

Doležel, S. (1972) Monoaminergic innervation of the aorta. Folia morpholog. 20, 14-20.

Doležel, S. (1973) Über die Variabilität der adrenergen innervation der grassen Gefässe. Acta Anat. 85, 123-132.

Doležel, S., Gerová, M., and Gero, J. (1973) Sympathetic constriction and monoaminergic innervation of large arteries. Folia morphol., 21, 364-367.

Edvinsson, L., Lindvall, M., Nielsen, K.C., and Owman, Ch. (1973) Are brain vessels innervated also by central (non-sympathetic) adrenergic neurones? Brain Res., 63, 496-499.

Evinsson, L., Nielsen, K.C., Owman, Ch., and Sporrong, B. (1972) Cholinergic mechanisms in pial vessels. Histochemistry, electron microscopy and pharmacology. Z. Zellforsch., 134, 311-325.

Ehinger, B., Falck, B., and Sporrong, B. (1967) Adrenergic fibres to the heart and to peripheral vessels. Bibl. anat., 8, 35-45.

Ellison, J.P. (1974) The adrenergic cardiac nerves of the cat. Am. J. Anat., 139, 209-226.

Euler, U.S. Von (1938) Action of adrenaline, acetylcholine and other substances on nerve-free vessels (human placenta). J. Physiol. (Lond.), 93, 129-143.

Fillenz, M. (1967) Innervation of blood vessels of lung and spleen. Bibl. anat., 8, 56-59.

Folkow, B., and Neil, E. (1971) Circulation, Oxford University Press, New York.

Forrester, T. (1972) A quantitative estimation of adenosine triphosphate released from human forearm muscle during sustained exercise. J. Physiol. (Lond.) 220, 26-27P.

Friend, D.S., and Gilula, N.B. (1972) Variations in "tight" and "gap" junctions in mammalian tissues. J. Cell Biol., 53, 758-776.

Fry, G.N., Devine, C.E., and Burnstock, G. (1977) Freeze-fracture studies of nexuses between smooth muscle cells; close relationship to sarcoplasmic reticulum. J. Cell Biol., 72, 26-34.

Furness, J.B. (1973) Arrangement of blood vessels and their relation with adrenergic nerves in the rat mesentery. J. Anat., 115, 347-364.

Furness, J.B., and Burnstock, G. (1975) Role of circulating catecholamines in the gastrointestinal tract. In: Handbook of Physiology, Section 7, Endocrinology, Vol 6. Eds. H. Blaschko and A.D. Smith, Am. Physiol. Soc., Washington. pp. 515-536.

Furness, J.B., McLean, J.R., and Burnstock, G. (1970) Distribution of adrenergic nerves and changes in neuromuscular transmission in the mouse vas deferens during postnatal development. Devel. Biol., 21, 491-505.

Furness, J.B., and Marshall, J.M. (1974) Correlation of the directly observed responses of mesenteric vessels of the rat to nerve stimulation and noradrenaline with the distribution of adrenergic nerves. J. Physiol. (Lond.), 239, 75-88.

Furshpan, E.J., and Potter, D.D. (1968) Low resistance junctions between cells in embryos and tissue culture. In: Current Topics in Developmental Biology (Ed. A.A. Moscona). Associated Press, N.Y., Vol. 3, 95.

Gabella, G. (1972) Intercellular junctions between circular and longitudinal intestinal muscle layers. Z. Zellforsch, 125, 191-199.

Gabella, G. (1974) The sphincter pupillae of the guinea-pig: structure of muscle cells, intercellular relations and density of innervation. Proc. Roy. Soc. Lond. B., 186, 369-386.

Gerová, M., Gero, J., Doležel, S., and Konečny, M. (1974) Postnatal development of sympathetic control in canine femoral artery. Physiol. Bohemoslov. 23, 289-295.

Gillespie, J.S., and Rae, R.M. (1972) Constrictor and compliance responses of some arteries to nerve or drug stimulation. J. Physiol., 223, 109-130.

Graham, J.M., and Keatinge, W.R. (1972) Differences in sensitivity to vasoconstrictor drugs within the wall of the sheep carotid artery. J. Physiol. (Lond.) 221, 477-492.

Grigor'eva, T.A. (1962) The Innervation of Blood Vessels. Pergamon Press (Oxford).

Grillo, M.A. (1966) Electron microscopy of sympathetic tissues. Pharmacol Rev., 18 387-399.

Hagen, E., and Wittkowski, W. (1969) Licht und electronenmikroskopische Untersuchung zur Innervation der Piagefasse. Z. Zellforsch, 95, 429-444.

Hartman, B.K., Zide, D., and Udenfriend, S. (1972) The use of dopamine-β-hydroxylase as a marker for the central noradrenergic nervous system in rat brain. Proc. Nat. Acad. Sci. U.S.A., 69, 2722-2726.

Hassler, O. (1962) Physiological intimal cushions in the large cerebral arteries of young individuals. 1. Morphological structure and possible significance for the circulation. Acta path. Microbiol. scand., 55, 19-27.

Henderson, R.M. (1975) Types of cell contacts in arterial smooth muscle. Experientia, 31, 103-105.

Henderson, R.M., Duchon, G., and Daniel, E.E. (1971) Cell contacts in duodenal smooth muscle layers. Am. J. Physiol. 221, 564-574.

Hillarp. N.A. (1959) On the histochemical demonstration of adrenergic nerves with the osmic acid-sodium iodide technique. Acta anat. (Basel), 38, 379-384.

Hökfelt, T. (1963) In vitro studies on central and peripheral monoamine neurons at the ultrastructural level. Z. Zellforsch, 91, 1-74.

Hökfelt, T. (1971) Ultrastructural localization of intraneuronal monoamines - some aspects of methodology. Progr. Brain Res., 34, 213-222.

Holman, M.E. (1969) Electrophysiology of vascular smooth muscle. Ergebn Physiol., 61, 137-145.

Holman, M.E., Kasby, C.G., Suthers, M.B., and Wilson, J.A.F. (1968) Some properties of the smooth muscle of rabbit portal vein. J. Physiol. (Lond.), 196, 111-132.

Holman, M.E., and McLean, J. (1967) Innervation of sheep mesenteric veins. J. Physiol. (Lond.), 190, 55-69.

Hudlická, O. (1973) Muscle Blood Flow. Its Relation to Muscle Metabolism and Function, Swets and Zeitlinger, B.V., Amsterdam.

Hughes, J., and Vane, J.R. (1967) An analysis of the responses of the isolated portal vein of the rabbit to electrical stimulation and to drugs. Brit. J. Pharmacol., 30, 46-66.

Hughes, J., and Vane, J.R. (1970) Relaxations of the isolated portal vein of the rabbit induced by nicotine and electrical stimulation. Brit. J. Pharmacol. 39, 476.

Hung, K.S., and Loosli, C.G. (1977) Electron microscopic studies of innervation of pulmonary veins of mouse. Act. Anatom., 97, 97-102.

Iwayama, T., and Furness, J.B. (1971) Enhancement of the granulation of adrenergic storage vesicles in drug-free solution. J. Cell Biol., 48, 699-703.

Iwayama, T., Furness, J.B., and Burnstock, G. (1970) Dual adrenergic and cholinergic innervation of the cerebral arteries of the rat. An ultrastructural study. Circ Res., 26, 635-646.

Jellinger, K. (1974) Intimal cushions in ciliary arteries of the dog. Experientia, 30, 188-189.

Johansson, B., Ljung, B., Malmfors, T., and Olson, L. (1970) Prejunctional and supersensitivity in the rat portal vein as related to its pattern of innervation. Acta physiol. scand., Suppl., 349, 5-16.

Katzman, R., Björklund, A., Owman, C., Stenevi, U., and West, K.A. (1971) Evidence for regeneration axon sprouting of central catecholamine neurons in the rat mesencephalon following electrolytic lesions. Brain Res., 25, 579-596.

Keatinge, W.R. (1966) Electrical and mechanical responses of arteries to stimulation of sympathetic nerves. J. Physiol. Lond.), 185, 701-715.

Keatinge, W.R., and Torrie, C. (1975) Response of inner and outer muscle of sheep carotid artery to noradrenaline and to activation of sympathetic nerves. J. Physiol. (Lond.), 244, 77-78P.

Kiernan, J.A. (1972a) The involvement of mast cells in vasodilatation due to axon reflexes in injured skin. Quart. J. exp. Physiol., 57, 311-317.

Kiernan, J.A. (1972b) Effects of known and suspected neurotransmitter substances and of some nucleotides on mast cells. Experientia, 28, 653-655.

Kirby, S., and Burnstock, G. (1969) Comparative pharmacological studies of isolated spiral strips of large arteries from lower vertebrates. Comp. Biochem Physiol., 28 307-319.

Lever, J.D., and Esterhuizen, A.C. (1961) Fine structure of the arteriolar nerves in the guinea-pig pancreas. Nature, 192, 566-567.

Lever, J.D., Graham, J.D., and Spriggs, T.L.B. (1967) Electron microscopy of nerves in relation to the arteriolar wall. Bibl. anat., 8, 51-55.

Lever, J.D., Spriggs, T.L.B., and Graham, J.D.P. (1968) A formol-fluorescence, fine-structural and autoradiographic study of the adrenergic innervation of the vascular tree in the intact and sympathectomized pancreas of the cat. J. Anat., 103, 15-34.

Lorez, H.P., Kuhn, H., and Tranzer, J.P. (1973) The adrenergic innervation of the renal artery and vein of the rat. A fluorescence histochemical and electron microscopical study. Z. Zellforsch., 138, 261-272.

Lowenstein, W. (1973) Membrane junctions in growth and differentiation. Fed. Proc., 32, 60-64.

McLean, J.R. and Burnstock, G. (1966) Histochemical localization of catecholamines in the urinary bladder of the toad (Bufo marinus). J. Histochem. Cytochem., 14, 538-548.

McNutt, N.S., and Weinstein, R.S. (1973) Membrane ultrastructure at mammalian intercellular junction. Progr. Biophys. Mol. Biol., 26, 45-101.

Matthews, M.A., and Gardner, D.L. (1966) The fine structure of the mesenteric arteries of the rat. Angiology, 17, 902-931.

Mekata, F. (1971) Electrophysiological studies of the smooth muscle cell membrane of the rabbit common carotid artery. J. Gen. Physiol., 57, 738-751.

Mekata, F. (1974) Current spread in the smooth muscle of the rabbit aorta. J. Physiol., 242, 143-155.

Merrillees, N.C. (1968) The nervous environment of individual smooth muscle cells of the guinea-pig vas deferens. J. Cell Biol., 37, 794-817.

Moffat, D.B., and Creasey, M. (1971) The fine structure of the intraarterial cushions at the origins of the justamedullary afferent arterioles in the rat kidney. J. Anat., 110, 409-419.

Mohri, K., Ohgushi, N., Ikeda, M., Yamamoto, K., and Tsunekawa, K. (1969) Histochemical demonstration of adrenergic fibers in smooth muscle layer of media of arteries supplying abdominal organs. Arch. Jap. Chir., 38, 236.

Motavkin, P.A., and Palaschchenko, L.D. (1973) Cholinergic and adrenergic nerves in the extramedullary arteries of the spinal cord. Acta. morph. Acad. Sci. hung., 21, 227-238.

Mott, J.C. (1961) The stability of the cardiovascular system. In: Somatic Stability in the Newly Born. A Ciba Foundation Symposium. (Eds. G.E.W. Wolstenholme and M. O'Connor). pp. 192-214, Little Brown Co., Boston.

Nedergaard, O.A., and Bevan, J.A. (1971) Neuronal and extraneuronal uptake of adrenergic transmitter in the blood vessel. In: Physiology and Pharmacology of Vascular Neuroeffector Systems, (Eds. J.A. Bevan, R.F. Furchgott, R.A. Maxwell, and A.P. Somlyo). pp. 22-34. Karger, Basel.

Nielsen, K.C., and Owman, Ch. (1967) Adrenergic innervation of pial arteries related to the circle of Willis in the cat. Brain Res., 6, 773-776.

Nielsen, K.C., Owman, Ch., and Sporong, B. (1971) Ultrastructure of the autonomic innervation apparatus in the main pial arteries of rats and cats. Brain Res., 27, 25-32.

Norberg, K.A. (1967) Transmitter histochemistry of the sympathetic adrenergic nervous system. Brain Res., 5, 125-170.

Norberg, K.A., and Hamberger, B. (1964) The sympathetic adrenergic neuron: some characteristics revealed by histochemical studies on the intraneuronal distribution of the transmitter. Acta physiol. scand., Suppl. 238, 1-42.

Olley, P.M., Bodach, E., Heaton, J., and Coceani, F. (1975) Further evidence implicating E-type prostaglandins in the patency of the lamb ductus arteriosus. Europ. J. Pharmac. 34, 247-250.

Owman, Ch., Edvinsson, L., Falck, B., and Nielsen, K.C. (1974) Amine mechanisms in brain vessels, with particular regard to autonomic innervation and blood-brain barrier. In: Pathology of Cerebral Microcirculation. (Ed. J. Cervós-Navarro). W. de Gruyter and Co., Berlin.

Paddle, B.M., and Burnstock, G. (1974) Release of ATP from perfused heart during coronary vasodilation. Blood Vessels, 2 110-119.

Pegram, B.L., Bevan, J.A., Su, C., and Ljung, B. (1973) The neuromuscular mechanism of veins. Proc. West. Pharmacol Soc., 16, 49-52.

Powis, D.A. (1974) Comparison of the effects of stimulation of the sympathetic vasomotor nerves and the adrenal medullary catecholamines on the hind limb blood vessels of the rabbit. J. Physiol. (Lond.) 240, 135-151.

Prosser, C.L., Burnstock, G., and Kahn, J. (1960) Conduction in smooth muscle: Comparative structural properties. Am. J. Physiol. (Lond.), 240, 135-151.

Reale, E., and Luziano, L. (1966) Elektronenmikroskopische beobachtungen an stellen der astabgabe der arterien. Angiologica, 3, 226-239.

Rennels, M.L., and Nelson, E. (1975) Capillary innervation in the mammalian central nervous system: an electron microscopic demonstration (1). Am. J. Anat., 144, 233-241.

Revel, J.P., Olson, W., and Karnovsky, M.J. (1967) A twenty-Angström gap junction with hexagonal array of subunits in smooth muscle J. Cell Biol., 35, 112A.

Rhodin, J.A.G. (1962) Fine structure of vascular walls in mammals, with special reference to smooth muscle component. Physiol. Rev., 42, (Suppl. 5), 48-81.

Rhodin, J.A.G. (1967) The ultrastructure of mammalian arterioles and precapillary sphincters. J. Ultrastr. Res., 18, 181-223.

Rickenbacher, J., and Ruflin, G. (1974). Zur Entwicklung der innervation der Extremitätengefässe beim Hühnchen. VASA, 3, 5-9.

Robinson, P.M., McLean, J.R., and Burnstock, G. (1971) Ultrastructural identification of non-adrenergic inhibitory nerve fibres. J. Pharmacol. exp. Ther., 179, 149-160.

Rosenblueth, W.I. (1976) Richness of sympathetic innervation - comparison of cerebral and extracerebral blood vessels. Stroke, 7, 270-271.

Samarasinghe, D.D. (1965) The innervation of the cerebral arteries in the rat: an electron microscope study. J. Anat., 99, 815-828.

Schenk, E.A., and El Badawi, A. (1968) Dual innervation of arteries and arterioles: a histochemical study. Z. Zellforsch., 91, 170-177.

Shibata, S., Hattori, K., Sakurai, I., Mori, J., and Fujiwara, M. (1971) Adrenergic innervation and cocaine-induced potentiation of adrenergic responses of aortic strips from young and old rabbits. J. Pharmacol., 177, 621-632.

Siefert, K. (1963) Electronmicroscope studies of the rabbit aorta. Z. Zellforsch., 60, 293-312.

Somlyo, A.P., and Somlyo, A.V. (1968) Vascular smooth muscle: 1. Normal structure, pathology, biochemistry and biophysics. Pharmacol. Rev., 20, 197-272.

Somlyo A.V., Woo, C., and Somlyo, A.P. (1965) Responses of nerve-free vessels to vasoactive amines and polypeptides. Am. J. Physiol., 208, 748-753.

Speden, R.N. (1970) Excitation of vascular smooth muscle. In: Smooth Muscle, (Eds. E. Bülbring, A. Brading, A. Jones, and T. Tomita), pp. 558-588. Edward Arnold, London.

Spiteri, M., Nguyen, J., and Anh, H. (1966) Ultrastructure du muscle lisse des artères du cordon ombilical humain. Path. Biol. (Paris), 14, 348-357.

Su, Ch. (1975) Purinergic nerves in blood vessels. Folio Pharmac. Jap., 71, 1.

Su, Ch., and Lee, T.J.F. (1976) Regional variation of adrenergic and nonadrenergic nerves in blood vessels. In: Vascular Neuroeffector Mechanisms, (Eds. J.A. Bevan, G. Burnstock, B. Johansson, R.A. Maxwell, and O.A. Nedergaard) pp. 33-40. Karger, Basel.

Su, Ch., and Sum, C. (1974) The uptake and release of adenine derivatives in the rabbit portal vein. Proc. West. Pharmacol Soc., 17, 122-124.

Takayanagi, T., Rennels, M.L., and Nelson, E. (1972) An electron microscopic study of intimal cushions in intracranial arteries of the cat. Am. J. Anat., 133, 415-438.

Tapp, R.L. (1969) A response of arteriolar smooth muscle cells to injury. Brit. J. exp. Path., 50, 356-360.

Thoenen, H., Tranzer, J.P., Hürliman, A., and Haefely, W. (1966) Untersuchungen zur Frage eines cholinergischen Gliedes in der postganglionären sympathetischen Transmission. Helv. physiol, pharmac. Acta, 24, 229-246.

Tranzer, J.P., and Thoenen, H. (1968a) An electron microscopic study of selective, acute degeneration of sympathetic nerve terminals after administration of 6-hydroxydopamine. Experientia, 24, 155-156.

Tranzer, J.P., and Thoenen, H. (1968b) Various types of amine-storing vesicles in peripheral adrenergic nerve terminals. Experientia, 24, 484-486.

Tranzer, J.P., Thoenen, H., and Snipes, R.L. (1969) Recent developments on the ultrastructural aspect of adrenergic nerve endings in various experimental conditions. Progr. Brain Res., 31, 33-46.

Trendelenburg, U. (1972) Factors influencing the concentration of catecholamines at the receptors. In: Catecholamines, (Eds. H. Blaschko and E. Muscholl), Springer, Heidelberg.

Tsunekawa, K., Mohri, K., Ideda, M., Ohgushi, N., and Fujiwara, M. (1967) Histochemical demonstration of adrenergic fibres in the smooth muscle layer of the pedal artery in dog. Experientia, 23, 842-843.

Uehara, Y., and Burnstock, G. (1970) Demonstration of 'gap junctions' between smooth muscle cells. J. Cell Biol., 44, 215-217.

Uvnäs, B. (1966) Cholinergic vasodilator nerves. Fed. Proc., 25, 1618-1622.

Van Orden, L.S. III, Bloom, F.E., Barnett, R.J., and Giarman, N.J. (1966) Histochemical and functional relationships of catecholamine in adrenergic nerve endings. 1. Participation of granular vesicles. J. Pharmac. exp. Ther., 154, 185-199.

Westfall, T.C. (1977) Local regulation of adrenergic neurotransmission. Physiol. Rev., 57, 659-728.
White, F.N., Ikeda, M., and Elsner, R.W. (1973) Adrenergic innervation of large arteries in the seal. Comp. Gen. Pharmacol., 4, 271-276.
Whittaker, V.P., Michaelson, I.A., and Kirkland, R.J. (1964) The separation of synaptic vesicles from nerve-ending particles ('synaptosomes'). Biochem. J., 90, 293-303.
Wyse, D.G. (1973) Inactivation of exogenous and neural noradrenaline by elastic and muscular arteries. Canad. J. Physiol. Pharmacol., 51, 164-168.
Yohro, T., and Burnstock, G. (1973a) Fine structure of "intimal cushions" at branching sites in coronary arteries of vertebrates. A scanning and transmission electron microscopic study. Z. Anat. Entw-Gesch., 140, 187-202.
Yohro, T., and Burnstock, G. (1973b) Filament bundles and contractility of endothelial cells in coronary arteries. Z. Zellforsch., 138, 85-95.
Zelander, T., Ekholm, R., and Edlund, Y. (1962) The ultrastructural organization of the rat exocrine pancreas. 111. Intralobular vessels and nerves. J. Ultrastruct. Res., 7, 84-101.

THE BLOOD SUPPLY TO NERVES

W. Pallie, M.B., D. Phil. (Oxon)

Chairman and Professor, Department of Anatomy, McMaster University Medical Centre, 1200 Main Street West, Hamilton, Ontario, Canada, L8S 4H9

By the turn of the 19th century, the gross morphological features of blood vessels to nerves had been recognized and attention had been drawn to their functional significance. Since detailed reviews in regard to this period are available (Adams, 1942; Blunt, 1957), only a brief introductory sketch is presented here in order to set a background to more recent information and interpretations that follow.

In the earliest literature, the blood supply to peripheral nerves and its functional significance had been noted (van der Spiegel, 1627; Ruysch, 1701; von Haller, 1752; Schmidel, 1755; Isenflamm and Doerffler, 1768). Later, further descriptions were added to the literature. Thus, Hyrtl (1859 and 1864) confirmed that all nerves are supplied by arteries that lead into capillary plexuses which, in turn, return blood via veins. Further, he reported that a continuous longitudinal anastomosis formed of ascending and descending branches ran within the nerve, constituting an important potential collateral circulation, a fact confirmed by others (Holl, 1880; Zuckerkandl, 1885). Indeed, Ranvier (1878) had described both a perifascicular and intrafascicular plexus of vessels.

Tonakoff (1898) affirmed that nerves received arteries from their nearest source. He also observed that when an artery crosses a nerve which is running by itself, it supplied it except when a distinct fascial plane existed. Thus, the transverse cervical artery is non-contributory to the brachial plexus. With the exception of the main "artery" to the sciatic nerve ("arteria ischiadi") and that to the median nerve ("arteria mediana") which may be called small arteries, the vessels supplying the nerves are strictly "arterioles." Tonakoff (1898) assessed the size of such vessels to be up to 0.5 mm diameter, while Bartholdy (1897) stated a higher maximum of 1.0 mm diameter.

Cranial nerves, both in their intra- and extracranial course, were said to be vascularized in the same manner as spinal nerves (Bartholdy, 1897). In regard to veins, it was believed that intraneural anastomoses existed in the same manner as arteries (Bichat, 1830). It was further stated that veins that emerged from nerves preferably drained into muscular veins, even if they had to pass through deep fascia to do so (Quènu and Lejars, 1890 and 1892).

It is convenient to categorize the various aspects pertaining to the vasculature of nerves.

[A]-EXTRANEURAL ("EXTRINSIC") ARTERIES OF NERVE

It is an established fact that all nerve trunks receive arterial supplies derived from branches of closely neighboring arteries. The origin of these arteriae nervorum were earlier subdivided into two categories. Firstly, "arteriae nutritae," preferably called "direct" branches that stem from main or named arteries; these seem to be the more common and are predominant in the case of large nerve trunks in the limbs. Secondly, the "arteriae comites," preferably called "indirect" branches that invariably arise from muscular or cutaneous branches and are the less common variety (Tonakoff, 1907; Sunderland, 1945 b). These extraneural parts of the arteriae nervorum are therefore short, being only 0.63 to 1.27 cm in length and only rarely up to 2.54 cm as in the example of the branch from the radial artery to the median nerve in the forearm (Sunderland, 1945 b).

In the case of large nerve trunks, neurovascular bundles are contained in fascial planes and ensheathed together, making it easy for the nerve to receive "direct" vasa nervorum that are short and conveniently supplied. In other instances, the nearest artery or arteriole will be the source of supply. For this reason departures from normal arterial patterns compound the variability in the source of arteriae nervorum. However, in usual circumstances, it is possible to identify the usual source vessels along the length of the nerves, though again, these are not regularly or predictably spaced, varying as they do from subject to subject and being asymmetrical even on the two sides of one individual (Bartholdy, 1897; Sunderland, 1945 a, b, and c). Deriving from this is the fact that the numbers of arteria nervorum are variable. Indeed, larger nerves do not necessarily receive a greater number of vessels, though generally larger nerve trunks receive larger arteriae nervorum, again not in strict proportion. Sunderland (1945 a, b, c) and Ramage (1927) have made counts of these vessels and the variability is quite apparent in these enumerations. Arteriae nervorum vary from 1 mm diameter to arterioles fine enough to be missed by the naked eye. By quantitation of extraneural vasa nervorum alone it has not been possible to assess or compare the degree of vascularity of nerve trunks or segments thereof (Sunderland, 1968, Blunt, 1957).

From a survey of the literature, it does not appear that independent arteriae nervorum approaching a nerve have extraneural anastomotic connections. However, a single nutrient artery may divide to supply more than one nerve trunk, particularly in plexuses of nerves (Day, 1964; Abdullah and Bowden, 1959).

It is claimed that these "extrinsic" or "extraneural" segments of arteriae nervorum possess a mesentery of "mesoneurium" (Smith, 1966 a and b; Nobel, 1968) which permits a nerve trunk to be mobile on a relatively fixed nerve bed. Sunderland (1968) stated that these vessels were straight or gently curved and tortuous only exceptionally, while Lundborg (1970) described tortuous vessels that permit some traction with impunity, though he himself was unable to identify a mesoneurium as such. We must remember that the epineurium is a collagenous connective tissue with little regular organization, though somewhat condensed to delineate the external layer of the nerve trunk and to separate it from the connective tissue and structures adjacent to it. Thus, extraneural arteries running closely paraneural are "ensheathed" by loose connective tissue. Further, while limited mobility of the nerve in its bed is possible, particularly side to side and perhaps partly dependent on the length of the regional vasa nervorum, we must recall that one of the methods that has been used to devascularize a nerve has been to exert traction on the nerve in a linear direction (Roberts, 1948). Indeed, in rabbit tibial nerve it has been reported from "in vivo" microscopic observations (Lundborg, 1970) that this is the case; stepwise, tortuous vessels were uncoiled, stretched sufficient to straighten them and then obstruct flow, and presumably, to rupture beyond this limit.

The patterns of behavior of "extrinsic" (extraneural) vasa vasorum have been analyzed (Sunderland, 1968) and all manner of forms are seen (Fig. 1A). The angle made with the parent vessel is generally large (70° or over), and is usual where a large vessel gives off a small caliber branch, an arrangement that encourages "plasma skimming" into the minor channel. It is also worthy of note that the extraneural anastomotic arrangement shown (Fig. 1A) is not a prevalent pattern.

Counts of arteriae nervorum have been made along nerves and sections of nerves, in order to compare their distribution in particular segments, and in comparisons from side to side and subject to subject (Sunderland, 1968; Ramage, 1927). It would appear that variability is the general rule. Thus a median nerve above the elbow

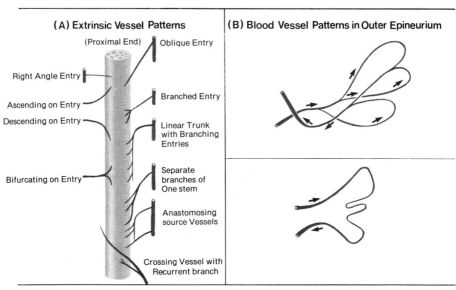
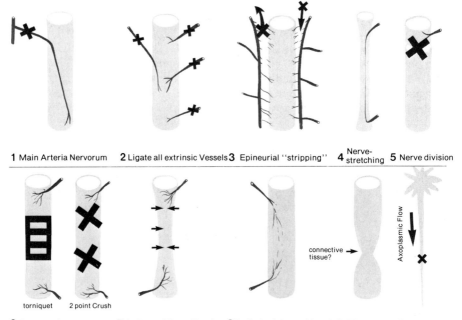

Fig. 1.

←Fig. 1 - Vasa Nervorum

    1A - Extrinsic vessel patterns. The manner in which extrinsic vessels enter a nerve trunk has been analyzed and ten different patterns are indicated in this composite diagram.
    1B - Two examples of epineural plexus patterns (modified from Lundborg and Brannemark, 1968).
    1C - Diagrams indicating the types of lesion that may compromise the blood supply to nerve trunks.

---

may have from as many as 14 to as few as 4 nutrient branches, while in the forearm alone the count ranges from 1 to 11 (Sunderland, 1968). Indeed, lengths of up to 8 cm of nerve devoid of extrinsic arteriae nervorum can be found, and the segments of ulnar and median nerve in the arm are sometimes examples of this.

(1) - Cranial Nerves

It is generally assumed that cranial nerves are no differently supplied than spinal nerves (Bartholdy, 1897; Grigorowsky, 1928), though it is claimed in particular instances that this may be more specialized and of clinical significance (Ashton, 1940). Some specific instances will be taken to illustrate features of the vascularization of cranial nerves and ganglia.

Optic nerve - This is not strictly a nerve but a "tract" of the central nervous system. Descriptions of its blood supply date back into the old literature (Haller, 1754; Zinn, 1755).

The optic chiasma is surrounded by pia mater and a plexus of vessels in the latter derived from many adjacent arteries that provide chiasmal twigs. Of these, the lateral chiasmal artery (Hughes, 1958; Blunt, 1956) and the anterior superior hypophyseal artery, both from the internal carotid, are the principal supply. The central chiasmal arteries are variable and minor (Fig. 2A).

The intra-cranial part of the optic nerve has no axial blood supply and is supplied by penetrating branches of the pial plexus that ensheaths the nerve. The major arterial twigs that supply this section are small branches from the internal carotid that pass to the inferior aspect of the nerve (Francois and Neetens, 1954) and of one of these is the recurrent branch from the anterior superior hypophyseal artery. Recurrent branches from the opthalmic artery also supply this portion (Behr, 1935; Hughes, 1958).

The intracanalicular part is a short continuation of the cranial segment and receives one to three branches (rarely more) from the opthalmic artery in the optic canal, or at the entry into the orbit or sometimes in the orbit (Hayreh, 1972).

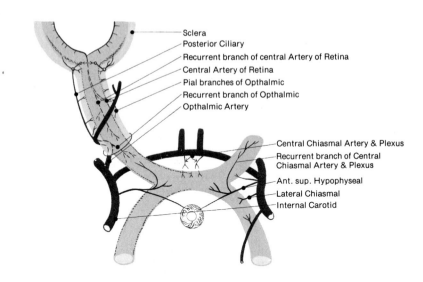
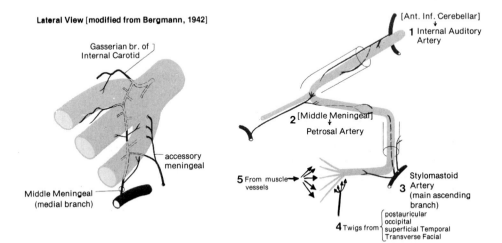

Fig. 2.

←Fig. 2 - Vasa Nervorum of Cranial Nerves

    2A - Diagram of the arterial supply of the optic nerve and chiasma.
    2B - Arterial supply of the trigeminal nerve and ganglion.
    2C - Arterial supply of the facial nerve. The numbers identify segments of the facial nerve supplied by different sources.

The intra-orbital part of the optic nerve -

(a) The posterior segment (of the intra-orbital part) is supplied by a "peripheral centripetal" system of vessels. These pial vessels (usually 1-3) come from the opthalmic artery directly or its branches; namely the muscular, posterior ciliary, or lacrimal, and rarely other branches. From the relationship of the opthalmic artery to the nerve (usually crossing superiorly to the medial side), these branches are rare on the lateral side of this segment of the optic nerve (Hayreh, 1972).

Previous literature has described an "axial" system in the nerve, and a "central artery of the optic nerve" has been described as a vessel arising from the opthalmic artery entering the nerve and running its length, with anterior and posterior divisions occupying the length of the optic nerve (Wolff, 1954; Hughes, 1958, and Wybar, 1956). This has been denied by Hayreh (1972) and the recent work of Francôis and Neetens (1972) who have reported in a study of over 100 human optic nerves. However, it has been conceded that a small recurrent branch from the central artery of the retina may be seen in the center of the nerve in this part (Hayreh, 1968; Singh and Dass, 1960).

(b) The anterior segment (of the intra-orbital part) This is supplied by an "axial centrifugal system" and a "peripheral centripetal system." The former is supplied by the central artery of the retina, which having entered the nerve (usually from below), lies centrally in it, accompanied by a comparison vein. In its course intraneurally, the artery supplies small branches to the nerve (Wybar, 1956).

The peripheral centrifugal system receives twigs from three sets of sources. (1) The central artery of the retina supplies pial branches. According to Singh and Dass (1960), these are present all around the nerve--inferiorly (100%), lateral (50%), medial (42%), and superior (18%). (2) Pial branches are also derived from branches of the thalmic artery e.g., ciliary arteries. (3) Recurrent branches from the circle of Zinn and peripapillary choroid vessels (Hyreh, 1963, and Singh and Dass, 1960).

According to Francôis and Neetens (1956), the vasa vasorum to the central retinal artery arise from the peripheral system of vessels. These authors have constructed a complex plan of hydrodynamics as it relates to the angioarchitecture of the anterior optic nerve (Francôis and Neetens, 1972).

Sympathetic fibers have been described accompanying the optic nerve (the "nerve of Tiedemann") confirmed by Krause (1875). Fukuda (1970) claimed that fibers are demonstrated by fluorescence and silver impregnation in the wall of the central retinal artery.

Oculomotor Nerve

The oculomotor nerve runs intracranially in close relation to and below the posterior cerebral artery, looping forward beneath this vessel below the interpeduncular fossa. The nerve is said to be particularly well vascularized, an observation made in the early literature (Sanger, 1880) and confirmed more recently (Plaut and Dreyfuss, 1940). Spontaneous intraneural hemorrhage has been reported in this nerve, and this having proceeded to its rupture and even fatal subarachnoid hemorrhage is on record (Plaut and Dreyfuss, 1940).

The trigeminal ganglion and nerve have drawn some attention in regard to their vascularization (Fig. 2B). The most conspicuous artery comes from the internal carotid and was recorded by early investigators (Tanasesco, 1905; Stohr, 1928). A more recent study (Bergman, 1942) cites three possible sources of supply:

(1) The internal carotid by a branch termed the "arteriae sinus cavernosi." This twig reaches the medial side of the ganglion and contributes largely to supplying the intracranial portions of the opthalmic and maxillary division of the nerve. It often forms the superior limb of an anastomotic loop with (2) and/or (3) below. (Fig. 2B).

(2) The middle meningeal artery reaching the outer side of the ganglion first, and forming a loop of supply.

(3) The accessory meningeal artery (or "arteria meningae parva") entering the cranium through the foramen ovale.

The middle meningeal and accessory meningeal arteries are "dural arteries," and constitute the inferior limb of the anastomotic loop that supplies the ganglion, though the loop may be incomplete or absent. However, a study of the accessory meningeal artery that reported the origin of this vessel equally, frequently from the middle meningea or maxillary arteries, also reported that the accessory meningeal is often purely of extracranial distribution (Baumel and Beard, 1961). Bergman (1942) also noted the asymmetry of arrangement on the two sides and recorded that the meningeal branch of the ascending pharyngeal artery, entering the cranium via the foramen lacerum, may very occasionally contribute to the blood supply of the ganglion. A description of the venous drainage is also given in that report.

Facial (and auditory) Nerve

The blood supply of the facial nerve has been described by many investigators (Bartholdy, 1897; Tobin, 1943; Guerrier, 1951; and Blunt, 1959 a). Its blood supply may be considered in five successive locations, though anastomotic connections generally link them (Fig. 2C).

(1) The anterior inferior cerebellar artery supplies direct branches to the auditory and facial nerves as they pass from the brain stem toward the temporal bone. Intracranially and into the auditory meatus, the internal auditory artery (there is a companion vein here) supplies both this nerve and the eighth nerve, and all four of these structures enter the internal auditory canal. Occasionally, the looping anterior cerebellar artery gives off anterior vestibular and vestibulo-cochlear branches directly, and these then supply the facial and auditory nerve trunks along their length within the canal, in the former case falling short of the geniculate ganglion.

(2) The geniculate ganglion and the adjacent part of the facial nerve in the facial canal are vascularized by the petrosal branch of the middle meningeal artery. This branch enters through the temporal bone in the company of the greater superficial petrosal nerve in the hiatus for that nerve (called the "hiatus Fallopii"). This petrosal artery supplies the geniculate ganglion (of which it is the principal supply) and goes on to enter the distal aspect of the facial nerve, and in this horizontal portion of the nerve, anastomoses with the stylomastoid artery. (The petrosal artery may be double, and sometimes the second branch is from the accessory meningeal.)

(3) The stylomastoid artery may arise from the occipital (twice as frequently [Blunt, 1959 a]) or the posterior auricular artery. The main ascending branch passing on the medial side of the nerve in the vertical part of the canal, goes on to anastomose with the petrosal artery.

(4) The extrapetrous portion immediately after leaving the stylomastoid foramen, receives twigs from the stylomastoid artery. As the facial nerve passes into the parotid gland, it receives twigs to each main branch from the nearest of several arteries in each respective case--posterior auricular, occipital, superficial temporal, and transverse facial arteries (Tobin, 1943).

(5) Near the terminal part of each subdivision of the facial nerve in the facial musculature, branches to muscles from the terminal branches of the facial artery (sub-mental, superior and inferior labial, etc.) supply indirect branches to the nerves, as do smaller arterioles in the face (Tobin, 1943).

There is no particular indication that any part of the facial nerve is poorly supplied, and further examination of intraneural vasculature (Blunt, 1959) substantiates this.

The Vagal Ganglia

These have been studied (Bartholdy, 1897; Grigorowsky, 1928; Quènu and Lejars, 1892). The supply comes from branches from the superior laryngeal artery which is importantly identified (Quènu and Lejars, 1892). Three or four branches also supply it from the ascending pharyngeal artery or its posterior meningeal branch.

(2) - Vascularization of Nerves of the Upper Extremity

Information in regard to aspects of the vascularization of nerves of the upper extremity are contained in the literature (Tonakoff, 1898; Bartholdy, 1897; Quènu and Lejars, 1890; Bourguet, 1913, Ramage, 1927; Sunderland, 1945 a; Bergman and Alexander, 1941; and Abdullah and Bowden, 1959).

The brachial plexus itself is supplied at its origin by spinal branches from the vertebral artery, reinforced by segmental branches from the ascending cervical, deep cervical, and superior intercostal arteries that enter the intervertebral foramina (Fig. 3A). Nearly each root has its own radicular artery, and cervical ganglia get direct branches from the vertebral artery. Indirect supplies to the roots may arise indirectly from vessels passing outward from the spinal cord surface (Tonakoff, 1898). The vertebral arteries are often unequal on the two sides, and may be compromised in varying degrees by osteoarthritis of the spine (Biemond, 1951; Hutchinson and Yates, 1956).

Fig. 3. →

3A - Blood supply of the brachial plexus in the neck. The "source" vessel is the subclavian artery; branches of the vertebral, ascending cervical, deep cervical, and superior intercostal arteries are the direct supply.
3B - Blood supply of the superior cervical sympathetic ganglion. The "source" vessel is the external carotid artery, the chief artery of direct supply usually being the ascending pharyngeal branch.

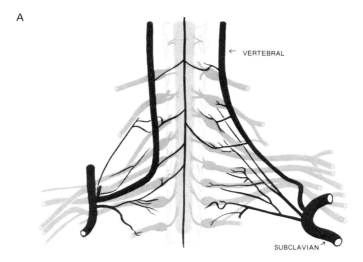

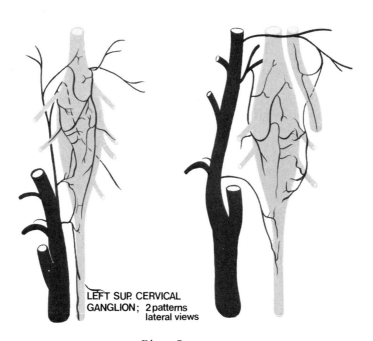

Fig. 3.

Though the ganglion cells are well supplied as compared to the fiber elements, this supply is claimed to be less rich than the grey matter elsewhere (Dunning and Wolf, 1937). The observation that six to seven vessels may enter some of the larger ganglia of the plexus (Anserow, 1925) has not been confirmed. The seventh cervical ganglion and nerve root were found to have the most variable supply (Tonakoff, 1898; Bergman and Alexander, 1941). It is often the bulkiest nerve root, and lies in the region of the cervical spine where excursions are maximal (Abdulla and Bowden, 1959). It has been pointed out that shearing stresses between nerve roots and the relatively fixed and adherent dura may entail repeated risk of minor vascular injury (Frykholm, 1951; Brain et al., 1952).

The trunks of the brachial plexus receive small direct branches from the ascending cervical, deep cervical, superior intercostal, and occasionally from the subclavian itself. The cords of the plexus, in turn, receive small direct branches from the subclavian, axillary, and subscapular arteries. Reportedly, the suprascapular and transverse cervical arteries supply no vasa nervorum to the brachial plexus (Abdulla and Bowden, 1959; Tonakoff, 1898, and Ramage, 1927). This is cited as the exception, where an artery running near a nerve does not supply it, since a distinct fascial plane intervenes (Tonakoff, 1898).

The ulnar nerve - is generally supplied by the following arteries in a proximo-distal direction along the nerve: axillary, (brachial), ulnar collateral, (supratrochlear), posterior ulnar recurrent, and the ulnar arteries; those in parenthesis are less common sources of supply, while the ulnar collateral artery is often a chief source. The segment of nerve in the medial epicondylar area is, fairly constantly, well supplied, usually by the posterior ulnar recurrent anastomosis, and this anchors the nerve trunk here, a circumstance that requires division of these nutrient vessels in the event that the nerve requires to be mobilized as in anterior transposition in symptom-producing valgus deformities of the elbow. Occasionally, no branches are seen to enter the nerve between the axilla and the elbow (Sunderland, 1968). On the other hand, some of the largest arteriae nervorum of the upper extremity were seen in the upper portions of this nerve (Ramage, 1927).

In the forearm, direct branches from the ulnar artery constitute the major supply to the ulnar nerve, and these serve to attach the nerve to this artery in the distal part of the forearm, in particular. Toward the wrist, muscular branches may cross the nerve before recurrent branches are supplied to it.

A total as high as 29 extrinsic branches, or as little as 7, have been counted in arm and forearm supplying the ulnar nerve (Sunderland, 1968). Other figures give an average of 3.8 in the arm (range 13 to 3) and 7.16 in forearm (range 19 to 2) Ramage (1927). (The ranges are from Sunderland, 1968.)

The median nerve -- In the axilla and upper arm, the axillary artery supplies the nerve directly, on occasion assisted by either of the acromiothoracic, lateral thoracic, or subscapular branches. Special mention has been made of a large nutrient artery (usually from the axillary) to the fork of the roots of this nerve (Robinson, 1910; Tonakoff, 1907) and is usually present (20 of 37 specimens, Sunderland, 1945 c). The brachial artery, and in the lower arm occasionally the ulnar collateral and supratrochlear arteries, are the source of supply.

In the cubital fossa, one or more nutrient arteries arise from among the brachial, division of radial and ulnar, common interosseous, anterior interosseous, and anterior ulnar recurrent arteries, all of which are in close proximity here.

In the forearm, when present, the "median artery" ("arteria nervi mediana") is a large arteria nervorum to the nerve. In 27 cases when it was present, in only 17 did it terminate in the nerve, while in the rest it did not enter the nerve, but terminated instead in flexor muscles of the forearm and hand; in the same series, it was absent in 10 cases (Sunderland, 1968). In fact, the forearm segment of the median nerve is said to receive no constant branch, but derives them from the following arteries: ulnar, radial, common interosseous, anterior interosseous, and the median arteries, and occasionally from their muscular branches. Particularly toward the wrist, the muscular branches that may supply vasa nervorum first cross the nerve before sending recurrent branches back to it.

At the wrist, vessels enter it above and below the flexor retinaculum (Smith, 1966 a), the distal arteriae nervorum arising from the superficial palmar arch and running in a proximal direction (Blunt, 1959 b; Smith, 1966 a).

The totals for extrinsic vasa nervorum to the median nerve range from 19 to 6 (Sunderland, 1968). In the arm the average is 3.8 (range 13 to 4) and in the forearm 6.4 (range 11 to 1) (Ramage, 1927; Sunderland, 1968 for ranges).

The Radial Nerve

In the first part of its course before entering the radial groove the radial nerve is supplied by the axillary, brachial, and profunda brachii arteries or their branches. Only rarely the subscapular artery supplies it with direct branches, though 1 to 3 twigs which it supplies to the brachial plexus pass into the radial nerve intraneurally.

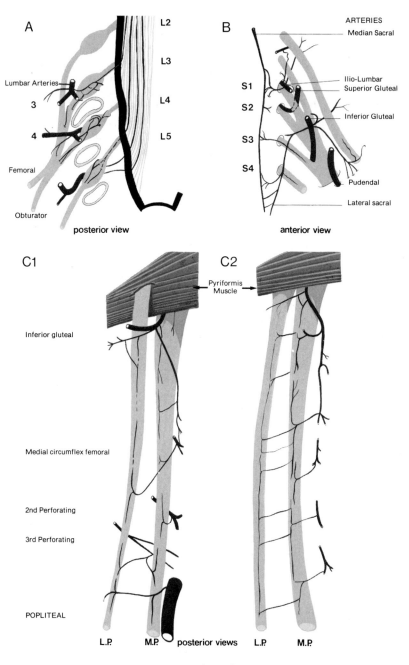

Fig. 4.

In the radial groove of the humerus, the radial nerve seems to have a constancy of supply and receives a total of some 2 to 9 branches to the principal trunk itself. The nerve tends to have loosely arranged strands when it lies here, and the arteriae nervorum are distributed through these patterned strands.

In the intermuscular furrow of the brachialis, brachioradialis and extensor carpi radialis longus muscles, muscular arteries supply the nerve. A total of 14 to 4 extrinsic nutrient vessels supplying the radial nerve have been counted.

The superficial continuation of the radial nerve receives branches from the anterior radial recurrent artery (1-4 branches) and the radialartery (1-4 direct; 1-5 indirect).

The posterior interosseous nerve is supplied by the anterior radial recurrent artery (1-3 branches), an anastomotic artery between the former vessel, the posterior interosseous artery within the supinator muscle, and by the posterior interosseous artery (1-4 branches) near its terminal portion. Occasionally, this part of the nerve also receives a supply from the anterior interosseous artery after it pierces the interosseous membrane. The nerve may have any number of branches from a total of 6 to 1 in an individual case (Sunderland, 1968).

In regard to arteriae nervorum of the upper extremity nerves, it is claimed that in the upper arm segments, large vessels lie in the peripheral part of the nerve (on the surface of the epineurium), while in the forearm, rather large arteriae nervorum run within (interfascicular) the nerve trunks (Ramage, 1927; Sunderland, 1945 a). The exception is the median artery, which when present, usually descends on the surface of the median nerve in the forearm.

(3) - Vascularization of Nerves of the Lower Limb

The literature in regard to the vascular supply of nerves of the lower extremity is rather limited (Haller, 1756; Bartholdy, 1897; Tonakoff, 1898; Hovelacque, 1927 and Day, 1964). In general terms, the pattern is similiar to the upper extremity. In the lower extremity, the nearest vessels supply the nerves, and all of them are as one would expect, indirectly, branches from the abdominal aorta (Fig. 4).

← Fig. 4 - Blood supply of the lumbosacral plexus.

    4A - Lumbar plexus (posterior view).
    4B - Sacral nerves and plexus (anterior view).
    4C - Two forms of sciatic nerve vascularization.

The lumbar roots and ganglia are supplied in a segmental pattern, L_1 to L_4 being supplied by lumbar arteries via the spinal branches. The fifth lumbar, probably due to the position of the aorta at its bifurcation, may be supplied differently on each side. On the left, the fifth lumbar or iliolumbar artery, while on the right, the superior gluteal is a third option of supply.

The sacral roots and ganglia are principally supplied by the lateral sacral arteries in segmental fashion. The other sources of supply are vessels of a non-segmental nature, and tend to be distributed over a group of successive segments; they are the inferior gluteal (rather rarely), ilio-lumbar, median sacral, and internal iliac arteries. There is a "variability of arteries supplying the fifth lumbar to the third sacral segment"--seeming to be directly or indirectly due to the abdominal aorta terminating as it does (Day, 1964).

The sciatic nerve

This is the largest nerve in the body and has the largest arteria nervorum that supplies it, namely the "arteria nervi ischiadi." It arises from the inferior gluteal artery and gives from 1 to 6 large branches. The old literature contains descriptions (Haller, 1756; Hyrtl, 1864; Quènu and Lejars, 1892; and Bartholdy, 1897). In the thigh, 1 to 6 arteriae nervorum to the sciatic nerve arise from the perforator arteries and their anastomoses, and these give rise to form 1 to 6 branches. The vessels that supply the trunk are in order: (rarely the superior gluteal or pudendal), inferior gluteal, "cruciate" anastomotic arteries, perforators and their anastomotic chain, genicular, and muscular arteries. The lateral popliteal may additionally receive twig(s) from the circumflex fibular artery. As a rule, the intraneural vascular plexus of the sciatic nerve in the buttock and thigh contains several arterial channels. The lateral and medial popliteal nerves may separate out from the sciatic nerve at any level, even within the plevis. The vascular connection of the two parts may be discrete in large measure or shared (Fig. 4C) (Hofmann, 1903). The lateral popliteal nerve has a vulnerable position as it approaches the fibular neck. In addition, it is claimed that the nerve has less supporting (adipose) tissue in its epineurial sheath (cp. medial popliteal), and the arteriae nervorum in 88% cases were superficially located and exposed.

The tibial nerve is supplied by the anterior tibial artery by 2 to 13 branches, only 1 or 2 being indirect. In the malleolar region the perforating peroneal and anterior malleolar arteries are sources of supply.

The peroneal nerve (posterior tibial) is supplied by the posterior tibial artery that, mostly directly, provides 2 to 11 branches.

The peroneal artery supplies the lower part of the nerve. In 8 of 40 specimens this nerve was penetrated by a main arterial channel (Sunderland, 1945 and 1968).

(4) - Blood Supply of Ganglia of Somatic Nerves

This has been described above in part, with some cranial nerves, and in the description of the plexuses. The literature contains some descriptions (Adamkiewicz, 1886; Tonakoff, 1898; Anserow, 1925; Grigorowsky, 1928; Bergman and Alexander, 1941).

Spinal ganglia are largely supplied on a strictly segmental plan by the spinal branch of the corresponding (inter)segmental artery (the veins, however, do not drain segmentally). The supply is "direct" from the radicular artery, but it is claimed that "indirect" supply from the channels on the surface of the spinal cord surface running toward the ganglion exists, and vessels may arise from the anterior or posterior spinal arteries, but especially along the posterior nerve roots. Larger ganglia (especially those of the plexuses) may receive several (up to 6 or 7) nutrient arteries that may branch before entering the ganglion (Tonakoff, 1898; Anserow, 1925).

Though the vascularity of ganglia are not described in quantitative terms, that they are well vascularized (as compared to nerve) (Blunt, 1959; Ishikawa, 1959) is recognized. Dilatations of capillaries and pre-capillaries exist in spinal ganglia, and it is claimed these are somehow related to atrophy of ganglion cells. Bergman and Alexander (1941) claim that after the third decade, reduction of posterior root fibers and loss of ganglion cells, with mild posterior column degeneration, accounts for the sensory loss, most discernible as loss of vibration sense in later life.

(5) - Sympathetic Ganglia

A review (Patterson, 1950) of the available literature points to the scantiness of information contained (Bartholdy, 1897; Delamere and Tanasesco, 1906; De Souza, 1932; Anserow, 1925).

The cervical ganglia are described by Patterson (1950). The superior cervical ganglia receives twigs principally from the ascending pharyngeal artery; additionally, the superior thyroid arteries and more rarely, the occipital and internal carotid arteries. (Fig. 3B) The rest of the cervical chain is supplied by the inferior thyroid. If the median ganglion is located in a low position in the neck, the thyrocervical and costocervical may supply it as they do the inferior cervical or stellate ganglion.

The paravertebral sympathetic ganglia in the rest of the chain are supplied by intersegmental arteries. The branches subdivide before entering the ganglion and provide an "intraganglionic plexus," a "periganglionic plexus," and may pass through to the other side of the ganglion. From these, branches pass to related nerves, e.g., splanchnic nerves, rami communicantes (De Souza, 1932). The lumbar ganglia, similiarly, are vascularized by the lumbar arteries as well as unnamed retroperitoneal branches of the aorta and common iliac arteries (Patterson, 1950). Nonidez (1942) reported A-V anastomoses in some sympathetic ganglia of dogs, but this has not been reported in other animals or in man.

(6) - Gangliform Enlargements of Nerves

Gangliform swellings on nerves are reportedly found normally in some sites on particular nerves, e.g., the teres (minor) branch of the axillary nerve, the median nerve in the hand, the posterior interosseous nerve on the dorsum of the wrist, and the deep peroneal nerve at the ankle. However, it has been suggested that fusiform swellings on nerves appear as a response to "trauma" or "chronic" friction, producing an increase of connective tissue (Chang et al., 1963; Gitlin, 1957; Daniell, 1954; Sunderland and Bradley, 1952). The increase of connective tissue appears to result from narrowing and obliteration of the arteriae nervorum as the result of endothelial proliferation and fibrosis of the tunica media. Associated with this may be axonal degeneration (Sunderland, 1968).

Age changes have been found in peripheral nerves, the most conspicuous of which is an increase in connective tissue in all sheaths (Cottrell, 1940).

[B] - INTRANEURAL ("INTRINSIC") ARTERIES OF NERVE

The general histological arrangement in a nerve trunk is well known (Fig. 5). However, since the blood vessels that are distributed within it have particular relationships in it, it is necessary to highlight some features of the supporting connective tissue in the course of this description.

The connective tissue contribution of a nerve trunk is said to vary from as little as 22% (ulnar nerve at elbow) to as much as 88% (sciatic nerve in gluteal region). It is claimed that nerve trunks containing a greater number of funiculi tend to have more connective tissue, as do nerve trunks at joint lines (Sunderland, 1968). Moreover, specific nerves and segments of them, with specific peculiarities in this regard, have been identified (Sunderland, 1968).

The epineurium is a connective tissue coat of collagen and a few elastic fibrils arranged in a largely longitudinally plane. The outer layer is somewhat condensed, separating the nerve trunk from adjacent tissues, and this layer has been called the "conjunctiva nervorum." Extrinsic blood vessels to the nerve that enter this

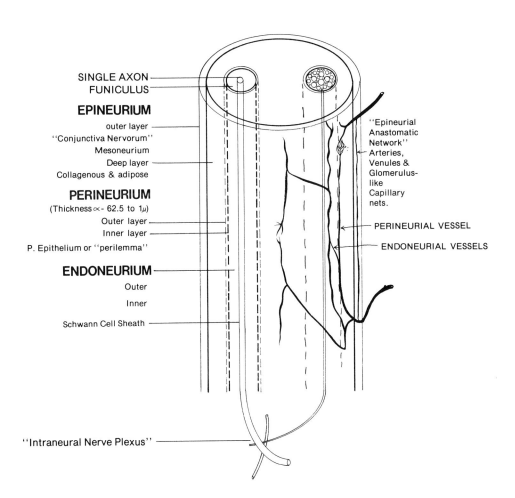

Fig. 5. - Diagram of nerve trunk, sheaths and vessels

The location of vessels in relation to epi-, peri- and endoneurium is indicated.

from outside are described as having a "mesoneurium" of this connective tissue (Smith, 1966; Noble, 1968). Adipose tissue is intermingled with collagen in the deeper part of the epineurium, and this is usually in larger trunks. It is also claimed that in general obesity, fat content in nerve trunks is also increased (Sunderland, 1968).

The extrinsic arteriae nervorum have been described previously, and their behavior on contact with the nerve trunk indicated (Fig. 1A). In some regions, the largest arterioles derived from them tend to be located on the peripheral surface of the nerves, such as in the upper arm. This is also true for the median nerve in the forearm. (Generally, they lie more internally within the nerve trunk of large nerves.)

An epineurial anastomotic network exists on the surface of the nerve trunk, fed by arterioles that enter to supply the nerve. Lundborg and Brannemek (1968) describe "glomerulus-like" patterns of capillaries in the outer epineurium of rabbit nerves, and arteriolo-venular shunts have been found in it (Lundborg, 1970).

The epineurium thus contains a "superficial anastomotic network" of nutrient arterioles that enter from without and form a capillary plexus with "glomerulus-like" patterns in this layer, leading away into venules on the surface (Fig. 1B). Some lymphatic capillaries lie in the epineurium alone, but do not enter the substance of the nerve trunk (Sunderland, 1968).

The perineurium surround each separate fascicle, and in contrast to the epineurium, has an outer layer of regular organization of flattened fibroblasts, alternating with fine collagen fibers in circular, oblique, and longitudinal orientation within this sheath. Other claims (Glees, 1943) stated that fibrils are arranged in double spirals, crossing at 45° and interdigitating with longitudinal fibers to form a compact network. The fibroblasts, surrounded by dense intercellular material, are claimed to form junctional complexes with each other, and from tracer experiment results, it is claimed that these limit the diffusion of large molecules across the sheath (Waggener and Beggs, 1967). The internal layer of perineurium is claimed to be special--smooth surfaced, and lined by 1 or 2 layers of flattened polygonal mesothelial cells to form an epithelium, "the perineural epithelium" (Shanthaveerappa and Bourne, 1962; Rohlick and Weiss, 1955) or the "perilemma". Muscle spindles, too, are ensheathed by the extension of perineural epithelium, and these epithelial cells usually intervene between capillaries and spindle contents (Banks and James, 1973). This is the morphological basis of the "diffusion barrier" (e.g., to phosphate ions [Causey and Palmer, 1953]) which maintains an environment in the peripheral nervous system that is almost akin to that of the central nervous system

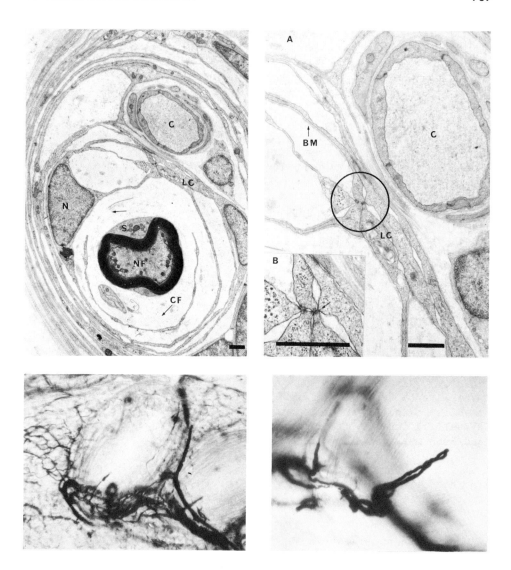

Fig. 6. - Blood vessels in nerve terminals

The vascular loop that enters the "core" of the pacinian corpuscle is shown in the two photographs in the lower half of the plate. In the upper portion, a transverse section through the myelinate axon shows the capillary (C) separated by a 4.5μ space from axon. This interval at high magnification (x 20,000) shows lamellar cells (LC) and nuclei concentrically arranged (inset x 40,000), together with collagen fibrils (CF) scattered in the interspaces.

(Shanta and Bourne, 1968). In the pacinian corpuscle, too, the terminal axon is "insulated" from the blood vessel that enters at one of its poles (Fig. 6). This vascular loop that supplies the corpuscle is walled off by "tight junctions" of cells that form the lamellations of this structure (Pallie, Nishii, and Oura, 1970; Nishii, Oura, and Pallie, 1969).

The perineurial cells are distinguishable from epineural fibroblasts by the presence of basement membranes and uniform intercellular contacts (Waggner et al., 1965). They also show pinocytotic vesicles, indicating a function of active transport.

It has been shown that it is the inner layers of the perineurium that are impermeable, since trypan blue, Evans blue, labelled serum albumin (Waksman, 1961), horseradish peroxidase, and lanthanum hydroxide (Olson and Reese, 1971), all pass into epineurium and enter the perineurium via "open" endothelial junctions, but no farther. "Tight junctions" exist in vessels in the inner perineurium and endoneurium (Waggner et al., 1965; Liebermann, 1968). It has been pointed out that nerves may pass inflammatory sites or even abscesses without being affected (Shanta and Bourne, 1963).

The "perineurial interfascicular network of blood vessels" form a continuous longitudinal anastomosis, and consists of arterioles that lead to pre-capillaries and capillaries, to become returning venules. Repeated division of these vessels, and oblique and spiralling interconnections between the nerve fasciculi, create a complex mesh in which the circulation flows in varied directions as has been visualized and diagramed in detail in the rabbit nerve (Lundborg and Branemark, 1968).

In the human fetal sural nerve, no blood vessels were seen within the perineural fascicles before the 18-week stage, and it was stated therefrom that "the endoneurial fibroblasts presumably penetrate the fascicles, together with the blood vessels" (Ochoa, 1971).

The endoneurium is mostly collagenous tissue that forms fine intrafunicular septa, incompletely subdividing the nerve fascicles. There are no lymphatic capillaries here, but fluid distribution in endoneurial spaces has been demonstrated moving proximo-distally (Weiss et al., 1945), by radioactive studies which favor their continuity with the subarachnoid space, or the subdural space (Sunderland (review), 1968). The endoneurium lining the external aspect of the "endoneurial tubes" forms a support to these cylindrical structures of schwan cell sheath that contains the axon and myelin sheath. In this tissue is supported the "endoneurial vascular network."

Ranvier (1878) has described "perifascicular" and "intrafascicular" plexuses of vessels which had been identified as important potential collateral circulatory pathways (Hyrtl, 1859 and 1864; Holl, 1880, and Zuckerkandl, 1885). Buitink (1934) stated that accessory longitudinal anastomoses may also be connected to a main one in a ladder-like fashion of cross-anastomosis, while yet more complex arrangements were described (Petrovits and Szabo, 1939). Adams (1943) stated that the vessels here are only capillaries, but precapillaries with muscular walls have been seen (Sunderland, 1945 a).

Fusiform dilatations in the transversely placed vessels of rabbit nerves were called "sinusoids" (Roberts, 1950) and thought to be regulatory shunts to control vascularity of nerves. In the dog sciatic nerve they have been labelled "micropools" (Nobel, 1969).

Capillary density has been measured as an area in nerve cross-sections in the cat and the capillary proportion given as 0.9% (Ishikawa, 1959). Sunderland (1945) has attempted to assess vascularity on the basis of counts of extrinsic as well as intrinsic vessels. He states that the number, size, and position of interfascicular vessels vary from level to level (also indicating that counts suggest that the number of interfascicular venules exceeds the number of arterioles). Nevertheless, the number of capillaries per unit area of fiber appears relatively constant, irrespective of the nerve or any segment thereof.

As indicated previously, endoneurial vessels have tight junctions in contrast to epineurium and outer perineurial vessels. It has been shown that these endoneurial vessels, unlike those of cerebral vessels, have their permeability increased by histamine and 5-hydroxytryptamine (Olsson, 1966) as well as in other pathological conditions (Olson, 1968).

The venous return from nerves has been referred to briefly above, there seemingly being little attention paid to it exclusively. In general, they resemble arteries in forming longitudinal intraneural anastomoses, but in large nerves, especially, veins leave independently of the arteries. Quènu and Lejars (1892 and 1894) stated that veins preferably drain into muscular veins, even if they have to pass deep fascia to do so. Though these workers claimed that, therefore, they do not drain into principal venous trunks directly, this has been declaimed (Sunderland, 1945 a). Arteriolovenular shunts have been described in the epineurial plexus as well as deeper in the nerve trunk (Lundborg, 1970). It has also been observed that among interfascicular vessels, venous channels appear to predominate (Sunderland, 1968). Indeed, it is remarked that endoneurial vessels appear to have wide lumina and some muscle, resembling venules, and these predominate over fine endothelium lined capillaries (Weddell, 1974). There are species differences in regard to endoneurial vessels, and these are evident from reports in the literature. In the guinea pig, large veins enter the endoneurium and subdivide into smaller, curving, and longitudinal venous type vessels before anastomosing with the capillary network. In this animal, unlike the rabbit, venules predominate and endoneurial arterioles were found in only one of every three or ten fascicles (Waksman, 1961).

Innervation of Vasorum Nervi

Vessels have been described in all positions in a nerve trunk cross-section, and some of these possess a tunica media, imparting contractile properties to these vessels. Postganglionic, non-medullated nervi nervorum, which are excitor fibers, were described by Sappey (1889), and these behaved in their pharmacological reactions like cerebral vessels (Bulbring and Burn, 1939). Adrenergic nerves in epi- and perineural vessels have been seen by fluorescence microscopy and reported in rat sciatic nerve (Olsson, 1965 and Dahlstrom, 1969) and sympathetic stimulation reported to reduce blood flow. In vivo studies in the rabbit have shown that, on the application of warmth, vessels and capillaries previously not visualized open up, suggesting the presence of a large reserve blood supply (Lundborg, 1970).

While the optic nerve and retina are strictly not "peripheral" nerve, the accompanying non-medullated plexus of nerves that run with the central retinal artery (the so-called nerve of Tiedemann (1824) has been recorded a century ago (Krause, 1875).

[C] - FUNCTIONAL CONSIDERATIONS OF VASA NERVORUM

There are complex and unresolved issues in regard to the precise part played by vasa nervorum, physiologically, as well as the degree of impunity with which pathological and surgical occurrences may affect them. A review is presented here with a view to illustrate perspectives in this field.

Several means have been employed to examine the role of vasa vasorum, and among them are the methods sought to compromise their function in order to examine effects of this (Fig. 1C). Ligation of the main artery of supply was carried out (Okada, 1905), in this instance the inferior gluteal artery of the rabbit, and this was followed by degeneration in the sciatic nerve. Employing animal sciatic nerve devascularization, conduction block has been demonstrated (Bentley and Schlapp, 1943), while devascularization of the rabbit sciatic nerve at one end produced marked alterations in recovery from fatigue in high frequency stimulation (Causey and Stratmann, 1953). The ligation of all nutrient arteries (as pointed out [Lundborg and Branemark, 1968] this is not easy to do) produced patchy areas of pathological change. It is claimed that injury to both the nutrient arteries as well as the longitudinal epineural vessels was required to cause true nerve fiber degeneration (Adams, 1943, and Durward, 1948).

Linear stretching of the nerve has been used effectively (Roberts, 1948) and indeed, it has been shown in vivo, under direct vision, that the circulation in the epineurial vessels can be arrested thereby (Lundborg, 1970). Edema of the nerve simulating the "carpal tunnel syndrome," produced in animal experiments, was shown to decrease conduction velocity (Weisel et al., 1964; Anderson, 1966), while exposure of the rat sciatic nerve to irradiation produced paresis eleven days later, and severe vasa nervorum damage was observed (Bergstrom, 1965).

A survey of the clinical literature is revealing in regard to the opinions that have been given to the importance of vasa nervorum, the extrinsic supply in particular. The older German and French literature records ischemia as a cause of peripheral neuritis (Joffroy and Achard, 1889; Dutil and Lamy, 1893; Oppenheim, 1889; Lapinsky, 1899; Desplats and Baillet, 1911; Foerster, 1912 and 1913; Gallavardin et al., 1925; Coste et al., 1933; and Schlesinger, 1933). Quènu and Lejars (1890; 1892 and 1894) reported difficulty of speech and respiration in post-thyroidectomy cases to be due to ischemia of the vagus and recurrent laryngeal nerves that were exclusively supplied by the thyroid arteries. Several other specific clinical entites have been cited as causative factors of peripheral neuropathy--arteriosclerosis (Priestly, 1932), thrombo angiitis obliterans (Meleney and Miller, 1925; Goldsmith and Brown, 1935; Barker, 1938; Dry and Hines, 1941), periarteritis nodosa (Lovshin and Kernohan, 1948; Kernohan and Woltman, 1938), diabetic neuropathy (Denny-Brown et al., 1945), Volkmann's ischemia (Holmes et al., 1944), and some forms of facial (Bell's) palsy (Holmes et al., 1944; McGovern et al., 1966; Oppenheim, 1893; Woltman and Wilder, 1929; Seddon, 1956; and Raff et al., 1968). Changes in senescence, with vascular changes, have been claimed to cause neuritis, and reviewing the older literature, Cottrell (1940) reported vascular pathology in vasa nervorum even proceeding to vascular obliteration, together with increased connective tissue in the endoperineurium and replacement fibrosis of axons, Richards (1951), reviewing the physiological and clinical evidence, stated that short periods of ischemia followed by relief, lead to rapid and complete recovery, while prolonged or several neural ischemia causes structural damage that may persist.

Reverting to the experimental evidence that has been documented, varying shades of opinion have been expressed. Several observations have been cited to adduce the ability of a nerve to survive deprivation of extrinsic vessels through a considerable length of its course. Thus, the ulnar nerve has been freed from the axilla to the wrist, reportedly with impunity (Bristow, 1941). The same nerve freed from axilla to forearm, on being sectioned, may be found to bleed (Bateman, 1962), and indeed even when freed along a length of 15 cms, the cut free end was found to bleed (Sunderland, 1968).

Indeed, it is claimed that in vivo examination of the microcirculation failed to reveal any effect of mobilization of 6 to 8 cms in rabbit experiments (Lundborg, 1970). Further, evidence supporting these views comes from work reporting that there was no statistical difference in the regeneration of nerves deprived of blood supply, as compared with a crushed nerve (Bacsich and Wyburn, 1945). However, in this latter regard we must recall that a collateral circulation may develop shortly after devascularization caused by section of a nerve (Edshage, 1964). Changes have been seen as soon as 48 hours after the trauma, though previously it was claimed that a collateral circulation was only satisfactory five days later, to become really effective 20 days later (Blunt and Stratton, 1956 a).

In contrast, reports claimed that in some circumstances damage to a major nutrient artery is sometimes followed by the development of ischemic lesions (Parks, 1945, Seddon and Holmes, 1945; Woollard and Davis, 1950; Okada, 1905). In fact, it had been shown that devascularization of the facial nerve in situ causes it to lose its excitability in 15 to 30 minutes (Mayer, 1878; Frolich and Tait, 1904). Further, the work of Bulling and Burn, (1939) demonstrated that ischemia need not be necessarily complete in order to produce significant effects on transmission of the nerve impulse.

Thus, it has become necessary to modulate general comments on the functional aspects of vasa nervorum, and to remember the considerable variability of the extrinsic pattern in particular. It is possible to state that the blood supply of a nerve is of a regional nature (the extent of the region varying from nerve to nerve and site to site). However, it is frequently the case that even where only one vessel may appear externally to supply a long segment of nerve, considerable overlap of supply exists from adjacent vessels, so as to prevent significant segmental ischemia from exclusion of a single nutrient artery, or even more than just one. However, the possibility that regional vascular dominance may be exerted by some vessels supplying a nerve is real. Smith (1966) has weighed the evidence and states that "*Reconstructive surgeons might be among the first to doubt that any pedicle with a 20:1 length width ratio, whether it be skin, nerve or tendon, would have adequate collateral circulation to its tip.*" Thus, he goes on to state that "*when nerves are isolated . . . over distances greater than 6 to 8 cms, the adjacent circulation will not be maintained through anastomotic channels. A large portion of any nerve, freed at surgery, must, on the basis of these observations function as a free graft until its blood supply is restored by the ingrowth of new blood vessels.*" Such effective proliferative changes have been seen and recorded (Torraca, 1920; Edshage, 1964).

An aspect of vasa nervorum involvement of practical interest is the question of causality in regard to compression effects on a nerve segment. Thus, in conditions of complete ischemia of a limb, a nerve ceases to conduct in 30 minutes at 25°C (Bentley and Schlapp, 1943). It is generally conceded that a tourniquet applied for 1½ hours and removed, permits recovery. In the case of nerve, this recovery commences in 30 seconds and is complete in 5 to 6 minutes (Bentley and Schlapp, 1943). However, the application of a tourniquet for even 30 minutes may produce effects that last up to 3 weeks (Eckhoff, 1931; Spiegel and Lewin, 1945). These are mostly motor in the milder cases (Richards 1951). Nevertheless, complete though slower recovery after 4 hours of ischemia is on record (Bentley and Schlapp, 1943).

These effects are claimed by some to be due to ischemia of the compressed nerve segment (Lewis et al., 1931; Denney-Brown and Brenner, 1944), while others claim that local deformation of nerve fibers is the significant factor, with ischemia playing only a subordinate role (Bishop et al., 1933; Bentley and Schlapp, 1943). In direct nerve compression such as "Saturday night paralysis" and "peroneal palsy," the circulatory effect, if any, is strictly localized. Here again, sensation is very little if at all affected, and there is retention of electrical excitability of the nerve and muscle below the block ("Erb's paradoxical phenomenon"). However, in chronic compression, where a habitual crossed legged posture caused foot drop, recovery was delayed as long as three months (Dunning, 1944). Clinical effects of major arterial obstruction (embolization) produce effects that lack uniformity (Geflan and Tarlov, 1956), making interpretations difficult.

It is also of interest to consider the several "entrapment syndromes" of nerve that have been reported--"tarsal tunnel syndrome," the lateral cutaneous nerve of the thigh at the inguinal ligament, "peroneal entrapment," "supinator syndrome" the subcostal nerve under the lateral arcuate ligament (Sidley, 1972; Watson-Jones, 1972; Applegate, 1973), and the "cubital tunnel syndrome" (Wadsworth, 1974). Generally, symptoms are attributed to direct nerve fiber compression, with no reference to circulation being compromised. However, the question has been raised as to what provokes the thickening of the surrounding fibrous tissue to such a degree in entrapments of nerve as to cause compression (Watson-Jones, 1972).

It is useful to recall that the experiments of Lundborg (1970) point to the considerable modification of neural microcirculation that is possible, as evidenced by warming a nerve trunk observed under direct vision. Further, the effects of cold applied to nerve trunks led Sinclair and Hinshaw (1951) to point to the effect on vasomotion in nerve. The theory of referred cardiac pain on the basis of spastic vaso-neuropathy (Roberts, 1948) thereby finds support. Another aspect of ischemic nerve phenomenon that has not found wide support is worth recording. According to Richards (1951) *"It seems not inconceivable that the pains of ischaemic neuritis are due to the presence of multiple artificial synapses in more than one nerve, which have been produced by the defective blood supply of nerve trunks."* The mechanism of the artificial synapse has been invoked by Doupe, Cullen, and Chance (1944) to account for the pain of causalgia. Their view is that *"as a result of either disease or injury, defective insulation occurs in the nerve trunk and, thus, impulses which are constantly passing down sympathetic nerve fibres are able to stimulate afferent pain fibres."*

Recent studies of peripheral nerve changes in amyloid neuropathy (Thomas and King 1974) point to the uncertain causation of nerve degeneration in these cases. Kernohan and Woltman (1942) viewed the ischemic changes as important, while Dyck and Lambert (1969) argued against this on the basis that early loss of unmyelinate axons that are relatively resistant to ischemia is contrary to that view. Thomas and King (1974) indicated that the concomitant lesion in dorsal root ganglia in the disease, confounds this issue further.

In human fetal nerves no blood vessels were seen before 18 weeks development (sural nerve) so that the inference was made that *"the metabolic needs of these sensory axons are probably met by their cell bodies in the dorsal root Ganglion"* (Ochoa, 1971). The metabolic processes within the cyton and its peripheral processes have been further complicated by the fact that besides the centrifugal axoplasmic flow that is recognized, centripetal flow also occurs as evidenced by the passage of horseradish peroxidase upstream in axons. What part the neural vasculature specifically contributes is incompletely understood, and is only conveniently expressed by suggesting "supportive nurture." It is indeed difficult to believe that for effective function, the cell body is solely responsible, unaided by neural circulation, for the nutrition of the entire length of such long axons that run the entire length of major peripheral nerve trunks.

REFERENCES

Abdulla, A., and Bowden, R.E.M. (1959) "The blood supply of the brachial plexus." Proc. Roy. Soc. Med., 53, 203-205.
Adamkiewicz, A. (1886) "Die Blutkreislauf der Ganglienzelle." Berlin.
Adams, W.E. (1942) "The Blood Supply of nerves. 1. Historical review." J. Anat., 76, 323-339.
Adams, W.E. (1943) "The blood supply of nerves. 2. The effects of exclusion of its regional sources of supply on the sciatic nerve of the rabbit." J. Anat., 77, 243-250.
Anderson, A. (1966) "Bestamning av overledningstiden i. n. medianus hos hanin peritendinost oedem i carpal tunneln." Nord. Med., 75, 492.

Anserow, N.I. (1925) "Die arterien der Ruckenmarks der intervertebralganglien und der sympathische ganglien." Anat Ber., 3, 348-349.

Applegate, W.Y. (1973) "Abdominal cutaneous nerve entrapment syndrome." Ann. Fam. Phys., 8, 132.

Asherton, N. (1940), Laryng, J., 55, 531. (quoted by Adams, 1942).

Bacsich, P., and Wyburn, G.M. (1945) "The vascular pattern of peripheral nerve during repair after experimental crush injury." J. Anat., 79, 9-14.

Barker, N.W. (1938) "Lesions in peripheral nerves in thromboangitis obliterans." Arch. Int. Med., 62, 271.

Banks, R.W., and James, N.T. (1973) "The blood supply of rabbit muscle spindles." J. Anat., 114, 7-12.

Bartholdy, K. (1897) "Die Arterien der Nerven." (Jena) Morph. Arb., 7, 393-458.

Bateman, J.E. (1962) "Trauma to nerves in limbs." W. B. Saunders, Phil., 1962.

Bentley, F.H., and Schlapp, W. (1942) "Experiments on the blood supply of nerves." J. Physiol., Lond., 102, 62-71.

Bergman, L. (1942) "Studies on the blood vessels of the human gasserian ganglion." Anat. Rec., 82, 609-630.

Bergman, L., and Alexander, L. (1941) "Vascular supply of the spinal ganglia." Arch. Neurol. Psychiat. (Chicago), 46, 761-782.

Bergstrom, R. (1965) in Swedish cancer year book 1963-65 (quoted by Lundborg and Branemark, 1968).

Bichat, M.F.X. (1830) in General Anatomy Vol, 1, p. 168, Lond. (quoted by Sunderland, 1968).

Biedmond, A. (1951) "Thrombosis of the basilar artery and the vascularization of the brain stem." Brain, 74, 300-317.

Bishop, G.H., Heinbecker, and O'Leary (1933) "Function of non-myelinated fibres of dorsal roots." Amer. J. Physiol., 106, 647.

Blunt, M.J. (1956) "Implications of the vascular anatomy of the optic nerve and chiasma." Proc. Roy. Soc. Med., 49, 433-439.

Blunt, M.J. (1957) "Functional and clinical implications of the vascular anatomy of nerves." Post-grad. Med. J., 33, 68-72.

Blunt, M.J. (1959a) "The blood supply of the facial nerve." J. Anat., 88, Pt. 4, 520-526.

Blunt, M.J., (1959b) "The vascular anatomy of the median nerve in the forearm and hand." J. Anat., 93, 15-22.

Blunt, M.J. (1960) "Ischaemic degeneration of nerve fibres." Arch. Neurol., 2, 528-536.

Blunt, M.J., and Stratton, K. (1956) "The development of a compensatory collateral circulation to nerve trunk." J. Anat., 90, 508-514.

Blunt, M.J., and Stratton, K. (1956) "The immediate effects of ligature of vasa nervorum." J. Anat., 90, 204-216.

Bourguet, M. (1913) "Des vasa nervorum." Compt. Rend. Soc. des Biol., 74, 656.

Brain, W.R., Northfield, D., and Wilkinson, M. (1952) "The neurological manifestations of cervical spondylitis." Brain, 75, 187-225.

Buitink, A.B. (1934) "Uber die Vaskularization der Nerven der unteren Extremitat." Anat. Anz., 79, 11-19.

Bulbring, E., and Burn, J.H. (1939) "Vascular changes affecting the transmission of nerve impulses." J. Physiol., 97, 250-264.

Causey, G., and Palmer, E. (1953) "The epineural sheath of a nerve as a barrier to the diffusion of phosphate ions." J. Anat., 87, 30.

Causey G., and Stratmann, C.J. (1953) "The relative importance of the blood supply and the continuity of the axon in recovery after prolonged stimulation of mammalian nerve." J. Physiol. (Lond.), 120, 373.

Chang, K.F.S., Low, W.D., Chan, S.T., Chuang, A., and Poon, K.T. (1963) "Enlargement of the ulnar nerve behind the medial epicondyle." Anat. Rec., 145, 149-153.

Coste, F., Bolgert, M., and De Bray, C. (1933) Bull. Soc. Med. Hosp., Paris, 43, 1026, (quoted by R. L. Richards (1951).

Cottrell, L. (1940) "Histological variations with age in apparently normal peripheral nerve trunks." Arch. Neurol. Psychiat. (Chicago), 43, 1138-1150.

Dahlstrom (1969) quoted as personal communication by Lundborg (1970).

Day, M.H. (1964) "The blood supply of the lumbar and sacral plexuses in the human foetus." J. Anat., 98, 1., 105-116.

Delamers, G., and Tanasesco, J.C. (1906) "Etudes sur les arteres du sympathique cephalique, cervical, thoracic et abdominal.", J. Anat. (Paris), 42, 98-107.

Denny-Brown, D., Adams, R.D., Brenner, C., and Doherty, M. (1945) "Pathology of injury in nerve induced by cold." J. Neuropath., 4, 305-323.

Denny-Brown, D., and Brenner, C. (1944) "Paralysis of nerve induced by direct pressure and by tourniquet." Arch. Neurol. Psychiat. (Chicago), 51, 1-26.

Desplats, R., and Baillet, A. (1911) Arch. Mal. Coeur., 4, 481. (quoted by R. L. Richards, 1951).

De Souza, O.M. (1932) "Contribution a l'etude de la vascularization du system nerveux-organobegetatif." Ann. Anat. Path. Med. Clur., 9, 975-997.

Doupe, J., Cullen, C.H., and Chance, G.Q. (1944) "Post traumatic pain and the causalgic syndrome." J. Neurol. Neurosurg. & Psychiat., 7, 33-48.

Dunning, H.S. (1944) "Injury to the peroneal nerve due to crossing the legs." Arch. Neurol. Psychiat., 51, 179-181.

Dunning, H.S., and Wolff, H.G. (1937) "The relative vascularity of the central nervous system and peripheral nervous system of the cat in relation to function." J. Comp. Neurol., 67, 433.

Durward, A. (1948) "The blood supply of nerves." Postgrad. Med. J., 24, 11-14.

Dutil, A., and Lamy, H. (1893) Arch. Med. Exp., 5, 102 (quoted by Richards, 1951).
Dyck, P.J., and Lambert, E.H. (1969) "Dissociated sensation in amyliodosis" Arch. Neurol. (Chicago), 20, 490-507.
Eckhoff, N.L. (1931) "Torniquet paralysis, a plan for the extended use of the pneumatic torniquet." Lancet, 2, 343.
Edshage, S. (1964) "Peripheral nerve suture." Acta. chir. scand. suppl., 331, 3-104.
Foerster, O. (1912) Veh. Ges. dtsch. Nervenarz, 6, 134. (quoted by Richards, 1951).
Foerster, O. (1913) Wein, Med. Wschr., 63, 313. (quoted by Richards, 1951).
Francôis, J., and Neetens, A. (1956) "Vascularization of the optic pathway." Brit. J. Opthal., 40, 45-52.
Francôis, J., and Neetens, A. (1972) in "The optic nerve." ed. J. Slantery Cant, Henry Kimpton, London.
Frykholm, R. (1951) Acta. chir. scand., 101, 345., and 102, 93 (suppl 10).
Fukuda, M. (1970) Jap. J. Opthalmology, 14, 91.
Gallavardin, L., Laroyenne, L., and Ravault, P. (1925) Lyon. Med., 136, 144. (quoted by R. L. Richards, 1951).
Geflan, S., and Tarlov, I.M. (1956) "Physiology of spinal cord, nerve root and peripheral nerve compression." Amer. J. Physiol., 185, 217-229.
Gitlin, G. (1957) "Concerning the gangliform enlargement ("pseudoganglion") on the nerve to the teres minor muscle." J. Anat. 91, 466.
Glees, P. (1943) "Observations on the structure of the connective tissue of the connective tissue sheaths of cutaneous nerves." J. Anat., 77, 153.
Grigorowsky, I.M. (1928) "Zur Anatomie der die Kopfnerven emahrenden Arterien." Zet. ges. Anat. I. Z. Anat. Entingesch., 87, 728.
Guerrier, Y. (1951) "Les Arteres du Nerf Facial." Montpellier med., 39, No. 5, 83-95.
Haller, A. von (1754) "Arteria oculi historia et tabulae arterium oculi." Gottingen.
Haller, A. von (1756) "Icones Anatomicae" Gottingen. A. Vandenhoeck.
Hayreh, S.S. (1963) "The central artery of the retina." Brit. J. Opthal., 47, 651-663.
Hayreh, S.S. (1969) "Blood supply of the optic nerve head and its role in optic atrophy, glaucoma and oedema of the optic disc." Brit. J. Opthal., 53, 721-748.
Hayreh, S.S. (1972) in "The optic nerve." ed. J. S. Cant, Henry Kimpton, London, 1972.
Hofmann, M. (1903) Arch. Klin. Chir., 69, 677. (cited by Day, 1964).
Holmes, W., Highet, W.B., and Seddon, H.J. (1944) "Ischaemic nerve lesions occurring in Volkmann's ischaemia." Brit. J. Surg., 32, 259-275.

Holl, M. (1880) "Verrenkung des linken Elibogengelenkes mit Zerreissung der A. ulnains und der N. medianus und ulnans." Heilung: Collateralkreislauf. Medsche Jb., 10, 151.

Hovelacque, A. (1927) "Anatomie des Nerfs Craniens et Rachidiens et du grand sympathique chez l'homme." Paris: Doin.

Hughes, B. (1958) "Blood supply of the optic nerves and chiasma and its clinical signficance." Brit. J. Opthal., 42, 106-125.

Hutchinson, E.C., and Yates, P.O. (1956) "The cervical portion of the vertebral artery -- a clinico-pathological study." Brain, 79, 319-331.

Hyrtl, J. (1859) Oesterr. Z. prakt. Heilk. (cited by Tonakow, 1907).

Hyrtl, J. (1864) Denkschr. Akael. Wiss., Wein, 23, (cited by Tonakow, 1907) (Anat. Anz., 30, 471).

Isenflamm, J.F., and Doerffler, J.R. (1768) "De vasis nervorum." Erlangen.

Ishikawa, H. (1959) "A metrological study of the blood capillary density of the Rhombencephalon, spinal cord and the peripheral nerves in the cat." Fukoka. Acta. Med., 50, No. 11, 4275.

Joffroy, A., and Achard, C. (1889) Arch. Med. exp., 1, 229. (quoted by R. L. Richards, 1951).

Kernohan, J.W., and Woltman, H.W. (1942) "Amyloid neuritis." Arch. Neurol. and Psychiat. (Chicago), 47, 132-140.

Krause, W. (1875) v. Graefs. Arch. Opthal., 21, (1), 296. (quoted in "System of Opthalmology" Vol. 11, ed. Duke-Elder, Henry Kimpton, London, 1961).

Leibermann, A.R. (1968) "The connective tissue elements of the mammalian Nodose ganglion." Z. Zeliforsch., 89 95.

Lewis, T., Pickering, and Rotschild (1931) "Centripetal paralysis arising out of arrested blood flow to limb." Heart, 16, 1.

Lundborg, G. (1970) "Ischaemic nerve injury." Scand. J. Plastic surg. suppl., 6, 3-113.

Lundborg, G., and Branemark, P.I. (1968) "Microvascular structure and function of peripheral nerves." Advances Microcirc., 1, 65-88.

McGovern, F.H., Thomson, E., and Nelson, L. (1966) "The experimental production of ischaemic facial paralysis." Laryngoscope, 76, 1338-1352.

Nishii, K., Oura, C., and Pallie, W. (1969) "Fine structure of Pacinian corpuscles in the mesentery of the cat." J. cell Biol., 43, 539-55.

Nobel, W. (1969) "Observations on the microcirculation of peripheral nerves." Fifth European Conference on Microcirculation, Gothenber, 1968. Bibl. Anat. (Basel), 10, 316-320. S. Kager, New York.

Nonidez, J.F. (1942) "Arterio venous anastomoses in the sympathetic chain ganglia of the dog." Anat. Rec., 82, 593-608.

Ochoa, J. (1971) "The sural nerve of the human foetus: electron microscopic observations and counts of axons." J. Anat., 108, 231-245.

Okada, E. (1905) Arb. neurol. Iust. Univ. Wein, 12, 59. (cited by Adams, 1942).
Olsson, Y. (1966) "Studies on vascular permeability in peripheral nerves." Acta. Neuropath.(Berlin). 7, 1.
Olsson, Y. (1968) "Topographical differences in the vascular permeability of the peripheral nervous system." Acta. Neuropath. (Berlin), 11, 103.
Olsson, Y., and Reese, T.S. (1971) "Permeability of vasa nervorum and perineurium in mouse sciatic nerve studied by fluorescence and electron microscopy." J. Neuropath. Exp. Neurol., 30, 105-119.
Oppenheim, H. (1893) "Uber die seirde Form der multiplen neuritis." (Berlin) Klin. Wchnschr, 25, 589-592.
Pallie W., Nishii, K., and Oura, C. (1970) "The pacinian corpuscle, its vascular supply and inner core." Acta. Anat., 77, 505-520.
Parkes, A.R. (1945) "Traumatic ischaemia of peripheral nerves with some observations on Volkmann's ischaemic contractive." Brit. J. Surg., 32, 403.
Patterson, E.L. (1950) "Sources of arterial blood supply to the superior and middle cervical sympathetic ganglia and the ganglion intermediaire." J. Anat., 84, 329-341.
Petrovitz, L., and Szabo, Z. (1939) Anat. Anz., 88, 392.
Plaut, A., and Dreyfuss, M. (1940) "Spontaneous haemorrhage into oculomotor nerve with rupture of nerve and fatal sub-arachnoid haemorrhage." Arch. Neurol. Psychiat. (Chicago), 43, 564-571.
Priestley, J.B. (1932) "Histopathological characteristics of peripheral nerves in amputated extrametres of patients with arteriosclerosis." J. Nerve. Ment. Dis., 75, 137.
Quènu, J., and Lejars, F. (1890) "Les arteres et les veines des nerfs." C. r. hebd. Seanc. Acad. Sci., Paris, III, 608.
Quènu, J., and Lejars, F. (1892) "Etude anatomique sur les vaisseaux sanguins des nerfs." Arch. Neurol., Paris, 23, 1.
Raff, M., Sangalang, V., and Asbury, A.R. (1968) "Ischaemic neuropathy multiplex associated with diabetes mellitus." Arch. Neurol., 18, 487-499.
Ramage, D. (1927) "The blood supply to the peripheral nerves of the superior extremity." J. Anat., 61, 198-205.
Ranvier, M.L. (1878) "System Nerveux, 1." (cited by M. J. Blunt, 1957).
Ruysch, F. (1701) "The saurus Anatomicus Primus." Amsterdam (cited by Blunt, 1957).
Richards, R.L. (1951) "Ischaemic lesions of peripheral nerves: a review." J. Neurol. Neurosurg. Psychiat., 14, 76-87.
Roberts, J.T. (1948) "The effect of occlusive arterial diseases of the extremities on the blood supply of nerves. Experimental and clinical studies on the role of the vasa nervorum." Amer. Heart. J., 35, 369-392.

Roberts, J.T. (1950) "Blood vessels of nerves seen in living animals with quartz rod illumination and their response to drugs including cortisone." Fed. Proc., 9, 106-107.

Robinson, R. (1910) C. R. Acad. Sci., Paris, 151, 532 (quoted by Adams, 1942).

Röhlich, P., and Weiss, M. (1955) "Studies on the histology and permeability of the peripheral nervous barrier." Acta. morph. hung., 5, 335.

Sanger, M. (1880) Arch. Psychiat. Nervenkr., 10, 158 (cited by Plaut and Dreyfuss, 1940).

Sappey, M.P.C. (1889) "Traite d'Anatomie descriptive," 3, 224. Paris.

Schlesinger, H. (1895) Neurol. Centralb., 14, 578 (cited by Cottrell, 1940).

Schlesinger, H. (1933) Wein. med. Wschr., 83, 98.

Schmidel, C.C. (1755) (cited by Blunt, 1957).

Seddon, H.J. (1956) "Wolkmann's contracture: treatment by excision of infarct." J. Bone & Jt. Surg., 38-B, 152-174.

Seddon, H.J., and Holmes, W. (1945) "Ischaemic damage in the peripheral stump of a divided nerve." Brit. J. Surg., 32, 389.

Shantha, T.R., and Bourne, G.H. (1968) "The structure and function of nervous tissue." Acad. Press, N.Y., 1968.

Shanthaveerappa, T.R., and Bourne, G.H. (1962) "The 'perineural epithelium', a metabolically active, continuous, protoplasmic cell barrier surrounding peripheral nerve fasciculi." J. Anat., 96, 527-537.

Sidey, J.D. (1972) "Trapped nerves." Brit. Med. J., 3, 647.

Sinclair, D.C., and Hinshaw, J.R. (1951) "Sensory changes in nerve blocks induced by cooling." Brain, 74, 318-335.

Singh, S., and Dass, R. (1960) "The central artery of the retina." Brit. J. Opthal., 44, 280-299.

Smith, J.W. (1966a) "Factors influencing nerve repair. 1. Blood supply of peripheral nerves." Arch. Surg., 93, 335-341.

Smith, J.W. (1966b) "Factors influencing nerve repair. 2. Collateral circulation of peripheral nerves." Arch. Surg., 93, 433-437.

Spiegel, I.J., and Lewin, P. (1945) "Tourniquet paralysis." J. Amer. Med. Assn., 129, 432.

Spiegel, van der. A. (1627) "De Humani Corpous Fabrica." Venice (quoted by Blunt, 1957).

Sunderland, S. (1945a) "Blood supply of the nerves of the upper limb." Arch. Neurol. Psychiat. (Chicago), 53, 91.

Sunderland, S. (1945b) "Blood supply of the sciatic nerve and its branches." Arch. Neurol. Psychiat. (Chicago), 54, 283.

Sunderland, S. (1945c) "Blood supply of peripheral nerves. Practical considerations." Arch. Neurol. Psychiat., 54, 280.

Sunderland, S. (1968) "Nerve and nerve injuries." E. & S. Livingstone, Ltd., Edin and Lond.

Stohr, P. (Jnr.) (1928) in Mollendorf's Handbuch der microscopischen Anatomie der Menschen, 4, 1.213. Berlin.

Tanasesco. J.C. (1905) Bull. Mem. Soc. Anat., 80, 834. (quoted by Bergmann, 1942).

Thomas, P.K., and King, R.H.M. (1974) "Peripheral nerve changes in amyloid neuropathy." Brain, 97, 395-406.

Tobin, C.E. (1943) "Injection method to demonstrate blood supply of nerves." Anat. Rec., 87, 341-344.

Tonakoff, V.N. (1898) "Die Arterien der Intervertebral-ganglien und der Cerebrospinal nerven des Menschen." Int. Mschr. Anat. Physiol., 15, 353-300.

Tonakoff, V.N. (1907) "Die nervenbegleitenden gefasnetze beim Embryo und die Arteriae nutriciae nervorum biem Erwachsen." Anat. Anz., 30, 471.

Torraca, L. (1920) "La circolazione sanguigne dei nervi isolati." Chir. Organi. Mov., 4, 279-295.

Wadsworth, G.T. (1974) "The cubital tunnel syndrome and the external compression syndrome." Anaesth. Analges. (Cleve), 53, 303-308.

Waggener, J.D., and Beggs, J. (1967) "The membranous coverings of neural tissue: an E. M. Study." J. Neuropath. and Expt. Neurol., 26, 412.

Waggener, J.D., Bunn, S.M., and Beggs, J. (1965) "The diffusion of ferritin within the peripheral nerve sheath: An electron microscope study." J. Neuropath. & Expt. Neurol., 24, 430-443.

Waksman, B.H. (1961) "Experimental study of diptheria polyneuritis in the rabbit and guinea-pig. III. The blood nerve barrier in the rabbit." J. Neuropath. & Expt. Neurol., 20, 35-77.

Weddell, G. (1974) Personal communications.

Weiss, P., Wang, H., Taylor, A.C., and Edds, MacV. (Jr.) (1945) "Proximodistal fluid connection in the endoneurial spaces of peripheral nerves demonstrated by colored and radioactive (isotope) tracers." Amer. J. Physiol., 143, 521-540.

Woltman, H.W., and Wilder, R.M. (1929) "Diabetes Mellitus - pathological changes in the spinal cord and peripheral nerves." Arch. Int. Med., 44, 576-603.

Woodall, B., and Davis, C. (Jr.) (1950) "Changes in the arteriae nervorum in peripheral nerve injuries in man." J. Neuropath. Exp. Neurol., 9, 335-343.

Zuckerkandl, O. (1885) "Zwei Falle von Collateral Kreislauf." Medsche Jb., 15, 273.

Zinn, J.G. (1755) "Descriptio anatomica oculi humani." Vandenhoeck, Gottingen.

INDEX

ATP
    effect on vasomotor control, 739
ATP Hydrolysis
    in cardiac contraction, 590
Acetycholine
    effect of on renal circulation, 677
    effect of on cardiac pacemakers, 590
Actin
    in cardiac contraction, 590
Adrenal glands
    circulation of, 407
Adrenergic
    innervation of arteries, 733-741
Afferent arteriole
    in non-human mammals, 646 649
    in kidney, 634, 640-642
    resistance in, 672
Aglomerular shunting, 664
Aldosterone
    effect of on renal circulation, 678
Alpha adrenergic receptors, 599-600
Anesthesia
    effect of on circulation, 607
    effect of on prostaglandin synthesis, 672
    effect of on renal circulation, 674
Angioblasts, 31
Angiotensin (see also Renin angiotensin system)
    effect of on vasomotor control, 739
    response of renal vessels to, 648
Angiotensin I, 684-685

Angiotensin II
    effect of on renal circulation, 678, 682, 686, 687, 701-703
    relation to aldosterone secretion, 688
Aorta
    collaterals of, 382-385
    development of, 100, 174-201
    discovery of, 2
    distribution of, 382-389
    elastic distensibility of, 195
    embryogenesis of, 52-58
    intimal cushions in, 200
    lipid desposits in, 52, 201
    mean pressure in, 593
    microscopic structure of, 40, 175-201
    nexuses in, 737
    rigidity of in newborns and infants, 197-199
    structure of, 174-201
    structure of during fetal life, 176-181
    structure of in newborns, 181-183
Aortic arch arteries, 51-53
Aortic arches
    in fetal circulation, 49, 51
Aortic isthmus, 154
    in fetal circulation, 109
Aortic sac
    in arterial embryogenesis, 51
Arcuate arteries
    in non-human mammals, 649
    in kidney, 634
Aretaeus of Cappadocia, 16
Aristotle, 1,2,3,7
Arterial anastomoses
    innervation of, 751
Arterial blood flow
    mechanisms of, 589-593
    regulation of, 587-630

Arterial circulation
    of extremities, 425, 485
Arterial embryology, 49-81
Arterial pressure (see also
    blood pressure)
    in fetal circulation, 111
    mechanisms of, 589-593
    regulation of, 587-630
Arterial wall
    nutrition of, 329-332
    tension of, 100
Arteries
    early distinction from
        veins, 2
    innervation of, 599, 600,
        674, 729-767
Arteriogenesis, 21
Arterioles
    effect on resistance, 594
    nexuses in, 737
Arteriovenous fistula, 528
    blood flow through, 497,
        498
Arteriovenous shunts
    in extremities, 467, 468
Atherosclerosis
    in iliac arteries, 301
    relation to arterial
        stenosis, 494
Atresia of the pulmonary ostium,
    155
Atrial pressures, left
    in fetal circulation, 115
Atrial septa
    in fetal circulation, 37a
Atrioventricular valves
    in fetal circulation, 44,
        45
Atrium
    in fetal circulation, 42-
        44
Atrium, right
    mean pressures in, 593
Auditory nerve
    blood supply of, 777

Autonomic nervous control
    of arteries, 599, 600,
        674, 731-741
    of fetal circulation, 117
    of veins, 743, 754
Autoregulation, 597, 598
    of cerebral circulation,
        543
Axelrod, J., 588
Axillary artery
    anatomy of, 431-435
Basement membrane
    of glomerular capillaries,
        643
Basilar artery, 244
Bayliss effect, 513
Beta adrenergic receptors,
    599-600, 739
    (see receptors, beta
    adrenergic)
Bile ducts
    blood supply of, 398
Bladder
    blood supply of, 413
Blood flow, microvascular
    flying-spot television,
        573
    high-speed cinematography,
        573
Blood flow in systemic vessels
    constant, 15, 557
    distribution of, 16, 17,
        558, 559
    laminar, 15, 16, 557, 558
    linear velocity of, 559
    linear volume of, 559
    measurement of, 18-32,
        560-578
    pulsatile, 15, 557
    Techniques of measurement,
        18-32, 560-578
    brain, 575-576
    by indicator dilution,
        563-567
    doppler, 569-571
    electromagnetic, 567-
        569

hepatic, 576
history of, 560, 561
microvascular, 573
muscle, 576
renal, 576
thermal dilution, 26, 27, 571-572
ultrasonic, 569-571
with Fick principle, 560-561
turbulent, 557, 558
velocity of, 557
volume of, 557
Blood flow, tissue
differential pressure, 577
drag, 577
Techniques of measurement, 573-578
thermal methods, 574-575
tissue clearance, 575
transit time, 577
venous occlusion plethysmography, 573-574
Blood islands
in arteriogenesis, 21, 29, 31
Blood loss
effect of on circulation, 614, 615
Blood pressure
in systemic circulation, 537-556
internal vs. transmural, 538-540
static vs. dynamic, 540
directional, 540
measurement of, 537-586
in retinal arteries, 556
in manifold capillaries, 556
in mesenteric vessels, 556
in ear vessels, 556
Bovet, Daniel, 588
Bowman's capsule, 642
Brödel's line, 632

Bradykinin
effect of on circulation, 600
effect of on umbilical vessels, 146
effect of on vasomotor control, 739
Brachial artery
anatomy of, 435-438
transitonal zones in, 98
Brachial plexus
blood supply of, 778
Bulbus cordis, 45
Calcifications
in arteries, 104, 105
in cerebral arteries, 265-270
in coronary arteries in infancy, 231, 232
in extremity arteries, 332
in iliac arteries, 278-289
Calcium ions
in cardiac contraction, 590
"Candelabra" anastomoses, 514
Capillaries
innervation of, 743, 753
Cardiac output
in infancy, 147
measurement of, 560, 572, 573
Cardiac pacemakers, 590
Cardiogenic plate, 31, 33
Cardiovascular control, 621-624
Carotid arteries, 232-243
lipid deposits in, 234
musculo-elastic layer in, 237
nexuses in, 737
post natal development of, 359
sensory neurons in, 741
Carotid sinus, 237-243
lipid deposits in, 237-241
transitional zones in, 98

Carotid siphon, 243, 251, 258-265, 326
Catecholamines
    effect of on arteries, 731-739
    effect of on cardiac pacemaker, 590
    effect on ductus arteriosus, 158
    effect of on pulmonary circulation, 337
    effect of on renal circulation, 674, 678, 699-700, 701-703
Celiac artery
    communication with superior mesenteric artery, 390, 391
Celsus, 5
Cerebral arteries, 243-248, 251-276
    intimal cushions in, 271, 274
    lipid deposits in, 270, 271
    medial gaps in, 275
    post natal development of, 359
    sensory neurons in, 741
Cerebral blood flow, 248-250
Cerebral circulation
    autoregulation of, 513, 543
    chemical regulation of, 513
    effect of hyperventilation on, 513
    innervation of, 753
    local vasodilators of, 513
    "steal" syndromes in, 514
    shunting in, 513
Cerebral oxygen consumption, 250, 512
Cerebral vessels
    embryogenesis of, 68, 69
Circle of Willis, 243, 251, 491
Chemoreceptors, 603, 611-614
Cholinergic nerves, 739, 741
Cholinergic receptors, 599, 600

Chromaffin cells
    extra-adrenal, 742
Coarctation of the aorta, 155
Cold
    effect of on cardiac pacemaker, 590
Collagen fibers
    distensibility of, 96
Collateral circulation
    anatomy of, 488-493
    biology of, 487-535
    cervical, 522
    chemical effects on development of, 500-503, 528, 529
    clinical significance of, 503, 530
    coronary, 498, 502-503, 509
    gastrointestinal, 521-522
    in brain, 509-516
    in extremities, 522-530
    in kidney, 519-521, 649
    in thrombophlebitis, 503
    in viscera, 516-523
    lower extremity, 462-463
    of femoral artery, 453
    of upper extremity, 443-445
    physiology of, 493-501
    reactivity of, 501, 502
    venous, 529-530
Colon
    blood supply of, 401, 402
Common iliac artery
    anatomy of, 445-448
Compliance in iliac arteries, 300
Conduction system
    in fetal circulation, 48
Congenital cardiac abnormalities
    atresia of pulmonary ostium, 155
    coarctation of aorta, 155
    left heart syndrome in, 152
    tricuspid atresia, 152, 161
    patent ductus arteriosus, 160
    pulmonary stenosis, 152, 160

Congestive heart failure
    renal circulation in, 668
Constant flow, 557
Contraceptive pills
    relation to renin activity, 683
Conus ridges
    in fetal circulation, 45
Coronary atherosclerosis
    in infancy, 229, 230
Coronary arteries
    atherosclerosis of in infancy, 230
    calcification of in infancy, 231, 232
    development of, 201-213
    gaps in internal elastic
    innervation of, 753
    membrane of, 327-332
    in fetal circulation, 109
    post natal remodeling, 359
    sensory neurons in, 741
    structure of, 201-230
    variation in distribution of, 204
Coronary arteriography, 508
Coronary artery pressure
    effect of collateral circulation on, 506
Corticosteroids
    effect of on renal circulation, 701
Cournand, Andre, 560, 588
Cranial nerves
    blood supply of, 773-778
Crista dividens
    in fetal circulation, 109
"Crista reuniens," 154
"Critical vessels"
    definition of, 99, 173
Cutaneous arteries
    anatomy of, 475-484
Dale, Sir Henry, 588
Depolarizing current
    in cardiac conducting tissue, 590

Diabetes
    renal vascular histology in, 640
Dive reflex, 603
Dopamine
    effect on vasomotor control, 739
Dorsal mesocardium, 33
Dorsalis pedis artery
    anatomy of, 461-462
Ductus arteriosus
    closing of, at birth, 81-83, 120, 146, 158-171
    effect of catecholamines, 158
    effect of prostaglandins, 161
    hematomas in, 166
    in fetal circulation, 109 155-158
    innervation of, 754
    "intimal mounds," 164
    necrosis of, 167, 169
    persistent patency of, 160
    smooth musculature of, 172
    structure of, 149-152
    thrombosis of, 168
Ductus venosus, 78
    course of, 107
    closure of, 82, 84
    in fetal circulation, 114
    structure of, 172
Duodenum
    circulation of, 392, 393
Ear artery
    innervation of, 745
Efferent arterioles
    of kidney, 634
    resistance in, 672
Einthoven, Willem, 588
Elastic arteries
    characteristics of, 397
    innervation of, 745
Elastic sheets, secondary
    relation to endothelium, 103

Elastic tendons
  function of in smooth muscle, 98
Elastic tissue
  presence of in blood vessels, 96
Electrocardiogram
  during closure of the ductus arteriosus, 120
  during fetal life, 111
Electronic coupling, 737
Electrophysiology
  of cardiac contraction, 590
Embryology, 21, 29, 31, 52-58, 66-70, 425, 426, 445
Emotional stress
  effect of on circulation, 615, 675
Empedocles, 1
Endocardial cushions
  in fetal circulation, 45
Endocardial tubes, 31
Endoneurium, 790
Endothelial lining
  role of in elasticity of vessels, 97
Epineurium, 771, 786
Erasistratus, 2, 3, 4, 5, 19
Erectile tissue
  innervation of, 754
Estrogens
  effect of on renal circulation, 701
  relation to renin activity, 683
Exercise
  effect of anticipation on circulation, 617
External iliac artery
  anatomy of, 450, 451
  (see extremity arteries, 301-324)
  transitional zones in, 98
Extracerebral (pial) arteries, innervation of, 751

Extremities
  arterial circulation of, 425-485
Extremity arteries, 301-324
  calcifications in, 332
  circular gaps in, 106
  embroygenesis of, 425, 426, 445
  microscopic structure, 313-324
  spindles in, 305
  transitional zones, 98, 303
Facial nerve
  blood supply of, 777
Femoral artery
  anatomy of, 451-454
Fetal circulation, 35-48, 107-111, 358
  autonomic nervous control of, 117
  characteristics of, 107-119, 358
Fetal nerves
  blood supply of, 796
Fetus
  circulation of, 107-119, 355
Fibro-muscular dysplasia
  in renal vessels, 640
Fick principle, 560-563
Filtration fraction, 672, 674
5-hydroxytryptamine (5-HT), (see serotonin)
Foramen ovale
  closure of, 85, 149
  in fetal circulation, 109-117
  premature closure of, 152
Forssman, Werner, 588
Fourier analysis
  in blood pressure measurement, 545, 578
Galen, 6-20
Gall bladder
  blood supply of, 398
"Gap junctions," 731, 736-739

Genitalia, external
    blood supply of, 417-418
Glomerular capillaries
    permeability of, 674
Glomeruli
    embryogenesis of, 59
Glomerulus, 642, 646, 656
Glycogen
    in fetal myocardial cells, 117
Greco-Roman medicine
    the arteries in, 1-20
Hales, Stephen, 542, 560, 587
Hamilton, W.F., 542
Harvey, William, 4
Heart failure
    effect of on circulation, 614, 615
Heart tube, undivided, 31-35
Heart volume,
    in infancy, 148
Hemodynamic ohm or hohm, 578
Herophilus, 2, 3, 4, 17
Hepatic artery
    collaterals of, 398, 399
    distribution of, 394-398
Hepatic circulation
    in fetal life, 34, 77, 78
Hepatic cirrhosis
    circulation in, 519
"Hepatorenal syndrome"
    relation to renin activity, 683
Heymans, C., 588
Histamine
    effect of on vasomotor control, 739
Histaminergic fibers, 600
Historical perspective, 1-20
Hoboken's valves, 126, 127
Hormones
    in development of collateral vessels, 499
Hunter's canal, 451
Hydraulic pressure
    effect on inferior vena cava pressure, 538

Hydronephrosis
    renal artery in, 633
Hydrostatic pressure
    effect on inferior vena cava pressure, 538
Hypertension
    renal circulation in, 521
    renal vascular histology in, 640
Hypothalamus
    innervation of capillaries of, 753
Hypothyroidism
    effect of on cardiac pacemaker, 590
Hypovolemia
    circulatory pressures in, 594
Iliac arteries, 276-301
    atherosclerosis in, 301
    calcification in, 103, 278-289
    circular gaps in, 104
    during fetal development, 287-289
    internal elastic membrane in, 281-287
    post natal development of, 359, 360
    relation to umbilical arteries, 276, 289, 291
Indomethacin
    effect on prostaglandin synthesis, 692, 695
Infarct size
    relation of collateral channels to, 506
Inferior thyroid artery
    anatomy of, 430
Inferior vena cava pressure
    effect of hydraulic pressure in, 538
    effect of hydrostatic pressure on, 538
    relation to intra thoracic pressure, 538
    relation to portal pressure, 538

Innervation of blood vessels
    development of, 754-755
Innervation of vascular smooth
    muscle, 598-614
Innominate artery
    embryogenesis of, 425
Input impedance, 578
Insulin
    effect of on renal
        circulation, 701-703
Interlobular arteries
    in non-human mammals, 649
    in kidney, 634, 640
Intermittment claudication, 529
Internal elastic membrane
    circular gaps in, 104
Internal elastic membrane
    gaps in, in coronary
        arteries, 327-329
    gaps in, in muscular
        arteries, 324-326
    in iliac arteries, 281-286
    "splitting" of, 103
Internal iliac artery
    anatomy of, 448-450
    effect of ligation of, 389
Internal mammary artery
    anatomy of, 429-430
Interventricular foramen
    in fetal circulation, 45
Interventricular septum
    in fetal circulation, 37,
        45
Intestinal angina, 522
Intestine
    blood supply of, 399-404
    collaterals of, 404-406
"Intimal cushions"
    innervation of in renal
        thyroid and ciliary
        arteries, 640, 751
    innervation of, 743
    in aorta, 200
    in cerebral arteries, 260,
        274

"Intimal mounds"
    in ductus arteriosus, 164
Intracerebral arteries
    innervation of, 751
"Intracerebral steal," 514
Intraneural arteries, 786-792
Intrathoracic pressure
    effect on systemic
        circulation, 538
Intravenous infusion
    effect of on circulation,
        614-615
Juxtaglomerular apparatus, 634
    640-642, 656
Kallikrein-Kinin system, 698-699
Katz, B., 588
Kety, Seymour, 561
Kidney
    anatomy of circulation,
        631-660
    circulation of, 407-411
    physiology of circulation,
        661-727
Kininases, 699
Kinins
    effect of on circulation,
        599-600
    effect of on renal circu-
        lation, 678, 698, 699,
        701-703
Korotkoff, Nicolai Sergeevich,
    542
Korotkoff sounds, 552
Krogh, August, 560
Lamellar unit
    of aorta, 175
Laminar flow, 557, 558
Laplace, law of, 95, 498, 594
Leeuwenhoeck, Antony van, 587
Left atrium
    pressures in, during fetal
        life, 115
Left heart syndrome, 152
Linear velocity
    of blood flow, 559
Linear volume
    of blood flow, 559

Lipid deposits
  in carotid arteries, 234–243
  in carotid sinus, 241
  in cerebral arteries, 270
  in vertebral arteries, 224–248
Liver
  circulation of, 394, 398, 516–519
Local vasodilators
  in cerebral circulation, 513
Locus coeruleus
  in innervation of intra-cerebral arteries, 751
Loewi, Otto, 588
Ludwig, Karl Friedrich Wilhelm, 560
Lumbar roots
  blood supply of, 784
Luminal diameter
  effect of on resistance, 595
Luminal retraction pattern in muscular arteries, 304–313
Lymphatics, renal, 676–677
Macula densa, 641, 656
Malpighi, 587
Manometers
  anaeroid, 544
  arterial loop, 550
  capacitance, 548
  catheter tip, 548–550
  Hamilton, 546
  inductive, 548
  mercury, 543, 544
  micro-, 550
  optical, 546, 547
  random-scale, 532
  saline, 543
  sphygmo, 551–553
  strain gauge, 547
  ultrasound, 553
Mechanoreceptors, 597, 598, 601–603
  endothelial, 496, 497
  high pressure, 601, 609, 610
  low pressure, 601, 602, 608, 609
  ventricular, 602, 603, 610, 611
Medial gaps
  in cerebral arteries, 275
Median nerve
  blood supply of, 781
Mesangium, 642
Mesenteric arteries
  innervation of, 748
  nexuses in, 737
  transitional zones in, 98
Mesenteric circulation
  innervation of, 753
"Mesoneurium," 771
Monckeberg sclerosis, 332
Mucocutaneous lymph node syndrome (MCLS) 232
Multipapillary kidney, 646
Muscular artery, 463–475
  innervation of, 748–751
  secondary elastic sheets in, 103
Muscular exercise
  effect of on circulation, 617–620
Musculo-elastic junctions
  types of in arterial wall, 98
Musculo-venous pumps
  effect of on blood flow, 593
Myocardial cells
  glycogen content during gestation, 117
Myocardial ischemia
  in development of collateral circulation, 504, 505
Myosin
  in cardiac contraction, 590
"Nexuses," 731, 736–739
Neogenic collaterals, 502, 503
Neonatal circulation, 147–161
  arterial pressures in, 149
  pulmonary vascular resistance, 149, 150

Nerve growth factor (NGF), 755
Nerves
    blood supply of, 769-803
Neuromuscular junctions
    autonomic, 731, 736-739
Noradrenaline (NA)
    as neurotransmitter, 731-739
Oculomotor nerve
    blood supply of, 776
Omphalomesenteric artery, 58
Operant conditioning
    effect of on circulation, 604
Optic chiasma
    blood supply of, 773
Optic nerve
    blood supply of, 773
Ostium secundum
    in fetal circulation, 115
Ovaries
    blood supply of, 417
PCO_2
    in development of collateral vessels, 499
PO_2
    in development of collateral vessels, 499
Pancreas
    circulation of, 392, 393
Parasympathetic activity
    effect of on cardiac pacemaker, 590
"Paravascular" nerves, 748
Pelvic arteries
    anatomy of, 634
Pelvic viscera
    blood supply of, 413-419
Penis
    blood supply of, 415, 416
Perfusion pressure
    relation to blood flow, 596
Pericytes
    in kidney, 641
Perineurium, 788
Peripheral resistance
    regulation of, 754
Peripheral resistance unit, 577
"Perivascular" nerves, 748
Peroneal artery
    anatomy of, 460-461
Peroneal nerve
    blood supply of, 784, 785
pH
    in development of collateral vessels, 499
Placenta
    circulation of, 77, 82
    synthesis of prostaglandins in, 697
Placental circulation
    cessation at birth, 120, 121
Plantar arteries
    anatomy of, 462
Plato, 3
Poiseuille, Jean Leonard Marie, 542
Popliteal arteries
    innervation of, 748
    (See extremity arteries, 301-324)
    anatomy of, 454-457
Portal hypertension
    circulation in, 519
Portal pressure
    effect on systemic circulation, 538
Portal vein
    nexuses in, 737
    smooth muscle in, 98
Posterior tibial artery
    anatomy of, 459-460
Postphlebitic syndrome, 530
Posture
    effect of on circulation, 615
Praxogoras, 2, 4, 17, 19
Precapillary arterioles
    innervation of, 743
    mesenteric, 736

Precapillary sphincters
  innervation of, 743
  nexuses in, 737
Pre-circulatory system, 29-31
Pregnancy
  circulatory adaptations in, 419
Pressure gradients
  in systemic circulation, 594
Primordial circulation, 31-35
Prostaglandins
  effect of on circulation, 600
  effect of on ductus arteriosus, 161
  effect of on renal circulation, 675, 678, 689-695, 701-703
  effect of on umbilical vessels, 146
  effect of on vasomotor control, 739
  synthesis of, 689-697
Prostate
  blood supply of, 415, 416
Psychic stimuli
  effect of on circulation, 615, 675
Psychoactive drugs
  effect of on circulation, 604
Pulmonary aeration at birth, 342
Pulmonary artery
  distensibility of, 347
  embryogenesis of, 53, 70
  nexuses in, 737
  structure of, 332-351
Pulmonary atresia, 161
Pulmonary circulation, 332-351
  effect of catecholamines, 337
  in fetal life, 37
  in fetus, 337
  in infancy, 342
Pulmonary ostium
  atresia of, 155

Pulmonary stenosis, 152, 160, 161
Pulmonary surfactant, 342
Pulmonary trunk, 98
  elastic and muscular elements in, 98
Pulmonary vascular resistance, 149, 152
  in fetal lung, 335
Pulmonary veins
  in fetal circulation, 43, 44
Pulsatile flow, 557
Purinergic nerves, 600, 739, 742
Purkinje cells
  in fetal circulation, 48
  in cardiac conduction, 590
Radial artery
  anatomy of, 438-440
Radial nerve
  blood supply of, 781, 782
Receptors
  alpha adrenergic, 599-600
  beta adrenergic, 599-600, 739
Rectum
  blood supply of, 402-404
Reinnervation of denervated vessels, 751
Renal arteries
  anatomy of, 632-635
  collaterals of, 411-412
  distribution of, 407-411
  innervation of, 748
  nexuses in, 737
  transitional zones in, 98
Renal circulation
  anatomy of, 631-660
  autoregulation, 675-676
  comparative studies, 646-656
  glomerular, 634, 672-674
  histology of, 640-646
  hormonal control of, 678-704

    in aging, 677–678
    medullary, 635–639, 662, 650–653
    neural control of, 674–675
    resistance in, 672
    sympathetic involvement in, 676–677
    variability of, 662
    Techniques of measurement:
      electromagnetic flowmeter, 665
      inert gas washout, 664–666
      measurement of, 664–668, 670–672
      PAH clearance, 665
      radioactive albumin, 670–671

Renal failure
    effect of on circulation, 614–615

Renal function
    relation to blood flow, 668

Renal oxygen consumption, 662

Renal pelvis
    blood supply of, 651

Renal vein
    anatomy of, 635

Renin
    inhibition of, 688
    production of, 634–641, 683
    relation to renal circulation, 521, 678–683

Renin-angiotensin system, 678–689
    activation of, 674, 675
    effect of on renal circulation, 686–689, 701–703

Renin release
    control of, 680–683

Renovascular hypertension
    histology in, 640, 653–654

Reserpine
    effect of on renin secretion, 683

Resistance
    estimation of, 577, 578
    in systemic circulation, 537–586

Resistance vessels, 595

Respiratory movements
    effect of on blood flow, 593

Retinal artery pressure
    measurement of, 556

Retinal vasculature
    embryogenesis of, 66–69

Richards, D.W., 588

Right atrium
    pressures in during fetal life, 115

Right ventricular output
    in fetal circulation, 109

Sacral roots
    blood supply of, 784

Salt loading
    effect of on renal circulation, 677

Saphenous artery
    innervation of, 748

Sarcoplasmic reticulum
    in cardiac contraction, 590

Scarpa's triangle, 451

Sciatic nerve
    blood supply of, 784

Seminal vesicles
    blood supply of, 415–416

Sensory neurons
    vascular, 741, 742

Septation of the heart, 35–37

Septum primum, 39

Septum secundum, 39

Serotonin
    effect of on umbilical vessels, 146
    effect on vasomotor control, 600, 739

Sinoatrial node, 590

Servo-control mechanisms, 606

Shear stress
    in development of collateral vessels, 497

INDEX

Shunting of blood
    in cerebral circulation, 513
Sine-wave generator, 544, 545
Sinus venosus
    in fetal circulation, 41, 42
Small intestine
    blood supply of, 399-401
Smooth musculature
    function of, 97
Somatic ganglia
    blood supply of, 785
Sphygmomanometry, 551-553
Spindles
    in extremity arteries, 305
Spiral valve, 37
Spleen
    circulation of, 393
Starling, Ernest Henry, 589, 591
"Steal" syndromes
    in cerebral circulation, 514
    in subclavian artery, 514
Stomach
    circulation of, 391, 392
Streeter horizons, 29-35
Stretch receptors
    pulmonary, 603
Subclavian artery
    anatomy of, 426-431
Subclavian steal syndrome, 514
Submaxillary gland
    renin production in, 680
Superior mesenteric artery
    communication with celiac artery, 390
Suprarenal glands
    (see adrenal glands)
Sympathetic ganglia
    blood supply of, 785, 786
Sympathetic innervation
    of arteries, 731-739
    effect of on renal circulation, 674

Syncope
    effect of on circulation, 614-615
Systemic circulation
    in fetal life, 37
Systemic resistance
    mechanisms of, 593-600
    regulation of, 587-630
Telemetry
    in monitoring circulation, 624
Temperature
    effect of on circulation, 620-621
Testes
    blood supply of, 414-415
Thoracic aorta
    innervation of, 745
Thrombophlebitis
    collateral circulation in, 503
Thyroid hormones
    effect of on renal circulation, 701
Tibial nerve
    blood supply of, 784
Tonin, 685, 686
Transitional zones, 98, 301
"Transmural" pressure
    components responsible for, 96
Transplanted tissue
    collateral circulation in, 502, 503
Tricuspid atresia, 152, 160
Trigeminal ganglion
    blood supply of, 776
"Trueta shunt" 638
Truncus arteriosus, 35
Turbulence
    effect of on arterial wall, 498
Turbulent flow, 557, 558
Ulnar artery
    anatomy of, 440-443
Ulnar nerve
    blood supply of, 780

Umbilical arteries
    changes in during closure, 127, 138, 144, 171
    characteristics of, 66
    effect of bradykinin on, 146
    effect of catacholamines on, 736
    effect of a single, 289-291
    relation to iliac arteries, 276, 289, 291
    smooth musculature of, 172
    structure of, 127-145
    tension of, 123
Umbilical blood flow
    in fetal circulation, 117-119
    regulation of, 145, 146
Umbilical cord
    nerve supply of, 147
Umbilical vein
    characteristics of, 66
    closure of, at birth, 122, 123, 148
    smooth musculature of, 172
    structure of, 123
    tension of, 123
Unipapillary kidney, 646
Units of measurement of flow, 559
Ureter
    blood supply of, 411
Uterine arteries
    innervation of, 748
Uterine blood flow
    regulation of, 680
Uterus
    blood supply of, 417, 418
    synthesis of prostaglandins in, 697
Vagal ganglia
    blood supply of, 778
Vagina
    blood supply of, 417-418
Valves
    venous, 529
Valvula foraminis ovalis
    during fetal life, 115

Vasa nervorum
    functions of, 792-796
    innervation of, 792
Vasa recta, 637, 640, 650
Vasa vasorum
    embryogenesis of, 61, 70
    innervation of, 748
Vascular arch
    in embryogenesis of extremity arteries, 425
Vascular smooth muscle
    nexuses in, 737
Vasoactive agents
    in study of collateral channels, 501, 502
Vasodilatation
    neural control of, 736, 739
Vasovagal attacks, 615
Veins
    early distinction from arteries, 2
    innervation of, 743, 754
Vena cava
    discovery of, 2
Vena cava inferior
    relation to fetal circulation, 107
    smooth muscle in, 98
Venous collaterals, 529, 530
"Venous gangrene," 530
Venous pressure (see also blood pressure)
    umbilical, 117
Venous valves
    embryogenesis of, 61
Ventricle
    in fetal circulation, 44-47
Venules
    innervation of, 743
Vertebral artery
    anatomy of, 243, 429
    lipid deposits in, 244-248
    post natal developments of, 359
    vertebral syphon, 244

Vertebro-basilar junction
    embryogenesis of, 62
Vertebro-basilar system, 258
Via dextra
    in fetal circulation, 109, 115
Via sinistra
    in fetal circulation, 109
Visceral ischemia, 522
Voight's method, 270
Von Euler, Ulf S., 588
Von Kossa staining technique, 270
Walking tolerance
    relation to collateral circulation, 525
Wedge pressure, 556
Wiggers, Carl, 542
Windkessel effect, 174, 195